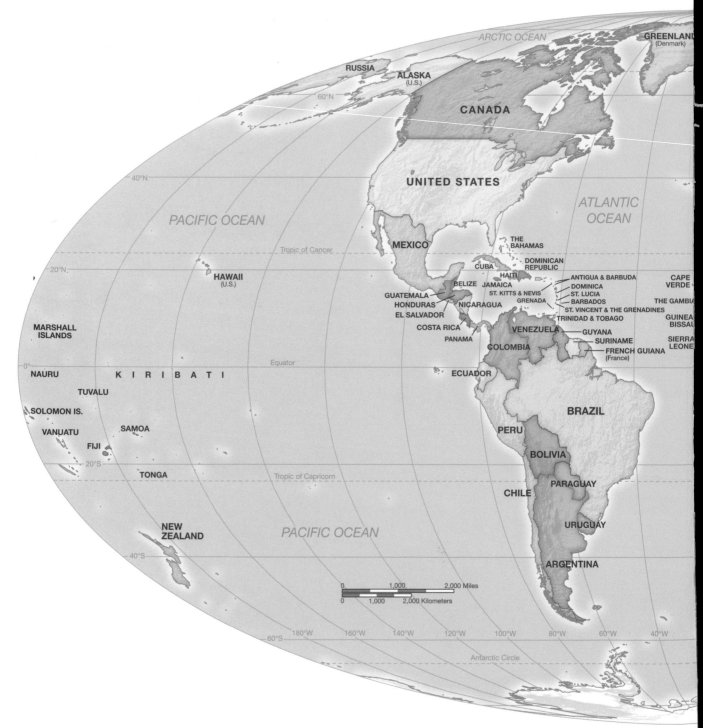

World States

See page 497 of the Appendix for an explanation of this projection

PLACES AND REGIONS IN GLOBAL CONTEXT

Human Geography

PLACES AND REGIONS IN GLOBAL CONTEXT

Human Geography

Fifth Edition

Paul L. Knox

Virginia Tech

Sallie A. Marston

University of Arizona

Prentice Hall

New York   Boston   San Francisco   London
Toronto   Sydney   Tokyo   Singapore   Madrid
Mexico City   Munich   Paris   Cape Town   Hong Kong   Montreal

Library of Congress Cataloging-in-Publication Data

Knox, Paul L.
    Places and regions in global context : human geography / Paul L. Knox, Sallie A. Marston. — 5th ed.
        p. cm.
    Includes bibliographical references and index.
    ISBN-13: 978-0-321-58002-3 (alk. paper)
    ISBN-10: 0-321-58002-8 (alk. paper)
    1. Human geography.   I. Marston, Sallie A.   II. Title.
    GF41.K56 2010
    304.2—dc22

                                                    2008047098

Acquisitions Editor: Christian Botting
Editorial Project Manager: Tim Flem
Editor in Chief, Science: Nicole Folchetti
Editorial Assistant: Haifa Mahabir
Marketing Manager: Amy Porubsky
Managing Editor, Chemistry and Geosciences: Gina M.
    Cheselka
Production Project Manager: Wendy Perez
Art Director: Suzanne Behnke
Interior Design: 511 Design
Cover Design: Suzanne Behnke

Art Project Manager: Connie Long
Art Studio: Spatial Graphics
Senior Operations Supervisor: Alan Fischer
Senior Media Producer: Angela Bernhardt
Media Production Manager: Natasha Wolfe
Photo Research Manger: Elaine Soares
Photo Researcher: Clare Maxwell
Composition/Full Service: GGS Book Services, A Division of
    Premedia Global Inc.
Production Editor, Full Service: Karpagam Jagadeesan
Cover Photo Credit: Sajjad Hussain/AFP/Getty Images

Printed in the United States of America
10   9   8   7   6   5   4   3   2   1

ISBN-13: 978-0-321-58002-3

ISBN-10:   0-321-58002-8

Prentice Hall
is an imprint of

www.pearsonhighered.com

Brief Contents

Contents

CHAPTER 2   THE CHANGING GLOBAL CONTEXT   41

CHAPTER 7   THE GEOGRAPHY OF ECONOMIC DEVELOPMENT   249

CHAPTER 8   AGRICULTURE AND FOOD PRODUCTION   297

List of Maps

Preface

A nation, like a person, has a mind—a mind that must be kept informed and alert, that must know itself, that understands the hopes and needs of its neighbors—all the other nations that live within the narrowing circle of the world.

Franklin Roosevelt, *Third Inaugural Address,*
Monday, January 20, 1941

Most people have an understanding of what their own lives are like and a good deal about their own neighborhood and perhaps even something of the larger city and state in which they live. Yet, even as the countries and regions of the world become more interconnected, most of us still know very little about the lives of people on the other side of our country, or in other societies, or about the ways in which the lives of those people connect to our own. In order to change the world, to make it a better place in which to live for all people, we need to understand not just our little corner of it, but the whole of it, the broad sweep of human geography that constitutes the larger world of which our small corners are just a part.

This book provides an introduction to human geography that will help young men and women to understand critically the world in which they live. To study human geography, to put it simply, is to study the dynamic and complex relationships between peoples and the worlds they inhabit. Our book gives students the basic geographical tools and concepts needed to understand the complexity of places and regions and to appreciate the interconnections between their own lives and those of people in different parts of the world—to make the world a better place.

OBJECTIVE AND APPROACH

The objective of the book is to introduce the study of human geography by providing not only a body of knowledge about the creation of places and regions but also an understanding of the interdependence of places and regions in a globalizing world. The approach is aimed at establishing an intellectual foundation that will enable a lifelong and life-sustaining geographical imagination: an essential tool for today's students in order to confront tomorrow's global, national, regional, and local challenges.

The book takes a fresh approach to human geography, reflecting the major changes that have recently been impressed on global, regional, and local landscapes. These changes include the globalization of industry and the related rapid rise of China and India as economic powerhouses, the upwelling of ethnic regionalisms on the heels of decolonization and the formation of new states, the movement of peoples around the world in search of better lives, the physical restructuring of cities, the transformation of traditional agricultural practices throughout much of the world, global environmental change and the movement for sustainability, the eruptions of war and the struggles for peace, and the emerging trend toward transnational political and economic organizations. The approach used in *Places and Regions in Global Context* provides access not only to the new ideas, concepts, and theories that address these changes but also to the fundamentals of human geography: the principles, concepts, theoretical frameworks, and basic knowledge that are necessary to more specialized studies.

The most distinctive feature of this approach is that it emphasizes the interdependence of both places and processes in different parts of the globe. In overall terms, this approach is designed to provide an understanding of relationships between global processes and the local places in which they unfold. It follows that one of the chief organizing principles is how globalization frames the social and cultural construction of particular places and regions.

This approach has several advantages. For example:

- It captures aspects of human geography that are among the most compelling in the contemporary world—the geographical bases of cultural diversity and their impacts on everyday life.

- It encompasses the salient aspects of new emphases in academic human geography—the new emphasis on sustainability and its role in the social construction of spaces and places.

- It makes for an easier marriage between topical and regional material by emphasizing how processes link them—technological innovation and the varying ways technology is adopted and modified by people in particular places.

- It facilitates meaningful comparisons between places in different parts of the world—how the core-generated industrialization of agriculture shapes gender relations in households both in the core and the periphery.

In short, the textbook is designed to focus on geographical processes and to provide an understanding of the interdependence among places and regions without losing sight of their individuality and uniqueness.

Several important themes are woven into each chapter, integrating them into the overall approach:

- the relationship between global processes and their local manifestations;

- the interdependence of people and places, especially the interactive relationships between core regions and peripheral regions;
- the continuing transformation of the political economy of the world system, and of nations, regions, cities, and localities;
- the social and cultural differences that are embedded in human geographies (especially the differences that relate to race, ethnicity, gender, age, and class).

CHAPTER ORGANIZATION

The organization of the book is innovative in several ways. First, the chapters are organized so that the conceptual framework—why geography matters in a globalizing world—is laid out in Chapters 1 and 2 and then deployed in thematic chapters (Chapters 3 through 11). The concluding chapter, Chapter 12, provides a coherent summary of the main points of the text by showing how future geographies may unfold, given what is known about present geographical processes and trends. Second, the conceptual framework of the book requires the inclusion of two introductory chapters rather than the usual one. The first describes the basics of a geographic perspective; the second explains the value of the globalization approach.

Third, the distinctive chapter ordering within the book follows the logic of moving from less complex to more complex systems of human social and economic organization, always highlighting the interaction between people and the world around them. The first thematic chapter (Chapter 3) focuses on human population. Its early placement in the book reflects the central importance of people in understanding geography. Chapter 4 deals with the relationship between people and the environment as it is mediated by technology. This chapter explores environmental problems and establishes a central theme: that all human geographical issues are about how people negotiate their environment—whether the natural or the built environment.

The chapter on nature, society, and technology is followed by Chapter 5 on cultural geography. The intention in positioning the cultural chapter here is to signal that culture is the primary medium through which people operate and understand their place in the world. In Chapter 6, the impact of cultural processes on the landscape is explored, together with the ways in which landscape shapes cultural processes.

In Chapter 7, the book begins the move toward more complex concepts and systems of human organization by concentrating on economic development. The focus of Chapter 8 is agriculture. The placement of agriculture after economic development reflects the overall emphasis on globalization. This chapter shows how processes of globalization and economic development have led to the industrialization of agriculture at the expense of more traditional agricultural systems and practices.

The final three thematic chapters cover political geography (Chapter 9), urbanization (Chapter 10), and city structure (Chapter 11). Devoting two chapters to urban geography, rather than a more conventional single chapter, is an important indication of how globalization increasingly leads to urbanization of the world's people and places. The final chapter, on future geographies (Chapter 12), gives a sense of how a geographic perspective might be applied to the problems and opportunities to be faced in the twenty-first century.

Features

To signal the freshness of the approach, the pedagogy of the book employs a unique cartography program, three different boxed features—"Visualizing Geography," "Geography Matters," and "Window on the World"— as well as more familiar pedagogical devices such as chapter overviews and end-of-chapter exercises.

Geography Matters Geography Matters boxes examine one of the key concepts of the chapter, providing an extended example of its meaning and implications through both visual illustration and text. The Geography Matters features demonstrate to students that the focus of human geography is on real-world problems.

Visualizing Geography Visualizing Geography boxes treat key concepts of the chapter through photographic essays. This feature helps students recognize that the visual landscape contains readily accessible evidence about the impact of globalization on people and places.

Window on the World Window on the World boxes take a key concept and explore its application in a particular location. This feature allows students to appreciate the relevance of geographic concepts to world events and brings far-flung places closer to their comprehension.

Pedagogical Structure Within Chapters Each chapter opens with a brief vignette that introduces the theme of the chapter and illustrates why a geographical approach is important. A list of the main points that will be covered in the chapter follows this vignette. Throughout each chapter, key terms are printed in boldface as they are introduced. These key terms are listed alphabetically at the end of the chapter, together with their location in the text. Figures with extensive captions are provided to integrate illustration with text.

At the end of each chapter, there are five useful devices to help students review. First comes a conclusion that summarizes the overarching themes and concepts of the chapter. Next the main points of the chapter are listed again, but this time they are expanded to include a summary of the text discussion of each main point. Then there is a comprehensive list of key terms for the chapter, followed by a number of suggested additional readings on the topic of the chapter. Each chapter concludes with two sets of exercises, some Internet-based (On the Internet) and some more traditional (Unplugged). Both sets of exercises require students to put into practice several of the key concepts of a chapter.

NEW TO THE FIFTH EDITION

The fifth edition of *Places and Regions in Global Context* represents a thorough revision. Every part of the book was examined carefully with the dual goals of keeping topics and data current and improving the clarity of the text and the graphics. The fifth edition of the book incorporates a comprehensive updating of all of the data, maps, photographs, and illustrative examples. In the text, we have added or expanded upon quite a few topics, including the global financial crisis; the credit crunch and mortgage foreclosures; climate change and issues of sustainability; global interdependence and food supplies; contract farming; water supply problems; oil and energy; geopolitics; political ecology; religion; biotechnology and agricultural systems; fast food and slow food; virtual social networks; global film and music; big-box retailing and global commodity chains; the "experience economy"; and the landscapes of the polycentric metropolis. These changes are designed to ensure that we offer the most up-to-date coverage of the field of human geography.

The pedagogical structure of the chapters has not been modified in any substantial way over the fourth edition. The beginning of each chapter continues to feature a section on the main points that will be covered. These main points are revisited at the end of each chapter to reinforce the most important points and themes from the chapter. There are end-of-chapter exercises, and the Web exercises and the book's website have been significantly upgraded and expanded. Lastly, each chapter includes an updated list of suggested further readings.

SUPPLEMENTS

The book includes a complete supplements program for both students and teachers.

For the Student
Study Guide (0321615433) The study guide includes chapter notes, key terms, 8–10 review questions

and 10–15 activities per chapter, and references the text-specific Web site throughout.

Online Study Guide (www.mygeoscienceplace.com)
This innovative online resource center is organized by text chapter and follows a four-step process that prepares students for exams. The site features Key Concepts, chapter-specific Human Geography Videos, Concept Review questions, Thinking Spatially mapping exercises, Critical Thinking essay questions, and Web Links to support and enhance students' study of human geography.

myeBook (www.mygeoscienceplace.com) Integrated
with the Online Study Guide, myeBook is a full-featured electronic version of the book that allows students to add personal notes and highlights to the pages and see any comments made by the instructor. The search function allows for full-text searching of book content.

Goode's World Atlas Prentice Hall and Rand McNally are pleased to announce that Pearson Prentice Hall is distributing the *Goode's World Atlas*—the atlas most widely used by college professors. It features nearly 250 pages of maps, from definitive physical and political maps to thematic maps that illustrate the spatial aspects of many important topics. The 22nd edition includes 160 pages of new, digitally-produced reference maps, as well as new thematic maps on global climate change, sea level rise, CO_2 emissions, polar ice fluctuations, deforestation, extreme weather events, infectious diseases, water resources, and energy production. Pearson Prentice Hall offers the atlas at a dramatically reduced price with *Places and Regions in Global Context, Fifth Edition*. See your local Pearson representative for details.

For the Instructor
Instructor Resource Center on DVD-ROM (0321590139) *The Instructor Resource Center on DVD-ROM* provides high-quality electronic versions of photos and illustrations from the book, as well as customizable PowerPoint™ lecture presentations, Classroom Response System questions in PowerPoint, and the *Instructor Resource Manual* and *Test Bank* in MS Word and TestGen formats. The DVD-ROM includes all of the illustrations and photos from the text presentation-ready JPEG files. For easy reference and identification, all resources are organized by chapter.

Instructor Resource Manual Available as chapter-by-chapter MS Word files on the *Instructor Resource Center on DVD-ROM*, the *Instructor Resource Manual* is intended as a resource for both new and experienced teachers. The IRM includes a variety of lecture outlines, additional source materials, teaching tips, advice on how to integrate visual supplements, answers to the end-of-chapter exercises, and various other ideas for the classroom.

Test Bank   An extensive array of chapter-by-chapter multiple-choice test questions is available in MS Word and TestGen formats on the *Instructor Resource Center on DVD-ROM.*

Overhead Transparencies (0321590120)   The transparencies feature 150 illustrations from the text, all enlarged for excellent classroom visibility.

Online Course Management Systems   Pearson Prentice Hall offers content specific to *Places and Regions in Global Context, Fifth Edition,* ready for easy importation into the BlackBoard, WebCT, and CourseCompass course management system platforms. Each of these platforms lets you easily post your syllabus, communicate with students online or off-line, administer quizzes, and record student results and track their progress. Please contact your local Pearson Prentice Hall representative for details.

CONCLUSION

The idea for this book evolved from conversations between the authors and colleagues about how to teach human geography in colleges and universities. Our intent was to find a way not only to capture the exciting changes that are rewriting the world's landscapes and reorganizing the spatial relationships between people but also to demonstrate convincingly why the study of geography matters. Our aim was to show why a geographical imagination is important, how it can lead to an understanding of the world and its constituent places and regions, and how it has practical relevance in many spheres of life.

The first edition of this book was written at the culmination of a significant period of reform in geographic education. One important outcome of this reform was the inclusion of geography as a core subject in Goals 2000: Educate America Act (Public Law 103–227). Another was the publication of a set of national geography standards for K–12 education (*Geography for Life,* published by National Geographic Research and Education for the American Geographical Society, the Association of American Geographers, the National Council for Geographic Education, and the National Geographic Society, 1994). This fifth edition builds on these reforms, offering a fresh and compelling approach to college-level geography.

Acknowledgments

We are indebted to many people for their assistance, advice, and constructive criticism in the course of preparing this book. Among those who provided comments on various drafts of the four editions of this book are the following professors:

Christopher A. Airriess *(Ball State University)*
Stuart Aitken *(University of California at San Diego)*
Kevin Archer *(University of South Florida)*
Sarah Bednarz *(Texas A&M University)*
Brian J. L. Berry *(University of Texas at Dallas)*
Brian W. Blouet *(College of William and Mary)*
George O. Brown, Jr. *(Boston College)*
Michael P. Brown *(University of Washington)*
Henry W. Bullamore *(Frostburg State University)*
Edmunds V. Bunske *(University of Delaware)*
Craig Campbell *(Youngstown State University)*
Dylan Clark *(University of Colorado)*
David B. Cole *(University of Northern Colorado)*
Jerry Crampton *(George Mason University)*
Christine Dando *(University of Nebraska, Omaha)*
Fiona M. Davidson *(University of Arkansas)*
Benjamin Dixon *(SUNY College at Oneonta)*
Vernon Domingo *(Bridgewater State College)*
Patricia Ehrkamp *(Miami University)*
Nancy Ettlinger *(The Ohio State University)*
Paul B. Frederic *(University of Maine)*
Kurtis G. Fuelhart *(Shippensburg University)*
Gary Fuller *(University of Hawaii at Manoa)*
Wilbert Gesler *(University of North Carolina)*
Melissa Gilbert *(Temple University)*
Jeffrey Allman Gritzner *(University of Montana)*
David Gwynn *(Michigan State University)*
Douglas Heffington *(Middle Tennessee State University)*
Andrew Herod *(University of Georgia)*
Peter Hugill *(Texas A&M University)*
David Icenogle *(Auburn University)*
Mary Jacob *(Mount Holyoke College)*
Wendy Jepson *(Texas A&M University)*
Douglas L. Johnson *(Clark University)*
Colleen E. Keen *(Minnesota State University)*
Paul Kelley *(University of Nebraska, Lincoln)*
Thomas Klak *(Miami University)*
James Kus *(California State University, Fresno)*
David Lanegran *(Macalester College)*
James Lindberg *(University of Iowa)*
John C. Lowe *(George Washington University)*
James McCarthy *(Penn State University)*
John Milbauer *(Northeastern State University)*
Byron Miller *(University of Cincinnati)*
Roger Miller *(University of Minnesota)*
Don Mitchell *(Syracuse University)*
Wendy Mitteager *(SUNY College at Oneonta)*
Woodrow W. Nichols, Jr. *(North Carolina Central University)*
Richard Pillsbury *(Georgia State University)*
James Proctor *(University of California at Santa Barbara)*

Mark Purcell *(University of Washington)*
Jeffrey Richetto *(University of Alabama)*
Andrew Schoolmaster *(University of North Texas)*
David Schul *(The Ohio State University, Marion)*
Alex Standish *(Rutgers University)*
Debra Straussfogel *(University of New Hampshire)*
Johnathan Walker *(James Madison University)*
Gerald R. Webster *(University of Alabama)*
Joseph S. Wood *(George Mason University)*
Wilbur Zelinsky *(Penn State University)*

In addition, Michael Wishart (World Bank Photo Library), Earthaline Harried (Photography Division, U.S. Department of Agriculture), and researcher Clare Maxwell were especially helpful in our photo gathering. Special thanks go to our project manager, Tim Flem, and to Ian Shaw at the University of Arizona for his excellent, swift, and insightful research assistance, and his patience. We want to extend our deepest gratitude, however, to our former publisher, Daniel Kaveney, who was always supportive, always alive with new ideas, and always thoroughly engaged in the project from ideas to marketing.

Finally, a number of colleagues gave generously of their time and expertise in guiding our thoughts, making valuable suggestions, and providing materials: Alejandro A. Alonso (University of Southern California), Michael Bonine (University of Arizona), Michael Brown (University of Washington), Martin Cadwallader (University of Wisconsin), Judith Carney (University of California, Los Angeles), Sarah Elwood (University of Washington), Kim Elmore (Centers for Disease Control), Efiong Etuk (Virginia Tech), Antonio Luna Garcia (Universitas de Pompeu Fabra), Hannes Gerhardt (University of Arizona), John Paul Jones, III (University of Arizona), Miranda Joseph (University of Arizona), Cindi Katz (City University of New York), Diana Liverman (Oxford University), Elaine Mariolle (University of Arizona), Beth Mitchneck (University of Arizona), Mark Nichter (University of Arizona), Mimi Nichter (University of Arizona), Asli Ceylan Oner (Virginia Tech), Natalie Oswin (University of British Columbia), Gearóid Ó Tuathail (Virginia Tech), Lynn Patterson (Kennesaw State University), Leland Pederson (University of Arizona), Paul Robbins (University of Arizona), Dereka Rushbrook (University of Arizona), Neil Smith (City University of New York), Matthew Sparke (University of Washington), Dick Walker (University of California, Berkeley), Marv Waterstone (University of Arizona), Michael Watts (University of California, Berkeley), Clyde Woods (University of Maryland), and Keith Woodward (University of Arizona).

Paul L. Knox
Sallie A. Marston

About the Authors

Paul L. Knox

Paul Knox received his Ph.D. in Geography from the University of Sheffield, England. In 1985, after teaching in the United Kingdom for several years, he moved to the United States to take up a position as professor of urban affairs and planning at Virginia Tech. His teaching centers on urban and regional development, with an emphasis on comparative study. He has received the university's award for teaching excellence. He has written several books on aspects of economic geography, social geography, and urbanization, serves on the editorial board of several scientific journals, and is co-editor of a series of books on world cities. In 2008, Professor Knox received the Distinguished Scholarship Award of the Association of American Geographers. He is currently a University Distinguished Professor in the College of Architecture and Urban Studies at Virginia Tech.

Sallie A. Marston

Sallie Marston received her Ph.D. in Geography from the University of Colorado, Boulder. She is a full professor in the Department of Geography and Regional Development at the University of Arizona. Her undergraduate teaching focuses on political and cultural geography and the American urban landscape. She is the recipient of the College of Social and Behavioral Sciences' Outstanding Teaching Award and the co-editor of five books and author or co-author of over fifty journal articles and book chapters. She serves on the editorial board of several scientific journals. In 1994/1995, she served as Interim Director of Women's Studies and the Southwest Institute for Research on Women.

Geography Matters

1

In today's world, where places are increasingly interdependent, it is important to know something about human geography and to understand how places affect, and are affected by, one another. Consider, for example, some of the prominent news stories of 2008. At first glance, they were a mixture of achievements, disputes, and disasters that might seem to have little to do with geography, apart from the international flavor of the coverage. There was a major crisis among banking and financial institutions around the world, accompanied by precipitous drops in stock markets. There were demonstrations and riots over food shortages in Burkina Faso, Cameroon, Mexico, Mozambique, North Korea, and Pakistan; the price of oil fluctuated sharply, reaching almost $150 a barrel at one time; Tata Group, India's biggest industrial conglomerate, acquired Jaguar and Land Rover from Ford and launched the world's cheapest car, the Tata Nano, for the domestic market in India; meanwhile in the U.S. domestic market there was a surge in demand for vehicles powered by alternative energy sources such as biofuels and electricity. Meanwhile, dominating headlines around the world, there was a continual stream of reports on the wars in Iraq and Afghanistan. Off the front pages we read of the startling economic growth of China and India, of concerns over the effects of global warming, and of a rash of foreclosures on suburban homes in U.S. cities.

Most of these stories did have important geographical dimensions. Stories about economic development, regional territorial disputes, ethnic conflict, and global warming, for example, all have strong geographical elements. Look a little closer, and we find important linkages among many of them. The problem of food shortages in many countries, for example, is partly the result of increasing food consumption in other parts of the world, especially in booming China and India, where many have stopped growing their own food and now have the cash to buy a lot more of it. Increasing meat consumption has helped drive up demand for feed grain, and this in turn has driven up the price of grain everywhere. Speculators in international commodity markets have joined in, further accelerating price rises. Another key linkage concerns energy prices: The increase in oil prices has pushed up fertilizer prices, as well as the cost of trucking food from farms to local markets and shipping it abroad. The popularity of biofuels as an alternative to hydrocarbons is further straining food supplies, especially in the United States, where generous federal subsidies for ethanol have lured thousands of farmers away from growing crops for food. Then there is climate change. Harvests in many countries have been seriously disrupted by extreme weather (including, in 2007–2008, prolonged droughts in Australia and southern Africa, floods in West Africa, an exceptionally deep and widespread frost in China, and record-breaking warmth in northern Europe).

Human geography is about recognizing and understanding this kind of interdependence among places and regions, without losing sight of the uniqueness

Main Points

- Geography matters because it is specific places that provide the settings for people's daily lives. In these settings important events happen, and from them significant changes diffuse.

- Places and regions are highly *interdependent*, each playing specialized roles in complex and ever-changing networks of interaction and change.

- Some of the most important aspects of the interdependence between geographic scales are provided by the relationships between the global and the local.

- Human geography provides ways of understanding places, regions, and spatial relationships as the products of a series of interrelated forces that stem from nature, culture, and individual human action.

- The first law of geography is that "everything is related to everything else, but near things are more related than are distant things."

- Distance is one aspect of this law, but connectivity is also important, because contact and interaction are dependent on channels of communication and transportation.

Ho Chi Minh City, Vietnam, during Chinese New Year

1

of specific places. *Places* are specific geographic settings with distinctive physical, social, and cultural attributes. *Regions* are territories that encompass many places, all or most of which share attributes different from the attributes of places elsewhere. Basic tools and fundamental concepts enable geographers to study the world in this way. Geographers learn about the world by finding out where things are and why they are there. Maps and mapping, of course, play a key role in how geographers analyze and portray the world. Maps are also important tools for introducing geographers' ideas about the way that places and regions are made and altered.

WHY PLACES MATTER

An appreciation of the diversity and variety of peoples and places is a theme that runs through all of *human geography,* the study of the spatial organization of human activity and of people's relationships with their environments. This theme is inherently interesting to nearly all of us. *National Geographic* magazine has become a national institution by drawing on the wonder and fascination of the seemingly endless variety of landscapes and communities around the world. Almost 8 million households, representing about 30 million regular readers, subscribe to this magazine for its intriguing descriptions and striking photographs. Millions more read it occasionally in offices, lobbies, or waiting rooms.

Yet many Americans often seem content to confine their interest in geography to the pages of glossy magazines, to television documentaries, or to vacations. It has become part of the conventional wisdom—both in the United States and around the world—that many Americans have little real appreciation or understanding of people and places beyond their own daily routines. This is perhaps putting it too mildly. Surveys have revealed widespread ignorance among a high proportion of Americans, not only of the fundamentals of the world's geography but also of the diversity and variety within the United States itself. One Gallup poll found that 70 percent of young adults in the United States could not find their own country on a map of the world. In a survey of more than 3,000 young adults in Canada, France, Germany, Great Britain, Italy, Japan, Mexico, Sweden, and the United States in 2002, Americans came in next to last in terms of geographic literacy.[1] Top scorers were young adults in Sweden, Germany, and Italy. Neither wars nor natural disasters appear to have compelled majorities of young adults to absorb knowledge about international places in the news. A follow-up survey in 2006[2] found that over 60 percent of young Americans could not find Iraq on a map of the Middle East, despite near-constant news coverage since the U.S. invasion of March 2003. Three-quarters of young men and women do not know that a majority of Indonesia's population is Muslim (making it the largest Muslim country in the world), despite the prominence of this religion in global news today. Also striking is young Americans' ignorance of how the United States fits into the wider world. Majorities overestimate the total size of the U.S. population and fail to understand how much larger the population of China is. Three-quarters believe English is the most commonly spoken native language in the world, rather than Mandarin Chinese. This lack of geographical knowledge carries over to people's perceptions of the United States and its role in the world. A poll in 2004 found, for example, that most Americans believe that the United States spends more than 20 percent of its budget on aid to poor countries; the actual figure has consistently been less than 0.2 percent of the country's gross national income, putting it last among rich industrial nations. So although most people in the United States are fascinated by different places, relatively few have a systematic knowledge of them. Fewer still understand how different places came to be the way they are or why places matter in the broader scheme of things. This lack of understanding is important because geographic knowledge can take us far beyond a simple glimpse of the inherently interesting variety of peoples and places.

The importance of geography as a subject of study is becoming more widely recognized, however, as people everywhere struggle to understand a world that is increasingly characterized by instant global communications, rapidly changing international relationships, unexpected local changes, and growing evidence of environmental degradation. Many more schools now require courses in geography than just a decade ago, and the College Board has added the subject to its Advanced Placement program (**Figure 1.1**). Meanwhile, many employers are coming to realize the value of employees with expertise in geographical analysis and an understanding of the uniqueness, influence, and interdependence of places.

The Influence and Meaning of Places

Places are dynamic, with changing properties and fluid boundaries that are the product of the interplay of a wide variety of environmental and human factors. This dynamism and complexity is what makes places so fascinating for readers of *National Geographic.* It is also what makes places so important in shaping people's lives and in influencing the pace and direction of change. Places provide the settings for people's daily lives and their social relations (patterns of interaction among family members, at work, in social life, in leisure activities, and in political activity). It is in these settings that people

[1] Roper ASW, *2002 Global Geographic Literacy Survey.* Washington, DC: National Geographic–Roper ASW, 2002.

[2] Roper Public Affairs, *2006 Global Geographic Literacy Survey.* Washington, DC: National Geographic Education Foundation–Roper Public Affairs, 2006.

Figure 1.1 Geographic literacy High school students check on their exam results.

learn who and what they are, how they should think and behave, and what life is likely to hold for them.

Places also exert a strong influence, for better or worse, on people's physical well-being, opportunities, and lifestyle choices. Living in a small town dominated by petrochemical industries, for example, means a higher probability than elsewhere of being exposed to air and water pollution, having a limited range of job opportunities, and having a relatively narrow range of lifestyle options because of a lack of amenities such as theaters, specialized stores and restaurants, and recreational facilities. Living in a central neighborhood of a large metropolitan area, on the other hand, usually means having a wider range of job opportunities and a greater choice of lifestyle options because of the variety of amenities accessible within a short distance. But it also means, among other things, living with a relatively high exposure to crime.

Places also contribute to people's collective memory and become powerful emotional and cultural symbols. Think of the evocative power for most Americans of places like Times Square and the site of the former World Trade Center in New York; the Mall in Washington, D.C.; Hollywood Boulevard in Los Angeles; and Graceland in Memphis. And for many people, ordinary places have special meaning: a childhood neighborhood, a college campus, a baseball stadium, or a family vacation spot. This layering of meanings reflects the way that places are *socially constructed*—given different meanings by different groups for different purposes. Places exist and are constructed by their inhabitants from a subjective point of view. The meanings given to a place may be so strong that they become a central part of the identity of the people experiencing them. People's **identity** is the sense that they make of themselves through their subjective feelings based on their everyday experiences and wider social relations. At the same time, though, the same places will likely be constructed rather differently by

outsiders. Your own neighborhood, for example, centered on yourself and your home, is probably heavily laden with personal meaning and sentiment. But your neighborhood may well be viewed very differently, perhaps unsympathetically, by outsiders. This distinction is useful in considering the importance of understanding spaces and places from the viewpoint of the insider—the person who normally lives in and uses a particular place—as well as from the viewpoint of outsiders (including geographers).

Finally, places are the sites of innovation and change, of resistance and conflict. The unique characteristics of specific places can provide the preconditions for new agricultural practices (such as the development of seed agriculture and the use of plow and draft animals that sparked the first agricultural revolution in the Middle East in prehistoric times—see Chapter 8); new modes of economic organization (such as the high-tech revolution that began in Silicon Valley in the late twentieth century); new cultural practices (the punk movement that began in disadvantaged British housing projects, for instance); and new lifestyles (for example, the "hippie" lifestyle that began in San Francisco in the late 1960s). It is in specific locales that important events happen, and it is from them that significant changes spread.

Nevertheless, the influence of places is by no means limited to the occasional innovative change. Because of their distinctive characteristics, places always modify and sometimes resist the imprint of even the broadest economic, cultural, and political trends. Consider, for example, the way that a global cultural trend—rock 'n' roll—was modified in Jamaica to produce reggae, while in Iran and North Korea rock 'n' roll has been resisted by the authorities, with the result that it has acquired an altogether different kind of value and meaning for the citizens of those countries. Similarly, Indian communities in London developed Bhangra—a "world beat" composite of traditional Punjabi music

Figure 1.2 Argentine dance spreads to China Popular trends are easily spread around the world. Here, tango dancers dance in front of the North Gate to the Forbidden City in Beijing during a tango marathon popularizing South America's most famous dance, which originated in Argentina at the end of the nineteenth century.

Mumbai (Bombay) movie scores, and Western disco (**Figure 1.2**). Cross-fertilization with local music cultures in New York and Los Angeles has produced Bhangra rap.

To take a very different illustration, think of the ways some communities have declared themselves "nuclear free" zones: places where nuclear weapons and nuclear reactors are unwelcome or even banned by local laws. By establishing such zones, individual communities are seeking to challenge national trends toward using nuclear energy and maintaining nuclear arms. They are, to borrow a phrase, "thinking globally and acting locally." Similarly, some communities have established "GM-free" zones, taking a stance against genetically modified crops and food (**Figure 1.3**). In adopting such strategies, they hope to influence thinking in other communities so that eventually their challenge could result in a reversal of established trends.

In summary, places are settings for social interaction that, among other things,

- structure the daily routines of people's economic and social life;
- provide both opportunities and constraints in terms of people's long-term social well-being;
- provide a context in which everyday, commonsense knowledge and experience are gathered;
- provide a setting for processes of socialization; and
- provide an arena for contesting social norms.

The Interdependence of Places

Places, then, have an importance of their own. Yet at the same time most places are *interdependent*,

filling specialized roles in complex and ever-changing geographies. The social and economic relations that lend distinctiveness to individual places also operate between places. Some of the social relations that help to shape a particular place stretch beyond it. It is this stretching of social and economic relations across space that connects places and the people who live in them with other places and people.

Figure 1.3 Acting locally The town of Überlingen, Germany, has established itself as a "GM-free" zone. Shown here is Cornelia Wiethaler, who initiated the movement to ban genetically modified crops and food from the town.

Consider, for example, the way that Manhattan, New York, operates as a specialized global center of corporate management, business, and financial services while relying on thousands of other places to satisfy its needs. For labor it draws on analysts and managers from the nation's business schools, blue- and pink-collar workers from neighboring boroughs, and skilled professional immigrants from around the world. For food it draws on fruits and vegetables from Florida, dairy produce from upstate New York, and specialty foods from Europe, the Caribbean, and Asia. For energy it draws on coal from southwest Virginia to fuel its power stations. And for consumer goods it draws on specialized manufacturing settings all over the world.

This interdependence means that individual places are tied into wider processes of change that are reflected in broader geographical patterns. New York's attraction for business-school graduates, for example, is reflected in the overall pattern of migration flows that cumulatively affect the size and composition of labor markets around the country: New York's gain is somewhere else's loss. An important goal for human geographers—and a central theme of this book—is to recognize these wider processes and broad geographical patterns without losing sight of the uniqueness of specific places. This means that we have to recognize another kind of interdependence: the interdependence that exists *between different geographic scales*.

Geography and Interdependence

Different aspects of human geography are understood best, and analyzed most effectively, at different spatial levels. At the same time, these different aspects are interrelated and interdependent, so that geographers have to be able to relate things at one level to things at another.

It is useful to think of geographical levels as conceptual devices for thinking about how processes might play out across the globe or how they affect local settings. In this sense, these levels or scales represent a tangible partitioning of space within which different processes (economic, social, political, etc.) might occur. This partitioning, in turn, often consolidates dominant patterns of geographical organization, at least until some major disruptive change occurs. The Industrial Revolution, for example, changed not only the character of economic development (from agrarian to manufacturing) but also the level at which industrial production and consumption were organized (from local to national and international). At any particular moment we can identify a sequence of specific levels that represent significant confluences of geographical processes. In today's world, the large scale is represented by international, or world, regions, large but relatively homogeneous territories with distinctive economic, cultural, and demographic characteristics (**Figure 1.4**).

World regions are large-scale geographic divisions based on continental and physiographic settings that contain major groupings of peoples with broadly similar

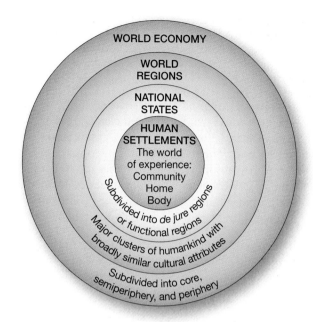

Figure 1.4 Spatial levels Geographic phenomena may be identified, analyzed, and understood at different levels. This diagram shows some of the principal ones that are commonly the focus of geographic research. (Adapted from S. Marston, P. Knox, and D. Liverman, *World Regions in Global Context*. Upper Saddle River, NJ: Prentice Hall, 2002, p. 18.)

cultural attributes. Examples include Europe, Latin America, and South Asia. These regions are constantly evolving as natural resources and technologies create opportunities and constraints to which particular cultures and societies respond.

Superimposed on these regions, sometimes with only approximate fit, are the *de jure* territories of national states. **States** are independent political units with territorial boundaries that are recognized by other states (see the discussion of states, nations, and national states in Chapter 9). *De jure* simply means legally recognized. Territories delimited by formal, legally recognized boundaries—national states, provinces, states, counties, municipalities, special districts, and so on—are known as *de jure* spaces or regions. Because of the inherent power of national governments, especially in relation to the flows of goods, money, and information that underpin reality, national states represent a geographic scale that is often very important. National states tend to be established to fit economic reality as closely as possible at the time of their foundation. Once their boundaries are set, however, principles of national sovereignty mean that these boundaries tend to become regarded by their inhabitants as natural and immutable. However, when economic circumstances change, national states may feel the need to adjust their boundaries or seek other means of accommodating themselves to economic reality, such as joining supranational organizations.

Supranational organizations are collections of individual states with a common goal that may be economic and/or political in nature and that diminish, to some extent, individual state sovereignty in favor of the group

1.1 GEOGRAPHY MATTERS

Making a Difference: The Power of Geography

The study of geography has become essential for understanding a world that is more complex and faster changing than ever before. Through an appreciation of the diversity and variety of the world's peoples and places, geography provides real opportunities not only for contributing to local, national, and global development but also for understanding and promoting multicultural, international, and feminist perspectives on the world.

The Importance of a Geographic Education

In the United States, a decade of debate about geographic education has resulted in a widespread acceptance that being literate in geography is essential in equipping citizens to earn a decent living, to enjoy the richness of life, and to participate responsibly in local, national, and international affairs. In response to the inclusion of geography as a core subject in the Goals 2000: Educate America Act (Public Law 103–227), a major report on the goals of geographic education was produced jointly by the American Geographical Society, the Association of American Geographers, the National Council for Geographic Education, and the National Geographic Society.[1] The report emphasizes the importance of being geographically informed; understanding that geography is the study of people, places, and environments from a spatial perspective; and appreciating the interdependent worlds in which we live:

> The power and beauty of geography allow us to see, understand, and appreciate the web of relationships between people, places, and environments.

> At the everyday level, for example, a geographically informed person can appreciate the locational dynamics of street vendors and pedestrian traffic or fast-food outlets and automobile traffic; the routing strategies of school buses in urban areas and of backpackers in wilderness areas; the land-use strategies of farmers and of real estate developers.

> At a more expanded spatial scale, that same person can appreciate the dynamic links between severe storms and property damage or between summer thunderstorms and flash floods; the use of irrigation systems to compensate for lack of precipitation . . . ; the seasonal movement of migrant laborers in search of work and of vacationers in search of sunshine and warmth.

At a global level, the geographically informed person can appreciate the connections between cyclical drought and human starvation in the Sahel or between the Chernobyl nuclear disaster and the long-term consequences to human health and economic activities throughout eastern and northwestern Europe; the restructuring of human migration and trade patterns as the European Union becomes increasingly integrated or as the Pacific Rim nations develop a commonality of economic and political interests; and the uncertainties associated with the possible effects of global warming on human society or the destruction of tropical rain forests on global climate.[2]

Most people want to understand the intrinsic nature of the world in which they live. Geography enables them to understand where they are both literally and figuratively. Geography provides knowledge of Earth's physical and human systems and of the interdependency of living things and physical environments. That knowledge, in turn, provides a basis for people to cooperate in the best interests of our planet. Geography also captures the imagination: It stimulates curiosity about the world and the world's diverse inhabitants and places, as well as about local, regional, and global issues. By better understanding the world, people can overcome parochialism and ethnocentrism. Last but not least, geography has utilitarian value in the modern world. As the interconnectedness of the world accelerates, the practical need for geographic knowledge becomes more critical. With a strong grasp of geography, people are better equipped to solve issues not only at the local level but also at the global level.[3]

Geographers at Work

Geography, then, is very much an applied discipline as well as a means of understanding the world. Geographers employed in business, industry, and government are able to use geographic theories and techniques to understand and solve a wide variety of specific problems. A great deal of the research undertaken by geography professors also has an applied focus. As a result, geography is able to make a direct and significant contribution to society. Because of the broad nature of the field, these contributions cover every aspect of human activity and every scale from the local to the global. A number of examples reflect this:

---

[1]Geography Education Standards Project, *Geography for Life: National Geography Standards 1994.* Washington, DC: National Geographic Research and Exploration, 1994.

[2]*Geography for Life,* p. 29.
[3]*Geography for Life,* pp. 23–24.

■ *International Affairs.* Geographers' knowledge and understanding of regional histories and geographies, along with their ability to analyze the interdependence of places and regions, enables them to effectively contribute to discussions of international policy. Geographers' work within governmental agencies, corporations, and nonprofit institutions in shaping international strategies is especially important in the context of accelerating globalization.

■ *Location of Public Facilities.* Geographers use specialized techniques to analyze the location patterns of particular population groups, to analyze transportation networks, and to analyze patterns of geographic accessibility to alternative sites. Such analysis enables geographers to determine the most effective locations for new public facilities, such as clinics, emergency rooms, social centers for the elderly, and shelters for the homeless.

■ *Marketing and Location of Industry.* Similar techniques are used in determining the most efficient, or most profitable, location for new factories, stores, and offices. Geographical research is also used to analyze the changing geography of supply and demand, allowing industry to determine whether, and where, to relocate. Basic techniques of geographic analysis are also used in geodemographic research, an important aspect of marketing.

■ *Geography and the Law.* Geographical analysis is increasingly required in helping to resolve complex social and environmental issues. One important example is the issue of racial segregation and the implications of residential segregation for policies such as school busing. Another example is the issue of property development and the implications of environmental hazards such as flooding, coastal erosion, toxic waste dumps, and earthquake fault zones for policies, codes, and regulations affecting development. A third is the issue of drawing up geographical boundaries for political units in ways that ensure equal representation by population size and racial and political composition—even as populations are changing between one election and another.

■ *Disease Ecology.* By analyzing social and environmental aspects of human diseases, geographers are able to shed light on the causes of disease, to predict the spread of particular outbreaks, and to suggest ways in which the incidence of disease might be controlled.

■ *Urban and Regional Planning.* Urban and regional planning adopts a systematic, creative approach to address and resolve physical, social, and economic problems of neighborhoods, cities, suburbs, metropolitan areas, and larger regions. Planners work directly on preserving and enhancing the quality of life in communities, protecting the environment, promoting equitable economic opportunity, and managing growth and change of all kinds. Planning has roots in engineering, law, architecture, social welfare, and government, but geography, because of its focus on the interdependence of peoples and places, offers the best preparation for specialized professional training in urban and regional planning.

■ *Economic Development.* Geographers' ability to understand the interdependence of places and to analyze the unique economic, environmental, cultural, and political attributes of specific regions enables them to contribute effectively to strategies and policies aimed at economic development. Geographers are involved in applied research and policy formulation concerning economic development all over the world, addressing the problems not only of individual places and regions but also of the entire world economy.

■ *Security.* Geographers' knowledge and understanding of geopolitics, political geography, demographics, medical geography, and cultural geography, together with an appreciation of the interdependent relationships among local, regional, and global systems, provide a sound basis for work in many areas of security. Knowing how places and regions "work" also means knowing about their vulnerability to potential security risks. A regional specialization may also be very relevant to certain occupations, while proficiency in using and interpreting GIS applications is fundamental to most aspects of security.

These are just a few examples of how geography is critical in today's world. The career choices for geography majors are diverse, challenging, and exciting. The single most popular choice for geography majors is, in fact, a career in marketing for retailing or industrial companies. Another popular choice is an administrative, managerial, or analytical post in local, state, or federal government. Most geography graduates are able to find careers in which they have the opportunity to make a positive contribution to the world through their analytical skills. These careers include cartography, geographic information systems (GIS), laboratory analysis, private consulting, urban and regional planning, international development, teaching, and management in private industry.

interests of the membership. Examples of supranational organizations include the European Union (EU), the North American Free Trade Association (NAFTA), and the Association of South East Asian Nations (ASEAN).

Within most national states and all international regions are smaller, functional regions. This geographical scale is constructed around specific resources and industries, with their networks of producers, suppliers, distributors, and ancillary activities and their associated social, cultural, and political identities. These are the classic functional regions of traditional regional geography: the American Corn Belt, the Argentine pampas, the Scottish coalfields, Japan's Pacific Corridor, and the Urals manufacturing region in Russia.

For most people, however, the realm of experience is encompassed by human settlements. This is the level that is constructed around the way people's lives are organized through their work, consumption, and recreation. It is also roughly coincident with another important aspect of geographical organization, *de jure* territories: local municipalities that provide the framework for public administration and the means for "collective consumption" of certain goods and services (public transport, education, public housing, recreational amenities, and so forth).

Within the realm of experience, there are other significant levels (Figure 1.4). Of these, the community is the most important but also the most difficult to pin down. It is the location of social interaction, of personal relationships and daily routine, that depends a great deal on the economic, social, and cultural attributes of local populations. Much more sharply defined is the home, which is an important geographic site insofar as it constitutes the physical setting for the structure and dynamics of family and household. It also reflects, in its spatial organization, the differential status accorded to men and women and to the young and the elderly.

Finally, the body and the self represent the most detailed level that geographers deal with. The body is of interest to geographers because it represents the intimate location where differences are ultimately defined—not only through physical attributes (hair, skin, facial features) but also through the socially constructed attributes of the body, such as norms of personal space, preferred bodily styles, and acceptable uses of bodies. Particularly important is the way that, in many cultures, the body is seen as less relevant to men. That is, men are regarded as able to transcend their bodies, while women are regarded as limited by bodily functions (for example, menstruation, pregnancy, and childbirth).

The result is that differential geographies are created and experienced by men and women—women's worlds and men's worlds. The self is of interest because it represents the operational scale for cognition, perception, imagination, free will, and behavior. The self has become an important scale of analysis for geographers because of the need to understand the interrelationships between nature, culture, and individual human agency in shaping places and regions.

Perhaps the most important conclusion that we can draw from this discussion is that while certain phenomena can be identified and understood best at specific spatial scales or levels, the reality of geography is that social, cultural, political, and economic phenomena are very fluid, constantly being constructed, reinforced, undermined, and rebuilt. Similarly, although certain scales represent manifestations of powerful real-world processes, the real world has to be understood, ultimately, as the product of interdependent phenomena at a variety of spatial scales.

In today's world some of the most important aspects of the interdependence between geographic levels of analysis are provided by the relationships between the *global* and the *local* scales. The study of human geography shows not only how global trends influence local outcomes but also how events in particular localities can influence patterns and trends elsewhere.

New York again can illustrate both sides of this relationship. In New York's stock exchanges and financial markets, brokers and clients must, in their own interest, take a global view. Their collective decisions influence stock prices, currency rates, and interest rates around the world, but they also often have very direct outcomes at the local level. Factories in certain localities may be closed and workers laid off because changed currency rates make their products too expensive to export successfully; elsewhere, new jobs may be created because the same change in currency rates gives a local economy an advantage within the global marketplace. On the other hand, local events can reverberate through New York's stock exchanges and financial markets with global effects. Political instability or natural disasters in a region that produces a key commodity, for instance, can result in changes in global pricing and affect the stock prices of many companies.

Interdependence as a Two-Way Process

One of the most important tenets of human geography is that places are not just distinctive outcomes of geographical processes; they are part of the processes themselves. Think of any city neighborhood, with its distinctive mix of buildings and people. This mix is the product of a combination of processes, including real estate development, the dynamics of the city's housing market, the successive occupancy of residential and commercial buildings by particular groups who move in and then out of the neighborhood, the services and upkeep provided by the city, and so on. Over time, these processes result in a distinctive physical environment with an equally distinctive population profile, social atmosphere, image, and reputation. Yet these neighborhood characteristics exert a strong influence in turn on the continuing processes of real estate redevelopment, housing market dynamics, and migration in and out of the neighborhood.

Places, then, are dynamic phenomena. They are created by people responding to the opportunities and constraints presented by their environments. As people live and work in places, they gradually impose themselves on their environment, modifying and adjusting it to suit their

> It is man who reveals a country's individuality by molding it to his own use. He establishes a connection between unrelated features, substituting for the random effects of local circumstances a systematic cooperation of forces. Only then does a country acquire a specific character, differentiating it from others, till at length it becomes, as it were, a medal struck in the likeness of a people.
>
> Paul Vidal de la Blache *Tableau de la Geographie de la France,* 1903, p. viii

Figure 1.5 Place-making People develop patterns of living that are attuned to the opportunities and constraints of the local physical environment. When this happens, distinctive regional landscapes are produced. These photographs and the map extract show part of the Chang Jiang (Yangtze) delta region of central China. In this watery region there has long been interplay between humans and the natural environment, and the result is a distinctive landscape with tens of thousands of kilometers of canals, irrigation ditches, and drainage channels, and thousands of ponds in rectilinear patterns. This "land of rice and fish" is extremely productive, the humid subtropical climate allowing for a triple-cropping system (two crops of rice, plus one of winter wheat or barley) in many locales.

needs and to express their values. At the same time, they gradually accommodate both to their physical environment and to the people around them. There is thus a continuous two-way process in which people create and modify places while at the same time being influenced by them (**Figure 1.5**).

Place-making is always incomplete and ongoing, and it occurs simultaneously at different scales. Processes of geographic change are constantly modifying and reshaping places, and places are constantly coping with change. It is important for geographers to be sensitive to this kind of interdependence without falling into the trap of over-generalization, or losing sight of the diversity and variety that constitute the heart of human geography. It is equally important not to fall into the trap of *singularity,* or treating places and regions as separate entities, the focus of study in and of themselves.

INTERDEPENDENCE IN A GLOBALIZING WORLD

As a subject of scientific observation and study, geography has made important contributions both to the understanding of the world and to its development. As we move further into the Information Age, geography continues to contribute to the understanding of a world that is more complex and fast-changing than ever before. With such an understanding, it is possible not only to appreciate the diversity and variety of the world's peoples and places but also to be aware of their relationships to one another and to be able to make positive contributions to local, national, and global development.

Today, in a world that is experiencing rapid changes in economic, cultural, and political life, geographic knowledge is especially important and useful. In a fast-changing

The Global Credit Crunch

The failure of several major private financial institutions in October 2008 prompted panic among international financial markets. Suddenly, the world economy was facing the prospect of recession, and millions of households in affluent countries were facing the very real prospect of the widespread loss of jobs, savings, and pension funds. The government of the United States, along with those of other leading economies, intervened with hundreds of billions of dollars of support for private financial institutions in an attempt to prop up the international financial system and, with it, their national, regional, and local economies. In several cases, including the United States, this involved the partial nationalization of financial institutions: a dramatic reversal of the neoliberal philosophy that had come to dominate the political economy of most countries over the previous 30 years.

How could this have happened? Part of the explanation lies in the steady increase in debt that had been fuelling every aspect of the world economy and, in particular, the American economy, since the late 1970s. Consumer spending had been increasingly financed by credit card debt; a housing boom had been financed by an expanded and aggressive mortgage market; and wars in Iraq and Afghanistan had been financed by U.S. government borrowing from overseas. By mid-2008, private debt in America had reached $41 trillion, almost three times the country's annual Gross Domestic Product; the external debts of the United States meanwhile had reached $13.7 trillion. All sorts of financial instruments had emerged in the speculative free-market climate engendered by neoliberalism: securitization, derivatives, hedge funds, collateralized debt obligations, mortgage-backed securities, and so on. Soon, banks and financial institutions in other countries were

joining in, profiting immensely from a global credit binge. Everyone, it seemed, was borrowing from everyone else in an international financial system that had become extremely complex, increasingly leveraged and decreasingly regulated.

Taking advantage of the relaxed controls on U.S. financial institutions resulting from neoliberal policies, mortgage lenders had been selling their mortgages on bond markets and to investment banks in order to fund the soaring demand for housing. But this also led to abuses, as mortgage lenders no longer had an incentive to check carefully on borrowers. Many lenders began to offer mortgages that included "subprime" lending to borrowers with poor credit histories and weak documentation of income. But it appears that few in the financial services industries fully understood the complexities of the booming mortgage market, while the various risk-assessment agencies were seriously at fault in underestimating the risks associated with these loans. Eventually, when interest rates increased, many households began to default on their monthly payments and, as the bad loans added up, mortgage lenders found themselves, in turn, in financial trouble. The first signs of the crisis came in 2007, when credit losses associated with the "subprime" mortgage business in the United States led to several large investment firms getting into difficulty.

Because the entire world economy has become so interdependent, the problem quickly spread. The internationalization of finance (see Chapter 2) meant that the banks in other countries were caught up in the complex web of loans, and by the autumn of 2008 the banking industry had been thrown into crisis worldwide. As credit markets seized up, manufacturers and other businesses found it difficult to get the credit they

world, when our fortunes and ideas are increasingly bound up with those of other peoples in other places, the study of geography provides an understanding of the crucial interdependencies that underpin everyone's lives. One of the central themes throughout this book is the *interdependence* of people and places.

Another central theme of this book is **globalization.** Globalization is the increasing interconnectedness of different parts of the world through common processes of economic, environmental, political, and cultural change. A world economy has been in existence for several centuries, and with it there has developed a comprehensive framework of sovereign national states and an international system of production and exchange. This system has been reorganized several times. Each time it has been reorganized, however, major changes have resulted, not

only in world geography but also in the character and fortunes of individual places.

Recently, there has been a pronounced change in both the pace and the nature of globalization, leading to a highly interdependent world. New telecommunication technologies, corporate strategies, and institutional frameworks have combined to create a dynamic new framework for real-world geographies. New information technologies have helped create a frenetic international financial system, while transnational corporations are now able to transfer their production activities from one part of the world to another in response to changing market conditions and changing transportation and communications technologies (see Chapter 7). This locational flexibility has meant that a high degree of functional integration now exists between economic activities that are increasingly dispersed so that needed in order to keep going. Understandably, investors big and small were shaken; stock markets collapsed (**Figure 1.A**). Consumer confidence also plummeted, prompting retailers to cut back on orders. The sophisticated flexible production system of global commodity chains (see Chapter 7) meant that the impact was almost instantaneous.

Within days of the U.S. stock-market crash in mid-October 2008, workers in Chinese toy factories found themselves unemployed, along with garment workers in Thailand. Meanwhile, in Iceland, the entire economy collapsed almost immediately, a result of its banks having been overextended in their search for profit from the international credit binge. When the Icelandic banks collapsed, their overseas clients—local authorities, trusts, charities, and nonprofit organizations as well as small corporate investors, mostly in Britain, Germany, and other European countries—collectively lost billions of dollars of savings and investments. Similar domino effects of the financial crisis were taking place around the world. European governments led a coordinated attempt to prop up the global financial system and provide a continuing flow of the credit needed to keep the global economy from falling into a deep depression.

The long-term success of these efforts remains to be seen, but it is clear that a major transition is imminent, not just in the organization of international finance and the operation of the global economy, but also in the broader political economy of Western countries, where decades of neoliberal policies have come to an abrupt halt with central governments having to nationalize corporate assets, and where there remains a great deal of grassroots resentment at the bailout of financial corporations while ordinary households have lost savings and had homes repossessed.

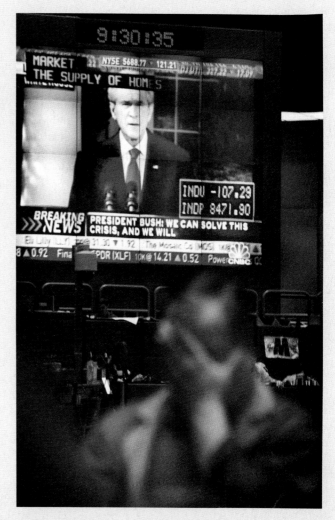

Figure 1.A A trader in the Dow Jones Industrial Average stock index futures pit at the Chicago Board of Trade reacts to trading as a monitor above the trading floor broadcasts U.S. President George W. Bush addressing the nation about the financial crisis October 10, 2008, in Chicago, Illinois.

products, markets, and organizations are both spread and linked across the globe. Governments, in their attempts to adjust to this new situation, have sought new ways of dealing with the consequences of globalization, including new international political and economic alliances such as NAFTA and the European Union (see Chapter 9).

The interdependence associated with globalization operates in a multitude of ways. In many cases, interdependence seems very unequal in nature—as in the example of a transnational firm based in one country taking advantage of low-cost labor in another. In some cases, interdependence can be seen to be to mutual advantage—as in countries that share the costs and responsibilities of transborder resource management. In almost every case, however, the outcomes of the increased geographic interdependence associated with globalization are very much open to interpretation.

Who "wins" and who "loses" depends very much on one's perspective and the geographic scale.

Perspectives on Globalization and Interdependence

An important aspect of globalization is the widespread perception that the world, through economic and technological forces, is increasingly becoming one shared political and economic space, with events in one region having repercussions for all others, whether near or far. Given that economic and technological forces are breaking down the barriers within and between near and distant places, will the most recent phase of globalization strengthen some regional connections and weaken others, or make regions

11

altogether irrelevant? Alternatively, will globalization enable some regions—core regions, for instance—to create even greater differences of wealth and power than already exist in the world system? Understanding what the experts believe about globalization will help us to get a better understanding of the complex interdependence between the global and the local.

The number of books published on globalization and its impact on the world's regions has grown at a tremendous rate over the last 15 years. Check out the shelves in any good bookstore and you will find hundreds of them, with whole sections devoted to globalization within the categories of political science, sociology, geography, economics, media studies, and business management. It is possible to group the main participants in the contemporary debates about globalization into three general camps: the hyperglobalists, the skeptics, and the transformationalists. Although these three viewpoints do not exhaust the range of the debates, they do provide a clear sense of the issues about which the experts on globalization agree and disagree.

The Hyperglobalist View

At one extreme is the view that open markets and free trade and investment across global markets allow more and more people to share in the prosperity of a growing world economy. Economic and political interdependence, meanwhile, creates shared interests that help prevent conflict and foster support for common values. Democracy and human rights, it is asserted, will spread to billions of people in the wake of neoliberal policies that promote open markets and free trade.

Neoliberal policies are economic policies that are predicated on a minimalist role for the state, assuming the desirability of free markets as the ideal condition not only for economic organization but also for political and social life. Hyperglobalists believe that the current phase of globalization signals the beginning of the end for the nation-state and the "denationalization" of economies. By economic denationalization they mean that national boundaries will become irrelevant with respect to economic processes and that national governments will not control their once geographically bounded economies but will instead facilitate connections among and between different parts of the world through supranational organizations such as NAFTA and the EU.

The wider implication of the hyperglobalist position is that the world will become borderless as national governments become increasingly meaningless or function merely as facilitators of global capital flows and investments. Hyperglobalizers believe that the nation-state, the primary political and economic unit of contemporary world society, will eventually be replaced by institutions of global governance in which individuals claim transnational allegiances that are founded upon a commitment to neoliberal principles of free trade and economic integration. Politically, the global spread of liberal democracy will reinforce the emergence of a global civilization with its own mechanisms of global governance, replacing the outmoded nation-state with global institutions like the International Monetary Fund (IMF) and the World Trade Organization (WTO).

The Skeptical View

A second broad argument within the globalization literature belongs to the skeptics, those who believe that contemporary levels of global economic integration represent nothing particularly new and that much of the talk about globalization is exaggerated. The skeptics look to the nineteenth century and draw on statistical evidence of world flows of trade, labor, and investment to fortify their position. They argue that contemporary economic integration is actually much less significant than it was in the late nineteenth century, when nearly all countries shared a common monetary system known as the gold standard. The skeptics are also dismissive of the idea that the nation-state is in decline. They argue that national governments are essential to the regulation of international economic activity and that the continuing liberalization of the world economy can only be facilitated by the regulatory power of national governments.

The skeptics assert that their analysis of nineteenth-century economic patterns demonstrates that we are today witnessing not globalization, but, rather, "regionalization," as the world economy is increasingly dominated by three major regional financial and trading blocs: Europe, North America, and East Asia (effectively, Japan). The skeptics understand regionalization and globalization to be contradictory tendencies. They believe that because of the dominance of these three major regional blocs, the world is actually less integrated than it once was, because Europe, North America, and Japan control the world economy and limit the participation of other regions in that economy.

The Transformationalist View

According to the transformationalist view, contemporary processes of globalization are historically unprecedented, as governments and peoples across the globe confront the absence of any clear distinction between the global and the local, between domestic affairs and international affairs. Like the hyperglobalists, this group understands globalization as a profound transformative force that is changing societies, economies, institutions of government—in short, the world order. In contrast to the hyperglobalists and the skeptics, however, the transformationalists make no claims about the future trajectory of globalization, nor do they see present globalization as a pale version of a more "globalized" nineteenth century. Instead, they see globalization as a long-term historical process that is underlain by crises and contradictions that are likely to shape it in all sorts of unpredictable ways. Moreover, unlike the skeptics, the transformationalists believe that the historically unprecedented contemporary patterns of economic, military, technological, ecological, migratory, political, and cultural flows have functionally linked all parts of the world into a larger global system in which countries and regions are increasingly interdependent

In this book we adhere generally to the transformationalist position. We suggest that we are all heading toward a world where places and regions will experience a wide range of internal changes at the same time that the strength of their connections with other parts of the world will increase. The complex interdependence of the contemporary world was underscored by the global financial crisis of October 2008 (see Box 1.2, "The Global Credit Crunch"), though it remains to be seen just how the ramifications of the crisis will affect individual places and regions.

What is perhaps most unsettling about the transformationalist view of globalization is the anticipated increase in disparities in wealth. Transformationalists believe that globalization is leading to increasing social stratification, in which some states and societies are more tightly connected to the global order while others are becoming increasingly marginalized. They contend that there is no evidence to sustain the hyperglobalist claim that the new global social structure is tending toward a global civilization where equality among individuals will eventually prevail. They argue the opposite: that the world will increasingly consist of a three-tiered system—comprising the elites, the embattled, and the marginalized—that cuts across national, regional, and local boundaries (see Chapter 12). Within nations, disparities of wealth—already striking in many countries—will increase, just as they will between nations (see Box 1.3, "Worlds Apart"). Meanwhile, the increasing interdependence among places and regions raises some key issues in relation to the environment, health, and security.

Key Issues in a Globalizing World

Environmental Issues   The sheer scale and capacity of the world economy means that humans are now capable of altering the environment at the global scale. The "footprint" of humankind extends to more than four-fifths of Earth's surface (**Figure 1.6**). Many of the important issues facing modern society are the consequences—intended and unintended—of human modifications of the physical environment.

Humans have altered the balance of nature in ways that have brought economic prosperity to some areas and created environmental dilemmas and crises in others. For example, clearing land for settlement, mining, and agriculture provides livelihoods and homes for some but alters physical systems and transforms human populations, wildlife, and vegetation. The inevitable by-products—garbage, air

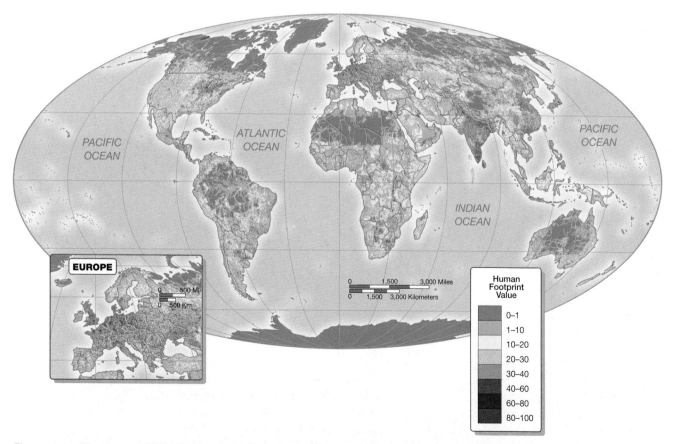

Figure 1.6   The human "footprint"   This map, prepared by a team of scientists from the New York–based Wildlife Conservation Society and Columbia University's Center for International Science Information Network (CIESIN), shows the extent and intensity of human influence on the land, reflecting population density, agricultural use, access from roads and waterways, electrical power infrastructure, and urbanization. The lower the number, the lesser the overall degree of human influence. (*Source:* www.wcs.org/humanfootprint/)

Worlds Apart

Meet Paul Rust and his family, who live and enjoy life in Zug, Switzerland, the richest canton in the world's richest country. And meet Hussein Sormolo and his family, who live in Addis Ababa, capital of the world's poorest country, Ethiopia.

Hussein Sormolo left the village where he was born for the big city in 1978. He left his eight brothers and seven sisters behind, as the land that the family farmed was being forcibly collectivized by a new regime. Hussein, aged 16, traveled the 100 miles north to the city in the back of a truck. A kinsman from the same village took him in until he found a job in a bakery. Paul Rust left his village when he was 17 and also ended up in a bakery. The two men are similar in other ways. Both are friendly, hospitable, and generous and love their families. Both work hard. Both like to watch the news. Both are active worshippers, without being religious dogmatists.

Yet their lives are different. Hussein lives with his wife, sons, and daughters in a leaky shack of corrugated asbestos and steel in the Nefas Silk district of Addis Ababa (**Figure 1.B**). Paul lives with his wife in a six-room house (not counting the ground-floor apartment where his son Martin lives with his girlfriend) overlooking the lustrous green waters and steep wooded slopes of Lake Aegeri in Zug (**Figure 1.C**).

The income difference is huge, of course. Hussein supports his wife and three younger children on wages of about $280 a year (more than twice the average income in Ethiopia). Paul and his wife, Hedi, draw about $68,000 between them each year from their bakery, though the Rusts are not affluent by Swiss standards (the average income per head in Zug is about $50,000).

It is the rainy season in Addis. Fat raindrops drum against, and often through, the rusting grooves of the corrugated roofs of the houses in Nefas Silk. Nights can be chilly and dank. From Debre Zeit road, the busy main north–south street lined with small businesses including the bakery where Hussein works, it's a 10-minute walk to the alley where he lives. Inside the Hussein shack, a single bare light bulb always burns. There is little natural light: There are no glass windows, and the openings punched in the asbestos walls are covered to keep out drafts. Hussein pays his neighbor 18 birr (about $2) a month, almost a tenth of his 200-birr salary, to sublet his electricity supply for the bulb in the shack. The family has no other electrical appliances, apart from a battery-operated radio. Neither Hussein, his wife Rukia, nor his eldest daughter, Fate, 17, who is lucky enough to be at school, has ever used a computer, taken a photograph, or made a phone call. Hussein and Rukia have a pair of shoes each. They buy new ones every two years. They have no savings and the family doesn't take holidays.

Except for feast days, the family eats the same dish every meal—a grey, spongy, limp bread called injera,

Figure 1.B   Hussein Sormolo and his family in Addis Ababa, Ethiopia.

spread out like a cloth, and a spicy vegetable stew. Meat, fish, cheese, and eggs are luxuries. They only buy fruit when one of the children is sick. Just under a quarter of the family income is spent on cooking charcoal and cans of water. In a country where only a quarter of the people in the countryside have access to safe drinking water, Hussein's family is lucky. There is a standpipe around the corner with reasonably clean water. That's about where their luck ends. With their neighbors, they used to have a toilet for 26 people. Now they have no toilet at all.

The Rust house, not counting the apartment, has three toilets, one each in the bathroom and two shower rooms of the four-story building. On the balconies under its broad, dark, solid eaves are cascades of red flowers. The well-used furnishings inside are not ostentatious, but the building is roomy and comfortable. From the top, there is a loft, four bedrooms, two living rooms, a kitchen, an office, a small wine cellar, a work room, garage parking for three cars (Paul, Hedi, and Martin Rust have one each) with room for another five on the forecourt. The house has its own elevator.

Paul and Hedi are going on vacation for two weeks in Austria this month, and usually take a week at Easter. Each has a mobile phone. The home office has computers and Internet access. They have a TV, a VCR, and a dishwasher. They eat what they want, although their tastes are plain—meat with several vegetables, salad, sometimes a little wine.

Switzerland is a rich country landlocked by other rich countries. Ethiopia is a poor country landlocked by other poor countries. Unlike other African nations, Ethiopia was not a European colony, but its people have endured regular European military incursions, proxy superpower duels, and local wars that have exacerbated the ravages of famine and disease. Famines in the 1970s, 1980s, and 1990s killed 1.3 million people. Through the 1970s and 1980s, the country was embroiled in ideological and ethnic civil war. HIV/AIDS has infected 3 million Ethiopians and kills 300,000 a year.

Hussein knows little about Switzerland. "I heard about Switzerland on the radio but I don't know. I heard it was a rich country, they help poor countries," he said.

Paul thought he could find Ethiopia on the map. Switzerland is not as aloof from the world as it was, he points out: They joined the boycott of apartheid South Africa. He said his brother helped build a dairy in Nepal 20 years ago. His church had adopted a village in Romania, giving it money for a new church and a school. The talk turned to immigration. "The really poor people, they can't come to Switzerland, they need money to get here," said Andrea. "We work, and have our life, we have our own problems," said Andrea. "So we don't think very often of other people's problems. It's a little bit selfish."

Based on an article by James Meek. *The Guardian*, 22 August 2002.

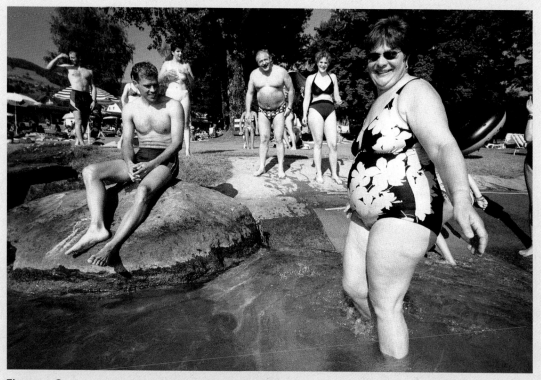

Figure 1.C Paul Rust and his family in Zug, Switzerland.

and water pollution, hazardous wastes, and so forth—place enormous demands on the capacity of physical systems to absorb and accommodate them.

Climate change as a result of human activity—in particular, the burning of fossil fuels, agriculture, and deforestation that cause emissions of carbon dioxide (CO_2) and other "greenhouse" gases—also has profound implications for environmental quality. Without concerted action to reduce greenhouse gas emissions, the global average surface temperature is likely to rise by a further 1.8–4.0°C this century. Even the lower end of this range would take the temperature increase since preindustrial times above 2°C, the threshold beyond which irreversible and possibly catastrophic changes become far more likely. Projected global warming this century is likely to trigger serious consequences for humanity and other life forms, including a rise in sea levels of between 18 and 59 centimeters, which will endanger coastal areas and small islands, and a greater frequency and severity of extreme weather events.

In addition to the specter of global warming, we also face the reality of serious global environmental degradation through deforestation, desertification, acid rain, loss of genetic diversity, smog, soil erosion, groundwater decline, and the pollution of rivers, lakes, and oceans.

Lake Baykal, in Russia (**Figure 1.7**), provides a salutary example. It is a place of incredible beauty—"the Pearl of Siberia"—that has long been emblematic of the pristine wilderness of the region. The lake holds 20 percent of the world's freshwater and is home to 2,500 species, many of them found nowhere else, such as the world's only exclusively freshwater seal.

The lake has warmed 1.21°C (2.18°F) since 1946 due to climate change, almost 3 times faster than global air temperatures. The lake's purity and unique ecosystem have meanwhile been compromised by environmental mismanagement. The first evidence of this was in the 1950s, when the lake's commercial fish populations nose-dived, partly as a result of overfishing and partly as a result of the construction of the Irkutsk dam (which raised the level of the lake and destroyed many of the shallow-water feeding grounds used by the fish). Then, in the 1960s, increasing levels of pollution were carried into the lake by the Selenga River, which supplies about half of the water that flows into the lake. The Selenga rises in mountain ranges to the south but collects human and industrial waste, much of it untreated, from several large cities before entering Lake Baykal. The Selenga and other rivers also began to carry increasing amounts of agricultural chemicals, such as DDT and PCBs into the lake.

Meanwhile, the purity of the lake's waters caught the attention of Soviet economic planners, who saw the lake as a good location for factories that needed plentiful supplies of pure water. The huge Baikalsk Pulp & Paper Mill was opened in the early 1960s to produce high-quality cellulose for the Russian defense industry. The mill pumps 140,000 tons of waste—including deadly dioxins—into the lake every day, along with 23 tons of pollutants into the atmosphere; over the past 40 years, the mill has spewed over a billion tons of waste into the lake. Since 1989 the mill has been partially privatized and now makes pulp for low-quality paper rather than cellulose. When thousands of the lake's freshwater seals began dying in 1997, the lake's fragile ecology came under international scrutiny, and in 1998 the lake was designated a World Heritage Site by UNESCO, the U.N. cultural agency. Nevertheless, it remains to be seen whether Russia can solve its environmental problems at a time when its economy is in disarray.

Environmental issues such as these point to the importance of sustainability. **Sustainability** is about the interdependence of the economy, the environment, and social well-being. This is often couched in terms of the "three Es" of sustainable development, referring to the environment, the economy, and equity in society (**Figure 1.8**). The oft-quoted definition of sustainable development from the Brundtland

Figure 1.7 Lake Baykal Lake Baykal is the world's deepest lake, at 1,615 meters (5,300 feet—over a mile), and contains about 20 percent of all the freshwater on Earth—more than North America's five Great Lakes combined. Lake Baykal is also an unusually ancient lake. Most of the world's large lakes are less than 20,000 years old, but evidence from the 7-kilometer-thick sediment at the bottom of the lake shows it to have been in existence for at least 25 million years, perhaps even 50 million. It has a unique ecology, with over 2,500 recorded plant and animal species, 75 percent of which are found nowhere else. These include the nerpa, Baykal's freshwater seal, separated by more than 3,000 kilometers (1,864 miles) from its nearest relative, the Arctic ringed seal. Ecologists have no understanding of how seals ever got to the lake or how they adapted to freshwater. Although Lake Baykal is now on UNESCO's list of World Natural Heritage sites, pollution from numerous paper mills along the lake shore continues to worsen.

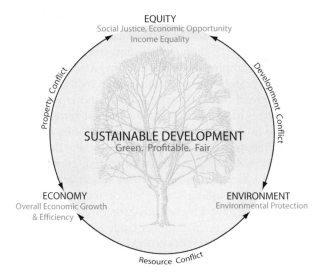

EQUITY
Social Justice, Economic Opportunity
Income Equality

Property Conflict

Development Conflict

SUSTAINABLE DEVELOPMENT
Green. Profitable. Fair

ECONOMY
Overall Economic Growth
& Efficiency

ENVIRONMENT
Environmental Protection

Resource Conflict

Figure 1.8   Sustainability There are thee key aspects of sustainability in the long run—the physical environment, equity, and efficiency—but there are also tensions between each of these.

Report, which examined the issues at the international scale, is "development that meets the needs of the present without compromising the ability of future generations to meet their own needs."[3] In fact, this is a relatively small part of the re-

[3]World Commision on Environment Development, *Our Common Future* (Bruntland Report), Oxford, Oxford University Press, 1987, p. 40.

port's concept of sustainability, which extends to reviving economic growth; meeting essential needs for jobs, food, energy, water, and sanitation; ensuring a balance between population and resources; conserving and enhancing the resource base; reorienting technology and managing risk; merging the environment and economics in decision making; and reorienting international economic relations.

Health Issues The increased intensity of international trade and travel has also heightened the risk and speed of the spread of disease. An example of the health risks associated with increasing interdependency was the outbreak of Severe Acute Respiratory Syndrome (SARS) in China in November 2002, which quickly spread around the world, causing widespread panic and serious disruption to business and tourism in East and Southeast Asia (**Figure 1.9**). By the time the outbreak had been contained four months later, almost 3,000 cases (including over 1,400 deaths) had been recorded, some as far afield as Brazil, Canada, Ireland, Romania, Spain, South Africa, Switzerland, the United Kingdom, and the United States.

Health care professionals are now concerned that a new strain of influenza virus is likely to result in an influenza pandemic. A **pandemic** is an epidemic that spreads rapidly around the world with high rates of illness and death. Although people are exposed to different strains of the influenza virus every year, entirely new flu strains develop several times each century. Because no one has had a chance to develop immunity to the new flu

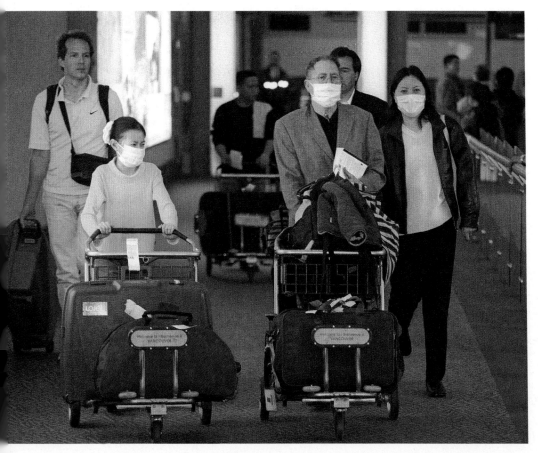

Figure 1.9 The outbreak of Severe Acute Respiratory Syndrome (SARS) in the winter of 2002 caused widespread disruption to international patterns of business and tourism for several months, underlining the increasing vulnerability of populations to disease epidemics as a result of global interdependency.

strain, it can spread rapidly and widely—and especially so in today's globalized and highly interconnected world. Similarly, there is serious concern about the possibility of epidemics in the human population resulting from zoonotic diseases (diseases originating with other species, e.g., anthrax, avian flu, ebola, West Nile virus). The most significant international health issue so far, however, has been the spread of HIV/AIDS (human immunodeficiency virus or acquired immunodeficiency syndrome).

Over the past quarter century, HIV/AIDS quickly spread around the world from a single hearth area. Medical geographers have concluded that the human immunodeficiency virus (HIV), which causes acquired immunodeficiency syndrome, or AIDS, spread in a hierarchical diffusion pattern from a hearth area in Central Africa in the late 1970s (**Figure 1.10**). The virus initially appeared almost simultaneously in the major metropolitan areas of North and South America, the Caribbean, and Europe. These areas then acted as localized diffusion poles for the virus, which next spread to major metropolitan areas in Asia and Oceania and to larger provincial cities in North and South America, the Caribbean, and Europe. Next in this cascading pattern of diffusion were provincial cities in Asia and Oceania and small towns in North and South America, the Caribbean, and Europe.

Today, sub-Saharan Africa is more severely affected by HIV/AIDS than any other part of the world, with the United Nations reporting between 25.0 and 28.2 million people infected in 2003—between 50 and 75 percent of the worldwide total (**Figure 1.11**). The infection rate is estimated at 8 percent of all adults compared with a 1 percent world rate, and more than 15 million Africans have lost their lives to HIV/AIDS since it was identified in 1981. It has become the main cause of death in Africa, killing more people than malaria and warfare combined.

The geography of HIV/AIDS in Africa varies by country, by regions within countries, and by social groups, and the spread of the disease is linked in many ways to the processes of globalization. The highest rates of infection were in eastern Africa in the early 1980s but have now shifted to southern Africa, especially Botswana, Zambia, and Zimbabwe, where more than 20 percent of adults are infected. Urban dwellers who have multiple sex partners, including young office workers and migrant workers, have a higher infection rate, as do women who work in the commercial sex trade and the wives and children of migrant workers. Migrant workers have taken the disease back to their home areas. The incidence is generally lower in rural areas, except along major truck routes and in areas where there are a lot of soldiers.

The deaths of skilled farm laborers have resulted in a decline in agricultural output, and many young professionals critical to the region's future have left their jobs because they have contracted HIV/AIDS. Major industries and companies in southern Africa, such as diamond mines and banks, estimate that absence and loss of employees to HIV/AIDS is costing them more than 5 percent of their profits.

Some countries have had success in combating HIV/ AIDS. Uganda and Senegal have promoted aggressive and successful HIV/AIDS education and prevention campaigns and have cut infection rates in half. International agreements with drug companies in combination with

Figure 1.10 Diffusion of the HIV virus The probable early diffusion of HIV/AIDS. (Adapted from M. Smallman-Raynor, A. Cliff, and P. Haggett, London International Atlas of AIDS. Oxford: Blackwell Reference, 1992, Fig. 4.1(c), p. 146.)

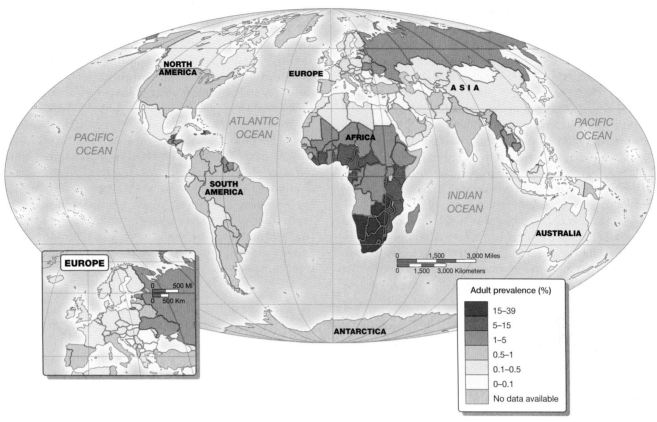

Figure 1.11   The geography of HIV/AIDS In 2003, between 35 and 42 million people were living with HIV/AIDS. More than half of these were in sub-Saharan Africa. (*Source:* http://www.who.int/hiv/facts/hiv2003/en/)

new assistance programs from the World Bank, charities, and donor countries are helping to bring down the cost of drugs.

Security Issues   International terrorism can also be attributed in part to globalization. While terrorism has a long history, it is only recently that terrorist attacks have

spilled beyond the sites of local conflict. In targeting the World Trade Center in New York in September 2001 (**Figure 1.12**), al-Qaeda terrorists were striving to tear down a potent symbol not merely of the economic might of the United States but also of the values of **capitalism** (a form of economic and social organization characterized by the profit motive and the control of the means of

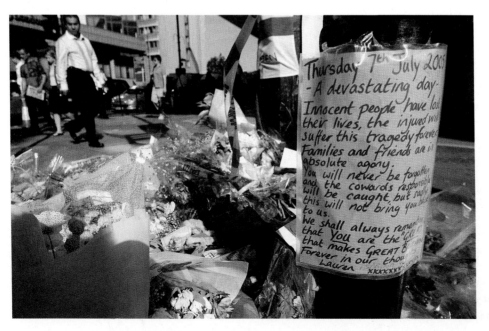

Figure 1.12   International terrorism Floral tributes lie outside Edgware Road underground station in London, England, after al-Qaeda bombers killed 49 people and injured 700 during morning rush hour terrorist attacks that were targeted at London's transport links on July 7, 2005.

production, distribution, and exchange of goods by private ownership) and Western materialism. Provoking an aggressive response from the United States would, it was hoped, prompt an outburst of anti-Americanism around the world. The al-Qaeda terrorists felt that they had a reasonable shot at triggering a new world disorder, pitching the Middle East into anarchy, launching a cultural war against the West, and sowing discord among peoples already uneasy about U.S. cultural and economic hegemony and among governments already unsettled by the U.S. government's increasing tendency to pursue narrow self-interest at the expense of transnational issues of economic inequality, geopolitics, and environmental sustainability.

More generally, as sociologist Ulrich Beck has pointed out, the high degree of interdependence that is now embedded in a globalizing and highly interconnected world has brought about all sorts of security issues. Beck argues that the contemporary world is characterized by the global—or at least transnational—production of risk. In traditional societies the risks faced by individuals and groups were associated mostly with hazards generated by nature (disease, flood, famine, and the like), along with socially determined hazards such as invasion and conquest and regressive forms of thought and culture. The industrial societies of the nineteenth and twentieth centuries, with more powerful technologies and weaponry, faced still more hazards, but they were mostly local and regional in nature. Contemporary society, Beck points out, is characterized by the endemic production of potentially catastrophic risks, so that we now face another set of hazards, many of them uncontrollable and with a global reach. Examples include climate change as a result of human activity; the spread of weapons of mass destruction (i.e., nuclear and biological warfare); the risk of accidents involving ionizing radiation or contamination by radioactivity from nuclear fuel or nuclear waste; the risk of epidemics in the human population resulting from zoonotic diseases; and the risk of epidemic disease in food animals—as in the devastating outbreak of foot-and-mouth disease, which affects cattle and sheep, in parts of northwestern Europe in 2001. The initial outbreak in England was a result of contaminated animal feed imported illegally, probably from South Africa. The spread of the disease within England was a result of the long-distance trade in animals that characterizes contemporary agriculture.

Overall, Beck argues, we are moving toward a **risk society,** in which the significance of wealth distribution is being eclipsed by the distribution of risk and in which politics is increasingly about avoiding hazards. As a result, knowledge—especially scientific knowledge—becomes increasingly important as a source of power, while science itself becomes increasingly politicized—as, for example, in the case of global warming.

Geography in a Globalizing World

All this adds up to an intensified global connectedness and the beginnings of the world as an interdependent system. Or, to be more precise, this is how it adds up for the 800 million or so of the world's people who are directly tied to global systems of production and consumption and who have access to global networks of communication and knowledge. All of us in this globalizing world are in the middle of a major reorganization of the world economy and a radical change in our relationships to other people and places.

At first glance it might seem that globalization will render geography obsolete—especially in the more developed parts of the world. High-tech communications and the global marketing of standardized products seem as if they might soon wash away the distinctiveness of people and places, permanently diminishing the importance of differences between places. Far from it. The new mobility of money, labor, products, and ideas actually increases the significance of place in some very important ways:

- *The more universal the diffusion of material culture and lifestyles, the more valuable regional and ethnic identities become.* One example of this is the way that the French government has actively resisted the Americanization of the French language and culture by banning the use of English words and phrases and by subsidizing France's domestic movie industry.

- *The faster the information highway takes people into cyberspace, the more they feel the need for a subjective setting—a specific place or community—they can call their own.* Examples can be found in the private, master-planned, residential developments that have sprung up around every U.S. metropolitan area. Unlike most previous suburban developments, each of these master-planned projects has been carefully designed to create a sense of community and identity for its residents.

- *The greater the reach of transnational corporations, the more easily they are able to respond to place-to-place variations in labor markets and consumer markets and the more often and more radically that economic geography has to be reorganized.* Global patterns of production are constantly being reorganized as transnational corporations seek to take advantage of geographical differences between places and regions and as workers and consumers in specific places and regions react to the consequences of globalization.

- *The greater the integration of transnational governments and institutions, the more sensitive people have become to local cleavages of race, ethnicity, and religion.* An example is the resurgence of nationalism and regionalism, as in the emergence of the Lega Nord (the Northern League) party in Italy in the 1990s. Lega Nord is a federalist political party whose supporters in northern Lombardy and rural northeastern Italy want to distance themselves from what they view as a distinctively different culture and society in the Italian South.

For some places and regions, globalization is a central reality; for others, it is still a marginal influence. While some places and regions have become more closely interconnected and interdependent as a result of globalization, others have been bypassed or excluded. In short, there is no one experience of globalization. All in all, the reality is that globalization is variously embraced, resisted, subverted, and exploited as it makes contact with specific cultures and settings. In the process, places are modified or reconstructed rather than destroyed or homogenized.

STUDYING HUMAN GEOGRAPHY

The study of geography involves the study of Earth as created by natural forces and modified by human action. This, of course, covers an enormous amount of subject matter. There are two main branches of geography: physical and human. **Physical geography** deals with Earth's natural processes and their outcomes. It is concerned, for example, with climate, weather patterns, landforms, soil formation, and plant and animal ecology. **Human geography** deals with the spatial organization of human activities and with people's relationships to their environments. This focus necessarily involves looking at natural physical environments insofar as they influence, and are influenced by, human activity. This means that the study of human geography must cover a wide variety of phenomena. These include, for example, agricultural production and food security, population change, the ecology of human diseases, resource management, environmental pollution, regional planning, and the symbolism of places and landscapes.

Regional geography combines elements of both physical and human geography. Regional geography is concerned with the way that unique combinations of environmental and human factors produce territories with distinctive landscapes and cultural attributes. The concept of **region** is used by geographers to apply to larger-size territories that encompass many places, all or most of which have similar attributes distinct from the attributes of other places.

What is distinctive about the study of human geography is not so much the phenomena that are studied as the way they are approached. The contribution of human geography is to reveal, in relation to a wide spectrum of natural, social, economic, political, and cultural phenomena, *how and why geographical relationships are important.* Thus, for example, human geographers are interested not only in patterns of agricultural production but also in the geographical relationships and interdependencies that are both causes and effects of such patterns. To put it in concrete terms, geographers are interested not only in what specialized agricultural subregions, such as the dairy farming area of Jutland, Denmark, are like (just what and how the region produces its agricultural output, what makes its landscapes and culture distinctive, and so on) but also in the role of subregions such as Jutland in national and international agro-food systems (their interdependence with producers, distributors, and consumers in other places and regions—see Chapter 8).

Basic Tools

In general terms, the basic tools employed in geography are similar to those in other disciplines. Like other social scientists, human geographers usually begin with observation. Information must be collected and data recorded. This can involve many different methods and tools. Fieldwork (surveying, asking questions, using scientific instruments to measure and record things), laboratory experiments, and archival searches all are used by human geographers to gather information about geographical relationships. Geographers also use **remote sensing**, the collection of information about parts of Earth's surface by means of aerial photography or satellite imagery designed to record data on visible, infrared, and microwave sensor systems (**Figure 1.13**). For example, agricultural productivity can be monitored by remotely sensed images of crops, and energy efficiency can be monitored by remotely sensed levels of heat loss from buildings.

Once data have been obtained through some form of observation, the next important step is to portray and describe them through *visualization* or *representation.* This can involve a variety of tools, including written descriptions, charts, diagrams, tables, mathematical formulas, and maps. Visualization and representation are important activities because they allow large amounts of information to be explored, summarized, and presented to others. They are nearly always a first step in the analysis of geographical relationships, and they are important in conveying the findings and conclusions of geographic research.

At the heart of geographic research, as with other kinds of research, is the *analysis* of data. The objective of analysis, whether of quantitative or qualitative data, is to discover patterns and establish relationships so that hypotheses can be established and models can be built. Models, in this sense, are abstractions of reality that help explain the real world. They require tools that allow us to generalize about things. Once again, we find that geographers are like other social scientists in that they utilize a wide range of analytical tools, including conceptual and linguistic devices, maps, charts, and mathematical equations.

In many ways, therefore, the tools and methods of human geographers are parallel to those used in other sciences, especially the social sciences. In addition, geographers increasingly use some of the tools and methods of the humanities—interpretive analysis and inductive reasoning, for example—together with ethnographic research and textual analysis. The most distinctive tools in the geographer's kit bag are, of course, maps and **geographic information systems (GIS)**. Geographic information systems involve an organized set of computer hardware, software, and spatially coded data that is designed to capture, store, update, manipulate, and display geographically referenced information (see Appendix 1, "Maps and Geographic

Landsat satellite images are digital images captured from spectral bands both visible and invisible to the human eye. Different kinds of vegetation cover, soils, and built environments are reflected by different colors in the processed image. This Landsat image is of part of southern Florida (Miami Beach is to the right of the image).

Aerial photographs can be helpful in explaining what would otherwise require expensive surveys and detailed cartography. They are especially useful in working with multidisciplinary teams. This example shows a remnant wetland adjacent to urban development in Muskegon River Flats, Michigan.

This spectacular image is the most detailed true-color image of the entire Earth to date. Using a collection of satellite-based observations, scientists and visualizers stitched together months of observations of the land surface, oceans, sea ice, and clouds into a seamless, true-color mosaic of every square kilometer (0.386 square mile) of the planet. Much of the information contained in this image came from a single remote-sensing device—NASA's Moderate Resolution Imaging Spectroradiometer, or MODIS—flying over 700 km above the Earth on board the Terra satellite.

Figure 1.13 Remotely sensed images Remotely sensed images can provide new ways of seeing the world, as well as unique sources of data on all sorts of environmental conditions. Such images can help explain problems and processes that would otherwise require expensive surveys and detailed cartography. They can be used to identify change in the environment and to monitor and analyze the rate of change. Examples of such applications have included studies of the deforestation of the Amazon rain forest, urban encroachment onto farmland, water pollution, and bottlenecks in highway systems. (*Source:* Top left photo courtesy of Spaceimaging.com; top right photo courtesy Joel Sartore www.joelsartore.com.)

Information Systems"). As we have seen, maps can be used not only to describe data but also to serve as important sources of data and tools for analysis. Because of their central importance to geographers, they can also be objects of study in their own right.

Spatial Analysis

The study of human geography is easily distinguished by its fundamental concepts. The study of many geographic phenomena can be approached in terms of their arrangement as points, lines, areas, or surfaces on a map. This is known as **spatial analysis**. *Location, distance, space, accessibility,* and *spatial interaction* are five concepts that are key to spatial analysis. Although these concepts may be familiar from everyday language, they require some elaboration.

Location Location is often nominal, or expressed solely in terms of the names given to regions and places. We speak, for example, of Washington, D.C. or of Georgetown, a location within Washington, D.C. Location can also be used as an absolute concept, whereby locations are fixed mathematically through coordinates of latitude and longitude (**Figure 1.14**).

Latitude refers to the angular distance of a point on Earth's surface, measured in degrees, minutes, and seconds

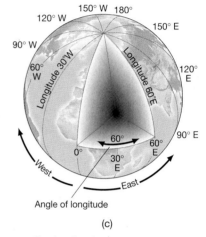

The prime meridian at the Royal Observatory in Greenwich, England—The observatory was founded by Charles II in 1675 with the task of setting standards for time, distance, latitude, and longitude—the key components of navigation.

Figure 1.14 Latitude and longitude Lines of latitude and longitude provide a grid that covers Earth, allowing any point on Earth's surface to be accurately referenced. Latitude is measured in angular distance (that is, degrees and minutes) north or south of the equator, as shown in (a). Longitude is measured in the same way, but east and west from the prime meridian, a line around Earth's surface that passes through both poles (North and South) and the Royal Observatory in Greenwich, just to the east of central London, in England. Locations are always stated with latitudinal measurements first. The location of Paris, France, for example, is 48°51'N and 2°20'E as shown in (b). (a) and (c), adapted from R. W. Christopherson, *Geosystems: An Introduction to Physical Geography*, 2nd ed., © 1994, pp. 13 and 15. (b), adapted from E. F. Bergman, *Human Geography: Cultures, Connections, and Landscapes*, © 1995, Figs. 1–10 and 1–13.

north or south from the equator, which is assigned a value of 0°. Lines of latitude around the globe run parallel to the equator, which is why they are sometimes referred to as *parallels*. **Longitude** refers to the angular distance of a point on Earth's surface, measured in degrees, minutes, and seconds east or west from the *prime meridian* (the line that passes through both poles and through Greenwich, England, which is assigned a value of 0°). Lines of **longitude**, called *meridians*, run from the North Pole (latitude 90° north) to the South Pole (latitude 90° south). Georgetown's coordinates are precisely 38°55'N, 77°00'E.

Thanks to the **Global Positioning System (GPS)**, it is very easy to determine the latitude and longitude of any given point. The Global Positioning System consists of 21 satellites (plus 3 spares) that orbit Earth on precisely predictable paths, broadcasting highly accurate time and locational information. The GPS is owned by the U.S. government, but the information transmitted by the satellites is freely available to everyone around the world. All that is needed is a GPS receiver. Basic receivers cost less than $50 and can relay latitude, longitude, and height to within 100 meters day or night, in all weather conditions, in any part of the world. Current production models of many automobiles are now equipped with GPS-based navigational systems. The most precise GPS receivers, costing thousands of dollars, are accurate to within a centimeter. The GPS has dramatically increased the accuracy and efficiency of collecting spatial data. In combination with GIS and remote sensing, GPS has revolutionized mapmaking and spatial analysis.

Location can also be *relative*, fixed in terms of site or situation. **Site** refers to the physical attributes of a location: its terrain, its soil, vegetation, and water sources, for example. **Situation** refers to the location of a place relative to other places and human activities: its accessibility to routeways, for example, or its nearness to population centers (**Figure 1.15**). Washington, D.C. has a low-lying riverbank site and is situated at the head of navigation of the Potomac River, on the eastern seaboard of the United States.

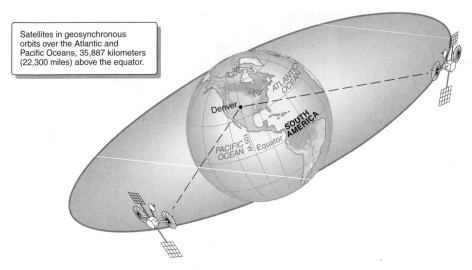

Satellites in geosynchronous orbits over the Atlantic and Pacific Oceans, 35,887 kilometers (22,300 miles) above the equator.

Figure 1.15   The importance of site and situation The location of telecommunications activities in Denver, Colorado, provides a good example of the significance of the geographic concepts of site (the physical attributes of a location) and situation (the location of a place relative to other places and human activities). Denver has become a major center for cable television, with the headquarters of giant cable companies such as Tele-Communications and DirecTV, an industrywide research lab, and a cluster of specialized support companies that together employ over 3,000 people. Denver's site, 1.6 kilometers (1 mile) above sea level, is important because it gives commercial transmitters and receivers a better "view" of communications satellites. Its situation, on the 105th meridian and equidistant between the telecommunications satellites that are in geostationary orbit over the Pacific and Atlantic oceans, allows it to send cable programming directly not just to the whole of the Americas but also to Europe, the Middle East, India, Japan, and Australia—to every continent, in fact, except Antarctica. This is important because it avoids "double-hop" transmission (in which a signal goes up to a satellite, then down, then up and down again), which increases costs and decreases picture quality. Before the location of telecommunications facilities in Denver, places east or west of the 105th meridian would have to double-hop some of their transmissions because satellite dishes would not have a clear "view" of both the Pacific and Atlantic telecommunications satellites.

Finally, location also has a *cognitive* dimension, in that people have cognitive images of places and regions, compiled from their own knowledge, experiences, and impressions. **Cognitive images** (sometimes referred to as *mental maps*) are psychological representations of locations that spring from people's individual ideas and impressions of these locations. These representations can be based on people's direct experiences, on written or visual representations of actual locations, on hearsay, on people's imaginations, or on a combination of these sources. Location in these cognitive images is fluid, depending on people's changing information and perceptions of the principal landmarks in their environment.

Some things may not be located in a person's cognitive image at all. **Figure 1.16** shows one person's cognitive image of Washington, D.C. Georgetown is given a location within this mental map, even though it is some distance from the residence of the person who sketched her image of the city. Less well-known and less distinctive places do not appear on this particular image.

Distance   Distance is also useful as an *absolute* physical measure, whose units we may count in kilometers or miles, and as a *relative* measure, expressed in terms of time, effort, or cost. It can take more or less time, for

example, to travel 10 kilometers from point A to point B than it does to travel 10 kilometers from point A to point C. Similarly, it can cost more or less. Geographers also have to recognize that distance can sometimes be in the eye of the beholder. It can seem longer or shorter, more or less pleasant, going from A to B as compared to going from A to C. This is **cognitive distance**, the distance that people perceive to exist in a given situation. Cognitive distance is based on people's personal judgments about the degree of spatial separation between points.

The importance of distance as a fundamental factor in determining real-world relationships is a central theme in geography. It was once described as the "first law" of geography: "Everything is related to everything else, but near things are more related than distant things." Waldo Tobler, the geographer from the University of California, Santa Barbara, who put it this way, is one of many who have investigated the friction of distance, the deterrent or inhibiting effect of distance on human activity. The **friction of distance** is a reflection of the time and cost of overcoming distance.

What these geographers have established is that these effects are not uniform—that is, they are not directly proportional to distance itself. This is true whether distance is measured in absolute terms (i.e., kilometers) or in

Figure 1.16 One person's cognitive image of Washington, D.C. This sketch was drawn by Rasheda DuPree, an urban affairs major at Virginia Tech, as part of a class exercise in recalling locations within students' hometowns. Rasheda has included many of the District's most prominent landmarks and some of its distinctive districts, including Georgetown. In contrast, there are no recorded locations in the city's southeastern quarter (marked by a skull and crossbones in Rasheda's sketch) or in the eastern outskirts (marked as "the burbs").

relative terms (i.e., time- or cost-based measures). What happens is that the deterrent effects of extra distance tend to lessen as greater distances are involved. Thus, for example, while there is a big deterrent effect in having to travel 2 kilometers rather than 1 to get to a grocery store, the deterrent effect of the same extra distance (1 kilometer) after already traveling 10 kilometers is relatively small.

This sort of relationship creates what geographers call a distance-decay function. A **distance-decay function** describes the rate at which a particular activity or phenomenon diminishes with increasing distance. A typical distance-decay function is described by the graph in **Figure 1.17**, which shows the effects of distance on people's willingness to travel for free medical care.

Distance-decay functions reflect people's behavioral response to opportunities and constraints in time and space. As such, they reflect the **utility** of particular locations to people. The utility of a specific place or location

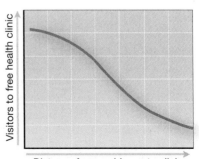

Figure 1.17 The friction of distance The effects of distance on people's behavior can be charted on graphs like this. The farther people have to travel, the less likely they are to do so. In this example, we can see clearly the deterrent effects of distance on people's attendance at a free health clinic.

refers to its usefulness to a particular person or group. In practice, utility is thought of in different ways by different people in different situations. The behavior of private firms and their agents or employees, for example, is most often guided by a bottom-line notion of utility that relates to dollar costs or profits. The same individuals will probably use a different notion of utility when it comes to their own lifestyle and the decisions they make in pursuing it. Prestige, convenience, or feelings of personal safety, communality, or happiness may well modify or override financial costs as the measure of utility. The business manager of a supermarket chain, for instance, will almost certainly decide on the utility of potential locations for a new store by weighing criteria based on the projected costs and revenues for each potential site. In deciding on the utility of potential locations in which to retire, however, that same manager will almost certainly weigh criteria based not only on costs but also on a wide range of quality-of-life aspects of potential retirement places.

The unifying theme here is that however place utility is thought of, people in most circumstances tend to *seek to maximize the net utility of location*. The supermarket chain's business manager, for example, will seek to find the location for the chain's new store that is most likely to yield the greatest profit. Upon retirement, he or she will choose to live in the place that represents the best trade-off among housing costs, cost of living, and quality of life. Seeking to maximize the net utility of location means that a great deal of human activity is influenced by what University of Washington geographer Richard Morrill once called the "nearness principle." According to this principle—a more explicit version of Tobler's first law—people will seek to

■ maximize the overall utility of places at minimum effort;

■ maximize connections between places at minimum cost; and

■ locate related activities as close together as possible.

The result is that patterns of behavior, locational decisions, and interrelations between people and places come to take on fairly predictable, organized patterns.

Space   Like distance, space can be measured in absolute, relative, and cognitive terms. **Table 1.1** lists the concepts human geographers use in talking about space in these various terms. Absolute space is a mathematical space described through points, lines, areas, planes, and configurations whose relationships can be fixed precisely through mathematical reasoning. Several ways of analyzing space mathematically are of use to geographers. The conventional way is to view space as a container, defined by rectangular coordinates and measured in absolute units of distance (kilometers or miles, for example). Other mathematical conceptions of space that geographers sometimes find useful also exist, however. One is **topological space**, defined as the connections between, or connectivity of, particular points in space (**Figure 1.18**). Topological space is measured not in terms of conventional measures of distance, but by the nature and degree of connectivity between locations.

Relative measurements of space can take the form of socioeconomic space or of experiential or cultural space (see Table 1.1). Socioeconomic space can be described in terms of sites and situations, routes, regions, and distribution patterns. In these terms, spatial relationships have to be fixed through measures of time, cost, profit, and production, as well as through physical distance. Experiential or cultural space is the space of groups of people with common ties, and it is described through the places, territories, and settings whose attributes carry special meaning for these particular groups. Finally, **cognitive space** is defined and measured in terms of people's values, feelings, beliefs, and perceptions about locations, districts, and regions. Cognitive space can be described, therefore, in terms of behavioral space—landmarks, paths, environments, and spatial layouts.

Accessibility   Given that people tend to pursue the nearness principle, the concept of accessibility is very important. **Accessibility** is generally defined by geographers in terms of relative location: the opportunity for contact or interaction from a given point or location in relation to other locations. It implies proximity, or nearness, to

TABLE 1.1 Different Kinds of Spaces Analyzed by Human Geographers

Absolute Space: Mathematical Space	Relative Space: Socioeconomic Space	Relative Space: Experiential/Cultural Space	Cognitive Space: Behavioral Space
Points	Sites	Places	Landmarks
Lines	Situations	Ways	Paths
Areas	Routes	Territories	Districts
Planes	Regions	Domains	Environments
Configurations	Distributions	Worlds	Spatial Layouts

Source: Based on H. Couclelis, "Location, Place, Region and Space," in R. Abler et al., *Geography's Inner Worlds.* New Brunswick, NJ: Rutgers University Press, 1992, Table 10.1, p. 231.

Figure 1.18 Topological space Some dimensions of space and aspects of spatial organization do not lend themselves to description simply in terms of distance. In other words, the connectivity of people and places is often important: whether they are linked, how they are linked, and so on. These attributes of connectivity define topological space. The map of the Metro system in Milan, Italy, is a topological map, showing how specific points are joined within a particular network. The most important aspects of networks of any kind, from the geographer's viewpoint, are their connectivity attributes. These attributes determine the flows of people and things (goods, information) and the centrality of places. As most Milanese know, the Metro system gives Duomo a very high degree of connectivity because trains on both the M1 and M3 lines stop there. Duomo is therefore relatively central within the "space of flows" of passenger traffic in central Milan. Piazza Missori—nearby in absolute terms—is much less central, however, and much less the focus of passenger flows.

something. Because it is a fundamental influence on the utility of locations, distance is an important influence on people's behavior. Distance is one aspect of accessibility, but it is by no means the only important aspect.

Connectivity is also an important aspect of accessibility because contact and interaction are dependent on channels of communication and transportation: streets, highways, telephone lines, and wavebands, for example. Effective accessibility is thus a function not only of distance but also of the configuration of networks of communication and transportation. Commercial airline networks provide many striking examples of this. Cities that operate as airline hubs are much more accessible than cities that are served by fewer flights and fewer airlines. Charlotte, NC, for example (a U.S. Airways hub), is more accessible from Albany, NY than from Richmond, VA, even though Richmond is 400 kilometers (248 miles) closer to Albany than Charlotte. To get to Richmond from Albany, travelers must fly to Charlotte or another hub and change—a journey that takes longer and often costs more.

Accessibility is often a function of economic, cultural, and social factors. In other words, relative concepts and measures of distance are often as important as absolute distance. A nearby facility, such as a health care clinic, is accessible to us only if we can actually afford the cost of getting there, if it seems close according to our own standards of distance, if we can afford to use the facility, if we feel that it is socially and culturally acceptable for us to use it, and so on. To take another example, a day care center may be located just a few blocks from a single-parent family, but the center is not truly accessible if it opens after the parent has to be at work or if the parent feels that the staff, children, or other parents at the center are from an incompatible social or cultural group.

Spatial Interaction Interdependence between places and regions can be sustained only through movement and flows. Geographers use the term **spatial interaction** as shorthand for all kinds of movement and flows involving human activity. Freight shipments, commuting, shopping trips, telecommunications, electronic cash transfers, migration, and vacation travel are all examples of spatial interaction. The fundamental principles of spatial interaction can be reduced to four basic concepts: complementarity, transferability, intervening opportunities, and diffusion.

- *Complementarity.* A precondition for interdependence between places is complementarity. For any kind of spatial interaction to occur between two places, there must be a demand in one place and a supply that matches, or complements, it in the other. This complementarity can be the result of several factors. One important factor is the variation in physical environments and resource endowments from place to place. For example, a heavy flow of vacation travel from Swedish cities to Mediterranean resorts is largely a function of climatic complementarity. To take another example, the flow of crude oil from Saudi Arabia (with vast oil reserves) to Japan (with none) is a function of complementarity in natural resource endowments.

A second factor contributing to complementarity is the international division of labor that derives from the evolution of the world's economic systems. The more developed countries of the world have sought to establish overseas suppliers for their food, raw materials, and exotic produce, allowing the more developed countries to specialize in more profitable manufacturing and knowledge-based industries (see Chapter 2). Through a combination of colonialism, imperialism, and sheer economic dominance on the part of these more developed countries, less powerful countries have found themselves with economies that directly complement the needs of the more developed countries. Among the many flows resulting from this complementarity are shipments of sugar from Barbados to the United Kingdom, bananas from Costa Rica and Honduras to the United States, palm oil from Cameroon to France, automobiles from France to Algeria, school textbooks from the United Kingdom to Kenya, and investment capital from the United States to most of the less developed countries.

A third contributory factor to complementarity is the operation of principles of specialization and **economies of scale**. Places, regions, and countries can derive economic advantages from the efficiencies created through specialization, which allows for larger-scale operations. Economies of scale are cost advantages to manufacturers that accrue from high-volume production, since the average cost of production falls with increasing output (**Figure 1.19**). Among other things, fixed costs (for example, the cost of renting or buying factory space, which will be the same—fixed—whatever the level of output from the factory) can be spread over higher levels of output so that the average cost of production falls. Economic specialization results in complementarities, which in turn contribute to patterns of spatial interaction. One example is the specialization of Israeli farmers in high-value fruit and vegetable crops for export to the European Union, which in return exports grains and root crops to Israel.

■ *Transferability.* Another precondition for interdependence between places is *transferability*, which depends on the frictional or deterrent effects of distance. Transferability is a function of two things: the costs of moving a particular item, measured in real money and/or time, and the ability of the item to bear these costs. If, for example, the costs of moving a product from one place to another make it too expensive to sell successfully at its destination, then that product does not have transferability between those places.

Transferability varies between places, between kinds of items, and between modes of transportation and communication. The transferability of coal, for example, is much greater between places that are connected by rail or by navigable waterways than between places connected only by highways. This is because it is much cheaper to move heavy, bulky

Figure 1.19 Economies of scale In many manufacturing enterprises, the higher the volume of production, the lower the average cost of producing each unit. This is partly because high-volume production allows for specialization and division of labor, which can increase efficiency and hence lower costs. It is also partly because most manufacturing activities have significant fixed costs (such as product design and the cost of renting or buying factory space) that must be paid for irrespective of the volume of production, so that the larger the output, the lower the fixed cost per unit. These savings are known as economies of scale.

materials by rail, barge, or ship. The transferability of fruit and salad crops, on the other hand, depends more on the speed of transportation and the availability of specialized refrigerated vehicles so the fruits and vegetables stay fresh. While the transferability of money capital is much greater by telecommunications than it is by surface transportation, it is also higher between places where banks are equipped to deal routinely with electronic transfers. Computer microchips have high transferability because they are easy to handle, and transport costs are a small proportion of their value. Computer monitors, on the other hand, have lower transferability because of their fragility and their relatively lower value by weight and volume.

Transferability also varies over time, as successive innovations in transport and communications technologies and successive waves of **infrastructure** development (canals, railways, harbor installations, roads, bridges, and so on) alter the geography of transport costs. New technologies and new or extended infrastructures have the effect of altering the transferability of particular things between particular places. *As a result, the spatial organization of many different activities is continually changing and readjusting.* The consequent tendency toward a shrinking world gives rise to the concept of **time-space convergence**, the rate at which places move closer together in travel or communication time or costs. Time-space convergence results from a decrease in the friction of distance as new technologies and infrastructure improvements successively reduce travel and communications time between places. Such space-adjusting technologies have, in general, brought places closer together over time (**Figure 1.20**). Overland travel between New York and Boston, for example, has been reduced from 3.5 days (in 1800) to 5 hours (in the 2000s) as the railroad displaced stagecoaches and was in turn displaced by interstate automobile travel. Other important space-adjusting innovations include air travel and air cargo; telegraphic, telephonic, and satellite communications systems; national postal services, package delivery services, and facsimile (fax) machines; and modems, fiber-optic networks, and electronic-mail software.

What is most significant about the latest developments in transport and communication is that they are not only global in scope but also are able to penetrate to local scales. As this penetration occurs, some places that are distant in kilometers are becoming close together, while some that are close in terms of absolute space are becoming more distant in terms of their ability to reach one another electronically (**Figure 1.21**). Much depends on the mode of communication—the extent to which people in different places are "plugged in" to new technologies. Older wire cable can carry only small amounts of information; microwave channels are good for person-to-person communication but depend on line of sight; telecommunications satellites are excellent for reaching remote areas but involve

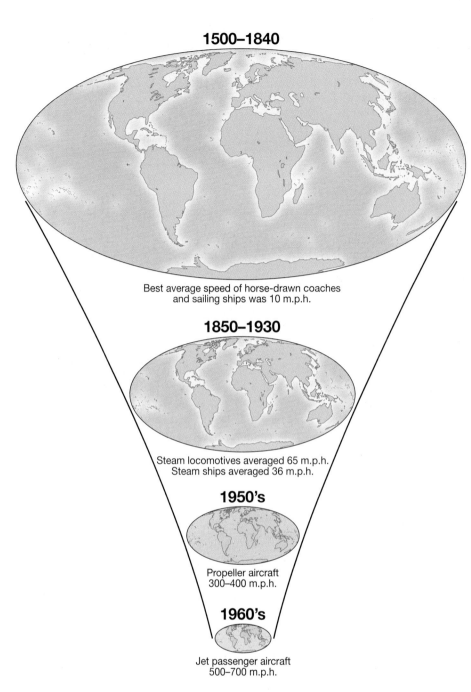

1500–1840

Best average speed of horse-drawn coaches
and sailing ships was 10 m.p.h.

1850–1930

Steam locomotives averaged 65 m.p.h.
Steam ships averaged 36 m.p.h.

1950's

Propeller aircraft
300–400 m.p.h.

1960's

Jet passenger aircraft
500–700 m.p.h.

Figure 1.20 Time-space convergence The effects of changing transportation technologies in "shrinking" the world in terms of travel time.

significant capital costs for users, while fiber-optic cable is excellent for areas of high-population density but not feasible for more remote, rural areas. The shrinking of space has important implications for people's everyday conceptions of space and distance and for their level of knowledge about other places.

■ *Intervening Opportunity.* While complementarity and transferability are preconditions for spatial interaction, intervening opportunities are more important in determining the *volume* and *pattern* of movements and flows. Intervening opportunities are simply alternative origins and/or destinations. Such opportunities are not necessarily situated directly between two points or even along a route between them. Thus, to take one of our previous examples, for Swedish families considering a Mediterranean vacation in Greece, resorts in Spain, southern France, and Italy are all likely to be intervening opportunities because they can probably be reached more quickly and cheaply than resorts in Greece.

The size and relative importance of alternative destinations are also important aspects of the concept of intervening opportunity. For our Swedish families, Spanish resorts probably offer the greatest intervening opportunity because they contain the largest aggregate number of hotel rooms and vacation apartments. We can therefore state the principle of intervening opportunity as follows: Spatial interaction between an origin and a destination will be proportional to the number of opportunities at that destination and inversely proportional to the number of opportunities at alternative destinations.

Figure 1.21 **Brooklyn, New York** In early 2008, more than 80 percent of American adults had regular access to the Internet. Increasing reliance on the Internet has led to the growth of Internet cafés, catering mostly to people's need for connectivity when they are away from home and/or work.

■ *Spatial Diffusion.* Disease outbreaks, technological innovations, political movements, and new musical fads all originate in specific places and subsequently spread to other places and regions. The way that things spread through space and over time—**spatial diffusion**—is one of the most important aspects of spatial interaction and is crucial to an understanding of geographic change.

Diffusion seldom occurs in an apparently random way, jumping unpredictably all over the map. Rather, it occurs as a function of statistical probability, which is often based on fundamental geographic principles of distance and movement. The diffusion of a contagious disease, for example, is a function of the probability of physical contact, modified by variations in individual resistance to the disease. The result is typically a "wave" of diffusion that describes an S-curve, with a slow buildup, rapid spread, and final leveling off (**Figure 1.22**).

It is possible to recognize several spatial tendencies in patterns of diffusion. In *expansion diffusion* (also called *contagion diffusion*—**Figure 1.23a**), a phenomenon spreads because of the proximity of carriers, or agents of change, who are fixed in their location. An example would be the diffusion of an agricultural innovation, such as the use of hybrid seed stock, among members of a local farming community. With *hierarchical diffusion* (also known as cascade diffusion), a phenomenon can be diffused from one location to another without necessarily spreading to

people or places in between (**Figure 1.23b**). An example would be the spread of a fashion trend from large metropolitan areas to successively smaller cities, towns, and rural settlements. Many patterns and processes of diffusion reflect both expansion and hierarchical diffusion (**Figure 1.23c**), as different aspects of human interaction come into play in different

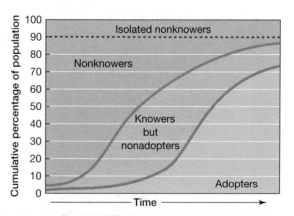

Figure 1.22 **Spatial diffusion** The spatial diffusion of many phenomena tends to follow an S-curve of slow buildup, rapid spread, and leveling off. In the case of the diffusion of an innovation, for example, it usually takes a while for enough potential adopters to get to know about the innovation, and even longer for a critical mass of them to adopt it. After that the innovation spreads quite rapidly, until most of the potential adopters have been exposed to it. (Adapted from D. J. Walmsley and G. J. Lewis, *Human Geography: Behavioral Approaches.* London: Longman, 1984, Fig. 5.3, p. 52.)

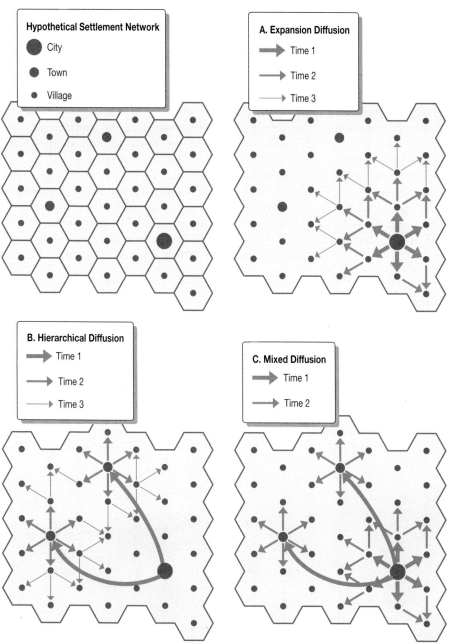

Figure 1.23 Patterns of spatial diffusion (a) Expansion diffusion (for example, the spread of an innovative agricultural practice, such as the use of hybrid seed stock, across a rural region); (b) hierarchical diffusion (the spread of a fashion trend from large metropolitan areas to smaller cities and towns); (c) mixed diffusion (the spread of a contagious disease across a region). (Adapted from E. K. Cromley and S. L. McLafferty, *GIS and Public Health.* New York: Guilford Press, p. 193.)

geographic settings. The diffusion of outbreaks of communicable diseases, for example, usually involves a combination of expansion and hierarchical diffusion, as in the case of HIV/AIDS (see p. 18).

Regional Analysis

Not all geographic phenomena are most effectively understood through spatial analysis. Geographers also seek to understand the complex relationships between peoples and places in terms of the similarities and differences among and between them and the identities and qualities associated with them. Here the key concepts are *regionalization*, *landscape*, and *sense of place.*

Regionalization The geographer's equivalent of *scientific classification* is **regionalization**, with individual places or areal units being the objects of classification.

The purpose of regionalization is to identify regions of one kind or another. There are several ways in which individual areal units can be assigned to regions. One is that of *logical division*, or "classification from above." This involves partitioning a universal set of areal units into successively larger numbers of regions, using more specific criteria at every stage. Thus, a classification of countries might be achieved by first differentiating between rich and poor countries, then dividing both rich and poor countries into those that have a trade surplus and those that have a deficit, and so on. A second way in which individual areal units can be assigned to regions is that of *grouping*, or "classification from below." This involves searching for regularities or significant relationships among areal units and grouping them in successively smaller numbers of classes, using a broader measure of similarity at each stage.

An implicit assumption in this type of classification of areal units into regions is that each unit is homogeneous with respect to the attribute or attributes under consideration. Where this assumption holds true, the result of regional classification is a set of formal regions. **Formal regions** are groups of areal units that have a high degree of homogeneity in terms of particular distinguishing features (such as religious adherence or household income). Few phenomena, however, exhibit such homogeneity over large areal units. For this reason, geographers also recognize **functional regions** (sometimes referred to as nodal regions)—regions within which, while there may be some variability in certain attributes (again, for example, religious adherence and income), there is an overall coherence to the structure and dynamics of economic, political, and social organization.

The concept of functional regions allows us to recognize that the coherence and distinctive characteristics of a region are often stronger in some places than in others. This point is illustrated by geographer Donald Meinig's *core-domain-sphere* model, which he set out in his classic essay on the Mormon region of the United States (**Figure 1.24**). In the core of a region the distinctive attributes are very clear; in the domain they are dominant but not to the point of exclusivity; in the sphere they are present but not dominant.

Finally, people's own conceptions of place, region, and identity may resonate with or against these boundaries to generate strong feelings of regionalism and sectionalism that feed back into the processes of place-making and regional differentiation. **Regionalism** is a term used to describe situations in which different religious or ethnic

Figure 1.24 The Mormon culture region Cultural attributes often gradually shade from one region to another, rather than having a single, clear-cut boundary. Geographer Donald Meinig's work on the Mormon culture region of the United States identified a "core" region, which exhibits all the attributes of Mormon culture; a "domain," where not all these attributes may be present (or may be less intense); and a "sphere," where some attributes of Mormon culture are present but often as a minority with the attributes of another culture region. (Adapted from D. Meinig, "The Mormon Culture Region: Strategies and Patterns in the Geography of the American West," *Annals, Association of American Geographers, 55,* 1965, pp. 191–220.)

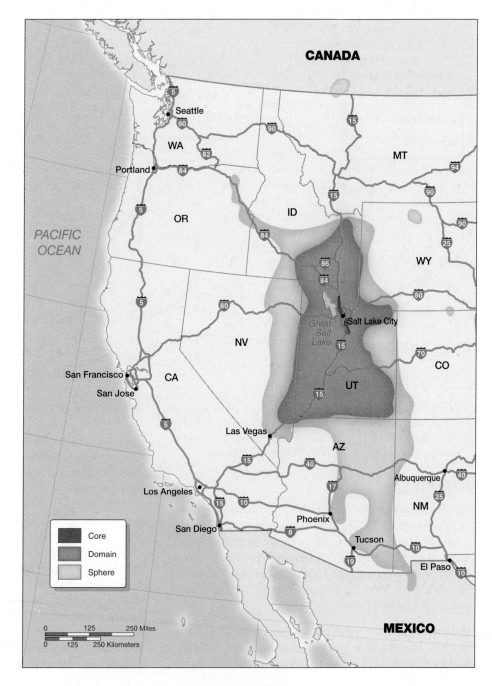

groups with distinctive identities coexist within the same state boundaries, often concentrated within a particular region and sharing strong feelings of collective identity. If such feelings develop into an extreme devotion to regional interests and customs, the condition is known as **sectionalism**. Regionalism often involves ethnic groups whose aims include autonomy from a national state and the development of their own political power (see Chapter 9). In certain cases, enclaves of ethnic minorities are claimed by the government of a country other than the one in which they reside. Such was the case, for example, of Serbian enclaves in Croatia, claimed by nationalist Serbs. The assertion by the government of a country that a minority living outside its formal borders belongs to it historically and culturally is known as **irredentism**. In some circumstances, as with Serbia's claims on Serbian enclaves in Croatia in the early 1990s, irredentism can lead to war.

Landscape Geographers think of landscape as a comprehensive product of human action such that every landscape is a complex repository of society. It is a collection of evidence about our character and experience, our struggles and triumphs as humans. To understand better the meaning of landscape, geographers have developed different categories of landscape types based on the elements contained within them. **Ordinary landscapes** (or vernacular landscapes, as they are sometimes called) are the everyday landscapes that people create in the course of their lives together. From crowded city centers to leafy suburbs and quiet rural villages, these are landscapes that are lived in and changed and that in turn influence and change the perceptions, values, and behaviors of the people who live and work in them.

Symbolic landscapes, by contrast, represent particular values or aspirations that the builders and financiers of those landscapes want to impart to a larger public. For example, the neoclassical architecture of the buildings of the federal government in Washington, D.C., along with the streets, parks, and monuments of the capital, constitute a symbolic landscape intended to communicate a sense of power but also of democracy in its imitation of the Greek city-state. Some landscapes become powerfully symbolic of national identity. Nation-building depends heavily on stories of golden ages, enduring traditions, heroic deeds, shared hardships, and dramatic destinies, all located in traditional (or promised) homelands with hallowed sites and scenery. Landscapes thus become a way of picturing a nation. With the creation of modern Italy during the *Risorgimento* ("revival through unification"—1815–1861), for example, the classical Tuscan landscape (**Figure 1.25**) became emblematic of Italy itself and has attracted landscape painters, romantic poets, and novelists ever since. Similarly, the West of Ireland (**Figure 1.26**) came to symbolize the whole of Ireland to Irish nationalists in the early twentieth century—partly because it was seen as the region least affected by British colonization, but also because its bare and rugged landscape seemed to contrast so strikingly with the more bucolic rural landscapes (**Figure 1.27**) by which England was popularly imagined.

Geographers now recognize that there are many layers of meaning embedded in the landscape, meanings that can be expressed and understood differently by different social groups at different times. Landscapes reflect the lives of ordinary people as well as the more powerful, and they reflect their dreams and ideas as well as their material lives. The messages embedded in landscapes can be read as signs about values, beliefs, and practices, though not every reader will take the same message from a particular landscape (just as people may differ in their interpretation of a passage from a book). In short, landscapes both produce and communicate meaning, and one of our tasks as geographers is to interpret those meanings.

Figure 1.25 The power of place In some countries, particular landscapes have become powerfully symbolic of national identity. In Italy it is the classical landscape of Tuscany, with its scattered farms and villas, elegant cypress trees, silvery-green olive trees, and rolling fields with a rich mixture of cereals, vegetables, fruit trees, and vines.

Figure 1.26 The national landscape of Ireland The rugged landscapes of the West of Ireland have come to symbolize the whole country to many people, both within Ireland and beyond.

Sense of Place The experience of everyday routines in familiar settings allows people to derive a pool of shared meanings. Often this carries over into people's attitudes and feelings about themselves and their locality. When this happens, the result is a self-conscious sense of place. The concept of a **sense of place** refers to the feelings evoked among people as a result of the experiences and memories they associate with a place and to the symbolism they attach to that place. It can also refer to the character of a place as seen by outsiders: its distinctive physical characteristics and/or its inhabitants.

For *insiders*, this sense of place develops through shared dress codes, speech patterns, public comportment,

and so on. A crucial concept here is that of the **lifeworld**, the taken-for-granted pattern and context for everyday living through which people conduct their day-to-day lives without having to make it an object of conscious attention. People's experience of everyday routines in familiar settings leads to a pool of shared meanings. People become familiar with one another's vocabulary, speech patterns, dress codes, gestures, and humor as a result of routine encounters and shared experiences in bars and pubs, cafes and restaurants, shops and street markets, and parks. This is known as **intersubjectivity**: shared meanings that are derived from the lived experience of everyday practice. Elements of daily rhythms (such as mid-morning grocery shopping with a

Figure 1.27 Castle Combe, Wiltshire, England The well-ordered and picturesque landscape of the southern parts of rural England have long been taken to be emblematic of England as a whole, and of the values and ideals of its people—even though urban and industrial development, together with modern agricultural practices, have brought about significant changes to both landscapes and society.

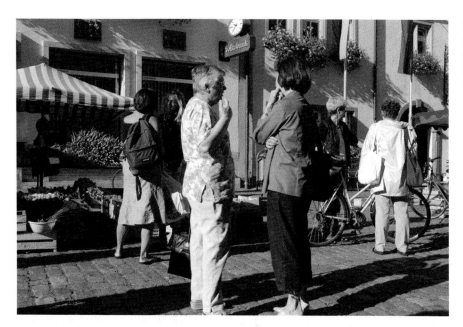

Figure 1.28 Intersubjectivity Routine encounters such as this, in Waldkirch, Germany, help to develop a sense of community and a sense of place among residents.

stop for coffee, the *aperitivo* en route from work to home, and the after-dinner stroll) are all critical to the density of routine encounters and shared experiences that underpin the intersubjectivity that is the basis for a sense of place within a community (**Figure 1.28**). The same is true of elements of weekly rhythms, such as street markets and farmers' markets, and of seasonal rhythms, such as festivals.

These rhythms, in turn, depend on certain kinds of spaces and places: not only streets, squares, and public open spaces but also "third places" (after home, first, and workplace, second): the sidewalk cafés, pubs, post offices, drug stores, corner stores, and family-run *trattoria* that are the loci of routine activities and sociocultural transactions.

Third places accommodate "characters," "regulars," and newcomers, as well as routine patrons and, like public spaces, facilitate casual encounters as well as settings for sustained conversations. The nature and frequency of routine encounters and shared experiences depend a great deal on the attributes of these spaces and places.

A sense of place also develops through familiarity with the history and symbolism of particular elements of the physical environment—a local mountain or lake, the birthplace of someone notable, the location of some particularly well-known event, or the expression of community identity through community art (**Figure 1.29**). Sometimes it is deliberately fostered by the construction

Figure 1.29 Community art Community art can provide an important element in the creation of a sense of place for members of local communities. This example is from the Mission district in San Francisco.

of symbolic structures such as monuments and statues. Often it is a natural outcome of people's familiarity with one another and their surroundings. Because of this consequent sense of place, insiders feel at home and "in place."

For *outsiders*, a sense of place can be evoked only if local landmarks, ways of life, and so on are distinctive enough to evoke a significant common meaning for people who have no direct experience of them. Central London, for example, is a setting that carries a strong sense of place to outsiders who have a sense of familiarity with the riverside panoramas, busy streets, and distinctive monuments and historic buildings that together symbolize the heart of the city.

Developing a Geographical Imagination

A **geographical imagination** allows us to understand changing patterns, processes, and relationships among people, places, and regions. Developing this capacity is increasingly important as the pace of change around the world increases to unprecedented levels. Whereas much of the world had remained virtually unchanged for decades, even centuries, the Industrial Revolution and long-distance, high-speed transportation and communications brought a rapid series of rearrangements to the countryside and to towns and cities in many parts of the world. Today, with a globalized economy and global telecommunications and transportation networks, places have become much more interdependent, and still more of the world is exposed to increasingly urgent imperatives to change.

It is often useful to think of places and regions as representing the cumulative legacy of successive periods of change. Following this approach, we can look for superimposed layers of development (**Figure 1.30**). We can show how some patterns and relationships last, while others are modified or obliterated. We can show how different places bear the imprint of different kinds of change, perhaps in different sequences and with different outcomes. To do so, we must be able to identify the kinds of changes that are most significant.

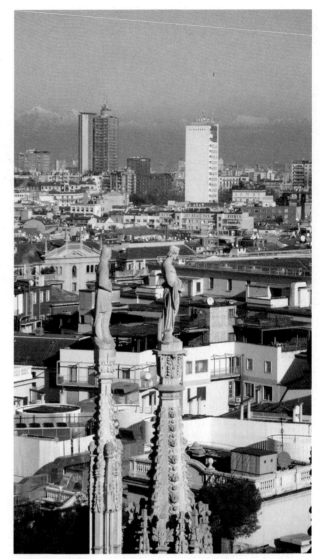

Figure 1.30   Places as the cumulative legacy of change To understand places properly, we have to see them as the cumulative legacy, not just of buildings and structures from different periods of the past but also of the laws, institutions, customs, and so on that developed within each of these periods. This photograph of Milan, Italy, is a very striking example, with modern urban development interlayered with surviving fragments of Medieval, Renaissance, and nineteenth-century development.

Recognizing the General and the Unique   We can prepare our geographical imagination to deal with an important aspect of spatial change by making a distinction between the *general* and the *unique*. This distinction helps account for geographical diversity and variety because it provides a way of understanding how and why one kind of change can result in a variety of spatial outcomes: It is because the *general effects* of a particular change always involve some degree of modification as they are played out in different environments, giving rise to *unique outcomes*.

Although we can usually identify some general outcomes of major episodes of change, there are almost always some unique outcomes, too. Let us take two related exam-

ples. The Industrial Revolution of nineteenth-century Europe provides a good example of a major period of change. A few of the general spatial outcomes were increased urbanization, regional specialization in production, and increased interregional and international trade. At one level, places could be said to have become increasingly alike: generic coalfield regions, industrial towns, ports, downtowns, worker housing, and suburbs.

It is clear, however, that these general outcomes were mediated by the different physical, economic, cultural, and social attributes of different places. Beneath the dramatic overall changes in the geography of Europe, new layers of diversity and variety also existed. Industrial

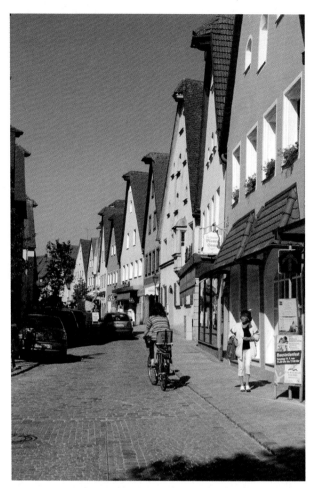

Figure 1.31 Hersbrück, Germany Hersbrück was once an important and prosperous regional center on an overland trade route—the "Golden Road"—between Nuremberg and Prague. After 1806, when Napoleon redrew the political map of Europe, the reorganization of the European economy, together with the onset of the Industrial Revolution, left Hersbrück somewhat isolated and economically disadvantaged. Hersbrück was never drawn into the industrial development of Germany and is not well connected to the transportation infrastructure of canals, railways, or major highways.

towns developed their own distinctive character as a result of their manufacturing specialties, their politics, the personalities and objectives of their leaders, and the reactions and responses of their residents. Downtowns were differentiated from one another as the general forces of commerce and land economics played out across different physical sites and within different patterns of land ownership. They also were differentiated as local socioeconomic and political factors gave rise to different expressions of urban design. Meanwhile, some places came to be distinctive because they were almost entirely bypassed by this period of change, their characteristics making them unsuited to the new economic and spatial order (**Figure 1.31**).

The second example of general and unique outcomes of change is the introduction of the railroad, one of the specific changes involved in the Industrial Revolution. In general terms, the railroad contributed to time-space convergence, to the reorganization of industry into larger market areas, to an increase in interregional and international trade, and to the interconnectedness of urban systems. Other unique outcomes, however, have also contributed to distinctive regional geographies. In Britain, the railroad was introduced to an environment that was partially industrialized and densely settled, and one of the main outcomes was that the increased efficiencies provided by the railroad helped to turn Britain's economy into a highly integrated and intensively urbanized national economy. In Spain, however, the railroad was introduced to an environment that was less urbanized and industrialized and less able to afford the costs of railroad construction. The result was that the relatively few Spanish towns connected by the railroads gained a massive comparative advantage. This situation laid the foundation for a modern space-economy that was much less integrated than Britain's, with an urban system dominated by just a few towns and cities.

CONCLUSION

Human geography is the systematic study of the location of peoples and human activities across Earth's surface and of their relationships to one another. An understanding of human geography is important both from an intellectual point of view (that is, understanding the world around us) and a practical point of view (for example, contributing to environmental quality, human rights, social justice, business efficiency, political analysis, and government policymaking).

While modern ideas about the study of human geography developed from intellectual roots going back to the classical scholarship of ancient Greece, as the world itself has changed, our ways of thinking about it have also changed. What is distinctive about the study of human geography today is not so much the phenomena that are studied as the approach taken by the investigator. The contribution of human geography is to

reveal, in relation to economic, social, cultural, and political phenomena, how and why geographical relationships matter in terms of cause and effect.

Geography matters because it is in specific places that people learn who and what they are and how they should think and behave. Places are also a strong influence, for better or worse, on people's physical well-being, their opportunities, and their lifestyle choices. Places also contribute to peoples' collective memory and become powerful emotional and cultural symbols. Places are the sites of innovation and change, of resistance and conflict.

To investigate specific places, however, we must be able to frame our studies of them within the compass of the entire globe. This is important for two reasons. First, the world consists of a complex mosaic of places and regions that are interrelated and interdependent in many ways. Second, place-making forces—especially economic, cultural, and political forces that influence the distribution of human activities and the character of places—are increasingly operating at global and international scales. The interdependence of places and regions means that individual places are tied into wider processes of change that are reflected in broader geographical patterns. An important issue for human geographers is to recognize these wider processes and broad geographical patterns without losing sight of the uniqueness of specific places.

This global perspective leads to the following principles:

- Each place, each region, is largely the product of forces that are both local and global in origin.

- Each is ultimately linked to many other places and regions through these same forces.

- The individual character of places and regions cannot be accounted for by general processes alone. Some local outcomes are the product of unusual circumstances or special local factors.

MAIN POINTS REVISITED

- **Geography matters because it is specific places that provide the settings for people's daily lives.**

  Places are settings for social interaction that, among other things, structure the daily routines of people's economic and social lives; provide both opportunities for—and constraints on—people's long-term social well-being; establish a context in which everyday commonsense knowledge and experience are gathered; provide a setting for processes of socialization; and provide an arena for contesting social norms.

- **Places and regions are highly *interdependent*, each filling specialized roles in complex and ever-changing networks of interaction and change.**

  Individual places are tied into wider processes of change that are reflected in broader geographical patterns. An important issue for human geographers is to recognize these wider processes and broad geographical patterns without losing sight of the individuality and uniqueness of specific places. Processes of geographic change are constantly modifying and reshaping places, and places are constantly coping with change.

- **In today's world some of the most important aspects of the interdependence between geographic scales are provided by the relationships between the global and the local.**

  The study of human geography shows not only how global trends influence local outcomes but also how events in particular localities can come to influence patterns and trends elsewhere. With an understanding of these trends and outcomes, it is possible to appreciate the diversity and variety of the world's peoples and places and also to be aware of their relationships to one another and to be able to make positive contributions to local, national, and global development.

- **Human geography provides ways of understanding places, regions, and spatial relationships as the products of a series of interrelated forces that stem from nature, culture, and individual human action.**

  Places are dynamic phenomena. They are created by people responding to the opportunities and constraints presented by their environments. As people live and work in places, they gradually impose themselves on their environment, modifying and adjusting it to suit their needs and express their values. At the same time, people gradually accommodate both to their physical environment and to the people around them. There is thus a continuous two-way process in which people create and modify places while at the same time being influenced by the settings in which they live and work. Places are not just distinctive outcomes of geographical processes; they are part of the processes themselves.

- **The first law of geography is "Everything is related to everything else, but near things are more related than are distant things."**

  A great deal of human activity is influenced by the "nearness principle," according to which people tend to seek to maximize the overall utility of places at minimum effort, to maximize connections between places at minimum cost, and to locate related activities as close together as possible. In doing so, people are responding to the friction of distance, the deterrent or inhibiting effect of distance on human activity. A distance-decay function describes the rate at which a particular activity or phenomenon diminishes with increasing distance from a given point.

- **Connectivity is also important, because contact and interaction are dependent on channels of communication and transportation: streets, highways, and telephone lines, for example.**

  Interdependence between places and regions can be sustained only through movement and flows. Accessibility and spatial interaction are two of the fundamental concepts that distinguish the study of human geography. The fundamental principles of spatial interaction can be reduced to four basic concepts: complementarity, transferability, intervening opportunities, and diffusion.

KEY TERMS

accessibility (p. 26)
capitalism (p. 18)
cognitive distance (p. 24)
cognitive image (p. 24)
cognitive space (p. 26)
distance-decay function (p. 25)
economies of scale (p. 28)
formal region (p. 32)

friction of distance (p. 24)
functional region (p. 32)
geodemographic research (p. 7)
geographical imagination (p. 35)
geographic information systems (GIS) (p. 21)

Global Positioning System (GPS) (p. 23)
globalization (p. 10)
human geography (p. 21)
identity (p. 3)
infrastructure (p. 28)
intersubjectivity (p. 34)
irredentism (p. 33)

latitude (p. 22)
lifeworld (p. 34)
longitude (p. 23)
neoliberal policies (p. 12)
ordinary landscapes (p. 33)
physical geography (p. 21)
place (p. 2)
region (p. 20)

regional geography (p. 21)
regionalism (p. 32)
regionalization (p. 31)
remote sensing (p. 21)
risk society (p. 19)
sectionalism (p. 33)

sense of place (p. 34)
site (p. 23)
situation (p. 23)
social relations (p. 5)
spatial analysis (p. 22)
spatial diffusion (p. 30)

spatial interaction (p. 27)
states (p. 5)
supranational organization
(p. 5)
symbolic landscapes (p. 33)
time-space convergence (p. 28)

topological space
(p. 26)
utility (p. 25)
world region
(p. 5)

ADDITIONAL READING

Agnew, J., D. Livingstone, and A. Rogers (eds.), *Human Geography: An Essential Anthology*. Cambridge, MA: Blackwell, 1996.

Beck, U., *Risk Society: Towards a New Modernity*. Thousand Oaks, CA: Sage, 1992.

Boyd, A., *An Atlas of World Affairs*, 11th ed. New York: Routledge, 1998.

Cloke, P., P. Crang, and M. Goodwin, *Envisioning Human Geographies*. London: Edward Arnold, 2004.

Dorling, D., and D. Fairbairn, *Mapping: Ways of Seeing the World*. London: Addison Wesley Longman, 1997.

Giddens, A., *Runaway World: How Globalization Is Shaping Our Lives*. New York: Routledge, 2000.

Gould, P., *Becoming a Geographer*. Syracuse: Syracuse University Press, 1999.

Harvey, D. W., *Explanation in Geography*. London: Edward Arnold, 1969.

Hubbard, P., *Thinking Geographically*. Continuum International Publishing, 2002.

Johnston, R. J., and J. D. Sidaway, *Geography and Geographers: Anglo-American Human Geography Since 1945*, 6th ed. London: Edward Arnold, 2004.

Livingstone, D. N., *The Geographical Tradition: Episodes in the History of a Contested Enterprise*. Oxford: Blackwell, 1993.

Phillips, M., *Contested Worlds: An Introduction to Human Geography*. Aldershot, UK: Ashgate, 2005.

Unwin, T., *The Place of Geography*. Upper Saddle River, NJ: Prentice Hall, 1996.

Warf, B. (ed.), *Encyclopedia of Human Geography*. Thousand Oaks, CA: Sage, 2005.

Wilford, J. N., *The Mapmakers*. New York: Vintage Books, 2000.

EXERCISES

At the end of each chapter, you will find exercises and activities based on using the Internet, along with some that do not require access to the Internet. This book has its own home page on the Internet, where you will find additional resources—maps, photographs, data—as well as exercises and activities that relate to each chapter. You will also find an evaluation checklist and suggestion form that you can mail to the authors electronically.

On the Internet

The Companion Website at www.mygeoscienceplace.com contains a number of tools designed to reinforce the geographic concepts presented in this chapter. Structured within a four-step learning path, these online resources include: chapter Key Concepts; a fully-searchable myeBook version of the chapter, which allows you to add notes and highlights; RSS news feeds from various media agencies; links to additional web resources; Human Geography Videos; a digital Glossary; Concept Review questions; Thinking Spatially mapping exercises; and Critical Thinking essay questions.

Unplugged

1. Consider geographical interdependence from the point of view of your own life. Take an inventory of your clothes, noting, where possible, the location each garment was manufactured. Where did you buy the garments? Was the store part of a regional, national, or international chain? What can you find out about the materials used in the garments? Where were they made?

2. Describe, as exactly and concisely as possible, the site (see p. 23) of your campus. Then describe its situation (see p. 23). Can you think of any reasons why the campus is sited and situated where it is? Would there be a better location? If so, why?

3. Choose a local landscape, one with which you are familiar, and write a short essay (500 words, or two double-spaced typed pages) on how the landscape has evolved over time. Note especially any evidence that physical environmental conditions have shaped any of the human elements in the landscape, together with any evidence of people having modified the physical landscape.

4. Figure 1.25 through Figure 1.27 show several examples of landscapes that have acquired a strong symbolic value because of the buildings, events, people, or histories with which they are associated. List five additional landscapes that have strong symbolic value to a large number of citizens of your own state or country, and state in 25 words or less why each setting has acquired such value.

The Changing Global Context

2

The story is told of a little Japanese girl who arrives in Los Angeles, sees a McDonald's restaurant, tugs her mother's sleeve, and says, "Look, Mother, they have McDonald's in this country, too."

It has become a cliché about the twenty-first century that everywhere will come to look like everywhere else, with the same McDonald's, Pizza Huts, and Starbucks; the same television programming with Hollywood movies and TV series; and the same malls selling the same Nike shoes, Philips electronics, IKEA furniture, and GAP clothing. Another cliché is that instantaneous global telecommunications, satellite television, and the Internet will soon overthrow all but the last vestiges of geographical differentiation in human affairs. Large corporations, according to this view, will no longer have strong ties to their home country, scattering their activities around the world in search of low-cost, low-tax locations. Employees will work as effectively from home, car, or beach as they could in the offices that need no longer exist. Events halfway across the world will be seen, heard, and felt with the same immediacy as events across town. National differences and regional cultures will dissolve, the cliché has it, as a global marketplace brings a uniform dispersion of people, tastes, and ideas.

Such developments are, in fact, highly unlikely. Even in the Information Age, geography will still matter and may well become more important than ever. Places and regions will undoubtedly change as a result of the new global context of the Information Age. But geography will still matter because of several factors: transport costs, different resource endowments, fundamental principles of spatial organization, people's territorial impulses, the resilience of local cultures, and the legacy of the past.

In this chapter we take a long-term, big-picture perspective on changing human geographies, emphasizing the continuing interdependence among places and regions. We show how geographical divisions of labor have evolved with the growth of a worldwide system of trade and politics and with the changing opportunities provided by successive technology systems. As a result of this evolution, the world is now structured around a series of core regions, semiperipheral regions, and peripheral regions; and globalization seems to be intensifying many of the differences among places and regions, rather than diminishing them.

Main Points

- The modern world-system has evolved through several distinctive stages, each of which has left its legacy in different ways on particular places, depending on their changing role within the world-system.

- At the end of the eighteenth century, the new technologies of the Industrial Revolution brought about the emergence of a global economic system that reached into almost every part of the world and into virtually every aspect of people's lives.

- Places and regions are part of a world-system that has been created as a result of processes of private economic competition and political competition between national states.

- Today, the world-system is highly structured and is characterized by three tiers: core regions, semiperipheral regions, and peripheral regions.

- The growth and internal colonization of the core regions could take place only with the foodstuffs, raw materials, and markets provided by the colonization of the periphery.

- Within each of the world's major regions, the successive technological innovations have transformed regional geographies.

- Globalization has intensified the differences between the core and the periphery and has contributed to the emergence of a digital divide and an increasing division between a fast world (about 15 percent of the world's population) and a slow world (the remaining 85 percent) with contrasting lifestyles and levels of living.

THE PREMODERN WORLD

The essential foundation for an informed human geography is an ability to understand places and regions as components of a constantly changing global system. In this sense, all geography is historical geography. Built into every place and each region is the legacy of a sequence of major changes in world geography. We can best understand the consequences of these changes for specific places and regions by thinking of the world as an evolving, competitive, political-economic system that has developed through successive stages of geographic expansion and integration. This evolution has affected the roles of individual places in different ways. It has also affected the nature of the interdependence among places. It explains why places and regions have come to be distinctive and how this distinctiveness has formed the basis of geographic variability. To understand the sequence of major changes in world geography, we need to begin with the hearth areas of the first agricultural revolution.

Hearth Areas

Systematically differentiated human geographies began with minisystems. A **minisystem** is a society with a single cultural base and a *reciprocal* social economy. That is, each individual specializes in particular tasks (tending animals, cooking, or making pottery, for example), freely giving any excess product to others, who reciprocate by giving up the surplus product of their own specialization. Such societies are found only in subsistence-based economies. Because they do not have (or need) an extensive physical infrastructure, minisystems are limited in geographic scale. Before the first agricultural revolution, in prehistoric times, minisystems had been based on hunting-and-gathering societies that were finely tuned to local physical environments.

The first agricultural revolution was a transition from hunter-gatherer minisystems to agricultural-based minisystems that were both more extensive and more stable. The transition began in the Proto-Neolithic (or early Stone Age) period, between 9000 and 7000 B.C., and was based on a series of technological preconditions: the use of fire to process food, the use of grindstones to mill grains, and the development of improved tools to prepare and store food. One key breakthrough was the evolution and diffusion of a system of slash-and-burn agriculture (also known as "swidden" cultivation—see Chapter 8). **Slash-and-burn** is a system of cultivation in which plants are harvested close to the ground, the stubble left to dry for a period, and then ignited, the burned stubble providing fertilizer for the soil. In Neolithic times (7000 to 5500 B.C.), this system involved planting familiar species of wild cereals or tubers on scorched land. Slash-and-burn did not require special tools or weeding, provided that cultivation was switched to a new plot after a few crops had been taken from the old one. Another key breakthrough was the domestication of cattle and sheep, a technique that had become established in a few regions by Neolithic times.

As cultural geographer Carl Sauer pointed out in his book *Agricultural Origins and Dispersals* (1952), these agricultural breakthroughs could take place only in certain geographic settings: where natural food supplies were plentiful; where the terrain was diversified (thus offering a variety of habitats and species); where soils were rich and relatively easy to till; and where there was no need for large-scale irrigation or drainage. Archaeological evidence suggests that the breakthroughs took place independently in several agricultural hearth areas and that agricultural practices diffused slowly outward from each (**Figures 2.1 and 2.2**). **Hearth areas** are geographic settings where new practices have developed and from which they have subsequently spread. The

Figure 2.1   Old World hearth areas The first agricultural revolution took place independently in several hearth areas, where hunter–gatherer communities began to experiment with locally available plants and animals in ways that eventually led to their domestication. From these hearth areas, improved strains of crops, domesticated animals, and new farming techniques diffused slowly outward. Farming supported larger populations than were possible with hunting and gathering, and the extra labor allowed the development of other specializations, such as pottery making and jewelry.

ANATOLIA
Fertile Crescent

Agricultural economy; 7000 B.C.

Direction of spread

Figure 2.2   New World hearth areas The agricultural hearth area in Central America eventually gave rise to the Aztec and Mayan world-empires; the hearth areas of the Andes in South America developed into the Inca world-empire.

main agricultural hearth areas were situated in four broad regions:

- *In the Middle East:* in the so-called Fertile Crescent around the foothills of the Zagros Mountains (parts of present-day Iran and Iraq), along the floodplains of the Tigris River and Euphrates River, around the Dead Sea Valley (Jordan and Israel), and on the Anatolian Plateau (Turkey).

- *In South Asia:* along the floodplains of the Ganga (Ganges), Brahmaputra, Indus, and Irawaddy rivers (Assam, Bangladesh, Burma, and northern India).

- *In China:* along the floodplain of the Huang He (Yellow) River.

- *In the Americas:* in Mesoamerica (the middle belt of the Americas that extends north to the North American Southwest and south to the Isthmus of Panama) around Tamaulipas and the Tehuacán Valley (Mexico) and in Arizona and New Mexico, and along the western slopes of the Andes in South America.

This transition to food-producing minisystems had several important implications for the long-term evolution of the world's geographies:

1. It allowed much higher population densities and encouraged the proliferation of settled villages.

2. It brought about a change in social organization, from loose communal systems to systems that were more highly organized on the basis of kinship. Kin groups provided a natural way of assigning rights over land and resources and of organizing patterns of land use.

3. It allowed some specialization in nonagricultural crafts, such as pottery, woven textiles, jewelry, and weaponry.

4. This specialization led to a fourth development: the beginnings of barter and trade between communities, sometimes over substantial distances.

Most minisystems vanished a long time ago, although some remnants have survived to provide material for Discovery Channel and National Geographic "Life in a Time Warp" specials (**Figure 2.3**). Examples of these residual and fast-disappearing minisystems are the bushmen of the Kalahari, the hill tribes of Papua New Guinea, and the

Figure 2.3   A remnant minisystem An Amazon tribal group, photographed in May 2008. Minisystems were rooted in subsistence-based social economies that were organized around reciprocity, each person specializing in certain tasks and freely giving any surplus to others, who in turn passed on the surplus from their own work. Today few of the world's remnant minisystems are "pure," unaffected by contact with the rest of the world. This group is believed to be related to the Tano and Aruak tribes of the Peru-Brazil border region. Brazil's National Indian Foundation believes there may be as many as 68 "uncontacted" groups around Brazil, although only 24 have been officially confirmed. Anthropologists say almost all of these tribes know about Western civilization and have sporadic contact with prospectors, rubber tappers, and loggers, but choose to turn their backs on civilization, usually because they have been attacked.

tribes of the Amazon rain forest. They contribute powerfully to regional differentiation and sense of place in a few enclaves around the world, but their most important contribution to contemporary human geographies is that they provide a stark counterpoint to the landscapes and practices of the contemporary world-system. A **world-system** is an *interdependent* system of countries linked by political and economic competition. The term, which was coined by historian Immanuel Wallerstein, is hyphenated to emphasize the interdependence of places and regions around the world.

The Growth of Early Empires

The higher population densities, changes in social organization, craft production, and trade brought about by the first agricultural revolution provided the preconditions for the emergence of several world-empires. A **world-empire** is a group of minisystems that have been absorbed into a common political system while retaining their fundamental cultural differences. The social economy of world-empires can be characterized as redistributive-tributary. That is, wealth is appropriated from producer classes by an elite class in the form of taxes or tribute. This redistribution of wealth is most often achieved through military coercion, religious persuasion, or a combination of the two. The best-known world-empires were the largest and longest lasting of the ancient civilizations—Egypt, Greece, China, Byzantium, and Rome (**Figure 2.4a**).

Urbanization These world-empires brought important new elements to the evolution of the world's geographies. One was the emergence of *urbanization* (see Chapter 10). Towns and cities became essential as centers of administration, as military garrisons, and as theological centers for the ruling classes, who were able to use a combination of military and theological authority to hold their empires together. As long as these early world-empires were successful, they gave rise not only to monumental capital cities but also to a whole series of smaller settlements, which acted as intermediate centers in the flow of tribute and taxes from colonized territories.

The most successful world-empires, such as the Greek and Roman, established quite extensive urban systems. In general, the settlements in these urban systems were not very large—typically ranging from a few thousand inhabitants to about 20,000. The seats of empire grew quite large, however. The Mesopotamian city of Ur, in present-day Iraq, for example, has been estimated to have reached a population of around 200,000 by 2100 B.C., and Thebes, the capital of Egypt, is thought to have had more than 200,000 inhabitants in 1600 B.C. Athens and Corinth, the largest cities of ancient Greece, had populations between 50,000 and 100,000 by 400 B.C. Rome at the height of the Roman Empire (around A.D. 200) may have had as many as a million inhabitants. The most impressive thing about these cities, though, was not so much their size as their degree of sophistication: elaborately laid out, with paved streets, piped water, sewage systems, massive monuments, grand public buildings, and impressive city walls.

Colonization Another important contribution of world-empires to evolving world geographies was **colonization**, the physical settlement in a new territory of people from a colonizing state. In part, this was an indirect consequence of the operation of the **law of diminishing returns**. This law refers to the tendency for productivity to decline after a certain point with the continued addition of capital and/or labor to a given resource base. World-empires could support growing populations only if overall levels of productivity could be increased. While some productivity gains could be achieved through better agricultural practices, harder work, and improvements in farm technology, a fixed resource base meant that as populations grew, overall levels of productivity fell. For each additional person working the land, the gain in production per worker was less. The usual response of empire builders was to enlarge the resource base by colonizing nearby land. This colonization had immediate spatial consequences in terms of establishing dominant/subordinate spatial relationships between hearths and colonies. It was also important in establishing hierarchies of settlements and creating improved transportation networks. The military underpinnings of colonization also meant that new towns and cities came to be carefully sited for strategic and defensive reasons.

The legacy of these important changes is still apparent in today's landscapes. The clearest examples are in Europe, where the Roman world-empire colonized an extensive territory that was controlled through a highly developed system of towns and connecting roads. Most Western European cities that are important today had their origin as Roman settlements. In quite a few it is possible to trace the original street layouts. In some it is possible to glimpse remnants of defensive city walls, paved streets, aqueducts, viaducts, arenas, sewage systems, baths, and public buildings (**Figure 2.4b,c**). In the modern countryside, meanwhile, we can still read the legacy of the Roman world-empire in arrow-straight roads built by Roman engineers and maintained and improved by successive generations. These early world-empires were also significant in developing a base of geographic knowledge (see Box 2.1, "Early Geographic Knowledge").

Some world-empires were exceptional in that they were based on a particularly strong central state, with totalitarian rulers who were able to organize large-scale communal land-improvement schemes using forced labor. These world-empires were found in China, India, the Middle East, Central America, and the Andean region of South America. Their dependency on large-scale land-improvement schemes (particularly irrigation and drainage schemes) as the basis for agricultural productivity has led some scholars to characterize them as *hydraulic societies*. Today their legacy can be seen in the

(a)

Figure 2.4   Greek colonies and the extent of the Roman empire (a) This map shows the distribution of the Greek poleis (city-states) and Carthaginian colonies and the spread of the Roman empire from 218 B.C. to A.D. 117.
(b) The Roman world-empire relied on a highly developed infrastructure of roads, settlements, and utilities. Roman road building was an important aspect of empire building, and many of the roads that were built by Romans became established as major routes throughout Europe. Wherever they could, Roman surveyors laid out roads in straight lines. This photo shows part of a Roman road that crosses the moorlands outside Manchester, England.
(c) The Pont du Gard aqueduct near Remoulins, France. Completed in 19 B.C., the whole project is a testament to the surveying and engineering skills of the Romans. (*Source:* Adapted from R. King et al., *The Mediterranean.* London: Arnold, 1977, pp. 59 and 64.)

(b)

(c)

2.1 GEOGRAPHY MATTERS

Early Geographic Knowledge

Several of the early world-empires were important in developing the intellectual significance and utility of geographic knowledge. Greek scholars developed the idea that places embody fundamental relationships between people and the natural environment and that the study of geography provides the best way of addressing the *interdependencies* between places and between people and nature. The Greeks were also among the first to appreciate the practical importance and utility of geographic knowledge, not least in politics, business, and trade. The word *geography* is in fact derived from the Greek language, the literal translation meaning "earth-writing" or "earth-describing." As Greek civilization developed, descriptive geographical writing came to be an essential tool for recording information about sea and land routes and for preparing colonists and merchants for the challenges and opportunities they would encounter in faraway places.

As Greek civilization blossomed, other aspects of geography also came to be valued. One important strand of scholarship, drawing on mathematics and astronomy, was concerned with the measurement of Earth and its accurate representation. Another important strand was philosophical. It was from this philosophical strand that the foundations of human geography were established. Strabo's 17-volume *Geography*, written between 8 B.C.

and A.D. 18, was particularly influential. Strabo systematically described places in order to address what he saw as the distinctive local relationships between nature and society. Today this sort of descriptive approach to geographical differentiation is known as *chorology* (or *chorography*) and is often referred to as the *regional approach* to geography. The problem with chorology is that in focusing on the distinctiveness of individual parts of Earth's surface (that is, distinctive places or homogeneous regions), it can lose sight not only of the whole but also of the relationships between places and regions. This problem was recognized by another Greek scholar, Ptolemy (ca. A.D. 90–168), whose eight-volume *Guide to Geography* was concerned with developing a comprehensive view of the world (see **Figure 2.A**).

Both Strabo and Ptolemy were writing at the height of the Roman Empire, and their work was part of the mainstream of classical scholarship. The Romans, however, were less interested than the Greeks in the scholastic and philosophical aspects of geography, though they did appreciate geographical knowledge as an aid to conquest, colonization, and political control. With the decline and fall of the Roman Empire, however, geography and geographical knowledge were neglected.

Figure 2.A   Ptolemy's map of the world Ptolemy began his *Guide to Geography* with an explanation of how to construct a globe, together with its parallels and meridians, and then showed how to project the world onto a plane surface. His map of the world stood for centuries as the basis for cartography. This example, published in 1482 by Leinhart Holle in Ulm, Germany, was typical of the basic map of the world in use at the time that Cristóbal Colón (Columbus) was considering the feasibility of going to China by sailing west. Unfortunately, Ptolemy had disregarded Eratosthenes' earlier calculations on the circumference of the Earth. As a result, Ptolemy's painstaking work, apparently so precise, contained fundamental inaccuracies that led generations of philosophers, monarchs, and explorers to underestimate the size of the globe. When Columbus did sail west from Spain in search of Asia, it was partly as a result of reading Ptolemy, whose miscalculations had led to an exaggerated conception of the size of Asia.

landscapes of terraced fields that have been maintained for generations in places like Sikkim, India, and East Java, Indonesia.

The Geography of the Premodern World

Figure 2.5 shows the generalized framework of human geographies in the Old World as they existed around A.D. 1400. The following characteristics of this period are important:

1. Harsher environments in continental interiors were still characterized by isolated, subsistence-level, kin-ordered hunting-and-gathering minisystems.

2. The dry belt of steppes and desert margins stretching across the Old World from the western Sahara to Mongolia was a continuous zone of kin-ordered pastoral minisystems.

3. The hearths of sedentary agricultural production extended in a discontinuous arc from Morocco to China, with two main outliers, in the central Andes and in Mesoamerica.

The dominant centers of global civilization were China, northern India (both of them hydraulic variants of world-empires), and the Ottoman Empire of the eastern Mediterranean. They were all linked by the Silk Road, a series of overland trade routes between China and Mediterranean Europe (**Figure 2.6**).

By A.D. 1400, other important world-empires had developed in Southeast Asia, in Muslim city-states of coastal North Africa, in the grasslands of West Africa, around the gold and copper mines of East Africa, and in the feudal kingdoms and merchant towns of Europe. Over time, all of these more developed realms were interconnected through trade, which meant that several emerging centers of capitalism existed. Port cities were particularly important, and among the leading centers were the city-state of Venice; the Hanseatic League of independent city-states in northwestern Europe (including Bergen, Bremen, Danzig, Hamburg, Lübeck, Riga, Stockholm, and Tallinn and affiliated trading outposts in other cities, including Antwerp, Bruges, London, Turku, and Novgorod); and Cairo, Calicut, Canton, and Malacca in North Africa and Asia. Traders in these port cities began to organize the production of agricultural

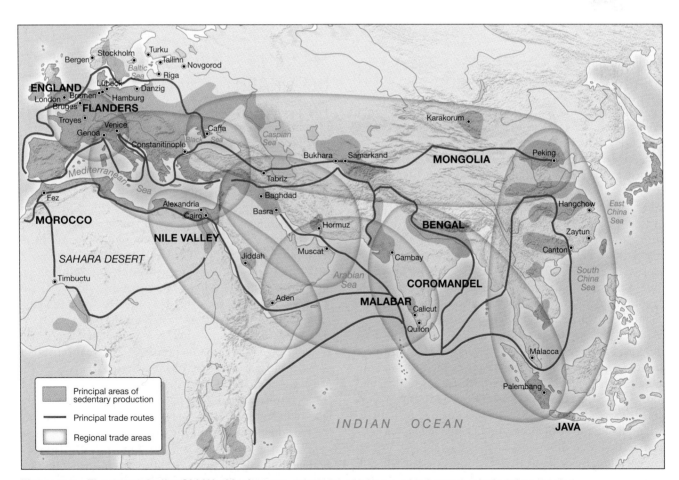

Figure 2.5   The precapitalist Old World, circa A.D. 1400 Principal areas of sedentary agricultural production are shaded. Some long-distance trade took place from one region to another, but for the most part it was limited to a series of overlapping regional circuits of trade. (*Source:* Adapted from R. Peet, *Global Capitalism: Theories of Societal Development.* New York: Routledge, 1991; J. Abu-Lughod, *Before European Hegemony: The World-System* A.D. *1200–1350.* New York: Oxford University Press, 1989; and E. R. Wolf, *Europe and the People Without History.* Berkeley: University of California Press, 1983.)

Figure 2.6   The Silk Road This map shows the trade routes of the Silk Road as they existed between 112 B.C. and A.D. 100. From Roman times until Portuguese navigators found their way around Africa and established seaborne trade routes, the Silk Road provided the main East–West trade route between Europe and China. This shifting trail of caravan tracks facilitated the exchange of silk, spices, and porcelain from the East and gold, precious stones, and Venetian glass from the West. The ancient cities of Samarkand, Bukhara, and Khiva stood along the Silk Road, places of glory and wealth that astonished Western travelers such as Marco Polo in the thirteenth century. These cities were East–West meeting places for philosophy, knowledge, and religion, and in their prime they were known for producing scholars in mathematics, music, architecture, and astronomy, such as Al Khoresm (780–847), Al Biruni (973–1048), and Ibn Sind (980–1037). The cities' prosperity was marked by impressive feats of Islamic architecture.

specialties, textiles, and craft products in their respective hinterlands.

The **hinterland** of a town or city is its sphere of economic influence—the tributary area from which it collects products to be exported and through which it distributes imports. By the fifteenth century, several regions of budding capitalism existed: northern Italy, Flanders, southern England, the Baltic Sea region, the Nile Valley, Malabar, Coromandel, Bengal, northern Java, and southeast coastal China (Figure 2.5).

Between roughly A.D. 500 and A.D. 1400, geographic knowledge was preserved and expanded by Chinese and Islamic scholars. Chinese maps of the world from the same period were more accurate than those of European cartographers because the Chinese were able to draw on information brought back by Imperial China's admirals, who successfully navigated much of the Pacific and Indian oceans. They recognized, for example, that Africa was a southward-pointing triangle, whereas on European and Arabic maps of the time it was always represented as pointing eastward.

With the rise of Islamic power in the Middle East and the Mediterranean in the seventh and eighth centuries A.D., centers of scholarship emerged throughout these regions, including Baghdad, Damascus, Cairo, and Granada, Spain. Here, surviving Greek and Roman texts were translated into Arabic by scholars such as Al-Battani, Al-Farghani, and Al-Khwarazmi. These Islamic scholars were also able to draw on Chinese geographical writing and cartography brought back by traders along the Silk Road. The requirement that the Islamic religious faithful undertake at least one pilgrimage to Mecca created a demand for travel guidebooks. It also brought scholars from all over the Arab world into contact with one another, stimulating considerable debate over different philosophical views of the world and of people's relationship with nature.

MAPPING A NEW WORLD GEOGRAPHY

The modern world-system had its origins in parts of fifteenth-century Europe, when exploration beyond European shores began to be seen as an important way of opening up new opportunities for trade and economic expansion. By the sixteenth century, new techniques of shipbuilding and navigation had begun to bind more and more places and regions together through trade and political competition. As a result, more and more peoples around the world became exposed to one another's technologies and ideas. Their different resources, social structures, and cultural systems resulted in quite different pathways of development, however. Some societies were incorporated into the new, European-based international economic system faster than others; some resisted incorporation; and some sought alternative systems of economic and political organization. Some parts of the world were barely penetrated, if at all, by the European world-system. Australia and New Zealand, for example, were discovered by Europeans only in the late eighteenth century. Regions not yet absorbed into the world-system are called **external arenas**.

With the emergence of this modern world-system at the beginning of the sixteenth century, a whole new geography began to emerge. Although several regions of budding capitalist production existed, and although imperial China could boast of sophisticated achievements in science, technology, and navigation, it was European merchant capitalism that reshaped the world. Several

Figure 2.7 Cotton plantation Plantation agriculture dominated the many territories acquired by European settlers.

factors motivated European overseas expansion. A relatively high-density population and a limited amount of cultivable land meant that a continuous struggle occurred to provide enough food. Meanwhile, the desire for overseas expansion was intensified both by competition among a large number of small monarchies and by inheritance laws that produced large numbers of impoverished aristocrats with little or no land of their own. Many of these were eager to set out for adventure and profit.

Added to these motivating factors were the enabling factors of innovations in shipbuilding, navigation, and gunnery. In the mid-1400s, for example, the Portuguese developed a cannon-armed ship—the caravel—that could sail anywhere, defend itself against pirates, pose a threat to those who were initially unwilling to trade, and carry enough goods to be profitable. The quadrant (1450) and the astrolabe (1480) enabled accurate navigation and mapping of ocean currents, prevailing winds, and trade routes (see Box 2.2, "Geography and Exploration"). Naval power enabled the Portuguese and the Spanish to enrich their economies with capital from gold and silver plundered from the Americas.

Equipped with better maps and navigation techniques, Europeans not only sent adventurers in search of gold and silver but also to commandeer land, decide on its use, and exploit coerced labor to produce high-value crops (such as sugar, cocoa, cotton, and indigo) on **plantations**, large landholdings that usually specialize in the production of one particular crop for market (**Figure 2.7**). Those regions whose populations were resistant to European disease and

which also had high population densities, a good resource base, and strong states were able to keep Europeans at arm's length. For the most part, these regions were in South and East Asia. Their dealings with Europeans were conducted through a series of coastal trading stations. Textiles were an important commodity, as reflected by English language:

> The word "satin" comes from the name of an unknown city in China that Arab traders called Zaitun. "Khaki" is the Hindi word for dusty. The word "calico" comes from India's southwestern coastal city of Calicut; "chintz," from the Hindi name for a printed calico; "cashmere," from Kashmir. "Percale" comes from the Farsi word *pargalah*. Another Farsi derivative is "seersucker," whose bands of alternating smooth and puckered fabric prompted a name that literally means milk and sugar. Still another Farsi borrowing is "taffeta," which comes from the Farsi for "spun." The coarse cloth we call "muslin" is named for Mosul—the town in Iraq—while "damask" is a short form of Damascus. Cotton takes its name from *qutun*, the Arabic name of the fiber.[1]

Within Europe, meanwhile, innovations in business and finance (banking, loan systems, credit transfers, commercial insurance, and courier services, for example)

[1] B. Wallach, *Understanding the Cultural Landscape*. New York: Guilford Press, 2005, p. 148.

Geography and Exploration

The Portuguese, under the sponsorship of Dom Henrique (known as Prince Henry the Navigator), established a school of navigation and cartography and began to explore the Atlantic and the coast of Africa. **Cartography** is the name given to the system of practical and theoretical knowledge about making distinctive visual representations of Earth's surface in the form of maps. **Figure 2.B** shows the key voyages of discovery. Portuguese explorer

Departure from Lisbon for Brazil, the East Indies, and America. Illustration from "Americae Tertia Pars" (1592).

Figure 2.B The European Age of Discovery The European voyages of discovery can be traced to Portugal's Prince Henry the Navigator (1394–1460), who set up a school of navigation and financed numerous expeditions with the objective of circumnavigating Africa in order to establish a profitable sea route for spices from India. The knowledge of winds, ocean currents, natural harbors, and watering places built up by Henry's captains was an essential foundation for the subsequent voyages of Cristóbal Colón (Columbus), da Gama, de Magalhães (Magellan), and others. The end of the European Age of Discovery was marked by Captain James Cook's voyages to the Pacific.

Bartholomeu Dias reached the Cape of Good Hope (the southern tip of Africa) in 1488. In 1492, Cristóbal Colón (Christopher Columbus), from Genoa, Italy, sailed to Hispaniola (the island that is now Haiti and the Dominican Republic) under the sponsorship of the Castillian (Spanish) monarchy. Six years later, Vasco da Gama reached India; two years after that, Pedro Cabral crossed the Atlantic from Portugal to Brazil. A small fleet of Portuguese ships reached China in 1513, and the first circumnavigation of the globe was completed in 1522 by Juan Sebastián del Cano, a survivor of the expedition led by Portuguese navigator Fernando de Magalhães (better known to English speakers as Magellan) in 1519. Portuguese successes inspired other countries to attempt their own voyages of discovery, all of them in pursuit of commercial advantage and economic gain. These explorations led to an invaluable body of knowledge about ocean currents, wind patterns, coastlines, peoples, and resources.

Geographical knowledge acquired during this Age of Discovery was crucial to the expansion of European political and economic power in the sixteenth century. In societies that were becoming more commercially oriented and profit conscious, geographical knowledge became a valuable commodity in itself. Information about regions and places was a first step toward controlling and influencing them, and this in turn was an important step toward amassing wealth and power. At the same time, every region began to be opened up to the influence of other regions because of the economic and political competition that was unleashed by geographical discovery. As the New World was being affected by European colonists, missionaries, and adventurers, the countries of the Old World found themselves pitched into competition with one another for overseas resources. Meanwhile, new crops, like corn and potatoes, introduced to Europe from the New World, profoundly affected local economies and ways of life.

The growth of a commercial world economy meant that objectivity in cartography and geographical writing became essential. Navigation, political boundaries, property rights, and rights of movement all depended on accuracy and impartiality. Success in commerce depended on clarity and reliability in describing the opportunities and dangers presented by one region or another. International rivalries required sophisticated understandings of the relationships among nations, regions, and places. Geography became a key area of knowledge. The historical period in Europe known as the Renaissance (from the mid-fourteenth to the mid-seventeenth centuries) saw an explosion of systematic map-making and the development of new **map projections** (see Appendix 1) and geographical descriptions (**Figure 2.C**).

Figure 2.C   Dutch master Johannes Vermeer's painting *The Geographer* (1668–1669) In Renaissance Europe, the study of geography not only contributed to the growth of scientific knowledge but also helped to support European overseas expansion. Vermeer's geographer is surrounded with accurately rendered cartographic objects, including a wall chart of the seacoasts of Europe, published by Willem Blau in 1658, and a globe made by Jodocus Hondius in 1618.

Throughout the seventeenth and eighteenth centuries, the body of geographic knowledge increased steadily as more and more of the world was explored, using increasingly sophisticated techniques of survey and measurement. Some of the most important advances in scientific cartographic techniques took place in France, where Louis XIV appointed an official cartographer, Nicolas Sanson, to produce a set of accurate maps of French territory. This was achieved through innovative techniques of measurement. Levels, surveying instruments, and telescopes, together with improved mathematical methods, helped to establish an accurate framework of known points throughout France. These points formed the basis of a system of triangulation, allowing French cartographers to produce advanced and innovative maps that became the model for cartographers in other countries. In 1747, under the direction of the Cassini family, the first of a series of 182 sheet maps designed to cover the whole of France was published. Engraved on copper plate and printed at a scale of 1:86,400 (1 inch to 2,400 yards), these maps showed towns in plan view; used a variety of symbols to represent important landmarks such as churches and mills; used a series of short parallel lines (hachures) instead of perspective drawings to show relief; and used tints to show various categories of land cover, such as forest and marsh.

helped to increase savings, investment, and commercial activity. European merchants and manufacturers also became adept at **import substitution**—copying and making goods previously available only by trading. The result was the emergence of Western Europe as the core region of a world-system that had penetrated and incorporated significant portions of the rest of the world.

For Europe this overseas expansion stimulated still further improvements in technology. These included new developments in nautical mapmaking, naval artillery, shipbuilding, and the use of sail. The whole experience of overseas expansion also provided a great practical school for entrepreneurship and investment. In this way, the self-propelling growth of merchant capitalism was intensified and consolidated.

For the periphery, European overseas expansion meant dependency (as it has ever since for many of the world's peripheral regions). At worst, territory was forcibly occupied and labor systematically exploited. At best, local traders were displaced by Europeans, who imposed their own terms of economic exchange. Europeans soon destroyed most of the Muslim shipping trade in the Indian Ocean, for example, and went on to capture a large share of the oceangoing trade within Asia, selling Japanese copper to China and India, Persian carpets to India, Indian cotton textiles to Japan, and so on.

As revolutionary as these changes were, however, they were constrained by a technology that rested on wind and water power, on wooden ships and structures, and on wood for fuel. Grain mills, for example, were built of wood and powered by water or wind. They could generate only modest amounts of power and only at sites determined by physical geography, not human choice. Within the relatively small European landmass, wood for structural use and for fuel competed for acreage with food and textile fibers.

More important, however, was the size and strength of timber, which imposed structural limits on the size of buildings, the diameter of waterwheels, the span of bridges, and so on. In particular, it imposed limits on the size and design of ships, which in turn imposed limits on the volume and velocity of world trade. The expense and relative inefficiency of horse- or ox-drawn wagons for overland transportation also meant that for a long time the European world-system could penetrate into continental interiors only along major rivers.

After 300 years of evolution, roughly between 1450 and 1750, the world-system had incorporated only parts of the world. The principal spheres of European influence were Mediterranean North Africa, Portuguese and Spanish colonies in Central and South America, Indian ports and trading colonies, the East Indies, African and Chinese ports, the Greater Caribbean, and British and French territories in North America. The rest of the world functioned more or less as before, with slow-changing geographies based on modified minisystems and world-empires that were only partially and intermittently penetrated by market trading.

Core and Periphery in the New World System

With the new production and transportation technologies of the Industrial Revolution (from the late 1700s), capitalism truly became a global system that reached into virtually every part of the inhabited world and into virtually every aspect of people's lives. Human geographies were recast again, this time with a more interdependent dynamic. New production technologies, based on more efficient energy sources, helped raise levels of productivity and create new and better products that stimulated demand, increased profits, and created a pool of capital for further investment. New transportation technologies triggered successive phases of geographic expansion, allowing for internal development as well as for external colonization and **imperialism** (the deliberate exercise of military power and economic influence by powerful states in order to advance and secure their national interests—see Chapter 7). As a result, human geography came to be reinterpreted by new generations of scholars (see Box 2.3, "The Foundations of Modern Geography").

Since the seventeenth century, the world-system has been consolidated with stronger economic ties between countries. It has also been extended, with all the world's countries eventually becoming involved to some extent in the interdependence of the capitalist system. Although there have been some instances of resistance and adaptation, the overall result is that a highly structured relationship between places and regions has emerged. This relationship has been organized around three tiers: *core, semiperipheral,* and *peripheral* regions. These broad geographic divisions have evolved—and are still evolving—through a combination of processes of private economic competition and competition among states.

The **core regions** of the world-system at any given time are those that dominate trade, control the most advanced technologies, and have high levels of productivity within diversified economies. As a result, they enjoy relatively high per capita incomes. The first core regions of the world-system were the trading hubs of Holland and England, joined soon afterward by France (**Figure 2.8**). The continuing success of core regions depends on their dominance and exploitation of other regions. This dominance in turn depends on the participation of these other regions within the world-system. Initially, such participation was achieved by military enforcement, then by European colonialism.

Colonialism involves the establishment and maintenance of political and legal domination by a state over a separate and alien society. This domination usually involves some colonization (that is, the physical settlement of people from the colonizing state) and always results in economic exploitation by the colonizing state. After World War II, the sheer economic and political influence of the core regions was sufficient to maintain their dominance without political and legal control, and colonialism was gradually phased out.

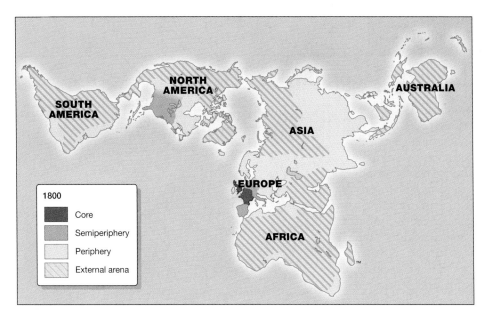

Figure 2.8 The world-system core, semiperiphery, and periphery in 1800 (*Source:* Map projection, Buckminster Fuller Institute and Dymaxion Map Design, Santa Barbara, CA. The word Dymaxion and the Fuller Projection™ Map design are trademarks of the Buckminster Fuller Institute, Santa Barbara, California, © 1938, 1967 & 1992. All rights reserved.)

Regions that have remained economically and politically unsuccessful throughout this process of incorporation into the world-system are peripheral. **Peripheral regions** are characterized by dependent and disadvantageous trading relationships, by primitive or obsolescent technologies, and by undeveloped or narrowly specialized economies with low levels of productivity.

Transitional between core regions and peripheral regions are semiperipheral regions. **Semiperipheral regions** are able to exploit peripheral regions but are themselves exploited and dominated by core regions. They consist mostly of countries that were once peripheral. This semiperipheral category underlines the fact that neither peripheral status nor core status is necessarily permanent. The United States and Japan both achieved core status after having been peripheral; Spain and Portugal, part of the original core in the sixteenth century, became semiperipheral in the nineteenth century but are now once more part of the core. Quite a few countries, including Brazil, India, Mexico, South Korea, and Taiwan, have become semiperipheral after first having been incorporated into the periphery of the world-system and then developing a successful manufacturing sector that moved them into semiperipheral status.

An important determinant of these changes in status is the effectiveness of states in ensuring the international competitiveness of their domestic producers. They can do this in several ways: by manipulating markets (protecting domestic manufacturers by charging taxes on imports, for example); by regulating their economies (enacting laws that help to establish stable labor markets, for example); and by creating physical and social infrastructures (spending public funds on road systems, ports, educational systems, and so on). Because some states are more successful than others in pursuing these strategies, the hierarchy of three geographical tiers is not rigid. Rather, it is fluid, providing a continually changing framework for geographical transformation within individual places and regions.

The colonization and imperialism that accompanied the expansion of the world-system was closely tied to the evolution of world leadership cycles. **Leadership cycles** are periods of international power established by individual states through economic, political, and military competition. In the long term, success in the world-system depends on economic strength and competitiveness, which brings political influence and pays for military strength. With a combination of economic, political, and military power, individual states can dominate the world-system, setting the terms for many economic and cultural practices and imposing their particular ideology by virtue of their preeminence. This kind of dominance is known as hegemony.

Hegemony refers to domination over the world economy, exercised—through a combination of economic, military, financial, and cultural means—by one national state in a particular historical epoch. Over the long run, the costs of maintaining this kind of power and influence tend to weaken the hegemon. This phase of the cycle is known as imperial overstretch. It is followed by another period of competitive struggle, which brings the possibility of a new dominant world power (see Box 2.4, "World Leadership Cycles").

The Industrialization of the World's Core Regions

Beginning in the late eighteenth century, a series of technological innovations in power and energy, transportation, and manufacturing processes resulted in some crucial changes in patterns of economic development. Each of these major clusters of technological innovations created new demands for natural resources as well as labor forces and markets. The result was that each major cluster of technological innovations—called technology

2.3 GEOGRAPHY MATTERS

The Foundations of Modern Geography

The foundations of geography as a formal academic discipline can be traced to the writings of a few key scholars in the eighteenth and nineteenth centuries. These included Immanuel Kant, Alexander von Humboldt, Karl Ritter, and Friedrich Ratzel. Their contribution, in general terms, was to move geography away from straightforward descriptions of parts of Earth's surface toward explanations and generalizations about the relationships of different phenomena within and among particular places.

Writing in the latter part of the eighteenth century, Immanuel Kant (**Figure 2.D**) distinguished between specific fields of knowledge represented by disciplines such as physics and biology, and two general fields of knowledge: geography, which classified things according to space (that is, regions), and history, which classified things according to time (that is, periods or epochs). Kant's interpretation of the field of geography was influential. He recognized the importance of commercial geography, theological geography, moral geography (concerning people's customs and ways of life), mathematical geography (concerning the Earth's shape and motion), and political geography—and saw them all as being heavily influenced by the underlying physical geography.

German geographer Alexander von Humboldt (1769–1859) set about the task of collecting and analyzing data about the relationships between the spatial distribution of rocks, plants, and animals. He traveled for five years (1799–1804) in South America, collecting data in order to identify relationships and make generalizations. Von Humboldt emphasized the mutual causality among species and their physical environment: the interdependence of people, plants, and animals on one another within specific physical settings. In this way, von Humboldt showed how people, like other species, have to adapt to their environment and how their behaviors also affect the environment around them.

O. B. Bosio dis.   *F. Birck inc.*

Emanuele Kant.

Figure 2.D   Immanuel Kant German philosopher Immanuel Kant (1724–1804) is regarded as one of the most influential thinkers of the Enlightenment in Europe. His belief in the intellectual importance of geography was an important element in establishing the subject as a formal discipline.

systems—tended to favor different regions and different kinds of places. **Technology systems** are clusters of interrelated energy, transportation, and production technologies that dominate economic activity for several decades at a time—until a new cluster of improved technologies evolves. What is especially remarkable about technology systems is that they have come along at about 50-year intervals. Since the beginning of the Industrial Revolution, we can identify four of them:

1790–1840: early mechanization based on water power and steam engines; development of cotton

textiles and ironworking; development of river transport systems, canals, and turnpike roads.

1840–1890: exploitation of coal-powered steam engines; steel products; railroads; world shipping; and machine tools.

1890–1950: exploitation of the internal combustion engine; oil and plastics; electrical and heavy engineering; aircraft; radio and telecommunications.

1950–1990: exploitation of nuclear power, aerospace, electronics, and petrochemicals; development of limited-access highways and global air routes.

German geographer Karl Ritter (1779–1859) was the founder of the tradition of regional geography. Ritter's 20-volume *Erdkunde*, published between 1817 and 1859, was a monumental work of comparison and classification. Like von Humboldt, Ritter saw geography as an integrative science, able to reveal the interdependencies between people and nature. Ritter's approach was to establish a framework for scientific comparisons and generalizations by dividing the continents into broad physical units and then subdividing these into coherent regions with distinctive attributes.

By the mid-nineteenth century, thriving geographical societies had been established in a number of cities, including Berlin, London, Frankfurt, Moscow, New York, and Paris. By 1899, there were 62 geographical societies worldwide, and university chairs of geography had been created in many of the most prestigious universities around the world. It must be said, however, that geography was seen at first in narrow terms, as the discipline of exploration. Because the importance of geography was linked so clearly to European commercial and political ambitions, ways of geographic thinking also changed. Places and regions tended to be portrayed from a distinctly European point of view and from the perspective of particular national, commercial, and religious interests. Geography mattered, but mainly as an instrument of colonialism.

One result of geography's involvement with colonialism and imperialism was that the discipline fostered ethnocentrism and masculinism. **Ethnocentrism** is the attitude that one's own race and culture are superior to those of others. **Masculinism** is the assumption that the world is, and should be, shaped mainly by men for men. These trends became more and more explicit as European dominance increased, reaching a peak in the late nineteenth century at the height of European geopolitical influence.

We should also note that most of the geographic writing in the nineteenth century was strongly influenced by environmental determinism. **Environmental determinism** is a doctrine holding that human activities are controlled by the environment. It rests on a belief that the physical attributes of geographical settings are the root not only of people's physical differences (skin color, stature, and facial features, for example) but also of differences in people's economic vitality, cultural activities, and social structures.

One of the most influential of the pioneer academic geographers who propagated this belief was Friedrich Ratzel (1844–1904). Ratzel, a German, was strongly influenced by Charles Darwin's theories about species' adaptation to environmental conditions and competition for living space. (Darwin's *On the Origin of Species* was published in 1859.) Ratzel's own ideas were carried to the United States by Ellen Churchill Semple, a former student of his who went on to teach at the University of Chicago and, later, Clark University. One of the best-known U.S. proponents of Ratzel's ideas was Ellsworth Huntington, who taught geography at Yale University between 1907 and 1917. Ratzel argued that civilization and successful economic development are largely the result of invigorating climates, which he defined as temperate climates with marked seasonal variations and varied weather but without prolonged extremes of heat, humidity, or cold. Huntington used this assertion to provide a rationale for the domination of northwestern Europe and the northeastern United States within the world economy.

Environmental determinists thus tend to think in terms of the influence of the physical environment on people rather than the other way around. The idea that peoples' social and economic development and behavior are fundamentally shaped by their physical environment lasted well into the twentieth century, though geographers now regard it as simplistic.

A fifth technology system, still incomplete, began to take shape in the 1980s with a series of innovations that are now being commercially exploited:

1990 onward: exploitation of solar energy, robotics, microelectronics, biotechnology, nanotechnology, advanced materials (fine chemicals and thermoplastics, for example), and information technology (digital telecommunications and geographic information systems, for example).

Each of these technology systems has rewritten the geography of economic development as it has shifted the balance of advantages between regions (**Figure 2.9**). From the mid-1800s, industrial development spread to new regions, whose growth then became interdependent with the fortunes of others through a complex web of production and trade.

The Industrial Revolution began in England toward the end of the eighteenth century and eventually resulted not only in the complete reorganization of the geography of the original European core of the world-system but also in an extension of the world-system core to the United States and Japan.

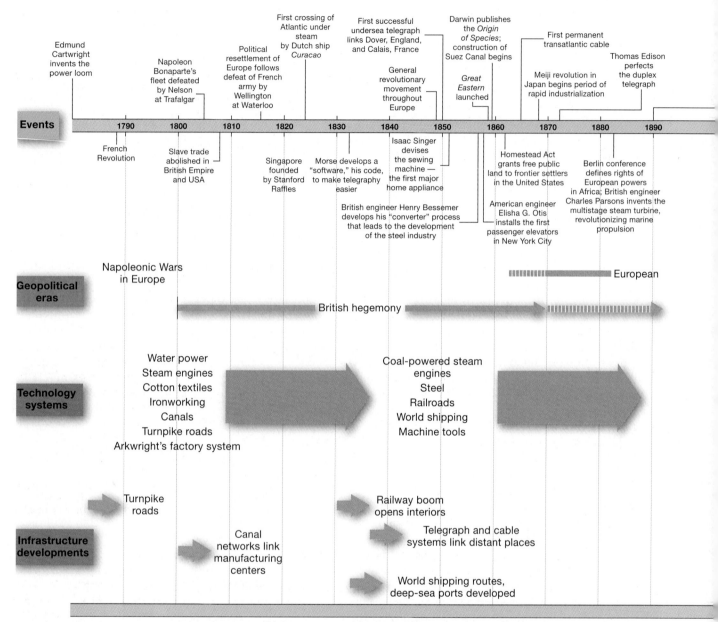

Figure 2.9 Technological change and economic development The Industrial Revolution at the end of the eighteenth century was driven by a technology system based on water power and steam engines, cotton textiles, ironworking, river transport systems, and canals. Since then there have been several more technology systems, each opening new geographic frontiers and rewriting the geography of economic development while shifting the balance of advantages between regions. Overall, the opportunities for development created by each new technology system have been associated with distinctive economic epochs and long-term fluctuations in the overall rate of change of prices in the economy.

Europe: Three Waves of Industrialization

In Europe, three distinctive waves of industrialization occurred. The first, between about 1790 and 1850, was based on the initial cluster of industrial technologies (steam engines, cotton textiles, and ironworking) and was very localized (**Figure 2.10**). It was limited to a few regions in Britain where industrial entrepreneurs and workforces had first exploited key innovations and the availability of key resources (coal, iron ore, water). Although these regions shared the common impetus of certain key innovations, each of them retained its own technological traditions and

industrial style. From the start, then, industrialization was a regional-scale phenomenon.

The second wave of industrialization, between about 1850 and 1870, involved the diffusion of industrialization to most of the rest of Britain and to parts of northwest Europe, particularly the coalfields of northern France, Belgium, and Germany (Figure 2.10). This second wave also brought a certain amount of reorganization to the first-wave industrial regions as a new cluster of technologies (steel, machine tools, railroads, steamships) brought new opportunities, new locational

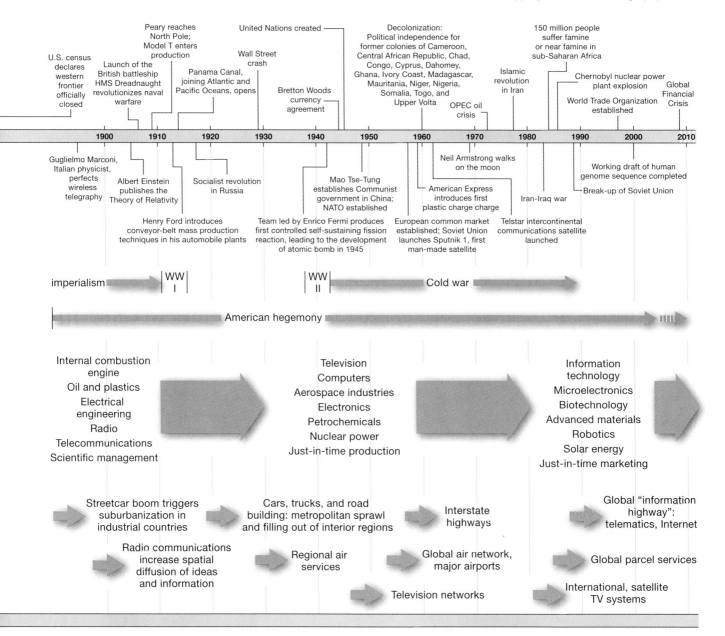

Figure 2.9 Continued

requirements, new business structures, and new forms of societal organization. New opportunities were created as railroads and steamships made more places accessible, bringing their resources and their markets into the sphere of industrialization. New materials and new technologies (steel, machine tools) created opportunities to manufacture and market new products. These new activities brought some significant changes to the logic of industrial location.

The importance of railway networks, for example, attracted industry away from smaller towns on the canal systems toward larger towns with good rail connections. The importance of steamships for coastal and international trade attracted industry to larger ports. At the same time, the importance of steel produced concentrations of heavy industry in places with nearby supplies

of coal, iron ore, and limestone. The scale of industry increased as new technologies and improved transportation made larger markets accessible to firms. Local, family firms became small companies that were regional in scope. Small companies grew to become powerful firms serving national markets, and specialized business, legal, and financial services emerged within larger cities to assist them. The growth of new occupations transformed the structure of social classes, and this transformation in turn came to be reflected in the politics and landscapes of industrial regions.

The third wave of industrialization, between 1870 and 1914, saw a further reorganization of the geography of Europe as yet another cluster of technologies (including electricity, electrical engineering, and telecommunications) imposed different needs and created new

Figure 2.10   The spread of industrialization in Europe
European industrialization began with the emergence of small industrial regions in several parts of Britain, where early industrialization drew on local mineral resources, water power, and early industrial technologies. As new rounds of industrial and transportation technologies emerged, industrialization spread to other regions with the right locational attributes: access to raw materials and energy sources, good communications, and large labor markets.

opportunities. During this period, industrialization spread for the first time to more remote parts of Britain, France, and Germany and to most of the Netherlands, southern Scandinavia, northern Italy, eastern Austria, and Catalonia, Spain. The overall result was to create a core within a core. Within the world-system core, processes of industrialization, modernization, and urbanization had forged a core of prosperity centered on the "Golden Triangle" stretching between London, Paris, and Berlin.

The United States Manufacturing Belt   By the end of the nineteenth century, the core of the world-system had itself extended to include the United States and Japan (**Figure 2.11**). The United States, politically independent just before the onset of the Industrial Revolution, was able to make the transition from the periphery to the core because of several favorable circumstances. Its vast natural resources of land and minerals provided the raw materials for a wide range of industries that could grow and organize without being hemmed in and fragmented by political boundaries. Its population, growing quickly through immigration, provided a large and expanding market and a cheap and industrious labor force. Its cultural and trading links with Europe provided business contacts, technological know-how, and access to capital (especially British capital) for investment in a basic infrastructure of canals, railways, docks, warehouses, and factories.

As in Europe, industrialization developed around preexisting centers of industrialization and population and was shaped by the resource needs and market opportunities of successive clusters of technology. America's industrial strength was established at the beginning of the twentieth century with the development of a new cluster of technologies that included the

Figure 2.11   The world-system core, semiperiphery, and periphery in 1900
(*Source:* Map projection, Buckminster Fuller Institute and Dymaxion Map Design, Santa Barbara, CA. The word Dymaxion and the Fuller Projection™ Map design are trademarks of the Buckminster Fuller Institute, Santa Barbara, California, © 1938, 1967 & 1992. All rights reserved.)

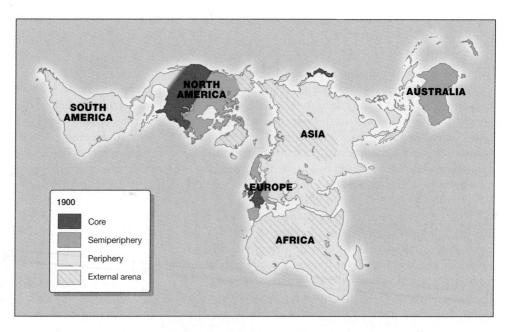

internal combustion engine, oil and plastics, electrical engineering, and radio and telecommunications (see Chapter 7). The outcome was a distinctive manufacturing region, known as the Manufacturing Belt, that by 1920 stretched from Boston and Baltimore in the East to Milwaukee and St. Louis in the West (**Figure 2.12**)—another core within a core.

Japan's First Economic Miracle Japan remained a feudal world-empire for a long time before vaulting into the core of the world-system. In 1868, a revolution deposed the feudal Tokugawa regime and restored an old imperial dynasty—the Meiji—whose backers deliberately set out to modernize the country as quickly as possible, whatever the cost. Under the slogan "National Wealth and Military Strength," the government took an unprecedented role, strong and deliberate, in creating capitalist enterprises and establishing a modern infrastructure. The costs were borne chiefly by a domestic workforce (whose rewards for hard work were meager) and by the victims of Japanese military aggression in Korea, Formosa (now Taiwan), and Manchuria.

Just as Japanese industry was becoming established with a base in textiles and shipbuilding, World War I provided a timely opportunity to expand productive capacity. With much European and American industry diverted to supply war materials, Japanese textile manufacturers were able to expand into Asian and Latin American markets. Meanwhile, with a great deal of merchant shipping destroyed by the hostilities, Japanese shipping industries took the opportunity to expand their merchant fleet. The profits from this commercial activity paid for the rebirth of the Japanese navy, which by 1918 had a dozen battleships and battle cruisers, with 16 more under construction. The United States at the time had only 14 battleships, with 3 under construction. Within 50 years of the Meiji revolution, Japan had joined the core of the world-system.

Internal Development of the Core Regions

Within the world's core regions, the transformation of regional geographies hinged on successive innovations in transport technology. These innovations opened up agrarian interiors and intensified interregional trading networks. Farmers were able to mechanize their equipment, and manufacturing companies were able to take over more resources and more markets.

The first phase of this internal geographic expansion and regional integration was based on an old technology: the canal (**Figure 2.13**). Merchant trade and the beginnings of industrialization in both Britain and France

Figure 2.12 The Manufacturing Belt of the United States The cities of this region, already thriving industrial centers that were well connected through the early railroad system, were ideally placed to take advantage of a series of crucial shifts between 1880 and 1920: a general upsurge in demand for consumer goods, the increased efficiency of the telegraph system and postal services, advances in manufacturing technologies, and the opening up of the national market through the extension of the railroad system. Individual cities began to specialize in their industrial products (e.g., grain milling in Minneapolis, brewing in Milwaukee and St. Louis, coach building and furniture in Cincinnati, agricultural machinery in Springfield, Illinois), producing in volume for the national market rather than local ones. This specialization in turn required an increase in commodity flows among the cities of the Manufacturing Belt, which bound the region more closely together.

Figure 2.13 Canal systems The canal systems that opened up the interiors of Europe and North America in the eighteenth century were initially dependent on horse power. Early examples carried horse-drawn barges. Later, barges were able to utilize steam- and oil-powered engines, extending the importance of larger canals and navigable rivers. Barges are still an important mode of transport for bulky goods. This photograph shows part of the Burgundy canal in France.

World Leadership Cycles

The modern world-system has so far experienced five full leadership cycles. In much simplified terms, and omitting for the moment any details of the consequent geographic changes, they have involved dominance by Portugal, the Netherlands, Great Britain (twice), and the United States.

Portuguese dominance was established through initial advantages derived from Atlantic exploration, trade, and plunder. The Treaty of Tordesillas, arbitrated by Pope Alexander VI in 1494, consolidated these advantages by limiting direct competition with Portugal's chief rival, Spain. The treaty allowed Portugal to lay claim to any territory to the east of a line drawn north–south, 370 leagues (1,850 kilometers, or 1,110 miles) west of Cape Verde (that is, between 48° and 49° West), which included the eastern coast of Brazil (**Figure 2.E**). Spain was allowed to claim the less accessible lands to the west of this line. Subsequently allied through royal marriage, Spain and Portugal sought to head off competition from England and Holland through imposing a global world-empire based on plunder rather than trade. By 1550, Portugal had established many important outposts around the world: in Brazil, Madagascar, India, the East Indies, and in parts of West, Southwest, and East Africa. The Portuguese attempt to establish a global world-empire was thwarted by English naval superiority, but it was the Dutch whose superior merchant economy gave them the resources to become the next hegemon.

Dutch dominance began with the failure of the Portuguese-backed Spanish Armada (against England) in 1588. The foundation of Dutch economic prosperity lay in the fishing and shipping industries. Dutch ports and Dutch shipping dominated European trade, and the Dutch government coordinated Dutch merchants' long-distance trade through the Dutch East India Company and the Dutch West India Company (see **Figure 2.F**). The hegemony of the United Provinces of the Netherlands (the so-called Dutch Republic) continued until the 1660s, when both the English and the French were able to mount a serious challenge. A series of three Anglo-Dutch wars (1652–1654, 1664–1667, and 1672–1678) helped to settle the issue in favor of England, which consolidated its power soon afterwards through the Act of Union (1707) with Scotland.

British dominance was sustained, in spite of a relatively poor domestic resource base, by overseas trade and colonization, backed up by a powerful navy. In the 1770s, this dominance was shaken by an alliance between France and the American revolutionaries, and there followed a period of struggle between Britain and France that came to involve much of Europe. A series of Napoleonic Wars marked the competitive phase of the leadership cycle, which was brought to a close by decisive victories by Vice-Admiral Lord Horatio Nelson at Trafalgar (1805; see **Figure 2.G**) and Lord Wellington at Waterloo (1815).

Figure 2.E Portugal's claims to the New World The Treaty of Tordesillas (1494) gave Portugal control over what were the more accessible parts of the New World. This map, drawn in 1502 by Alberto Contino, shows Portuguese territories marked by flags with a blue interior and a red border. Spanish territories, farther west, are marked by the red, gold, and black flag of Spain.

Figure 2.F Merchant vessels of the Dutch and English East India Companies The English East India Company was established in 1600 as a monopolistic trading company and agent of British imperialism in India. In response, the Dutch East India Company was founded in 1602 by the government of the United Provinces of the Netherlands in order to protect its trading interests in the Far East. The company was almost a branch of the government. It was granted a trade monopoly in the waters of the Indian and Pacific oceans, together with the rights to conclude treaties with local rulers, to build forts and maintain armed forces, and to carry on administrative functions through officials who were required to take an oath of loyalty to the Dutch government.

By then, the Industrial Revolution had given Britain an economic and military edge that proved decisive.

The second cycle of British dominance was based on the economic advantages of early industrialization, which allowed for an unprecedented degree of incorporation of the world under British imperial and economic hegemony. But from the 1860s onward, imperial overstretch and increasing economic competition from the United States and Germany put an end to hegemony and marked the beginning of a period of struggle. The German challenge culminated in the Great War of 1914–1918 (World War I), which left Germany defeated, Britain weakened, and the United States strengthened.

The **United States** was economically dominant within the world-system by 1920 but did not achieve hegemonic power because of a failure of political will. It took a second World War, prompted by the resurgence of Germany and the sudden rise of Japan, to force the United States out of its isolationism. After World War II, the United States was unquestionably the hegemonic power. Its economic and cultural dominance within the world-system has not yet been seriously challenged, though its political and military superiority was in question for several decades as the Soviet Union sought a noncapitalist path to modernization and power. This challenge disappeared in 1989, but by then the economic foundation for U.S. hegemony had come under serious threat because of the resurgence of Japanese and European industry and the globalization of economic activity through transnational corporations (see **Figure 2.H**). With the instability following the global credit crisis of 2008, (see Box 1.2, "The Global Credit Crunch"), the economic dominance of the United States is now under serious threat.

(a)

(b)

Figure 2.G   British Hegemony extended (a) The Battle of Trafalgar. On the afternoon of October 21, 1805, a few miles offshore from the Spanish port of Cádiz in the Bay of Trafalgar, 27 line-of-battle sailing ships commanded by Vice-Admiral Lord Horatio Nelson annihilated a combined French and Spanish fleet of 33 ships of the line under French Admiral Pierre Villeneuve. Nelson was mortally wounded in the action, but his victory that afternoon established a naval supremacy that enabled Britain to build and maintain a global empire, and to enjoy a leadership cycle of such unchallenged dominance that the following 100 years came to be known as the *Pax Brittanica*.
(b) The Battle of Waterloo was fought near the town of Waterloo in Belgium on June 18, 1815. Napoleon Bonaparte of France, having escaped from prison and assembled an army of 74,000 with 250 guns, resolved to attack the British army before other powers could come to their assistance. He was met by the Duke of Wellington with 23,000 British and 44,000 allied troops and 160 guns. Napoleon was defeated in what Wellington described as "the nearest-run thing you ever saw in your life." Britain and European allies restored Louis XVIII to the French throne, while Napoleon was exiled to the island of Saint Helena in the South Atlantic.

Figure 2.H   The Gulf War All hegemonic powers must protect the economic foundations of their power. They must also resist challenges to their political, cultural, and ideological dominance. Occasionally, this calls for military intervention. The Gulf War, fought in 1991, was an example of this sort of intervention. It was prompted by the invasion of Kuwait in 1990 by Iraq, where President Saddam Hussein had consolidated popularity through an aggressive stance toward American culture and business interests. The invasion of Kuwait not only threatened to disrupt the supply of crude oil to the world economy but also posed a threat to the U.S.-sponsored regional balance of power in the Middle East. On January 15, 1991, the United States, with 500,000 troops and massive air power, led a 100-hour war against Iraq, driving Iraqi troops out of Kuwait and destroying Iraq's military infrastructure.

were underpinned by extensive navigation systems that joined one river system to another. By 1790, France had just over 1,000 kilometers (620 miles) of canals and canalized rivers; Britain had nearly 3,600 kilometers (2,230 miles). The Industrial Revolution provided both the need and the capital for a spate of additional canal building that began to integrate and extend emerging industrial regions.

In Britain, 2,000 more kilometers (1,240 miles) of canals were built between 1790 and 1810. In France, which did not industrialize as early as Britain, 1,600 additional kilometers (990 miles) were built between 1830 and 1850. In the United States, the key event was the opening of the Erie Canal in 1825, which enabled New York, a colonial gateway port, to reorient itself toward the nation's growing interior. The Erie Canal was so profitable that it set off a "canal fever" that resulted in the construction of some 2,000 kilometers (1,240 miles) of navigable waterways in the next 25 years. This canal system helped to bind the emergent Manufacturing Belt together.

The scale of the United States was such that a network of canals was a viable proposition only in more densely settled areas. The effective colonization of the interior could not take place until the development of steam-powered transportation—first riverboats and then railroads. The first steamboats were developed in the early 1800s, offering the possibility of opening up the vast interior by way of the Mississippi and its tributaries (**Figure 2.14**). By 1830, the technology and design of steamboats had been perfected and navigable channels had been established. The heyday of the river steamboat was between 1830 and 1850. During this period, vast acreages of the U.S. interior were opened up to commercial, industrialized agriculture—especially cotton production for export to British textile manufacturers. At the same time, river ports such as New Orleans, St. Louis, Cincinnati, and Louisville grew rapidly, extending the frontier of industrialization and modernization.

Railroads and Regional Economic Integration

By 1860, the railroads had taken over the task of internal development, further extending the frontier of settlement and industrialization and intensifying the use of previously developed regions. The railroad originated in Britain, where George Stephenson engineered the world's first commercial railroad, a 20-kilometer (12.4-mile) line between Stockton and Darlington that was opened in 1825. The famous *Rocket*, the first locomotive for commercial passenger trains, was designed mainly by Stephenson's son Robert for the

Figure 2.14 Steamboats The interior of continental North America was first opened up as a result of steamboat trade along the Mississippi and its tributaries. Cotton and other plantation crops were shipped north to manufacturing cities, grain crops were shipped south, and both grain and cotton were exported. This illustration shows the levee at New Orleans, drawn in 1884.

Liverpool & Manchester line, which opened four years later. The economic success of this line sparked two railroad-building booms that eventually created a highly efficient transportation network for Britain's manufacturing industry.

In other core countries, where sufficient capital existed to license (or copy) the locomotive technology and install the track, railroad systems led to the first full stage of economic and political integration. The first passenger-train services averaged little more than 20 to 35 kilometers per hour (15–20 mph), but locomotive technology changed rapidly and made it easier, faster, and cheaper to conquer the vast interior distances of the United States. Between 1830 and 1845, the United States created the world's largest rail system—some 5,458 kilometers (3,688 miles) compared to Britain's 3,083 kilometers (2,069 miles), Germany's 2,956 kilometers (1,997 miles), France's 817 kilometers (552 miles), and Belgium's 508 kilometers (343 miles).

While the railroads integrated the economies of entire countries and allowed vast territories to be colonized, they also brought some important regional and local restructuring and differentiation. In the United States, for example, the railroads led to the consolidation of the Manufacturing Belt. They also contributed to the mushrooming of Chicago as the focal point for railroads that extended the Manufacturing Belt's dominance over the West and South (**Figure 2.15**). This reorientation of

Figure 2.15 Railroads (a) The Illinois Central Railroad issued a promotional poster in the 1860s celebrating the role it envisaged—that its trunk line from Chicago to New Orleans would significantly affect the settling of the American interior. Note the deliberate juxtaposition of the high technology of the railroad locomotive and the telegraph line, contrasting with the obsolescent technologies of the stagecoach (inset top right) and barge (inset bottom right). Even the oceangoing ship is shown with masts and sails to diminish its status in comparison to the railroad locomotive. (b) The poster issued 50 years later, on the other hand, shows the railroad locomotive in the company of several important new technologies: aircraft, automobiles and trucks, and electric streetlights.

the nation's transportation system effectively ended the role of the cotton regions of the South as outliers of the British trading system. Instead, they became outliers of the U.S. Manufacturing Belt, supplying factories in New England and the Mid-Atlantic Piedmont. This left New Orleans, which had thrived on cotton exports, to cope with an abrupt end to its phenomenal growth.

The railroads reorganized the geography of America's cities and opened up its interior regions. Railway stations reordered land-use patterns in central business districts, where new hotels and department stores competed for sites near the station's entrance. Farther away, the railroad lines cut swathes through the urban fabric, separating neighborhoods from one another and establishing the basis for a radial framework for the physical development of the city.

Tractors, Trucks, Road Building, and Spatial Reorganization

In the twentieth century, the internal combustion engine powered further rounds of internal development, integration, and intensification. The replacement of horse-drawn farm implements with lightweight tractors powered by internal combustion engines, beginning in the 1910s, amounted to a major revolution in agriculture (see Chapter 8). Productivity was increased, the frontiers of cultivable land were extended, and vast amounts of labor were released for industrial work in cities. The result was a parallel revolution in the geographies of both rural and urban areas.

The development of trucks in the 1910s and 1920s suddenly released factories from locating near railroads, canals, and waterfronts. Trucking allowed goods to be moved farther, faster, and cheaper than before. As a result, trucking made it feasible to locate factories on inexpensive land on city fringes and in smaller towns and peripheral regions where labor was cheaper. It also increased the market area of individual factories and reduced the need for large product inventories.

This decentralization of industry, in conjunction with the availability of buses, private automobiles, and massive road-building programs, brought about another phase of spatial reorganization. The outcomes of this phase were the specialized and highly integrated regions and urban systems of the modern core of the world-system. This integration was not simply a question of their being interconnected through highway systems. It also involved close economic linkages among manufacturers, suppliers, and distributors—linkages that enabled places and regions to specialize and develop economic advantages (see Chapter 7).

Organizing the Periphery

Parallel to the internal development of core regions were changes in the geographies of the periphery of the world-system. Indeed, the growth and internal development of the core regions simply could not have taken place without the foodstuffs, raw materials, and markets provided

by the colonization of the periphery and the incorporation of more and more territory into the sphere of industrial capitalism.

As soon as the Industrial Revolution had gathered momentum in the early nineteenth century, the industrial core nations embarked on the penetration of the world's inland midcontinental grassland zones in order to exploit them for grain or stock production. This led to the settlement, through the emigration of European peoples, of the temperate prairies and pampas of the Americas; the veld in southern Africa; the Murray-Darling Plain in Australia; and the Canterbury Plain in New Zealand. At the same time, as the demand for tropical plantation products increased, most of the tropical world came under the political and economic control—direct or indirect—of one or another of the industrial core nations. In the second half of the nineteenth century, and especially after 1870, there was a vast increase in the number of colonies and the number of people under colonial rule.

The International Division of Labor

The fundamental logic behind all this colonization was economic: the need for an extended arena for trade, an arena that could supply foodstuffs and raw materials in return for the industrial goods of the core. The outcome was an international division of labor, driven by the needs of the core and imposed through its economic and military strength. This **division of labor** involved the specialization of different people, regions, and countries in certain kinds of economic activities. In particular, colonies began to specialize in the production of commodities meeting certain criteria:

- where an established demand existed in the industrial core (for foodstuffs and industrial raw materials, for example);
- where colonies held a **comparative advantage** in specializations that did not duplicate or compete with the domestic suppliers within core countries (tropical agricultural products like cocoa and rubber, for example, simply could not be grown in core countries).

The result was that colonial economies were founded on narrow specializations that were oriented to and dependent upon the needs of core countries. Examples of these specializations were many: bananas in Central America; cotton in India; coffee in Brazil, Java, and Kenya; copper in Chile; cocoa in Ghana; jute in East Pakistan (now Bangladesh); palm oil in West Africa; rubber in Malaya (now Malaysia) and Sumatra; sugar in the Caribbean islands; tea in Ceylon (now Sri Lanka); tin in Bolivia; and bauxite in Guyana and Surinam. Most of these specializations persist today. Thus, for example, 45 of the 55 countries in sub-Saharan Africa still depend on just three products—tea, cocoa, and coffee—for more than half of their export earnings.

This new global economic geography took some time to establish, and the details of its pattern and timing

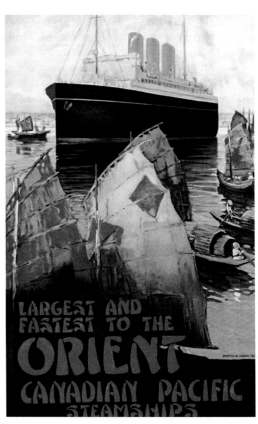

Figure 2.16 Major steamship routes in 1920 The shipping routes reflect (1) the transatlantic trade between the bipolar core regions of the world-system at the time and (2) the colonial and imperial relations between the world's core economies and the periphery. Transoceanic shipping boomed with the development of steam-turbine engines for merchant vessels and with the construction of shipping canals, such as the Kiel Canal, the Suez Canal, the St. Lawrence Seaway, and the Panama Canal. When the 82 kilometers (51 miles) of the Panama Canal opened in 1914, shipping could move between the Atlantic and the Pacific without having to go around South America, saving thousands of kilometers of steaming. The poster dates from 1924.

were heavily influenced by technological innovations. The incorporation of the temperate grasslands into the commercial orbit of the core countries, for example, involved successive changes in regional landscapes as critical innovations—such as barbed wire, the railroad, and refrigeration—were introduced.

The single most important innovation stimulating the international division of labor, however, was the development of metal-hulled, oceangoing steamships. This development was cumulative, with improvements in engines, boilers, transmission systems, fuel systems, and construction materials adding up to produce dramatic improvements in carrying capacity, speed, range,

and reliability. The construction of the Suez Canal (opened in 1869) and the Panama Canal (opened in 1914) was also critical, providing shorter and less hazardous routes between core countries and colonial ports of call. By the eve of World War I the world economy was effectively integrated by a system of regularly scheduled steamship trading routes (**Figure 2.16**). This integration, in turn, was supported by the second most important innovation stimulating the international division of labor: a network of telegraph communications (**Figure 2.17**) that enabled businesses to monitor and coordinate supply and demand across vast distances on an hourly basis.

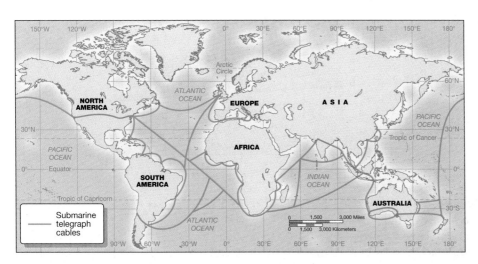

Figure 2.17 The international telegraph network in 1900 For Britain, submarine telegraph cables were the nervous system of its empire. Of the global network of 246,000 kilometers (152,860 miles) of submarine cable, Britain had laid 169,000 kilometers (105,015 miles).

The international division of labor brought about a substantial increase in trade and a huge surge in the overall size of the capitalist world economy. The peripheral regions of the world contributed a great deal to this growth. By 1913, Africa and Asia provided more *exports* to the world economy than either North America or the British Isles. Asia alone was *importing* almost as much, by value, as North America. The industrializing countries of the core bought increasing amounts of foodstuffs and raw materials from the periphery, financed by profits from the export of machinery and manufactured goods. Britain, the dominant power of the period, drew on a trading empire that was truly global (**Figure 2.18**).

Patterns of international trade and interdependence became increasingly complex. Britain used its capital to invest not just in peripheral regions but also in profitable industries in other core countries, especially the United States. At the same time, these other core countries were able to export cheap manufactured items to Britain. Britain financed the purchase of these goods, together with imports of food from its dominion states (Canada, South Africa, Australia, and New Zealand) and colonies, through the export of its own manufactured goods to peripheral countries. India and China, with large domestic markets, were especially important. Thus there developed a widening circle of exchange and dependence, with constantly switching patterns of trade and investment.

Imperialism: Imposing New Geographies on the World

The incorporation of the periphery was by no means entirely motivated by this basic logic of free trade and investment. Although Britain was the dominant power in the late nineteenth century, several other European countries (notably Germany, France, and the Netherlands), together with the United States—and later Japan—were competing for global influence. This competition developed into a scramble for territorial and commercial domination. The core countries engaged in preemptive geographic expansionism in order to protect their established interests and to limit the opportunities of others. They also wanted to secure as much of the world as possible—through a combination of military oversight, administrative control, and economic regulation—in order to ensure stable and profitable environments for their traders and investors. This combination of circumstances defined a new era of imperialism.

In the final quarter of the nineteenth century, a surge of imperialism brought a competitive form of colonialism that resulted in a scramble for territory. Africa, more than any other peripheral region, was given an entirely new geography. It was carved up into a patchwork of European colonies and protectorates in just 34 years, between 1880 and 1914, with little regard for either physical geography or the preexisting human geographies of minisystems and world-empires. Whereas European interest had previously focused on coastal trading stations and garrison ports, it now extended to the entire continent.

Within just a few years the whole of Africa became incorporated into the modern world-system, with a geography that consisted of a hierarchy of three kinds of spaces. One consisted of regions and localities organized by European colonial administrators and European investors to produce commodities for the world market.

Figure 2.18   The British Empire in the late 1800s Protected by the all-powerful Royal Navy, the British merchant navy established a web of commerce that collected food for British industrial workers and raw materials for its industries, much of it from colonies and dependencies appropriated by imperial might and developed by British capital. So successful was the trading empire that Britain also became the hub of trade for other states. (*Source:* Adapted from P. Hugill, *World Trade Since 1431.* Baltimore: Johns Hopkins University Press, 1993, p. 136.)

A second consisted of zones of production for local markets, where peasant farmers produced food for consumption by laborers engaged in commercial mining and agriculture. The third consisted of widespread regions of subsistence agriculture whose connection with the world-system was as a source of labor for the commercial regions.

Meanwhile, the major powers jostled and squabbled over small Pacific islands that had suddenly become valuable as strategic coaling stations for their navies and merchant fleets. Resistance from indigenous peoples was quickly brushed aside by imperial navies with ironclad steamers, high-explosive guns, and troops with rifles and cannons. European weaponry was so superior that Otto von Bismarck, the founder and first chancellor (1871–1890) of the German empire, referred to these conflicts as "sporting wars." Between 1870 and 1900, European countries added almost 22 million square kilometers (10 million square miles) and 150 million people to their spheres of control—20 percent of the Earth's land surface and 10 percent of its population.

The imprint of imperialism and colonization on the geographies of the newly incorporated peripheries of the world-system was immediate and profound. The periphery was rendered almost entirely dependent on European and North American capital, shipping, managerial expertise, financial services, and news and communications. Consequently, it also became dependent on European cultural products: language, education, science, religion, architecture, and planning. All of these influences were etched into the landscapes of the periphery in a variety of ways as new places were created, old places were remade, and regions were reorganized.

The Third World and Neocolonialism

The imperial world order began to disintegrate shortly after World War II, however. The United States emerged as the new hegemonic power, the dominant state within the world-system core, which came to be called the First World. The Soviet Union and China, opting for alternative, socialist paths of development for themselves and their satellite countries, were seen as a Second World, withdrawn from the capitalist world economy. Their pursuit of alternative political economies was based on radically different values.

By the 1950s, many of the old European colonies began to seek political independence. Some of the early independence struggles were very bloody because the colonial powers were initially reluctant to withdraw from colonies where strategic resources or large numbers of European settlers were involved. In Kenya, for example, a militant nationalist movement known as the Mau Mau launched a campaign of terrorism, sabotage, and assassination against British colonists in the early 1950s. They killed more than 2,000 white settlers between 1952 and 1956; in return, 11,000 Mau Mau rebels were killed by the colonial army and 20,000 put into detention camps by the colonial administration. By the early 1960s, however, the process of decolonization had become relatively smooth. (In 1962, Jomo Kenyatta, who had been jailed as a Mau Mau leader in 1953, became prime minister of a newly independent Kenya.) The periphery of the world-system now consisted of a Third World of politically independent states, some of which adopted a policy of nonalignment, vis-à-vis the geopolitics of the First and Second Worlds. They were nevertheless still highly dependent, in economic terms, on the world's core countries.

As newly independent peripheral states struggled to be free of their economic dependence through industrialization, modernization, and trade from the 1960s onward, the capitalist world-system became increasingly integrated and interdependent. The old imperial patterns of international trade broke down and were replaced by more complex patterns. Nevertheless, the newly independent states were still influenced by many of the old colonial links and legacies that remained intact. The result was a neocolonial pattern of international development. **Neocolonialism** refers to economic and political strategies by which powerful states in core economies indirectly maintain or extend their influence over other areas or people. Instead of formal, direct rule (colonialism), controls are exerted through such strategies as international financial regulations, commercial relations, and covert intelligence operations. Because of neocolonialism, the human geographies of peripheral countries continued to be heavily shaped by the linguistic, cultural, political, and institutional influence of the former colonial powers and by the investment and trading activities of their firms.

At about the same time, a new form of imperialism was emerging. This was the *commercial imperialism* of giant corporations. These corporations had grown within the core countries through the elimination of smaller firms by mergers and takeovers. By the 1960s, quite a few of them had become so big that they were *transnational* in scope, having established overseas subsidiaries, taken over foreign competitors, or simply bought into profitable foreign businesses.

These **transnational corporations** have investments and activities that span international boundaries, with subsidiary companies, factories, offices, or facilities in several countries. By 2006, over 78,000 transnational corporations were operating, 90 percent of which were headquartered in the core states. Among them these corporations control about 780,000 foreign affiliates and account for the equivalent of 10 percent of world GDP and one-third of world exports.

Transnational corporations have been portrayed as imperialist by some geographers because of their ability and willingness to exercise their considerable power in ways that adversely affect peripheral states. They have certainly been central to a major new phase of geographical restructuring that has been under way for the last 30 years or so. This phase has been distinctive because

Figure 2.19 The world-system core, semiperiphery, and periphery in 2005 (*Source:* Map projection, Buckminster Fuller Institute and Dymaxion Map Design, Santa Barbara, CA. The word Dymaxion and the Fuller Projection™ Map design are trademarks of the Buckminster Fuller Institute, Santa Barbara, California, © 1938, 1967 & 1992. All rights reserved.)

an unprecedented amount of economic, political, social, and cultural activity has spilled beyond the geographic and institutional boundaries of states. It is a phase of *globalization,* a much fuller integration of the economies of the worldwide system of states and a much greater interdependence of individual places and regions from every part of the world-system (**Figure 2.19**).

The current phase of globalization also involves a distinctive new geopolitical element that has been described as the "new imperialism": the imperialism of the United States, the world's only effective superpower. Although Americans do not like to think of their country as territorially aggressive or exploitative, the "war on terror" and invasion of Afghanistan and Iraq following the al-Qaeda attacks in 2001 are widely interpreted elsewhere in the world as comprising an exercise in imperialism, motivated in large part by a desire for military control over global oil resources. This interpretation of the United States as the instigator of a new imperialism has been reinforced by the country's military threats against Iran and North Korea, its gulag-style detention of prisoners without trial, its deployment of special forces around the globe, its use of "extraordinary rendition" (the apprehension or kidnapping of suspects followed by their transfer to countries known to employ

harsh interrogation techniques or torture), and its unilateral rejection of international environmental treaties and international aid agreements. What is less widely discussed in the opinion pages of the world's newspapers is that this new imperialism is also seen by some academics as the result of a highly competitive global economic environment in which the United States is no longer able to achieve superiority through innovation, product design, productivity, and marketing and so has had to resort to military intervention.

CONTEMPORARY GLOBALIZATION

Globalization has been under way since the inception of the modern world-system in the sixteenth century. The basic framework for globalization has been in place since the nineteenth century, when the competitive system of states fostered the emergence of international agencies and institutions, global networks of communication, a standardized system of global time, international competitions and prizes, international law, and internationally shared notions of citizenship and human rights. Global connections today, though, differ in at least four important ways from those in the past.

First, they function at much greater *speed* than ever before. Second, globalization operates on a much larger *scale*, leaving few people unaffected and making its influence felt in even the most remote places. Third, the *scope* of global connections is much broader and has *multiple dimensions*: economic, technological, political, legal, social, and cultural, among others. Fourth, the interactions and interdependencies among numerous global actors have created a new level of *complexity* for the relationships between places and regions.

The most distinctive feature of globalization over the past 30 years or so is a decisive increase in the proportion of the world's economic and cultural activities that are transnational in scope. This increase is linked to a significant shift in the *nature* of international economic activity. Flows of goods, capital, and information that take place within and between transnational corporations are becoming more important than imports and exports between countries. At the same time, all these flows and activity have helped to spread new values around the world. These new values range from changing consumer lifestyle preferences to increased concern about global resources, global environmental change, and famine relief.

The contemporary world economy is constituted through the myriad commodity chains that crisscross global space. **Commodity chains** are networks of labor and production processes that originate in the extraction or production of raw materials and whose end result is the delivery and consumption of a finished commodity (see Box 2.5, "Commodity Chains"). These networks often span countries and continents, linking into vast global assembly lines the production and supply of raw materials, the processing of raw materials, the production of components, the assembly of finished products, and the distribution of finished products. As we shall see in Chapter 7, these global assembly lines are increasingly important in shaping places and regions.

The issues raised under the heading "globalization" are often controversial because the term has different meanings for different people. Most broadly, globalization is the expansion and intensification of linkages and flows of capital, people, goods, ideas, and cultures across national borders. To some this process implies a serious decline in the importance of local communities and national governments. Globalization has produced a more complex system of interdependent states in which *trans*national rules and organizations have gained influence. States pursuing their national interests are still a major force, but corporations and international nongovernmental organizations (INGOs) can now critically influence world politics. No clear hierarchy of interests common to all states exists, and the major powers are reluctant to use military force except under extreme provocation. World society therefore contains many centers of power and has no single power hierarchy. As power disperses and

goals diverge, a new pattern of complex interdependence is emerging.

Globalization also has important cultural dimensions (see Chapter 5). One is quite simply the diffusion around the world of all sorts of cultural forms, practices, and artifacts that had previously been confined to specific places or regions. Examples include "ethnic" and regional cuisine, "world" music, Caribbean carnivals, and "charismatic" Christian sects. Another dimension of cultural globalization derives from consumer culture: everything that is sold in international markets, from sneakers, replica soccer shirts, and automobiles to movies and rock concert tours. This had led some observers to believe that globalization is producing a new set of universally shared images, practices, and values: literally, a global culture.

Most observers, though, regard such notions of a single global culture as overly simplistic. They point to other dimensions of globalization such as the fusion of different regional and international cultural elements that creates a hybridization of culture. Transgressions, adaptations, and subversions of "conventional" or "traditional" cultures originate in many different places and regions and then spread, resulting in a series of open-ended "global" cultures. Cultural anthropologist Arjun Appadurai has described five kinds of cultural flows that contribute to these global cultures:

Ethnoscapes: produced by flows of people, including tourists, immigrants, refugees, exiles, and guest workers;

Technoscapes: resulting from the diffusion of goods, technologies, and architectural styles;

Finanscapes: produced by rapid flows of money in currency markets and stock exchanges;

Mediascapes: images of the world produced by news agencies, magazines, television, and film;

Ideoscapes: resulting from the diffusion of ideas and ideologies, concepts of human rights, democracy, welfare, and so on.

Meanwhile, a more integrated global system has also increased awareness of a set of common problems—climate change, pollution, disease, crime, poverty, and inequality—that many see as a consequence of globalization. The globalization of the contemporary world—its causes and effects on specific aspects of human geographies at different spatial scales—is a recurring theme through the rest of this book. For the moment, we need only note in broad outline its principal causes and outcomes.

The Causes and Consequences of Globalization

The globalization of the past quarter century has been caused by four important and interrelated factors: a new international division of labor, an internationalization of

2.5 GEOGRAPHY MATTERS

Commodity Chains

A commodity chain is a network of labor and production processes whose end result is a finished commodity. Global commodity chains link the progression of a commodity from design through procurement of raw materials and production to import or export to the point of sale, distribution for sale, marketing, and advertising. They are often entirely internal to the global operations of transnational corporations.

Advances in telecommunications, management techniques, transportation, informatics, finance, and other services to industry have made possible the segmentation of corporate production lines, as well as services, across multiple settings. Manufacturing companies can design a product in one country, have it produced by contractors in various countries continents apart, sell the product with its brand name by telephone or Internet almost anywhere in the world, and have other contractors deliver it. The services involved—design, sales, financing, and delivery—can be undertaken without the various actors ever meeting face to face (except for the carrier services). Advances in technology and management have also permitted the reproduction and standardization of services and products on a global basis. Certain patented services, such as fast-food restaurants (McDonald's, Kentucky Fried Chicken, Burger King), rely on computer-regulated technology to deliver a standard service and product over time and geographic space.

Almost every mass-marketed manufactured product involves a complex commodity chain. Take, for example, the manufacture of Lee Cooper jeans (**Figure 2.I**). Designed in the United States, advertised globally, retailed in stores across Europe, the United States, and the metropolises of semiperipheral and peripheral countries, their manufacture draws on labor and products from around the world. The final stages of the commodity chain of one particular pair of jeans sold in a large discount store of a provincial British city were "in a van that came up the A12 [road] from Lee Cooper's warehouse at Staples Corner, just at the bottom of the M1 [highway] in North London. . . . Before that they came through the Channel Tunnel in a lorry from similar warehouses in Amiens, France and before that, by boat and train from Tunis in Tunisia . . .[1]"

There are three broad types of global commodity chains. The first is producer-driven, in which large, often transnational, corporations coordinate production networks. A good example of this is the U.S. pharmaceutical industry. Research and development of drugs are conducted in the United States; materials and components are produced in a vertically integrated global production line (i.e., a commodity chain); the drugs are assembled and marketed out of a semiperipheral site (in particular, Puerto Rico and Ireland), where companies enjoy certain competitive advantages not only in the industrial segment of the assembly but in marketing, financial, and management services.

A second type of commodity chain is consumer-driven, where large retailers, brand-name merchandisers, and trading companies influence decentralized production networks in a variety of exporting countries, often in the periphery. A good example is the case of Lee Cooper jeans described here. Another good example is the discount chain-store company Wal-Mart, which contracts directly with producers in low-wage countries (China, in particular) for the bulk of its merchandise (which is sold in the United States to the accompaniment of advertising that invokes community-oriented and even patriotic themes).

The third type, the marketing-driven commodity chain, represents a hybrid of the first two types. It involves the production of inexpensive consumer goods—such as colas, beers, breakfast cereals, candies, cigarettes, and infant formula—that are global commodities and carry global brands yet are often manufactured in the periphery and semiperiphery for consumption in those regions. These commodities take their globalized status not only from their recipes and production techniques but, even more important, from their globally contrived cultural identities.

Different forms of commodity chains provide varying opportunities for firms and national economies to enter into and improve their position within the global division of labor. They are an important dimension of the complex transnationalization of economic space. They are also an important dimension of the complex currents of cultural globalization. In addition to facilitating the standardization of products and services (for those who can afford them) around the world, commodity chains also reflect the inequalities that are inscribed into the global economy. If core-country consumers stop to think about the origins of many of the products they consume, they may recognize what is happening "down the line," where poverty wages and grim working and living conditions are a precondition for the beginning of many commodity chains. In contrast, those who work on the farms and plantations and in the workshops and factories at the beginning of commodity chains are acutely aware—thanks to contemporary media—of the dramatically more affluent lifestyles of those who will eventually consume the fruits of their labor.

[1]Abrams, F. and J. Astill, "Story of the Blues," *The Guardian.* May 29, 2001, p. 2; quoted in L. Crewe, "Unravelling Fashion's Commodity Chains." In A. Hughes and S. Reimer (eds.), *Geographies of Commodity Chains.* London: Routledge, 2004, p. 201.

Figure 2.1   The making of a pair of Lee Cooper jeans (*Source:* Adapted from A. Hughes and S. Reimer (eds.), *Geographies of Commodity Chains.* New York: Routledge, 2004.)

finance, a new technology system, and a homogenization of international consumer markets.

A New International Division of Labor

The new international division of labor has involved three main changes. First, the United States has declined as an industrial producer, relative to the development of Japan, the resurgence of Europe, and the spectacular growth of China as industrial producers. Second, manufacturing production has been decentralized from these core regions to some semiperipheral and peripheral countries. In 2006, U.S.-based companies employed about 31 million workers overseas, 80 percent of whom were in manufacturing jobs. An important reason for this trend has been the prospect of keeping production costs low by exploiting the huge differential in wage rates around the world:

> Nike pays USD $16 million (13 million Euro) a year to the Brazilian national football team and Adidas pays USD $1.8 million (1.5 million Euro) per year to French player Zinedine Zidane. Meanwhile the Asian workers who make the football boots and other sports gear worn by players are paid as little as 47 cents Euro per hour—3.76 Euro for a standard working day. Shopping at their cheapest local markets, women producing brand-name sportswear in Indonesia need to work 3.75 hours to earn enough to purchase 1.5 kg of uncooked chicken, which for some is all the meat they can afford for a month.[2]

In Indonesia, apparel workers were paid US$0.15 an hour and expected to work up to 70 hours a week.

A third result of the new international division of labor is that new specializations have emerged within the core regions of the world-system: high-tech manufacturing and **producer services** (that is, services such as information services, insurance, and market research that enhance the productivity or efficiency of other firms' activities or that enable them to maintain specialized roles). One significant reflection of this new international division of labor is that global trade has grown much more rapidly over the past 30 years than global production—a clear indication of the increased economic integration of the world-system.

The Internationalization of Finance

The second factor contributing to today's globalization is the internationalization of finance: the emergence of global banking and globally integrated financial markets. The pivotal moment was a "system shock" to the international economy that occurred in the mid-1970s. World financial markets, swollen with U.S. dollars by the U.S.

[2]Offside! Oxfam International, "Labour Rights and Sportswear Production in Asia," 2006, p. 2. www.oxfam.org.au/campaigns/labour/06report, accessed May 7, 2008.

government's deficit budgeting and by huge currency reserves held by OPEC (Organization of Petroleum Exporting Countries) after they had orchestrated a four-fold increase in the price of crude oil, quickly evolved into a new and sophisticated system of international finance, with new patterns of investment and disinvestment.

These changes are, of course, tied in to the new international division of labor. In particular, they are a consequence of massive increases in levels of international direct investment. Between 1975 and 2006, private capital flows increased from less than 5 percent of world GDP to just under 25 percent. Until the early 1970s, U.S.-based transnational corporations accounted for about two-thirds of the total outflows of foreign direct investment (FDI), and about four-fifths of this was directed toward Canada and the more advanced industrial national economies of Western Europe. By 2006, U.S. transnational corporations' share of the total had dropped to less than half, while direct foreign investment by Japanese, Canadian, and German corporations increased significantly. Meanwhile, another source of foreign direct investment had begun to show up: transnational corporations based in semiperipheral countries. Several of the top 100 corporations in the world are now based in these newly industrializing countries (NICs). Sinopec Corp., a Chinese petroleum and chemical corporation, is bigger than Boeing, and Samsung, the South Korean electronics firm, is bigger than Japan's Toshiba, Germany's BMW, and the U.S. conglomerate of Procter and Gamble.

Along with these changes in the sources of investment have been changes in destination. Core countries now absorb most of the inflows of capital: Between 1998 and 2004 the United States, the European Union, and Japan accounted for three-quarters of global inflows of foreign direct investment. They also accounted for 85 percent of outflows, so that the dominant outcome of globalization is a circular movement of FDI between North America, Western Europe, and Japan, at the expense of peripheral regions. Nevertheless, as we shall see in Chapter 7, the globalization of economic activity has also brought significant flows of capital into NICs such as Argentina, Brazil, China (including Hong Kong), Malaysia, Mexico, Singapore, and South Korea. As noted in Chapter 1, all of these changes have been fostered by neoliberal policies and institutions such as the World Trade Organization and World Economic Forum.

Meanwhile, the capacity of computers and information systems to deal very quickly with changing international conditions has added a speculative component to the internationalization of finance. All in all, about $2 trillion worth of currencies are exchanged every day. The volume and complexity of international investment and financial trading has created a need for banks and financial institutions that can handle investments on a large scale, across great distances, quickly and efficiently. The nerve centers of the new system are located in just a few places—London, Frankfurt, New York, and Tokyo, in particular. Their activities are interconnected around the clock for almost

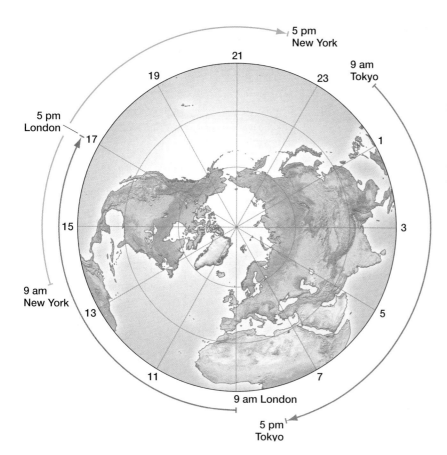

Figure 2.20 24-hour trading between major financial markets Office hours in the two most important financial centers—New York and London—overlap one another because the two cities are situated in broadly separated time zones. While these markets are both closed, Tokyo offices are open. This means that between them the world's three major financial centers span the globe with almost 24-hour trading in currencies, stocks, and other financial instruments.

24 hours (**Figure 2.20**), and their networks penetrate into every part of the globe. As noted in Chapter 1, it was this interconnectedness and complexity that contributed to the global financial crisis of October 2008.

A New Technology System The third factor contributing to globalization is a new technology system (see p. 54) based on a combination of innovations, including solar energy, robotics, microelectronics, biotechnology, and digital telecommunications and information systems. This new technology system has required the geographical reorganization of the core economies. It has also extended the global reach of finance and industry and permitted a more flexible approach to investment and trade. Especially important in this regard have been new and improved technologies in transport and communications—the integration of shipping, railroad, and highway systems through containerization (**Figure 2.21**); the introduction of wide-bodied cargo jets; and the development of fiber-optic networks, communications satellites, and electronic mail and information retrieval systems. Finally, many of these telecommunications technologies have also introduced a wider geographical scope and faster pace to many aspects of political, social, and cultural change, as we shall see in subsequent chapters.

Global Consumer Markets The fourth factor in globalization has been the growth of consumer markets. Among the more affluent populations of the world,

similar social processes create parallel trends in consumer taste. A new and materialistic international culture has taken root, in which people save less, borrow more, defer parenthood, and indulge in affordable luxuries that are marketed as symbols of style and distinctiveness. This culture is easily transmitted through the new telecommunications media, and it has been an important basis for transnational corporations' global branding and marketing of "world products" (German luxury automobiles, Swiss watches, British raincoats, French wines, American soft drinks, Italian shoes and designer clothes, and Japanese consumer electronics, for example). Nine of the ten most trusted brands in India in 2007, for example, were products of British or American transnational corporations: Colgate, Vicks, Lux, Nokia, Britannia, Dettol, Lifebuoy, Pepsodent, and Pond's. Tata Tea was the only Indian brand in the top-ten list.

This materialistic international culture is reinforced through other aspects of globalization, including the internationalization of television, especially CNN, MTV, Fox Broadcasting, Sky, Star Television, and the syndication of TV movies and light entertainment series. The number of television sets per 1,000 people worldwide doubled between 1980 and 2000, and multimedia industries have been booming. The global market for popular cultural products carried by these media is becoming concentrated, however. At the core of the entertainment industry—film, music, and television—there is a growing dominance of U.S. products, and many countries

Figure 2.21 The impact of containerization on world trade Containerization revolutionized long-distance transport because it did away with the slow, expensive, and unreliable business of loading and unloading ships with manual labor. Before containerization, ships spent one day in port for every day at sea; after containerization, they spent a day in port for every ten days at sea. By 1965, an international standard for containers had been adopted, making it possible to transfer goods directly from ship to rail to road and allowing for a highly integrated global transport infrastructure. The average container ship today holds 4,000 20-foot containers, but some are able to carry 6,000 or 8,000. Containerization requires a heavy investment in both vessels and dockside handling equipment, however. As a result, container traffic has quickly become concentrated in a few ports that handle high-volume transatlantic and transpacific trade. This photograph shows the Burchardkai Terminal in Hamburg, Germany.

have seen their homegrown industries wither. Hollywood obtains more than 50 percent of its revenues from overseas, up from just 30 percent in 1980. Movies made in the United States account for about 50 percent of the market in Japan, 70 percent in Europe, and 85 percent in Latin America. Similarly, U.S. television series have become increasingly prominent in the programming of other countries.

Globalization and Core–Periphery Differences

The single most dramatic outcome of the globalization that has resulted from all these changes is the consolidation of the core of the world-system. The core is now a close-knit triad of the geographic centers of North America, the European Union, and Japan. These three geographic centers are connected through three main circuits, or flows, of investment, trade, and communication: between Europe and North America, between Europe and the Far East, and among the regions of the

Pacific Rim. **Figure 2.22,** for example, shows just how dominant North America has become in accounting for flows of international telephonic communication. As we shall see in Chapter 7, this consolidation of the core of the world-system is having some profound effects on economic geography. Within the core regions, for example, a new hierarchy of regional economic specialization has been imposed by the locational strategies of transnational corporations and international financial institutions.

Globalization, although incorporating more of the world more completely into the capitalist world-system, has intensified the differences between the core and the periphery. According to the United Nations Development Program, the gap between the poorest fifth of the world's population and the wealthiest fifth increased more than threefold between 1965 and 2005. Per capita GDP in the world's 20 richest countries is now 40 times higher than in the world's poorest 20 countries. Some parts of the periphery have almost slid off the economic map. In 55 countries, per capita income actually fell during the 1990s. In sub-Saharan Africa, economic output fell by one-third during the 1980s and stayed low during

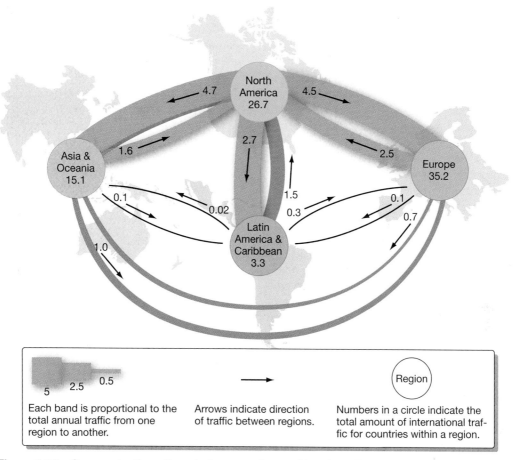

Figure 2.22 **Communications flows between major world regions** This diagram shows the flows, in billions of minutes of telecommunications traffic over public telephone networks, between major regions. (*Source:* Adapted from G. C. Staple (ed.), *TeleGeography 1999.* Washington, DC: TeleGeography Inc., 1999, Fig. 4, p. 255.)

the 1990s, so that people's standard of living there is now, on average, lower than it was in the early 1960s. In 2008, the fifth of the world's population living in the highest-income countries had:

- 75 percent of world income (the bottom fifth had just 1 percent);
- 83 percent of world export markets (the bottom fifth had just 1 percent);
- 76 percent of world telephone lines, today's basic means of communication (the bottom fifth had just 1.5 percent).

While 3 billion people around the world struggle to live on less than US$2.00 a day, the world's billionaires— fewer than 500 people—were worth $3.5 trillion (over 7 percent of world GDP). The GDP of the 41 Heavily Indebted Poor Countries (567 million people) is less than the wealth of the world's 7 richest individuals combined. OECD countries (the Organization for Economic Cooperation and Development, an association of 30 industrialized countries), with 19 percent of global population, control more than 75 percent of global trade in goods and services and consume 86 percent of the world's

goods. More than 850 million people are still clinically malnourished and permanently hungry, including more than 140 million in India, 220 million in China, and 200 million in sub-Saharan Africa.

Such enormous differences lead many people to question the equity of the geographical consequences of globalization. The concept of **spatial justice** is important here, because it requires us to consider the distribution of society's benefits and burdens at different spatial scales, taking into account both variations in people's need and in their contribution to the production of wealth and social well-being.

In particular, many people, nations, and ethnic groups around the world feel marginalized, exploited, and neglected as a result of the quickening pace of transformation that has come with globalization. Across much of the peripheral world, the perception of injustice has been brewing for a long time. Resentment at past colonial and imperial exploitation has been compounded as the more affluent places and regions of the world have become increasingly dependent on the cheap labor and resources of the periphery and as transnational businesses have displaced the traditional economic and social practices of peripheral and semiperipheral regions under the banner of modernization. Thinking about spatial justice is an important aspect of the "geographical imagination" described in Chapter 1, and is a recurring theme in the remainder of this book.

Meanwhile, some other important and controversial issues are tied to economic globalization: the depletion of the ozone layer, for example, together with threats to biodiversity and marine life. Carbon emissions, an important factor in global climate change, tend to reflect core-periphery patterns of economic development. On a per capita basis, the United States generates roughly 7 times more carbon emissions than China and 9 times more than India.

Of equal concern to many are the various local outcomes of the operation of the international economy: resource depletion and environmental degradation in some regions, ecotourism in others, industrialization in still others, and so on. Globalization has also been accompanied by increased international flows of narcotics and laundered money, a more extensive diffusion of communicable diseases like tuberculosis and HIV/AIDS, and greater freedom of movement for terrorists and weapons of mass destruction. These unwelcome consequences of globalization pose huge challenges to an international community more interconnected—and therefore more vulnerable—than ever before.

Jihad vs. McWorld

Related to core-periphery and fast-world/slow-world contrasts and interdependencies are broad cultural differences that have been characterized by political scientist Benjamin Barber as "Jihad vs. McWorld." In his characterization, McWorld is shorthand for the pop culture and shallow materialism that is part of Western, capitalist modernization. "Jihad" is shorthand for cultural values that are underpinned by religious fundamentalism, traditional tribal allegiances, and opposition to Western materialism. (Note that the term *jihad* properly refers to a struggle waged as a religious duty on behalf of Islam.) Neither "Jihad" nor "McWorld" is conducive to a healthy democracy or civil society, argues Barber, while tensions between the two make for potentially volatile situations.

At the heart of these tensions is a marked disillusionment with the West, especially within traditional Islamic societies. Across much of the world, modernization is now taken to mean Westernization and, more specifically, Americanization (**Figure 2.23**). While most Americans think of modernization as necessary and good, many other people see it as having resulted in their exploitation and humiliation. In most peripheral countries, only a minority can enjoy Western-style consumerism and the impoverished majority is acutely aware of the affluence of the core countries. While the gap between rich and poor countries has actually been widening for several decades, the U.S. aid budget—already low compared to the aid budgets of other developed countries—has been declining substantially. The United States, therefore, tends to be seen as a swaggering superpower, rigging the world-system in its own interests but doing relatively little by way of economic or humanitarian aid.

In the Arab world, the sense of grievance at such injustice, along with a sense of dismay at the way modernization is undermining Islamic values and traditions, has been intensified by opposition to U.S. policy toward the Middle East, which is seen as being cynically

Figure 2.23 Westernization
Turkish Moslems chat near an illuminated billboard advertising a Turkish Internet company at Istanbul airport.

geared to U.S. oil interests and support for Israel. In addition, there is widespread resentment at the injustices experienced by Palestinians, at the suffering of Iraqi civilians resulting from the U.S. invasion of Iraq, at the presence of U.S. troops in Saudi Arabia, and at the repressive and corrupt nature of U.S.-backed Persian Gulf governments.

Meanwhile, the world's core countries, acting on the confident assumption that people everywhere want Western-style modernization, have until recently failed to recognize the cultural resentment, the sense of injustice, and the genuine rejection of modernization that exists in many parts of the world. The terrorist attacks on the Pentagon and on New York's World Trade Center in September, 2001, brought an end to such assumptions. In the aftermath of the attacks it became clear that resentment toward the United States and its economic and foreign policies has bred widespread hostility and even hatred.

Mobilization Against Globalization

Economic globalization is in many ways still in its beginning stages, but already it has brought a great deal of change to the economic, cultural, and political geography of places and regions throughout the world. A great deal of this change has been progressive, bringing increased overall levels of economic well-being, a strengthening of free enterprise and democracy, and an enriched flow of products, ideas, and culture among and between places and regions. Inevitably, though, as with previous epochs of economic change, economic globalization has also brought problems as some places and regions—the U.S. Manufacturing Belt, northern England, central Scotland, and northeast France, for example—have experienced disinvestment in order that capital could be made available for more profitable investments elsewhere. Economic globalization has also undercut the power of national and local governments to regulate economic affairs and has erased a great deal of local diversity because of the economic success of global products: the "McDonaldization of everywhere."

Fundamental geographic differences—in climate, resources, culture, and so on—mean that economic globalization is variously embraced, modified, or resisted in different parts of the world. Indeed, a countermovement has emerged, a "mobilization against globalization," that could well affect the whole dynamic of economic globalization as it is played out over the next decade or two.

One form of this mobilization is exemplified by the efforts of activists who use the legal system in order to resist what they see as the undesirable local outcomes of transnational business practices. In the United States, activists have resurrected a 1789 alien tort law that was originally designed to provide redress for foreigners against sea pirates and slavers. The law had already been used to pursue individuals such as Ferdinand Marcos, the late dictator of the Philippines. In the late 1990s, activists pursued transnational corporations, accusing them of helping to suppress human rights. For example, four U.S. retailers and clothing manufacturers charged with unethical labor practices in a $1 billion alien tort suit that had been filed on behalf of some 50,000 garment workers in Saipan agreed to settle in 1999. While admitting no liability, Nordstrom, Gymboree, Cutter & Buck, and J. Crew agreed to pay $1.25 million into a fund to support the independent monitoring of their overseas suppliers.

A second form of mobilization is old-fashioned popular protest (**Figure 2.24**). French farmers, for example, regularly employ tactics such as blocking streets with tractors, with produce, with farmyard manure,

(a)   (b)

Figure 2.24 Antiglobalization demonstrations Globalization often leads to the downward convergence of wages and environmental standards, an undermining of democratic governance, and a general recoding of nearly all aspects of life to the language and logic of global markets. (a) Antiglobalization demonstration in the old town, Bern, Switzerland, prior to the 2008 meeting of the World Economic Forum (WEF) in the Swiss Alpine resort of Davos. (b) French farmers dump imported Chilean and Argentinian apples and melons before the headquarters of the Mutualite Sociale Agricole (MSA) in Marseille, 2005.

or with farm animals to protest trade-liberalization policies. A third form of mobilization is to counter the prevailing trend with alternatives, as in the case of Mecca Cola.

Fourth, and perhaps most significant in terms of future cultural struggles between local interests and transnational business interests, mobilization can be organized by coalitions of nongovernmental organizations (NGOs). This form of mobilization against globalization became much more powerful in the 1990s as a result of the Internet. Groups such as Kenya's Consumers' Information Network, Ecuador's Acción Ecologica, and Trinidad and Tobago's Caribbean Association for Feminist Research and Action are linked through scores of Web sites, listserves, and discussion groups to U.S., European, and Asian counterparts. NGOs set the agenda for the Earth Summit in Rio de Janeiro in 1992 and lobbied governments to attend; they publicized the Chiapas rebellion in Mexico in 1994, thereby preventing the Mexican government from suppressing it violently. In 1997, a loose alliance of 350 NGOs from 23 countries set out to ban land mines; they soon persuaded 122 nations to sign on to a treaty. In 1998, another NGO alliance, this time estimated to number 600 groups in nearly 70 countries, blocked a painstakingly negotiated treaty on international multilateral investment.

In 1999, more than 775 NGOs registered with the World Trade Organization (WTO) and took more than 2,000 observers to the WTO summit in Seattle, Washington. They also helped to organize some 70,000 protesters who took part in the most extensive teach-ins and demonstrations in the United States since the Vietnam War. Riot police fired tear gas, pepper spray, and rubber bullets at peaceful protesters blockading the WTO meeting, while roving gangs of anarchists smashed windows, overturned newspaper stands, and attacked cars. When demonstrators refused to disperse, the city declared a state of civil emergency. The WTO meetings themselves, delayed by the protests, ended in collapse. In 2001, a young man was killed by paramilitary police during antiglobalization riots in Genoa, Italy. Since then, economic summit meetings have been rescheduled to meet in hard-to-reach locations in an attempt to curb the effectiveness of demonstrations.

These demonstrations highlight some of the central issues that surround economic globalization. Economic globalization depends on free trade, but should the abolition of economic protectionism be accompanied by the abolition of social and environmental protection? The WTO's mandate is the "harmonization" of safety and environmental standards among member nations as well as the removal of tariffs and other barriers to free trade. Most people support free trade, but not if it harms public health, and not if it is based on child labor.

There have been several examples of how the free trade principles embodied in the WTO can erode national environmental standards. In 1997, Venezuela and Brazil, on behalf of their gasoline producers, challenged U.S. Environmental Protection Agency (EPA) regulations on gasoline quality, which were designed to ensure minimal levels of pollution. The WTO ruled in favor of Venezuela. The EPA subsequently had to change its regulations, leaving it with a weakened ability to enforce federal air-quality standards. When the European Union in 1998 banned U.S. beef raised with the assistance of hormone injections, the United States took the case to the WTO. The WTO ruled that the European Union's act was illegal under international trading rules, even if the Europeans were distrustful of the hormone-injected meat. Another WTO ruling in 1998 undermined the U.S. Endangered Species Act. The United States has attempted to protect endangered sea turtles from extinction by requiring that shrimp boats install devices that allow the turtles to escape their nets. The law applied to all shrimp sold in the United States, but the WTO ruled that this was a restraint on other countries' ability to trade freely. In 2003, the U.S. government, urged by biotechnology giants like Monsanto, Aventis, DuPont, and Dow Chemical and big agricultural groups such as the National Corn Growers Association, filed a formal case before the WTO, challenging the European Union's regulatory system on genetically modified foods. The WTO ruled in favor of the challenge in 2006, setting a broad precedent that inhibits the ability of WTO member states to set food safety, public health, and environmental health measures related to genetic engineering.

CONCLUSION

Places and regions everywhere carry the legacy of a sequence of major changes in world geography. The evolution of world geography can be traced from the prehistoric hearths of agricultural development and human settlement, through the trading systems of the precapitalist, pre-industrial world, to the foundations of the geography of the modern world. These foundations were built on industrialization, the colonization of many parts of the world, and the spread of an international market economy. Today these foundations can be seen in the

geography of the Information Age, a geography that now provides a global context for places and regions.

Today's world is highly integrated. Places and regions have become increasingly interdependent, linked through complex and rapidly changing commodity chains that are orchestrated by transnational corporations. Using new technology systems that allow for instantaneous global telecommunications and flexible patterns of investment and production, these corporations span the fast world of the core and the slow worlds of the periphery.

This integration does tend to blur some national and regional differences as the global marketplace brings about a dispersion of people, tastes, and ideas. The overall result, though, has been an intensification of differences between the core and the periphery. Within this new global context, local differences in resource endowments remain, and people's territorial impulses endure. Many local cultures continue to be resilient or adaptive. Fundamental principles of spatial organization also continue to operate. All this ensures that even as the world-system becomes more and more integrated, places and regions continue to be made and remade.

At first glance the emergence of globalization—with its transnational architectural styles, dress codes, retail chains, and popular culture and its ubiquitous immigrants, business visitors, and tourists—seems as if it might inevitably impose a sense of placelessness and dislocation, a loss of territorial identity, and an erosion of the distinctive sense of place associated with certain localities. Yet the common experiences associated with globalization are still modified by local geographies. The structures and flows of globalization are variously embraced, resisted, subverted, and exploited as they make contact with specific places and specific communities. In the process, places and regions are reconstructed rather than effaced. Often this involves deliberate attempts by the residents of a particular area to create or re-create territorial identity and a sense of place. Human geographies change, but they don't disappear.

MAIN POINTS REVISITED

- **The evolution of the modern world-system has exhibited distinctive stages, each of which has left its legacy in different ways on particular places, depending on their changing role within the world-system.**

  The modern world-system was first established over a long period that began in the late fifteenth century. More and more peoples around the world became exposed to one another's technologies and ideas over the next five centuries. Their different resources, social structures, and cultural systems resulted in quite different pathways of development, however. Some societies were incorporated into the new, European-based international economic system faster than others; some resisted incorporation; and some sought alternative systems of economic and political organization.

- **At the end of the eighteenth century, the new technologies of the Industrial Revolution brought about the emergence of a global economic system that reached into almost every part of the world and into virtually every aspect of people's lives.**

  New transportation technologies triggered successive phases of geographic expansion, allowing for an intensive period of external colonization and imperialism. The core of the world-system grew to include the United States and Japan, while most of the rest of the world was systematically incorporated into the capitalist world-system as a dependent periphery.

- **Places and regions are part of a world-system that has been created as a result of processes of private economic competition and political competition between national states.**

  Each place and region carries out its own particular role within the competitive world-system. Because of these different roles, places and regions are dependent on one another. The development of each place affects, and is affected by, the development of many other places.

- **Today the world-system is highly structured and is characterized by three tiers: core regions, semiperipheral regions, and peripheral regions.**

  The core regions of the world-system are those that dominate trade, control the most advanced technologies, and have high levels of productivity within diversified economies. Peripheral regions are characterized by dependent and disadvantageous trading relationships, by primitive or obsolescent technologies, and by undeveloped or narrowly specialized economies with low levels of productivity. Semiperipheral regions are able to exploit peripheral regions but are themselves exploited and dominated by the core regions. This three-tiered system is fluid, providing a continually changing framework for geographical transformation within individual places and regions.

- **The growth and internal colonization of the core regions could only take place with the foodstuffs, raw materials, and markets provided by the colonization of the periphery.**

  In the eighteenth and nineteenth centuries, the industrial core nations embarked on the inland penetration of the world's midcontinental grassland zones in order to exploit them for grain or stock production. At the same time, as the demand for tropical plantation products increased, most of the tropical world came under the political and economic control—direct or indirect—of one or another of the industrial core nations. For these peripheral regions, European overseas expansion meant political and economic dependency.

- **Within each of the world's major regions, successive technological innovations have transformed regional geographies.**

  Each new system of production and transportation technologies has helped raise levels of productivity and create new and better products that have stimulated demand, increased profits, and created a pool of capital for further investment. This investment, however, has taken place in new or restructured geographic settings.

- **Globalization has intensified the differences between the core and the periphery and has contributed to the emergence of a digital divide and an increasing division between a fast world (about 15 percent of the world's population) and a slow world (about 85 percent of the world's population) with contrasting lifestyles and levels of living.**

  The leading edge of the fast world is the Internet, which is now the world's single most important mechanism for the transmission of scientific and academic knowledge. Today, flows of goods, capital, and information that take place within and between transnational corporations are becoming more important than imports and exports between countries. At the same time, all these flows have helped to spread new values—from consumer lifestyle preferences to altruistic concerns with global resources, global environmental change, and famine relief—around the "fast" world.

KEY TERMS

British dominance (p. 60)
cartography (p. 50)
colonialism (p. 52)
colonization (p. 44)
commodity chain (p. 69)
comparative advantage (p. 64)
core regions (p. 52)
division of labor (p. 64)
Dutch dominance (p. 60)
environmental determinism (p. 55)

ethnocentrism (p. 55)
external arena (p. 48)
hearth areas (p. 42)
hegemony (p. 53)
hinterland (p. 48)
imperialism (p. 52)
import substitution (p. 52)
law of diminishing returns (p. 44)
leadership cycles (p. 53)
map projections (p. 51)

masculinism (p. 55)
minisystem (p. 42)
neocolonialism (p. 67)
peripheral regions (p. 53)
plantation (p. 49)
Portuguese dominance (p. 60)
producer services (p. 72)
semiperipheral regions (p. 53)
slash-and-burn (p. 42)

spatial justice (p. 75)
technology systems (p. 54)
transnational corporations (p. 67)
United States (p. 61)
world-empire (p. 44)
world-system (p. 44)

ADDITIONAL READING

Amoore, L. (ed.), *Global Resistance Reader.* London: Routledge, 2005.

Appadurai, A., "Disjuncture and Difference in the Global Cultural Economy," in M. Featherstone (ed.), *Global Culture: Nationalism, Globalization, and Modernity.* London: Sage, 1990, pp. 295–310.

Barber, B. R., *Jihad vs. McWorld: How Globalism and Tribalism Are Reshaping the World.* New York: Ballantine Books, 1995.

Daniels, P., and W. F. Lever (eds.), *The Global Economy in Transition.* New York: Addison Wesley Longman, 1996.

Diamond, J., *Collapse: How Societies Choose to Fail or Succeed.* New York: Viking, 2005.

Diamond, J., *Guns, Germs, and Steel.* New York: W. W. Norton, 1997.

Giddens, A., *Runaway World: How Globalization is Reshaping Our Lives.* New York: Routledge, 2000.

Harvey, D., *The New Imperialism.* Oxford: Oxford University Press, 2003.

Harvey, D., *Spaces of Global Capitalism.* New York: Verso, 2006.

Haywood, J., *Historical Atlas of the 19th Century World, 1783–1914.* New York: Barnes and Noble, 1998.

Held, D., A. McGrew, D. Goldblatt, and J. Perraton, *Global Transformations.* Malden, MA: Blackwell, 1999.

Held, D., and A. McGrew (eds.), *The Global Transformations Reader.* Malden, MA: Blackwell, 2000.

Hopkins, A. G. (ed.), *Globalization in World History.* London: Pimlico, 2002.

Hughes, A., and S. Reimer (eds.), *Geographies of Commodity Chains.* New York: Routledge, 2004.

Hugill, P., *World Trade Since 1431.* Baltimore: Johns Hopkins University Press, 1993.

Hutton, W., and A. Giddens (eds.), *Global Capitalism.* New York: New Press, 2000.

Johnston, R. J., P. J. Taylor, and M. Watts (eds.), *Geographies of Global Change: Remapping the World.* Cambridge, MA: Blackwell, 2002.

Knox, P. L., J. Agnew, and L. McCarthy, *The Geography of the World Economy,* 5th ed. New York: Routledge, 2008.

Milanovic, B., *Worlds Apart: Measuring International and Global Inequality.* Princeton, NJ: Princeton University Press, 2005.

O'Loughlin, J., L. Staeheli, and E. Greenburg (eds.), *Globalization and Its Outcomes.* New York: Guilford Press, 2004.

Ritzer, G., *The Globalization of Nothing.* Thousand Oaks, CA: Pine Forge Press, 2004.

Royal Geographical Society, *Atlas of Exploration.* New York: Oxford University Press, 1997.

Smith, N., *The Endgame of Globalization.* London: Routledge, 2004.

Urry, J., *Global Complexity.* Cambridge, MA: Polity Press, 2003.

Wallach, B., *Understanding the Cultural Landscape.* New York: Guilford Press, 2005. See especially Part II ("Historical Developments").

Wang, J. D. Oliver, T. Nottebottom, and B. Slack, *Ports, Cities and Global Supply Chains.* Burlington, VT: Ashgate, 2007.

EXERCISES

On the Internet

The Companion Website at www.mygeoscienceplace.com contains a number of tools designed to reinforce the geographic concepts presented in this chapter. Structured within a four-step learning path, these online resources include: chapter Key Concepts; a fully-searchable myeBook version of the chapter, which allows you to add notes and highlights; RSS news feeds from various media agencies; links to additional web resources; Human Geography Videos; a digital Glossary; Concept Review questions; Thinking Spatially mapping exercises; and Critical Thinking essay questions.

Unplugged

1. The present-day core regions of the world-system, shown in Figure 2.19 (p. 68), are those that dominate trade, control the most advanced technologies, have high levels of productivity within diversified economies, and enjoy relatively high per capita incomes. What statistical evidence can you find in your local library to support this characterization? (*Hint:* Look for data in annual compilations of statistics published in annual reviews such as the Encyclopaedia Britannica's *Yearbook* and in the annual reports of organizations such as the United Nations Development Programme (UNDP), the World Bank, and the World Resources Institute.)

2. The idea of an international division of labor is based on the observation that different countries tend to specialize in the production or manufacture of particular commodities, goods, or services. In what product or products do the following countries specialize: Bolivia, Ghana, Guinea, Libya, Namibia, Peru, and Zambia? You will find the data you need in a good statistical yearbook, such as the Encyclopaedia Britannica's *Yearbook,* or in a world reference atlas or economic atlas.

Geographies of Population

3

In the spring of 2008, food protests began erupting from Haiti to Bangladesh to Egypt and Mozambique in more than 37 countries across the globe in response to the soaring costs of basic grains like wheat and rice. In a period of two months, the global price of rice jumped 75 percent and over a twelve-month period, the price of wheat increased 120 percent. Rice farmers in Thailand began staying awake in shifts at night to guard their fields from thieves. In Peru, shortages of wheat flour prompted the military to make bread with potato flour, a native crop. Though no riots have been reported in core countries and regions, such as the United States and in Europe, more and more core poor families are struggling to feed themselves as the price of food skyrockets and less food is available through community food banks and other nonprofit outlets. Surging oil prices—which then boost related costs like fertilizer and transportation—combined with a drop in grain production due to droughts in Australia and the Ukraine, the falling U.S. dollar, and a growing global biofuels policy to promote ethanol (a corn-based product) in place of oil are the main factors in draining global foodstocks and stimulating the rising cost of food. And to add further concern, the United Nations Food and Agricultural Organization expects food prices to stay high through 2013.

The reason the price of grains is so important to population geography is that food shortages lead to malnutrition, specifically **undernutrition**, which is the result of an insufficient intake of food or the inability to eat or absorb food due to disease. Undernutrition includes being underweight for one's age, too short for one's age (stunted), dangerously thin (wasted), and deficient in vitamins and minerals (micronutrient malnutrition). When individuals are undernourished, they can no longer maintain natural bodily capacities, such as growth, resisting infections and recovering from disease, learning and physical work, and pregnancy and lactation in women. Persistent undernutrition also ultimately leads to lowered life expectancy and higher mortality rates.

For population geographers, however, knowing that mortality, or death from starvation and hunger-related illnesses, is on the rise in the periphery is not enough. They also want to know where, specifically, increased mortality from starvation is occurring, why it is occurring, and what the consequences of its occurrence—environmental, economic, political, or otherwise—are for the remaining population. And they want to know about fertility, how undernourishment affects birthrates, for instance; and about migration and why and to where people are migrating.

In this chapter we examine population distribution and structure, as well as the dynamics of population growth and change, with a special focus on spatial variations and implications. In short, we look at the locations of population clusters, the numbers of men and women and old and young, and the different ways in which these dynamics combine to create overall population growth or decline. We examine as well population movements and the models and concepts that population experts have developed to better understand the potential problems posed by human populations.

Main Points

- Population geographers depend on a wide array of data sources to assess the geography of populations. Chief among these sources are censuses; other sources include vital records and public health statistics.

- Population geographers are largely concerned with the same sorts of questions that other population experts study, but they also investigate "the why of where." *Why* do particular aspects of population growth and change (and problems) occur *where* they do, and what are the implications of these factors for the future of places?

- Two of the most important factors that make up population dynamics are birth and death. These variables may be examined in simple or complex ways, but in either case the reasons for the behavior of these variables are as important as the numbers themselves.

- A third crucial force in population change is the movement of populations. The forces that push populations from particular locations as well as those that pull them to move to new areas are key to understanding new settlement patterns. Population migration may not always be a matter of choice.

- Perhaps the most pressing question facing scholars, policymakers, and other interested individuals was articulated at the U.N. Millennium Summit in 2000: How can the global economy provide the world's growing population with enough food and safe drinking water, and a sustainable environment, so that all the world's people have the basic necessities for enjoying happy, healthy, and satisfying lives?

Protest over the soaring cost of food, Jakarta, Indonesia, 2008.

THE DEMOGRAPHER'S TOOLBOX

Demography, or the study of the characteristics of human populations, is an interdisciplinary undertaking. Geographers study population to understand the areal distribution of Earth's peoples. They are also interested in the reasons for, and the consequences of, the distribution of populations from the international to the local level. While historians study the evolution of demographic patterns, and sociologists study the social dynamics of human populations, geographers focus on the spatial patterns of human populations, the implications of such patterns, and the reasons for them. Using many of the same tools and methods of analysis as other population experts, geographers think of population in terms of the places that populations inhabit. They also consider populations in terms of the way that places are shaped by populations and in turn shape the populations that occupy them.

Censuses and Vital Records

Population experts rely on a wide array of instruments and institutions to carry out their work. Government entities, schools, and hospitals collect information on births, deaths, marriages, migration, and other aspects of population change. The most widely known instrument for assessing the state of the population is the census, a survey the ancient Romans developed to obtain information for tax collection.

At a simple level, a **census** is a straightforward count of the number of people in a country, region, or city. Censuses, however, are not usually so simple. Most are also directed at gathering other information about people, such as previous residences, marital status, and income. Many of the developed countries of the Western Hemisphere comprehensively assess the characteristics of their national populations every 10 years. In the United States, for example, the Bureau of the Census has surveyed the population every 10 years since 1790. The information gathered is used to apportion seats in the U.S. House of Representatives, as well as to redistribute federal tax funds and other revenues to states, counties, and cities.

In addition to the census, population experts employ other data sources to assess population characteristics. One such source is **vital records**, which report births, deaths, marriages, divorces, and the incidence of certain infectious diseases. These data are collected, and records of them are kept by local, county, state, and other levels of government. Schools, hospitals, police departments, prisons, other public agencies such as the Immigration and Naturalization Service in the United States and international organizations like the World Health Organization also collect demographic statistics that are useful to population experts.

Limitations of the Census

Census enumerations are an extremely expensive and labor-intensive undertaking for any governmental jurisdiction, and as a result, they occur rather infrequently—usually no more than every 5 or 10 years. Historically, the United States has undertaken a population census every 10 years, but in 1985 it introduced a quinquennial (every 5 years) census to augment the decennial (every 10 years) census. Most prominent among the reasons for initiating an additional, quinquennial census was that data collected every 10 years quickly became obsolete.

In many peripheral and semiperipheral countries, governments are not always able to finance a decennial census such as the comprehensive surveys undertaken in more developed countries like France or Germany. The small country of Andorra in southwestern Europe conducted its first comprehensive census in 1948; the next in 1954; followed by its last one in 1975, over 30 years ago! In southeast Asia, Cambodia conducted its first complete census in 1962 and did not conduct another until 1998. Liberia conducted its first census in 24 years in late March 2008. Pop stars were enlisted and billboards about the census were erected by the government to remind villagers to stay home for three days and be counted. As the median age of the population is 15, there were lots of people who had never before experienced a census and had no idea how to participate.

The incompatibility of enumeration dates makes comparisons among, between, and within countries quite difficult. For example, the United States conducted its last decennial census on April 1, 2000. China conducted a comparable decennial census on November 1, 2000, its fifth nationwide census. The seven-month difference in collection times between the two censuses, however, makes comparisons difficult. Demographers compare the populations of countries all the time, but they can do so only in very general terms because processes such as international migration and the difference caused by gaps between census collection dates can have a significant impact on the actual numbers.

Self-identification of ethnicity or race by indicators such as chief language spoken at home or skin color mask any number of other, often more meaningful, cultural attachments. A problematic category in the last several U.S. censuses has been "Hispanic," a term used to refer to residents whose cultural heritage derives from a Spanish-speaking country. Hispanics have in common ancestral ties to Latin America or Spain, but this apparent commonality obscures a great deal of cultural differentiation because a self-identified Hispanic can be Mexican American, Puerto Rican, Cuban American, Spanish, or someone from all but a couple of the South or Central American countries.

Possibly even more complicated is the "racial" category available on the 2000 census (**Figure 3.1**). The category is intended to allow individuals who feel they do

Person 1

Your answers are important! Every person in the Census counts.

1 **What is this person's name?** *Print the name of Person 1 from page 2.*

Last Name

First Name   MI

2 **What is this person's telephone number?** *We may contact this person if we don't understand an answer.*

Area Code + Number

3 **What is this person's sex?** *Mark* ☒ *ONE box.*
☐ Male
☐ Female

4 **What is this person's age and what is this person's date of birth?**

Age on April 1, 2000

Print numbers in boxes.
Month   Day   Year of birth

→ NOTE: Please answer BOTH Questions 5 and 6.

5 **Is this person Spanish/Hispanic/Latino?** *Mark* ☒ *the* **"No"** *box if* **not** *Spanish/Hispanic/Latino.*
☐ **No,** not Spanish/Hispanic/Latino
☐ Yes, Mexican, Mexican Am., Chicano
☐ Yes, Puerto Rican
☐ Yes, Cuban
☐ Yes, other Spanish/Hispanic/Latino — *Print group.* ↘

6 **What is this person's race?** *Mark* ☒ *one or* **more races** *to indicate what this person considers himself/herself to be.*
☐ White
☐ Black, African Am., or Negro
☐ American Indian or Alaska Native — *Print name of enrolled or principal tribe.* ↘

☐ Asian Indian
☐ Chinese
☐ Filipino
☐ Japanese
☐ Korean
☐ Vietnamese
☐ Other Asian — *Print race.* ↘

☐ Native Hawaiian
☐ Guamanian or Chamorro
☐ Samoan
☐ Other Pacific Islander — *Print race.* ↙

☐ Some other race — *Print race.* ↘

7 **What is this person's marital status?**
☐ Now married
☐ Widowed
☐ Divorced
☐ Separated
☐ Never married

8 **a. At any time since February 1, 2000, has this person attended regular school or college?** *Include only nursery school or preschool, kindergarten, elementary school, and schooling which leads to a high school diploma or a college degree.*
☐ No, has not attended since February 1 → *Skip to 9*
☐ Yes, public school, public college
☐ Yes, private school, private college

2043

Form D-2

Figure 3.1   Demographic questions from the U.S. Census Shown here is a page from the 2000 U.S. census. In an attempt to get more accurate information about race, the questions allow respondents to choose more than one racial category to describe their own racial identity.

not truly belong to one racial category—because their parents or grandparents were not from the same racial category—not to have to choose between White and Asian, for instance. The appearance of the category on the census form reflects the complex, polyglot world of U.S. demographics. But it is also a confusing category, and few people actually chose it. It is not even clear that those who did choose it really understood the Census Bureau's racial categories.

The bottom line is that no census is entirely comprehensive (or comprehensible). All censuses tend to underrepresent nonmainstream kinds of households as well as homeless individuals. For instance, more than 8 million people were not counted in the U.S. 1990 census, most of whom were Hispanics or African Americans. As a result of the 1990 undercount, many municipalities complained to the U.S. Bureau of the Census. The city of Inglewood in metropolitan Los Angeles estimated that 13,400 individuals were not found by the enumerators, leading to an undercount of 10.9 percent, the largest of any city in the country. Because federal revenue-sharing formulas as well as the apportionment of U.S. House seats are tied to population numbers, urban political officials from many of America's large and medium-size cities complained extensively about the numbers generated from the 1990 census.

The designers of the 2000 census suggested a correction for the 1990 undercounting problem by proposing a sampling plan. The aim of the plan was to sample statistically for overcounted as well as undercounted individuals so that minority residents would be acknowledged and counted. Thus, although every individual would not actually be counted through the submission of a census form, all would be accounted for by statistically manipulating a sample of the counted population.

In a case brought before a federal district court by former Republican Speaker of the U.S. House of Representatives Newt Gingrich, the sampling plan was defeated, and it was charged that undercounting occurred again in the 2000 census. After several reevaluations of the data, however, the Census Bureau determined that, in fact, there was not an undercount of the total population but rather a net overcount of 1.3 million persons. When looking at specific populations, the Census Bureau found that not only had non-Hispanic Whites been overcounted, so had children between the ages of 0 and 9. It was actually Native Hawaiian and Other Pacific Islanders and American Indians and Alaskan Natives who had been underercounted.

The counting of the U.S. population has a real impact on peoples' lives. Since billions of dollars in federal grants for education, transportation, and health care are distributed on the basis of population, undercounted communities can lose significant financial support for local schools, roads, and health care services, while overcounted communities can reap more grants than warranted.

POPULATION DISTRIBUTION AND STRUCTURE

Because human geographers explore the interrelationships and interdependencies between people and places, they are interested in demography. Population geographers bring to demography a special perspective—the spatial perspective—that emphasizes description and explanation of the "where" of population distribution, patterns, and processes.

For instance, the seemingly simple fact that as of early 2007 the world was inhabited by 6,600 billion people is one that geographers like to think about in a more complex way. While this number is undeniably phenomenal and increasing with each passing second, the most important aspect of this number for geographers is its uneven spatial expression from region to region and from place to place. Equally important are the implications and impacts of these differences. Looking at population numbers, geographers ask themselves two questions: Where are these populations concentrated, and what are the causes and consequences of such a population distribution?

Population Distribution

At a basic level, many geographic reasons exist for the distribution of populations throughout the globe. As the world population density map demonstrates (**Figure 3.2**), some areas of the world are very heavily inhabited, others only sparsely. Some areas contain no people whatsoever. Overall, Bangladesh and the Netherlands, for example, have high population densities. India, however, displays a different pattern, with especially high population concentrations along the coasts and rivers but a relatively low population density in its western sector. The sparsest inland population concentrations in the world occur in the densely forested frontier regions of the Amazon. Thus, environmental and physical factors are important influences on population distributions and concentrations.

Degree of accessibility, topography, soil fertility, climate and weather, water availability and quality, and type and availability of other natural resources are some of the factors that shape population distribution. Other factors are also crucial—first and foremost are a country's political and economic experiences and characteristics. For example, the high population concentrations along Brazil's Atlantic coast date back to the trade patterns set up during Portuguese colonial control in the sixteenth and seventeenth centuries. Another important factor is culture as expressed in religion, tradition, or historical experience. Cities like Medina and Mecca, in the Middle East, comprise important population concentrations in no small part because they are Islamic sacred sites.

Table 3.1 shows population estimates in terms of traditional continental distributions. From the table it is clear that Asia is far and away the most populous

Figure 3.2   World population density, 2006 As this map shows, the world's population is not uniformly distributed across the globe. Such maps are useful in understanding the relationships between population distribution and the national contexts within which they occur. China and India are the largest countries in the world with respect to population; the core countries have much smaller populations. (Adapted from World Bank, *World Development Indicators 2007*, World Bank: Washington, DC.)

continent, including nearly 61 percent of the world's inhabitants in more than 40 countries. In Asia, China and India alone constitute 37 percent of the continental population total and 21 percent of the world total. Running a distant second and third are Africa, with 14 percent of the world's population, and Europe, with 11.5 percent.

The population clusters that take shape have a number of physical similarities. Almost all of the world's inhabitants live on 10 percent of the land. Most live near the edges of landmasses, near the oceans or seas, or along rivers with easy access to a navigable waterway.

Approximately 90 percent live north of the equator, where the largest proportion of the total land area (63 percent) is located. Finally, most of the world's population lives in temperate, low-lying areas with fertile soils.

Population numbers are significant not only on a global scale but also at other levels. Population concentrations within countries, regions, and even metropolitan areas are also important. For example, much of the population of North Africa is distributed along the coastal areas where most of the large cities are. Egypt, however, is an exception. While it does possess a significant coastal population, a large part of its population is

TABLE 3.1 World Population Estimates by Continents

Continent	Number of Inhabitants (in millions)	% of Total Population
Africa	944	14
Asia	4,010	60.5
Oceania	35	0.5
Europe	733	11.5
North America	335	5
Latin America and Caribbean	569	8.5
Total	6,626	100

Source: Population Reference Bureau Web site, www.prb.org. *World Population Data Sheet, 2007.*

88 CHAP

Population distribution of Egypt, 2004 Egypt's population distribution is closely linked to the proximity of water. In the north, the population clusters along the Mediterranean, and in the interior, along the banks of the Nile River. (*Source:* Gridded Population of the World [GPW], Version 2. Palisades, NY: CIESIN, Columbia University. Available at http://sedac.ciesen.org/plu/gpw/index.html?main.html@2)

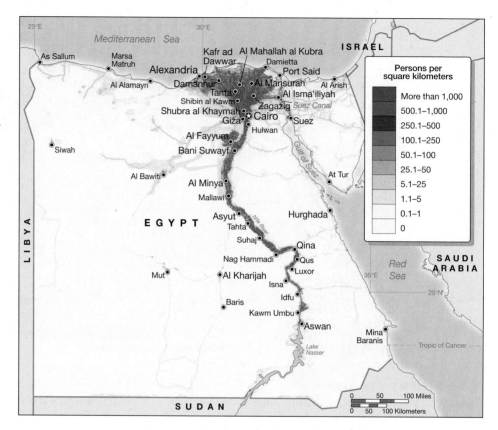

Population Density and Composition

Another way to explore population is in terms of density, a numerical measure of the relationship between the number of people and some other unit of interest expressed as a ratio. For example, crude density is probably the most common measurement of population density. **Crude density**, also called **arithmetic density**, is the total number of people divided by the total land area. The metropolitan area of Melbourne, Australia represents a classic low-density urban settlement predicated on a quarter-acre home and garden enabled by the widespread use of automobiles. Melbourne, the second most populous city in Australia, has a population density of approximately 1,566 persons per square kilometer (4,056 per square mile; **Figure 3.4**). (Since the 1990s, Melbourne has begun to address this sprawling urban form through infill.) By comparison, the city of Tokyo, the most populous city in the world, containing 12,790,000 residents as of 2007, is a classic high-density urban settlement with a population density of 5,848 persons per square kilometer (15,281 persons per square mile; **Figure 3.5**).

The limitation of the crude density ratio—and hence the reason for its "crudeness"—is that it is one-dimensional. It tells us very little about the opportunities and obstacles that the relationship between people and land

also distributed narrowly in a line running through the interior of the country along the banks of the Nile River, which marks the essential importance of water to this desert population (**Figure 3.3**).

contains. For that, we need other tools for exploring population density, such as nutritional density or agricultural density. **Nutritional density** is the ratio between the total population and the amount of land under cultivation in a given unit of area. **Agricultural density** is the ratio between the number of agriculturists—people earning their living or subsistence from working the land—per unit of farmable land in a specific area. **Figure 3.6** is a map of the ratio of population to the number of physicians, an indicator of the quality of a country's health care.

In addition to exploring patterns of distribution and density, population geographers also examine population in terms of its composition—that is, the subgroups that constitute it. Understanding population composition enables geographers to gather important information about population dynamics. For example, knowing the composition of a population in terms of the total number of males and females, number and proportion of senior citizens and children, and number and proportion of people active in the workforce provides valuable insights into the ways in which the population behaves now and how it might behave in the future.

Facing unique challenges are countries with populations that contain a high proportion of old people—a situation most core countries will soon be facing as their "baby boom" generation ages. The **baby boom** generation includes those individuals born between 1946 and 1964. A considerable amount of a country's resources and energies will be necessary to meet the needs of a large number of people who may no longer be contributing in any significant fashion to the creation of the

**Figure 3.4
Population density,
Melbourne,
Australia** The urban
landscape of metropoli-
tan Melbourne is low-
profile and widely spread
predicated on the early
availability of cheap land
and the automobile dur-
ing a critical period of
its growth. Melbourne is
largely a regional urban
hub, whereas Tokyo's
importance in the world
economic system is
more national and
international.

wealth necessary for their maintenance. There might also be a need to import workers to supplement the relatively small working-age population.

Knowing the number of women of childbearing age in a population, along with other information about their status and opportunities, can provide valuable information about the future growth potential of that population. For example, populations in core countries like Denmark, which has a small number of women of childbearing age relative to the total population size but with high levels of education, socioeconomic security, and wide opportunities for work outside the home, will generally grow very slowly if at all. Populations in peripheral countries like Kenya, on the other hand, where a large number of women of

childbearing age have low levels of education and socio-economic security and relatively few employment opportunities, will continue to experience relatively high rates of population growth, barring unforeseen changes. The variety that exists within a country's population very much shapes the opportunities and challenges it must confront nationally, regionally, and locally.

Understanding population composition, then, not only can tell us much about the future demographics of regions but is also quite useful in the present. For example, businesses in core countries use population composition data to make marketing decisions and to decide where to locate their operations. For many years, these businesses used laborious computer models to help target their markets. With

Figure 3.5   Population density, Tokyo, Japan Urban form in Tokyo is the result of many factors. Most important in terms of its population size is its role as a central node in the world system of cities. Such a prominent role is built upon both a varied and extensive population base. The urban economy is complex, requiring a wide range of labor skills from professional and managerial to low-skilled, service-oriented workers. It is also the case that Japan is a small country, geographically, with a large population and the cities there tend to be particularly large.

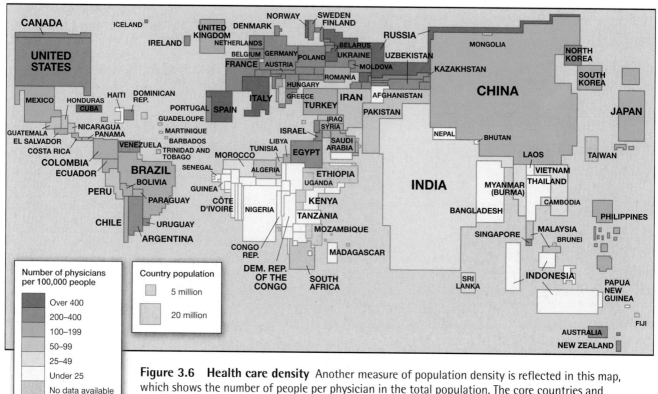

Figure 3.6   Health care density Another measure of population density is reflected in this map, which shows the number of people per physician in the total population. The core countries and China have the highest ratio of doctors to overall population. Most of the continent of Africa, except South Africa and Egypt, Libya, and Morocco in North Africa, has the lowest ratio, reflecting another dimension of core-periphery inequality. (*Source:* Adapted from H. Veregin (ed.), *Goode's World Atlas,* 21st ed. Chicago, IL: Rand McNally & Co, 2005, p. 37.)

the recent development of geographic information systems (GIS), however, this process has been greatly simplified. Some examples of assessing the location and composition of particular populations, known as **geodemographic analysis,** are provided in Box 3.1, "GIS Applications and Google Earth," which discusses private marketing applications as well as community-oriented ones.

Age-Sex Pyramids

Areal distributions are not the only way that demographers portray population distributions. To display variations within particular subgroups of a population or with respect to certain of its descriptive aspects, such as births or deaths, demographers also often use bar graphs displayed both horizontally and vertically. The most common way for demographers to graphically represent the composition of the population is to construct an **age-sex pyramid,** which is a representation of the population based on its composition according to age and sex. An age-sex pyramid is actually a bar graph displayed horizontally. Ordinarily, males are portrayed on the left side of the vertical axis and females on the right. Age categories are ordered sequentially from the youngest at the bottom of the pyramid to the oldest at the top. By moving up or down the pyramid, one can compare the opposing horizontal bars in order to assess differences in frequencies for each age group.

Age-sex pyramids also allow demographers to identify changes in the age and sex composition of populations. For example, an age-sex pyramid depicting Germany's 2000 population clearly revealed the impact of the two world wars, especially with regard to the loss of large numbers of males of military age and the deficit of births during those periods (**Figure 3.7**). Demographers call population groups like these cohorts. A **cohort** is a group of individuals who share a common *temporal* demographic experience. A cohort is not necessarily based only on age, however. Cohorts may be defined based on criteria such as time of marriage or time of graduation.

In addition to revealing the demographic implications of war or other significant events, age-sex pyramids can provide information necessary to assess the potential impacts that growing or declining populations might have. As illustrated in **Figure 3.8**, the shape of an age-sex pyramid varies depending on the proportion of people in each age cohort. The pyramid for the peripheral countries, shown in Figure 3.8a, reveals that many dependent children, ages 0 to 14, exist relative to the rest of the population. The considerable narrowing of the pyramid toward the top indicates that the population has been growing very rapidly in recent years. The shape of this pyramid is typical of peripheral countries with high birthrates and low death rates.

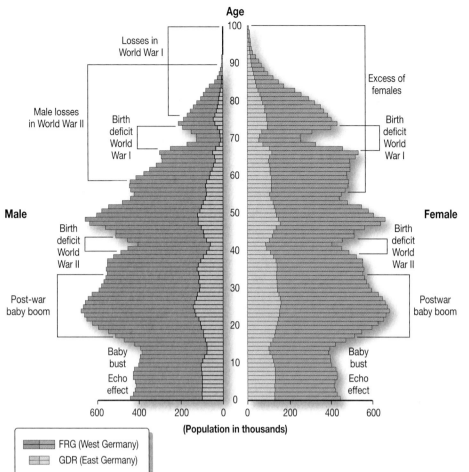

Figure 3.7 Population of Germany, by age and sex, 2000 Germany's population profile is that of a wealthy core country that has passed through the postwar baby boom and currently possesses a low birthrate. It is also the profile of a country whose population has experienced the ravages of two world wars. (*Source:* Adapted from J. McFalls Jr., "Population: A Lively Introduction," 5th ed., *Population Bulletin, 62* (1), 2007, p. 20.)

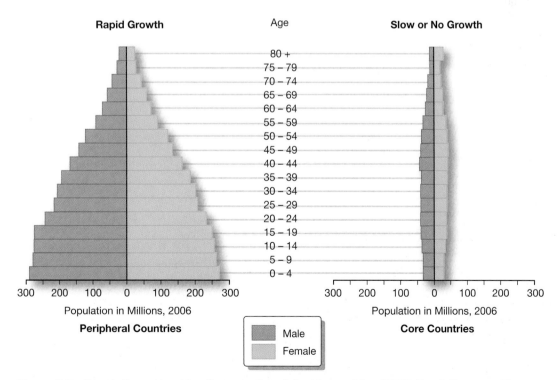

Figure 3.8 Population pyramids of core and peripheral countries, 2006 Population pyramids vary with the age and gender structure of the population being depicted. We can derive important information about the population growth rates of different countries over time by analyzing changes in the numbers of people in each sex and age category.

3.1 GEOGRAPHY MATTERS

GIS Applications and Google Earth

Geographic information systems, or GIS, possess the powerful ability to reveal spatial patterns in demographic data, such as the occurrence of crime and the distribution of air pollution or patterns brought about by economic or political practices, such as retail location analysis or land-use planning. Recognizing the potential commercial applications, marketing firms and in-house site-location departments have readily incorporated GIS into their daily operations. Prior to using GIS for marketing, businesses made crude estimates of the location and behavior of their clientele. By using GIS to analyze demographic data, marketing companies are now able to provide businesses with maps identifying areas of potential customers. With this information, businesses can engage in promotional "sales blitzes," targeting specific customers identified from GIS analysis.

The key to using GIS effectively in marketing is the ability to link demographic data to particular locations. This is known as *georeferencing*. Demographic data are linked to ZIP code areas and telephone prefix zones. Then the data are analyzed to determine whether meaningful spatial relationships exist among specified areal units, based on given demographic data. For example, the residences of people with a particular level of education may cluster around a few contiguous ZIP code regions.

Demographic data such as income level, education level, home value, age, and number of household residents provide marketing firms with information that can be linked to certain consumer goods. For example, people who have high income and high education levels tend to own luxury cars. Luxury-car dealerships would want to target such households. A GIS could produce a map that would show the general location of such households, and advertisements could then be distributed throughout this area. Shopping-center developers analyze demographic data with a GIS to evaluate mall location opportunities. In addition to the demographic data—which include variables such as population density, income, and education—developer databases also often contain information about the characteristics of competing shopping centers, as well as tenant and ownership data.

Another marketing area in which GIS is used is locational analysis—especially determining where to locate a business. The success of businesses such as gasoline stations and fast-food restaurants depends on locational factors such as traffic density. GIS is also aptly suited for modeling traffic patterns, which are frequently determined by studying demographic data

that indicate how far, how often, and in which direction people travel to work. Many GIS systems have models that can be used to predict traffic patterns. By identifying areas with high traffic volumes, businesses can choose strategic locations. Other conditions, such as zoning restrictions and proximity to competition, can also be factored into a GIS to narrow the number of potential retail sites.

The Arizona Republic, the premier newspaper for the Phoenix metropolitan area, uses GIS to more carefully target advertising for its business customers and subscribers. For example, one drug store wanted to reach customers within a certain number of blocks from the store, avoiding those areas where large transportation arterials would prevent people from accessing the store. A grocery store chain wanted different versions of advertising for its different store locations. *The Arizona Republic* advertising salespeople used their in-house GIS to determine the number of subscribers and nonsubscribers in an area, as well as other variables, including income, ethnicity, age, education, as well as car or pet ownership among a host of other variables. Their database incorporates existing government-generated data—such as U.S. census data; 3- and 5-digit ZIP codes; highway data; and state, county, and city street maps—as well as specific subscriber information collected by the newspaper, including carrier routes and subscriber profiles (**Figure 3.A**) to produce more specifically focused advertising.

It should be pointed out, however, that the potential for invasion of privacy through the coupling of different sorts of data sets is very real. For instance, when you shop in grocery stores with electronic inventory systems and pay for your groceries by credit card or check, a record of your purchases can be joined with your name. Such a coupling allows marketers to target you based on information on sales and manufacturers' coupons. It also, potentially, gives them access to information about what prescription drugs you are taking, how much alcohol you consume, what sorts of magazines you buy, or which videos you rent. Thus, while GIS and the massive data sets that support it open up new opportunities for marketing, the possibility also exists for the erosion of privacy.

It is also possible to use GIS in all sorts of noncommercial sectors, including government as well as private endeavors. For instance, public participation GIS, also known as PPGIS, is a way that communities, especially poor and minority communities, can use the technology to offer their own solutions to local issues and problems, from crime to conservation.

Figure 3.A   Potential advertising coverage Maps like this help *Arizona Republic* salespeople to link store addresses with carrier routes in order to more precisely target advertising.

Two Georgia planners, for example, have employed a PPGIS model to help Atlanta's inner-city poor neighborhoods—among the poorest in the country—by teaching them how to use GIS in order to produce knowledge that will help them make more informed decisions. One of the ways that professional planners helped poor Atlanta neighborhoods was through a project aimed at addressing the decline in housing conditions. Through the efforts of the planners and some volunteers, selected neighborhood residents were trained to identify building code violations. Next, the residents were taught how to enter the code violation data into a GIS program so that maps showing the distribution of the violations could be generated. The maps were used to persuade city officials to take action to improve code enforcement.

A more popular and widely accessible medium for accessing spatial data by way of home computers, laptops, and wireless handheld devices like i-Phones and BlackBerries is Google Earth, which enables users to go anywhere on Earth by way of satellite imagery, maps, terrain models, and 3D buildings. Google Earth's Sketch Up free software enables its users to re-create ancient cities or view details of the damage created by Hurricane Katrina along the U.S. Gulf Coast. And Street View enables users to see the streets, sidewalks, and buildings (where imagery is available) of places all over the world and explore landscapes and take virtual trips around neighborhoods. Google Earth and its related software like Sketch Up and Street View enable people who are far away to get a more intimate view of another place. Michael Jones, chief technical officer of Google Earth, believes that the more you can connect with other people and places across the world, the better you will understand them.

As with GIS, skeptics believe that Google Earth is a Trojan horse—a surveillance mechanism in disguise. There are indeed components of this threat and other visual technologies that expose aspects of our lives that had previously been more private. But Google Earth, as geographers John Paul Jones III and Paul Kingsbury argue, also offers unique opportunities for pleasure and enjoyment as the world becomes exposed to us in new ways.

Source: http://www. esri. com/library/fliers/pdfs/cs-arizona-republic. pdf, accessed May 27, 2008; D. S. Sawici, and P. Burke, "The Atlanta Project: Reflections on PPGIS Practice," in *Community Participation and Geographic Information Systems*, W. J. Craig, T. M. Harris, and D. Weiner (eds.), New York: Taylor and Francis, 2002, pp. 89–100; M. Wagner, "The View from Google Earth: An Interview with Michael Jones," Geospatial Solutions, http://gismap.geospatial-solutions.com/gssgismap/article/articleDetail.jsp?id=362082, accessed May 22, 2008; and P. Kingsbury and J. P. Jones III, "Beyond Apollo and Adorno: Dionysus and Walter Benjamin on Google Earth," *Geoforum*, forthcoming.

Serious implications are associated with this type of pyramid, however. First, in the absence of high productivity and wealth, resources are increasingly stretched to their limit to accommodate even elemental schooling, nutrition, and health care for the growing number of dependent children. Furthermore, when these children reach working age, a large number of jobs will have to be created to enable them to support themselves and their families. Also, as they form their own families, the sheer number of women of childbearing age will almost guarantee that the population explosion will continue. This will be true unless strong measures are taken, such as intensive and well-funded birth-control campaigns, improved education and outside opportunities for women, and modifications of cultural norms that place a high value on large family size.

In contrast, the pyramid for the core countries (Figure 3.8b) illustrates the typical shape for a country experiencing a slow rate of growth. Most countries in the core are experiencing birthrates that are at or below replacement level. Thus, the pyramid is very columnar, hardly a pyramid at all. People are equally distributed among the cohorts, though the base is perceptibly narrower. Many of the core European countries, such as Denmark, Spain, and Germany, as well as the United States, can expect the same demographic challenges, including an extremely large elderly population. In all these countries, however, high levels of production and wealth, combined with low birthrates, translate into a generally greater capacity to provide not only high levels of health, education, and nutrition but also jobs, as those children grow up and join the workforce. Whether these opportunities are equitably distributed among individual members of the population remains to be seen.

Table 3.2 and Figure 3.9 show the present state and potential future impact of the baby boom cohort, the largest population cohort in U.S. demographic history. Figure 3.9 provides a series of pyramids that illustrate how the configuration changes as the boomers age. The narrower column of younger people rising below the boomer cohort in these pyramids reveals the biggest problem facing this population: a significantly smaller cohort moving into its main productive years having to support a growing cohort of aging and decreasingly productive boomers (**Figure 3.10**). Congressional fights over Social Security and Medicare funding are only the tip of the iceberg with regard to this problem, however (see Box 3.2, "The Baby Boom and the Aging of the Population").

Construction and analysis of age-sex pyramids need not be restricted to the national level. **Figure 3.11**, for example, shows pyramids for three census tracts within the same Minnesota county. They show that at the local level, variations exist in the geography of population composition that are as informative as those constructed at the national level. The proportional representation of each cohort, by age and sex, raises provocative questions about why such spatialized differences exist and what the implications of the distributions might be for social services, marketing strategies, and a host of other applications. For example, a census tract with a large number of children may lead decision makers to push for greater opportunities for organized sports and perhaps a children's health care facility. In contrast, a growing concentration of older people might prompt the opening of a senior citizen center and a geriatric care facility.

A critical aspect of the population pyramid is the **dependency ratio**, which is a measure of the economic impact of the young and old on the more economically productive members of the population. In order to assess this relation of dependency in a particular population, demographers divide the total population into three age cohorts, sometimes further dividing those cohorts by sex. The **youth cohort** consists of those members of the population who are less than 15 years of age and generally considered to be too young to be fully active in the labor force. The **middle cohort** consists of those members of the population aged 15 to 64, who are considered economically active and productive. Finally, the **old-age cohort** consists of those members of the population aged 65 and older, who are considered beyond their economically active and productive years. By dividing the population into these three groups, it is possible to obtain a measure of the dependence of the young and old upon the economically active, and the impact of the dependent population upon the independent.

TABLE 3.2 Baby Boomer Population Structure

While the baby boom has demographically dominated the last half of the twentieth century, its influence will begin to wane in the first half of the new one.

Year	% of Population Who Are Baby Boomers	Age
1990	30	25–44
2000	20	35–59
2020	15	55–79
2040	7	75–85+

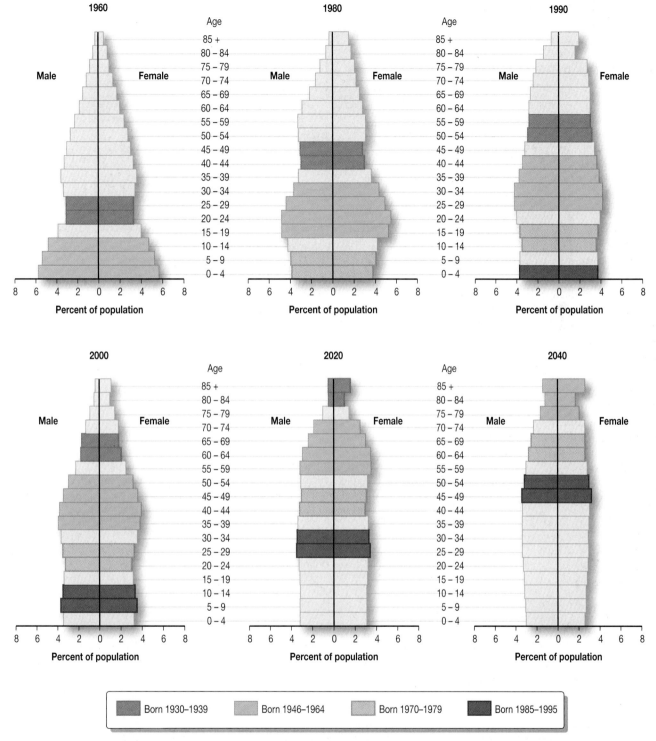

Figure 3.9   Population pyramids of U.S. baby boomers, 1960–2040 Population pyramids are not only based on past or present statistics but may also be constructed based on estimates of the future. In this series of pyramids, we can clearly see the progression of the baby boomers, shown in blue, up the pyramid as their cohort ages. Note how the "pyramid" becomes a column by the year 2040 as birthrates remain below death rates for each cohort. Note also the significantly higher number of women compared with men in the oldest age group for this same year, reflecting the statistical tendency for women to live longer than men. (Adapted from L. Bouvier and C. DeVita, "The Baby Boom: Entering MidLife," *Population Bulletin, 46* (3), 1991.)

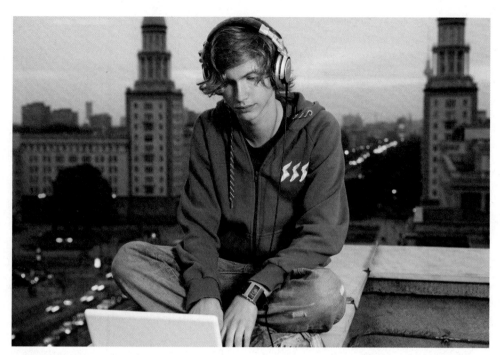

Figure 3.10   The Net Generation Members of the Net Generation, people who are currently in their late teens to late twenties, are faced with the awesome burden of having to help support a huge, aging, baby boom population. The story is more complex than that, however, as both generations are likely to have different effects on culture and the economy. A sharp rise in the number of wealthy older people will likely have an impact on the demand for several types of consumer goods or services, such as health care, travel, and luxury products. They have already manifested their desire to delay the aging process by whatever means they can. As millions of them move into their sixties, they will almost certainly avail themselves of such cosmetic procedures as Botox shots to erase wrinkles, while others may opt for skin treatments at health spas. At the same time, the Net Generation is likely to continue to drive demand around increasingly sophisticated personal technology and demonstrate a desire for social inclusion, innovation, and "living in the moment" experiences. Their impact on politics in the United States has just begun to be tested in the 2008 presidential elections.

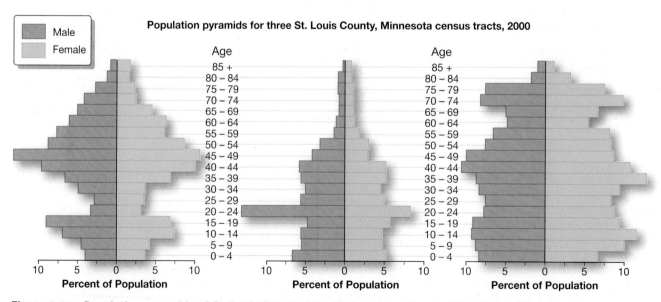

Figure 3.11   Population pyramids of St. Louis County, Minnesota, census tracts, 2000 Population pyramids are as useful for analyzing populations within a single region as for an entire country. Here we see how age-sex profiles can also vary within different census tracts of the same city. The tracts show that even within a city, variation in populations can be substantial. Information like this can be very valuable in decision-making and policymaking at state, regional, and local levels as well as for marketers who can target different populations through direct mailings. (Minnesota Department of Administration, http://www.lmic.state.mn.us/datanetweb/php/census2000/PopReport.php?year=2000&final=1&area=T&subtop=P02&tract=1600&tractname=16&cnty=69&cntyname, accessed May 31, 2008.)

3.2 GEOGRAPHY MATTERS

The Baby Boom and the Aging of the Population

The baby boom generation—those individuals born between 1946 and 1964 in the United States—has been described as "one of the most powerful and enduring demographic influences on this nation." It has certainly been one of the most studied, analyzed, vilified, and sanctified. The case of the baby boom generation provides a useful way of understanding the complex factors that shape demographic change.

Why so many births between 1946 and 1964? Most demographers cannot give a definitive answer to this question. In fact, in the 1940s, population experts were predicting that the American population would stop growing. And although increases in births following a war are expected, the extended surge in births that followed the end of the Second World War came as a surprise to population experts, policymakers, government officials, and most others. Demographic transition theory would have led demographers to anticipate declining births as urbanization and "modernization" accelerated after the war. Sociological theories and predictions based on past trends were not very useful, however, in anticipating the unprecedented phenomenon of the baby boom. To understand it, then, we need to examine a whole host of factors.

Demographic Factors

Demographers insist that the baby boom not be seen as a direct or indirect result of the end of the war. In fact, although marriages (and divorces) did increase dramatically after the war, the accompanying rise in birthrates accounts only for the time from 1946 to 1947. By 1950, birthrates had actually fallen slightly (**Figure 3.B**). In 1951, the birthrates once again began to climb and continued to be high until 1964.

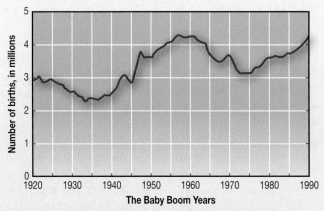

Figure 3.B   U.S. baby boom crude birthrate The chart very clearly shows the big increase in births following the Second World War. (*Source:* Adapted from L. Bouvier and C. DeVita, "The Baby Boom: Entering MidLife," *Population Bulletin, 46* (3), 1991.)

Political and Economic Factors

It is significant that the end of the Second World War witnessed a phenomenal expansion in the U.S. economy. Initially stimulated by the war, it was extended by important transformations in transportation and technology. Additionally, government had expanded and created programs for education and housing that helped returning veterans start married life with property and the opportunity for improving their economic status by attending college. Transportation—especially through the construction of the interstate highway system—helped to fuel suburbanization, which meant growth in the building industry, in automobile manufacturing, and in the manufacturing of durable goods for the home. Not surprisingly, demand for labor was high in a growing economy, and young people, most of whom had attained a higher level of education than the previous generation, were able to obtain good jobs with relatively high wages, decent benefits, and reasonable prospects for promotion.

The Aging of the Population

It is important to recognize that something like the baby boom occurred not only in the United States but also in most of the core countries of the world and even in parts of the periphery. The result is that there is a very large cohort of individuals worldwide, currently in their early forties to mid- to late fifties, that has had and will continue to have tremendous impacts on the rest of the population, especially as they enter their sixties, seventies, and eighties. In fact, the most fundamental demographic transformation of the twenty-first century is the aging of the population (**Figure 3.C**). In the United States, this is a result not only of the aging of the baby boom cohort but also of longer life expectancies and falling fertility rates, that are also falling all across the globe. In the year 2005, the median age of the world's populations was 28, but by 2050 it is expected to increase to 38.

The aging of the population is currently most pronounced in the core countries, where in 1998 the number of older persons (60 and over) actually exceeded the number of children (under 15) for the first time. As a result, for the core countries the median age of the population will rise from 37.5 in 2000 to 45.6 in 2050, in many cases resulting in decreasing national populations more generally. **Table 3.A** lists some of the core countries and regions that are expected to experience overall population decline. In most cases, populations that are getting both older and smaller are also likely to experience a decrease in the ratio of persons of working age (15–64) to older persons (65 and older); that is, there will be fewer working-age people to support the needs of the aging population. What this means is that

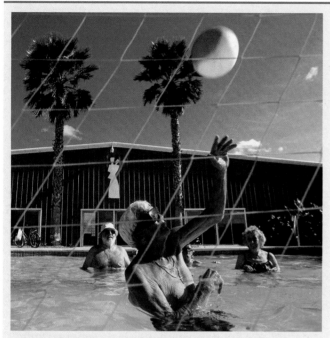

Figure 3.C   Elderly people in the United States In the core countries, men and women are living longer through lifestyle changes and improved health care.

for countries like Spain, Italy, Japan, and Germany, just to name a few, the prospect of smaller and older populations raises some very serious concerns about employment, economic growth, health care, pensions, and social support services. One possible "solution" to the problem of the aging of the core's population that could have immediate effects is increasing immigration opportunities for young migrants from countries and regions that are experiencing rapid population growth.

And yet, although the aging of the population has been slower in peripheral countries because of high twentieth-century birthrates, the next 50 years will see the median age rise there as well. By 2050 the median age in countries of the periphery is expected to be 36.7 years, up from 24.4 years in 2000. This is actually a bigger jump than in the core. East and Southeast Asia are the regions whose populations are aging most rapidly, with Africa aging the least.

In the United States, the aging of the population is expected to have increasing impacts on the economy, politics, society, and culture. A significant difference of opinion exists among population experts on the future impacts of aging baby boomers. Some see all these highly educated workers as continuing to stimulate the economy in terms of improved productivity and increasing innovations. Others worry that this cohort will eventually place unfulfillable demands on the nation's social service system. Certainly the debates about health care and Social Security reform are related to the increasing demands that this group is beginning to place on the health care and social service system. In addition, the pressures

generated by its sheer size will certainly continue well into the first half of the twenty-first century. For example, by 2030 the oldest surviving boomers will be age 84 and the youngest will be 65. At that point, a projected 65 million people in the United States will be over the age of 65, compared with roughly 30 million today.

The Impacts on Younger Americans

Demographers, in addition to identifying the existence of the baby boom generation, have also demonstrated the presence of a "baby bust" generation. Two ways exist to determine the extent of this cohort. One is through annual birthrates; if we use this measure, this generation includes those individuals born between 1965 and 1980, when births were below the baby boom average of 4 million births per year. A second way to determine the extent of the baby bust cohort is in terms of the total fertility rate. Between 1947 and 1964, the TFR ranged between 3.1 and 3.7, well above replacement level. In 1965, however, the TFR dropped below replacement level for the first time in U.S. history, and in 1976 it hit an all-time low of 1.7. Between 1977 and 1987, the TFR remained pretty stable at 1.8, and in 1990 it climbed to 2.09—close to replacement level, marking the end of the baby bust generation.

The term *baby bust* refers to the demographic features of a cohort; cultural terms have also been offered to describe this group as well as groups that have followed it. Those persons born between 1965 and 1975 have also been called Generation X. The cultural label placed on Generation Xers is "slackers," because some of the most famous of this birth cohort, such as Kurt Cobain of the band Nirvana, were associated with grunge music and the antiestablishment attitude it stood for. Those individuals born after the Xers—between 1982 and 1994—are sometimes called Generation Y, the Echo Boomers, or the iGeneration or NetGeneration. As their dates make clear, this group is not part of the baby bust cohort. Importantly, they represent a small rise, or a "boomlet," in birthrates that "echoes" the baby boom, though on a much smaller scale than what happened in the late 1940s, 1950s, and early 1960s. They also represent the first generation born "digital," that is, into a society wired into the Internet.

Generations X and Y will, of course, be the ones that have to "mop up" after the flood of boomers moves up their career ladders toward retirement and beyond. The tremendous size of the baby boom cohort is likely to affect the career and job mobility of the busters, as well as labor costs, which will affect taxes, health care, and other benefit costs. These older workers will be more costly for their employers as their salaries and benefits increase in line with their seniority.

The actual impact that the baby boom generation will have into the first half of the twenty-first century remains to be seen. What is clear, however, is that it carries the future of the nation's economic and social security along with it. Because of its sheer size, the cohort will continue to influence public policy through its political participation at the ballot box and as national, regional, and local leaders and officials. Most important for the years to come, however, is that boomers will force the issues of generational equity and the allocation of scarce resources across all cohorts to the center of political and ethical debate.

Source: Adapted from L. Bouvier and C. De Vita, "The Baby Boom: Entering Midlife," *Population Bulletin 46* (3), 1991; and K. Kinsella and V. A. Velkoff, *An Aging World: 2001*, U.S. Census Bureau, Series P95/01-1, Washington, DC, U.S. Government Printing Office, 2001.

TABLE 3.A Countries Where Population is Expected to Decline, 2007 and 2050

Country or area*	Population (thousands)		Population Change		Percent 65 years or older		Change in proportion 65 years or older (percent)
	2000	2050	(thousands)	(percent)	2000	2050	
Austria	8,211	7,094	−1,117	−14	15	30	106
Belarus	10,236	8,330	−1,907	−19	14	25	86
Belgium	10,161	8,918	−1,243	−12	17	28	65
Bosnia and Herzegovina	3,972	3,767	−205	−5	10	27	171
Bulgaria	8,225	5,673	−2,552	−31	16	30	88
China, Hong Kong SAR[a]	6,927	6,664	−263	−4	11	33	217
Croatia	4,473	3,673	−800	−18	15	26	77
Cuba	11,201	11,095	−105	−1	10	27	176
Czech Republic	10,244	7,829	−2,415	−24	14	33	144
Denmark	5,293	4,793	−500	−9	15	24	59
Estonia	1,396	927	−469	−34	14	29	107
Finland	5,176	4,898	−278	−5	15	26	72
Germany	82,220	73,303	−8,917	−11	16	28	73
Greece	10,645	8,233	−2,412	−23	18	34	92
Hungary	10,036	7,488	−2,548	−25	15	28	92
Italy	57,298	41,197	−16,101	−28	18	35	92
Japan	126,714	104,921	−21,793	−17	17	32	86
Latvia	2,357	1,628	−728	−31	14	27	86
Lithuania	3,670	2,967	−704	−19	13	27	102
Luxembourg	431	430	−1	0	14	27	84
Netherlands	15,786	14,156	−1,629	−10	14	28	104
Poland	38,765	36,256	−2,509	−6	12	26	118
Portugal	9,875	8,137	−1,738	−18	16	31	99
Romania	22,327	16,419	−5,908	−26	13	31	131
Russian Federation	146,934	121,256	−25,678	−17	13	25	100
Slovakia	5,387	4,836	−551	−10	11	27	139
Slovenia	1,986	1,487	−499	−25	14	32	131
Spain	39,630	30,226	−9,404	−24	17	37	117
Sweden	8,910	8,661	−249	−3	17	27	53
Switzerland	7,386	6,745	−641	−9	15	30	104
Ukraine	50,456	39,302	−11,154	−22	14	27	91
United Kingdom	58,830	56,667	−2,163	−4	16	25	56
Yugoslavia	10,640	10,548	−92	−1	13	23	73

[a]As of July 1, 1997 Hong Kong became a Special Administrative Region (SAR) or China.

*Countries or areas with 150,000 persons or more in 1995.

Source: U.N. Population Division, *World Population Prospects: The 1998 Revision.*

POPULATION DYNAMICS AND PROCESSES

In order to arrive at a different understanding of population growth and change, experts look at two significant factors: fertility and mortality. Birth and death rates, as they are also known, are important indicators of a region's level of development and its place within the world economy.

Birth, or Fertility, Rates

The **crude birthrate (CBR)** is the ratio of the number of live births in a single year for every thousand people in the population. The crude birthrate is indeed crude, because it measures the birthrate in terms of the total population and not with respect to a particular age-specific group or cohort. For example, as of 2005, the CBR of the entire U.S. population was 14.0, while the birthrate for Asian American women aged 20 to 24 was 72.0 and for Hispanic women in the same age category 184.6. Clearly, differences exist when we look at specific groups and especially at age and sex cohorts at their reproductive peak.

Although the level of economic development is a very important factor shaping the CBR, other, often equally important, influences also affect CBR. In particular, it may be heavily affected by the demographic structure of the population, as graphically suggested by age-sex pyramids. In addition, an area's CBR is influenced by women's educational achievement, religion, social customs, and diet and health, as well as by politics and civil unrest. Most demographers also believe that the availability of birth-control methods is critically important to a country's or region's birthrate (**Figure 3.12**). A world map of the CBR (**Figure 3.13**) shows high levels of fertility in most of the periphery of the world economy and low levels of fertility in the core. The highest birthrates occur in Africa, the poorest region in the world.

The crude birthrate is only one indicator of fertility and in fact is somewhat limited in its usefulness, telling very little about the potential for future fertility levels. Two other indicators formulated by population experts—the total fertility rate and the doubling time—provide more insight into the potential of a population. The **total fertility rate (TFR)** is a measure of the average number of children a woman will have throughout the years that demographers have identified as her childbearing years, approximately ages 15 through 49 (**Table 3.3**). Whereas the CBR indicates the number of births in a given year, the TFR is a more predictive measure that attempts to portray what birthrates will be among a particular cohort of women over time. A population with a TFR of slightly higher than 2 has achieved replacement-level fertility. This means that birthrates and death rates are approximately balanced and there is stability in the population.

TABLE 3.3 Total Fertility Rates for Selected Countries, 2007

Country	Total Fertility Rate (TFR)
Afghanistan	6.8
Argentina	2.5
China	1.6
India	2.9
Angola	6.8
Mauritius	1.7
Spain	1.4
Russia	1.3
United States	2.1

Source: Population Reference Bureau, *World Population Data Sheet,* 2007.

Figure 3.12 Birth–control promotion in India Controlling fertility has been an objective of population policy at the international level for several decades. For many years, however, birth-control programs were rejected by peripheral countries because having children was regarded as an economically sound decision. Birth-control programs coupled with improved educational and economic opportunities for women have proved to be far more effective than birth-control policies alone. But in India, issues of ethnicity complicate things because one ethnic group is fearful that if it limits its births, it will soon be outnumbered by another ethnic group.

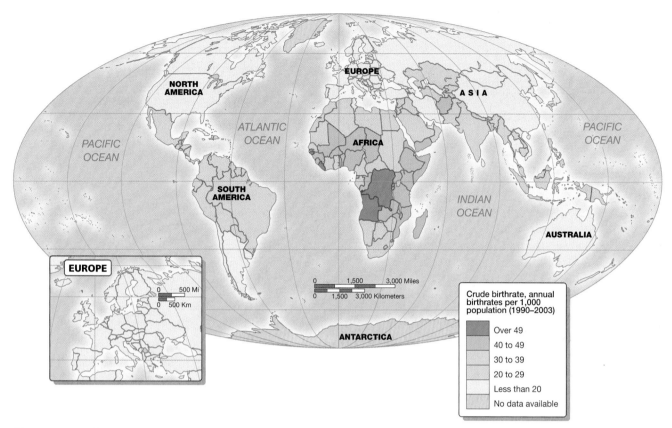

Figure 3.13 World crude birthrates, 2007 Crude birthrates and crude death rates are often indicators of the levels of economic development in individual countries. Compare, for example, Australia, a core country, which offers a stark contrast to statistics for Ethiopia, a very poor and underdeveloped peripheral country. (*Data source:* Population Reference Bureau, *World Population Data Sheet, 2007*.)

Closely related to the TFR is the doubling time of the population. The **doubling time**, as the name suggests, is a measure of how long it will take the population of an area to grow to twice its current size. For example, a country whose population increases at 1.8 percent per year will have doubled in about 40 years. In fact, world population is currently increasing at this rate. By contrast, a country whose population is increasing 3.18 percent annually will double in only 22 years—the doubling time for Kenya. Birthrates and the population dynamics we can project from them, however, tell us only part of the story of the potential of the population for growth. We must also know the death, or mortality, rates.

Death, or Mortality, Rates

Countering birthrates and shaping overall population numbers and composition is the **crude death rate (CDR)**, the ratio of the number of deaths in one year to every thousand people in the population. As with crude birthrates, crude death rates often roughly reflect levels of economic development—countries with low birthrates generally have low death rates (**Figure 3.14**).

Although often associated with economic development, CDR is also significantly influenced by other factors. A demographic structure with more men and elderly people, for example, usually means higher death rates.

Other important influences on mortality include health care availability, social class, occupation, and even place of residence. Poorer groups in the population have higher death rates than the middle class. In the United States, coal miners have higher death rates than schoolteachers. Also in the United States, urban areas often have higher death rates than rural areas. The difference between the CBR and CDR is the rate of **natural increase**—the surplus of births over deaths—or the rate of **natural decrease**—the deficit of births relative to deaths (**Figure 3.15**).

Death rates can be measured for both sex and age cohorts, and one of the most common measures is the **infant mortality rate**. This figure reflects the annual number of deaths of infants less than 1 year of age compared to the total number of live births for that same year. The figure is usually expressed as number of deaths during the first year of life per 1,000 live births.

The infant mortality rate has been used by researchers as an important indicator both of the adequacy of a country's health care system and of the general population's access to health care. Global patterns show that while infant mortality rates are high in the peripheral countries of Africa and Asia, they are low in the more developed countries of Europe and North America (**Figure 3.16**). Generally, the core's low rates reflect adequate maternal nutrition and the wider availability of health care resources and personnel.

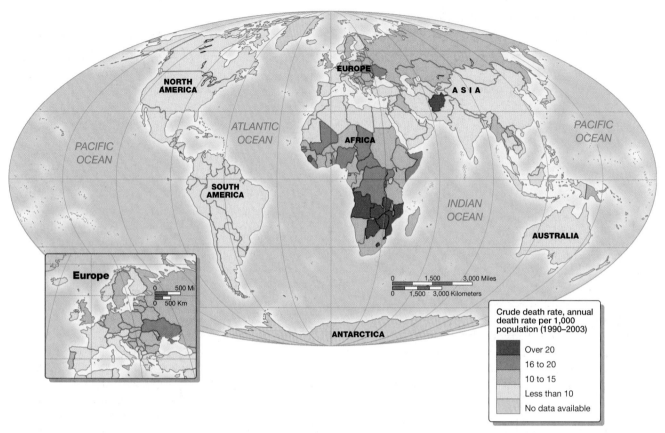

Figure 3.14   World crude death rates, 2007 The global pattern of crude death rates varies from crude birthrates. Most apparent is that the difference between highest and lowest crude death rates is relatively smaller than is the case for crude birthrates, reflecting the impact of factors related to the middle phases of the demographic transition. (*Data source:* Population Reference Bureau, *World Population Data Sheet, 2007.*)

When patterns are examined below the global scale, at the level of countries, regions, and cities, infant mortality rates are not uniform. In the United States, for example, African Americans, as well as other ethnic minorities in urban and rural areas, suffer infant mortality rates that are twice as high as the national average. In east central Europe, Bulgaria has a 9.7-per-thousand infant mortality rate, while the Czech Republic has a rate of 3.3 per thousand. In Israel, the infant mortality rate per thousand is 3.9 and in the neighboring Palestinian Territories it is 25. The point is that global patterns often mask regional, local, and even neighborhood variation in mortality rates for both infants and other population cohorts.

Related to infant mortality and the crude death rate is **life expectancy,** the average number of years an infant newborn can expect to live. Not surprisingly, life expectancy varies considerably from country to country, region to region, and even from place to place within cities and among different classes and racial and ethnic groups. In the United States, an infant born in 2005 can expect to live more than 78 years. If we begin to specify the characteristics of that infant by sex and race, however, variation emerges. An African American male born in 2005 has a life expectancy of 69.5 years, while an Anglo-American female born the same year can expect to live, on average, 81 years.

Another key factor influencing life expectancy is epidemics, which can quickly and radically alter population numbers and composition. In our times, epidemics can spread rapidly over great distances, largely because people and other disease carriers can now travel from one place to another very rapidly. Epidemics can have profound effects, from the international to the local level, and reflect the increasing interdependence of a shrinking globe. They may affect different population groups in different ways and, depending on the quantity and quality of health and nutritional care available, may have a greater or lesser impact on different localities.

One of the most widespread epidemics of modern times is HIV/AIDS (human immunodeficiency virus/ acquired immunodeficiency syndrome). The disease is now a serious problem in regions ranging from Southeast Asia to sub-Saharan Africa and also affects certain populations in many core countries of Europe as well as the United States. To date, the Islamic countries of the Middle East and North Africa, as well as South America, seem to be least affected, though in the last several years there has been a marked increase in the spread of HIV/AIDS in the Middle East and North Africa, especially in Djibouti, Somalia, and the Sudan. Eastern Europe—especially the Russian Federation and Ukraine—is experiencing the fastest-growing epidemic in the world, with the number of

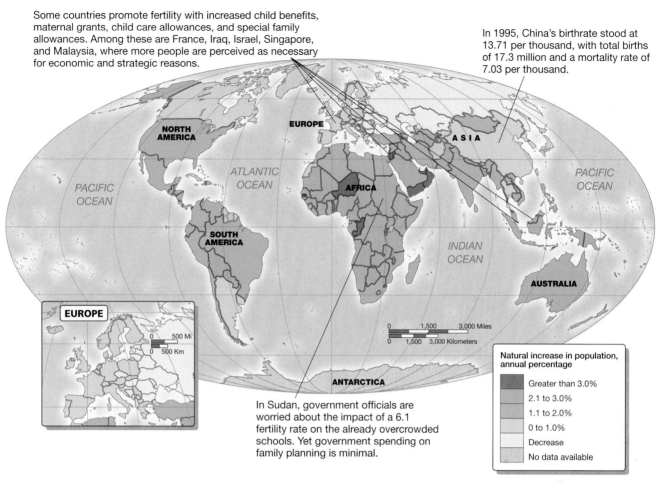

Some countries promote fertility with increased child benefits, maternal grants, child care allowances, and special family allowances. Among these are France, Iraq, Israel, Singapore, and Malaysia, where more people are perceived as necessary for economic and strategic reasons.

In 1995, China's birthrate stood at 13.71 per thousand, with total births of 17.3 million and a mortality rate of 7.03 per thousand.

In Sudan, government officials are worried about the impact of a 6.1 fertility rate on the already overcrowded schools. Yet government spending on family planning is minimal.

Natural increase in population, annual percentage

Greater than 3.0%
2.1 to 3.0%
1.1 to 2.0%
0 to 1.0%
Decrease
No data available

Figure 3.15 World rates of natural increase, 2007 As the map shows, rates of natural increase are highest in sub-Saharan Africa, the Near East, and parts of Asia, as well as parts of South and Central America. Europe, the United States, and Canada, as well as Australia and parts of central Asia and Russia, have slow to stable rates of natural increase. (*Data source:* Population Reference Bureau, *World Data Sheet, 2007.*)

new HIV infections rising steeply. In 2005, there were estimated to be approximately 1 million people living with HIV/AIDS in the Russian Federation; Ukraine possesses the highest rate of increase in newly reported cases. Transmission in this region appears to occur largely through high rates of intravenous drug use among young people and through risky practices among sex workers.

Medical geographers have made important contributions to the study of the diffusion of HIV/AIDS. **Medical geography** is a subarea of the discipline that specializes in understanding the spatial aspects of health and illness. This spatial perspective includes disease mapping as well as the distribution and diffusion of health and illness. The importance of medical geography lies in the fact that the landscape of disease that we are seeing today is changing: Diseases we thought were eradicated, such as tuberculosis, are coming back; diseases that were previously unknown are emerging, such as ebola; and diseases that have maintained geographic limitations, such as dengue hemorrhagic fever, are spreading.

The most common health issue addressed by medical geographers is HIV/AIDS. With respect to this disease, medical geographers ask: Where did AIDS originate? Why

did it originate there? How did it spread to surrounding areas? How is the HIV/AIDS crisis related to other things? How has the epidemiology of HIV/AIDS changed over time? In the United States, for example, HIV/AIDS first arose largely among male homosexuals and intravenous drug users who shared needles. Geographically, early concentrations of AIDS occurred in places with high concentrations of these two subpopulations, such as San Francisco, New York, Los Angeles, and southern Florida. Since first appearing, the disease has hierarchically diffused to other groups, often manifesting itself differently in each group, including hemophiliacs and other recipients of blood transfusions. It has had perhaps the most severe impact in inner-city areas but has cropped up in every region of the United States, increasingly appearing in the male and female heterosexual population. Especially alarming is the increasing incidence of HIV/AIDS among heterosexual teenagers and 20-year-olds in this country, a group largely untouched by HIV/AIDS previously.

The pattern of the disease is different, however, in central Africa, where it is overwhelmingly associated with heterosexual, nondrug users and affects both sexes equally. The geographical diffusion of HIV/AIDS in Africa

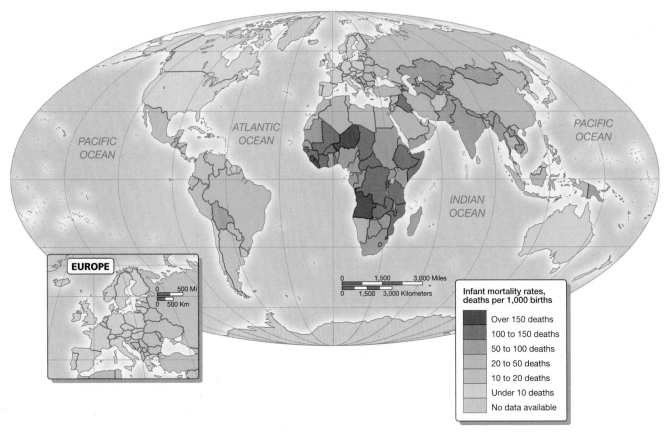

Figure 3.16 World infant mortality rates, 2007 The geography of poverty underlies the patterns shown in this map and allows us to analyze the linkages between population variables and social conditions. Infant mortality rates seem generally to parallel crude death rates, with sub-Saharan Africa generally reporting the highest rates. These rates reflect a number of factors, including inadequate or completely absent maternal health care as well as poor nutrition for infants. (*Data source:* Population Reference Bureau, *World Data Sheet, 2007.*)

has occurred along roads, rivers, and coastlines, all major transportation routes associated with regional marketing systems. The central African nations, including Congo (formerly Zaire), Zambia, Uganda, Rwanda, and the Central African Republic, have been especially hard-hit. For sub-Saharan Africa it is estimated that approximately

22.5 million people are infected with the virus (**Figure 3.17**). The impact is worst in urban areas, though no area has been immune to the disease's spread. High death rates in these countries—1.6 million in 2007—reflect in no small part the ravages of HIV/AIDS. That these countries continue to experience high growth rates, however, is a

Figure 3.17 Adults and children living with and dying from HIV/ AIDS, 2007 HIV/AIDS infections are concentrated in the periphery and semiperiphery, with most of the people with HIV/AIDS living in Asia and sub-Saharan Africa. Compare the number of adults and children living with HIV/AIDS in Africa with those in North America or Europe. Deaths from HIV/AIDS have also been highest in Africa. Campaigns in some African countries, especially in Zimbabwe and Uganda, have helped to stem the number of new cases and ultimately the number of deaths. (Adapted from UNAIDS, AIDS Epidemic Update 07, http://data.unaids. org/pub/EPISlides/2007/2007_epiupdate_en. pdf, p. 7, accessed May 31, 2008.)

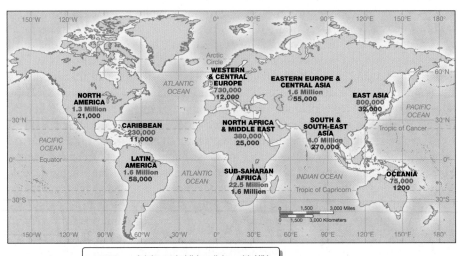

reflection of the huge proportion of the population that is in the peak years of biological reproduction.

Although no cure yet exists, some countries have successfully slowed the spread of HIV/AIDS. Finland, for example, has significantly inhibited the diffusion of HIV through an intensive public relations campaign and the provision of top-notch health services to all its citizens. Few of the sub-Saharan African countries, mired in poverty and with inadequate health care systems, political inefficiency, and often civil unrest, are able to approach these levels of activity, although it appears as though Uganda, which has extremely high rates of HIV/AIDS, may be experiencing a decline in new cases. In the summer of 2000, activists in Durban, South Africa, were effective in forcing several large pharmaceutical companies to slash the price of HIV/AIDS drugs for victims of the disease; by 2004, the number of Africans on treatment had doubled. The downside of this important success is that as the numbers in treatment expand over the next few years, the virus will develop resistance to the current drugs and the need for new drugs will grow. World Trade Organization intellectual property rules mean that countries like India that now produce cheap generic versions of the drugs will not be able to legally provide affordable generic versions of the new patented drugs and the high cost of treatment will resume.

Many Southeast Asian countries also face difficult economic constraints in attempting to slow the diffusion of the disease. In Cambodia and Thailand, however, prompt large-scale prevention programs and cheaper, government-subsidized drug treatments have begun to stem the spread of the epidemic. After peaking at around 140,000 cases in 1991, the number of new HIV cases dropped to an estimated 21,000 in 2003, due in large part to lower drug prices and very aggressive and more widespread antiretroviral treatment. However, as the United States pushes for a Free Trade Agreement with Thailand that will require stricter patent rules, the price of new and improved HIV/AIDS drugs could increase by as much as 90 percent.

Demographic Transition Theory

Many demographers believe that fertility and mortality rates are directly tied to the level of economic development of a country, region, or place. Pointing to the history of demographic change in core countries, they contend that many of the economic, political, social, and technological transformations associated with industrialization and urbanization lead to a demographic transition. A **demographic transition** is a model of population change in which high birth and death rates are replaced by low birth and death rates. Once a society moves from a preindustrial economic base to an industrial one, population growth slows. According to the demographic transition model, the slowing of population growth is attributable to improved economic production and higher standards of living brought about by better health care, education, and sanitation.

As **Figure 3.18** illustrates, the high birth and death rates of the preindustrial phase (Phase 1) are replaced by the low birth and death rates of the industrial phase (Phase 4) only after passing through the critical transitional phase of steady birthrates and falling death rates (Phase 2) and then more moderate rates (Phase 3) of natural increase (increase through birth, not migration) and growth. This transitional phase of rapid population growth is the direct result of early and steep declines in mortality at the same time that fertility remains at levels characteristic of a place that has not yet been industrialized.

Some demographers have observed that many peripheral and semiperipheral countries appear to be stalled in the transitional phase, or caught in a "demographic trap." **Figure 3.19** illustrates the disparity between birth and death rates for core and peripheral countries. Despite a sharp decline in mortality rates, most peripheral countries retain relatively high fertility rates.

The reason for this lag in declining fertility rates relative to mortality rates is that while societies have developed new and more effective methods for fighting infectious diseases, social attitudes about the desirability of large

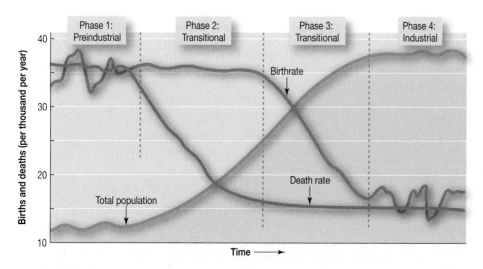

Figure 3.18 Demographic transition model The transition from a stable population based on high birth and death rates to one based on low birth and death rates progresses in clearly defined stages, as illustrated by this graph. With basic information about a country's birth and death rates and total population, it is possible to identify that country's position within the demographic transition process. Population experts disagree about the usefulness of the model, however. Many insist it is applicable only to the demographic history of core countries.

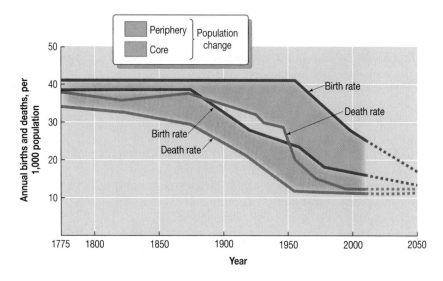

Figure 3.19 World trends in birth and death rates, 1775–2050 The graph is an illustration of the impact of affluence on reproductive choices. It also portrays an optimistic view of the future, where for peripheral regions, birthrates will continue to fall as we move closer to the midpoint of the next century. For the core, the projection is that birth and death rates will stay low. (Adapted from T. Allen and A. Thomas, *Poverty and Development in the 1990s.* London: Oxford University Press, 1992.)

families are only recently changing. Historical trends in birth and death rates and natural increase are shown in **Table 3.4** for Scotland and England. Over a roughly 50-year period (1870–1920), both places were able to reduce their rates of natural increase by nearly one-third. In the fourth phase of postindustrial development (not shown in the table), birth and death rates both have stabilized at a low level, which means that population growth rates are very slow and birthrates are more likely to oscillate than death rates. Whereas England and Scotland, early industrializers, passed through the demographic transition over a period of about 150 years beginning in the mid-nineteenth century, most peripheral and semiperipheral countries have yet to complete the transition.

Although the demographic transition model is based on actual birth and death statistics, many population geographers and other population experts question its generalizability to all places and all times. In fact, although the model adequately describes the history of population change in the core countries, it appears less useful for explaining the demographic history of countries and regions in the periphery. Its significance for understanding population geography has therefore been contested. Among other

criticisms is that industrialization—which, according to the theory, is central to moving from Phase 2 to Phases 3 and 4—is seldom domestically generated in the peripheral countries. Instead, foreign investment seems to drive peripheral industrialization. As a result, the features of demographic change, such as higher living standards, witnessed in the core countries, where industrialization was largely a result of internal capital investment, have not been as dramatic in many peripheral countries. Other critics of the demographic transition model point to several factors undermining a demographic transition fueled by economic growth: the shortages of skilled laborers, the absence of advanced educational opportunities for all members of the population (especially women), and limits on technological advances. In other words, while a demographic transition may be a characteristic experience of the core regions of the globe, its applicability to the periphery needs to be premised on a different set of factors.

Whether or not the demographic transition model, derived from core countries' experiences, is directly relevant to the periphery, it is becoming increasingly clear that some sort of demographic transformation is occurring globally, resulting in a decrease in the TFR from Europe and North

TABLE 3.4 Birth and Death Rates for England and Scotland, 1870–1920
The CBRs and CDRs for England and Scotland for the period between 1870 and 1920 illustrate countries moving between Phase 2 and Phase 3 of the demographic transition, in which death rates are lower than birthrates. England and Scotland are clear examples of the way in which the demographic transition has been theorized to operate in the core countries. During the 50-year period covered in the table, both countries were completing their transformation as key industrial regions.

	1870	1880	1890	1900	1910	1920
Crude Birthrate						
England	35.2	35.4	32.5	29.9	27.3	22.7
Scotland	35	34.8	32.3	30.2	28.4	24
Crude Death Rate						
England	22.5	21.4	19.2	18.2	15.4	14.6
Scotland	22.1	21.6	19.2	18.5	16.6	15.3

Source: M. Anderson and D. Morse, "High Fertility, High Emigration, and Low Nuptuality: Adjustment Processes in Scotland's Demographic Experience, 1861–1914, Part I *Population Studies, 47,* 1993, p.8.

America to East Asia and the Caribbean, with Africa and parts of the Middle East being exceptions.

POPULATION MOVEMENT AND MIGRATION

In addition to the population dynamics of death and re-production, the third critical influence on population is the movement of people from place to place. Individuals may make far-reaching, international or intraregional moves, or they may simply move from one part of a city to another. For the most part, mobility and migration reflect the interdependence of the world-system. For example, global shifts in industrial investment result in local adjustments to those shifts as populations move or remain in place in response to the creation or disappearance of employment opportunities.

Mobility and Migration

One way to describe such movement is by the term **mobility**, the ability to move from one place to another, either permanently or temporarily. Mobility may be used to describe a wide array of human movement, ranging from a journey to work (for example, a daily commute from suburb to city or suburb to suburb) to an ocean-spanning permanent move.

The second way to describe population movement is in terms of **migration**, a long-distance move to a new location. Migration involves a permanent or temporary change of residence from one neighborhood or settlement (administrative unit) to another. Moving from a particular location is defined as **emigration**, also known as out-migration. Moving to a particular location is defined as **immigration** or in-migration. Thus, a Russian who moves to Israel emigrates from Russia and immigrates to Israel. This type of move, from one country to another, is termed **international migration**. Moves may also occur within a particular country or region, in which case they are called **internal migration**. Both permanent and temporary changes of residence may occur for many reasons but most often involve a desire for economic betterment or an escape from adverse political conditions, such as war or oppression.

Governments are concerned about keeping track of migration numbers and rates, as well as the characteristics of the migrant populations, because these factors can have profound consequences for political, economic, and cultural conditions on national, regional, and local levels. For example, a peripheral country that has experienced substantial out-migration of highly trained professionals, such as Cuba, may find it difficult to provide needed services such as health care. Benefiting from lower labor costs are countries that have received large numbers of low-skilled in-migrants willing to work for extremely low wages, such as the United States, Germany, and France. These countries may also face considerable social stress in times of economic recession, when unemployed citizens begin to blame the immigrants for "stealing" their jobs or receiving welfare benefits.

Demographers have developed several calculations of migration rates. Calculation of the in-migration and out-migration rates provides the foundation for discovering gross and net migration rates for an area under study. **Gross migration** refers to the total number of migrants moving into and out of a place, region, or country. **Net migration** refers to the gain or loss in the total population of that area as a result of the migration. Migration is a particularly important concept because the total population of a country, region, or locality is dependent on migration activity as well as on birth and death rates.

Migration rates, however, provide only a small portion of the information needed to understand the dynamics of migration and its effects from the local to the national level. In general terms, migrants make their decisions to move based on push factors and pull factors. **Push factors** are events and conditions that impel an individual to move from a location. They include a wide variety of possible motives, from the idiosyncratic, such as an individual migrant's dissatisfaction with the amenities offered at home, to the dramatic, such as war, economic dislocation, or ecological deterioration. **Pull factors** are forces of attraction that influence migrants to move to a particular location. Again, factors drawing individual migrants to chosen destinations may range from the highly personal (such as a strong desire to live near the sea) to the very structural (such as strong economic growth and thus relatively lucrative job opportunities).

Usually the decision to migrate is a combination of both push and pull factors, and most migrations are voluntary. In **voluntary migration** an individual chooses to move. Where migration occurs against the individual's will, push factors can produce **forced migration**. Oftentimes, the decision to migrate is a mixed one reflecting both forced and voluntary factors. Forced migration (both internal and international) remains a critical problem in the contemporary world. The United Nations reported that a total of 32.9 million people were forced migrants or refugees at the end of 2006, nearly twice as many as were reported in 2004 . These migrants may be fleeing a region or country for many reasons, but some of the most common are war, famine (often war-induced), life-threatening environmental degradation or disaster, or governmental coercion or oppression. The terrorist attacks on the World Trade Center and the Pentagon on September 11 are believed to have caused large refugee movements not only in Central Asia, especially Afghanistan, but also globally. And while **refugees**—individuals who cross national boundaries to seek safety and asylum—are a significant global problem, **internally displaced persons (IDPs)**—the number of individuals who are uprooted within the boundaries of their own country because of conflict or human rights abuse—is also growing globally (see Box 3.3, "Internal Displacement").

3.3 GEOGRAPHY MATTERS

Internal Displacement

While the global humanitarian community has paid close attention to refugee populations for decades, the problem of internally displaced persons has only recently begun to draw international attention. Around the globe, in 2007 there were an estimated 26 million internally displaced persons (IDPs)—individuals who were uprooted within their own countries due to civil conflict or human rights violations, sometimes by their own governments (**Figure 3.D**). This number is actually twice as high as the global refugee population and it is frequently the case that the plight of IDPs is actually worse than that of the refugees. This is because their own governments are either unable or unwilling to provide them with the protection or assistance they have a right to expect. It is also the case that the international community is either unaware of them or has not secured the resources needed to help them (**Table 3.B**).

Two of the world's regions with the most difficult IDP challenges are Africa and Latin America. Africa possesses nearly 13 million IDPs due to rebel activities and intercommunal violence as well as some direct abuse by national governments. National security forces and government-backed militias have deliberately displaced large numbers of people in Zimbabwe, the Democratic Republic of Congo, and Côte d'Ivoire. In Liberia and Somalia, IDPs had virtually nowhere to go to escape attacks and find safe shelter; many were killed or died of hunger and disease. In Sudan, more than 2.5 million people have fled the western state of Darfur since 2003 as a response to attacks by government troops and raids by militias reportedly backed by the government (**Figure 3.E**). In 2008, there has still not been any kind of substantial intervention in the conflict besides humanitarian aid. This relative inaction is compounded by the UN's refusal to call the killing of an estimated 200,000–400,000 civilians by government, Janjaweed and rebel forces a "genocide" because, according to *The Report of the International Commission of Inquiry on Darfur to the United Nations Secretary-General*, ". . . the crucial element of genocidal intent appears to be missing." Some of the reasons for the inaction include a lack of cooperation from the

Figure 3.D   Internally displaced people worldwide, 2007 This map provides an indication of the geographical extent and numbers of people displaced from their livelihoods and homes but still living within their own countries, oftentimes in temporary camps provided by humanitarian organizations or the UN. As of December 2007, the Internal Displacement Monitoring Center estimated that 26 million people had been displaced within their own countries (*Source:* Adapted from IDMC, http://www.internal-displacement.org/8025708F004CE90B/(httpPages)/22FB1D4E2B196DAA802570BB005E787C?OpenDocument&count=1000, accessed May 31, 2008.)

TABLE 3.B Regional Distribution of IDPs, Refugees, Asylum Seekers, and Stateless Persons, 2006

Region (UN major area)	Refugees	Asylum-seekers	Returned refugees	IDPs Protected/assisted	Returned IDPs	Stateless persons	Others of concern	Total
Africa	2,607,600	244,100	312,200	5,373,000	1,043,900	100,100	71,700	9,752,600
Asia	4,537,800	90,100	408,900	3,879,100	811,600	5,026,900	156,500	14,910,900
Europe	1,612,400	244,000	12,500	542,200	8,700	679,000	331,700	3,430,500
Latin America and the Caribbean	40,600	16,200	100	3,000,000	–	–	485,600	3,542,500
Northern America	995,300	147,800	–	–	–	–	–	1,143,100
Oceania	84,000	1,700	–	–	–	–	–	85,700
Total	9,877,700	743,900	733,700	12,794,300	1,864,200	5,806,000	1,045,500	32,865,300

Source: UN High Commission for Refugees, *Statistical Yearbook*, 2006.

Sudanese government, fragmented rebel groups unable to cohere around peace plans, and Chinese intervention. (China has prevented Sudan from suffering from sanctions possibly because it buys about 60 percent of Sudan's oil and, in return, sells it weapons.)

In the South American country of Colombia, both guerrillas and paramilitaries continue to target civilian populations through arbitrary killings, looting, and destruction of property in order to depopulate rural areas for political and economic gains and to control or regain strategic territories, displacing some 2 million people as of early 2006. The war and the postconflict situation in Iraq has also had a dramatic impact on Iraqi citizens, displacing nearly 1.2 million people internally in addition to the approximately 2.2 million people who have left the country altogether. Other populations of concern to the UNHCR include Azerbaijan, Somalia,

Sri Lanka, Serbia and Montenegro, Liberia, Georgia, Boznia and Herzegovinia, the Russian Federation, and Afghanistan (**Figure 3.F**).

The U.S.-led "war on terror" has had a decidedly negative effect on the worldwide displacement crisis by encouraging governments to seek military solutions to conflicts. As a result, international humanitarian and human rights standards, including those relating to the protection of IDPs, have been undermined or ignored altogether in numerous places around the globe (**Table 3.C**). In fact, there is little doubt that the international antiterrorism campaign has enabled some nondemocratic governments to characterize armed opposition movements as "terrorists" and to present their counterinsurgency operations as part of the international "war on terror." This posture has earned these governments—many with a long history of instability,

Figure 3.E Darfur, Sudan, internally displaced persons Shown here are Sudanese women and children who have fled a village in the region of Darfur in order to escape the attacks of the Janjaweed tribal militia, sponsored by the Sudanese government, that have been occuring throughout the southern region of the country since early 2003. In mid-2005 there were nearly 2 million internally displaced Sudanese in camps like this one.

Figure 3.F Peace and return in the Balkans Pictured here is a Muslim man who has returned to Republika Srpska Bosnia with his family to rebuild their home and restore their lives. Since the 1990s wars have ended, about 1.8 million people have returned to the Balkans.

TABLE 3.C Countries Where the War on Terror Has Affected the Protection of IDPs

Afghanistan
Colombia
Indonesia
Iraq
Nepal
Israel/Palestinian Territories
Philippines
Russian Federation
Uganda

Sources: Global IDP Database, UNHCR, USCR, UNWRA.

military coups, and human rights violations—substantial military support, mainly from the United States. In Indonesia and the Philippines, for example, tens of thousands of people have been displaced because of counterinsurgency operations conducted under the banner of the "war on terror." Importantly, these military campaigns were ongoing before the "war on terror" was launched in 2001. And yet, by invoking the rhetoric of terror, they have, ironically, undermined the protection of civilians, causing them to flee their homes for safe havens elsewhere in the country.

Source: Adapted from *Internal Displacement: A Global Overview of Trends and Developments in 2006,* by the Global IDP Project of the Norwegian Refugee Council, Geneva, Switzerland, 2006.

International Voluntary Migration

Migration does not always involve force or even a permanent change of residence. Voluntary migration can occur for any number of reasons, such as high wage differentials between places, better experience and job opportunities elsewhere, family links abroad, or local underemployment or unemployment conditions (**Figure 3.20**). Temporary labor migration has long been an indispensable part of the world economic order and has at times been actively pursued by governments and companies alike. Individuals who migrate temporarily to take up jobs in other countries are known as **guest workers.** Sending workers abroad is an important economic strategy for many peripheral and semiperipheral countries; it lessens local unemployment and also enables workers to send substantial amounts of money to their families at home. This arrangement helps to supplement the workers' family income and supports the dominance of the core in global economic activities. It is not always an ideal situation, however, because economic downturns in the guest workers' host country may result in a large decrease in remittances received by the home country, thus further aggravating that country's economic situation.

It is also important to know the gender of temporary workers and the gender-based differences in the types of work performed. The Philippines, for example, has an Overseas Contract Worker (OCW) program that links foreign demand for workers to Philippine labor supply. The proportion of men and women OCWs is approximately equal, but the jobs they hold are gender-biased. Men receive most of the higher-level positions, while women are largely confined to the service and entertainment sectors. Some interesting geographical variations exist as well. Although constituting a small percentage of the total number of women working abroad, a large proportion of female OCWs who have professional positions (such as doctors and nurses) work in the United States and Canada. More typical of Filipina OCW experience, however, are the patterns in Hong Kong, Singapore, and Japan. In Hong Kong and Singapore, Filipinas are almost exclusively employed as domestic servants; in Japan, most work in the "entertainment industry," a term often synonymous with prostitution. These women in many ways have been transformed from individuals to commodities—many are even chosen by catalog. The OCW program has been criticized by feminist groups as well as

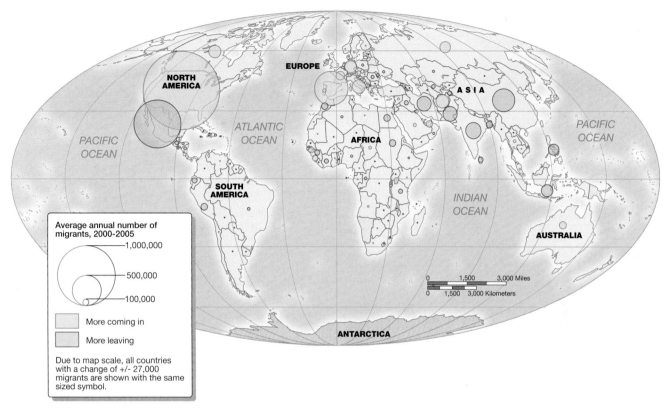

Figure 3.20 Global voluntary migration, 2005 This *New York Times* snapshot map shows those countries that are largely receiving and those that are largely sending migrants. In 2005, 190 million people, nearly 3 percent of the world's population, lived outside their country of origin (*Source: New York Times,* http://www. nytimes.com/ref/world/20070622_CAPEVERDE_GRAPHIC.html, accessed May 31, 2008.)

human rights organizations. Many regard the OCW treatment of women as a contemporary form of slavery.

An increasingly important category of international migrants are **transnational migrants**, so called because they set up homes and/or work in more than one nation-state. Sometimes these migrants are low-paid workers, as in the case of many Mexican migrants to the United States who take up jobs in the unskilled sectors of manufacturing, agriculture, and the service economy. These migrants characteristically possess domiciles in Mexican cities, towns, or villages as well as in the United States, maintaining regular contact between the two places through visits and phone calls and sending substantial sums of money to support family members in Mexico to improve their residences there.

Other transnational migrants occupy the higher end of the socioeconomic spectrum, such as the Hong Kong Chinese, who have established residences in Vancouver, Canada. Geographer Katharyne Mitchell has written extensively about this group of transnational migrants, who moved to Canada in the 1980s and 1990s as a result of Hong Kong's pending transfer from British to Chinese (PRC) control in 1997. A business immigration program, intended to encourage Canadian economic development, enabled prospective migrants with a personal net worth of C$500,000 (US$503,117) and a promise to invest C$350,000 (US$352,140) in a Cana-

dian venture to jump to the head of a long immigration waiting list and receive a visa. The Hong Kong Chinese who have migrated to Vancouver maintain extensive and substantial property and business investments in both places and after three years are eligible to become Canadian citizens, while maintaining their citizenship status in Hong Kong as well. Transnational migrants seek destinations in other areas of the world than North America, with substantial migrant labor communities also occuring in France, Germany, Saudi Arabia, and South Africa, to name a few.

International Forced Migration

Forced migration is a worldwide phenomenon (**Figure 3.21**). A long-standing example of international forced migration is the Lebanese and Kurdish people in the Middle East. As **Figure 3.22** shows, these two groups have been scattered widely throughout the globe because of war and civil strife. Over the last century, and because of the tensions existing in this multiethnic, multireligious society, Lebanon has lost a great deal of its population to forced migration. The 1975–1990 civil war and more recent violence in Lebanon have resulted in a particularly large number of migrations. The failure of the Kurds to establish an autonomous state in the early twentieth century led to their being split among Iran, Iraq, and

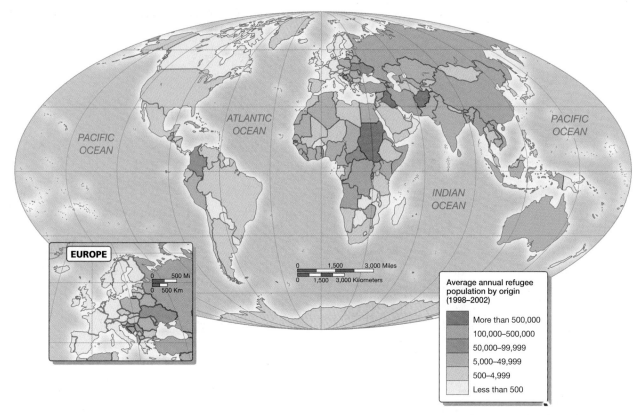

Figure 3.21 Refugee-sending countries, 2006 War is certainly the most compelling factor in forcing refugee migration. Shown are the sending countries, those whose internal situations propelled people to leave. What is perhaps most distressing about this graphic is that refugee populations have increased over the last decade, and almost exclusively so in the periphery. (*Data source:* UNHCR, http://www.unhcr.org/cgi-bin/texis/vtx/home/opendoc.pdf?id=478ce34a2&ttbl=STATISTICS, accessed May 31, 2008.)

Figure 3.22 Kurdish and Lebanese diaspora, 1990 The most significant diasporic populations of the Middle East and North Africa during modern times are the Palestinians, Lebanese, and Kurdish peoples. This map shows the scattering of the Lebanese and the Kurds (the dispersal of the Palestinians is shown in Figure 3.23). (Adapted from A. Segal, *An Atlas of International Migration.* London: Hans Zell, 1993, pp. 95 and 103.)

Turkey, with a small minority in Syria. Many Kurds have moved to different parts of the region or left it altogether because of military aggression, persecution, and the failure of repeated attempts to establish a Kurdish state.

Other prominent examples of international forced migration include the migration of Jews from Germany and Eastern Europe preceding the Second World War and the deportation of Armenians from the Ottoman Empire after the First World War. **Figure 3.23** shows the scattering of Palestinians from their homeland since the establishment of the state of Israel in 1948. The refugees have been fleeing ongoing violence as well as discrimination and land dispossession.

Recently—and especially since September 11— many European countries, faced with rising numbers of refugees seeking political asylum, have tightened their previously liberal asylum policies. Since September 11, the immigration issue has frequently been reconfigured as a security issue and national governments across Europe have voted to restrict the conditions under which asylum would be granted. This is despite the fact that many European countries are in need of immigrants to help counter labor shortages in skilled and unskilled jobs. In short, in the last several years national elections in which xenophobic politicians were either elected or garnered a sizable portion of the vote in normally liberal countries, such as the Netherlands, Switzerland and France, make it quite clear that Europeans are becoming increasingly resistant to absorbing refugee populations.

In other parts of the world, though, refugee populations are pouring across borders and creating very difficult conditions for national governments. Sudan, for example, is in the difficult position of having displaced several million of its own citizens as a result of civil war, while at the same time hosting over 750,000 refugees from neighboring countries. The situation provides a clear example of the openness of the borders that define many African states in contrast to the tight border security in most core countries. In some cases, refugee populations remain in a host country for long periods of time. In Jordan and Lebanon, for example, some of the Palestinian refugee camps are multigenerational, dating back to the 1948 Arab–Israeli war. High fertility rates in some of these refugee camps put considerable strain on the humanitarian aid resources of both international organizations and the host countries.

Internal Voluntary Migration

One way to understand the geographical patterns of migration is to think in terms of waves of migration. In the United States, for example, three important and overlapping waves of internal migration over the past two centuries altered the population geography of the country. As with a great deal of migration activity, these three major migrations were tied to broad-based political, economic, and social changes.

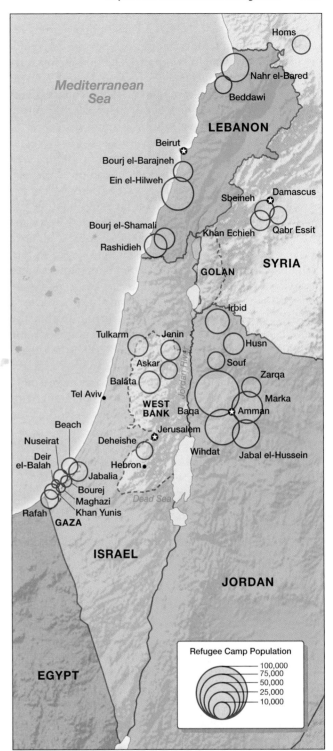

Figure 3.23 Palestinian refugees in the Middle East, 2008 This map shows the dispersion of Palestinian refugees—in camps and elsewhere—in the states around Israel. One of the biggest sticking points in negotiations between Israelis and Palestinians is the question of whether refugees will be allowed to return and, if so, where they will be allowed to settle, given that most of their land has been occupied by Israeli settlers. More than 200,000 Jews have settled on Palestinian land. The settlers include some of the most radical Jewish nationalists in the region, and although the most recent round of removal resulted in protests, the settlers were ultimately forced to move as sections of the West Bank were returned to Palestinian control. (Adapted from *The Guardian*, October 14, 2000, p. 5.)

The first internal migration wave began with colonization and increased steadily through the twentieth century. This wave had two parts. The first was characterized by the large movement of people from the settled eastern seaboard into the interior of the country. This westward expansion, which began spontaneously during the British colonial period—in blatant disregard of British restrictions on such expansion—became official settlement policy after the American Revolution. The federal government encouraged migration over the course of more than a century as part of the country's expansionist strategies. **Figure 3.24** illustrates how the demographic center of the United States moved westward in the years between 1790 and 2000 as more and more areas were settled, largely through transportation and communications innovations, capital investment, and large internal population movements. A second part consisted of massive rural-to-urban migration associated with industrialization, especially occurring during the mid-nineteenth to the early twentieth centuries. **Table 3.5** illustrates how, between 1860 and 2000, the United States was transformed from a rural to an urban society as industrialization created new jobs and as increasingly redundant numbers of agricultural workers (along with foreign immigrants) moved to urban areas to work in the manufacturing sector.

The second great migration wave, which began early in the 1940s and continued through the 1970s, was the massive and very rapid movement of mostly African Americans out of the rural South to cities in the South, North, and West. Although African Americans formed considerable populations in cities such as Chicago and New York before the onset of this wave, the mechanization of cotton picking pushed additional large numbers of these people out of the rural areas. Tenant cotton picking was a major source of livelihood for African Americans in the Deep South until mechanization reduced the number of jobs available. At the same time, pull factors attracted them to the large cities. In the early 1940s, large numbers of jobs in the defense-oriented manufacturing sector became available as other urban workers joined the war effort. This second wave of migration can be seen as part of a wider pattern of rural-to-urban migration among agricultural workers as industrialization spread globally.

After World War II a more important catalyst drove this migration: the increasing emphasis on high levels of mass consumption, which reoriented industry toward production of consumer goods and which in turn stimulated large increases in the demand for unskilled and semiskilled labor. The impact on the geography of racial distribution in the United States was profound.

The third internal migration wave began shortly after World War II and continues into the present. Following the end of the war and directly related to the impact of governmental defense policies and activities on the country's politics and economy, the region of the United States lying below the thirty-seventh parallel, also known as the Sunbelt, emerged. Between 1950 and 1990, this region, which includes 15 states and extends from North Carolina in the east to southern California in the west and Florida and Texas to the south, experienced a 97.9 percent increase in population. Beginning in the late 1970s and early 1980s, the West also began to grow dramatically. At the same time that the West and the South were booming, the Midwest and Northeast, known variously as the Snowbelt, Frostbelt, or Rustbelt, together grew by only

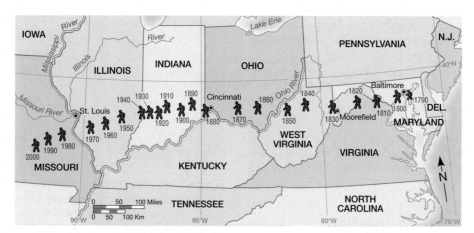

Figure 3.24 Changing demographic center of the United States, 1790–2000
Knowing where the demographic center of a country's population is located at given times allows us to track that population's movement, growth, distribution, and concentration. Here we can clearly perceive the expansion of the American frontier and see that although the East continued to be the most populated part of the country early on, the demographic center moved as more and more of the population dispersed to the West and the population of the West grew through natural increase. (Reprinted with permission of Prentice Hall, adapted from J. M. Rubenstein, *The Cultural Landscape: An Introduction to Human Geography*, 5th ed., © 1996, p. 122, and updated with data from the U.S. 2000 census.)

TABLE 3.5 United States Rural to Urban Population Change, 1860–2000

Year	1860	1870	1880	1890	1900	1910	1920	1930	1940	1950	1960	1970	1980	1990	2000
Total U.S. population (in millions)	31.4	38.5	50.1	62.9	76.2	92.2	106	123.2	132.1	151.3	179.3	203.2	226.5	248.7	281.4
Total urban population (in millions)	6.2	9.9	14.1	22.1	30.2	42	54.2	69.1	74.7	96.8	125.2	149.6	167	187	222.3
% Urban	19.8	25.7	28.2	35.1	39.6	45.6	52.2	58.1	58.5	64	69.9	73.6	73.7	75.2	78.9
% Increase in total population	35.6	22.6	30.2	25.5	21	21	15	16.2	7.3	14.5	18.5	13.4	11.4	9.8	13.1
% Increase in urban population	35.6	22.6	30.2	20.5	21	21	15	16.2	7.3	14.5	18.5	13.4	11.4	9.8	18.8

Source: United States Census Bureau.

33.3 percent. **Figure 3.25** is one illustration of the relative decline and disinvestment experienced in the Rustbelt during the 1970s and early 1980s. Since the late 1980s the Midwest and the Northeast have stabilized and even added population in most cases.

Preceding the decline of the Rustbelt and the rise of the Sunbelt has been the steady suburbanization of the U.S. population. **Suburbanization** is the growth of population along the fringes of large metropolitan areas.

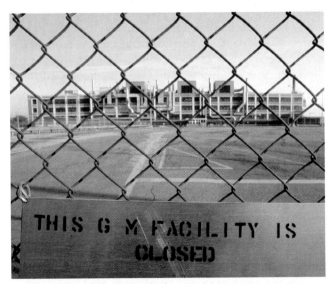

Figure 3.25   U.S. Rustbelt In the mid-twentieth century, the most important automobile manufacturers in the United States located their factories in the Midwest, where skilled labor and raw materials were readily available. By the late 1970s and 1980s, cities like Flint, Michigan, began losing their automobile employment base as companies, such as Ford, moved to suburban locations in the Midwest or out of the region either to the South or to foreign locations, such as Mexico, where labor is far cheaper and environmental laws less stringent.

The first evidence of U.S. suburbanization can be traced back to the late eighteenth century, when wealthy city dwellers began seeking more environmentally healthy and scenic residential locations. By the early twentieth century, residents fled to the suburbs to get away from the new immigrants and their increasing hold over urban political machines (**Figure 3.26**).

Between 1860 and 1920, the United States went from a primarily agricultural country to a primarily industrial one. In the United States, industrialization was consistently associated with urbanization: As the country industrialized, its population urbanized, with machines replacing human labor on the farm and former farmhands and others operating machines in factories.

The process really took on its own life, however, with the introduction of new transportation technologies—first with horse-drawn streetcars, then intraurban rail services, and finally automobiles. Each innovation in transportation enabled people to travel longer distances to and from work within the same or a shorter time period. They chose to move to the suburbs in massive numbers, not in the least because the suburbs were—and are—considered by many to be healthier and more desirable places to raise a family. It is important, however, to understand that deeper political and economic factors, beyond individual lifestyle choices, were also at work.

The most compelling explanation for the large-scale population shift characteristic of the third migration wave was the pull of economic opportunity. Rather than being invested in upgrading the aged and obsolescent urban industrial areas of the Northeast and Midwest, venture capital was invested in Sunbelt locations, where cheaper land, lower labor costs, and the absence of labor union organizations made the introduction of manufacturing and service-sector activity more profitable. By the 1990 census, there was a decrease in the amount of in-migration to the

Figure 3.26 Suburbanization, 2008 Suburban growth includes not only large tracts of residential development, as pictured here, but large retailers and strip malls, mostly oriented to automobile dependency. As energy costs continue to climb in the first decade of the twentieth century, some urbanists argue that the suburbs will experience problems of mobility and access.

Sunbelt, but the geography of the U.S. population at the end of the twentieth century was nevertheless almost diametrically opposed to that of 150 years ago. This new population distribution—largely suburban and more evenly spread across the country—illustrates the way in which political and economic transformations play an especially significant role in shaping individual choice and decision making.

In other parts of the world, rural-to-urban migration trends have changed population geographies in equally dramatic ways, although local manifestations have often been quite different from those experienced in the United States. Shantytowns ring all major cities of the periphery countries, posing almost insurmountable problems for city managers attempting to provide the most basic health, sanitation, educational, and occupational services. In some places, such as Brazil, an effort is ongoing to redirect these migrants to other locations. The Brazilian government constructed highways into the Amazon region, which has drawn many migrants into the frontier area. Nevertheless, pressures on major cities like Rio de Janeiro remain overwhelming.

Another approach aimed at redressing a severe population imbalance has been taken by Indonesia. In 2000, Indonesia's population surpassed 200 million people, more than 60 percent of whom live on the islands of Java and Bali (these two islands comprise just 7 percent of the nation's land base). Because of tremendous population growth, 2 million new jobseekers enter the market annually, causing the government to create programs aimed at inducing the people to move to less densely populated locations. These efforts date back to the turn of the nineteenth century, when the area was still under Dutch colonial rule. For many years they involved offering incentives to selected landless and jobless people to move to new agriculturally based settlements in the Outer Islands. Since 1984, however, facing a limited development budget, the government has turned to sponsoring investment in labor-intensive enterprises in established settlement areas. To date, over 1.5 million people have migrated, more than 60 percent of them to the large island of Sumatra.

Internal Forced Migration

One of the best-known forced migrations in the United States is the "Trail of Tears," a tragic episode in which nearly the entire Cherokee Nation was forced to leave their once treaty-protected Georgia homelands for Oklahoma. **Figure 3.27** illustrates this forced migration as well as the removal of four other eastern tribes to territory west of the Mississippi. Despite sustained legal resistance, approximately 16,000 Cherokees were forced to march across the continent in the early 1830s, suffering from drought, food scarcity, bitterly cold weather, and sickness along the way. By some estimates at least a quarter of the Cherokees died as a result of the removal.

Placed within the national and international context, the movement of Native American populations during the nineteenth century can be seen as a response to larger political and economic forces. European populations were participating in a massive migration to the United States, and the national economy was on the threshold of an urban-industrial revolution. The eastern Native American populations posed an obstacle to economic expansion, which itself was dependent upon geographic expansion. Growing Anglo-American prosperity, it was thought, had to be secured by taking Indian land.

Other recent examples of internal forced migration are provided by China and South Africa. In the late

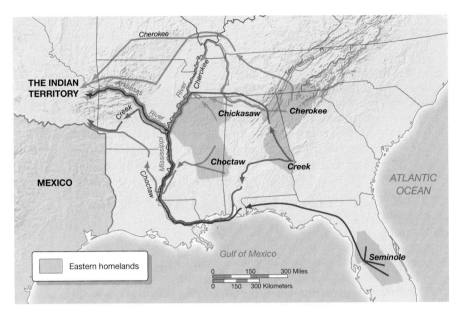

Figure 3.27   Trail of Tears, 1830s In the Indian Removal Act of 1830, lands held by eastern Native American tribes were exchanged for territory that the federal government had acquired west of the Mississippi, effectively eliminating Indian title to land previously protected by treaties. Without government protection of their rights to their land, members of the Choctaw, Chickasaw, Cherokee, Seminole, Creek, and other eastern tribes had to move west to the lands that had been acquired for them. The map shows the routes taken by these tribes as they vacated their ancestral lands for the federally designated Indian Territory in what is now Oklahoma. (Adapted from *Atlas of the North American Indian* by C. Waldman; maps and illustrations by M. Braun. Copyright © 1985 by C. Waldman. Reprinted by permission of Facts on File, Inc.)

1960s and 1970s, the government of China forcibly relocated 10 to 17 million of its citizens to rural communes in order to enforce Chinese Communist dogma and to ease pressures arising from high urban unemployment. The policy has since been disavowed, but the effects on an entire generation of young people were profound. Another example took place in South Africa between 1960 and 1980, when apartheid policies forced some 3.6 million blacks to relocate to government-created homelands, causing much suffering and dislocation. Indeed, civil war, ethnic conflict, famine, deteriorating economic conditions, and political repression have produced an extraordinary series of internal forced migrations in several sub-Saharan countries. These forced migrations, both internal and international in scope, become particularly significant in light of changing population. Forecasts predict that 80 percent of the world population increase in the next decade will take place within the poorest countries of the world. Many of these countries have some of the highest rates of forced migration. The combination bodes ill for improving these countries' prospects for economic and political improvement.

Today forced **eco-migration**—population movement caused by the degradation of land and essential natural resources—has created a new category of migrants and refugees. In Bangladesh, for example, floodplain settlement that began in the 1960s has led to severe losses of life and property and forced temporary relocation of huge numbers of people whenever severe flooding occurred. In Ethiopia the 1984–1985 famine was attributed not only to drought and pest infestation but also to governmental policies and actions that favored urban populations and rendered rural populations especially vulnerable to environmental stress. In the summer of 2002, 10 million Africans in Malawi, Zimbabwe, Lesotho, and Angola faced starvation, with hundreds of thousands having left their villages in hopes

of food and employment opportunities in urban areas in the region and beyond (**Figure 3.28**).

The tsunami that hit South Asia and parts of southern Africa on December 26, 2004 and killed anywhere from 228,000 to 310,000 people is an example of how natural disasters and tenuous environmental conditions have their greatest impact on poor people, in both the core and the periphery, whose only coping mechanism in the face of impending disaster or devastation is migration. A UN report indicates that as a result of the tsunami, 111,073 fishing vessels were destroyed or damaged; 36,235 engines were lost or damaged beyond repair; and 1.7 million units of fishing gear—such as nets, tackle, and similar equipment—were destroyed. Without the tools of their livelihood, the poor fishers along the southern coasts of

Figure 3.28   Acute malnutrition in Ethiopia, 2008 Another casuality of ecological catastrophe is Ethiopia where late rains, failure of crops and soaring food prices have led to severe food crisis and dislocation.

peripheral countries like India, Indonesia, Sri Lanka, the Maldives, Burma, and Thailand must find the resources to purchase or build new boats and fishing gear—which is unlikely, as most have no savings to fall back on—or move elsewhere in search of work.

POPULATION DEBATES AND POLICIES

One big question occupies the agenda of population experts studying world population trends today: How many people can Earth sustain without depleting or critically straining its resource base? The relationship between population and resources, which lies at the heart of this question, has been a point of debate among experts since the early nineteenth century.

Population and Resources

The debate about population and resources originated in the work of an English clergyman named Thomas Robert Malthus (1766–1834), whose theory of population relative to food supply established resources as the critical limiting condition upon population growth. Malthus's theory was published in 1798 in a famous pamphlet called *An Essay on the Principle of Population*. In this tract Malthus sets up two important postulates:

- Food is necessary to the existence of human beings.
- The passion between the sexes is necessary and constant.

It is important to put the work of Malthus into the historical context within which it was written. Revolutionary changes—prompted in large part by technological innovations—had occurred in English agriculture and industry and were eliminating traditional forms of employment faster than new ones could be created. This condition led to a fairly widespread belief among wealthy members of English society that a "surplus" of unnecessary workers existed in the population. The displaced agriculturists began to be a heavy burden on charity, and so-called poor laws were introduced to regulate begging and public behavior.

In his treatise, Malthus insisted that "the power of the population is indefinitely greater than the power of the earth to produce subsistence." He also believed that if one accepted this premise, a natural law would follow; that is, the population would inevitably exhaust food supplies. Malthus's response to this imbalance was to advocate for the creation of laws to limit human reproduction, especially among poor people.

Malthus was not without his critics, and influential thinkers such as William Godwin, Karl Marx, and Friedrich Engels disputed his premises and propositions. Godwin argued that "there is no evil under which the human species cannot labor, that man is not competent to cure." Marx and Engels were in general agreement that technological development and an equitable distribution of resources would solve what they saw as a fictitious imbalance between people and food.

The debate about the relationship between population and resources continues to this day, with provocative and compelling arguments for both positions being continually advanced. The geographer David Harvey, for example, has explored the population-resources issue in great detail. He has shown that by adopting Malthus's approach, only one outcome is possible—a doomsday conclusion about the limiting effect of resources on population growth. By following Marx's approach, however, quite different perspectives on, and solutions to, the population-resources issue can be generated. These solutions are based on human creativity and socially generated innovation, which allow people to overcome the limitations of their environment.

Neo-Malthusians—people today who share Malthus's perspective—predict a population doomsday. They believe that growing human populations the world over, with their potential to exhaust Earth's resources, pose the most dangerous threat to the environment. Although they acknowledge that the people of the core countries consume the vast majority of resources, they and others argue that only strict demographic controls everywhere will solve the problem, even if they require severely coercive tactics.

A more moderate approach argues that people's behaviors and governmental policies have a much greater impact on the condition of the environment and the state of natural resources than population size in and of itself. Proponents of this approach reject casting the population issue as a biological one in which an ever-growing population will inevitably create ecological catastrophe. They also reject framing it as an economic issue in which technological innovation and the sensitivities of the market will regulate population increases before a catastrophe can occur. Rather, they see the issue as a political one—one that governments have tended to avoid dealing with because they lack the will to redistribute wealth or the resources to reduce poverty, the latter condition being strongly correlated with high fertility.

The question of whether too many people exist for Earth to sustain bedeviled population policymakers and political leaders for most of the second half of the twentieth century. This concern led to the formation of international agencies that monitor and often attempt to influence population change. It also led to the organizing of a series of international conferences that attempted to establish globally applicable population policies. The underlying assumption of much of this policymaking, that has continued into the twenty-first century, is that countries and regions have a better chance of achieving improvement in their level of development if they can keep their population from outstripping the supply of resources and jobs.

Population Policies and Programs

Contemporary concerns about population—especially whether too many people exist for Earth to sustain—have led to the development of international and national policies and programs. A **population policy** is an official government strategy designed to affect any or all of several objectives, including the size, composition, and distribution of population. The implementation of a population policy takes the form of a population program. Whereas a policy identifies goals and objectives, a program is an instrument for meeting those goals and objectives.

Most of the international population policies of the last three decades have attempted to reduce the number of births worldwide. The main instruments developed to address rising fertility took the form of family-planning programs. The desire to limit fertility rates by the international population-planning community is a response to concerns about rapidly increasing global population—an increase that is more significant in the periphery and semiperiphery than in the core countries. Accompanying this situation of imbalanced population growth between the core and the periphery is gross social and economic inequality, as well as overall environmental degradation and destruction.

Figure 3.29 provides a picture of the recent history of world population growth by region and a reasonable projection of future growth. The difference between core and periphery is dramatically illustrated. In the summer of 2008 the world contained 6,677,563,921 billion people. This means that by the year 2050 the world is projected to contain nearly 9 billion people. In comparison, over the course of the entire nineteenth century, fewer than a billion people were added to the population. **Figure 3.30** shows a projection of world population in the year 2020.

The geography of projected population growth is noteworthy. Over the next century, population growth is predicted to occur almost exclusively in Africa, Asia, and Latin America, while Europe and North America will experience very low, and in some cases zero, population growth. The differing rates of natural increase listed in **Table 3.6** illustrate this point.

At the turn of the twentieth century, the core contained 32 countries with zero population growth. By contrast, the periphery contained 39 countries with rates of natural increase of 3.0 or more. A sustained rate of natural increase of 3.0 per year means a population will double in approximately 24 years. Such remarkable rates of population increase and their relationship to quality of life and environment have led to a concerted international response.

Since 1974, the United Nations has sponsored international conferences at 10-year intervals to develop population policy at the global level. Out of each conference emerged explicit population policies aimed at lowering fertility rates in the periphery and semiperiphery. The 1974 world conference was held in Bucharest, Romania, where two opposing approaches to lowering fertility were expressed. In fact, general disagreement emerged as to whether a population problem even existed. While the core countries were warning that a "population bomb" was ticking because the peripheral countries were producing too many babies, the peripheral countries accused the core of advocating genocide. They argued that it wasn't the periphery but the core that was degrading the environment with its industry and heavy consumption of natural resources, such as petroleum. The peripheral countries promoted the view that the most effective path to lowering fertility was through economic development, not the family-planning policies that the core was advocating. Ten years later, in Mexico City, the positions were reversed and peripheral countries were pressing for increased family-planning programs and the distribution of contraceptives, while core countries had adopted the perspective that "development is the best contraceptive."

The 1994 world population conference took place in Cairo, Egypt. Called the International Conference on Population and Development, the name itself reflects the changes in ideas and practices that 20 years of population

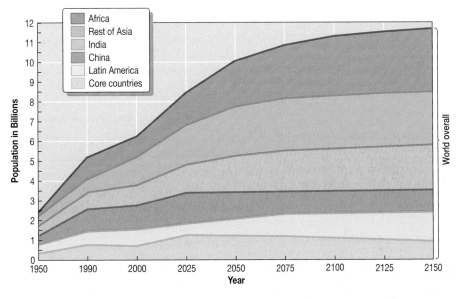

Figure 3.29 World population projection by region This graph represents a medium-variant projection, one that takes the middle position of three possible scenarios. In this projection, population continues to expand in the periphery, though in some regions more than others. Africa is projected to experience the greatest growth, followed by Asia (not including China), where growth is expected to level off by 2150. Less dramatic growth is expected to occur in Latin America, while in the core population, numbers remain constant or drop slightly. Though the total number of people in the world will be dramatically greater by 2150, the medium-variant forecast indicates a gradual leveling off of world population. (Adapted from I. Hauchler and P. Kennedy (eds.), *Global Trends: The World Almanac of Development and Peace*. New York: Continuum, 1994, p. 109.)

Figure 3.30 World population, 2020 This map provides a sense of how much the populations of various countries are expected to change by the year 2020. Although the populations of nearly all countries are expected to increase, it is clear that some populations will grow far more dramatically than others. Notice the substantial growth expected in Saudi Arabia and Afghanistan in contrast to the United States and Europe, where little, if any, population growth will occur. Italy, for example, is expected to lose population, while the Netherlands will grow by only 5 percent.

policy and programs had wrought. The Cairo conference, rather than focusing on the increase in global population growth rates, pointed to the fact that birthrates in almost every country on Earth are dropping—in many cases by significant amounts. Recognizing that a leveling off of the

population was possible in the foreseeable future, conference participants from both the core and the periphery agreed that efforts to bring down the growth rates must continue so that human populations will peak sooner rather than later. The policy that emerged from the Cairo

TABLE 3.6 Global Demographic Indicators, 2006 and 2007

This table illustrates the substantial population pressures emanating from the global periphery, which contained roughly four times the population of the core with a rate of natural increase also nearly four times as great. While Asian countries possess the largest proportion of the world's population, their rate of natural increase is considered moderate and their level of economic development—measured by their GNI PPP—higher than the average overall for the periphery. Most troubling to demographers, however, are the population dynamics of Africa, with a rate of natural increase of nearly 3 percent per year and a GNI PPP of only $2,550 per capita.

Region	Population Mid-2007 (millions)	Natural Increase (annual %)	Birthrate (per 1,000 pop.)	Death Rate (per 1,000 pop.)	Life Expectancy (years at birth)	GNI PPP Capita ($US 2006)
World	6625	1.2	22	9	68	9,940
More Developed	1221	0.1	11	10	77	29,680
Less Developed	5404	1.5	25	8	68	5,480
Africa	944	2.4	38	14	53	2,550
Asia	4010	1.2	19	7	68	6.630
Latin America and Caribbean	569	1.5	21	6	73	8,630
Europe	733	−0.1	10	11	75	22,690
North America	335	0.6	14	8	78	49,290

conference called for governments not simply to make family-planning programs available to all but also to take deliberate steps to reduce poverty and disease; improve educational opportunities, especially for girls and women; and work toward environmentally sustainable development. The Cairo meeting was the first of the three international population conferences in which nearly all the core and peripheral countries agreed on a plan for achieving population stabilization, a plan that would encourage freedom of choice in the matter of family size, population policies and programs, and development policies and practices. The goals established at the Cairo meeting have been embraced and extended in the most recent population conference held at the United Nations in 2000, the UN Millenium Summit. We discuss the summit in greater detail below.

All the world population conferences have recognized that the history, social and cultural practices, development level and goals, and political structures for countries and even regions within countries are highly variable and that one rigid and overarching policy to limit fertility will not work for all. Whereas some programs and approaches will be effective for some countries seeking to cut population numbers, they will be fruitless for others.

For instance, China's family-planning policy of one child per household appears to be effective in driving down the birthrate and damping the country's overall population growth. Because so little data are available on China, however, it is not clear whether the policy is operating in the same fashion throughout the country. For instance, some population experts believe an urban bias exists. That is, while the policy is adhered to in the cities, it is disregarded in the countryside. Interestingly, in response to the May 2008 Sichuan earthquake, with a projected death toll of 80,000—many of them children caught in collapsed school buildings—the Chinese government is suspending the policy and allowing couples whose children have died or are severely injured or ill, to have another child.

Family-planning regulations include authorizing the age of marriage, offering incentives to couples to have only one child, and mandating increasingly severe disincentives to couples who have larger families. In India, family-planning policies offering free contraceptives and family-planning counseling have resulted in lowering the birthrate. Other countries, such as Sri Lanka, Thailand, and Cuba, as well as the Indian state of Kerala, have lowered their birthrates not through regulation of family size, but by increasing access to social resources such as health care and education, particularly for women.

Indeed, it is now a widely accepted belief among demographers and policymakers that a close relationship exists between women's status and fertility. Women who have access to education and employment tend to have fewer children because they have less of a need for the economic security and social recognition that children are thought to provide (**Figure 3.31**). In Botswana, for instance, women with no formal education have, on average, 5.9 children, while those with four to six years of schooling have just 3.1 children. In Senegal, women with no education give birth to an average of 7 children. In contrast, the average number of children born to a woman with 10 years of education drops to 3.6. The numbers are comparable for Asia and South America.

More equality between men and women inside and outside the household is also believed to have a significant impact on reducing fertility. Giving both men and women choices about birth control, and educating them about the implications of such choices, appears to be especially successful in small island populations—such as those of Bali, Barbados, and Mauritius—with historically high population growth. In Mauritius in just 24 years (between 1962 and 1986) the introduction of voluntary constraints lowered the total fertility rate from 5.8 to 1.9. It is hardly any wonder, then, that the 1994 population conference placed such a clear and well-received emphasis on (1) rejecting coercive measures, including government sterilization quotas, that force people to violate their personal moral

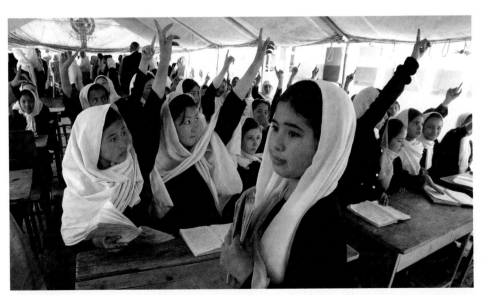

Figure 3.31 Educating girls in Afghanistan Improving the economic status of women is central to the success of controlling population growth. Access to education and employment security are seen as critical factors shaping a woman's decisions about how many children to have and when to have them.

codes and (2) improving the rights, opportunities, and economic status of girls and women as the most effective way of easing global population growth.

Ultimately, successful family-planning programs rely on the widespread availability of contraceptive commodities and practices. And while 40 years ago, contraception was socially and culturally unacceptable for many women in both core and peripheral countries, today the challenge is a different one: There is far more demand for it than there is availability. Ironically, family-planning and reproductive health programs are a victim of their own success. Having effectively increased knowledge about contraceptives over the past four decades, the demand for access to them has increased by over 50 percent.

By 2001, 92 percent of all countries supported family-planning programs and contraceptives, either directly through government facilities or indirectly through support of nongovernmental activities, such as nonprofit family-planning associations. Despite the pervasiveness of government support for contraceptive methods, however, some 200 million women currently do not have ready access to safe and effective modern means of contraception. The reasons why supply is not keeping up with demand are complex but they revolve around the increasing cost of contraception commodities themselves; the cost of providing clinicians, facilities, counselors, and education materials; the financial support of international donor organizations such as the U.S. Agency for International Development (USAID) (the single largest donor of contraceptives globally) and the UN Population Fund; and the U.S. policy of prohibiting USAID financial support for contraception to those countries where abortion is a legal means of terminating pregnancy.

Sustainable Development, Gender, and Population Issues

Globalization appears to lead to increasing levels of participation by women in the formal labor force, which is expected to lead to an increase in their economic and social value overall and a decline in birthrates globally. Large firms producing for export tend to employ women in assembly-line jobs because they can be hired for lower wages than men. And yet, while increasing labor-force participation has its benefits for women, it can also be based on implicit and explicit gender discrimination. Women constitute a large share of workers in informal subcontracting—often in the garment industry—at low wages and under poor conditions. Globalization is also associated with increasing levels of home work, telework and part-time work. This is a mixed blessing. Informal work arrangements can accommodate women's care obligations to the family, but such jobs are typically precarious and underpaid.

Recognizing the central importance of economic development to improving the lives of women, men, and children throughout the world, international agencies are increasingly turning to sustainable economic development as a way of limiting births and ensuring an improved quality of life through the reduction of poverty. A host of related international conferences have been held to address this theme, the most central one being the UN Millennium Summit. The goals, listed in **Table 3.7**, are aimed at reducing poverty by improving economic development through aid, trade, and debt relief and through the enhancement of democratic governance institutions and careful attention to the impact of development on the environment. The UN Development Program (UNDP) Millennium Development Goals are based on a partnership between core and peripheral countries. With respect to governance transformations, for instance, UNDP is working with an oil company and Amnesty International in Venezuela to provide the country's judges with a comprehensive understanding of human rights laws, regulations, and issues. With respect to the environment, UNDP is working with farmers in Ethiopia by supporting the planting and marketing of traditional crops and, in the process, strengthening the country's Biodiversity Research Institute and encouraging farmers to create biodiversity banks, while the crops make their incomes more secure.

The eight major MDGs reflect the neoliberal turn in international development. As discussed in Chapter 1, neoliberalism promotes a reduction in the role and budget of government, including reduced subsidies and the privatization of formerly publicly owned and operated concerns such as utilities. The goal of neoliberal development policies—such as the one being widely advanced by

---

TABLE 3.7 Millennium Development Goals (MDGs)

The goals and targets are based on the UN Millennium Declaration, and the UN General Assembly has approved them as part of the Secretary General's road map toward implementing the declaration. UNDP worked with other UN departments, funds, and programs, the World Bank, the International Monetary Fund, and the Organization for Economic Cooperation Development to identify over 40 quantifiable indicators to assess progress.

Goals and Targets

Goal 1: Eradicate extreme poverty and hunger

Goal 2: Achieve universal primary education

Goal 3: Promote gender equality and empower women

Goal 4: Reduce child mortality

Goal 5: Improve maternal health

Goal 6: Combat HIV/AIDS, malaria, and other diseases

Goal 7: Ensure environmental sustainability

Goal 8: Develop a Global Partnership for Development

Figure 3.32 UN World Summit, 2005 In September 2005, world leaders met to assess the level of progress made since the Millennium Summit. Development, human rights, and institutional reform were the focus of the meeting, which took place at the UN headquarters in New York City.

the UNDP—is to enable peripheral countries to achieve core economic standards of wealth and prosperity while recognizing that preexisting conditions will have to be taken into account to construct a place-specific development path. As the goals imply, enabling more sustainable economic development worldwide is seen as a way of also shaping population growth and the quality of life for population in the periphery. It is also a way of opening up new markets for core products and services and extending the capitalist world-system.

In 2005, world leaders met again as a follow-up summit meeting to the UN 2000 Millennium Summit. At the "World Summit," representatives (including many leaders) of the 191 (now 192) member states met in New York City for what the UN described as "a once-in-a-generation opportunity to take bold decisions in the areas of development, security, human rights and reform of the United Nations" (**Figure 3.32**). In addition to reporting on progress and discussing the challenges of meeting the MDGs, the World Summit leaders agreed to the formation of a UN Human Rights Council, which was launched in 2006, with the goal of strengthening the UN human rights machinery. A fact sheet derived from the summit can be found at http://www.un.org/summit2005/presskit/ fact_sheet.pdf [accessed September 3, 2008.]

CONCLUSION

The geography of population is directly connected to the complex forces that drive globalization. And since the fifteenth century, the distribution of the world's population has changed dramatically as the capitalist economy has expanded, bringing new and different peoples into contact with one another and setting into motion additional patterns of national and regional migrations.

When capitalism emerged in Europe in the fifteenth century, the world's population was experiencing high birthrates, high death rates, and relatively low levels of migration or mobility. Four hundred years later, birth, death, and migration rates vary—sometimes quite dramatically—from region to region, with core countries experiencing low death and birthrates and peripheral and semiperipheral countries generally experiencing high birthrates and fairly low death rates. Migration rates vary within and outside the core. These variations may be seen to reflect the level and intensity of political, economic, and cultural connectedness between core and periphery.

The example of formerly colonized peoples migrating to their ruling countries in search of work provides insights into the dynamic nature of the world economy. The same can be said of U.S. migrants who in the 1970s and 1980s steadily left their homes in the Northeast and Midwest to take advantage of the employment opportunities that were emerging in the Sunbelt: the Southeast, Southwest, and West. Both examples show the important role that people play in acting out the dynamics of geographic variety.

Death rates, birthrates, and migration rates are the central variables of population growth and change. These indicators tell us much about transforming regions and places as elements in a larger world-system. Globalization has created many new maps as it has unfolded; the changing geography of population is just one of them.

MAIN POINTS REVISITED

- Population geographers depend on a wide array of data sources to assess the geography of populations. Chief among these are censuses; other sources include vital records and public health statistics.

  Census collection methods vary from country to country and are often conducted in different years, making it difficult to compare different places.

- Population geographers are largely concerned with the same sorts of questions that other population experts address, but they also investigate "the why of where." *Why* do particular aspects of population growth and change (and problems) occur *where* they do, and what are the implications of these factors for the future of places?

  A geographic perspective is sensitive to the important influences of place and sees geographic factors as helping to explain population growth and change.

- Two of the most important factors that make up population dynamics are birth and death. These variables may be examined in simple or complex ways, but in either case the reasons for their variation are as important as the numbers themselves.

  Birth and death rates are fairly crude measures of population change, so population geographers also look at factors such as the particular experiences of certain age cohorts or

race cohorts and how those factors influence birth and death rates.

- A third crucial force in population change is the movement of populations. The forces that push populations from particular locations as well as those that pull them to move to new areas are key to understanding new settlement patterns. Population migration may not always be a matter of choice.

  Migration is one of the most important factors affecting the distribution of world population today. For some countries whose birthrates are especially low, migration is one way of reversing that trend.

- Perhaps the most pressing question facing scholars, policymakers, and other interested individuals was articulated at the UN Millennium Summit in 2000: How can the global economy provide the world's growing population with enough food and safe drinking water and a sustainable environment so that all the world's people have the basic necessities for enjoying happy, healthy, and satisfying lives?

  The most recent approach to limiting population growth is to look toward sustainable economic and environmentally sound development as a way of limiting births and ensuring an improved quality of life.

KEY TERMS

age-sex pyramid (p.90)
agricultural density (p. 88)
arithmetic density (p. 88)
baby boom (p. 88)
census (p. 84)
cohort (p. 90)
crude birthrate (CBR) (p. 100)
crude death rate (CDR) (p. 101)
crude density (p. 88)
demographic transition (p. 105)
demography (p. 84)

dependency ratio (p. 94)
doubling time (p. 101)
eco-migration (p. 117)
emigration (p. 107)
forced migration (p. 107)
geodemographic analysis (p. 90)
gross migration (p. 107)
guest workers (p. 110)
immigration (p. 107)
infant mortality rate (p. 101)
internally displaced persons (IDPs) (p. 107)

internal migration (p. 107)
international migration (p. 107)
life expectancy (p. 102)
medical geography (p. 103)
middle cohort (p. 94)
migration (p. 107)
mobility (p. 107)
natural decrease (p. 101)
natural increase (p. 101)
net migration (p. 107)
nutritional density (p. 88)
old-age cohort (p. 94)

population policy (p. 119)
pull factors (p. 107)
push factors (p. 107)
refugees (p. 107)
suburbanization (p. 115)
total fertility rate (TFR) (p. 100)
transnational migrant (p. 111)
undernutrition (p. 83)
vital records (p. 84)
voluntary migration (p. 107)
youth cohort (p. 94)

ADDITIONAL READING

Baldwin, D., *Chicago's New Negroes: Modernity, the Great Migration, & Black Urban Life.* Chapel Hill, NC: University of North Carolina Press, 2007.

Bernard, R. M., and B. R. Rice, *Sunbelt Cities: Politics and Growth Since World War II.* Austin, TX: University of Texas Press, 1983.

Birdsall, N., A. C. Kelley, and S. A. W. Sinding, *Population Matters: Demographic Change, Economic Growth and Poverty in the Developing World.* New York: Oxford University Press, 2001.

Bouvier, L., and C. DeVita, "The Baby Boom—Entering Midlife," *Population Bulletin* 46 (3), 1991, pp. 2–33.

Castles, S., and M. J. Miller, *The Age of Migration: International Population Movements in the Modern World,* 3rd ed. New York: Guilford Press, 2003.

Cernea, M. M., and C. McDowell, *Risks and Reconstruction: Experiences of Resettlers and Refugees.* Washington, DC: World Bank, 2000.

Citro, C. F., D. L. Cork, and J. L. Norwood (eds.), *The 2000 Census: Interim Assessment.* National Research Council, Committee on National Statistics. Washington, DC: National Academy Press, 2001.

Kinsella, K., and V. A. Velkoff, *An Aging World: 2001.* U.S. Census Bureau, Series P95/01-1. Washington, DC: U.S. Government Printing Office, 2001.

Klein, H. S., *A Population History of the United States.* Cambridge, UK: Cambridge University Press, 2004.

Kunstler, J. H., *The Long Emergency: Surviving the Converging Catastrophes of the Twenty-First Century.* New York: Atlantic Monthly Press, 2005.

Macunovich, D. J., *Birth Quake: The Baby Boom and Its Aftershocks.* Chicago: University of Chicago Press, 2002.

Mann, C. C., "How Many Is Too Many?" *Atlantic Monthly,* February 1993, pp. 47–67.

McFalls, J., Jr., "Population: A Lively Introduction," 5th ed., *Population Reference Bureau,* 62 (1), 2007, available at http://www.prb.org/pdf07/62.1LivelyIntroduction.pdf, accessed September 3, 2008.

McIntosh, A. C., and J. Finkle, "The Cairo Conference on Population and Development: A New Paradigm?" *Population and Development Review,* 21 (2), 1995, pp. 223–260.

Organization for Economic Cooperation and Development, *Policy Coherence for Development 2007: Migration and Developing Countries.* Paris: OECD, 2007.

Population Division of the Department of Economic and Social Affairs of the United Nations Secretariat, *World Population Prospects: The 2006 Revision, Highlights.* New York: United Nations, 2007.

Population Reference Bureau, "World Population Highlights," *Population Bulletin,* 62 (3), 2007.

Tichenor, D. J., *Dividing Lines: The Politics of Immigration Control in America.* Princeton, NJ: Princeton University Press, 2002.

UNAIDS, *2007 UNAIDS Annual Report: Knowing Your Epidemic.* Geneva: UNAIDS, 2008.

UNHCR, *Statistical Yearbook 2006: Trends in Displacement, Protection and Solutions.* Geneva: UNHCR, 2007.

UNHCR, *The State of the World's Refugees: Human Displacement in the New Millennium.* Oxford, England, and New York: Oxford University Press, 2006.

World Bank, *World Development Indicators 2008.* Washington, DC: World Bank, 2008.

World Summit Outcome: Resolution/Adopted by the General Assembly. New York: United Nations, 2005.

EXERCISES

On the Internet

The Companion Website at www.mygeoscienceplace.com contains a number of tools designed to reinforce the geographic concepts presented in this chapter. Structured within a four-step learning path, these online resources include: chapter Key Concepts; a fully-searchable myeBook version of the chapter, which allows you to add notes and highlights; RSS news feeds from various media agencies; links to additional web resources; Human Geography Videos; a digital Glossary; Concept Review questions; Thinking Spatially mapping exercises; and Critical Thinking essay questions.

Unplugged

1. The distribution of population is a result of many factors, such as employment opportunities, culture, water supply, climate, and other physical environment characteristics. Look at the distribution of population in your state or province. Is it evenly distributed, or are the majority of people found in only a few cities? What role do you think various factors have played in influencing where people live in your state or province? Can you think of other reasons for this distribution?

2. Immigration is an important factor contributing to the increase in the population of the United States. Chances are your great-grandparents, grandparents, parents, or even you immigrated to, or migrated within, your country of residence. Construct your family's immigration or migration history. What were some of the push and pull factors influencing your family's decision to immigrate to or migrate within your country?

3. During the 1990s, an interesting example of return migration occurred. Some areas of the United States that had previously lost large numbers of African Americans began to regain them. Compare data from the 1980 census with data from the 2000 census for the southeastern region of the United States. Which states experienced an increase of African American migrants? What are some of the characteristics of this migrant stream, and why might it have occurred? What impact might this have on electoral issues?

4. Every few years UNESCO publishes a data book on global refugee statistics indicating both sending and receiving countries, among other variables. Use your library's most recent data book and identify one country that has been a large sender of refugees and the country that has been the largest receiver of those refugees. The data book provides information not only on the numbers of refugees but on their age, gender, and other variables. Discuss some of the demographic implications for both countries if the refugee population was not to be allowed to return to the sending country.

Nature and Society

4

One of the most crucial elements in the relationship between people and their environment is water. We can't live without it and all of our economic practices depend on it too. For instance, it takes 150 gallons of water to produce one newspaper; 1,000 gallons for one quart of orange juice, 2,500 gallons for 1 pound of beef, and 40,000 gallons of water to produce 1 new car. The water embedded in the food or other products we consume is called **virtual water**. Without abundant water, we in the core could not live the high-quality lives we currently enjoy; without clean water, those in the periphery die. It's therefore important to question what our lives would be like if the price of water increased at the same rate as, for instance, gasoline prices did recently in the United States ($3.50 nationwide average, January 2008; $4.15 nationwide average, August 2008).

While this may seem a far-fetched question for those of us who enjoy very low water costs, in the periphery populations have been feeling the effects of escalating water prices for nearly a decade, and in some places far longer. The *2006 UN Human Development Report* states that 1.1 billion people in peripheral countries do not have adequate access to water, with 2.6 billion lacking basic sanitation. Almost two-thirds of the people lacking access to clean water survive on less than $2 dollars a day. Access to piped household water is unequal, with 85 percent of the wealthiest 20 percent of the population having adequate access, compared to a meager 20 percent of the poorest 20 percent of the population. Perhaps most disturbing is the statistic that 1.4 million children will die each year from lack of access to safe drinking water and sanitation.

It is becoming more and more clear that one of the most significant challenges of the twenty-first century will be how to ensure the wide accessibility of adequate supplies of quality drinking water for the world's people. While in the core most households take the availability of safe drinking water for granted, it is not the case in the periphery. In fact, experts around the world have begun to talk about a global "water crisis," which is occurring not because there is a shortage, but because water as a public good is rapidly turning into a privatized commodity. That is, the privatization of water (rather than the provision of water through publicly owned utilities)—touted as a way of bringing equitable access and efficiency to all water users—has instead led to increased prices and accessibility problems for poor and marginalized peoples not only in the global periphery but in peripheral areas of many core countries as well. Ten years ago, many governments—national as well as local—looked to the privatization of water as a way of unburdening themselves of a relatively expensive service in response to the public's demand for smaller government. Corporations—mostly

Main Points

- Nature and society constitute a complex relationship. It is our view that nature is both a physical realm and a social creation.

- Because in this text we regard nature as a social creation, it is important to understand the many views of nature in society today as well as the history of those views. The most prominent view of nature in Western culture is derived from the Judeo-Christian tradition, which is founded on a belief that humans should dominate nature.

- People's relationships with nature have developed over the course of human history, beginning with the early Stone Age. The early human history included people who revered nature, as well as those who abused it. Urbanization and industrialization have had extremely degrading impacts on the environment.

- The globalization of the political economy has meant that environmental problems are also global in scope. Deforestation, acid rain, and disappearing species affect us all. Many new ways of understanding nature have emerged in the last several decades in response to global environmental crises.

- Sustainability has recently become a predominant way of approaching global economic development and environmental change. In addition, new institutional frameworks, including conventions, protocols, and organizations, have emerged to promote global sustainability.

multinational ones—began buying up municipal water providers and offering water provision based on profit-and-loss considerations.

In response to the global water privatization movement, individuals as well as governments and nongovernmental organizations are increasingly arguing that access to safe and adequate supplies of water should be seen as a human right and that large corporations are not capable of guaranteeing that right. It is important to be aware that the conflict over who should provide safe water is occurring in cities and towns throughout the core as well as in the periphery. For instance, in the United States hundreds of cities and towns have been embroiled in battles with their local governments over the selling of the public utility to private water companies. Besides issues of access and public safety, many of those opposing water privatization argue that conflicts over water are really about fundamental questions of democracy. In particular, if water is no longer a public good, who will make the decisions that affect our future access to it, and who will be excluded? This question is an especially significant one as climate change is expected to have dramatic impacts on water quality and quantity throughout the world.

The relationship between people and the environment is perhaps the most central of all relationships within the discipline of geography. Indeed, geography consists of those who study natural systems, those who study human systems, and those who study the connections between them. In this chapter, we explore the ways that society has used technology to transform and adapt to nature, together with the impact of those technological adaptations on humans and the environment.

NATURE AS A CONCEPT

As discussed in Chapter 2, a simple model of the nature-society relation is that nature, through its awesome power and subtle expressions, limits or shapes society. This model is known as environmental determinism. A second model posits that society also shapes and controls nature, largely through technology and social institutions. This second model, explored in this chapter, emphasizes the complexity of nature-society interactions.

Interest in the relationship between nature and society has experienced a resurgence over the last two or three decades largely because the scope of environmental problems seems to have increased from those that are locally or regionally defined to those that have implications for the whole planet. The single most dramatic manifestation of this interest occurred in the summer of 1992, when more than 100 world leaders and 30,000 other participants attended the second Earth Summit in Rio de Janeiro (the first Earth Summit was held in Stockholm in 1972). The central focus of the agenda was to ensure a sustainable future for Earth by establishing treaties on global environmental issues, such as climate change and biodiversity. The signatories to the 1992 Earth Summit conventions created the Commission on Sustainable Development to monitor and report on implementation. A five-year review of Earth Summit progress was made in 1997 through a UN General Assembly meeting in special session. In July 2002, world leaders and other environmentally concerned individuals met for another Earth Summit, this time in Johannesburg, South Africa (**Figure 4.1**).

Some very important and significant changes have occurred since Rio. One change has been the emergence of international institutions to facilitate and monitor environmental improvements. Another has been real progress on the global phaseout of leaded gasoline. A third has been rising scientific and popular interest in global environmental issues. Perhaps most significant has been the proliferation of concern for sustainability, especially as it has moved away from a conference theme to become a set of practices being taken up by governments, corporations, and public and private organizations. Among these new practices and objectives are: environmental restoration,

Figure 4.1   2002 Earth Summit, Johannesburg, South Africa Global concern about environmental issues persists as the new century unfolds. Pictured here are attendees at the 2002 Earth Summit, with then UN Secretary General Kofi Annan at the podium.

green urban development, alternative energy technologies (such as wind, geothermal, and solar, as well as nuclear), sustainable tourism, "greening" the commodity supply chain, sustainable agriculture, biomass and biofuels promotion, green real estate development, and "greening" government practices, the emergence and growth of carbon markets, sustainable buildings, and "green marketing."

Renewed interest in the nature-society relationship is the result of the persistence, increasing number, and wider impact of environmental crises. This interest has led to attempts to rethink the relationship. For example, at the beginning of the twentieth century, Gifford Pinchot, as an influential citizen and later as director of the National Conservation Commission, advocated environmental conservation as the nation's forests and wild lands were increasingly given over to development. In 1962, Rachel L. Carson, in a groundbreaking book, *Silent Spring*, warned of the dangers of agricultural pesticides to ecosystems. Yet as the new century begins, the pesticide problem persists in both peripheral and core countries.

In the past, technology emerged as the apparent solution to most environmental problems, but today's technological progress seems often to aggravate rather than to solve such problems (**Figure 4.2**). As a result, researchers and activists have begun to ask different questions and abandon the assumption that technology is the *only*

Figure 4.3   Wangari Maathai In 2004, environmental activist, scientist, and assistant minister for environment and natural resources in Kenya became the first African woman to receive a Nobel Prize. Dr. Maathai received the prize for her work in the Green Belt Movement (GBM) in Kenya and the pan-GBM. The GBM is a grassroots organization involved in tree planting/biodiversity conservation, civic and environmental education, advocacy and networking, food security, and capacity building for women and girls. The aim of the organization is to address issues of sustainable development through environmental alternatives to technology.

Figure 4.2   Deformities in frogs A 2005 study by a Pennsylvania State University scientist published in the *Proceedings of the National Academy of Sciences* found that the combination of pesticide contamination and parasite infection has caused missing legs and extra legs in wood frogs in 43 states in the United States and 5 provinces in Canada. Both factors acting together are thought to be responsible for the mutations. It is believed that exposure to agricultural chemicals may weaken amphibian immune systems, making the frogs more vulnerable to parasitic infection leading to limb deformities. Worldwide, populations of amphibians have been decreasing at an alarming rate and scientists like this one believe that global climate change and other human-induced factors are responsible for this decline.

solution. For instance, 2004 Nobel Laureate and scientist Wangari Maathai has led a movement advocating tree planting as a way of addressing the linkage between environmental conservation and economic development (**Figure 4.3**). The Green Belt Movement, as it is known, works to conserve local biodiversity, prevent soil erosion, and increase the forest cover of Kenya. At present, Kenya's forest cover is less than 2 percent. The GBM is therefore fully engaged in a reforestation campaign aimed at preserving and improving local biological diversity. The purchase of seedlings for the campaign helps to generate income for local groups at the same time that reforestation is advanced.

The GBM and others like it have influenced environmental experts, including a number of geographers, to conceptualize nature not as something apart from humans but as inseparable from us. These experts believe that nature and questions about the environment need to be considered in conjunction with society, which shapes our attitudes toward nature and how we identify sources of and solutions to environmental problems. Such an approach—uniting nature and society as interactive components of a complex system—enables us to ask new questions and consider new alternatives to current practices with respect to nature.

But before asking new questions about nature, we examine the nature-society relationship by looking first at different approaches to it. We then examine how changing conceptions of nature have translated into very different uses of and adaptations to it. We conclude with an examination of sustainable development as a way of addressing global environmental problems and the new institutional frameworks and activist organizations that are emerging to promote sustainability.

Nature and Society Defined

The central concepts of this chapter—nature and society—have very specific meanings. Although we discuss the changing conceptions and understandings of nature in some detail, we hold to one basic conception here, that **nature** is a social creation as much as it is the physical universe that includes human beings. Therefore, understandings of nature are the product of different times and different needs. Nature is not only an object, it is a reflection of society in that philosophies, belief systems, and ideologies shape the way people think about and use nature. The relationship between nature and society is two-way: Society shapes people's understandings and uses of nature at the same time that nature shapes society. The amount of shaping by society is dependent to a large extent on the state of technology and the constraints on its use at any given time.

Society is the sum of the inventions, institutions, and relationships created and reproduced by human beings across particular places and times. Society's relationship with nature is just one of its relationships, and the social relationship to nature varies from place to place among different social groups.

The relationship between society and nature is usually mediated through technology. Knowledge, implements, arts, skills, and the sociocultural context all are components of technology. If we accept that all of these components are relevant to technology, then we can provide a definition that has three distinguishable, though equally important, aspects. **Technology** is defined as

- physical objects or artifacts (for example, the plow);
- activities or processes (for example, steelmaking);
- knowledge or know-how (for example, biological engineering).

This definition recognizes tools, applications, and understandings as critical components of the human production of technology. The manifestations and impacts of technology can be measured in terms of concepts, such as level of industrialization and per capita energy consumption.

The definitions provided in this section reflect current thinking on the relationship between society and nature. For centuries, humankind, in response to the constraints of the physical environment, has been as much influenced by prevailing ideas about nature as by its realities. In fact, prevailing ideas about nature have changed over time, as evidence from literature, art, religion, legal systems, and technological innovations makes abundantly clear.

A recent attempt to conceptualize the relationship between social and environmental changes has emerged from concern about global environmental changes. Based on the premise that individual societal changes can be both subtly and dramatically related to environmental changes, a formula for distinguishing the sources of social impacts on the environment has been advanced and is now widely used. The formula, known as $I = PAT$, relates human population pressures on environmental resources to the level of affluence and access to technology in a society. More specifically, the formula states that $I = PAT$, where I (impact on Earth's resources) is equal to P (population) times A (affluence, as measured by per capita income) times T (a technology factor). For example, the differential impact on the environment of two households' food consumption in two different regions in China would equal the number of people per household times the per capita income of the household times the type of technology and energy used in producing foodstuffs for that household (**Figure 4.4**).

Each of the variables in the formula—population, affluence, and technology—is complex. For example, with regard to population numbers, it is generally believed that fewer people on the planet will result in fewer direct pressures on resources. Some argue, however, that increased world population is quite desirable, since more people means more labor coupled with more potential for the emergence of innovation to solve present and future resource problems. Clearly, there is no simple answer to the question of how many people are too many.

Affluence also cannot simply be assessed in terms of "less is better." Certainly, increasing affluence—a measure of per capita consumption multiplied by the number of consumers and the environmental impacts of their technologies—is a drain on Earth's resources and a burden on Earth's ability to absorb waste. Yet how much affluence is too much is difficult to determine. Furthermore, evidence shows that the core countries, with high levels of affluence, are more effective than the poor countries of the periphery at protecting their environments. Unfortunately, core countries often do so by exporting their noxious industrial processes and waste products to peripheral countries. By exporting polluting industries and the jobs that go with them, however, core countries may also be contributing to increased affluence in the receiving countries. Given what we know about core countries, such a rise fosters a set of social values that ultimately leads to better protection of the environment in a new place. It is difficult to identify just when environmental consciousness goes from being a luxury to a necessity. The role of affluence in terms of environmental impacts is, in short, like population, difficult to assess.

Not surprisingly, the technology variable is no less complicated. Technologies affect the environment in three ways, through

(a)

(b)

Figure 4.4   Different levels of affluence in China As the $I = PAT$ formula suggests, the level of affluence of households plays an important role in their impact on the global environment. Pictured here are two families, one from the capitol city Beijing and the other from rural China. Both families are pictured in their dining rooms with a week's worth of food arrayed around them. Look closely for the differences in the arrays. The urban family, the Dongs (a), are upwardly mobile and aspiring to be middle class, and their weekly groceries reflect their higher level of affluence (far more processed food and even a couple of French baguettes). The Cuis (b) of Weitai Village, 60 miles east of Beijing, consume much larger amounts of local produce and freshly butchered meats than their urban counterparts. While both families possess refrigerator-freezers, the technology and energy required to produce, distribute, and market the processed foods the Dongs consume in a week indicate a greater impact on the environment than the Cuis mostly locally sourced and procured foods.

■ the harvesting of resources;

■ the emission of wastes in the manufacture of goods and services;

■ the emission of waste in the consumption of goods and services.

A technological innovation can shift demand from an existing resource to a newly discovered, more plentiful one. In addition, technology can sometimes be a solution and sometimes a problem. Both principles can be seen in the case of nuclear energy, widely regarded as cleaner and more efficient than coal or oil as energy sources. Producing this energy creates hazards, however, which scientists are still unable to prevent.

It is therefore clear that increases in human numbers, in levels of wealth, and in technological capacity are key components of social and economic progress whose impact on the environment has been extremely complex. In the last 100 years this complexity has increasingly come to be seen as a triple-barreled threat to the quality of the natural world and the availability and quality of environmental resources. Before we look more carefully at the specific impacts of populations, affluence, and technology on nature, we need to look first at how differing social attitudes toward nature shape the human behaviors that are a basis for $I = PAT$.

Nature–Society Interactions

The concept of *adaptation* to the natural environment is part of the geographical subfield of cultural ecology most closely associated with the work of Carl Sauer and his students. **Cultural ecology** is the study of how human society has adapted to environmental challenges such as aridity and steep slopes through technologies such as irrigation and terracing and the organization of people to construct and maintain these systems. These adaptations can be seen clearly in the landscape, such as the rice terraces of Southeast Asia or the canals and reservoirs of the southwestern United States. More recent adaptations include the use of biotechnology and agricultural chemicals to increase agricultural production and the development of new pharmaceuticals to cope with diseases.

Human adaptation has gone far beyond responses to natural constraints to produce widespread modification of environment and landscapes. In some cases, the human use of nature has resulted in environmental degradation or pollution. For example, overcultivation of steep slopes can result in erosion of the soil needed for subsequent agricultural production, and the use of agricultural chemicals has caused the contamination of adjacent rivers and lakes by chemicals that are toxic to fish and humans. The Industrial Revolution produced a dramatic growth in the emissions of waste material to land, water, and atmosphere and resulted in serious air pollution and health problems in many areas.

The massive transformation of nature by human activity led geographers such as Neil Smith and Margaret Fitzsimmons to claim that we can no longer talk about "natural" environments or untouched wilderness. They use the phrase *social production of nature* to describe the refashioning of landscapes and species by human activity, especially capitalist production and labor processes.

Geographers have played a major role in highlighting the global scope of this transformation in their discussions of the human dimensions of global environmental change, in which they explore the social causes and consequences of changes in global environmental conditions. Of particular concern are global patterns of fossil-fuel use and land-use change that are producing serious changes in climate and biodiversity through carbon dioxide–induced global warming or deforestation.

Global climate change is causing sea levels to rise as polar ice caps melt and has increased the strength and severity of violent storms. These storms will have dramatic impacts on people and the environment. Although it is not possible to state unequivocally that Cyclone Nargis, which devastated large parts of Burma in the spring of 2008, was a result of climate change, it is instructive to appreciate the impacts such extreme storms can have. According to the UN Office for the Coordination of Humanitarian Affairs, Nargis, one of the deadliest named cyclones of all times, is believed to have been directly or indirectly responsible for the deaths of anywhere from 100,000 to 220,000, displacing a futher 150,000 people, with an estimated $10 billion in property damage. In addition to personal and property impacts, it is also important to recognize that Nargis has had very detrimental short- and long-term impacts on Burmese fisheries, forests, and agriculture as well.

Warmer oceans surrender greater quantities of water as evaporation. Warmer surface temperatures and more-humid air masses intensify weather systems, resulting in fiercer cyclones and hurricanes. In the summer of 2005, the twin disasters associated with global warming—violent storms and flooding—came together in the United States as Hurricane Katrina bore down on a wide swathe of the Gulf Coast that extended from Pensacola, Florida, to New Orleans, Louisiana. The hurricane destroyed extensive sections of the built environment and caused the flooding of low-lying areas, especially Greater New Orleans—with a population of 1.4 million—where thousands died or were injured and more than 1 million people were displaced. The media, politicians, government officials, planners, and engineers labeled Hurricane Katrina the worst natural disaster the United States had ever experienced.

It is now largely accepted that the root causes of the calamity were far from a "natural" disaster. While there is no doubt that the force of the winds slamming the Gulf Coast were extreme, the fact that districts and parishes in and around New Orleans (where the most dramatic impacts occurred) flooded and so many people (who had not evacuated) died was not because of the hurricane but was the result of political and social factors that could have been avoided. The event is instructive because it provides insight into the ways that global climate change can have catastrophic effects because of the social as well as the environmental vulnerability of populations.

As with Hurricane Andrew, the violent storm that hit the Miami-Dade area in 1992, weak building codes and poor building code enforcement in New Orleans made the wood-frame housing stock especially susceptible to damage from high winds. And critically needed improvements to the sinking levees—which hold back the Mississippi River and Lake Ponchartrain and enable various districts of the city of New Orleans to sit nearly 5 feet below sea level—were repeatedly postponed by the federal government. The strong winds mostly brought down power and communication lines, leaving the region without electricity or phone service but the housing stock fairly in tact. But the storm surge that accompanied Katrina breached the levees and caused flooding to occur in over 80 percent of the city. Add to these technological vulnerabilities the fact that although the city's evacuation plan worked well for many, tens of thousands of people were too poor or disabled or fearful to find their way out of the city before the storm hit and were thus put directly in harm's way.

The argument that nature was to blame for the flooding and the tremendous loss of life in New Orleans, if not also in the rest of the affected Gulf region, is difficult to sustain. This is especially so when comparing Katrina's impact on the Gulf Coast of the United States in 2005 to Hurricane Ivan's impact on Cuba just a year before. Ivan was a Category 5 storm with 160-mile per hour winds when it hit the west coast of Cuba; Katrina was a Category 4 storm with winds of 125–140 miles per hour when it hit the Gulf Coast. Yet 1.5 million Cubans successfully evacuated to higher ground ahead of the storm. They also evacuated pets, TVs, and refrigerators, so people weren't reluctant to leave because their belongings would be unprotected. And though 20,000 houses were destroyed, no one died, no curfews were imposed, and no looting or violence took place. After Hurricane Ivan, the United Nations International Secretariat for Disaster Reduction cited Cuba as a model for hurricane preparation.

Atmospheric scientists have been arguing for over a decade that global warming is putting Atlantic and Gulf coastal cities in the United States in a very vulnerable position as higher-than-normal ocean temperatures help to intensify hurricanes and tropical storms. Many believe that Hurricane Katrina is not an example of a "perfect storm," where essential meteorological elements combine to produce an extreme and devastating event, but the perfect example of how global climate change in combination with human practices is transforming Earth's environment in dramatic and devastating ways.

During the twentieth century, global sea level rose by 20 centimeters (7.9 inches), and a recent report by Britain's Meteorological Office warned that flooding will increase more than nine-fold over the twenty-first century, with four-fifths of the increase coming in South and Southeast Asia. A rising sea level would be disastrous for some countries. About 70 percent of Bangladesh, for example, is at sea level, as is much of Egypt's most fertile land in the Nile delta. In contrast, farmers in much of Europe and North America would welcome a local rise in mean temperatures, since it would extend their options for the kinds of crops that they could profitably raise.

The causes and consequences of these global climate changes vary considerably by world region. For example, the industrial countries have higher carbon dioxide emissions. Increased carbon dioxide emissions are contributing to rising temperatures through the trapping of heat in Earth's atmosphere. In order to survive in many of the world's peripheral regions, the rural poor are often impelled to degrade and destroy their immediate environment by cutting down forests for fuelwood, leading to the destruction of forests, which help to cool Earth's surface. Thus, both the core and the periphery are contributing to the problem of global change in different but equally significant ways (**Figure 4.5**).

Population growth patterns and the changing geography of economic development allow us to predict with

some confidence that the air and water pollution generated by low-income countries will more than double in the next 15 years as they become more industrialized. Thus, environmental problems are becoming inseparable from processes of demographic change, economic development, and human welfare. In addition, regional environmental problems are becoming increasingly enmeshed in matters of national security and regional conflict. Before the 1980s, the principles of cultural ecology did not include the political dimensions of ecological questions, but since then cultural ecologists have moved away from a strict focus on particular cultural groups' relationship with the environment, placing that relationship within a wider context instead. The result is political ecology, the merging of political economy with cultural ecology. **Political ecology** stresses that human-environment relations can be adequately understood only by reference to the relationship of patterns of resource use to political and economic forces (see Box 4.1, "Cultural Ecology and Political Ecology").

U.S. Environmental Philosophies and Political Views of Nature

As mentioned, nature is a construct that is very much shaped by social ideas and values. As a result, different societies and different cultures have very different views of nature. In the contemporary world, views of nature are dominated by the Judeo-Christian tradition, which understands humans to be superior to nature, such that nature is something to be tamed or dominated. But other views of nature have emerged and have departed dramatically from the Judeo-Christian tradition, especially the environmental philosophies that became popular in the nineteenth and early twentieth centuries and the more radical political views of nature that gained prominence in the late twentieth century. We examine both of these approaches in this section in terms of changing U.S. philosophies and views of nature, which are largely representative of the core countries of the world economy.

Henry David Thoreau (1817–1862), an American naturalist and activist, perhaps best illustrates the Western incorporation of North American Indian conceptions of nature into ecological approaches that began emerging in the mid-nineteenth century in the United States. Thoreau lived and studied the natural world around the town of his birth, Concord, Massachusetts. He is most famous for his book *Walden*, which chronicles the two years he spent living and observing nature in solitude in a house he built at Walden Pond, 1½ miles from the village green of Concord. Thoreau represents a significant alternative to the "humans-over-nature" approach that characterized his times. Many people credit him to be the originator of a U.S. ecological philosophy.

Thoreau was impressed with the power of nature. He often described its unrestrained and sometimes explosive capacity, which he thought had the potential to overthrow human dominion if left unchecked. He also

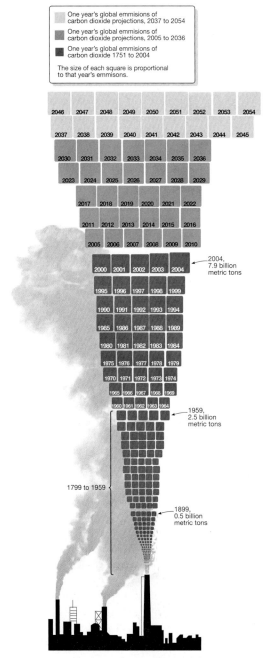

Figure 4.5 Global emissions of carbon dioxide Each square represents one year's global emissions of carbon dioxide, measured by the weight of carbon it contains. (*Source:* Adapted from *New York Times*, December 16, 2007. http://www.nytimes.com/interactive/2007/12/16/weekinreview/20071216_EMISSIONS_GRAPHIC.html, accessed June 17, 2008.)

emphasized the interrelatedness of the natural world, where birds depended upon worms, fish depended upon flies, and so on along the food chain (**Figure 4.6**). Most notably, however, Thoreau regarded the natural world as an antidote to the negative effects of technology on the landscape and the American character. Concord was just 20 miles west of Boston and an equal distance south of the booming mill towns of Lowell and Lawrence. Although he spent his life in a more or less rural setting, the Industrial Revolution was in full force all around Thoreau, and he was keenly aware of its impacts. In fact, Thoreau's approach to the natural world was very much a response to the impacts on nature of the early forces of globalization. His research into the animals and plants that surrounded Concord was an attempt to reconstruct the

landscape as it had existed before colonization and massive European immigration.

Thoreau's ideas embraced European notions of **romanticism**, a philosophy that emphasized the interdependence of humans and nature. In direct revolt against those who espoused a Judeo-Christian understanding of nature, the romantics believed that *all* creatures—human and otherwise—were infused with a divine presence that commanded respect and that humans were not exceptional in this scheme. Rather, their divinity issued from humble participation in the natural community.

A branch of American romanticism known as **transcendentalism** also influenced contemporary understandings of nature. Transcendentalism was espoused most eloquently by a Unitarian minister turned poet and

Figure 4.6   Generalized aquatic food chain Although a contemporary diagram, this illustration of the food chain in a Long Island estuary demonstrates the nineteenth-century naturalist's view. Carnivores, herbivores, and photosynthesizers are all seen to be part of a complex whole, such that elimination or injury to one element affects the entire system. Although most ecosystems have complex food chains containing numerous relationships among the different parts, one rule holds for all: The higher the entity is in a food chain, the fewer there are of it. This rule is illustrated in the bar graph at the top of the illustration, which shows that the photosynthesizers (those at the bottom) are numerous, whereas the carnivores (at the top) are much less so. The photosynthesizers are the most numerous because they are the most energy efficient. The farther you go up the chain from the photosynthesizers—who are also known as primary producers—the less energy efficient the animal is. (Adapted from C. Ponting, *A Green History of the World.* London: Sinclair-Stevenson, 1991, p. 13.)

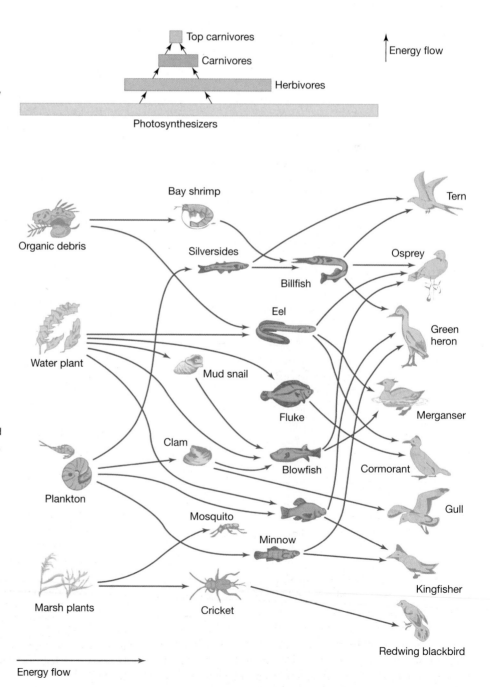

philosopher, Ralph Waldo Emerson, a neighbor and contemporary of Thoreau's. It encouraged people to attempt to rise above nature and the limitations of the body to the point where the spirit dominates the flesh, where a mystical and spiritual life replaces a primitive and savage one. Thoreau and Emerson are two of the most important influences on contemporary ideas about the human-nature relationship.

Another major influence on U.S. environmentalism derives from the writings of George Perkins Marsh, a native Vermonter who in 1864 wrote a treatise entitled *Man and Nature, or Physical Geography as Modified by Human Action* (heavily revised and republished in 1874 as *The Earth as Modified by Human Action*). As the first work to suggest that human beings are significant agents of environmental change, it is considered one of the most important advances in geography, ecology, and resource management in the nineteenth century. Marsh's ideas served as the foundation of the U.S. environmental movement in the twentieth century.

Early in the twentieth century, writers like Gifford Pinchot and politicians like Theodore Roosevelt drew on the ideas of Thoreau, Emerson, and Marsh to advocate the wise use of natural resources and the conservation of natural environments. Their view that nature should be conserved has persisted to the present. **Conservation** holds that natural resources should be used wisely and that humans should serve as stewards, not exploiters, of the natural world. Conservation implies responsibility to future generations as well as to the natural world itself in the utilization of resources. The writings of all these individuals eventually helped to inspire a wide range of environmental organizations, including the Environmental Defense Fund, World Watch Institute, Nature Conservancy, and the Sierra Club.

For those who espouse a more radical approach to nature, the conservation approach is seen as too passive to be truly effective in protecting the environment. Such individuals believe that conservation leaves intact the political and economic system that drives the exploitation of nature. They believe that nature is sacred and should be preserved, not used at all. This more extreme position, **preservation**, advocates that certain habitats, species, and resources should remain off-limits to human use, regardless of whether the use maintains or depletes the resource in question. The philosophy of groups such as the Earth Liberation Front (ELF) and Earth First! is closely aligned with the preservationist perspective. Earth First!, for example, unlike the Sierra Club, operates outside the bounds of mainstream institutional frameworks. Whereas the Sierra Club takes its opponents into the courtroom, Earth First! employs extralegal tactics—often called *ecoterrorist* tactics—such as driving spikes into trees to discourage logging. Both Earth First! and ELF are leaderless organizations that use direct-action economic sabotage to stop the destruction of the environment.

Focusing on environmental polluters and combining the strategies of both the Sierra Club and Earth First!, Greenpeace is yet another environmental organization. Greenpeace is global in its reach, meaning that both its membership and its areas of emphasis are international (**Figure 4.7**). (We talk more about global environmental organizations later in this chapter.) Borrowing from the tactics of both Earth First! and the Sierra Club, Greenpeace utilizes direct oppositional measures as well as formal international legal actions. In its membership—with the world headquarters in Amsterdam and regional offices in most major industrial countries—as well as in its objectives—halting environmental pollution worldwide—Greenpeace articulates the belief that places are interdependent and what happens in one part of the globe affects us all.

Organizations such as Earth First! and Greenpeace are practical illustrations of approaches to understanding human interactions with nature that have developed since the publication of *Silent Spring* over 40 years ago. These, as well as

Figure 4.7 Greenpeace protest in Europe The sign attached to the Brandenburg Gate in Berlin is displayed by Greenpeace, Europe, one of the many regional branches of the organization. Activists like these and in other environmental organizations throughout Europe and North America were among those who opposed the war in Iraq for humanitarian as well as environmental reasons. The Persian Gulf War (1991) became an environmental disaster as Iraqi troops released oil into the Persian Gulf and ignited 732 oil wells in Kuwait as they fled the country. These protesters and others like them would like to see a different energy source than oil fueling the global economy.

Cultural Ecology and Political Ecology

The impact of Spanish agricultural innovations on the culture of the indigenous people of the Central Andes region of South America (an area encompassing the mountainous portions of Peru, Bolivia, and Ecuador) presents an excellent case study in cultural ecology. The transformation of Andean culture began when Pizarro arrived in Peru from Spain in 1531 and set about vanquishing the politically, technologically, and culturally sophisticated Incas. The Spaniards brought with them not only domestic plants and animals (mainly by way of Nicaragua and Mexico) but also knowledge of tool fabrication and a strong sense of what was necessary for a "civilized" life.

By the 1590s, a bundle of Spanish cultural traits had been integrated into the Central Andean rural culture complex, creating a hybridized rural culture. The hybridized culture—and cultural landscape—combined a much simplified version of Spanish material life with important (though altered) Incan practices of crop growing, herding, agricultural technology, and settlement patterns. That this hybrid culture complex remains identifiable today, even after four centuries and in the face of contemporary globalizing forces, is due to the peasants' strong adherence to custom, geographic isolation, and poverty.

By 1620, the indigenous Andean people had lost 90 percent of their population and had been forced to make significant changes in their subsistence lifestyles (an illustration of demographic collapse, as discussed on pages 139–145). The Inca empire, with its large population base, had once engaged in intensive agricultural practices, including building and maintaining irrigation systems, terracing fields, and furrowing hillsides. With the severe drop in population and consequent loss of labor power, the survivors turned to pastoralism because herding requires less labor than intensive agriculture. Ultimately, it was the introduction of Old World domesticated animals that had the greatest impact on the Central Andes (**Figure 4.A**).

Cultural ecologists study the material practices (food production, shelter provision, levels of biological reproduction) as well as the nonmaterial practices (belief systems, traditions, social institutions) of cultural groups. Their aim is to understand how cultural processes affect groups' adaptation to the environment. Whereas the traditional approach to the cultural landscape focuses on human impacts on the landscape or its form or history, cultural ecologists seek to explain how cultural processes affect adaptation to the environment. Cultural adaptation involves the complex strategies human groups employ to live successfully as part of a natural system. Cultural ecologists recognize that people are components of complex ecosystems and that the way they manage and consume resources is shaped by cultural beliefs, practices, values, and traditions, as well as by larger institutions and power relationships.

The cultural ecology approach incorporates three key points:

- Cultural groups and the environment are interconnected by systemic interrelationships. Cultural ecologists examine how people manage resources through a range of strategies to comprehend how the environment shapes culture, and vice versa.

- Cultural behavior is examined as a function of the cultural group's relationship to the environment through both material and nonmaterial cultural

Figure 4.A Andean woman weaving Though sheep are not indigenous to the Andes, they have been widely adopted in this region since the colonial period. Sheep are well adapted to high altitudes and provide wool and meat. Shown here is a woman weaving, accompanied by another woman.

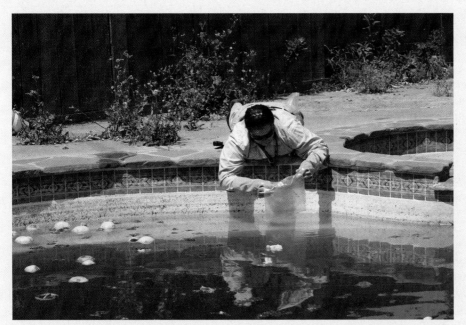

Figure 4.B Mosquitoes breed in abandoned pools With the number of foreclosed homes at record levels in 2008, government officials worry about the mounting number of swimming pools that are not being maintained as they become breeding sites for mosquitoes carrying the West Nile virus. Pictured here is a government technician releasing mosquito-eating fish at a foreclosed home in California.

elements. Such examinations are conducted through intensive fieldwork.

■ Most studies in cultural ecology investigate food production in rural and agricultural settings in the periphery in order to understand how change affects the relationship between cultural groups and the environment.[1]

Cultural ecologists look at food production, demographic change and its impacts on ecosystems, and ecological sustainability. The scale of analysis is not based on cultural areas or cultural regions, but on small groups' adaptive strategies to a particular place or setting.

In the Andean example above, cultural ecologists have been able to understand complex relationships between two cultural groups and their environment, showing how the groups' choices were shaped by both culture and environmental conditions. Some critics have argued, however, that cultural ecology leaves out other intervening influences of the relationship between culture and the environment: the impact of political and economic institutions and practices.

During the 1980s, cultural ecologists began moving away from a strict focus on a particular group's interactions with the environment, instead placing that relationship within a wider context. The result is political ecology, the merging of political economy with cultural ecology. Political ecology stresses that human-environment relations can be adequately understood only by reference to the relationship of patterns of resource use and environmental

practices to political and economic forces. Just as with the study of agriculture, industrialization, urbanization, and comparable geographical phenomena, this perspective requires an examination of the impact of the state and the market on the ways in which particular groups utilize their resource base.

Political ecology incorporates the same human-environment components analyzed by cultural ecologists. However, because political ecologists frame cultural ecology within the context of political and economic relationships, they go beyond what cultural ecologists seek to understand. The case of people and mosquitoes in Tucson, Arizona, is a good illustration of this point because it shows how government institutions and market forces and preferences shape the relationship between these two entities.

Given the arid conditions of the Sonoran Desert, is perhaps surprising that mosquitoes have been historically present in the Southwest of the United States.[2] The ways that humans have modified their desert environment with swimming pools, irrigated lawns, and shaded urban areas goes some way to explaining both the survival and expansion of mosquito populations in an otherwise unfavorable environment. The 2003 arrival of the potentially fatal West Nile virus triggered widespread public and institutional response in Arizona.

There are over 2,700 species of mosquito worldwide, collapsed into three principle genera, all of which are now present in Arizona: the *Aedes*, the *Anopheles*, and the *Culex*. All three genera have, in principle, different

[1]K. Butzer, "Cultural Ecology," in G. L. Gaile and C. J. Wilmot (eds.), *Geography in America*. Columbus, OH: Merrill Publishing Co., 1989, p. 192.

[2]P. Robbins, R. Farnsworth, and J. P. Jones III, "Insects and Institutions: Managing Emergent Hazards in the US Southwest," *Journal of Environmental Policy and Planning*, 10 (1), 2008, pp. 95–112.

geographies. *Aedes aegypti* are sometimes referred to as the "yellow fever mosquito," vectoring both yellow fever and dengue among other diseases. They are strong fliers and persistently bite mammals, especially humans, at dawn and early in the evening. They enjoy great success in urban environments, due to their predilection for stagnant water found in objects such as flower vases, discarded tires, swimming pools, buckets, gutters, and other household debris in the back yard. As such, they live intimately alongside humans. The *Anopheles* genus is found worldwide and often breeds in bodies of permanent freshwater. This genus includes the common malaria mosquito (*Anopheles quadrimaculatus*), which continues to ravage vast swathes of Africa. Finally, the *Culex*, like the *Aedes*, prefers to breed in standing water (**Figure 4.B**). The *Culex* is the primary vector of West Nile virus, a relatively high-profile disease that has a human case fatality rate of around 7 percent and is responsible for the majority of West Nile activity found in Arizona. The relationship between people and the environments they create that encourage the proliferation of mosquitoes explains the "ecology" part of political ecology. But the political part is also important.

Significantly, when West Nile virus arrived in Arizona, the city of Tucson did not have any specific agency designed to manage mosquitoes. Instead, responsibility fell on both the water department and the health department—and those staff had to become mosquito "experts" in a matter of months. The real difficulty facing these policymakers, epidemiologists, and vector control specialists has been just how to reconcile the unpredictable and emergent quality of mosquitoes with static and legislative spatial boundaries found in the city of Tucson. Put differently, the geography of the mosquito (which is constantly shifting with ecological conditions such as temperature and water sources) does not always align with the political geography of the city (with its jurisdictional boundaries, subdivisions, and other spatial features such as roads and washes). It is because a variety of actors are responsible for a variety of geographies, that managing mosquitoes is so complicated. Through modifying their environment, humans have created an oasis in the desert for mosquitoes. But it is through different government agencies at the state, county and city level—some of them at serious odds with one another in terms of their missions—that the problem needs to be managed.

other new approaches—environmental ethics, ecofeminism, deep ecology, and environmental justice—take the view that nature is as much a physical universe as it is a product of social thought. All provide different ways of understanding how society shapes our ideas about nature.

Environmental ethics is a philosophical perspective that prescribes moral principles as guidance for our treatment of nature. Environmental ethics insists that society has a moral obligation to treat nature according to the rules of moral behavior that exist for treating human beings. An aspect of environmental ethics that has caused a great deal of controversy is the perspective that animals, trees, rocks, and other elements of nature have rights in the same way that humans do. If the moral system of our society insists that humans are to have the right to a safe and happy life, then it is argued that the same rights should be extended to nonhuman nature.

Ecofeminism shares much of this philosophical perspective. Ecofeminists hold that patriarchy—a system of social ideas that values men more highly than women—is at the center of our present environmental malaise. Because patriarchy has equated women with nature, it has promoted the subordination and exploitation of both. The many varieties of ecofeminism range from nature-based spirituality oriented toward a goddess to more political approaches that emphasize resistance and opposition to the dominant masculine models that devalue what is not male. Some ecofeminists are also environmental ethicists. Not only a movement of the core, ecofeminism has also been widely embraced in the periphery, where women are primarily responsible for the health and welfare of their families in environments that are being rapidly degraded. The unifying objective in all of ecofeminism is to dismantle the patriarchal biases in Western culture and replace them with a perspective that values social, cultural, and biological diversity.

Deep ecology, which shares many points of view with ecofeminism, is an approach to nature revolving around two key components: self-realization and biospherical egalitarianism. Self-realization embraces the view that humans must learn to recognize that they are part of the nonhuman world, whereas biospherical egalitarianism insists that Earth, or the biosphere, is the central focus of all life and that all components of nature, human and nonhuman, deserve the same respect and treatment. Deep ecologists, like environmental ethicists, believe that there is no absolute divide between humanity and everything else and that a complex and diverse set of relations constitutes the universe. The belief that all things are internally related could enable society to treat the nonhuman world with respect and not simply as a source of raw materials for human use.

Activists in the **environmental justice** movement consider the pollution of their neighborhoods through such elements as nearby factories and hazardous-waste dumps to be the result of a structured and institutionalized inequality that is pervasive in both the capitalist core and periphery. They see their struggles as distinct from the more middle-class and mainstream struggles of groups such as the Sierra Club or even Earth First! and Greenpeace. Environmental justice activists view their struggles as rooted in their economic status. Thus, their struggles are not about quality-of-life issues, such as whether any forests will be left for recreation, but about sheer economic and physical survival. As a result, the questions raised by environmental justice activists involve the distribution of economic and political resources. Such questions are not easily resolved in courts of law, but speak to more complex issues such as the nature of racism and sexism and of capitalism as a class-based economic system.

The environmental justice movement is not restricted to the core. Indeed, poor people throughout the world are concerned that the negative impacts of economic development consistently affect them more than the rich. Furthermore, although most people support improving the lives of people throughout the globe who are currently living in poverty, many have serious reservations about trying to raise all living standards to the level that core regions and the elite enjoy. In fact, some globalization experts argue that some of the highest standards of living around the globe will have to be lowered in order to raise others up, because the widespread economic development that universally high standards would require is not environmentally sustainable. In short, because of the limitations set by Earth's resources—both in terms of the amount and quality of key resources—improvements in social well-being at the level now enjoyed by the core cannot be sustained over the long term and may not even be possible to begin with. As a result, discussions of the future of economic globalization are increasingly being accompanied by discussions about sustainability and the ways economic development might proceed without continuing to damage the environment. (We discuss sustainability in greater detail later in this chapter.)

All of these approaches to the environment attest to a growing concern regarding the effects of globalization. Acid rain, deforestation, the disappearance of species, nuclear accidents, and toxic waste have all been important stimuli for newly emerging philosophies about the preferred relationships between society and nature within a globalizing world. While none of these philosophies is a panacea, each has an important critique to offer. More than anything, however, each serves to remind us that environmental crises are not simple, and simple solutions will not suffice.

The Concept of Nature and the Rise of Science and Technology

Let us return to the point made earlier: that the most widespread ideas of nature current in Western thought—and ones that have persisted under different labels for thousands of years—are that humans are the center of all creation and that nature in all its wildness was meant to be dominated by humans. This Judeo-Christian belief insists that humans, made in the image of God, were set apart from nature and must be encouraged to control it.

While Christianity held that nature was to be dominated, that idea existed more in the religious and spiritual realm than in the political and social realm. In terms of the conduct of everyday life, it was not until the sixteenth century that Christian theology was conscripted to aid the goals of science. Before 1500 in Europe there existed a widely held view of Earth as a living entity such that human beings conducted their daily lives in an intimate relationship with the natural order of things. The prevailing image was that of an organism, which emphasized interdependence among human beings and between them and Earth. Yet even within this organic idea of nature, we can find two opposing conceptions. One was of a nurturing Earth that provided for human needs in a beneficent way; the other was of a violent and uncontrollable nature that could threaten human lives. In both views Earth and nature were regarded as female.

Francis Bacon (1561–1626) and Thomas Hobbes (1588–1679) were English philosophers who, as prominent promoters of science and technology, were influential in changing the prevailing organic view of nature. Borrowing from Christian theology, they advanced a view of nature as something subordinate to the human race. Bacon and Hobbes sought to rationalize benevolent nature as well as to dominate disorderly and chaotic nature largely through the application of new technologies that were emerging in anticipation of the Industrial Revolution, including pumps, windmills, chains, excavators, gears, ratchets, and wrenches, which aided directly in altering the land though mining, drainage, and timber harvesting.

THE TRANSFORMATION OF EARTH BY ANCIENT HUMANS

Although the previous discussion might suggest that Earth remained relatively unaffected by human action until well into the early modern period, the Paleolithic and Neolithic peoples altered the environment even without machines or elaborate tools. Considerable evidence exists that early humans were very active agents of change. People's perceptions of nature were usually quite influential in shaping their environmental behaviors, although many examples exist of contradictions between attitudes and actions. In this section we see that contemporary humans have inherited an environment that was significantly affected even by the practices of our very earliest ancestors.

Paleolithic Impacts

Although humans are thought to have first inhabited Earth approximately 6 million years ago, almost no evidence exists of how the very earliest hominids, as they are called, used the natural world around them to survive. What is known is that their numbers were not large and that they left little behind in the way of technology or art to help us understand their relationship to nature. The earliest evidence about the environmental relationships of our ancestors comes from the **Paleolithic period** (about 1.5 million years ago), a cultural period also known as the

Figure 4.8 The settlement of the world This map shows one theory of the direction and timing of movement of early humans. The constant search for food promoted such movement. The map represents over 1 million years of migration. (Adapted from C. Ponting, *A Green History of the World.* London: Sinclair-Stevenson, 1991, p. 25.)

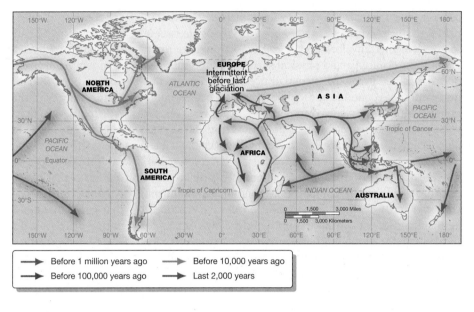

early Stone Age, because this was the period when chipped-stone tools first began to be used.

Hunters and gatherers living on the land in small groups, the early Stone Age people mainly foraged for wild food and killed animals and fish for their survival. Hunting under early Stone Age conditions could not support a growing population, however. It is estimated that on the African grassland, where humans are believed to have first evolved, only two people could survive on the vegetation and wildlife available within about a 2.5-square-kilometer (1-square-mile) area. To help ensure survival, early Stone Age people constantly moved over great distances, which ultimately made them a dispersed species, creating the foundation for the world's population distribution of today (**Figure 4.8**).

Evidence also exists of early Stone Age tools, as well as of the importance of hunting to human existence. The cave paintings of Vallon-pont-d'arc, France, for example, illustrate that hunting was the primary preoccupation of the Stone Age mind (**Figure 4.9**). Because these early peoples lived in small bands and moved frequently in wider and wider ranges, it is tempting to conclude that they had very little impact on their environment. It does appear, however, that Stone Age people frequently used the powerful tool of fire. They used it to attract game, to herd and hunt game, to deflect predators, to provide warmth, and to encourage the growth of vegetation that would attract grazing animals like antelope and deer.

The impact of frequent and widespread fire on the environment can be dramatic. Fire alters or destroys vegetation—from entire forests to vast grasslands (**Figure 4.10**). Fire can encourage the growth and survival of some species while eliminating others. When fire destroys the vegetation that anchors the soil, however, and is followed by heavy rains, it can lead to soil erosion and, in areas with a steep slope, to the total denudation of the landscape. The use of fire by early Stone Age peoples certainly had all of these impacts on the environment.

Archaeologists believe that many large North American animal species disappeared around 11,000 years ago, at the end of the Pleistocene—the geologic and climatic age immediately preceding the one in which we now live. Large, slow animals such as the mastodon, mammoth, cave bear, woolly rhino, and giant deer became extinct. These species constituted over two-thirds of the megafauna, or large animals, of the region. A great deal of controversy exists about why these megafauna became extinct and others did not. Climate change or large-scale natural disasters are not satisfying explanations because neither is particularly selective

Figure 4.9 Cave paintings Cave paintings are an important record of the imagination of early hunter-gatherers. Some regard these paintings as clear evidence of the development of forward thinking or anticipation among humans. This cave painting from Vallon-pont-d'Arc in southern France portrays large animals hunted for their meat and skins.

Figure 4.10 Fire and its impacts on the landscape Fires can have a devastating impact on the landscape. Especially hot fires can burn not only the surface vegetation but also the organic material in the soil. The loss of these organic materials and other nutrients hinders the regeneration of vegetation. Without vegetation to anchor the soil, heavy rains can cause massive soil erosion. Shown in the photograph is eastern Arizona, the site of the convergence of the Rodeo and Chediski fires in 2002.

such that megafauna would be eliminated but not other animals.

Another explanation for the extinction of the larger, slower megafauna is that early Stone Age peoples were directly responsible. Although it might seem implausible that primitive hunting techniques could have such a devastating impact, consider the point made earlier that 2.5 square kilometers (1 square mile) of vegetation and wildlife were adequate for the survival of only two people. As population sizes increased, more pressure was placed on animal populations to satisfy human food requirements. Those animals easiest to bring down were certainly the slowest and largest—the ones that could not escape quickly.

It is also the case that early Stone Age peoples had, over time, refined their killing technologies, particularly stone blades and spearheads. The double-edged **Clovis point,** for example, increased the likelihood of a kill rather than an injury to an animal. The Clovis point (**Figure 4.11**) is a flaked, bifaced projectile whose length is more than twice its width. Used to kill large animals such as bison, the point is so named because archaeologists found the first projectiles in conjunction with kill sites in Clovis, New Mexico. Some Paleolithic hunters used the natural landscape to trap or kill large numbers of animals. Driving animals into canyons where they could be contained, or over cliffs where they would be killed en masse, ensured that huge numbers could be eliminated at one time (**Figure 4.12**). Other Paleolithic groups hunted small game, such as rabbits, using traps and small projectiles.

Neolithic Peoples and Domestication

The credit for the development of agriculture—a technological triumph with respect to nature—goes to the Neolithic peoples, also known as the late Stone Age peoples. The divide between them and the Paleolithic peoples occurred about 10,000 years ago, but it is not known exactly when Neolithic peoples shifted from hunting and gathering to cultivating certain plants and taming and herding wild animals. We have termed that period the First

Agricultural Revolution (described in greater detail in Chapter 8). Climatically, we know that for many regions of the globe, this period coincided with the end of the last Ice Age, which means that spring slowly began to occur in places that had not experienced it for thousands upon thousands of years.

At this time, environmental conditions made possible the domestication of plants and animals, which also requires a sedentary lifestyle based in permanent settlements. The first domestication successes of the late Stone

Figure 4.11 The Clovis point Clovis points are named after Clovis, New Mexico, where the points were first discovered. The Clovis points in this photo were found in Southern Arizona, and they show the typical shape of the points, which were attached to spears and used to hunt game. (*Source:* Photo courtesy Arizona State Museum © Jerry Jacka Photography.)

Figure 4.12 Massive animal kills Paleolithic hunters appear to have used features of the landscape to aid them in hunting large game. Archaeologists believe that the mounds of skeletal remains of large animals found at various sites are evidence of this. It is not clear whether hunters and their kin were even able to consume all the animal flesh made available through such killing methods. It has been speculated that such gross killing methods may have led to the extinction of some species. (After Arthur Lidov/National Geographic Society Image Collection.)

Age peoples were with the most docile animals (herbivores) and the hardiest plants (those with large seeds and a tolerance to drought). Once early domestication was established, it became possible for small groups of Neolithic peoples to cease being nomads. As Chapter 2 discusses, permanent settlement enabled further refinements in domestication. Eventually, Neolithic people achieved a truly dramatic innovation—the breeding of plants and animals to produce desired genetic characteristics, such as disease resistance in plants.

The emergence of agriculture changed the course of human history and had important environmental impacts—both negative and positive. One negative impact was the simplification of ecosystems as the multiplicity of wild species began to be replaced by fewer cultivable crops. An **ecosystem** is a community of different species interacting with each other and with the larger physical environment that surrounds it. Along with the vast number of wild species lost has gone the opportunity to understand their benefit to humans and the wider ecosystem. More positively, however, increased crop yields through greater control over available foodstuffs helped to improve human health and eventually increased population growth.

It is from this early period of plant domestication that we also find widespread evidence of a growing appreciation of nature through ritual, religion, and art. Human beings depended upon rain, soil fertility, and an abundance of sunlight to produce a successful harvest, and reverential attitudes toward nature appear to have been pervasive. In many places in both the Old and the New Worlds, people worshiped the gods Earth, Sun, and Rain, and they sacrificed deer and bear to them in an attempt to ensure survival.

Early Settlements and Their Environmental Impacts

What is perhaps the most significant aspect of plant and animal domestication is that it eventually enabled a food surplus to be produced. It also permitted the formation of human settlements in which small groups—probably

craftspeople and political and religious elites—were able to live off the surplus without being responsible for its production. Eventually, growing numbers of people, bolstered by increasing surpluses, were able to settle in places where water was available and the land cultivable.

The invention of agricultural tools helped to further the domestication of plants and animals as well as multiply the early agriculturalists' impact on the landscape. Among the early tools that enabled humans a greater measure of control over nature were the sickle for harvesting wheat (**Figure 4.13**); the plow for preparing the soil; the yoke for harnessing draft animals, such as oxen, to pull the plow; and the wheel for grinding wheat, creating pottery, and later enabling transportation. The wheel was also used as a pulley to draw water. In Sumer and Assyria, for example, the wheel enabled the development of large-scale irrigation systems.

Irrigation is one of the most significant ways that humans have been able to alter the limits of their

Figure 4.13 Wheat and flint sickle blade Perhaps the most significant factor in the spread of agriculture from Mesopotamia was the occurrence of hybrid forms of wheat, one of many wild grasses found in the area. Even before a fertile hybrid emerged, however, people were harvesting the wild forms. Sickle blades made from flint and set into a horn handle were the most common harvesting tools.

environment. Throughout much of the world, in fact, agriculture could not occur without irrigation. And as agriculture has spread, irrigation has increased (**Table 4.1**). Following the success of the Fertile Crescent (an area, shaped roughly like a crescent, of fertile land in the Middle East where plants were domesticated and ancient civilization flourished), agriculture diffused and new settlements emerged. The food-producing minisystems of China, the Mediterranean, Mesoamerica, the Middle East, and Africa were sustained largely through irrigated agriculture. Yet despite the existence of a vast irrigation network and a whole social structure bound up with agricultural production and attendant activities, the cities of the Mesopotamian region collapsed around 4,000 years ago. There is no undisputed explanation for why this occurred, but many researchers believe—based on archaeological evidence—that it was due to environmental mismanagement. The irrigation works became clogged with salt, resulting in increasingly saline soils. To counteract the effect of salt on production, agriculturalists switched to barley, which is more salt-resistant than wheat, but the ultimate result was a significant drop in yields. **Siltation** (the buildup of sand and clay in a natural or artificial waterway) associated with **deforestation** (the removal of trees from a forested area without adequate replanting) also occurred, filling up the deltas for nearly 200 miles (322 kilometers). The canals eventually filled with salt and the soils became too saline to support cultivation.

TABLE 4.1 World Irrigated Area Since 1700

Date (A.D.)	Area (in thousands of square kilometers/square miles)
1700	50/19
1800	80/31
1900	480/185
1949	920/355
1959	1,490/575
1980	2,000/772
1981	2,130/822
1984	2,200/849
2000	2,740/1,057

Source: W. Meyer, *Human Impact on the Earth.* Cambridge: Cambridge University Press, 1990, p. 59. *Original source:* B. G. Rozanov, V. Targulian, and D. S. Orlov, "Soils," in *The Earth Transformed by Human Action*, B. L. Turner II, W. L. Clark, R. W. Kates, J. F. Richards, J. T. Matthews, and W. B. Meyer (eds.). Cambridge: Cambridge University Press,1990. Updated from FAOSTAT Agricultural Database, 2001. Irrigation. 10 July 2001. Web site: http://apps.fao.org/page/form?collection=Irrigation&Domain= Land&servlet=1&language=EN&hostname=apps.fao.org&version=default.

Poorly informed management may have led to the demise of the ancient Mesopotamian cities, but increasingly saline soils also currently plague agriculture in California and southwestern Arizona (**Figure 4.14**). And it was not only the Mesopotamians who made environmental

Figure 4.14 Irrigation system near El Centro, Southern California This photo illustrates a large irrigation system in one of the driest areas of the United States. Much of the remaining cultivable land in the United States lies in dry areas like the Southwest. Large-scale irrigation is required there to sustain agricultural productivity. Under these circumstances, the application of water to crops is an expensive undertaking—it has to be pumped some distance and is subject to rapid evaporation during the dry, hot season of summer. Irrigation in the Southwest also contributes to the depletion of groundwater supplies. The systems that supply such irrigation are expensive to build and maintain. Much of the water delivered to the agricultural sector in the U.S. Southwest is heavily subsidized by the federal government in order to protect agricultural producers from the negative impact that high water prices would have on the number of farms as well as overall productivity. The agricultural sector is often the largest water user among all sectors (including residential, commercial, and government) in this part of the country.

mistakes. Other early urban civilizations, such as the Mayans in Central America and the Anasazi of Canyon de Chelly in Arizona, are also thought to have collapsed due to environmental mismanagement of water.

In the following section, we examine the period of European expansion and globalization. Although many other important cultures and civilizations affected the environment in the intervening periods, the impacts of their technological developments were much the same as those we have already described. The period of European colonialism, however, had a profoundly different impact from that of preceding periods in extent, magnitude, and kind. Furthermore, it set the stage for the kinds of environmental problems contemporary society has inherited, perpetuated, and magnified.

EUROPEAN EXPANSION AND GLOBALIZATION

The history of European expansion provides a powerful example of how a society with new environmental attitudes was able to transform nature in radically new ways. These new attitudes drew from a newly emerging science and its contribution to technological innovation; the consolidation of the population around Judeo-Christian religious beliefs; and, most important, the development of a capitalist political and economic system.

Initially, European expansion was internal—largely contained within its continental boundaries. The most obvious reason for this expansion was population increase: from 36 million in 1000 to over 44 million in 1100, nearly 60 million in 1200, and about 80 million by 1300 (**Figure 4.15**). As population continued to increase, more land was brought under cultivation. In addition, more forest land was cleared for agriculture, animals killed for food, and minerals and other resources exploited for a variety of needs. Forests originally covered upward of 90 percent of Western and Central Europe. At the end of the period of internal expansion, however, around 1300, the forested area was only 20 percent.

The bubonic plague, also known as the Black Death, had temporarily slowed population growth by wiping out over a third of Europe's people in the mid-fourteenth century. By then, agricultural settlement had been extended to take up all readily available land and then some. In England, Italy, France, and the Netherlands, for example, marshes and fens had been drained and the sea pushed back or the water table lowered to reclaim and create new land for agriculture and settlement.

In the fifteenth century, Europe initiated its second phase of expansion—external—which not only changed the global political map but launched a period of environmental change that continues to this day. European external expansion—colonialism—was the response to several impulses, ranging from self-interest to altruism. Europeans were fast running out of land, and as we saw in Chapter 2, explorers were being dispatched by monarchs to conquer new territories, enlarge their empires, and collect

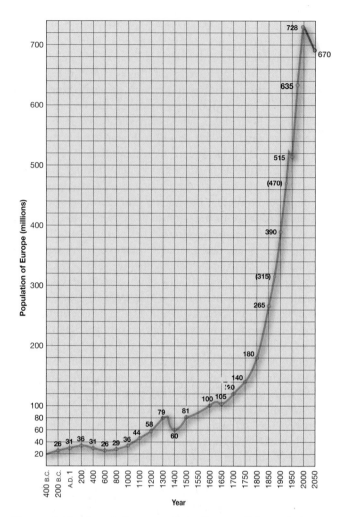

Figure 4.15 Population growth in Europe This graph shows growth and change in the European population from 400 B.C. to A.D. 2050. As is apparent from the graph, the growth in European population has been especially dramatic in the last 500 years as a result of capitalist globalization. The increase in human numbers at the beginning of the 1500s gave an important push to exploration and colonization beyond the confines of the continent. The dip in the graph from 1300 to 1500 is partially explained by the bubonic plague epidemic and food shortages. The dip around the mid twentieth century shows the effect of the two world wars. European population is expected to decline over the next four decades. (Adapted from C. McEvedy and R. Jones, *Atlas of World Population History.* London: Allen Lane, 1978, Fig. 1.2, p. 18.)

tax revenues from the monarch's new subjects. Many of these adventurous individuals were also searching for fame and fortune or avoiding religious persecution. Behind European external expansion was also the Christian impulse to bring new souls into the kingdom of God. Other forces behind European colonialism included the need to expand the emerging system of trade, which ultimately meant increased wealth and power for a new class of people—the merchants—as well as for the aristocracy.

Over the centuries, Europe came to control increasing areas of the globe. Two cases illustrate how the introduction of European people, ideologies, technologies, plant species, pathogens, and animals changed the environments into which they were introduced and also the societies they encountered.

Disease and Depopulation in the Spanish Colonies

Little disagreement exists among historians that the European colonization of the New World was eventually responsible for the greatest loss of human life in history. Moreover, very little disagreement exists that the primary factor responsible for that loss was disease. New World populations, isolated for millennia from the Old World, possessed immune systems that had never encountered some of the most common European diseases. **Virgin soil epidemics**—where the population at risk has no natural immunity or previous exposure to the disease within the lifetime of the oldest member of the group—were common in the so-called Columbian Exchange, though the exchange in this case was mostly one-way. The **Columbian Exchange** was the interaction between the Old World (Europe) and the New World (the Americas) initiated by the voyages of Columbus. For example, diseases such as smallpox, measles, chicken pox, whooping cough, typhus, typhoid fever, bubonic plague, cholera, scarlet fever, malaria, yellow fever, diphtheria, influenza, and others were unknown in the pre-Columbian New World.

Geographer W. George Lovell has examined the role disease played in the depopulation of some of Spain's New World colonies from the point of initial contact until the early seventeenth century, using several cases to illustrate his argument.[1] The first case is Hispaniola (present-day Haiti and the Dominican Republic), where Columbus's 1493 voyage probably brought influenza through the introduction of European pigs carrying swine fever. Subsequent voyages brought smallpox and other diseases, which eventually led to the extinction of the island's Arawak population.

In a second example, in Central Mexico, Lovell writes of Hernán Cortés's contact with the Aztec capital of Tenochtitlán in the first decades of the sixteenth century, which led to a devastating outbreak of smallpox among a virgin soil population. A native Aztec text provides a graphic description of the disease:

> While the Spaniards were in Tlaxcala, a great plague broke out here in Tenochtitlán. It began to spread during the thirteenth month [September 30–October 19, 1520] and lasted for seventy days, striking everywhere in the city and killing a vast number of our people. Sores erupted on our faces, our breasts, our bellies; we were covered with agonizing sores from head to foot.
>
> The illness was so dreadful that no one could walk or move. The sick were so utterly helpless that they could only lie on their beds like corpses, unable to move their limbs or even their heads. They could not lie face down or roll from one side to the other. If they did move their bodies, they screamed with pain.

> A great many died from this plague, and many others died of hunger. They could not get up and search for food, and everyone else was too sick to care for them, so they starved to death in their beds.[2]

In a third example, Lovell described the Jesuits' efforts to establish missions in northern Mexico during a slightly later period. Because these efforts gathered dispersed population groups into single locations, conditions for the outbreak of disease were created. Contact with Spanish conquistadors in advance of the missionaries had already reduced native populations by perhaps 30 to 50 percent. When groups were confined to smaller areas organized around a mission, mortality rates climbed to 90 percent. Eventually the disease was diffused beyond the initial area of contact as traders carried it across long-distance trade routes to the periphery of the Mayan empire in advance of the Spanish armies and missionaries. The Mayans were not defeated by European technological superiority, but by the ravages of a new disease against which they possessed no natural defenses.

Lovell provides similar descriptions of disease impacts in Mayan Guatemala and the Central Andes of South America that led to devastating depopulation. Scholars refer to the phenomenon of near genocide of native populations as **demographic collapse**. The ecological effect of the population decline caused by the high rates of mortality was the transformation of many regions from productive agriculture to abandoned land. Many of the Andean terraces, for example, were abandoned and dramatic soil erosion ensued. In contrast, large expanses of cleared land eventually returned to forest in areas such as the Yucatán in present-day Mexico.

Old World Plants and Animals in the New World

A second case study of the environmental effects of European colonization involves the introduction of Old World plants and animals into the New World, and vice versa. The introduction of exotic plants and animals into new ecosystems is called **ecological imperialism**, a term now widely used by geographers, ecologists, and other scholars of the environment. The interaction between the Old and the New Worlds resulted in both the intentional and unintentional introduction of new crops and animals on both sides of the Atlantic. Europeans brought from their homelands many plants and animals that were exotics, that is, unknown to American ecosystems. For example, the Spanish introduced wheat and sugarcane, as well as horses, cattle, and pigs.

These introductions altered the environment, particularly as the emphasis on select species led to a reduction in the variety of plants and animals that constituted local

[1]W. G. Lovell, "Heavy Shadows and Black Night: Disease and Depopulation in Colonial Spanish America," *Annals, Association of American Geographers*, 82, 1992, pp. 426–443.

[2]W. G. Lovell, 1992, p. 429, quoting from M. León-Portilla, *The Broken Spears: The Aztec Account of the Conquest of Mexico*. Boston: Beacon Press, 1962, pp. 92–93.

ecosystems. Inadvertent introductions of hardy exotic species included rats, weeds such as the dandelion and thistle, and birds such as starlings, which crowded out the less hardy indigenous species. As with the human population, the indigenous populations of plants, birds, and animals had few defenses against European plant and animal diseases and were sometimes seriously reduced or made extinct through contact.

Contact between the Old and the New Worlds was, however, an exchange—a two-way process—and New World crops and animals as well as pathogens were likewise introduced into the Old World, sometimes with devastating implications. Corn, potatoes, tobacco, cocoa, tomatoes, and cotton were all brought back to Europe; so was syphilis, which spread rapidly throughout the European population.

Contacts between Europe and the rest of the world, though frequently violent and exploitative, were not uniformly disastrous. There are certainly examples of beneficial contacts for both sides. The largely beneficial impacts of the Columbian Exchange were mostly knowledge-based or nutritional. Columbus's voyages (**Figure 4.16**) added dramatically to global knowledge, expanding understanding of geography, botany, zoology, and other rapidly growing sciences. It has been argued that the availability of New World gold bullion and silver enlarged Europe's capacity for trade and may even have

been indirectly responsible for creating the conditions that launched the Industrial Revolution.

The encounter also had significant nutritional impacts for both sides by bringing new plants to each. European colonization, although responsible for the extermination of hundreds of plant and animal species, was also responsible for increasing the types and amounts of foods available worldwide. It is estimated that the Columbian Exchange may have tripled the number of cultivable food plants in the New World. It certainly enabled new types of food to grow in abundance where they had never grown before, and it introduced animals as an important source of dietary protein. The advantages of having a large variety of food plants are several. For instance, if one crop fails, another more than likely will succeed because not all plants are subject to failure from the same set of environmental conditions.

The introduction of animals provided the New World with both additional sources of protein and additional animal power. Before the Columbian Exchange, the only important sources of animal energy were the llama and the dog. The introduction of the horse, the ox, and the ass created a virtual power revolution in the New World. These animals also provided fibers and, after death or slaughter, hides and bones to make various tools, utensils, and coverings. Most significant in its environmental impact, however, was the ox.

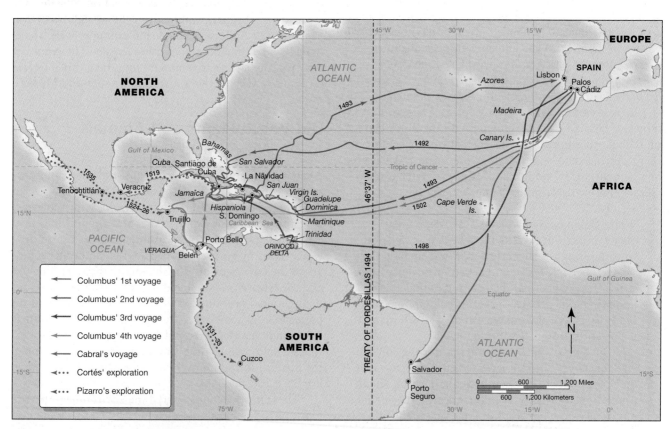

Figure 4.16   European voyages of exploration This maps shows the voyages and missions of Columbus, Pizarro, Cabral, and Cortés. Departing from Portugal and Spain, Columbus encountered several of the islands of the Caribbean as well as the coastal area of present-day Honduras and Venezuela. (Adapted from *The Penguin Atlas of the Diasporas,* by G. Chaliand and J. P. Rageau, translated by A. M. Berrett. Translation copyright © 1995 by G. Chaliand and J. P. Rageau).

Land that had escaped cultivation because the indigenous digging sticks and tools were unable to penetrate the heavy soil and matted root surface became available to an ox-drawn plow. The result was that the indigenous form of intensive agricultural production (small area, many laborers) was replaced by extensive production (large area, fewer laborers). This transformation, however, was not entirely without negative impacts, such as soil destabilization and erosion.

It is also important to explore the impact of native New World peoples on their environment. The popular image of indigenous peoples living in harmony with nature, having only a minimal impact on their environment, is flawed. In reality, different groups had very different impacts, and it is erroneous to conflate the thousands of groups into one romanticized caricature.

In New England, for example, prior to European contact, groups existed who hunted for wild game and gathered wild foods. More sedentary types also existed, living in permanent and semipermanent villages, clearing and planting small areas of land. Hunter-gatherers were mobile, moving with the seasons to obtain fish, migrating birds, deer, wild berries, and plants. Agriculturalists planted corn, squash, beans, and tobacco and used a wide range of other natural resources. The economy was a fairly simple one based on personal use or on barter (trading corn for fish, for example). The idea of a surplus was foreign here: People cultivated or exploited only as much land and resources as they needed to survive. Land and resources were shared, without concepts such as private property or land ownership. Fire was used to clear land for planting as well as for hunting. Although vegetation change did occur, it was minimal and not irreversible.

Indians of South and Central America altered their environment as well, though perhaps in more dramatic ways. The Aztecs of Mexico and the Incas of Peru had developed complex urban civilizations—many of them more sophisticated and highly populated than any European cities—dependent on dense populations employing intensive agricultural techniques on the urban outskirts (**Figure 4.17**). These groups were responsible for dramatic environmental modifications through cultivation techniques that included the irrigation of dry regions and the terracing of steep slopes. As we have seen, irrigation over several centuries results in the salinization of soils. In the lowland tropics, intensive agricultural practices resulted in widespread deforestation as people cut and set fire to patches of forest, planted crops, and then moved on when soil fertility declined. A surplus was key to the operations of both societies, as tribute by ordinary people to the political and religious elite was required in the form of food, animals, labor, or precious metals. The construction of the sizable Inca and Aztec empires required the production of large amounts of building materials in the form of wood and mortar. Concentrated populations and the demands of urbanization meant that widespread environmental degradation existed prior to European contact.

Figure 4.17   Tenochtitlán, circa 1500 This famous contemporary painting by Miguel Covarrubias of the capital city of the Aztecs, Tenochtitlán, illustrates the existence of dense social, cultural, and political activity in the core of the city with agricultural fields on the periphery, particularly to the north. Agricultural goods were also imported from the area surrounding the capital beyond the shores of the lake. With a population of two hundred thousand, it was roughly twice the size of Seville, then Spain's largest city. When Cortés came upon the capital he noted that ". . . in Spain there is nothing to compare with it."

HUMAN ACTION AND RECENT ENVIRONMENTAL CHANGE

No other transition in human history has had the impact on the natural world that industrialization has. When we couple industrialization with its frequent companion, urbanization, we have the two processes that, more than any others, have revolutionized human life and effected far-reaching ecological changes. The changes wrought by industrialization and urbanization have moved beyond a local or a regional level to affect the entire globe. In this section, we explore some of the dramatic contemporary environmental impacts that industrial technology and urbanization have produced. In doing so, we highlight the two issues most central to environmental geography today: energy-use and land-use change.

The Impact of Energy Needs on the Environment

Certainly the most central and significant technological breakthrough of the Industrial Revolution was the discovery and utilization of fossil fuels: coal, oil, and natural gas. Although the very first factories in Europe and the United States relied on waterpower to drive the machinery, hydrocarbon fuels provided a more constant, dependable, and effective source of power. A steady increase in power production and demand since the beginning of the Industrial Revolution has been paralleled, not surprisingly, by an increase in resource extraction and conversion.

At present, the world's population relies most heavily for its energy needs on nonrenewable energy resources that include fossil fuels and nuclear ones, as well as renewable resources such as solar, hydroelectric, wind, and geothermal power. Fossil fuels are derived from organic materials and are burned directly to produce heat. Nuclear energy originates with isotopes, that emit radiation. Most commercial nuclear energy is produced in reactors fueled by uranium. Renewable sources of energy, such as the sun, wind, water, and steam, are captured in various ways and used to drive pumps, machines, and electricity generators.

The largest proportion of the world's current consumption of energy resources, 34 percent, is from oil (see Box 4.2, "Peak Oil"); 25 percent is from coal; 21 percent from gas; 2 percent from hydropower (largely from dams); 7 percent from nuclear power; and 11 percent from combustible renewables and waste (which includes wood, charcoal, crop waste, and dung). The production and consumption of these available resources, however, are geographically uneven, as **Figure 4.18** shows. In 2007, 30 percent of the world's oil supplies were from the Middle East, and most of the coal was from the Northern Hemisphere, with China accounting for 41 percent of world production, followed by the United States at 19 percent and the former Soviet Union at 8

percent. Nuclear reactors are a phenomenon of the core regions of the world. For example, France has 59 nuclear reactors operated by government-owned Electricité de France, supplying the country with over 430 billion kilowatt hours per year of electricity, which is 78 percent of the total generated there.

The consumption side of energy also varies geographically. It has been estimated that current annual world energy consumption is equal to what it took about 1 million years to produce naturally. In one year, global energy consumption is equal to about 1.3 billion tons of coal. What is most remarkable is that this is 4 times what the global population consumed in 1950 and 20 times what it consumed in 1850. And as the $I = PAT$ formula suggests, the affluent core regions of the world far outstrip the peripheral regions in energy consumption. With nearly 4 times the population of the core regions, the peripheral regions account for less than one-third of global energy expenditures. Yet consumption of energy in the peripheral regions is rising quite rapidly as globalization spreads industries, energy-intensive consumer products such as automobiles, and energy-intensive agricultural practices into regions of the world where they were previously unknown. It is projected that within the next decade or so, the peripheral regions will become the dominant consumers of energy (**Figure 4.19**).

Most important for our discussion, however, is that every stage of the energy conversion process—from discovery to extraction, processing, and utilization—has an impact on the physical landscape. In the coal fields of the world, from the U.S. Appalachian Mountains to western Siberia, mining results in a loss of vegetation and topsoil, in erosion and water pollution, and in acid and toxic drainage. It also contributes to cancer and lung disease in coal miners. The burning of coal is associated with relatively high emissions of environmentally harmful gases, such as carbon dioxide and sulfur dioxide (**Figure 4.20**).

The burning of home heating oil, along with the use of petroleum products for fuel in internal combustion engines, launches harmful chemicals into Earth's atmosphere—causing air pollution and related health problems. The production and transport of oil have resulted in oil spills and substantial pollution to water and ecosystems. Media images of damage to seabirds and mammals after tankers have run aground and spilled oil have shown how immediate the environmental damage can be. Indeed, the oceans are acutely affected by the widespread use of oil for energy purposes. Thousands of tons of oil are spilled into the world's oceans each year from leaking ships, oil drilling, transporting oil, and natural seeps. Oil drilling can also have other profound environmental consequences in the form of well explosions and fires (**Figure 4.21**).

Natural gas is one of the least noxious of the hydrocarbon-based energy resources because it is converted relatively cleanly. Now supplying nearly one-quarter of global commercial energy, natural gas is predicted to be the fastest-growing energy source in the new century. Reserves are still being discovered, with Russia holding

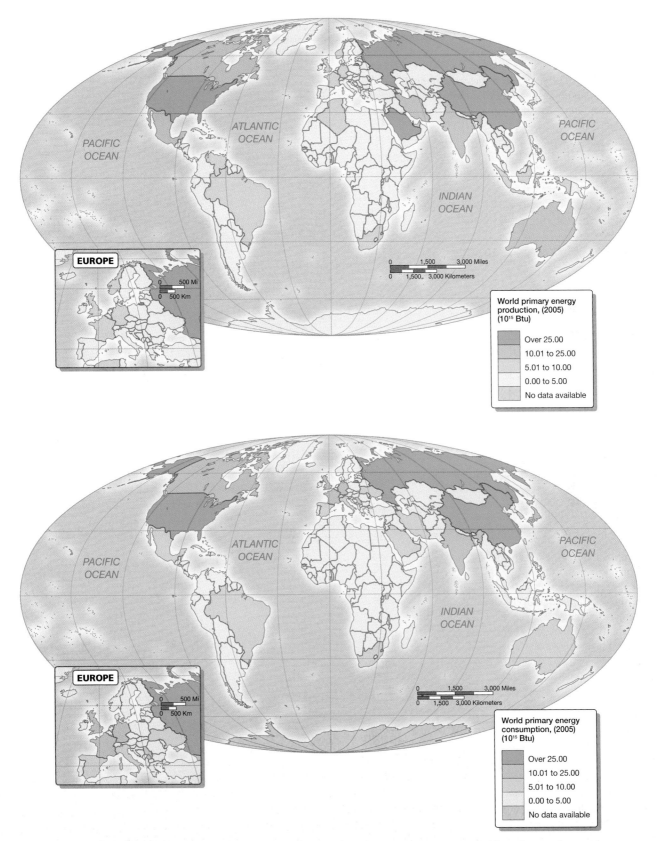

Figure 4.18   World production and consumption of energy, 2005 These paired maps provide a picture of the uneven distribution of the production and consumption of energy resources around the world. The United States is the largest producer and consumer of a range of energy resources. Notice that although the Middle East and North African countries as well as Nigeria are important producers of energy resources, their consumption (as well as that of the rest of the African continent, excluding South Africa) is very low. Japan produces a negligible amount of the total of world energy resources but consumes a relatively high share. (*Source:* Adapted from Energy Information Agency, *International Energy Annual 2005*, Tables E1 and F1. http://www.eia.doe.gov/iea/wepbtu.html, accessed June 19, 2008.)

Figure 4.19 Pottery furnace, China Black smoke billows from the chimneys of an unlicensed pottery furnace on the out-skirts of Wuhan in China's central province of Hubei. The workers use material like old tires and asphalt as fuel, which produces excessive pollutants. Chinese authorities are fighting an uphill battle to eliminate unlicensed factories throughout China that continue to contaminate the environment.

Figure 4.20 Coal mining This coal-mining operation in western Germany is typical of the surface-mining technologies used to exploit shallow deposits in most parts of the world. After deposits are located, the vegetation and over-burden (rocks and dirt overlaying the coal seam) are removed by bull-dozers and discarded as spoil (or waste material). With the coal seam exposed, heavy equipment (such as the type pictured) is used to mine the deposit. Some countries, such as the United States, require restoration of newly mined land-scapes. Sites exploited before these laws were introduced remain unre-stored, however. Successful restora-tion makes it difficult to tell that the area was once a mining site. Unfortunately, many such sites are in arid or semiarid areas where soil and climate prevent full restora-tion. In addition to substantial land disturbance, the mining and pro-cessing of coal resources often cause soil erosion as well as water and air pollution.

Figure 4.21   A tanker oil spill Although the international press is quick to report the disastrous oil spills into the world's oceans, most of the petroleum released into the environment every year comes from nonaccidental sources. The largest contributor to oil pollution of the oceans comes from the extraction process. A very small amount of oil released in oceans is due to major tanker accidents. Tanker spills, like the one shown here, are dramatic because huge amounts of oil end up affecting fairly small areas.

the largest amount—about one-third of the world's total (**Figure 4.22**). While regarded as a preferred alternative to oil and coal, natural gas is not produced or consumed without environmental impacts. The risk of explosions at natural-gas conversion facilities is significant; leakages and losses of gas from distribution systems contribute to the deterioration of Earth's atmosphere.

At the midpoint of the twentieth century, nuclear energy for civilian use was widely promoted as a clearly preferable alternative to fossil fuels. It was seen by many

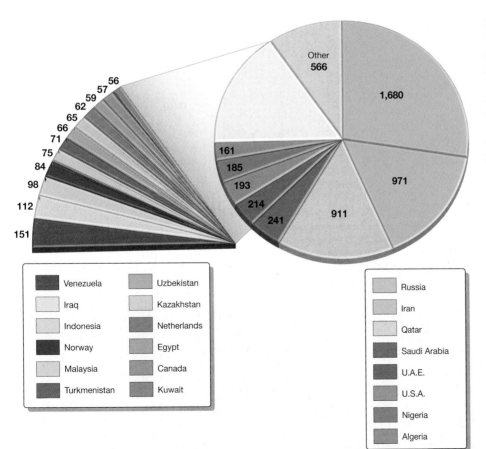

Figure 4.22   Global natural gas reserves Although this pie chart demonstrates that Russia has more gas reserves than any other country, energy experts are forecasting that Russia will have a severe shortfall by 2010. The reason for the looming gas shortfall is that over the past several years, Gazprom, Russia's state-owned natural gas monopoly, has not invested sufficiently, and lacks the technology to develop new gas fields to replace its rapidly depleting ones. The numbers on the pie chart correspond to trillion cubic feet of reserves. (*Source:* Adapted from Russia Profile.org. http://www.russiaprofile.org/resources/business/sectors/oil/index.wbp, accessed June 19, 2008.)

Peak Oil

By Ian G. R. Shaw

Any finite resource has a midway point between the growth and decline of its extraction from the ground. If we were to plot the rates of oil extraction on to a graph, it would loosely resemble a bell-shaped curve and we would label "Peak Oil" the name of the half-way point, or peak, of this imaginary graph. In this sense, Peak Oil is the point at which the highest global rate of oil production is reached. Many experts believe that if world consumption of petroleum-derived products is not slowed, the arrival of Peak Oil is likely to bring with it a worldwide energy crisis. Keep in mind, that in order to determine the point of Peak Oil it is necessary to know something about existing oil reserves. It is generally agreed that the peak of oil discovery was passed in the 1960s, and the world started using more oil than was contained in new fields in 1981. This means that the calculations that predict the point of Peak Oil are premised not only on current use data but also on the known global oil reserves (**Figure 4.C**).

Oil is the lifeblood of most economies, and the increasing dependence on oil in core, semiperipheral, and peripheral countries—especially rapidly developing and highly populous ones like China and India—means that Peak Oil will affect the global economy, demanding proactive mitigation measurements by businesses, governments, and individuals. Optimistic predictions of Peak Oil production place it somewhere in the next few decades, whereas pessimistic predictions indicate that Peak Oil has already occurred—or is imminent. For example, two well-respected scientific networks, the Association for the Study of Peak Oil and Gas (ASPO)

Figure 4.C Proven oil reserves by region, 2007 In order to reasonably project future oil production, it is necessary to have reliable data on known oil reserves. (*Source*: Adapted from "Worldwide Look at Reserves and Production," *Oil and Gas Journal, 104* (47), 2006, pp. 24–25.)

and Energy Watch Group (EWG), have projected fairly similar peaks. ASPO predicts Peak Oil will be reached in 2010, whereas EWG believes Peak Oil production already occurred in 2006. There are some others—mostly influential corporations and government entities—who believe that Peak Oil is a problematic theory that cannot be substantiated by the evidence (**Figure 4.D**). Whatever one's position on Peak Oil, the other aspect of oil production and consumption that will unarguably affect both the global economy and our individual practices is the rapidly escalating demand for and price of oil.

Figure 4.D Predictions of future oil production This figure illustrates Energy Watch Group's prediction for global oil supply through 2030. The prediction is dramatically different from the International Energy Agency's (IEA) "World Energy Outlook 2006," which is also plotted on the graph, in red. This divergence reflects not only differences in methodology but also contradictory attitudes to future ambiguity—with the IEA opting for a far more optimistic and "business-as-usual" scenario. Muddying these predictions further is the notorious difficulty of obtaining accurate estimations of oil reserves from corporations and governments. (*Source:* Adapted from Energy Watch Group, *Crude Oil: The Supply Outlook*, 2007, p. 12. http://www.energywatchgroup.org/fileadmin/global/pdf/EWG_Oilreport_10-2007.pdf, accessed June 24, 2008.)

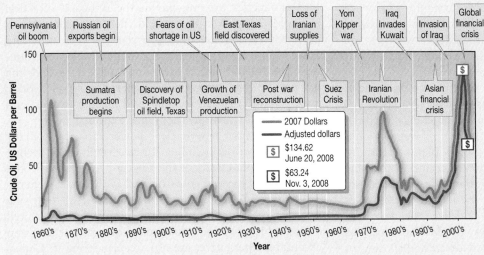

Figure 4.E Price of crude oil, 1986–2008 As this graph shows, the barrel price of oil was relatively stable until 2003. Production, consumption, and political factors all contributed to its recent escalation and decline. (*Source:* Adapted from M. Janofsky, "Forecast of $250 Oil Finds Few Believers," *Arizona Daily Star*, June 22, 2008. http://www.azstarnet.com/allheadlines/244867.php, accessed June 22, 2008.)

The forecasts in demand for oil must be considered alongside the increasing cost of oil. **Figure 4.E** illustrates the history of increases of the barrel price of oil over the last 22 years. Significantly, the rising price of oil occurred over a relatively short period of time—in June 2003 the price of a barrel of crude oil was around $30, compared to a record-breaking $130 in June 2008—a rise of over 333 percent which would begin a precipitous decline in advance of the global financial crises of October 2008. This drop in price illustrates how entwined oil production and worldwide economies are. The underlying causes of rising oil prices are highly contested. First, are the economic explanations: Record demand for oil is simply outstripping available supplies, leading to surging prices. Currently, world consumption of oil is at around 83 million barrels per day, but this is set to jump to 118 million barrels in 2030, with two-thirds of this rise coming from increased use in the transportation sector. The Organization of the Petroleum Exporting Countries (OPEC), comprised of 13 countries and responsible for two-thirds of the international oil supply, has also been negatively criticized by some for restricting the levels of supply. Many commentators also blame "speculators" for "betting" on rising prices of oil. We can add to our list of economic explanations the falling price of the U.S. dollar. Geopolitical explanations for rising prices are also important: There are fears about the political stability of oil-producing countries like Nigeria and Venezuela and mounting tensions between the United States and Iran. Wars and skirmishes can also drive the price of oil by compromising supply. The U.S.-led "War in Iraq" reduced excess production capacity in Iraq from over 6 million barrels per day in 2002 to less than 2 million in 2004–2005. Given these economic and geopolitical driving forces, combined with increased population growth, urbanization, and rising living standards throughout the world, what are some of the impacts and responses we are likely to see?

We can think about mitigation responses from governments and corporations, as well as individuals. The arrival of Peak Oil is likely to affect people and places differently—depending not so much on future policies and responses, but the responses made right *now*. This urgency is precipitated by the huge time lags involved with mitigation responses. The *Hirsch Report*, published in 2005 for the U.S. Department of Energy, investigated the likelihood of Peak Oil and the necessary mitigation responses. Crucially, if governments wait for Peak Oil production to arrive, the consequences will be severe, with liquid-fuel deficits for over two decades. Even a response 10 years in *advance* of Peak Oil would likely leave the world in deficit for over a decade. What is further cause for concern is the rapidity with which oil reserves can dwindle once the Peak is reached—for example, supplies in United Kingdom oilfields might more or less "crash" after the Peak, rather than follow the smoother bell-shaped predictions. All this points to the need to make a response *now* rather than later. Remember, Peak Oil is not just a momentary "energy crisis"—think of all the cars, trains, planes, and ships used every day. Right now, there are simply no alternatives for these modes of transportation. Large-scale agriculture will become increasingly expensive due to the huge energy inputs required. This may lead to the mushrooming of smaller, local and organic farms and produce. Perhaps we will also see the reshaping of the urban landscape in the United States and other core countries with the so-called "end of suburbia." Low-density housing tracts built around the use of the automobile will become increasingly costly places to reside, and we can expect to see migration back to inner cities, or at the very least, surging demand for public transportation. Of course, these responses are complemented by more traditional strategies such as switching to renewable energy sources. However, given the mounting social unrest targeted at rising gas prices in Western Europe and much of the world, it is likely that the transition through Peak Oil will be difficult, contested, and fraught with social and political upheaval.

Ian Shaw is a PhD student in the Department of Geography and Regional Development at the University of Arizona.

as the answer to the expanding energy needs of core countries, especially as the supply of uranium worldwide was thought to be more than adequate for centuries of use. Nuclear energy was also regarded as cleaner and more efficient than fossil fuels. Although nuclear war was a pervasive threat, and there were certainly critics of nuclear energy even in the early years of its development, the civilian "atomic age" was widely seen as a triumphant technological solution to the energy needs of an expanding global economic system. It was not until serious accidents at nuclear power plants began to occur—such as at Windscale in Britain and Three Mile Island in the United States—that the voices of concerned scientists and citizens began to be heard. These voices described, with incontestable evidence, the problems associated with nuclear energy production, such as ensuring nuclear reactor safety and safely disposing of nuclear waste (which remains radioactive for tens of thousands of years). After these accidents and the meltdown of the Chernobyl nuclear power plant in Russia in 1986, many core countries reduced or eliminated their reliance on nuclear energy. Sweden, for example, is set to eliminate entirely its reliance on nuclear power—from which a large portion of its historical electricity use was derived—by the year 2010.

Interestingly, while many core countries have moved away from nuclear energy because of the possibility of environmental disaster in the absence of fail-safe nuclear

reactors, a few semiperipheral—and especially populous—countries are moving in the opposite direction (**Figure 4.23**). Given that world energy consumption is predicted to increase by 57 percent between now and 2030, it is not surprising that rapidly developing countries like India, South Korea, and China have fledgling nuclear energy programs. Higher fossil-fuel prices, energy security concerns, improved reactor designs, and environmental considerations are expected to improve the prospects for new nuclear power capacity in many parts of the world including Russia, which is still feeling the devastating effects of the Chernobyl disaster. So far, no accidents have been associated with nuclear energy production in the periphery. And, because of the rising price of oil, some core countries, such as the United States, which had once abandoned nuclear energy as an acceptable alternative, are now reconsidering it. Generally speaking, though, core countries are expected to experience an overall decline in nuclear energy production over the next two decades.

Nuclear power problems are largely confined to the core, but the periphery is not without its energy-related environmental problems. Because a large proportion of populations in the periphery rely on wood for their energy needs, as the populations have grown, so has the demand for fuelwood. One of the most immediate environmental impacts of wood burning is air pollution, but the most alarming environmental problem is the rapid

Figure 4.23 World distribution of nuclear reactors, 2008 Most of the dependence on nuclear power is concentrated in core countries. South America and Africa together contain only four nuclear reactors. Whereas some semi-peripheral countries, such as India, are enthusiastic about increasing their nuclear energy production, core countries such as Sweden are phasing out dependence on nuclear power. Australia, where there is a very strong antinuclear movement, is one of the few core countries to have rejected nuclear power altogether. (Adapted from International Nuclear Safety Center Web site, Maps of Nuclear Power Reactors: World Map 2002, p. 1. http://www.insc.anl.gov/pwrmaps/map/world_map.html, accessed June 18, 2008.)

depletion of forest resources. With the other conventional sources of energy (coal, oil, and gas) being too costly or unavailable to most peripheral households, wood or other forms of biomass—any form of material that can be used as fuel such as animal wastes, livestock operation residues, and aquatic plants—is the only alternative. The demand for fuelwood has been so great in many peripheral regions that forest reserves are being rapidly used up (**Figure 4.24**).

Fuelwood depletion is extreme in the highland areas of Nepal, as well as in Andean Bolivia and Peru. The clearing of forests for fuelwood in these regions has led to serious steep-slope soil erosion. In sub-Saharan Africa, where 90 percent of the region's energy needs are supplied by wood, overcutting of the forests has resulted in denuded areas, especially around rapidly growing cities. And although wood gathering is usually associated with rural life, it is not uncommon for city dwellers to use wood to satisfy their household energy needs as well. For example, in Niamey, the capital of Niger, the zone of overcutting is gradually expanding as the city itself expands. It

is now estimated that city dwellers in Niamey must travel from 50 to 100 kilometers (31 to 62 miles) to gather wood. The same goes for inhabitants of Ouagadougou in Burkina Faso, where the average haul for wood is also over 50 kilometers. And fuelwood use in Asian and South American highland regions as well as in sub-Saharan Africa is expected to continue to increase substantially (34 percent by some estimates) over the next decade, at least.

Hydroelectric power was also once seen as a preferred alternative to the more obviously environmentally polluting and nonrenewable fossil-fuel sources. The wave of dam building that occurred throughout the world over the course of the twentieth century improved the overall availability, quality, cost, and dependability of energy (**Figure 4.25**). Unfortunately, however, dams built to provide hydroelectric power (as well as water for irrigation, navigation, and drinking) for the burgeoning cities of the core and to encourage economic development in the periphery and semiperiphery have also had profound negative environmental impacts. Among the most significant of these impacts are changes in downstream flow, evaporation, sediment transport and

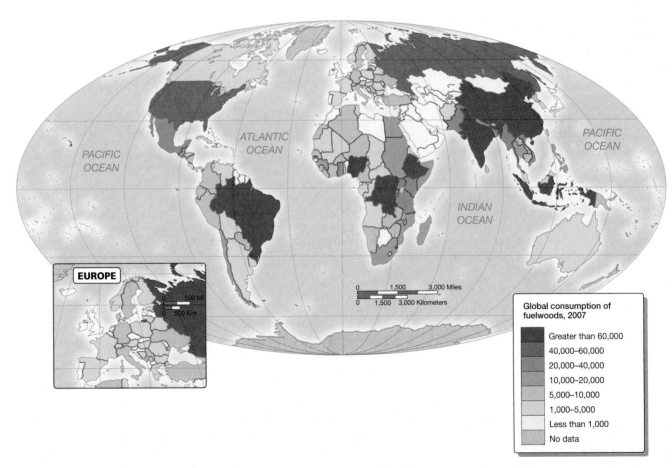

Figure 4.24   Global consumption of fuelwoods, 2007 Firewood, charcoal, and dung are considered traditional fuels, and although their availability is decreasing, dependence upon them is increasing. Dependence on traditional sources of fuel is especially high in the periphery, where, in Africa, for example, they are the most important energy source for cooking and heating. Wood and charcoal, although renewable sources, are replenished very slowly. Acute scarcity will be a certainty for most African households in the twenty-first century. (Data from UN Food and Agricultural Organization, *State of the World's Forests 2007*, Annex, Tab. 4, Rome, 2007. ftp://ftp.fao.org/docrep/fao/009/a0773e/a0773e10.pdf, accessed June 18, 2008.)

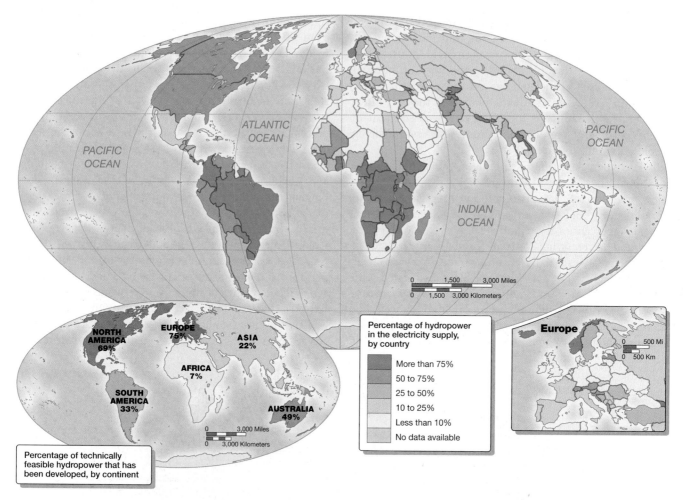

Figure 4.25   Percentage of hydropower in the electricity supply by country, 2002 Although the great dam-building era for core countries is now largely completed, many peripheral countries, in a bid to participate more actively in the world economy, are building dams. Only a few countries are almost exclusively dependent on the hydropower produced from dams. These include Norway, Nepal, Zambia, Ghana, Paraguay, and Costa Rica. Although the power produced by dams is environmentally benign, the construction of large dams can be extremely destructive to the environment and can dislocate large numbers of people. Still, given the increasing need for electricity by rapidly developing peripheral countries, hydropower is becoming a more attractive energy option for many of them because of the uncertain supply of oil in the future. The larger map shows the amount of hydropower that is currently available; the smaller one shows the potential for hydropower development, especially for peripheral regions.

deposition, mineral quality and soil moisture, channeling and bank scouring, and aquatic biota and flora, as well as conditions threatening to human health. Furthermore, the construction of dams dramatically alters the surrounding terrain, often with serious consequences. For example, clearance of the forest for dam construction often leads to large-scale flooding. The felled trees are usually left to decay in the impounded waters, which become increasingly acidic. The impounded waters can also incubate mosquitoes, which carry diseases such as malaria.

Despite these problems and in response to the rising cost of fossil fuels, the use of hydroelectricity is expected to continue to expand over the next two decades by 1.9 percent per year, particularly where its use is supported by government policies and incentives, such as in Asia and South America. Three Gorges Dam, as well as several other large-scale hydroelectric projects in China,

illustrates this trend (**Figure 4.26**). In contrast, most of the increase in core energy production is expected to be in the form of renewable resources, such as wind, solar, geothermal, municipal solid waste, and biomass.

One reason hydroelectric power continues to be appealing in the periphery is that it produces fewer atmospheric pollutants as compared to fossil fuels. Indeed, coal and gas power stations as well as factories, automobiles, and other forms of transportation are largely responsible for the increasingly acidic quality of Earth's atmosphere. Although it is true that people as well as other organisms naturally produce many gases, including oxygen and carbon dioxide, increasing levels of industrialization and motor vehicle use have destabilized the natural balance of such gases, leading to serious atmospheric pollution (see Box 4.3, "Global Climate Change"). Increasing the level of acids in the atmosphere

Figure 4.26   Three Gorges Dam, China In May 2006, the world's largest dam was completed in Yichang, central China's Hubei Province. The axis of the dam is 2,309 meters (1.4 miles), the longest in the world. Begun in 1993, the project proceeded with the backing of the communist leadership despite objections to its $22 billion cost and projected environmental and social impact. More than 1.3 million people have been relocated to make way for the dam and its reservoir; Environmentalists and engineers warned that the reservoir risks becoming polluted with waste from cities and towns upriver. One consequence of the dam has been the sharp increase in landslides. Studies by geologists have shown that water seeping out of the reservoir as well as huge pressure changes are weakening the banks. The shore of the reservoir has already collapsed in 91 places, a total of 36 kilometers (22 miles), killing scores of people. Although developing countries in Asia and South America are increasingly looking to dams for the provision of hydroelectricity, there is still a great deal of environmental opposition to them globally.

are sulfur dioxide, nitrogen oxides, and hydrocarbons, among other gases, which are released into the atmosphere from motor vehicle exhaust, industrial processes, and power generation (based on fossil fuels). If these gases reach sufficient concentrations and are not effectively dispersed in the atmosphere, acid rain can result.

Acid rain is the wet deposition of acids upon Earth through the natural cleansing properties of the atmosphere. Acid rain occurs as the water droplets in clouds absorb certain gases that later fall back to Earth as acid precipitation. The term *acid rain* also includes acid mists, acid fogs, and smog. The effects of acid rain are widespread. Throughout much of the Northern Hemisphere, for example, forests are being poisoned and killed and soils are becoming too acidic to support plant life. Lakes are becoming acidic in North America and Scandinavia. In urban areas, acid rain is corroding marble and limestone buildings, such as the Parthenon in Athens and St. Paul's Cathedral in London as well as other historic structures in Europe. **Figure 4.27** illustrates the global problem of acid emissions that come back to Earth as acid rain.

Before giving up all hope that the use of energy can ever be anything but detrimental to the environment, it is important to realize that alternatives exist to fossil fuels, hydroelectric power, and nuclear energy. Energy derived from the sun, the wind, Earth's interior (geothermal sources), and the tides has been found to be clean, profitable, and dependable. Japan, the United States, and Germany all have solar energy production facilities that have proved to be cheap and nonpolluting. Although contributing only small amounts to the overall energy supply, the production of energy from geothermal and wind sources has also been successful in a few locations around the globe. Italy, Germany, the United States, Mexico, Denmark, and the Philippines all derive some of their energy production from geothermal or wind sources.

And since, as was previously mentioned, most of the world's energy consumption is for transportation, it's important to mention the trend toward hybrid and electric vehicles which is restricted largely to the core, due to the high cost of this technology. Booz Allen Hamilton, a highly respected global strategy and technology marketing firm, predicts that by 2015, 80 percent of all cars sold in the United States will be hybrid, as consumers choose from more than 50 models. U.S. city and county governments are already converting

Figure 4.27 Global acid emissions Acid emissions affect various elements of the natural and the built environment. In some parts of the world, the damage to soils is especially severe. In others, acid emissions cause serious air pollution. Lakes and rivers are also affected by acid emissions, resulting in fish and other wildlife kills. Large amounts of acid-producing chemicals may be generated in one place but exported to another by prevailing winds. More than three-quarters of the acid deposition in Norway, Switzerland, Austria, Sweden, the Netherlands, and Finland is blown in from Western and Eastern Europe. Most industrialized countries have cut sulfur dioxide emissions to help mitigate the damage of acid rain on ecosystems. But the acid rain problem is not yet solved: Emissions of oxides of nitrogen have remained constant or are rising in North America and Europe. At the same time, acid rain is emerging as a major problem in the developing world, especially in parts of the Asia and the Pacific region where energy use has surged. It is likely that the acid emissions experienced in the core countries in the twentieth century will be repeated in the periphery in the twenty-first century. (Adapted from J. L. Allen, *Student Atlas of Environmental Issues.* Guilford, CT: Duskin/McGraw Hill: 1997, p. 45; World Resources Institute, "Acid Rain: Downpour in Asia," *World Resources 1998–1999*, 1998.)

their fleets to hybrids along with large corporations like Federal Express (FedEx). Hybrid and electric automobiles are also very popular in Europe and the United Kingdom.

Nonrenewable alternative sources of energy, such as fuel cells and cogeneration, are also beginning to appear and become more widely adopted. A fuel cell converts chemical energy directly into electricity by combining hydrogen and oxygen in a controlled reaction. Fuel cells emit no pollution, as the waste exhaust is water vapor and heat. One of the most promising uses of fuel cells is for road transportation. Cogeneration, also known as Combined Heat and Power or CHP, is the production of energy and heat in one single process for dual output streams. In conventional electricity generation, only about 35 percent of the fuel is actually converted to electricity while the rest is lost as waste heat. Since cogeneration produces both heat and electricity, it is able to achieve efficiency of upwards of 90 percent. It is the most efficient way to use fuel.

Monies to support the development of geothermal, wind, and tidal energy, as well as fuel cells and cogeneration have been scarce, however, due to the opposition of oil and gas companies, as well as other political factors, such as powerful oil and gas lobbies. While viable alternatives exist to traditional energy sources, the further development of these alternatives is likely to hinge on future political and economic factors.

Impacts of Land-Use Change on the Environment

In addition to industrial pollution and steadily increasing demands for energy, the environment is also being dramatically affected by pressures on the land. The clearing of land for fuel, farming, grazing, resource extraction, highway building, energy generation, and war all have significant impacts. Land may be classified into five categories: forest, cultivated land, grassland, wetland, and areas of settlement. Geographers understand land-use change as occurring in either of two ways: conversion or modification. *Conversion* is the wholesale transformation of land from one use to another (for example, the conversion of forest to settlement). *Modification* is an alteration of existing cover (for example, when a grassland is overlaid with railroad line or when a forest is thinned and not clear-cut). As human populations have increased and the need for land for settlement and cultivation has also increased, changes to the land have followed.

One of the most dramatic impacts of humans upon the environment is loss or alteration of forest cover, which has been cleared for millennia to make way for cultivation and settlement. Forests are cleared not only to obtain land to accommodate increases in human numbers but also to extract the vast timber resources they contain. The approximate chronology and estimated extent of the clearing of the world's forests are listed in **Table 4.2**, which shows that the forested area of the world has been reduced by about 8 million square kilometers (about 3 million square miles) since preagricultural times. Rapid clearance of the world's forests has occurred either through logging, settlement, and agricultural clearing or through fuelwood cutting around urban areas. **Figure 4.28** shows the global extent of deforestation in recent years.

The permanent clearing and destruction of forests, *deforestation*, is currently occurring most alarmingly in the world's rain forests. The UN Food and Agricultural Organization has estimated that rain forests globally are being destroyed at the rate of 0.40 hectare (1 acre) per second. Today rain forests cover less than 7 percent of the land surface, half of what they covered only a few thousand years ago.

Destruction of the rain forests, however, is not just about the loss of trees, a renewable resource that is being eliminated more quickly than it can be regenerated. It is also about the loss of the biological diversity of an ecosystem, which translates into the potential loss of biological compounds that may have great medical value. The destruction of rain forests can also lead to the destabilization of

TABLE 4.2 Estimated Area Cleared ($\times 1000$ km^2)

Region or Country	Pre-1650	1650–1749	1750–1849	1850–1978	1990–2000	2000–2005	Total Estimate
North America	6	80	380	641	0	1	1,108
Central America	15	30	40	200	+4*	3	288
Latin America and Carribean	12	100	170	637	42	45	1,006
Australia, New Zealand, and the South Pacific	4	5	6	362	4	4	385
Former USSR	61	150	260	575	–	–	1,046
Europe	176	54	186	18	+9	+7	481
Asia	732	190	606	1,220	8	10	2,766
Africa	126	24	72	469	44	40	775
Total Estimate	**1,132**	**633**	**1,720**	**4,185**	**85**	**96**	**7,855**

* "Plus" sign indicates a growth, not loss, of total forest area ($\times 1,000$ km^2).

Source: Adapted from B. L. Turner II, W. C. Clark, R. W. Kates, J. F. Richards, J. T. Mathews, and W. B. Meyer, *The Earth as Transformed by Human Action: Global and Regional Changes in the Biosphere over the Past 300 Years.* Cambridge: Cambridge University Press, 1990, p. 180; UN Food and Agricultural Organization, *State of the World's Forests, 2007*, pp. 109–115.

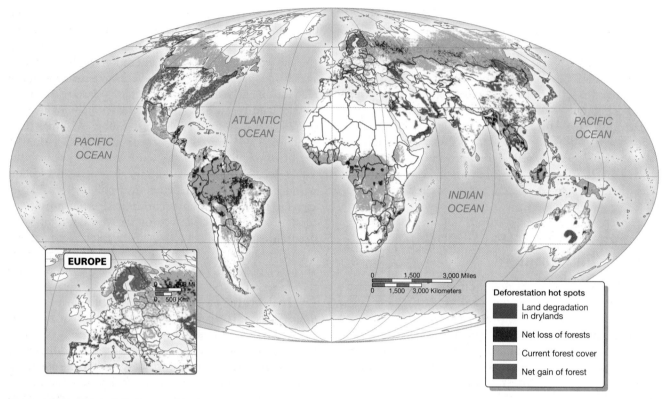

Figure 4.28 Global deforestation The world's forests are disappearing or being reduced or degraded every-where, but especially in tropical countries. Since agriculture emerged about 10,000 years ago, human activities have diminished the world's forest resources by about 25 percent. Whereas forests once occupied about one-third of Earth's surface, they now take up about one-quarter. Playing an important role in the global ecosystem, they fil-ter air and noise pollution, provide a habitat for wildlife, and slow down water runoff, helping to recharge streams and groundwater. They also influence climate at local, regional, and global levels. (*Source:* Adapted from E. B. da Silva, GeoScience Blog: The World of Geography, "World Deforestation," Map. http://images.wri.org/sdm-gene-02-deforestation.jpg, accessed June 24, 2008.)

the oxygen and carbon dioxide cycles of the forests, which may have long-term effects on global climate. Much of the destruction of the South American rain forests is the result of peripheral countries' attempts at economic develop-ment. **Figure 4.29** illustrates this point with reference to the Bolivian Amazon rain forest. The introduction of coca pro-duction has become an important source of revenue for farmers in the region and has led to the removal of small tracts of forest. The pressure of economic development persist in the region such that net rain forest and forest land more generally continue to decline in Central and South America at a faster rate than in the 1990s.

Great geographical variability exists with respect to human impacts on the world's forests. For most of the core regions, net clearance of the forests has been replaced by regeneration. Yet for most of the periphery, clearance has accelerated to such an extent that one esti-mate shows a 50 percent reduction in the amount of for-est cover since the early 1900s.

Cultivation is another important component of global land use, that we deal with extensively in Chapter 8. However, one or two points about the environmental impacts of cultivation are pertinent here. During the past 300 years the land devoted to cultivation has expanded globally by 450 percent. In 1700, the global stock of land in

cultivation took up an area about the size of Argentina. Today, it occupies an area roughly the size of the entire con-tinent of South America. While the most rapid expansion of cropland since the mid-twentieth century has occurred in the peripheral regions, the amount of cropland has either held steady or been reduced in core regions. The expansion of cropland in peripheral regions is partly a response to grow-ing populations and rising levels of consumption worldwide. It is also due to the globalization of agriculture (see Chapter 8), with some core-region production having been moved to peripheral regions. The reduction of cropland in some core regions is a result both of this globalization and of a more intensive use of cropland—utilizing more fertilizers, pesti-cides, and farm machinery—and new crop strains.

Grasslands are also used productively the world over, either as rangeland or pasture for livestock grazing. Most grasslands are found in arid and semiarid regions that are unsuitable for farming because of lack of water or poor soils (**Figure 4.30**). Some grasslands, however, occur in more rainy regions where tropical rain forests have been removed. Other grasslands occur at the mid-latitudes, such as the tall- and short-grass prairies of the central United States and Canada. Approximately 68 million square kilometers (26 million square miles) of the land surface is currently taken up by grasslands.

Figure 4.29 Coca growing in Bolivia The Bolivian portion of the Amazon remains relatively intact as compared with nearby Brazil. Two factors, however, may diminish its relatively pristine state. The first is the increased logging of hardwoods. The second is the production of coca for export as cocaine. Shown in this photograph are coca leaves drying on white cloths in a cleared-out area of the rain forest. Coca is Bolivia's chief export and accounts for over half a million jobs in a largely subsistence economy. In the Chapare region, over 100,000 hectares (247,100 acres) of coca are under production.

Human impacts on grasslands are largely of two sorts. The first is the clearing of grasslands for other uses, most frequently settlement. As the global demands for beef production have increased, so has the use of the world's grasslands intensified. Widespread overgrazing of grasslands has led to their acute degradation. In its most severe form, overgrazing has led to desertification.

Desertification is the spread of desert conditions resulting from deforestation, overgrazing, and poor agricultural practices, as well as reduced rainfall associated with climatic change. One of the most severe examples of desertification has been occurring in the Sahel region of Africa since the 1970s. The degradation of the grasslands bordering the Sahara Desert has not been a simple case of careless overgrazing by thoughtless herders, however. Severe drought, land decline, recurrent famine, and the breakdown of traditional systems for coping with disaster have all combined to create increased pressure on fragile resources, resulting in a loss of grass cover and extreme soil degradation. The factors behind the human impacts on the Sahelian grasslands are complex, but the fact remains that the grasslands have been severely degraded and the potential for their recovery is still unknown (**Figure 4.31**).

Land included in the wetland category covers swamps, marshes, bogs, peatlands, and the shore areas of lakes, rivers, oceans, and other water bodies. Wetlands can be associated either with saltwater or freshwater (**Figure 4.32**). Most of Earth's wetlands are associated with the latter.

Human impacts on wetland environments are numerous. The most widespread has been the draining or filling of wetlands and their conversion to other land uses, such as settlement or cultivation. One reliable estimate places the total area of the world's wetlands at about 8.5 million square kilometers (3.3 million square miles), with about 1.5 million square kilometers (0.6 million square miles) lost to drainage or filling. For example, Australia has lost all of its original 20,000 square kilometers (7,740 square miles) of wetlands to conversion. For the last 400 years or

Figure 4.30 African grasslands Also known as savannas, grasslands include scattered shrubs and isolated small trees and are normally found in areas with high-to-average or high average temperatures and low-to-moderate precipitation. They occur in an extensive belt on both sides of the equator. African tropical savannas, such as the one pictured here in Kenya, contain extensive herds of hoofed animals, including gazelles, giraffes, zebras, wildebeests, antelope, and elephants. Wild species are essential to the safari tourism industry in Africa and would die off without the nourishment that savannas provide.

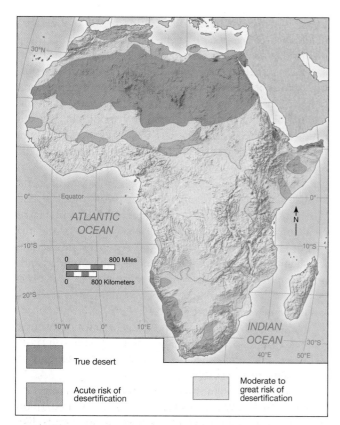

Figure 4.31 Desertification in sub-Saharan Africa
Desertification is a mounting problem in many parts of the world, but especially in Africa south of the Sahara Desert. Overgrazing on fragile arid and semiarid rangelands and deforestation without reforestation are thought to be the chief causes of desertification in this part of Africa.

century allowed speculators and real estate developers to extend significantly the central downtown area into the once marshy edges of San Francisco Bay. The Gold Rush in the Sierra Nevada sent millions of tons of sediment down the rivers into the bay, filling in marshland and reducing its nearshore depth. It is estimated that in 1850 the San Francisco Bay system, which includes San Pablo Bay as well as Suisun Bay, covered approximately 315 square kilometers (about 120 square miles). One hundred years later only about 125 square kilometers (about 50 square miles) remained. By the 1960s the conversion and modification of the wetlands (as well as the effects of pollution pouring directly into the bay) had so dramatically transformed water quality and the habitats of fish, fowl, and marine life that the viability of the ecosystem was seriously threatened. Since then, restoration actions have been put into place and parts of the Bay have returned to something approximating their former state, but large parts are heavily urbanized and cannot be restored.

The combustion of fossil fuels, the destruction of forest resources, the damming of watercourses, and the massive change in land-use patterns brought about by the pressures of globalization—industrialization being the most extreme phase—contribute to environmental problems of enormous proportions. It is now customary to speak of the accumulation of environmental problems we, as a human race, experience as global in dimension. Geographers and others use the term **global change** to describe the combination of political, economic, social, historical, and environmental problems with which human beings across Earth must currently contend. Very little, if anything, has escaped the embrace of globalization, least of all the environment.

In fact, no other period in human history has transformed the natural world as profoundly as the last 500 years. While we reap the benefits of a modern way of life, it is important to recognize that these benefits have not

so, people have regarded wetlands as nuisances, if not sources of disease. In core countries, technological innovation made modification and conversion of wetlands possible and profitable. In San Francisco, California, for example, the conversion of wetlands in the mid-nineteenth

Figure 4.32 Wetlands
Throughout the world, freshwater and saltwater wetlands help to control floods, preserve the water supply, and provide a natural environment for a wide variety of wildlife and a source of recreation for residents of nearby population centers. Shown here is the Yolo Basin Wildlife Area, with the skyline of Sacramento, California, in the background.

been without cost. Fortunately, the costs have not been accepted uncritically. Over the last two to three decades, responses to global environmental problems have been on the increase as local groups have mobilized internationally. In the next section, we examine popular and institutional responses to global environmental problems.

THE GLOBALIZATION OF THE ENVIRONMENT

Global Environmental Politics

The increasing importance of flows and connections—economic, political, social, and cultural—means that contemporary globalization has resulted in an increasingly shrinking world. In addition to allowing people and goods to travel farther faster and to receive and send information more quickly, the smaller world that globalization has made possible means that political action has also become global. It can now move beyond the confines of the state into the global political arena, where rapid communications enable complex supporting networks to be developed and deployed, facilitating interaction and decision making. A good example can be seen in the protests that occurred in Seattle, Washington, in 1999 at the World Trade Organization (WTO) meeting; in Genoa, Italy, in July 2001 at the Group of 8 (G8) summit; and in Bangkok in 2003 to protest the meeting of the Asian Pacific Economic Cooperation summit meeting. Telecommunications, and especially the Internet, enabled protest leaders to organize and deploy demonstrators from interested groups all over the world. Such protests reflect politics that attempts to match the political reach of institutions like the WTO, the International Monetary Fund (IMF), the World Bank, and the G8.

One indication of the expanding influence and geographical extent of popular political groups is the growth of environmental organizations whose purview and membership are global (**Figure 4.33**). These organizations have emerged in response to the global impact of such contemporary environmental problems as fisheries depletion, global warming, the increasing use of genetically modified

seeds, and the widespread decline in global biodiversity. Since the 1990s, these groups—ranging from lobbying organizations to nongovernmental organizations (NGOs) to direct-action organizations to political parties like the Green Party in Europe and drawing on distinctive traditions and varying levels of resources—have become an important international force. Although states tend to exclude global environmental organizations from formal political decision making, these groups have had important impacts on the creation of international environmental institutions and laws, such as the regulation of international waters and the control of marine pollution (the London Dumping Convention in 1972 and the United Nations Law of the Sea in 1982), as well as specific agreements on protection of such wildlife as polar bears and seals, trade in internationally determined rare species, and on the Antarctic ecosystem.

Most dramatic have been the major conventions signed on the international transport of hazardous waste materials (the Basel Convention in 1989), air pollution controls on chlorofluorocarbon (CFC) emissions (the Vienna and Montreal Protocols in 1985 and 1987), as well as a range of treaties regulating transboundary acid rain in North America and Europe.

Increasingly, agreements and conventions protecting biodiversity are being created, and not a moment too soon. The decline in the diversity of simple foodstuffs, like lettuce, potatoes, tomatoes, and squash, occurred most dramatically over the course of the twentieth century. For instance, in 1903 there were 13 known varieties of asparagus; by 1983 there was just one, a decline of 97.8 percent. There were 287 known varieties of carrots in 1903, but this has fallen to just 21, a fall of 92.7 percent. A decline in the diversity of foodstuffs means that different resistances to pests inherent in these different varieties have also declined, as have their different nutritional values and tastes. It should be pointed out, however, that a growing local food movement in the core countries has begun to recover some of these lost varieties.

Moreover, new sources of medicine may be lost not only because of deforestation in tropical forests but also because of the decline in indigenous languages, cultures, and traditions. The Convention on Biological Diversity that emerged from the Rio Summit in 1992 is attempting

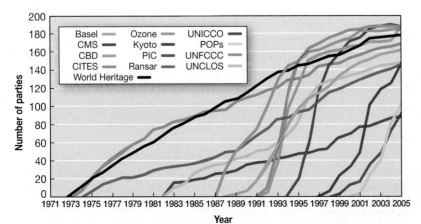

Figure 4.33 Growth of international environmental organizations Since the Second World War, there has been a dramatic increase in global environmental problems, as well as national, international, and voluntary organizations to assess and address those problems. (*Source:* Adapted from Millennium Ecosystem Assessment Board, *Living Beyond Our Means: Natural Assets and Human Well-being.* Washington, DC: Island Press, 2006, pp. 79 and 80.)

Global Climate Change

In 2007, the Intergovernmental Panel on Climate Change (IPCC) and former vice president Al Gore, Jr., were awarded the Nobel Peace Prize. In making its decision, the Nobel Committee determined that the appropriateness of the prize was due not only to the science involved but also to the urgency of the problem that Gore and the scientists were addressing. The Nobel Committee Web site provided this statement:

> Indications of changes in the earth's future climate must be treated with the utmost seriousness, and with the precautionary principle uppermost in our minds. Extensive climate changes may alter and threaten the living conditions of much of mankind. They may induce large-scale migration and lead to greater competition for the earth's resources. Such changes will place particularly heavy burdens on the world's most vulnerable countries. There may be increased danger of violent conflicts and wars, within and between states. (http://nobelprize.org/nobel_prizes/peace/laureates/2007/press. html, accessed June 22, 2008).

The issue of a human-induced (or anthropogenic) change in Earth's climate distributed *unevenly* across the planet has become a growing concern for scientists and policymakers since the 1980s. By 1992, at the Earth Summit in Rio de Janiero, the international community began seeking a way to strike a balance between increasing the pace of economic development without further threatening the global environment. The biggest potential threat to the global environment then and now is the impact that increased energy use will have on global climate. At the Rio Earth Summit, 167 nations ratified the UN Framework Convention on Climate Change with the aim of solving the problem of how to reduce the amount of **greenhouse gases (GHGs)**— any gas that absorbs infrared radiation in the atmosphere, including, but are not limited to, water vapor, carbon dioxide (CO_2), methane (CH_4), nitrous oxide (N_2O), chlorofluorocarbons (CFCs), and ozone (O_3)— that are generated by energy use. An equally critical aim is to ensure that the burden of protecting the environment is shared equitably across all nations.

In December 1997, these nations began to confront the problem of balancing global economic development and environmental protection more substantively by forging the Kyoto Protocol. The protocol marks the first time that an attempt has made to limit the amount of greenhouse gas emissions generated by core countries. The aim of the protocol is to cut the combined emissions of greenhouse gases from core countries by roughly 5 percent from their 1990 levels by 2012. (Core countries account for a disproportionate amount of CO_2 emissions). In late 2007, the climate community met again, this time in Bali, Indonesia, for the UN Climate Change Conference. Over 10,000 participants attended, including representatives of over 180 countries together with observers from intergovernmental and nongovernmental organizations and the media. The result of the conference was the adoption of the Bali Roadmap, which consists of a number of forward-looking commitments that represent the various tracks that are essential to reaching a secure climate future.

In the 10 years that have elapsed since the Kyoto meeting, there is mounting evidence that human-induced climate change is affecting a wide range of Earth systems and that the only way to address these changes is to begin to modify human behavior as individuals and as members of local, national, and international communities. Here we use the IPCC's definition of **climate change:** "a change in the state of the climate that can be identified (e.g., using statistical tests) by changes in the mean and/or the variability of its properties, and that persists for an extended period, typically decades or longer. It refers to any change in climate over time, whether due to natural variability or as a result of human activity." **Figure 4.F** provides a schematic understanding of the process of climate change as well as its effects as humans interact with the environment in complex ways.

The IPCC's *Fourth Assessment Report: Climate Change 2007* may be *the* most significant document produced in the new century. The 15-page *Synthesis Report: Summary for Policymakers* is an easily accessible and should be required reading for literate adults across the globe (http://www.ipcc.ch/pdf/assessment report/ar4/syr/ar4_syr_spm.pdf, accessed June 22, 2008). We provide verbatim text from the *Synthesis Report* here, beginning with scientific observations about climate change as well as some of its observed effects. We then selectively present a summary of what are believed to be the causes of climate change and end with impacts and possible mitigation efforts.

Observations of Climate Change

Eleven of the last twelve years (1995–2006) rank among the twelve warmest years in the instrumental record of global surface temperature (since 1850).

- The temperature increase is widespread over the globe and is greater at higher northern latitudes. Average Arctic temperatures have increased at almost twice the global average rate in the past 100 years. (See **Figure 4.G.**)

- Increases in sea level are consistent with warming. Global average sea level rose at an average rate of 1.8 [1.3 to 2.3] mm per year over 1961 to 2003 and at an average rate of about 3.1 [2.4 to 3.8] mm per year from 1993 to 2003.

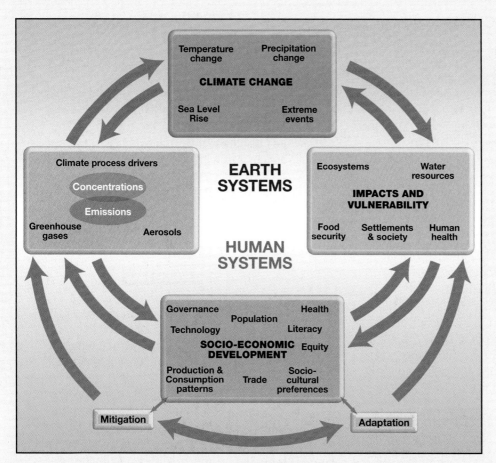

Figure 4.F Schematic framework of climate change drivers, impacts, and responses This diagram represents human-induced drivers, impacts of and responses to climate change, and their linkages. (*Source:* Adapted from IPCC, *Climate Change 2007, Synthesis Report*, p. 26. http:// www.ipcc.ch/pdf/ assessment-report/ar4/ syr/ar4_syr.pdf, accessed June 23, 2008.)

Figure 4.G Global and continental temperature change These graphs compare observed continental and global changes in surface temperature with results simulated by climate models using either natural or both natural and human-induced forcings. Lines are dashed where spatial coverage is less than 50 percent. (*Source:* Adapted from IPCC, *Climate Change 2007: Synthesis Report*, p. 40. http:// www.ipcc.ch/pdf/assessment-report/ ar4/syr/ar4_syr.pdf, accessed June 23, 2008.)

- Satellite data since 1978 show that annual average Arctic sea ice extent has shrunk by 2.7 [2.1 to 3.3] % per decade, with larger decreases in summer of 7.4 [5.0 to 9.8] % per decade. Mountain glaciers and snow cover on average have declined in both hemispheres.

- Trends from 1900 to 2005 have been observed in precipitation amount in many large regions. Over this period, precipitation increased significantly in eastern parts of North and South America, northern Europe and northern and central Asia whereas precipitation declined in the Sahel, the Mediterranean, southern Africa and parts of southern Asia. Globally, the area affected by drought has likely increased since the 1970s.

- Average Northern Hemisphere temperatures during the second half of the 20th century were very likely higher than during any other 50-year period in the last 500 years and likely the highest in at least the past 1300 years.

Observed Effects of Climate Change

There is *high confidence* that natural systems related to snow, ice, and frozen ground (including permafrost) are affected. Examples include: enlargement and increased numbers of glacial lakes; increasing ground instability in permafrost regions and rock; changes in some Arctic and Antarctic avalanches in mountain regions ecosystems, including those in sea-ice biomes, and predators at high levels of the food chain.

- There is *very high confidence*, based on more evidence from a wider range of species, that recent warming is strongly affecting terrestrial biological systems, including such changes as earlier timing of spring events, such as leaf-unfolding, bird migration, and egg-laying; and poleward and upward shifts in ranges in plant and animal species.

- There is high confidence, based on substantial new evidence, that observed changes in marine and freshwater biological systems are associated with rising water temperatures, as well as related changes in ice cover, salinity, oxygen levels, and circulation. These include: shifts in ranges and changes in algal, plankton and fish abundance in high-latitude oceans; increases in algal and zooplankton abundance in high-latitude and high-altitude lakes; and range changes and earlier fish migrations in rivers.

- Sea level rise and human development together are contributing to losses of coastal wetlands and mangroves and increasing damage from coastal flooding in many areas.

Causes of Climate Change (Natural and Anthropogenic)

The radiative forcing (warming influence) of the climate system is dominated by the long-lived GHGs: Global GHG emissions due to human activities have grown since pre-industrial times, with an increase of 70% between 1970 and 2004. (p. 36)

- Carbon dioxide (CO_2) is the most important anthropogenic GHG. Its annual emissions have grown between 1970 and 2004 by about 80%, from 21 to 38 gigatonnes (Gt), and represented 77% of total anthropogenic GHG emissions in 2004. The largest growth in GHG emissions between 1970 and 2004 has come from energy supply, transport, and industry, while residential and commercial buildings, forestry (including deforestation) and agriculture sectors have been growing at a lower rate.

- Changes in the atmospheric concentrations of GHGs and aerosols, land cover, and solar radiation alter the energy balance of the climate system and are drivers of climate change. They affect the absorption, scattering, and emission of radiation within the atmosphere and at the Earth's surface. The resulting positive or negative changes in energy balance due to these factors are expressed as radiative forcing, which is used to compare warming or cooling influences on global climate.

- Human activities result in emissions of four long-lived GHGs: CO_2, methane (CH_4), nitrous oxide (N_2O) and halocarbons (a group of gases containing fluorine, chlorine, or bromine). Atmospheric concentrations of GHGs increase when emissions are larger than removal processes.

- Global atmospheric concentrations of CO_2, CH_4, and N_2O have increased markedly as a result of human activities since 1750 and now far exceed pre-industrial values determined from ice cores spanning many thousands of years. The atmospheric concentrations of CO_2 and N_2O in 2005 exceed by far the natural range over the last 650,000 years.

- The global atmospheric concentration of CO_2 increased from a pre-industrial value of about 280 ppm to 379 ppm in 2005. The global atmospheric concentration of CH_4 has increased from a pre-industrial value of about 715 ppb to 1732 ppb in the early 1990s. The global atmospheric N_2O concentration increased from a pre-industrial value of about 270 ppb to 319 ppb in 2005.

■ There is very high confidence that the global average net effect of human activities since 1750 has been one of warming, with a radiative forcing of +1.6 [+0.6 to +2.4] W/m^2.

Some Projected Impacts

Continued GHG emissions at or above current rates would cause further warming and induce many changes in the global climate system during the twenty-first century that would very likely be larger than those observed during the twentieth century.

■ Snow cover area is projected to contract. Widespread increases in thaw depth are projected over most permafrost regions. Sea

TABLE 4.A Corporate and Government Responses to Climate Change

Sector	Key mitigation technologies and practices currently commercially available.	Policies, measures and instruments shown to be environmentally effective	Key constraints or opportunities (Constraints above; opportunities below)
Energy Supply	Improved supply and distribution efficiency; fuel switching from coal to gas; nuclear power; renewable heat and power (hydropower, solar, wind, geothermal and bioenergy); combined heat and power; early applications of carbon dioxide capture and storage (CCS) (e.g. storage of removed CO_2 from natural gas)	Reduction of fossil fuel subsidies; taxes or carbon charges on fossil fuels feed-in tariffs for renewable energy technologies; renewable energy obligations; producer subsidies	Resistance by vested interests may make them difficult to implement May be appropriate to create markets for low emissions technologies
Transport	More fuel-efficient vehicles; hybrid vehicles; cleaner diesel vehicles; biofuels; modal shifts from road transport to rail and public transport systems; non-motorised transport (cycling, walking); land-use and transport planning;	Mandatory fuel economy; biofuel blending and CO_2 standards for road transport taxes on vehicle purchase, registration, use and motor fuels; road and parking pricing	Effectiveness may drop with higher incomes Particularly appropriate for countries that are building up their transportation systems
Buildings	Efficient lighting and daylighting; more efficient electrical appliances and heating and cooling devices; improved cook stoves, improved insulation; passive and active solar design for heating and cooling; alternative refrigeration fluids, recovery and recycling of fluorinated gases	Appliance standards and labelling, building codes and certification, public sector leadership programmes, including procurement	Periodic revision of standards needed, Enforcement can be difficult Government purchasing can expand demand for energy-efficient products
Industry	More efficient end-use electrical equipment; heat and power recovery; material recycling and substitution; control of non-CO_2 gas emissions; and a wide array of process-specific technologies	Provision of benchmark information; performance standards; subsidies; tax credits, tradable permits, voluntary agreements*	Predictable allocation mechanisms and stable price signals important for investments May be appropriate to stimulate technology uptake
Agriculture	Improved crop and grazing land management to increase soil carbon storage; restoration of cultivated peaty soils and degraded lands; improved rice cultivation techniques and livestock and manure management to reduce CH_4 emissions; improved nitrogen fertiliser application techniques to reduce N_2O emissions; dedicated energy crops to replace fossil fuel use; improved energy efficiency	Financial incentives and regulations for improved land management; maintaining soil carbon content; efficient use of fertilisers and irrigation	May encourage synergy with sustainable development and with reducing vulnerability to climate change, thereby overcoming barriers to implementation
Forestry/ forests	Afforestation; reforestation; forest management; reduced deforestation; harvested wood product management; use of forestry products for bioenergy to replace fossil fuel use	Financial incentives (national and international) to increase forest area, to reduce deforestation and to maintain and manage forests; land-use regulation and enforcement	Constraints include lack of investment capital and land tenure issues Can help poverty alleviation
Waste	Landfill CH_4 recovery; waste incineration with energy recovery; composting of organic waste; controlled wastewater treatment; recycling and waste minimisation	Financial incentives for improved waste and wastewater management	Most effectively applied at national level with enforcement strategies May stimulate technology diffusion

Source: IPCC, *Climate Change 2007: Synthesis Report,* p. 60. http://www.ipcc.ch/pdf/assessment-report/ar4/syr/ar4_syr.pdf, accessed June 23, 2008.

ice is projected to shrink in both the Arctic and Antarctic.

- Based on a range of models, it is likely that future tropical cyclones (typhoons and hurricanes) will become more intense, with larger peak wind speeds and more heavy precipitation associated with ongoing increases of tropical sea-surface temperatures.

- Anthropogenic warming and sea level rise would continue for centuries due to the time scales associated with climate processes and feedbacks, even if GHG concentrations were to be stabilized.

- The resilience of many ecosystems is likely to be exceeded this century by an unprecedented combination of climate change, associated disturbances (e.g., flooding, drought, wildfire, insects, ocean acidification), and other global change drivers (e.g., landuse change, pollution, fragmentation of natural systems, overexploitation of resources).

- Crop productivity is projected to increase slightly at mid- to high latitudes for local mean temperature increases of up to 1 to 3°C depending on the crop, and then decrease beyond that in some regions. At lower latitudes, especially in seasonally dry and tropical regions, crop productivity is projected to decrease for even small local temperature increases (1 to 2°C), which would increase the risk of hunger.

- Coasts are projected to be exposed to increasing risks, including coastal erosion, due to climate change and sea level rise.

- The health status of millions of people is projected to be affected through, for example, increases in malnutrition; increased deaths, diseases, and injury due to extreme weather events; increased burden of diarrhoeal diseases; increased frequency of cardio-respiratory diseases due to higher concentrations of ground-level ozone in urban areas related to climate change; and the altered spatial distribution of some infectious diseases.

- Climate change is expected to exacerbate current stresses on water resources from population growth and economic and land-use change, including urbanization.

Mitigation Efforts

There are so many ways that individuals can begin to address the effects of climate change as described above, from using compact fluorescent light bulbs to taking shorter showers, to buying local produce and using reusable water bottles instead of disposable ones. More concentrated efforts are the responsibility of corporations, governments, and large organizations.

For example, one increasing popular strategy for reducing GHGs is emissions trading. Already an established practice, emissions trading allows governments to regulate the amount of emissions produced in aggregate by setting the overall cap, but allowing corporations the flexibility of determining how and where the emissions reductions will be achieved. Corporations that want to limit their emissions are allocated allowances, with each allowance representing a ton of the relevant emission, such as carbon dioxide. Corporations can then emit in excess of their allocation of allowances by purchasing allowances from the market. Similarly, a company that emits less than its allocation of allowances can sell its surplus allowances.

Emissions trading is believed to be a more desirable mitigation strategy than strict regulation because it gives companies the flexibility to meet emission reduction targets according to their own strategy; for example by reducing emissions on-site or by buying allowances from other companies who have excess allowances. Theoretically, the environmental outcome is not affected because the amount of allowances allocated is fixed. **Table 4.A** provides an overview of what some of these mitigating responses might look like and how our practices will have to change to make them possible.

to protect global biodiversity by preserving and protecting indigenous cultures and traditions, recognizing that many indigenous people have extensive knowledge of local plants and animals and their medicinal uses. As globalization homogenizes languages and draws more and more people into a capitalist market system, traditional knowledge and practices are being lost. The UN Environment Program devotes a great deal of its energies to biological and cultural diversity. Even the WTO has begun to recognize the value of indigenous knowledge and the promise of biodiversity through its advocacy of intellectual property rights of both corporations and indigenous peoples. For the latter, this means protection against "bioprospecting," the practice by which companies attempt to exploit indigenous knowledge of plants for commercial purposes without compensating those individuals who have developed detailed knowledge of their medicinal value.

Clearly, global environmental awareness is on the rise from both the conservative (such as the WTO) and progressive (such as programs devoted to preserving genetic diversity in seed strains) ends of the political spectrum. This increasing awareness is directly responsible for the staging of global environmental conferences like Rio in 1992, Kyoto in 1997, and Johannesburg in 2002, which have not only affected international laws but continue to shape the debates about and responses to environmental problems. Most recently, the terms of these debates have centered on the concept of sustainability.

Environmental Sustainability

The interdependence of economic, environmental, and social problems, often located within widely different political contexts, means that some parts of the world are ecological time bombs. The world currently faces a daunting list of environmental threats, including the destruction of tropical rain forests and consequent loss of biodiversity; widespread health-threatening pollution; the degradation of soil, water, and marine resources essential to food production; stratospheric ozone depletion; and acid rain. Most of these threats are greatest in the world's periphery, where daily environmental pollution and degradation amount to a catastrophe that will continue to unfold in slow motion in the coming years.

In the peripheral regions there is simply less money to cope with environmental threats. The poverty endemic to peripheral regions also adds to environmental stress. In order to survive, the rural poor are constantly impelled to degrade and destroy their immediate environment, cutting down forests for fuelwood and exhausting soils with overuse. In order to meet their debt repayments, governments feel compelled to generate export earnings by encouraging the harvesting of natural resources. In the cities of the periphery, poverty encompasses so many people in such concentrations as to generate its own vicious cycle of pollution, environmental degradation, and disease. Even climate change, an inherently global problem, seems to pose its greatest threats to poorer, peripheral regions.

A more benign relationship between nature and society has been proposed under the principle of *sustainable development*, a term that incorporates the ethic of intergenerational equity, with its obligation to preserve resources and landscapes for future generations (see Box 7.1, "Sustainable Development"). Geographers such as William Adams and Timothy O'Riordan conceive of sustainable development as including ecological, economic, and social measures to prevent environmental degradation while promoting economic growth and social equality. Sustainable development means that economic growth and change should occur only when the impacts on the environment are benign or manageable and the impacts (both costs and benefits) on society are fairly distributed across classes and regions. This means finding less-polluting technologies that use resources more efficiently and managing renewable resources (those that replenish themselves, such as water, fish, and forests) to ensure replacement and continued yield. In practice, sustainable development policies of major international institutions, such as the World Bank, have promoted reforestation, energy efficiency and conservation, and birth control and poverty programs to reduce the environmental impact of rural populations. At the same time, however, the expansion and globalization of the world economy has resulted in increases in resource use and inequality that contradict many of the goals of sustainable development.

CONCLUSION

The relationship between society and nature is very much mediated by institutions and practices, from technology to religious beliefs. In this chapter, we have seen how the nature-society relationship has changed over time and how the globalization of the capitalist world economy has had a more widespread impact on attitudes and practices than any cultural or economic system that preceded it.

Many ancient humans apparently displayed a reverential attitude toward the natural world, an attitude still evident among native populations in many parts of the New World as well as in Africa and Asia. With the emergence of Judaism and, later, Christianity, humans began adopting a more dominant attitude about nature. The expansion of European trade, followed by colonization and eventually industrialization, broadcast worldwide the belief that humans should take their place at the apex of the natural world. It is the Judeo-Christian attitude toward nature as it was taken up by the emergence of the capitalist economic system that is the most pervasive shaper of nature-society interactions today.

Besides exploring the history of ideas about nature and contemporary environmental philosophies and organizations in the United States, this chapter has also shown that society and nature are interdependent and that events in one part of the global environmental system affect conditions in the system elsewhere. Finally, this chapter has shown that events in the past have shaped the contemporary state of society and nature.

In short, as economies have globalized so has the environment. We can now speak of a global environment in which not only the people but also the physical environments where they live and work are linked in complex and essential ways. Along with the recognition of a globalized environment have come new ways of thinking about global economic development. Sustainable development, one of these new ways of thinking, has come to dominate the agenda of international institutions, as well as environmental organizations, as the new century begins.

MAIN POINTS REVISITED

- **Nature and society constitute a complex relationship. In our view, nature is both a physical realm and a social creation.**

  Recognizing the interactive relationship of nature and society requires us also to acknowledge that humans are not separate from nature but an integral part of it.

- **Because in this text we regard nature as a social creation, it is important to understand views of nature in society today as well as the history of those views. The most prominent view of nature in Western culture is derived from the Judeo-Christian tradition, one core belief of which is that nature exists to be dominated by humans.**

Environmental philosophies have emerged and spread in an attempt to counter this view of nature. A wide range of philosophies and approaches are circulating around the globe.

- **Humankind's relationships with nature have developed over the course of human history, beginning with the early Stone Age. The early history of humankind included people who revered nature as well as those who abused it. Urbanization and industrialization have had extremely degrading impacts on the environment.**

Although other societies have had substantial impacts on nature, the extent of the contemporary core society's impact on the environment is unprecedented. As peripheral countries aim to achieve the level of prosperity enjoyed in the core, their economic practices have similar environmental impacts. The result is that while core countries have begun to limit their negative environmental impacts, peripheral countries in many ways are just beginning to produce their own significant environmental problems.

- **The globalization of the political economy has meant that environmental problems are also global in scope. Deforestation, acid rain, and nuclear fallout affect us all. Many new** ways of understanding nature have emerged in the last several decades in response to global environmental crises.

Some of the most disturbing problems have to do with extensive land-use changes, such as deforestation, as well as with widespread air pollution from the burning of fossil fuels, which many scientists believe is leading to global climate change. In response to these global crises, many new ways of understanding nature have emerged in the last several decades, offering insight into our world as a complexly integrated natural system.

- **Sustainability is fast becoming the most significant approach to addressing global economic and environmental challenges. In addition, new institutional frameworks, including conventions, protocols, and organizations, are rapidly emerging to promote global sustainability.**

New institutions and organizations continue to emerge to demand accountability for a changing global economy that is creating new, more widespread, and oftentimes more disastrous environmental problems. In some important cases such as carbon trading, local food movements, and alternative energy conversion, promising positive changes are being made.

KEY TERMS

acid rain (p. 157)
Climate Change (p. 164)
Clovis point (p. 141)
Columbian Exchange (p. 145)
conservation (p. 135)
cultural ecology (p. 131)
deep ecology (p. 138)

deforestation (p. 143)
demographic collapse (p. 145)
desertification (p. 161)
ecofeminism (p. 138)
ecological imperialism (p. 145)
ecosystem (p. 142)
environmental ethics (p. 138)

environmental justice (p. 138)
global change (p. 162)
greenhouse gases (GHG) (p. 164)
nature (p. 130)
Paleolithic period (p. 139)
political ecology (p. 133)
preservation (p. 135)

romanticism (p. 134)
siltation (p. 143)
society (p. 130)
technology (p. 130)
transcendentalism (p. 134)
virgin soil epidemics (p. 145)
virtual water (p. 127)

ADDITIONAL READING

Allister, M. (ed.), *Eco-man: New Perspectives On Masculinity And Nature*. Charlottesville, NC: University of Virginia Press, 2004.

Bartsch, U., B. Muller, and A. Aaheim, *Fossil Fuels in a Changing Climate: Impacts of the Kyoto Protocol and Developing Country Participation*. New York: Oxford, 2000.

Bebbington, A. "Movements, Modernizations, and Markets: Indigenous Organizations and Agrarian Struggle in Ecuador." In R. Peet and M. Watts (eds.), *Liberation Ecologies: Environment, Development, Social Movements*. London: Routledge, 1996.

Burkett, P., *Marx and Nature: A Red and Green Perspective*. New York: St. Martin's Press, 1999.

Callicott, J. B., and M. P. Nelson (eds.), *The Great New Wilderness Debate*. Athens, GA: University of Georgia Press, 1998.

Campbell, C. J., *Oil Crisis*. Essex, UK: Multi-Science Publishing, 2005.

Centers for Disease Control and Prevention, 2005, West Nile Virus, Division of Vector Borne Infectious Diseases, http://www.cdc.gov/ncidod/dvbid/westnile/background.htm, accessed June 15, 2005.

Chasek, P. S., *The Global Environment in the 21st Century: Prospects for International Cooperation*. New York: United Nations, 1999.

Collingwood, R., *The Idea of Nature*. London: Oxford University Press, 1960.

Crosby, A. W., *The Columbian Exchange: Biological and Cultural Consequences of 1492*. Westport, CT: Greenwood Press, 1972.

Crosby, A. W., *Germs, Seeds, and Animals: Studies in Ecological History*. Armonk, NY: M. E. Sharpe, 1994.

de Jonge, E., *Spinoza and Deep Ecology: Challenging Traditional Approaches to Environmentalism*. Aldershot, England, and Burlington, VT: Ashgate, 2004.

Energy Information Adminstration, International Energy Outlook, Washington, DC, 2007. www.eia.doe.gov/oiaf/ieo/index.html, accessed June 13, 2008.

Energy Watch Group, *Crude Oil: The Outlook, 2007*. http://www.energywatchgroup.org/fileadmin/global/pdf/EWG_Oilreport_10-2007.pdf, accessed June 1, 2008.

Fagen, B., *The Long Summer: How Climate Changed Civilization*. New York: Basic Books, 2004.

Glacken, C., *Traces on the Rhodian Shore*. Berkeley: University of California Press, 1967.

Goudie, A., and D. J. Cuff, *Encyclopedia of Global Change, Environmental Change and Human Society*. New York: Oxford University Press, 2002.

Harvey, F., "World Carbon Trading Doubles," *Financial Times/FT.com*. http://www.ft.com/cms/s/0/94e2baa2-1c8f-1dd-8bfc-000077b07658,dwp_uuid=3c093daa-edc1-11db-8584-000b5df10621.html?nclick_check=1, accessed June 13, 2008.

Hirsch, R. L., *The Inevitable Peaking of World Oil Production*. The Atlantic Council of the United States Bulletin, Vol. reformat for consistency: XVI (in ital.) (3), 2005. http://www.acus.org/docs/051007-Hirsch_ World_Oil_Production. pdf, accessed June 16, 2008.

Hughes, J. D., *What is Environmental History?* Cambridge, UK, and Malden, MA: Polity Press, 2005.

Intergovernmental Panel on Climate Change, *Climate Change 2007: Synthesis Report.* http://www. ipcc. ch/pdf/ assessment-report/ ar4/ syr/ ar4_syr. pdf, accessed June 17, 2008.

International Energy Agency, *Oil Market Report, 2008.* http:// omrpublic.iea.org/currentissues/full.pdf, accessed June 10, 2008.

Lovejoy, T. E., and L. J. Hannah (eds.), *Climate Change and Biodiversity.* New Haven: Yale University Press, 2005.

Lovell, W. G., "Heavy Shadow and Black Night: Disease and Depopulation in Colonial Spanish America," *Annals of the Association of American Geographers* 82 (3), 1992, pp. 426–443.

Maathai, W., *The Green Belt Movement: Sharing the Approach and the Experience.* New York: Lantern Books, 2004.

Marsh, G. P., *Man and Nature.* New York: Scribner, 1864.

Meade, M. S., and R. J. Erickson, *Medical Geography*, 2nd ed. New York: Guildford, 2005.

Mills, R., *The Myth of the Oil Crisis: Overcoming the Challenges of Depletion, Geopolitics, and Global Warming.* New York: Praeger, 2008.

Peet, R., and M. Watts, *Liberation Ecologies: Environment, Development, Social Movements.* New York: Routledge, 2004.

Raven, P. H., *Nature and Human Society: The Quest for a Sustainable World: Proceedings of the '97 Forum on Biodiversity.* Washington, DC: National Academy Press, 2000.

Robbins, P., *Political Ecology: A Critical Introduction.* Oxford: Blackwell, 2004.

Sayer, J., and B. Campbell, *The Science of Sustainable Development: Local Livelihoods and Global Environment.* Cambridge, England: Cambridge University Press, 2003.

Smith, M., and L. Marx, *Does Technology Drive History? The Dilemma of Technological Determinism.* Cambridge, MA: MIT Press, 1995.

Thomas, W. L. (ed.), *Man's Role in Changing the Face of the Earth.* Chicago: University of Chicago Press, 1956.

Turner, B. L., II, et al., *The Earth Transformed by Human Action: Global and Regional Changes in the Biosphere over the Past 300 Years.* New York: Cambridge University Press, 1990.

Warren, K. J., *Ecofeminist Philosophy: A Western Perspective on What It Is and Why It Matters.* Lanham, MD: Rowman & Littlefield Publishers, 2000.

Wood, D., *Five Billion Years of Global Change: A History of the Land.* New York: Guildford, 2004.

Worldwatch Institute, *State of the World 2002: A Worldwatch Institute Report on Progress Toward a Sustainable Society.* New York: W.W. Norton & Co., 2002.

Worster, D., *Nature's Economy: A History of Ecological Ideas.* Cambridge, England: Cambridge University Press, 1977.

EXERCISES

On the Internet

The Companion Website at www.mygeoscienceplace.com contains a number of tools designed to reinforce the geographic concepts presented in this chapter. Structured within a four-step learning path, these online resources include: chapter Key Concepts; a fully-searchable myeBook version of the chapter, which allows you to add notes and highlights; RSS news feeds from various media agencies; links to additional web resources; Human Geography Videos; a digital Glossary; Concept Review questions; Thinking Spatially mapping exercises; and Critical Thinking essay questions.

Unplugged

1. Many communities have begun to produce an index of stress, which is a map of the toxic sites of a city or region. One way to plot a rudimentary map is to use the local phone book as a data source. Use the Yellow Pages to identify the addresses of environmentally harmful and potentially harmful businesses, such as dry-cleaning businesses, gas stations, automotive repair and car-care businesses, aerospace and electronic manufacturing companies, agricultural supply stores, and other such commercial enterprises where noxious chemicals may be produced, sold, or applied. Or consider your college campus and where its most toxic sites are. Your campus directory is the ideal place to start. Compile a map of these activities to begin to get a picture of your locale's geography of environmental stress.

2. Locate and read a natural history of the place where your college or university is located. What sorts of plants and animals dominated the landscape there during the Paleolithic period? Do any plants or animals continue to survive in altered or unaltered form from that period? What new plants have been introduced and how extensive are they? What sorts of tending and maintenance do these new species require and how have the changed the landscape?

3. Colleges and universities are large generators of waste, from plain-paper waste to biomedical and other sorts of wastes that can have significant environmental impacts. Identify how your college or university handles this waste stream and how you, as a member of the academic community, contribute to it. Where does the waste go when it leaves the university? Is it locally deposited? Does it go out of state? Remember to trace the stream of all the types of waste, not just the paper.

Cultural Geographies

5

Descend into the World of Warcraft and join thousands of mighty heroes in an online world of myth, magic and limitless adventure. Jagged snowy peaks, mountain fortresses, harsh winding canyons. Zeppelins flying over smoldering battlefields, epic sieges—an infinity of experiences await. So what are you waiting for?

http://www.blizzard.com/store/details.xml?id=110000034, accessed June 26, 2008.

There are over 10 million subscribers to the World of Warcraft (WOW), an online role-playing video game set in the fantasy Warcraft universe. Subscribers participate by assuming roles as Warcraft heroes who interrelate in a vast, challenging experience. Thousands of players are able to converge across the globe within the same virtual world where they band together or fight against each other in epic battles, competing for power and glory. This phenomenon of multiple players involved interactively in online play is known as MMORPG, which is an acronym for "Massively Multiplayer Online Role-Playing Game." At present there are 16 million subscribers to these online games where the Internet and a monthly subscription is the ticket to an alternative world that can be realistic or complete fantasy.

Millions more possess their own game consoles in order to have a dedicated platform for enjoying this form of entertainment with friends and family at home. According to the Entertainment Software Association, more than 67 percent of U.S. households play video games, and video game sales exceeded $9.5 billion in 2007. In Europe, 40 percent of the population play video games between 6 and 14 hours a week, making it one of the most popular forms of leisure activities there.

The games work through players who control a virtual avatar—for instance a warrior, a druid, or a member of the criminal underworld. In the case of WOW, avatars undertake quests in order to "level-up" (improve their relative standing) and win treasure from slain monsters. WOW epic quests go on for hours as players across the world "come together" to participate in complex battles that have no predetermined outcome.

Not suprisingly, the video game industry is highly lucrative and it is expanding at an unprecedented rate. In terms of annual sales, it now rivals the Hollywood film industry. Indeed in recent years the mainstream status of video games has been confirmed through huge releases, such as the highly anticipated and critically acclaimed *Grand Theft Auto IV*. Competitive video game events, once limited to a few networked computers in a basement, are now international competitions, such as the World Cyber Games. And although core-country players dominate the video game world, players come from every corner of the globe, including peripheral and semiperipheral countries, especially the rapidly developing ones like China, where video games are

Main Points

- Though culture is a central, complex concept in geography, it may be thought of as a way of life involving a particular set of skills, values, and meanings.

- Geographers are particularly concerned about how place and space shape culture and, conversely, how culture shapes place and space. They recognize that culture is dynamic and is contested and altered within larger social, political, and economic contexts.

- Like other fields of contemporary life, culture has been profoundly affected by globalization. Globalization, however, has not produced a homogenized culture so much as distinctive impacts in different societies and geographical areas as local cultures resist or modify or embrace global forces.

- Contemporary approaches in cultural geography seek to understand the role played by politics and the economy in establishing and perpetuating cultures, cultural landscapes, and global patterns of cultural traits and complexes.

- Cultural geography has been broadened to include analysis of gender, class, sexuality, race, ethnicity, and life-cycle stage, in recognition that important differences can exist within, as well as between, cultures.

- Globalization does not necessarily mean that the world is becoming more homogeneous. In some ways, globalization has made the local even more important than before. In other ways, it has meant that local cultures have indeed suffered, as in the case of the disappearance of indigenous languages. In still other cases, globalization has helped to foster hybrid forms that blend distant cultures in unique ways.

becoming more and more popular every day as avid players dress up like their favorite avatars.

It may seem strange to use the example of video games to talk about cultural geography, but the games provide an excellent entry point into the complexity of the globalization of culture and the new geographies that are being produced around developing technologies. For example, the fact that online players come from all over the world and join in one event demonstrates how culture and geography can be juxtaposed differently through technology. This conjunction serves to make clear the way that technology is changing the ways cultural practices come into contact across great distances. It also illustrates that even though online participants are converging around a single event, each one experiences the event differently as nationality, age, viewing venue, level of skill, gender, and a whole host of other variables influence the experience. In short, video games are just one way of exploring how the culture of technology-mediated experiences is shaped by geography and how geography shapes it.

In this chapter, we examine the many ways geographers have explored the concept of culture—including the globalization of culture—and the insights they have gained from these explorations. We ask: What counts as culture? How do geographers study it? How do we come to grips with the fact that U.S. cultural practices are being exported to the far reaches of the globe at the same time that our own cultural practices are being shaped by forces beyond our national borders?

CULTURE AS A GEOGRAPHICAL PROCESS

Geographers seek to understand the manifestations and impacts of culture on geography and of geography on culture. Whereas anthropologists are concerned with the ways in which culture is created and maintained by human groups, geographers are interested not only in how place and space shape culture but also the reverse—how culture shapes place and space.

Anthropologists, geographers, and other scholars who study culture, such as historians, sociologists, and political scientists, agree that culture is a complex concept. Over time, our understanding of culture has been changed and enriched. A simple understanding of culture is that it is a particular way of life, such as a set of skilled activities, values, and meanings surrounding a particular type of economic practice. Scholars also describe culture in terms of classical standards and aesthetic excellence in opera, ballet, or literature, for example.

The term *culture* is also used to describe the range of activities that characterize a particular group, such as working-class culture, corporate culture, or teenage culture. Although all these understandings of culture are accurate, for our purposes they are incomplete. Broadly speaking, **culture** is a shared set of meanings that is lived through the material and symbolic practices of everyday life. Our understanding in this book is that culture is not something that is necessarily tied to a place and thus a fact to be discovered. Rather, we regard the connections among people, places, and cultures to be social creations

that can be altered and are therefore always changing, sometimes in subtle and other times in more dramatic ways. The "shared set of meanings" can include values, beliefs, practices, and ideas about religion, language, family, gender, sexuality, and other important identities (**Figure 5.1**). These values, beliefs, ideas, and practices are, because of globalization, increasingly subject to reevaluation and redefinition and can be, and very frequently are, transformed from both within and outside a particular group.

In short, culture is a dynamic concept that revolves around and intersects with complex social, political, economic, and even historical factors. This understanding of culture is part of a longer, evolving tradition within geography and other disciplines, such as anthropology and sociology. We will look more closely at the development of the cultural tradition in geography in the following section, in which we discuss the debates surrounding culture within the discipline.

For much of the twentieth century, geographers, like anthropologists, have focused most of their attention on material culture, as opposed to its less tangible symbolic or spiritual manifestations. For example, they have explored the spatial extent of particular religious practices (the global distribution of Buddhism) and the physical expression of religiosity, such as the appearance of crosses along roadways where fatal traffic accidents have occurred (**Figure 5.2**). In the last 20 or so years, the near-exclusive focus on material cultural practices has changed—driven by the larger changes that are occurring in the world around us.

Figure 5.1   Maori men with traditional tattoos These men, indigenous peoples of New Zealand, are dancing a *haka*, a traditional war dance meant to display aggressiveness and fear-lessness. The dancers use facial contortions to deride their enemy. Notice as well the tattoos, known as *ta moko*, marks used as signs of identification, rank, geneaology, tribal history, eligibility to marry, beauty and ferocity.

As with agriculture, politics, and urbanization, glob-alization has also had complex effects on culture. Terms such as *world music* and *international television* are a reflection of the sense that the world has become a very small place indeed, and people everywhere are sharing aspects of the same culture through the widespread in-fluence of television, the Internet, and other media, such as radio. Yet, as pointed out in Chapter 2, although powerful homogenizing global forces are certainly at work, the world has not become so uniform that place no longer matters. With respect to culture, just the opposite is true. Place matters more than ever in the negotiation of global forces, as local forces confront globalization and translate it into unique place-specific forms. Box 5.1, "The Culture of Hip-Hop," illustrates the global nature of hip-hop music.

The place-based interactions occurring between cul-ture and global political and economic forces are at the

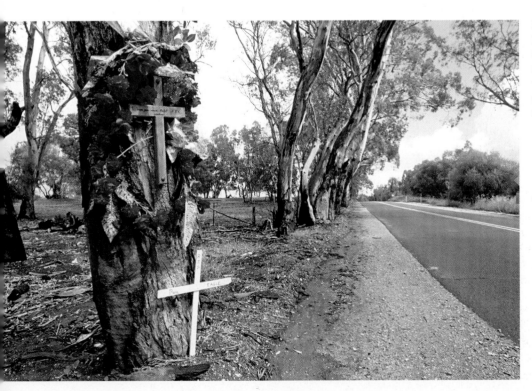

Figure 5.2   Road accident memorial Known in Mexico and Spain as *descansos*, these memorials are artistic expressions that mark the place where the soul has left the body in a fatal accident.

The Culture of Hip-Hop

Hip-hop is a popular manifestation of contemporary cultural practices among U.S. big-city and especially inner-city youth. Characterized to some extent by graffiti art, and earlier on by break dancing, hip-hop is understood globally through rap music and a distinctive idiomatic vocabulary. Like most nations, hip-hop has its forebears. These include boxer Muhammad Ali, Jamaican Rastafarian and reggae musician Bob Marley, Black Panther Huey Newton, and funksters James Brown and George Clinton. Also like most nations, the hip-hop nation has its origins in the Bronx, in New York City.

But it also has much older roots in the West African storytelling culture known as *griot*. Hip-hop in the United States enlarged upon those origins, and now hip-hop is both appreciated and produced on six continents. Hip-hop is a cultural practice that has truly globalized, not because its practitioners have migrated far and wide, but because its culture has migrated via telecommunications and the music and film industries. Hip-hop has become a cultural phenomenon that exists beyond geography in the music, the clothes, and the language of its practitioners.

Although hip-hop is predominantly black and male, hip-hop crosses all color lines and includes women and gays, though the latter two groups have been the targets of entrenched sexism and homophobia by a wide range of rappers. Its pioneers include white graffiti artists and Latinos who influenced break dancing as well as hip-hop DJ (disk jockey) and MC (rapper) styles.

Music is the heart and soul of hip-hop nation and the geography of U.S. hip-hop—where the most important music has come from—can be crudely divided into East Coast, West Coast, South Coast, and a region in and around Detroit where white rap-metal groups have become popular. The East Coast includes the five boroughs of New York, Long Island, Westchester County, New Jersey, and Philadelphia. The West Coast includes Los Angeles, Compton, Long Beach, Vallejo, and Oakland. The South Coast region is made up of Atlanta, New Orleans, Miami, and Memphis (**Figure 5.A**).

But just as hip-hop has broken out of its regional boundaries, it has transcended national boundaries as well. Hip-hop graffiti art can be found in urban areas as distant as Australia and South Africa. Rap music is as popular in the Philippines as it is in Paris. And individual DJs have had significant influences well beyond their old neighborhoods. For example, Afrika Bambaataa, a former gang member, organized the Universal Zulu Nation over 25 years ago. Bambaattaa, or "Bam," incorporated former gang members into a community-building group that has become a household name in hip-hop circles all around the world.

But hip-hop is also very clearly about the more local space of the neighborhood, or "the 'hood"—rap's dominant spatial trope—which is portrayed in all its complexity in songs, music videos, and hip-hop films like *Breakin'* (1984), *Beat Street* (1985), *Wild Style* (1982), and *8 Mile* (2002). While rap and dancing are central to hip-hop, the local context in which the story, music, or dance unfold is also critical. As hip-hop cultural theorist Murray Forman argues, "Virtually all of the early descriptions of hip-hop practices identify territory and the public sphere as significant factors, whether in visible artistic expression and appropriation of public space via

Figure 5.A The sources and diffusion of U.S. rap This map portrays the centers of rap music in the United States today, showing how rap, which began in African American inner-city neighborhoods in New York City in the late 1970s, moved westward and then southward. Most recently, a hybrid form of rap metal has emerged in the U.S. Midwest urban center of Detroit. The Detroit metropolitan area contains large numbers of African Americans and working-class whites who lost their jobs in the restructuring of the automobile industry in the 1970s and early 1980s. What the rap-metal genre confirms is that although hip-hop culture has its roots in the African American experience, it derives much of its power from issues of poverty and class.

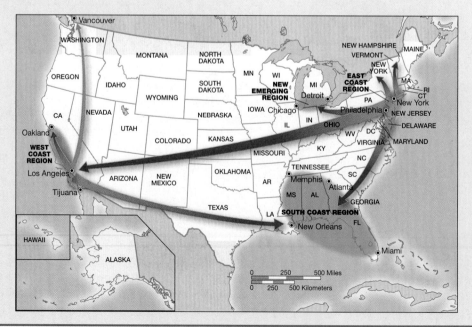

graffiti or b-boying [break dancing], the sonic impact of a pounding bass line, or the discursive articulation of urban geography in rap lyrics [and films]."[1]

Hip-hop is effectively about how space and place shape the identities of rappers in particular but also African Americans more generally, showing how race, space, and place come together to produce the contradiction of "home" as a locus of roots and the foundation of personal history but also as a site of devaluation vis-à-vis the dominant white society.

The most controversial variant of hip-hop is "gangsta rap" pioneered in the late 1980s by artists like Ice T and groups such as N.W.A. Today, gansta rap is one of the most popular and lucrative of all hip-hop genres. The element that has drawn the most condemnation from both the left and the right is gansta lyrics, which are said to glorify promiscuity, violence, misogyny (hatred of women), police brutality, gang culture, rape, drug dealing, and other acts of criminal behavior both major and minor. Gangsta artists respond, however, by arguing that theirs is not glorification, but merely a retelling of the inner-city life of young African Americans. Popular gangta rap artists today include Snoop Doggy Dog, 50 Cent, Dr. Dre, and Ice Cube, all of whom have achieved global success (**Figure 5.B**).

Cultural critic bell hooks turns the condemnation of gangsta rap on its head by pointing out that gangsta rap does not exist in a vacuum, but is an extension of white, male-dominated, capitalist society.[2] She argues that it is far easier to attack gansta rap than the culture that produces and reproduces it (through white, middle-class consumption of the music). Simultaneously, hooks points out that while we are repelled by the misogyny promoted by gangsta rappers, we fail to see it everywhere else—including within mainstream U.S. culture. For instance, who's to say which is more misogynist, *Sex and the City* or Snoop Dog?

Gangsta rap is a huge commercial success, with rappers like 50 Cent involved in a range of projects from promoting Reebok G-Unit sneakers to appearing in an episode of *The Simpsons*. He provided the voice-over as the protagonist in the video game *50 Cent: Bulletproof* and the sequel *50 Cent: Blood on the Sand*, which were released for PlayStation 2, Xbox, and the PlayStation

Figure 5.B Rapper 50 Cent Curtis James Jackson, III, is better known by his stage name 50 Cent. Born in Queens, New York, 50 Cent began drug dealing at the age of 12. He catapulted to stardom with his first multiplatinum CD in 2003, *Get Rich or Die Tryin'*. He grew up without a father and was raised by his mother, Sabrina Jackson, a cocaine dealer, who gave birth to him at the age of 15. Sabrina raised Jackson until he was 8, when she was murdered. Following his mother's death, he was raised by his grandparents, spending much of his youth selling drugs, experiencing several arrests, being shot nine times but surviving, and serving a prison term. His biography starkly reflects the controversial lyrics in his songs as well as those of other gangsta rappers.

Portable. He worked with Glacéau, a beverage company, to create a vitamin water drink called *Formula 50* and teamed up with Right Guard to launch a body spray called *Pure 50 RGX Body Spray* and a condom line called *Magic Stick Condoms*.

Hip-hop, as a youth-oriented cultural product that has become widely commercialized in core countries by multinational corporations, is still in many parts of the world, a fully homegrown phenomenon. Interestingly, rather than homogenizing local cultures, the global spread of hip-hop from the United States outward has tended to create hybrid and synthetic forms of music, adjusting to local cultures and struggles. What is most consistent about the rap music that

[1]M. Forman, "Ain't No Love in the Heart of the City: Hip-Hop, Space, and Place," in M. Forman and M. A. Neal (eds.), *That's the Joint! The Hip Hop Studies Reader.* New York: Routledge, 2004, p. 155.

[2] b. hooks, "Sexism and Misogyny: Who Takes the Rap?" *Z Magazine*, February 1994, pp. 26–29, http://race.eserver.org/misogyny.html, accessed June 29, 2008.

is being produced in the periphery is that it has become a focal point for the underprivileged around the world to unite and challenge the oppression of the status quo.

Social struggle seems to be the single strongest thread that weaves together the disparate sounds, cultures, and artists of global hip-hop. In 2006 on his own Web site, the British, Pakistani-born Muslim rapper Aki Nawaz, released, by anyone's standards, a phenomenally angry album. *All is War* (*The Benefits of G-had*) contains a track that uses the words of Osama Bin Laden and equates him as a hero on par with Che Guevara. Another re-creates a suicide bomber at work. Nawaz claims that the album is a rejection of the hypocrisy and immorality of the West.

In Sudan, Emmanuel Jal, a former child soldier in Sudan's People's Liberation Army, uses rap as a way of educating the world about the struggles that oppressed people face in Africa: political injustice, government terror, war, as well as the hope of freedom from these tyrannies.

Worldwide sales of rap music have been dropping since 2000. It is not clear whether this signals a death knell for rap or whether audiences have simply gotten tired of the U.S.-dominated scene or whether this music is often downloaded from the Web illegally and the sales therefore only reflect part of its popularity. Whatever the reason or reasons, it is doubtful that rap as a musical form and hip-hop as a cultural complex will disappear entirely any time soon.

heart of cultural geography today. **Cultural geography** focuses on the way in which space, place, and landscape shape culture at the same time that culture shapes space, place, and landscape. As such, cultural geography demarcates two important and interrelated parts. Culture is the ongoing process of producing a shared set of meanings and practices, while geography is the dynamic setting that groups operate in to shape those meanings and practices and in the process to form an identity and act. Geography in this definition can be as small as the body and as large as the globe.

A contemporary example of this two-way relationship between geography and culture can be seen in the widespread popularity of online social networks designed to connect friends, family, co-workers and people with similar interests. Two of the most popular online social networking Web sites are MySpace and Facebook, with around 80 and 110 million registered worldwide users, respectively.

These Web sites allow members to share photos, blogs, music and videos, and a host of other personal information, as well as organize events, join groups, or seek relationship partners. For example, a Facebook user creates an individual "profile" that shows his or her personal information and interests on-screen for other users to look at. Users can then communicate with each other by writing private messages or, more popularly, by writing on each others' "walls."

These virtual networks have revolutionized the way we communicate with each other, as friendship groups and circles obscure the boundaries between online and offline domains. Cultural scholars who study them believe that online social networks are not turning us into "virtual creatures" who exist only in the ether of the Internet, but rather into hybrid creatures with a foot in both the virtual and the real world. This hybridity enables us to be in more than one place at any one time and to connect to others whom we may never meet in person, but with whom we may form significant relationships or simply passing ones. And in those interactions we are changed, perhaps only in small ways but those ways can and often do add up to something quite substantial.

The virtual world of social networking sites enables users to create new geographies where new cultural practices prevail. Following the 2004 presidential elections in the United States, for example, they have become important locations for developing new political practices in support of candidates as well as causes. Examples include "Support the Monks," a group organized to protest state violence in Burma in 2008. Another is the "One Million Strong for Barack" group, which developed to support Barack Obama in his bid for presidency of the United States.

Privacy concerns remain a dominant issue with most social networking sites. Their open access means that anyone from stalkers to employers to marketing companies and the police can use the sites to gain access to members' personal information. Not surprisingly, because these networking sites allow users from all over the world to express their opinions and engage with other users, the same cultural fault lines that occur in everyday spaces are also drawn within cyberspace, though the connections can be far more extensive and immediate.

Before we proceed any further, it is important to discuss a significant difference between our view of culture and that of more conventional cultural geographers. Many introductory human geography texts divide culture into two major categories: folk and popular culture. **Folk** culture is seen by specialists as the traditional practices of small groups, especially rural people with a simple lifestyle (compared with modern, urban people), such as the Amish in Pennsylvania or the Roma (also known as Gypsies or Travellers) in Europe, who are seen as homogeneous in their belief systems and practices. **Popular culture,** by contrast, is viewed by some cultural geographers as the practices and meaning systems produced by large groups of people whose norms and tastes are often heterogeneous and change frequently, often in response to commercial products. Hip-hop would be seen by these theorists as an example of popular culture, as would the Cosplay young people in the chapter's opening photo.

In this text, we do not divide culture into categories. We see culture as an overarching process that is shaped by and shapes politics, the economy, and society and cannot be neatly demarcated by reference to the number of characteristics or degree of homogeneity of its practitioners. We see culture as something that can be enduring as well as newly created, but always influenced by a whole range of interactions as groups maintain, change, or even create traditions from the material of their everyday lives. For us, there is no purpose served in categorically differentiating between hip-hop and Hinduism, as both are significant expressions of culture and both are of interest to geographers.

BUILDING CULTURAL COMPLEXES

Geographers focus on the interactions between people and culture and among space, place, and landscape. One of the most influential geographers was Carl Sauer, who taught at the University of California, Berkeley, in the early to mid-twentieth century. Sauer was largely responsible for creating the "Berkeley school" of cultural geography. He was particularly interested in trying to understand the material expressions of culture by focusing on their manifestations in the landscape (**Figure 5.3**). This interest came to be embodied in the concept of the **cultural landscape,** a characteristic and tangible outcome of the complex interactions between a human group—with its own practices, preferences, values, and aspirations—and its natural environment. Sauer differentiated the cultural landscape from the natural landscape. He emphasized that the former was a "humanized" version of the latter, such that the activities of humans resulted in an identifiable and understandable alteration of the natural environment. **Figure 5.4** lists the differences between a natural and a cultural landscape.

For roughly five decades, interest in culture within the field of U.S. geography largely followed Sauer's important work. His approach to the cultural landscape was ecological, and his many published works reflect his attempts to understand the myriad ways that humans transform the surface of Earth. In his own words:

> The cultural landscape is fashioned from a natural landscape by a cultural group. Culture is the agent, the natural area is the medium, the cultural landscape is the result. Under the influence of a given culture, itself changing through time, the landscape undergoes development, passing through phases, and probably reaching ultimately the end of its cycle of development. With the introduction of a different—that is an alien—culture, a rejuvenation of the cultural landscape sets in, or a new landscape is superimposed on remnants of an older one.[1]

[1]C. Sauer, "The Morphology of Landscape," in J. Leighly (ed.), *Land and Life: Selections from the Writings of Carl Ortwin Sauer.* Berkeley, CA: University of California Press, 1964, pp. 315–350.

Figure 5.3   Masai village, Kenya The cultural landscape, as defined by Carl Sauer, reflected the way that cultural and environmental processes came together to create a unique product. Pictured here is a small village where herding is the major occupation. The village is enclosed by thorny brambles and branches harvested from the surrounding area. Within the enclosure, the dwellings are arranged in a unique circular pattern, with the animal pens in the middle of the settlement for easy observation by the residents.

Figure 5.4 Sauer's cultural landscape This figure summarizes the ways the natural and cultural landscapes are transformed. Physical and climatic factors shape the natural landscape. Cultural practices also have an important impact upon it. The results of cultural factors are cultural forms, such as population distributions and patterns and housing. Over time, people—through culture—reshape the natural landscape to meet their needs. (Adapted from C. Sauer, "The Morphology of Landscape." in J. Leighly (ed.), *Land and Life: Selections from the Writings of Carl Ortwin Sauer.* Berkeley and Los Angeles: University of California Press, 1963, pp. 315-350.)

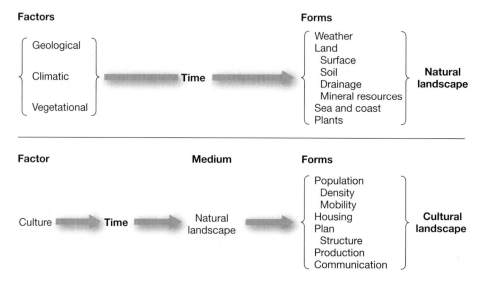

Factors

Geological
Climatic
Vegetational

→ **Time** →

Forms

Weather
Land
  Surface
  Soil
  Drainage
  Mineral resources
Sea and coast
Plants

Natural landscape

Factor

Culture → **Time** →

Medium

Natural landscape →

Forms

Population
Density
Mobility
Housing
Plan
  Structure
Production
Communication

Cultural landscape

In Europe, geographers interested in human interactions with the landscape took slightly different approaches. For example, in Great Britain the approach to understanding the human imprint on the landscape was termed *historical geography*, while in France it was conceptualized as *genre de vie*. **Historical geography**, very simply defined, is the geography of the past. Its most famous practitioner was H. C. Darby, who attempted to understand "cross sections" or sequences of evolution, especially of rural landscapes. **Genre de vie**, a key concept in Vidal de la Blache's approach to cultural geography in France, referred to a functionally organized way of life that was seen to be characteristic of a particular culture group. *Genre de vie* centered on the livelihood practices of a group that were seen to shape physical, social, and psychological bonds (**Figure 5.5**). Although emphasizing some landscape components over others or giving a larger or smaller role to the physical environment, all of these approaches placed the cultural landscape at the heart of their study of human-environment interactions.

H. C. Darby most successfully implemented his historical approach to cultural geography and landscape by developing a geography based on the *Domesday Book*. William the Conqueror ordered the *Domesday* compiled in 1085 so he could have a list of his spoils of war. The book provides a rich catalog of the ownership of nearly every tract of land in England, including the names of landholders in each county, the manors they held and their values, the names of their subtenants, the names of many boroughs and details

Figure 5.5 Market gardens in Corsica This image shows a rural setting in Corsica, an island nation in the Mediterranean, where commercial agriculture is being undertaken. Farming is a way of life—a *genre de vie*—that we can read from the landscape where extensive cultivated fields and isolated farmhouses constitute key elements.

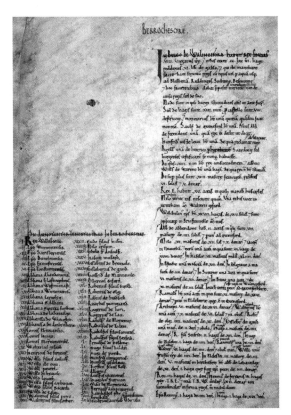

Figure 5.6   Domesday Book Pictured here is a page from *Great Domesday Book* that shows the accounts for Berkshire County in England. The listing begins with the lands belonging to the King followed by the lands of the bishops of Winchester, Salisbury, and Durham, then the lands of the Abbot of Abingdon, and lastly those of the great lay barons. *Domesday Book* reveals an elaborate feudal structure of landholding from the King down. Under this system land was supposedly held directly or indirectly by the King, who stood at the top of the feudal ladder. He granted land called fiefs to his chief barons, bishops, and abbots.

of their customs, the number of freemen, sokemen, unfree peasants and slaves on each manor, and the resources of each manor (**Figure 5.6**). For geographers like Darby, such data were invaluable for reconstructing the political, economic, and social forces that shaped past landscapes.

Vidal de la Blache emphasized the need to study small, homogeneous areas in order to uncover the close relationships that exist between people and their immediate surroundings. He constructed complex descriptions of preindustrial France that demonstrated how the various *genres de vie* emerged from the possibilities and constraints posed by local physical environments. Subsequently, he wrote about the changes in French regions brought on by industrialization, observing that regional homogeneity was no longer the unifying element. Instead, the increased mobility of people and goods produced new, more complex geographies where previously isolated *genres de vie* were integrated into a competitive industrial economic framework. Anticipating the widespread impacts of globalization, de la Blache also recognized how people in various places struggled to cope with the big changes transforming their lives.

Geographers also examine specific aspects of culture, ranging from single attributes to complex systems. One simple aspect of culture of interest to geographers is the idea of special traits, which include such things as distinctive styles of dress, dietary habits, and styles of architecture (**Figure 5.7**). A **cultural trait** is a single aspect of the complex of routine practices that constitute a particular cultural group. For example, Canon Law for Catholics proscribes the consumption of meat on Fridays. This avoidance may be said to be a cultural trait of Catholic people. Fasting is another aspect of religious law that is practiced not just by Catholics on certain holy days but also by Jews and Muslims. Fasting, too, is considered a cultural trait. Geographers are also interested in learning how cultural traits come together to form larger frameworks for living in the world. Ultimately, cultural traits are not necessarily unique to one group, and understanding them is only

Figure 5.7   Iroquois longhouse A longhouse is a long, narrow, single-room building built by peoples in various parts of the world, including Asia, Europe, and North America. Pictured here is an Iroquois longhouse. On average, a typical Iroquois longhouse was about 60 feet long by 18 feet wide by 18 feet high with doors at each end and no windows. The longhouses were meant to house up to 20 or more families at once.

Figure 5.8 A coming-of-age ceremony, South Korea The coming-of-age ceremony for girls in South Korea is held every May 15 in Seoul. There, young women who are turning 20 participate in dances and a citywide celebration that is meant to remind them of the responsibilities they face as adults. In South Korea, people who turn 60 are also ritually celebrated. This ritual is not so much a coming-of-age celebration as it is a coming-*to*-an age celebration. Called *hwan-gap*, this celebration is considered one of the larger celebrations in a person's life. It is significant because it marks the day on which an individual has completed a full zodiacal cycle. But, more importantly, it is also celebrated, because to live to be 60 is seen as a great accomplishment, because in the past most people in South Korea died before their 60th birthday.

one aspect of the complexity of culture. For instance, there are a number of cultural groups, including Hindus, Muslims, and Jews, who avoid pork in their diet.

Many cultures also recognize the passage from childhood into adulthood with a celebration or ceremony. Called **rites of passage,** these are acts, customs, practices, or procedures that recognize key transitions in human life—birth, menstruation, and other markers of adulthood such as sexual awakening and marriage. Such rites of passage are not uncommon among many of the world's cultures. Some non-Western cultures, for example, send adolescent boys away from the village to experience an ordeal—ritual scarring or circumcision, for example—or to meditate in extended isolation on the new roles they must assume as adults. After an extended absence from the social group, these youths return transformed and ready to lay down their previous childish occupations (**Figure 5.8**).

In Roman Catholicism, the passage of boys and girls into adulthood, traditionally around the age of 12, is celebrated by confirmation. In this religious ceremony, the confirmed chooses a new name to mark this important spiritual transition. Jews mark the passage of adolescent boys and girls into adulthood with separate religious ceremonies: a *bar mitzvah* for boys and a *bat mitzvah* for girls. Although marking the passage into adulthood is a trait of both religious groups, they do not exhibit the trait in exactly the same way. In fact, this and other traits always occur in combination with others. The combination of traits characteristic of a particular group is known as a **cultural complex.** The avoidance of pork, the celebration of *bar* and *bat mitzvahs,* and other dietary, religious, and social practices constitute the cultural complex of Judaism, although it is important to note that even within the cultural complex of Judaism, variation exists among regions and sects.

Another concept key to traditional approaches in cultural geography is the cultural region. Although a cultural region may be quite extensive or very narrowly described and even discontinuous in its extension, it is the

area within which a particular cultural system prevails. A **cultural region** is an area where certain cultural practices, beliefs, or values are more or less practiced by the majority of the inhabitants. For example, the state of Utah is considered to be a Mormon cultural region because the population of the state is dominated by people who practice the Mormon religion and presumably adhere to its beliefs and values.

Figure 5.9 illustrates the overall religious geography of the United States. The map shows that the southeastern part of the United States has a large Baptist population, whereas the West and Southwest, particularly California, Nevada, Arizona, and New Mexico, are dominated by Roman Catholics. As with the large concentration of Roman Catholics in the northeastern and Middle Atlantic states, these concentrations reflect the immigration of people from Catholic countries. In the case of the western/southwestern United States, these are largely populations of fairly recent immigrants from Mexico and other parts of Central America as well as those who occupied the region well before the Anglo population arrived. With respect to the Eastern seaboard, the dominance of Roman Catholicism reflects older immigration from Catholic Europe, as well as more recent immigration from the Caribbean Basin. In the upper central part of the United States, the scattered concentrations of Lutherans reflect previous immigration from the Scandinavian countries.

CULTURAL SYSTEMS

Broader than the cultural complex concept is the cultural system, a collection of interacting components that, taken together, shape a group's collective identity. A **cultural system** includes traits, territorial affiliation, and shared history, as well as other, more complex elements, such as language. In a cultural system it is possible for internal variations to exist in particular elements at the same time that broader similarities lend coherence. For example,

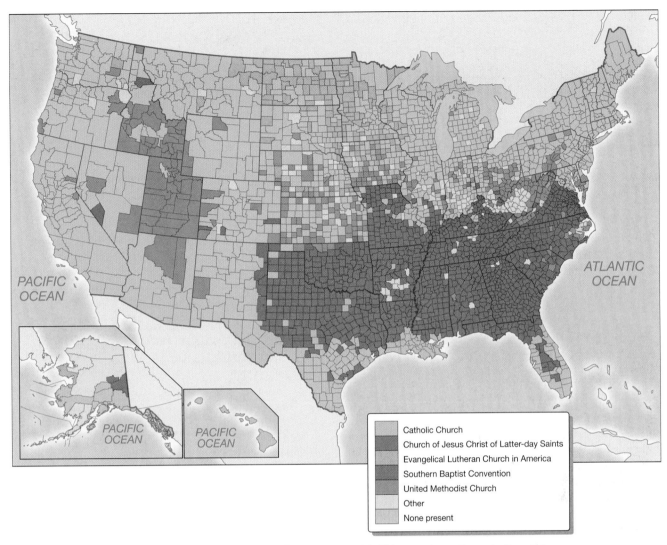

	Catholic Church
	Church of Jesus Christ of Latter-day Saints
	Evangelical Lutheran Church in America
	Southern Baptist Convention
	United Methodist Church
	Other
	None present

Figure 5.9   U.S. religious population distribution by county, 2000 This map shows the majority religion by county for the United States and illustrates the concept of cultural regions based on religion. It is important to remember, however, that at such a scale, it would be erroneous to assume too much homogeneity within these regions. In each aggregation of counties or region, there are likely to be substantial variations in belief systems and practices at the local level. (*Source:* Adapted from D. E. Jones, S. Doty, C. Grammich, J. E. Horsch, R. Houseal, M. Lynn, J. P. Marcum, K. M. Sanchagrin, and R. H. Taylor, *Religious Congregations and Membership in the United States 2000: An Enumeration by Region, State and County Based on Data Reported for 149 Religious Bodies.* Nashville, TN: Glenmary Research Center, 2000, p. 562.)

Christianity unites all Protestant religions (as well as Catholic ones), yet the practices of particular denominations—Lutherans, Episcopalians, and Quakers—vary. And while Mexicans, Bolivians, Cubans, and Chileans exhibit variations in pronunciation, pitch, stress, and other aspects of vocal expression, they all speak Spanish. This means they share a key element of a cultural system (which, for these nationalities, also includes Roman Catholicism and a Spanish colonial heritage).

Geography and Religion

Two key components of a cultural system for most of the world's people are religion and language. **Religion** is a belief system and a set of practices that recognizes the existence of a power higher than humankind. Although religious affiliation is on the decline in some parts of the world's core

regions, it still acts as a powerful shaper of daily life, from eating habits and dress codes to coming-of-age rituals and death ceremonies, to holiday celebrations and family practices, in both the core and the periphery. And, like language, religious beliefs and practices change as new interpretations are advanced or new spiritual influences are adopted.

The most important influence on religious change has been conversion from one set of beliefs to another. From the Arab invasions following Muhammad's death in 632, to the Christian Crusades of the Middle Ages, to the onset of globalization in the fifteenth century, religious missionizing—propagandizing and persuasion—as well as forceful and sometimes violent conversion have been key elements in changing geographies of religion. Especially in the 500 years since the onset of the Columbian Exchange, conversion of all sorts has escalated throughout the globe. In fact, since 1492, traditional religions have become dramatically

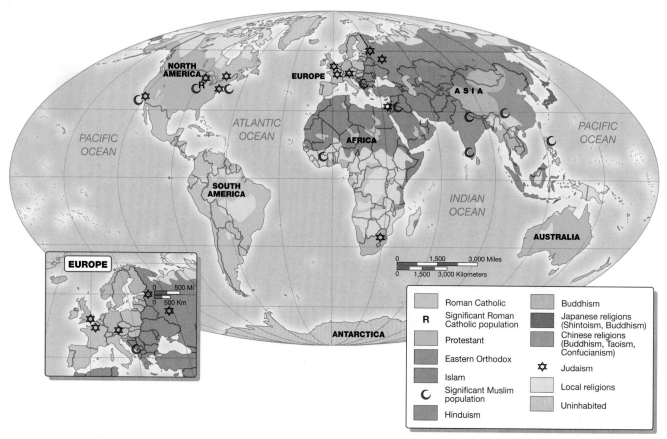

Figure 5.10   World distribution of major religions Most of the world's peoples are members of these religions. Not evident on this map are the local variations in practices, as well as the many smaller religions that are practiced worldwide.

dislocated from their sites of origin not only through missionizing and conversion but also by way of diaspora and emigration. Whereas missionizing and conversion are deliberate efforts to change the religious views of a person or peoples, diaspora and emigration involve the involuntary and voluntary movement of peoples who bring their religious beliefs and practices to their new locales.

Diaspora is a spatial dispersion of a previously homogeneous group. The processes of global political and economic change that led to the massive movement of the world's populations over the last five centuries have also meant the dislodging and spread of the world's many religions from their traditional sites of practice. Religious practices have become so spatially mixed that it is a challenge to present a map of the contemporary global distribution of religion that reveals more than it obscures. This is because the globe is too gross a level of resolution to portray the wide variation that exists among and within religious practices. **Figure 5.10** identifies the contemporary distribution of what religious scholars consider to be the world's major religions because they contain the largest number of practitioners. As with other global representations, the map is useful in that it helps to present a generalized picture.

Figure 5.11 identifies the source areas of four of the world's major religions and their diffusion from those sites over time. The map illustrates that the world's major religions originated and diffused from two fairly small areas of the globe. The first, where Hinduism and Buddhism (as well as Sikhism) originated, is an area of lowlands in the subcontinent of India drained by the Indus and Ganges rivers (Punjab on the map). The second area, where Christianity and Islam (as well as Judaism) originated, is in the deserts of the Middle East.

Hinduism was the first religion to emerge, among the peoples of the Indo-Gangetic Plain, about 4,000 years ago. Buddhism and Sikhism evolved from Hinduism as reform religions, with Buddhism appearing around 500 B.C. and Sikhism developing in the fifteenth century. It is not surprising that Hinduism helped to produce new religions, because India has long been an important cultural crossroads. As a result, ideas and practices originating in India spread rapidly, at the same time that other ideas and practices were being brought to India from far-flung places and then absorbed and translated to reflect Indian needs and values.

For example, Buddhism emerged as a branch of Hinduism in an area not far from the Punjab. At first a very

Figure 5.11 Origin areas and diffusion of four major religions
The world's major religions originated in a fairly small region of the world. Judaism and Christianity began in present-day Israel and Jordan. Islam emerged from western Arabia. Buddhism originated in India, and Hinduism in the Indus region of Pakistan. The source areas of the world's major religions are also the cultural hearth areas of agriculture, urbanization, and other key aspects of human development.

small group of practitioners surrounding Prince Siddhartha Gautama, the founder of the religion, were confined to northern India. Slowly and steadily, however, Buddhism dispersed to other parts of India and was carried by missionaries and traders to China (100 B.C.–A.D. 200), Korea and Japan (A.D. 300–500), Southeast Asia (A.D. 400–600), Tibet (A.D. 700), and Mongolia (A.D. 1500) (see **Figure 5.12**). And as it spread, Buddhism developed many regional forms, so that Tibetan Buddhism is distinct from Japanese Buddhism.

Religious beliefs are organized and codified, often based on the teachings and writings of one or more of its founders. And it is important to recognize that the world's religions contain all sorts of variation within them. For example, Christians may be Catholics or Protestants, but even within these large groups there exists a great deal of variation. The same is true for all religions.

Hinduism, the oldest religion, is codified in the Veda and the Upanishads, its sacred scriptures, but *Hinduism is a term that broadly describes a wide array of sects.* Hindus strive, through the practice of yoga, adherence to Vedic scriptures, and devotion to a personal guru, for release from repeated reincarnation. Hinduism's most central deities are the divine trinity, representing the cyclical nature of the universe: Brahma the creator, Vishnu the preserver, and Shiva the destroyer. Splitting off from Hinduism, Buddhism was founded by Gautama—

known as the Buddha (Enlightened One)—in southern Nepal. The central tenets of Buddhism are that meditation and the practice of good religious and moral behavior can lead to nirvana, the state of enlightenment. Before reaching nirvana, however, one must experience repeated lifetimes that are good or bad depending on one's actions (karma). There are four noble truths in Buddhism: (1) existence is a realm of suffering; (2) desire, along with the belief in the importance of one's self, causes suffering; (3) achievement of nirvana ends suffering; and (4) nirvana is attained only by meditation and by following the path of righteousness. Like Hinduism and Buddhism, Sikhism advocates the pursuit of salvation through disciplined, personal meditation in the name and message of God, who is regarded not as a person, but as the Universe itself.

Christianity, Islam, and Judaism all developed among the Semitic-speaking people of the deserts of the Middle East. And like the Indo-Gangetic religions, these three religions are related. Although Judaism is the oldest of the three, it is the least widespread. Judaism originated about 4,000 years ago, Christianity about 2,000 years ago, and Islam about 1,300 years ago. Judaism developed out of the cultures and beliefs of Bronze Age peoples and was the first monotheistic (belief in one God) religion. Although Judaism is the oldest monotheistic religion, and one that spread widely and rapidly, it

Figure 5.12　Spread of Buddhism This map illustrates the diffusion of Buddhism from its source area in India to China and then from China on to Korea and Japan. Commercial routes, like the Silk Road, were important vectors for the spread of the religion from India to China. Missionaries were responsible for the spread of Buddhism from China to Korea and Japan. (Adapted from C. Schirokauer, *A Brief History of Chinese and Japanese Civilizations* 2nd ed. Florence, KY: Wadsworth 2005.)

is numerically small because it does not seek converts. Christianity developed in Jerusalem among the disciples of Jesus; they proclaimed that he was the Messiah expected by the Jews. As it moved east and south from its hearth area, Christianity's diffusion was helped by missionizing and by imperial sponsorship. The diffusion of Christianity in Europe is illustrated in **Figure 5.13**. Although we discuss Islam and Islamism later in this chapter, it is important to point out here that Islam, like Christianity, was for centuries routinely spread by the force of conversion to save souls but also often for the purposes of political control.

Christianity is a religion based on the life and teachings of Jesus of Nazareth as written in the New Testament of the Holy Bible and is the world's largest religion, with 1 to 2 billion adherents. Christians believe that Jesus is the son of God and the Messiah, prophesied in the Old Testament, the holy scripture that is also significant to Judaism. As a central figure in all Christian religions, Jesus is revered as a teacher, the model of a pious life, the manifestation of God, and most importantly the saviour of humanity who suffered, died, and was resurrected in order

to bring about salvation from sin. Christians maintain that upon his death, Jesus ascended into heaven and that he will return to judge the living and the dead and grant life everlasting to those who have followed his teachings. Christians and Jews, along with Muslims, as well as Bahá'í and Druze followers, share the same source as Abrahamic religions, which means that they all claim a strong connection to the prophet Abraham and the source area of the Middle East.

Although sharing the same origin, Judaism is also based on the teachings of the prophets Isaac and Jacob as well as Abraham. Jews espouse belief in a God who leads his people by speaking through these prophets. They believe that God's word is revealed in the Hebrew Bible (or Old Testament), especially in that part known as the Torah, which contains a total of 613 biblical commandments, including the Ten Commandments, explicated in another part of the Hebrew Bible known as the Talmud. Jews believe that the human condition can be improved, that the letter and the spirit of the Torah must be followed, and that a Messiah will eventually bring the world to a state of paradise.

Figure 5.13 Spread of Christianity in Europe Christianity diffused through Europe largely because of missionary efforts. Monks and monasteries were especially important as hubs of diffusion in the larger network. Monasteries and other sorts of religious communities are indicated by dots on the map. The shaded areas indicate places where Christian converts dominated by the year 300. (Adapted from C. Park, *Sacred World.* New York: Routledge, 1994.)

While these, along with Islam, are the world's major religions, there are many others that figure prominently in the cultural lives of people around the world. Among them are Confucianism, Taoism, and Shintoism, Mormonism, Zorastrianism and Jansenism, as well as Voodoo, Rastafarianism, and animism, among many others. The point is that faith, a trusting belief in a transcendent reality or a supreme being, is a profoundly powerful force in guiding people's actions and attitudes and shaping the worlds in which they live. Faith is also a signficant element of globalization as disparate belief systems come into contact, in some cases causing tension and even violent conflict, as new religions are introduced among populations who are adherents of a diferent one.

An excellent illustration of the global forces behind the changing geography of religious faith is through Columbian contact with the New World. Before Columbus and later Europeans reached the continents of North and South America, the people living there practiced, for the most part, various forms of animism and related rituals. They viewed themselves holistically, as one part of the wider world of animate and inanimate nature. They used religious rituals and charms to guide and enhance the activities of everyday life, as well as the more extreme situations of warfare. Shamanism, in which spiritually gifted individuals are believed to possess the power to control preternatural forces, is one important aspect of the belief system that existed among Native

American populations at the time of European contact (**Figure 5.14**).

European contact with the New World was, from the beginning, accompanied by Christian missionizing efforts directed at changing the belief systems of the aboriginal peoples and converting them to what the missionizers believed to be "the one, true religion" (**Figure 5.15**). Religion, especially for the Spanish colonizing agents, was especially important in integrating Native Americans into the feudal system.

Perhaps what is most interesting about the present state of the geography of religion is how during the colonial period religious missionizing and conversion flowed from the core to the periphery. In the postcolonial period, however, an opposite trend has occurred (see Box 5.2, "Changing Religious Practices in Latin America and the Caribbean"). For example, the fastest-growing religion in the United States today is Islam, and it is in the core countries where Buddhism is making the greatest numbers of converts. The recently deceased Pope John Paul was one of the most widely traveled leaders in Roman Catholic history, but the same can be said for the Dalai Lama, who as the spiritual leader of Tibetan Buddhists and the head of the Tibetan government-in-exile, is a tireless world traveler for Tibetan Buddhism. Pope John Paul's efforts were mostly directed at maintaining Roman Catholic followers and attempting to dissuade their conversion to other religions, such as evangelicalism in the United States and Latin America. The Dalai Lama is promoting conversion

Figure 5.14   Pre-Columbian religions in North America   Before European contact, Native Americans had developed a range of religious practices. Religious traditions based on agrarian practices diffused from south to north, while those based on hunting diffused from north to south. (*Source:* Adapted from B. E. Carroll, *The Routledge Historical Atlas of Religion in America.* New York: Routledge, 2000, pp. 15–16.)

to Buddhism by carrying its message to new places, especially in the core, and advocating sovereignty for Tibet, a region once independent but currently controlled by China (**Figure 5.16**). It is important to be aware, however, that religious missionizing in the periphery by core practitioners has not ceased, as Box 5.2 makes clear.

One other impact of globalization upon religious change occurs by conversion through the electronic media. The rise of television evangelism, or *televangelism*—especially in the United States—has led to the conversion of large numbers of people to Christian fundamentalism, which is a term popularly used to describe strict adherence

Figure 5.15 San Xavier del Bac mission near Tucson, Arizona The influence of missionaries on native populations was profound, as indigenous rituals became mixed with Christian ones. Even today in missions such as San Xavier, native peoples practice Christianity along with components of their native belief system.

to Christian doctrines based on a literal interpretation of the Holy Bible. The term *fundamentalism* derives from a late nineteenth- and early twentieth-century trans-denominational Protestant movement that opposed the accommodation of Christian doctrine to modern scientific theory and philosophy, especially Darwinian theories about the origin of the universe. It has its origin in a series of pamphlets published between 1910 and 1915 entitled "The Fundamentals: A Testimony to the Truth." Televangelism occurred in the wake of the decline of traditional fundamentalist revivalism—the practice whereby itinerant preachers exhorted their hearers to accept forgiveness of personal sin through faith in Jesus Christ and to commit themselves to spiritual self-discipline and religious exercises such as prayer, Bible reading, and church support. Televangelism began to grow in the 1950s, through the

mass consumption of television sets in the United States. By the early 1990s, however, televangelism was in decline as scandal and corruption among Christian fundamentalist television ministers like James Baaker undermined the confidence of the faithful, leading many to return to more conventional worship. While televangelism persists but is less popular, church-based Christian fundamentalism is strong and growing stronger with a trend toward mega churches as popular sites of worship (**Figure 5.17**).

Geography and Language

Languages are another aspect of cultural systems that interest geographers. Language is an important focus for study because it is a central aspect of cultural identity. Without language, cultural accomplishments could

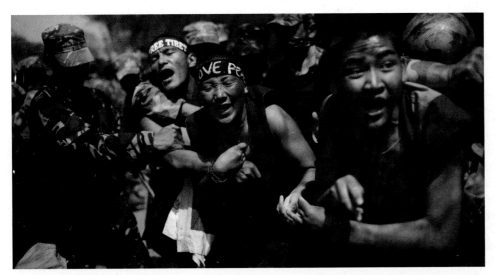

Figure 5.16 Tibetan monks protest in Nepal China's refusal to acknowledge an independent Tibet has sent the Dalai Lama on numerous international tours to broadcast the plight of Tibetans, who are experiencing extreme persecution. Pictured here are Tibetan monks in exile in Nepal who are protesting ouside the Chinese embassy there marking the anniversary of the failed 1959 uprising against China and demanding that China grant sovereignty to Tibet.

Figure 5.17 Megachurch, South Barrington, Illinois Willow Creek Community church is typical of the megachurch phenomenon that has come to characterize fundamental Christian sects in many places across the United States. A Sunday service at Willow Creek often draws 20,000 or more. According to a recent survey, there are 1,200 Protestant churches that claim more than 2,000 weekly worship attendees in the United States. There are megachurches in 45 out of 50 states; Texas has 174 (14 percent), California has 169 (13.7 percent), Florida 83 (6.7 percent), and Georgia 64 (5.2 percent). Houston and Dallas alone account for 56 megachurches, or 4.5 percent of the total.

not be transmitted from one generation to the next. And language itself reflects the ways that different groups regard and interpret the world around them. The distribution and diffusion of languages tell much about the changing history of human geography and the impact of globalization on culture (see Box 5.3, "Language and Ethnicity in Africa"). Before looking more closely at the geography of language and the impacts of globalization upon the changing distribution of languages, however, it is necessary to become familiar with some basic vocabulary.

Language is a way of communicating ideas or feelings by means of a conventionalized system of signs, gestures, marks, or articulate vocal sounds. In short, communication is symbolic, based on commonly understood meanings of signs or sounds. Within standard languages (also known as official languages because they are maintained by offices of government such as education and the courts), regional variations, known as **dialects,** exist. Dialects emerge and are distinguishable through differences in pronunciation, grammar, and vocabulary that are place-based in nature.

For the purposes of classification, languages are grouped into families, branches, and groups. A **language family** is a collection of individual languages believed to be related in their prehistorical origin. About 50 percent of the world's people speak a language that is in the Indo-European family. A **language branch** is a collection of languages that possesses a definite common origin but has split into individual languages. A **language group** is a collection of several individual languages that is part of a language branch, shares a common origin in the recent past, and has relatively similar grammar and vocabulary. Spanish, French, Portuguese, Italian, Romanian, and Catalan are a language *group,* classified

under the Romance *branch* as part of the Indo-European language *family.*

Traditional approaches in cultural geography have identified the source areas of the world's languages and the paths of diffusion of those languages from their places of origin. Carl Sauer identified the origins of certain cultural practices with the label "cultural hearth." **Cultural hearths** are the geographic origins or sources of innovations, ideas, or ideologies. Language hearths are a subset of cultural hearths; they are the source areas of languages. **Figures 5.18** and **5.19** show the locations of the world's indigenous languages and the hearth area of the Indo-European language. Compare them to see the spread of the latter throughout Europe and into North and South Asia.

In India, for example, in the most recent population census, people reported the use of over 1,600 languages, which the government collapsed into 200 related ones. These translate into a broad regional grouping of four major language families. The Indo-European family of languages, introduced by the Aryan herdsmen who migrated from Central Asia between 1500 and 500 B.C., is prevalent in the northern plains region, Sri Lanka, and the Maldives. This language family includes Hindi, Bengali, Punjabi, Bihari, and Urdu. Munda languages are spoken among the tribal hill peoples who still inhabit the more remote hill regions of peninsular India. Dravidian languages (which include Tamil, Telegu, Kanarese, and Malayalam) are spoken in southern India and the northern part of Sri Lanka. Finally, Tibeto-Burmese languages are scattered across the Himalayan region (**Figure 5.20**).

In India, the boundaries of many of the country's constituent states were established after partition on the basis of language. Overall, no single language is spoken

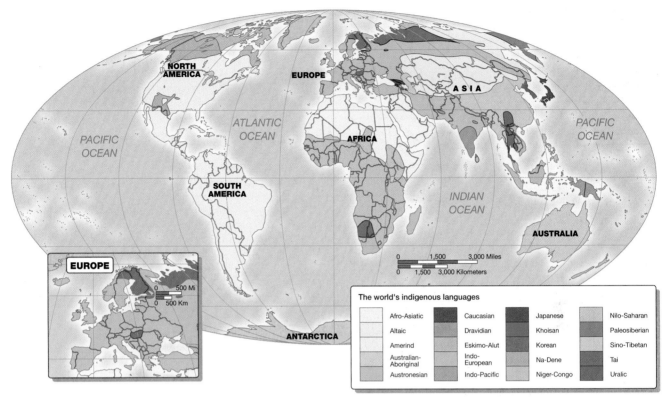

Figure 5.18   World distribution of major languages and major language families Classifying languages by family and mapping their occurrence across the globe provide insights about human geography. For example, we may discover interesting cultural linkages between seemingly disparate cultures widely separated in space and time. We may also begin to understand something about the nature of population movements across broad expanses of time and space. (Adapted from E. F. Bergman, *Human Geography: Cultures, Connections, and Landscapes,* Upper Saddle River, NJ: Prentice Hall, 1994; Western Hemisphere after J. H. Greenberg, *Language in the Americas.* Stanford, CA: Stanford University Press, 1987; Eastern Hemisphere after D. Crystal, *The Cambridge Encyclopedia of Language.* Cambridge: Cambridge University Press, 1997.)

or understood by more than 40 percent of the people. There have been efforts since India became independent to establish Hindi, the most prevalent language, as the national language, but this has been resisted by many of the states within India, whose political identity is now closely aligned with a different language. In terms of popular media and literature, there is a thriving Hindi and regional language press, while film and television are dominated by Hindi and Tamil, with some Telegu programming.

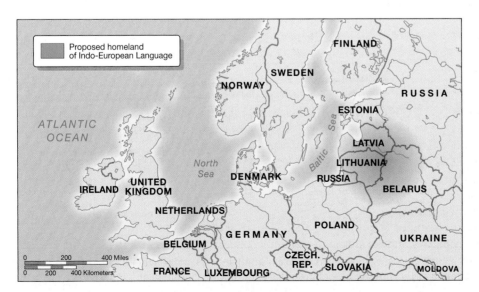

Figure 5.19   Source area of the Indo-European language family Although scholars are not certain where the Indo-European language arose, new evidence suggests that it blossomed in northeast central Europe, north of the Black Sea, in the fifth millennium B.C. As the language diffused west, east, and south, it mutated into the multiplicity of languages that today make up the Indo-European language family. (Adapted from C. Renfrew, *Archaeology and Language: The Puzzle of Indo-European Origins.* London: Jonathan Cape, 1987.)

Figure 5.20　Language map of India India's linguistic landscape is complex, with hundreds of distinct languages in use. This map provides an illustration of the intricate geography of language on the subcontinent.

Figure 5.21 African countries with extinct and threatened languages It is not absolutely certain how many languages are currently being spoken worldwide, but the estimates range between 4,200 and 5,600. While some languages are being created through the fusion of an indigenous language with a colonial language, such as English or Portuguese, indigenous languages are mostly dying out. Although only Africa is shown in this map, indigenous languages are dying out throughout the Americas and Asia as well.

English, spoken by fewer than 6 percent of the people, serves as the link language between India's states and regions. As in other former British colonies in South Asia, English is the language of higher education, the professions, and national business and government. Without English, there is little opportunity for economic or social mobility. Most children who do attend school are taught only their local language and so are inevitably restricted in their prospects. Guards, sweepers, cooks, or drivers who speak only Hindi or Urdu will likely do the same work all their lives. In contrast, those who can speak English—by definition the upper middle classes—are able to practice their profession or do business in any region of their country or in most parts of the world.

With globalization, the geography of language has become even more dynamic. As mentioned, the plethora of languages and dialects in a region like South Asia makes communication and commerce among different language speakers difficult. It may, furthermore, create problems for governing a population. That is why developing states created one national language to facilitate communication and enable the efficient conduct of state business. In general, where official languages are put into place, indigenous languages may eventually be lost (**Figure 5.21**). Yet the actual unfolding of globalizing forces—such as official languages—works differently in different places and in different times (**Figure 5.22**). The overall trend appears to be toward the loss of indigenous language (and other

Figure 5.22 The languages and dialects of France in 1789 On the eve of the French Revolution, language diversity in France was not so dissimilar from that of other European regions that were consolidating into states. A multiplicity of local languages and dialects prevailed before the emergence of a strong central state, and many governments created policies to eliminate them. Local languages made it difficult for states to collect taxes, enforce laws, and teach new citizens. (Adapted from D. Bell, "Lingua Populi, Lingua Dei," *American Historical Review.* 1995, p. 1406.)

5.2 GEOGRAPHY MATTERS

Changing Religious Practices in Latin America and the Caribbean

One of the main objectives of Spanish and Portuguese colonialism in Latin America and the Caribbean was the conversion of indigenous peoples to Catholicism. Some indigenous people fiercely resisted missionary efforts; others found ways to blend their own traditions with those of the Roman Catholic Church. The process of conversion was facilitated by the reported appearance of the Virgin Mary of Guadalupe to an Indian convert in Mexico on December 9, 1531, leaving behind her brown-skinned image on his mantle, and by the efforts of some priests to protect local communities from the Spanish government's efforts to obtain land, tribute, and labor by force.

The slave trade brought African religious traditions to Latin America and the Caribbean, and these eventually merged with indigenous and Catholic beliefs to form Candomble and Umbanda in Brazil, Voodoo in Haiti, and Santería in Cuba and other islands (**Figure 5.C**). Candomble and Umbanda are both sects of Santería, with rituals that involve dances, offerings of candles and flowers, sacrifice of animals such as chickens, and mediums and priests who use trances to communicate with spirits that include several Catholic saints. Voodoo (also spelled Voudou, Voudun, and Voudoun) rituals include drumming, prayer, and animal sacrifice to important spirits based on traditional African gods and Catholic saints and are led by priests who act as healers and protectors against witchcraft. Santería is closely connected to the Yoruba religion of West Africa and blends saints with African spirits associated with nature, using rituals similar to other Latin American religions.

In the 1970s, liberation theology emerged, a new form of Catholic practice that focused on the poor and

Figure 5.C   Santería altar This altar such as the gold cross illustrates how Santería combines animistic elements with Catholic religious symbols.

forms of culture), as **Figure 5.23** shows. Language is also an important means of challenging the political, economic, cultural, and social forces of globalization as they occur in places like Belgium (among the Flemish people), Spain (with the Basque separatist movement), Canada (around the Québecois movement, see Box 5.4, "Separatism in Québec") and other countries.

Culture and Society

Culture has an important impact on society through its influence on social organization. Social categories like kinship, gang, or generation, or some combination of categories, may figure more or less prominently in a group,

depending upon geography. Moreover, the salience of these social categories may change over time as the group interacts with people and forces outside of its boundaries.

For example, countries in the Middle East and North Africa have a complex social organization—as complex as that of any other region of the globe—shaped by cultural ties and meaning systems that highlight gender, tribe, nationality, kinship, and family. Global media technologies, such as satellite television and the Internet, are increasingly penetrating the region, however, and the potential for new social forms to emerge and old ones to be reconfigured is increasing. Generally, the predominant forms of social organization in the region have persisted for hundreds of years. It would be incorrect, however, to disadvantaged. It was informed by the perceived preference of Jesus for the poor and helpless and by the writings of Karl Marx and other revolutionaries on inequality and oppression. This new orientation to the poor was espoused by the Second Vatican Council, called by Pope John XXIII in 1962. Priests preached grassroots self-help to organized Christian-based communities and often spoke out against repression and authoritarianism. Some were murdered by powerful interests who saw liberation theology as revolutionary and communistic. For example, Archbishop Oscar Romero in El Salvador was shot to death while saying mass on March 24, 1980.

In recent decades, evangelical Protestant groups with fundamentalist Christian beliefs have grown and spread rapidly in Latin America and the Caribbean. **Figure 5.D** shows the current distribution of these groups. Their message of literacy, education, sobriety, frugality, and personal salvation has become very popular in many rural areas. The conversion of Latin Americans to Protestant faiths grows stronger each year. Today over 15 percent of Latin Americans belong to evangelical churches, with the Latin American Catholic Bishops Conference claiming that 8,000 Latin Americans convert to evangelical Christianity every day. In Brazil, where evangelical Protestantism is second only to Catholicism in terms of number of adherents, half a million Catholics leave the church per year. In Mexico, there has been a 10 percent decline in the Catholic population since the mid-twentieth century. Observers of evangelical missionizing in Latin America believe that fundamentalist missionaries and ministers are younger and more numerous than Catholic priests and therefore provide a higher minister-to-worshiper ratio than the Catholic priest-to-worshiper ratio. For example, in Mexico, each

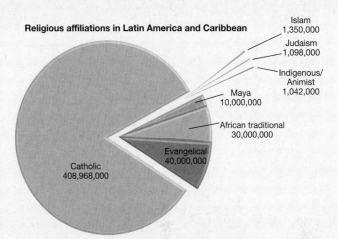

Figure 5.D   Latin America and Caribbean religions Although the majority of Latin Americans are Catholic, evangelical and African traditional religions, such as Santería, are also important to millions of people.

evangelical pastor serves 230, while each priest serves 8,600. Evangelical churches may also be more involved in indigenous communities than the Catholic churches.

There is also a strong thread of anticlericalism in many Latin American countries. This partly originated as a reaction to the authority given by the Pope to the Spanish crown, the control of land and labor by the missions, and the Catholic Church's alliance with the landowners and political leaders in colonial and postindependence conservative administrations. After the Mexican Revolution, for example, there was for many years a strict separation of Church and state, with priests forbidden to wear their clerical garb on the street.

assume that both subtle and dramatic changes within these forms have not already occurred.

Kinship is a form of social organization that is particularly central to the culture system of the Middle East and North Africa. **Kinship** is normally thought of as a relationship based on blood, marriage, or adoption. This definition needs to be expanded however, to include a *shared notion of relationship* among members of a group. The point is that not all kinship relations are understood by social groups to be exclusively based on biological or marriage ties. While biological ties, usually determined through the father, are important in the Middle East, they are not the only important ties that link individuals and families. In fact, though kinship is often expressed as a

"blood" tie among social groups throughout the Middle East, neighbors, friends, even individuals with common economic or political interests are often considered kin.

Kinship is such a valued relationship for expressing solidarity and connection that it is often used to assert a feeling of group closeness and as a basis for identity even where no "natural" or "blood" ties are present. This notion of kinship might also be seen displayed among fraternity "brothers," sorority "sisters," gang members, and even police officers and firefighters who feel a strong familial bond with co-workers.

In the Middle East and North Africa, kinship is even an important factor in shaping the spatial relationships of the home, as well as outside the home, determining

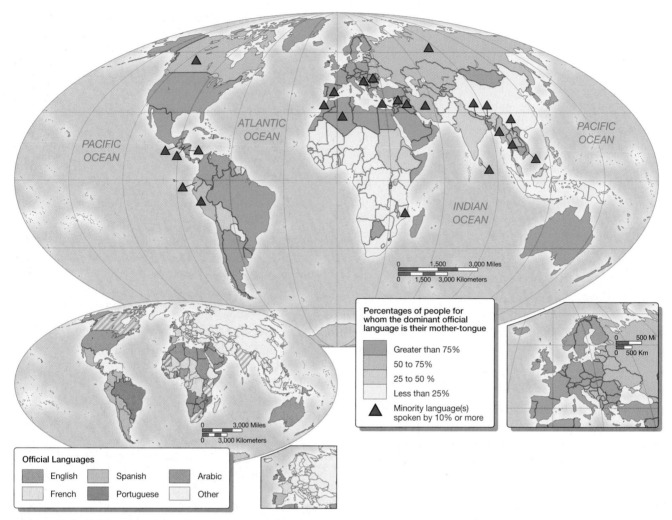

Figure 5.23 Tongue-tied Another way to examine the geography of language is to consider the proportion of people around the world whose mother tongue is not their country's official language. In places like China, where the dominant language is often not the mother tongue, simple communication may pose considerable challenges. While Mandarin is the official language, imposed by the imperial Mandarin bureaucracy, there are 6 major regional dialects of Mandarin as well as another 41 distinct languages spoken by ethnic minorities in China. In other cases, the disparity between the official language and the mother tongue is the result of colonization. For instance, in South Africa before apartheid was dismantled, there were two official languages, English and Afrikaans—both languages of the colonizers—but over 50 indigenous languages. Since the election of Nelson Mandela in the early 1990s, the number of official languages has grown to 11. The increase gives residents more access to public information and is intended to expand participation in civil society as well as preserve the cultural diversity of the country. (*Source:* Adapted from M. Kidron and R. Segal, *The New State of World Atlas,* 4th ed. New York: Simon and Schuster, 1991.)

who can interact with whom and under what circumstances. This is especially the case for the interaction of gender and kinship, where women's and men's access to public and private space are sharply differentiated.

The idea of the tribe is also central to understanding the sociopolitical organization of the Middle East and North Africa, as well as other regions of the world. Although tribally organized populations appear throughout the region, the tribe is not a widespread form of social organization. The term *tribe* is a highly contested concept and one that should be treated carefully. For instance, it is often seen as a negative label applied by the colonizers to suggest a primitive social organization throughout Africa. Where it is adopted in the Middle East and North Africa, however, tribe

is seen not as a primitive form of social organization, but rather as a valuable element in sustaining modern national identity. Generally speaking, a **tribe** is a form of social identity created by groups who share a set of ideas about collective loyalty and political action. Tribes are grounded in one or more expressions of social, political, and cultural identities created by individuals who share those identities.

The result of shared tribal identity is the formation of collective loyalties that result in primary allegiance to the tribe. External groups may recognize the existence of these self-defined tribal groups and seek to undermine or encourage their persistence. For instance, in early twentieth-century Iran, the state ruthlessly and systematically attempted to eliminate tribal affiliations. In contrast, during the European colonial period in Sudan and Morocco, tribes were seen as forms of social organization that might inhibit independence movements, so they were largely promoted and supported by the colonial state.

Throughout the Middle East and North Africa and beyond, globalization is contributing to the transformation of existing culture systems. While some groups accept these changes, others attempt to resist and actively protest them in a whole host of ways, from cultural protection legislation to acts of vandalism and even terrorism.

ISLAMIC CULTURAL NATIONALISM

The protection of regional languages as a way of resisting globalization is just one part of a larger movement that interests geographers and other scholars. The movement, known as **cultural nationalism,** is an effort to protect regional and national cultures from the homogenizing impact of globalization, especially from the penetrating influence of U.S. culture. **Figures 5.24** and **5.25** provide a picture of two widespread aspects of U.S. culture—films and television. Many other U.S. products also travel widely outside its borders (**Table 5.1**). Although many products of U.S. culture are welcomed abroad, many others are not. France, for example, has been fighting for years against the "Americanization" of its language, food, and films.

Nations can respond to the homogenizing forces of globalization and the spread of U.S. culture in any number of ways. Some groups attempt to seal themselves off from undesirable influences. Other groups attempt to legislate the flow of foreign ideas and values, as in some Muslim countries.

After Christianity, Islam possesses the next largest number of adherents worldwide—about 1 billion. The map in **Figure 5.26** shows the relative distribution of Muslims throughout Europe, Africa, and Asia; **Figure 5.27** shows the heartland of Islamic religious practice.

The Islamic world includes very different societies and regions, from Southeast Asia to Africa. Muslims comprise

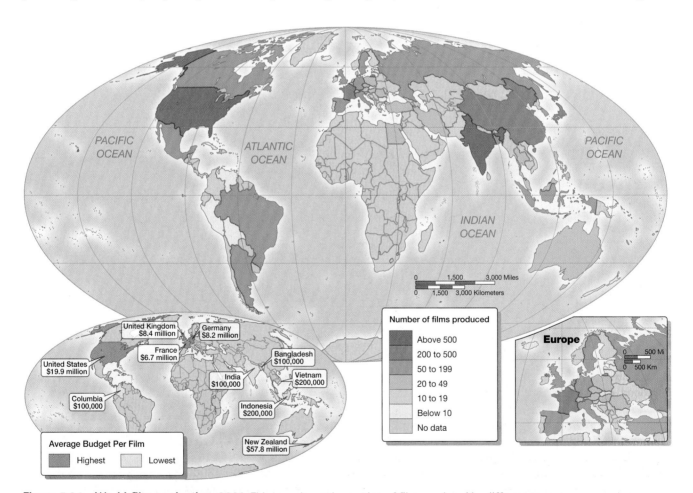

Figure 5.24   World film production, 2001 This map shows the number of films produced by different countries around the world. As expected, core countries are much more likely to be film producers. They also import fewer films as a result. For instance, the United States produces four times more films than it imports. In the semiperiphery, India, a country with an enormously popular film industry, produces eight times more films than it imports. Despite popular beliefs, the United States is not the largest film producer in the world. India far exceeds U.S. production and China and Hong Kong as well as the Philippines are also film production giants. (Adapted from UNESCO, *Survey on National Cinematography, 2001,* available at http://www.unesco.org/culture/industries/cinema/html_eng/prod.shtml. *World Culture Report,* 2000, table 4, http://www.unesco.org/culture/worldreport/html_eng/stat2/table4.pdf. All rights reserved.)

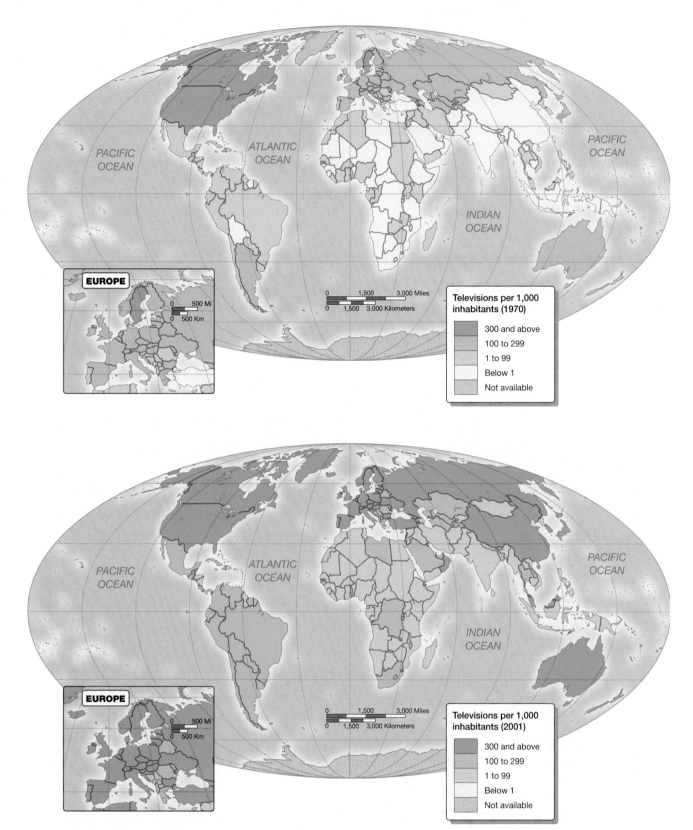

Figure 5.25 World distribution of TV sets and cable subscribers, 1970 and 2003 The ability to receive television broadcasts no matter who is exporting or producing them depends, of course, on access to a television set. In some areas of the periphery—for example, in some of the wealthier South American countries—there are over 300 television sets per 1,000 people. In much of sub-Saharan Africa, however, there are fewer than ten sets for every 1,000 people. In the core, in countries such as the United States and Japan, there is on average one color television set per household, as well as a high level of ownership of videocassette recorders. Cable subscription opens up a market to even more commercial television products. With respect to television programming exports, the core vastly overwhelms the periphery in the export of television programs, with the leaders being the United States and the United Kingdom. (Adapted from J. M. Rubenstein, *The Cultural Landscape: An Introduction to Human Geography*, 6th ed., © 1999. Update from World Bank, *World Development Indicators 2002*, World Bank: Washington, DC.)

TABLE 5.1 Most Popular Global Brands

This table shows the most popular brands across the globe. Of the twenty listed, sixteen are U.S. products.

2007 Rank	2006 Rank	Brand	Logo	Country of origin	Sector	2007 Brand Value ($m)	Change in brand value
1	1	Coca-Cola		US	Beverages	65,324	-3%
2	2	Microsoft		US	Computer Software	58,709	3%
3	3	IBM		US	Computer Services	57,091	2%
4	4	GE		US	Diversified	51,569	5%
5	6	Nokia		Finland	Consumer Electronics	33,696	12%
6	7	Toyota		Japan	Automotive	32,070	15%
7	5	Intel		US	Computer Hardware	30,954	-4%
8	9	McDonald's		US	Restaurants	29,398	7%
9	8	Disney		US	Media	29,210	5%
10	10	Mercedes		Germany	Automotive	23,568	8%
11	11	Citi		US	Financial Services	23,443	9%
12	13	Hewlett-Packard		US	Computer Hardware	22,197	9%
13	15	BMW		Germany	Automotive	21,612	10%
14	12	Marlboro		US	Tobacco	21,283	0%
15	14	American Express		US	Financial Services	20,827	6%
16	16	Gillette		US	Personal Care	20,415	4%
17	17	Louis Vuitton		France	Luxury	20,321	15%
18	18	Cisco		US	Computer Services	19,099	9%
19	19	Honda		Japan	Automotive	17,998	6%
20	24	Google		US	Internet Services	17,837	44%

Source: Interbrand, *All Brands are Not Created Equal: The Best Global Brands 2007*, p. 12, http://www.ourfishbowl.com/images/surveys/Interbrand_BGB_2007.pdf, accessed July 2, 2008.

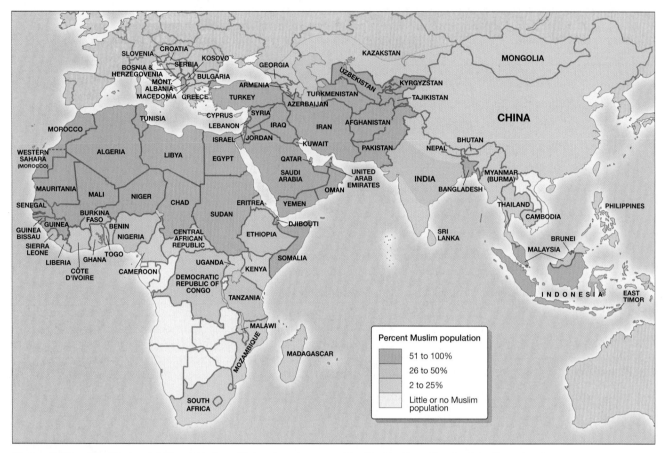

Figure 5.26 Muslim world The diffusion of Islam is quite extensive. Like the Spanish colonial effort, the rise and growth of Muslim colonization were accompanied by the diffusion of the colonizers' religion. The distribution of Islam in Africa, Southeast Asia, and South Asia that we see today testifies to the broad reach of Muslim cultural, colonial, and trade activities. (Adapted from D. Hiro, *Holy Wars.* London: Routledge, 1989.)

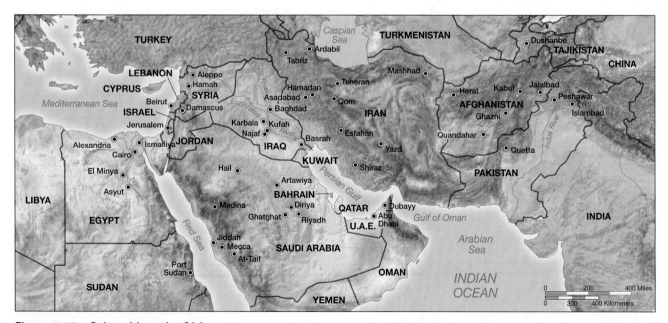

Figure 5.27 Cultural hearth of Islam Islam has reached into most regions of the world, but the heart of the Muslim culture remains the Middle East, the original cultural hearth. It is also in this area that Islamism is most militant. (Adapted from D. Hiro, *Holy Wars.* London: Routledge, 1989.)

5.3 GEOGRAPHY MATTERS

Language and Ethnicity in Africa

The geography of languages in Africa is remarkably complex, with more than 800 living languages, 40 of them spoken by more than 1 million people (**Figure 5.E**). The dominant indigenous languages, spoken by 10 million or more, are Hausa (the Sahel), Lingala (Central Africa), Swahili (East Africa), and Tswana-Sotho and Zulu (southern Africa). Hausa and Swahili are trade languages, spoken as second languages by many groups to facilitate trade. English, French, Portuguese, and Afrikaans are also spoken in regions of recent colonial control and education systems or white settlement, and Arabic is common in North Africa. Arabic has also strongly influenced Swahili along the east coast of Africa. Because most countries have no dominant indigenous African language, they have often chosen a European language for official business and school systems. The countries with the most coherent overlap between their territory and a dominant African language are Somalia (Somali), Botswana (Tswana), and Ethiopia (Amharic).

The indigenous languages of Africa have been grouped into larger language families, including the Afro-Asiatic languages of North Africa, including Somali, Amharic, and Tuareg; the Nilo-Saharan languages, which include Dinka, Turkana, and Nuer in East Africa; and the largest Niger-Congo group, which includes Hausa, Yoruba, Zulu, Swahili, and Kikuyu. A small family is the Khoisan languages spoken by the Bushmen of southern Africa, which have a distinctive "click" vocalization.

The multiplicity of languages and dialects reflects the large number of distinct cultural or ethnic groups in Africa. Groups are often led by chiefs, and some are monarchies with kings or queens. The largest ethnic groups in Africa are associated with certain dominant languages, such as Hausa, Yoruba, and Zulu, but almost all groups, however large or small, were either split geographically by colonial national boundaries or grouped with neighbors with whom they may have had no affinity.

Attempts to consolidate ethnic groups across boundaries and struggles for power between groups within countries are a major cause of conflict in contemporary Africa. For example, tensions between the Ibo and Yoruba in Nigeria led to civil war when the Ibo declared the independence of eastern Nigeria as Biafra in 1967. The conflict, which drew international attention and intervention because of starvation in Biafra and the presence of oil in the region, resulted in as many as a million deaths (mainly from hunger and disease) and lingering ethnic resentments after Nigeria was reunited. Another evident tension is between the ruling Kikuyu and other ethnic groups in Kenya, where opposition political parties have organized around ethnicity and threatened violence in the face of perceived election corruption and bias toward Kikuyu regions and individuals.

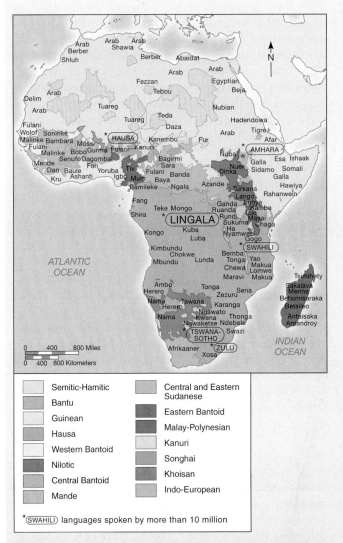

Figure 5.E Language map of Africa The cultural complexity of Africa is clearly demonstrated in the variety of indigenous languages shown in this map. Swahili and Hausa are spoken by millions as the trade languages of East and West Africa, respectively. (Adapted from S. Aryeetey–Attoh (ed.), *The Geography of Sub-Saharan Africa.* Upper Saddle River, NJ: Prentice Hall, 1997.)

over 85 percent of the populations of Afghanistan, Algeria, Bangladesh, Egypt, Indonesia, Iran, Iraq, Jordan, Pakistan, Saudi Arabia, Senegal, Tunisia, Turkey, and most of the newly independent republics of Central Asia and the Caucasus (including Azerbaijan, Turkmenistan, Uzbekistan, and Tajikistan). In Albania, Chad, Ethiopia, and Nigeria, Muslims make up 50 to 85 percent of the population. In India, Burma (Myanmar), Cambodia, China, Greece, Slovenia, Thailand, and the Philippines, significant Muslim minorities exist.

Islam is an Arabic term that means "submission," specifically submission to God's will. A **Muslim** is a member of the community of believers whose duty is obedience and submission to the will of God. As a revealed religion, Islam recognizes the prophets of the Old and New Testaments of the Holy Bible, but Muhammad is considered the last prophet and God's messenger on Earth. The Qur'an, the principal holy book of the Muslims, is considered the word of God as revealed to Muhammad by the Angel Gabriel beginning in about A.D. 610. There are two fundamental sources of Islamic doctrine and practice: the Qur'an and the Sunna. Muslims regard the Qur'an as directly spoken by God to Muhammad. The Sunna is not a written document, but a set of practical guidelines to behavior. It is effectively the body of traditions that are derived from the words and actions of the prophet Muhammad.

Islam holds that God has four fundamental functions—creation, sustenance, guidance, and judgment—the purpose of people is to serve God by worshiping him alone and adhering to an ethical social order. The actions of the individual, moreover, should be to the ultimate benefit of humanity, not the immediate pleasures or ambitions of the self. There are five primary obligations, known as the five pillars of Islam, that a Muslim must fulfill: repeating the profession of the faith ("There is no god but God; Muhammad is the messenger of God"); praying five times a day facing Mecca; giving alms or charitable donations; fasting from sunup until sundown during the holy month of Ramadan; and making at least one pilgrimmage, or **hajj,** to Mecca if financially and physically able.

The emergence and spread of Islam are linked to the commercial history of the Middle East and North Africa. The geographical origin of Islam is Mecca, in present-day Saudi Arabia. When Islam first emerged, Mecca, where Muhammad was born in A.D. 570, was an important node in the trade routes that at first connected Yemen and Syria and eventually linked the region to Europe and all of Asia. Today Mecca is the most important sacred city in the Islamic world, as well as an important commercial center. Eventually Medina also became an important sacred city because it was the place to which Muhammad fled when he was driven out of Mecca by angry merchants who felt his religious beliefs threatened their commercial practices.

Disagreement over the line of succession from the prophet Muhammad occurred shortly after his death in 632 and resulted in the split of Islam into two main sects, the Sunni and the Shi'i. The Sunni faction argued that the clergy should succeed Muhammad, while the Shi'i argued

that Muhammad's son Ali should succeed his father. Ali was killed, and the Sunnis became dominant. They remain the mainstream branch of Islam, but the pattern varies from one country to another. The majority of Iran's 60 million people follow Shi'i, the official state religion of the Islamic Republic of Iran, founded in 1979. The majority of Iraq's population is also Shi'i. It is also important to keep in mind that Islam is practiced differently in many different locales throughout the Middle East and North Africa and that Muslims who have migrated from the region—to Europe and the United States, for instance—are shaped by, and shape the practice of, Islam in the Middle East.

Perhaps one of the most widespread cultural counterforces to globalization has been the rise of Islamism, more popularly, although incorrectly, known as Islamic fundamentalism. Whereas *fundamentalism* is a general term that describes the desire to return to strict adherence to the fundamentals of a religious system, **Islamism** is an anticolonial, anti-imperial, and generally anticore political movement. In Muslim countries, Islamists resist core, especially Western, forces of globalization—namely modernization and secularization. Not all Muslims are Islamists, although Islamism is the most militant movement within Islam today.

The basic intent of Islamism is to create a model of society that protects the purity and centrality of Islamic precepts through the return to a universal Islamic state—a state that would be religiously and politically unified. Islamists object to modernization because they believe the corrupting influences of the core place the rights of the individual over the common good. They view the popularity of Western ideas as a move away from religion to a more secular (nonreligious) society. Islamists desire to maintain religious precepts at the center of state actions, such as introducing principles from the sacred law of Islam into state constitutions.

Another important aspect of the Islamist movement is the concept of **jihad,** which is a sacred struggle. When this struggle is violently directed against the enemies of Islam, jihad is understood to be a holy war. But jihad can also be a more peaceful struggle to establish Islam as a universal religion through the conversion of nonbelievers. One example of jihad today is the struggle of Shi'ite Muslims for social, political, and economic rights within Sunni-dominated Islamic states.

A heightened fear of Islamic militancy and the "War on Terror" more generally has resulted in increased security measures in most core countries. Many commentators believe that a "culture of fear" now permeates society in these places—with governments and media outlets routinely warning of impending threats and disaster. This fear has led to what many believe is an erosion of civil liberties—for example, the government of the United Kingdom recently passed a bill that would allow suspected terrorists to be detained for up to 42 days without charge. Related to this is the notion of a "surveillance society" where civilians are monitored constantly through closed-circuit television (CCTV) and other forms of surveillance, such as national identification cards. The United Kingdom is the most

(a)   (b)

Figure 5.28   Muslim women in Afghanistan and Turkey (a) The fundamentalist, theocratic Taliban regime required women to be extremely conservatively dressed when appearing in public. These strictures were dramatically relaxed when the new government took power in early 2002. Women may now exercise a great deal of choice with respect to how to dress in public. Some wear simple head scarves, while others, such as those shown here shopping at a street market in Mazar-e-Sharif, in the northern part of the country, still wear burquas. (b) Three young women stroll down a street in Istanbul. Although Turkey is 99 percent Muslim, it has a secular democratic government and a population with generally liberal attitudes toward the West, especially in urban areas. While some older women and women in the countryside prefer to appear in public with head coverings, much of the younger generation, especially middle- and upper-class urban women, have adopted Western styles of dress and overall appearance.

"spied upon" nation on Earth, with 20 percent of the world's CCTV cameras, and one of very few countries to be introducing national identification cards capable of holding various types of biometric data.

As popular media reports make clear, no other movement emanating from the periphery is as widespread and has had more of an impact politically, militarily, economically, and culturally than Islamism. Yet Islamism—a radical and sometimes militant movement—should not be regarded as synonymous with the practices of Islam, any more generally than Christian fundamentalism is with Christianity. Islam is not a monolithic religion, and even though all adherents accept the basic pillars, specific practices vary according to the different histories of countries, nations, and tribes. Some expressions of Islam allow for the existence and integration of Western styles of dress, food, music, and other aspects of culture, while others call for the complete elimination of Western influences (**Figure 5.28**).

CULTURE AND IDENTITY

In addition to exploring cultural forms, such as religion and language, and movements, such as cultural nationalism, geographers have increasingly begun to ask questions about other forms of identity. This interest largely has to do with the fact that certain long-established and some more recently self-conscious cultural groups have begun to use their identities to assert political, economic, social, and cultural claims.

Sexual Geographies

One identity that has captured the attention of cultural geographers is sexuality. **Sexuality** is a set of practices and identities that a given culture considers related to each other and to those things it considers sexual acts and desires. One of the earliest and most effective examples of the geographic study of sexuality is the examination of the spatial expression of prostitution. Research on prostitution in California found that sex work—as it is now more commonly known—is spatially differentiated based upon the target clientele as well as systems of surveillance. Female sex workers who service an upper-income clientele generally perform their work in the homes, hotels, and other private areas of upper- and middle-class society. Female sex workers serving a lower-class clientele tend to operate within more public spaces or bring clients to their homes or hotel rooms in the locales where they are "streetwalking."

Not surprisingly, female sex workers at the lower end of the economic spectrum are more likely to be subject to police surveillance and other forms of public scrutiny because their advertising and work activities are largely carried out in public places. In contrast, those oriented toward a wealthier clientele have, in most cases, little or no obvious public identity. The distinction should be clear with respect to geography. The "red-light districts" of most cities are widely known to residents. But sex workers who serve a higher-income clientele do not restrict themselves to segregated workspaces.

5.4 WINDOW ON THE WORLD

Separatism in Québec

The political movement in the province of Québec that seeks to separate Québec from the rest of Canada is premised on a deep desire to preserve and enrich Québecois culture. The movement can be traced back to the early history of colonial Canada, when the British defeated the French and secured possession of the entire Canadian territory in 1763. As the only "vanquished" European settlers in North America, the Québecois are undeniably unique.

Until Québec experienced its so-called Quiet Revolution in the 1960s, when a separatist political party—the Parti Québecois—emerged, French-Canadian nationalism was a largely conservative movement. With the French-Canadian family and Roman Catholic religion at its core, Québec oriented itself around clerical, rural, and agricultural values. Its nationalism was backward-looking and traditional. While numerous scholars have advanced any number of explanations for the rise and persistence of French nationalism in Québec, the most compelling one from our perspective is based on its peripheral status in the national and the global economy.

As the Canadian economy became increasingly integrated into the global economy and, more particularly, into the U.S. economy, Québec found itself less able than previously to benefit from U.S. hegemony. Largely because its identity was based on traditional ideas and practices, Québec was ill-equipped to participate in the revolutionary changes transpiring around it. As western Canada's economic power increased with the increasing importance of the energy sector, and as Toronto transformed itself based on the growing service sector, enabling it to become the financial and industrial capital of Canada, Québec was left behind.

Recognizing that these changes posed a major threat to the social, economic, and political survival of Québec's cultural identity, Québec's elites saw a way out of peripheralization by opting out of the Canadian national project altogether. Seizing control of the provincial government of Québec, the elites were able to translate their vision for a separate Québec by creating a polity that wished not simply to try to hold on to the old ways, but to achieve complete independence in order to determine their own path through the new economy.

The desire of the Québec separatists to opt out of their regional/provincial status and become an independent nation with a distinct identity became translated into provincial referenda on independence. In both 1980 and 1985, the separatists forced votes that, if they had passed, would have signaled the desire of residents of the province to separate politically from Canada. Both times the referenda failed. In 1995, another referendum was on the ballot asking voters if they wished to stay part of Canada; in this referendum, "yes" votes to remain in Canada exceeded "no" votes by a very small margin (**Figure 5.F**).

It is too soon to tell whether Québec will push once more for independence from Canada. There are signs that the province is actually becoming less interested in separation, given its increasing prosperity through incorporation into the global economy. Globalization and Québec's growing independent relationship with the United States have begun to influence its desire for nation-state status. Some scholars believe that rather than undermining Québecois national identity, NAFTA and Canada's increasing connections to Europe have enhanced it. In short, there is no clear answer at this

More typical of contemporary work on sexuality in geography is that which explores the spatial constraints on homosexuality and the ways in which lesbian, gay, bisexual, and transgendered (LGBT) people respond to and reshape them. Two particular areas of research have emerged: gay and lesbian consumerism, and the importance of the body as a site of performance of sexuality and gender identities.

One approach to understanding homosexuality, particularly as it is represented in contrast to heterosexual, is research and writing on the body and the ways in which individuals literally perform their sexual identities through clothing, attitude, body language, and grooming, such as hairstyles and bodily adornments. Research on the body has also been used to decode the social and

cultural practices that express other kinds of social identity, such as gender and class. But the work has been most fully developed in its application to sexuality.

The central theoretical position of this research is that sexual identities are learned and performed through standard, taken-for-granted practices that most people tend to take as "normal" or "natural." Using gender as a starting point and recognizing it as a key feature of all human identity that marks and defines us as "sexed" beings, researchers in this area contend that gender is not something we essentially are (because of a given set of physical characteristics), but something we do (something we enact by the way we present our bodily selves to the world). We may be born as biological females, but

Figure 5.F   Referendum vote on separation from Canada The distribution of the vote reflects the ethnic division of upper and lower Québec. While lower Québec is urbanized, with overwhelming numbers of French-speaking Canadians, upper Québec is largely rural and contains mostly non-Francophone communities.

point to the question of whether Québec will remain a distinct society within Canada or become a new nation-state. What is clear is that the context for making that decision is rapidly changing, as changes in the Canadian state brought about by globalization have begun to open up ways for Québec to maintain its distinct identity, to prosper economically, and to control more of its own destiny without having to secede from Canada.

In fact, a recent poll shows that for the first time since the 1990s, 50 percent of the residents of Québec support sovereignty for the province. At the same time, however, two-thirds of them indicated they were proud to be Canadian and most still wanted the province to remain part of Canada. In short, it is sovereignty but not separatism that increasingly characterizes the political desire of more and more Québec residents. It would appear that the question of Québec's relationship to Canada hinges on the attitudes of different generations, and time may change which of those attitudes will prevail. Those residents of Québec who opposed separation from Canada, or the "traditional federalists," were born before World War II and they are dying off. The baby boomers of the Quiet Revolution support sovereignty but wish to remain with Canada. It appears that today's young adults—the children of Bill 101, the landmark language law from 1977 that put French at the forefront of education—are likely to support independence from Canada. Time will tell whether the right conditions prevail for separation when this generation holds the reins of power.

we learn how to act as women. An illustration of the argument that sexuality is something that we perform through gendered acts is the gay skinhead or the lipstick lesbian. Generally, the skinhead is socially understood to be a hypermasculine, heterosexual male. But the gay skinhead complicates this construction by challenging the idea that all hypermasculine men are straight.

Research on sexuality and the body in geography recognizes that space is central to our understanding. Where gay or straight identities can be performed plays a central role in who occupies those spaces, how they are occupied, and even what sexual identity you might be performing if you occupy one space or another. But issues of sexuality and space go beyond the performance

of sexual identity to the emergence of new political cultures that are being constructed to protect the rights of lesbian, gay, and transgendered people, in particular national and international spaces (**Figure 5.29**).

With the increasing attention being paid to human rights issues around the globe, sexual rights have also been put on the agenda. For example, the Department of Reproductive Health and Research of the World Health Organization provides a Web site on sexual health that includes a set of working definitions intended to provide a foundation for discussion about sexual health around the world. These working definitions include the right of all persons, free of coercion, discrimination, and violence, to, among other things, choose their sexual partners, receive sexuality

Figure 5.29 Gay pride parade, São Paulo, Brazil Almost 2 million LGBT people and their supporters—many in lavish Carnival costumes and waving rainbow-colored flags—paraded in Brazil's biggest city in May 2005 to celebrate gay pride and call for the legalization of civil unions between homosexuals. In comparison, San Francisco's gay pride parade typically attracts tens of thousands of people, and the 2004 World gay pride day celebrations in Berlin attracted between 200,000 and 500,000 participants.

education, and pursue a satisfying, safe, and pleasurable sexual life. Although these definitions are provisional and not the official policy of the World Health Organization, they provide some insight into the transformations that are occurring around issues of sexuality—especially as they pertain to international human rights.

Ethnicity and the Use of Space

Ethnicity is another area in which geographers are exploring cultural identity. **Ethnicity** is a socially created system of rules about who belongs to a particular group based upon actual or perceived commonalities, such as language or religion. A geographic focus on ethnicity is an attempt to understand how it shapes and is shaped by space, and how ethnic groups use space with respect to mainstream culture. For cultural geographers, territory is also a basis for ethnic group cohesion (see Chapter 9 for more on

territory). For example, cultural groups—ethnically identified or otherwise—may be spatially segregated from the wider society in ghettos or ethnic enclaves (see Chapter 11). Or they may use space to declare their subjective interpretations about the world they live in and their place in it.

The use of the city streets by many different cultural groups demonstrates this point. Nineteenth-century immigrants to U.S. cities used the streets to broadcast their ideas about life in their adopted country. Ethnic parades in the nineteenth century—such as St. Patrick's Day for the Irish and Columbus Day for the Italians—were often very public declarations about the stresses that existed between and among classes, cultures, and generations.

An endless array of spaces besides city streets are also important to the making and remaking of ethnic identity, whether for a few hours or for generations. Ethnic restaurants are often places where individuals and groups sharing a common place of origin and cuisine come together to enjoy those foods as well as confirm their connections to each other and to a distant place. Even immigrant Web sites and blogs, where people far from their homeland can communicate with others from that same place, are virtual spaces where ethnic identity is made and remade through the words and ideas that are exchanged.

Geographers Maria Price and Courtney Whitworth have explored how Latino immigrants in Washington, DC, create vibrant, though temporary, ethnic spaces through immigrant soccer leagues. They write, "Leagues create a cultural space that is familiar, entertaining, practical, inexpensive, transnational, and ephemeral, where immigrants gather to reaffirm their sense of identity and belonging."[2] They also make it clear that it is not just that the soccer field itself constitutes a vibrant ethnic space, but it is more so that the social networks that enable the leagues to function are themselves a kind of "in-between" space that links the lived space of Washington, DC, with the remembered space of their home country. Like Web sites that link distant immigrants through discussions about the home country and the pleasures and difficulties of living abroad, the social networks that sustain soccer leagues also create a transnational space of ethnic identity.

Race and Place

Geographers also use prevailing ideas and practices with respect to race to understand the shaping of places.[3] **Race** is a problematic classification of human beings based on skin color and other physical characteristics. **Racialization**

[2]M. Price and C. Whitworth, "Soccer and Latino Cultural Space: Metropolitan Washington *Fútbol* Leagues," in *Hispanic Spaces, Latino Places: Community and Cultural Diversity in Contemporary America*, D. Arreola (ed.). 2004, pp. 167-186. Austin: University of Texas Press, 2004.

[3]Adapted from K. Anderson, "The Idea of Chinatown: The Power of Place and Institutional Practice in the Making of a Racial Category," *Annals of the Association of American Geographers* 77 (4), 1987, pp. 580–598.

Figure 5.30 Chinatown, San Francisco Chinatowns have been portrayed as places voluntarily chosen by immigrant Chinese. But they can also be seen as places of exclusion and racism created by the dominant society that prefers to segregate people of difference from the wider society.

is the practice of creating unequal castes where whiteness is considered the norm. Biologically speaking, however, no such thing as race exists within the human species. Yet consider the categories of race and place that correspond to "Chinese" and "Chinatown." Powerful Western ideas about Chinese as a racial category have enabled the emergence and perpetuation of Chinatowns in many North American cities (**Figure 5.30**). In this and other cases, the visible characteristics of hair, skin, and bone structure made race into a category of difference that was (and still is) widely accepted and often spatially expressed.

The mainstream approach to neighborhood is to see it as a spatial setting for systems of affiliation more or less chosen by people with similar skin color. Yet cultural geographers have begun to overturn this approach and to see neighborhoods as spaces that affirm the dominant society's sense of identity. For example, from the perspective of white society, nineteenth-century Chinatowns were the physical expression of what set the Chinese apart from whites. The distinguishing characteristics revolved around the way the Chinese looked, what they ate, their non-Christian religion, opium consumption, gambling habits, and other "strange" practices. Place—Chinatown—maintained and manifested differences between white and Chinese society. Furthermore, place continues to be a mechanism for creating and preserving local systems of racial classification and for containing geographical difference within defined geographical confines. The homelands of South Africa and the dismantling of apartheid there also illustrate the interaction of race and place, though on a much larger scale.

What remains unexamined in the kinds of perceptions and practices that create racialized places as different as Chinatown and the homelands of South Africa is the taken-for-grantedness of whiteness. Whiteness is seen to be the norm or the standard against which all other visible differences are compared. But whiteness, as geographers Owen Dwyer and John Paul Jones, III, have shown, is itself a category of difference that depends on visible distinctions and not biological ones and is always constructed in relation to other categories. Recently, researchers in the humanities and social sciences have begun to "denaturalize" whiteness by investigating the way it has been constructed as a social category in different periods and places as well as the ways that whiteness operates in particular sites such as classrooms, on the street, and in boardrooms. It is only when we begin to challenge the naturalness of whiteness and the spaces that enable it that we can truly begin to undo racist practices.

Gender and Other Identities

Gender is an identity that has received a great deal of attention from cultural geographers within the last three decades. **Gender** is a category reflecting the social differences between men and women. As with other forms of identity, it implies a socially created difference in power between groups. In the case of gender, the power difference gives an advantage to males over females and is not biologically determined, but socially and culturally created. Gender interacts with other forms of identity and can intensify power differences among and between groups. Furthermore, the implications of these differences are played out differently in different parts of the world.

For example, although gender differences play an important part in shaping social life for men and women in the Middle East, as elsewhere around the globe, there is no single Islamic, Christian, or Jewish notion of gender that operates exclusively in the region. Many of us have formed stereotypes about the restricted lives of Middle Eastern women because of the operation of rigid Islamic traditions. It is important to understand, however, that these are indeed stereotypes and do not capture the great variety in gender relations that exist in the Middle East and North Africa across lines of class, generation, level of education, and geography (urban versus rural origins), among other factors.

Figure 5.31 Indian women in the informal sector Many women in South Asia are self-employed as small vendors in the daily markets. Others do home-based work such as weaving and dyeing cloth or embroidering or sewing garments. Nearly all workers in the informal sector, whether male or female, lack any sort of social protection such as health or unemployment insurance.

What is pervasive throughout the Middle East and North Africa, as well as in many other societies throughout the world, is an ideological assumption that women should be subordinate to men. This view is largely held by both men and women in Middle Eastern and North African societies. Interestingly, it seems that men regard women's subordination as something natural, something that is effectively determined by biology. In contrast, most women in the Middle East tend to regard their subordination as something social, something that is the product of the society in which they live and operate, and therefore something that can be negotiated and manipulated. The gender systems that operate in a wide variety of contexts in the Middle East and North Africa are derived in large part from some of the same notions about men and women that inform gender systems in Western societies.

Another way to explore the implications of gender systems is to complicate them by introducing other identity elements, such as class. Among South Asia's poor, for example, women bear the greatest burden and the most suffering. South Asian societies are intensely patriarchal, though the form that patriarchy takes varies by region and class. The common denominator among the poor throughout South Asia is that women not only have the constant responsibilities of motherhood and domestic chores but also have to work long hours in informal-sector occupations (**Figure 5.31**). In many poor communities, 90 percent of all production occurs outside of formal employment, more than half of which is the result of women's efforts. In addition, women's property rights are curtailed, their public behavior is restricted, and their opportunities for education and participation in the waged labor force are severely limited.

Women's subservience to men is deeply ingrained within South Asian cultures, and it is manifest most clearly in the cultural practices attached to family life, such as the custom of providing a dowry to daughters at marriage. The preference for male children is reflected in the widespread (but illegal) practice of selective abortion and female infanticide. Within marriages, many (but by no means all) poor women are routinely neglected and maltreated. More extreme are the cases—usually reported only when they involve middle-class families—of "bride burning," whereby a husband or mother-in-law fakes the accidental death (kitchen fires are favored) or suicide of a bride whose parents had defaulted on their dowry payments. Several thousand such deaths are reported in India each year, and this is almost certainly only a fraction of the real incidence.

The picture for women in South Asia, as elsewhere, is not entirely negative, however, and one of the most significant developments has been the emergence of women's self-help movements. Perhaps the best known of these is the Grameen Bank, a grassroots organization formed to provide small loans to the rural poor in Bangladesh. Another is the Self-Employed Women's Association (SEWA) in India, which has made a major contribution to building self-confidence and self-reliance among poor working women by mobilizing and organizing them.

GLOBALIZATION AND CULTURAL CHANGE

The discussion of cultural geography in this chapter has raised one central question: How has globalization changed culture? We have seen that it affects different cultural groups differently and that different groups respond in different ways to these changes. With so much change occurring for so long, however, we must still ask ourselves what overall impact globalization is having on the multiplicity of culture groups that inhabit the globe.

Anyone who has ever traveled between major world cities—or, for that matter, anyone who has been attentive to the backdrops of movies, television news stories, and magazine photojournalism—will have noticed the many familiar aspects of contemporary life in settings that until recently were thought of as being quite different from one another. Airports, offices, and international hotels have become notoriously alike, and their similarities of architecture and interior design have become reinforced by the near-universal dress codes of the people who frequent them. The business suit, especially for males, has become the norm for office workers throughout much of the world. Jeans, T-shirts, and sneakers, meanwhile, have become the norm for young people, as well as those in lower-wage jobs. The same automobiles can be seen on the streets of cities throughout the world (though sometimes they are given different names by their manufacturers); the same popular music is played on local radio stations; and many of the movies shown in local theaters are the same (*Harry Potter and the Half-Blood Prince, Iron Man, Dark Knight*, for example). Some of the TV programming is also the same—not just the music videos on MTV, but CNN's news, the Internet, major international sports events, and drama series like *Family Guy, Grey's Anatomy*, and *CSI*. The same brand names also show up in stores and restaurants: Coca-Cola, Perrier, Carlsberg, Nestlé, Nike, Seiko, Sony, IBM, Nintendo, and Microsoft, to name a few. Everywhere there is Chinese food, pita bread, pizza, classical music, rock music, and jazz.

Americanization and Globalization

It is these commonalities that provide a sense of familiarity for core travelers abroad. From the point of view of cultural nationalism, the "lowest common denominator" of this familiarity is often seen as the culture of fast food and popular entertainment that emanates from the United States. Popular commentators have observed that cultures around the world are being Americanized, or "McDonaldized" (see Chapter 2), which represents the beginnings of a single global culture that will be based on material consumption, with the English language as its medium.

There is certainly some evidence to support this point of view, not least in the sheer numbers of people around the world who view *The Bachelorette*, drink Coca-Cola, and eat in McDonald's franchises or similar fast-food chains. Meanwhile, U.S. culture is increasingly embraced by local entrepreneurs around the world largely through consumer goods. It seems clear that U.S. products are consumed as much for their symbolism of a particular way of life as for their intrinsic value. McDonald's burgers, along with Coca-Cola, Hollywood movies, rock music, and NFL and NBA insignia, have become associated with a lifestyle package that features luxury, youth, fitness, beauty, and freedom.

It is important to recognize, however, that U.S. products often undergo changes when they travel across the globe. For instance, Revlon, a U.S.-based international public corporation and a leading mass-market producer of cosmetics, skin-care products, and fragrances, is very much aware of the need to vary its products for their consumption in non-U.S. markets. Not only are Revlon products manufactured and distributed in Asia, Africa, South America, and Europe, as well as in the United States, the company tailors its products and its advertising to the particularities of those markets. For example, Modi Revlon, a joint venture between Revlon and the Indian company Modi Mundipharma, provides a product line of color cosmetics that complement Indian women's skin tones and appeal to upper-income markets. The success of Modi Revlon, which has captured 80 percent of the Indian cosmetics market, has been remarkable.

The economic success of the U.S. entertainment industry has also helped reinforce the idea of an emerging global culture based on Americanization. Today the entertainment industry is the leading source of foreign income in the United States, with a trade surplus of over $25 billion. Similarly, the United States transmits much more than it receives in sheer volume of cultural products. Today the originals of over half of all the books translated in the world (more than 25,000 titles) are written in English, the majority of which are produced by U.S. publishers. In terms of international flows of everything from mail and phone calls to press-agency reports, television programs, radio shows, and movies, a disproportionately large share originates in the United States.

Neither the widespread consumption of U.S. and U.S.-style products nor the increasing familiarity of people around the world with global media and international brand names, however, adds up to the emergence of a single global culture. Rather, what is happening is that processes of globalization are exposing the world's inhabitants to a common set of products, symbols, myths, memories, events, cult figures, landscapes, and traditions. People living in Tokyo or Tucson, Turin or Timbuktu, may be perfectly familiar with these commonalities without necessarily using or responding to them in uniform ways. Equally, it is important to recognize that cultural flows take place in all directions, not just outward from the United States. Think, for example, of European fashions in U.S. stores; of Chinese, Indian, Italian, Mexican, and Thai restaurants in U.S. towns and cities; and of U.S. and European stores selling exotic craft goods from the periphery.

A Global Culture?

The answer to the question whether there is a global culture must therefore be "no," or at least there is no indisputable sign of it yet. While people around the world share an increasing familiarity with a common set of products, symbols, and events (many of which originate in the U.S.

Figure 5.32 World music Bands like Gogol Bordello combine punk with traditional Gypsy music to produce a sound that allows the two to come together not harmoniously, but chaotically. Gogol Bordello's objective is to produce what they call an "unstable hybrid" to demonstrate that culture is a living entity that is constantly confronted and challenged. The band sees its music not as celebrating multiculturalism, but as challenging listeners to understand contemporary alternative world music as "multi kontra culti," indicating that globalization is not conflict free.

culture of fast food and popular entertainment), these commonalties are configured in different ways in different places, rather than constituting a single global culture. The local interacts with the global, often producing hybrid cultures. Sometimes traditional, local cultures become the subject of global consumption; sometimes it is the other way around. This is illustrated very well by the case of two suqs (linear bazaars) in the traditional medieval city of Tunis, in North Africa. Both radiate from the great Zaytuna Mosque, which has always been the geographic focal point of the old city. Both are located near gateways from the medieval core to the French-built new city.

The older suq now specializes in Tunisian handicrafts, "traditional" goods, and so on. It has kept its exotic architecture and multicolored colonnades. The plaintive sound of the ancient nose flute and the whining of Arabic music provide background for the European tourists in their shorts and T-shirts, who amble in twos and threes, stopping to look and to buy. Few natives, except for sellers, are to be seen. The newer suq, formerly less important, is a bustling madhouse. It is packed with partially veiled women and younger Tunisian girls in blouses and skirts, with men in knee-length tunic/toga outfits or in a variety of pants and shirts, with children everywhere. Few foreigners can be seen. The background to the din is blaring rock-and-roll music, and piled high on the pushcarts that line the way are transistor radios, watches, blue jeans (some prewashed), rayon scarves, and Lux face and Omo laundry soaps.[4]

A second illustration of the absence of a homogenous global culture is **world music,** the musical genre defined largely in response to the sudden increase of non–English-language recordings released in the United

Kingdom and the United States in the 1980s. The term is one employed primarily by the media and record stores, and it includes such diverse sources as Tuvan throat singers and Malian griots. (Throat singing is a performance style originating in south-central Russia where the vocal sounds are projected from the throat, allowing two to four tones to be simultaneously produced by the singer. Griots were ancient praise singers of West Africa and also important advisors to royalty.)

There are at least two major views on the effect of globalization on indigenous musical productions. The first emphasizes how the Western music industry has enabled indigenous music to be more widely disseminated and therefore more widely known and appreciated. This position sees local roots mixing with Western popular musical styles, with a hybrid sound resulting. The second view worries that the influence of Western musical styles and the Western music industry have transformed indigenous musical productions to the point where their authenticity has been lost and global musical heterogeneity diminished. Despite their fundamental disagreement, holders of both positions recognize that world music has enabled cultural diversity to flourish and hope that indigenous performers will be able to resist the power of the Western music industry to homogenize their work at the same time that they rely on it to disseminate their music (**Figure 5.32**).

A final example of the variation that exists despite the globalization of commercial cultural products is the film industry. The commercial film industry has also become global; the sites of production vary and the products themselves are significantly aesthetically and practically different. Most Americans are aware of Bollywood films, made in India and distributed widely because of the large global diaspora of Indian peoples. No one would mistake a Bollywood film for one made in Hollywood largely because Indian audiences have different expectations for a filmic experience.

[4]J. Abu-Lughod, "Going Beyond Global Babble," in A. D. King (ed.), *Culture, Globalization, and the World-System.* Basingstoke, England: Macmillan, 1991, p. 132.

Figure 5.33   On the set of *This is Nollywood* Nollywood video films are popular the world over and rival Hollywood and Bollywood in numbers produced. The video films are unlike those of either of the other two leading film industries, following different production techniques and invoking different kinds of stories, ultimately resulting in a different aesthetic. Despite the name, Nollywood is *not* a cheap imitation of Hollywood, but a unique cultural form.

One of the least-well-known among Americans, however, but arguably more popular than either Hollywood or Bollywood, is Nollywood, the film industry centered in Lagos, Nigeria. Nollywood produces over 1,000 films a year and they are watched avidly not only by Nigerians but across the globe by the large diasporic African population (**Figure 5.33**). Unlike Hollywood films, with their high production values, Nollywood video films are made on inexpensive Chinese cameras with short production schedules and a minimum of editing so that within about a three-week period the video film has been shot and edited and is available for retail consumption. Americans watching the films might complain about the quality of the lighting and sound, among other things, but Africans appreciate the video films as reflections of their own lives and experiences. As with music, the world film industry translates into anything but homogeneity of the genre.

CONCLUSION

Culture is a complex and exceedingly important concept within the discipline of geography. A number of approaches exist to understanding culture. It may be understood through a range of elements and features, from single traits to complex systems. Cultural geography recognizes the complexity of culture and emphasizes the roles of space, place, and landscape and the ecological relationships between cultures and their environment. It distinguishes itself from other disciplinary approaches, providing unique insights that reveal how culture shapes the worlds we live in at the same time that the worlds we inhabit shape culture.

Two of the most universal forms of cultural identity are religion and language. Despite the secularization of many people in core countries, religion is still a powerful form of identity, and it has been used to mediate the impacts of globalization. Globalization has caused dramatic changes in the distribution of the world's religions, as well as interaction among and between religions. Perhaps most remarkable, religious conversion to religions of the periphery is now underway in the core.

While the number of languages that exist worldwide is threatened by globalization, some cultures have responded to the threat by providing special protection for regional languages.

The 500-year history of globalization has resulted in the steady erosion of many regional languages and heavy contact with and change in the languages that persist. It is not only religion and language, however, that are transformed by globalization; other forms of cultural expression such as art and film are as well.

Different groups in different parts of the world have begun to use cultural identities such as gender, race, ethnicity, and sexuality as a way of buffering the impacts of globalization on their lives. When the impacts of globalization are examined at the local level, some groups suffer more harm or reap more benefits than others. The unevenness of the impacts of globalization and the variety of responses to it indicate that the possibility of a monolithic global culture wiping out all differences is limited.

Finally, while globalization is undoubtedly reshaping the world and bringing different cultural groups closer together than they have ever been previously, there is no conclusive evidence that globalization leads to cultural homogenization. Instead, globalization seems to be a differential process, which means that it is deployed differently in different places and experienced and responded to differently by the people who live in those places.

MAIN POINTS REVISITED

- Though culture is a central, complex concept in geography, it may be thought of as a way of life involving a particular set of skills, values, and meanings.

  Culture includes styles of dress among youth as well as operatic arias and slang and ecclesiastical languages.

- Geographers are particularly concerned about how place and space shape culture and, conversely, how culture shapes place and space. They recognize that culture is dynamic and is contested and altered within larger social, political, and economic contexts.

  The places in which cultural practices are produced shape the cultural production as much as the cultural production shapes the places in which it occurs.

- Like other fields of contemporary life, culture has been profoundly affected by globalization. However, globalization has not produced a homogenized culture so much as distinctive impacts in different societies and geographical areas as global forces come to be modified by local cultures.

  Although U.S. culture, especially commercialized culture, is widely exported around the globe, foreign cultural practices affect the United States and other parts of the world as well. The French influence on Algeria, for instance, is much more pronounced than is that of the United States.

- Contemporary approaches in cultural geography seek to understand the role played by politics and the economy in establishing and perpetuating cultures, cultural landscapes, and global patterns of cultural traits and complexes.

  For example, the state often facilitates the import or export of cultural practices, such as movies or music, so that economic growth can be enhanced.

- Cultural geography has been broadened to include analysis of gender, class, sexuality, race, ethnicity, and life-cycle stage in recognition that important differences can exist within as well as between cultures.

  What geographers find important about these identities are the ways in which they are constructed in spaces and places and the ways those spaces and places shape the identities.

- Globalization does not necessarily mean that the world is becoming more homogeneous.

  In some ways, globalization has made the local even more important than before. In other ways, it has meant that local cultures have suffered, as in the case of the disappearance of indigenous languages. In still other cases, globalization has helped to foster hybrid forms that blend distant cultures in unique ways.

KEY TERMS

cultural complex (p. 182)
cultural geography (p. 178)
cultural hearths (p. 190)
cultural landscape (p. 179)
cultural nationalism (p. 197)
cultural region (p. 182)
cultural system (p. 182)
cultural trait (p. 181)
culture (p. 174)

dialects (p. 190)
diaspora (p. 184)
ethnicity (p. 206)
folk culture (p. 178)
gender (p. 207)
genre de vie (p. 180)
hajj (p. 202)
historical geography (p. 180)
Islam (p. 202)

Islamism (p. 202)
jihad (p. 202)
kinship (p. 195)
language (p. 190)
language branch (p. 190)
language family (p. 190)
language group (p. 190)
Muslim (p. 202)

popular culture (p. 179)
race (p. 206)
racialization (p. 206)
religion (p. 183)
rites of passage (p. 182)
sexuality (p. 203)
tribe (p. 196)
world music (p. 210)

ADDITIONAL READING

Altheide, D. L., *Terrorism and the Politics of Fear*. Lanham, MD: AltaMira Press, 2006.

Anderson, K., *Race and the Crisis of Humanism*. London: Routledge, 2006.

Barber, B. R., *Jihad vs. McWorld: Terrorism's Challenge to Democracy*. New York: Ballantine Books, 2001.

Bentahila, A., and E. E. Davies, "Language Mixing in Rai Music: Localisation or Globalisation?" *Language & Communication*, 22 (April 2, 2002), pp. 187–205.

Bonnett, A., "Constructions of 'Race,' Place and Discipline: Geographies of 'Racial' Identity and Racism," *Ethnic and Racial Studies*, 19, 1996, pp. 864–883.

Cronon, W., *Changes in the Land: Indians, Colonists, and the Ecology of New England*. New York: Hill and Wang, 2003.

Davis, S., *Parades and Power: Street Theatre in Nineteenth Century Philadelphia*. Philadelphia: Temple University Press, 1986.

Duncan, J. C., N. C. Johnson, and R. H. Schein (eds.), *A Companion to Cultural Geography*. Oxford: Blackwell, 2004.

Dwyer, O. J., and J. P. Jones, III, "White Socio-Spatial Epistemology," *Social and Cultural Geography*, 1 (2), 2000, pp. 209–221.

Forman, M., and M. A. Neal (eds.), *That's the Joint! The Hip-hop Studies Reader*. New York: Routledge, 2004.

Huq, R., *Beyond Subculture: Pop, Youth and identity in a Postcolonial World*. London: Routledge, 2005.

International Labor Office, *Women and Men in the Informal Economy: A Statistical Picture*. Geneva: International Labor Office, 2002.

Jackson, P., *Maps of Meaning: An Introduction to Cultural Geography*. London: Unwin Hyman, 1989.

Kachun, M. A., *Festivals of Freedom: Memory and Meaning in African American Emancipation Celebrations, 1808–1915*. Amherst, MA: University of Massachusetts Press, 2003.

Katz, C., and J. Monk, *Full Circles: Geographies of Women over the Life Course.* London: Routledge, 1993.

Kepel, G., *Jihad: The Trail of Political Islam.* Cambridge, MA: Harvard University Press, 2002.

Marston, S., "Making a Difference? Conflict over Irish Identity in the New York City St. Patrick's Day Parade," *Political Geography, 21*(3), 2001, pp. 373–392.

Mitchell, D., *Cultural Geography: A Critical Introduction.* Malden, MA: Blackwell, 2000.

Papademetriou, D. G., and D. W. Meyers, *Caught in the Middle: Border Communities in the Era of Globalization.* Washington, DC: Carnegie Endowment for International Peace, 2001.

Ramsey, G. P., *Race Music: Black Cultures from Bebop to Hip-hop.* Berkeley: University of California Press, 2003.

Ritzer, G., *McDonaldization: The Reader.* Thousand Oaks, CA: Pine Forge Press, 2002.

Robinson, J. M., "The Information Revolution—Culture and Sovereignty—A Canadian Perspective," *Canada—United States Law Journal, 24,* 1998, pp. 147–153.

Seago, A., "Where Hamburgers Sizzle on an Open Grill Night and Day: Global Pop Music," *American Studies, 41*(2/3, Summer/Fall 2000), pp. 119–136.

Stump, R:, *The Geography of Religion: Faith, Place and Space.* Landham, MD: Rowman and Littlefield, 2008.

Venkatesh, S. A., *Gang Leader for a Day: A Rogue Sociologist Takes to the Streets.* New York: Penguin Press, 2008.

EXERCISES

On the Internet

The Companion Website at www.mygeoscienceplace.com contains a number of tools designed to reinforce the geographic concepts presented in this chapter. Structured within a four-step learning path, these online resources include: chapter Key Concepts; a fully-searchable myeBook version of the chapter, which allows you to add notes and highlights; RSS news feeds from various media agencies; links to additional web resources; Human Geography Videos; a digital Glossary; Concept Review questions; Thinking Spatially mapping exercises; and Critical Thinking essay questions.

Unplugged

1. Using *Billboard Magazine* (the news magazine of the record industry), construct a historical geography of the top 20 singles over the last half century in order to determine how different regions of the country (or the globe!) have risen and fallen in terms of musical significance. You should also determine an appropriate interval for sampling—every three to five years is a generally accepted one. You may use the hometowns of the recording artists or the headquarters of the recording studios as your geographic variables. Once you have organized your data, you should be able to answer the following questions: How has the geography you have documented changed? What might be the reasons for these changes? What do these changes mean for the regions of the country (or globe) that have increased or decreased in terms of musical prominence?

2. Ethnic identity is often expressed spatially through the existence of neighborhoods or business areas dominated by members of a particular group. One way to explore the spatial expression of ethnicity in a place is to look at newspapers over time. In this exercise you are to look at ethnic change in a particular neighborhood over time. You can do this by using your library's holdings of local or regional newspapers. Examine change over at least a four-decade period. To do this, you must identify an area of the city in which you live or some other city for which your library has an extensive newspaper collection. You should trace the history of an area you know is now occupied by a specific ethnic group. How long has the group occupied that area? What aspects of the group's occupation of that area have changed over time (school, church, or sports activities, or the age of the households)? If different groups have occupied the area, what might be the reasons for the changes?

3. College and university campuses generate their own cultural practices and ideas that shape behaviors and attitudes in ways that may not be so obvious at first glance. For this exercise you are to observe a particular practice that occurs routinely at your college or university. For example, fraternity and sorority initiations are important rituals of college life, as are sports events and class discussions. Observe a particular cultural practice that is an ordinary part of your life at college. Who are the participants in this practice? What are their levels of importance? Are there gender, age, or status differences in those who carry out this practice? What are the time and space aspects of the practice? Who controls its production? What are the intended outcomes? How does the practice contribute to or detract from the maintenance of order in the larger culture?

4. Using your local library as your source, find a description of a coming-of-age ceremony for any part of the world other than the United States. Summarize that description and then compare it to one you have either experienced directly or observed in the United States. What are the differences and the similarities between your experience and the one you read about? What might be some of the reasons for these?

Interpreting Places and Landscapes

6

On 8/8/2008 at 8 pm Chinese standard time (because 8 is a special number in Chinese culture), the world was shown a slice of the new Beijing, when the Summer Olympics opening ceremonies commenced. The program at Beijing's iconic national stadium—known informally as the Bird's Nest—showcased more than 15,000 performers, largely from the People's Liberation Army song and dance troupes, who through various pieces honored the nation's ancient history and its rise to modern power. Behind the gorgeousness of the costumes, the awesome synchronicity of 2,000 percussionists, the ethereal beauty of the aerial artists, and the marvellous technological feats and computer graphics was a subtle message: the reignited greatness of China. On more than one occasion, broadcasters noted that in ten of the last eleven centuries (all except the twentieth) China possessed the largest economy in the world; and it expects to be a global powerhouse once again. Because of its vast territory and huge population, as well as its extensive natural resources and its drive to embrace capitalist growth, China's potential is unarguably enormous.

Although the spell-binding opening ceremony lasted a bit less than 4 hours, the transformed cityscape of Beijing is meant to be a more enduring reminder of China's ancient economic, political, and cultural greatness and, most importantly, its current ascendancy. In addition to the Bird's Nest, the Chinese government invested heavily in the construction of 31 new Beijing-based Olympic Games facilities, completely overhauled its transportation systems, and renovated an additional 59 training facilities. Beijing's airport underwent a major expansion, adding the new Terminal 3, the largest airport terminal in the world. Simultaneously, they opened the new Beijing South Railway Station and the 120-kilometer (74.5-mile) Beijing–Tianjin Intercity Rail, which possesses the world's fastest scheduled train service at 350 km/hr (217.5 mi/hr). As a result of all these and over 10,000 other construction projects throughout the city, the glittering modern and the enchanting ancient come together in Beijing's new landscape and, like the opening ceremonies, project for all to see the capital city's historic and gathering power.

As in all landscapes however, there is more to its material form than meets the eye. For instance, the Chinese government, at the same time that it claims the Bird's Nest is the world's largest steel structure, also boasts that it was built for only $423 million, a tenth of the price it would have cost had it been built in any western city. This is a building that at peak construction employed 17,000 workers. How did the government manage such a feat so cheaply? A report by Human Rights Watch in spring 2008 found that 90 percent of Beijing's vast army of construction workers are rural migrants who suffer a range of abusive conditions and discriminatory treatment from their employers and the government. For instance, they received meager wages, the payment of which was sometimes delayed for over a year: some workers, despite putting in 17-hour days, were never paid at all. A survey conducted by the Chinese government, also in

Main Points

- In addition to understanding how the environment shapes (and is shaped by) people, geographers seek to identify how it is perceived and understood by people.

- The language in which a landscape is written is a kind of code. The code includes signs that direct our attention toward certain features and away from others.

- The written code of landscape is also known as semiotics. Codes signify important information about landscapes, such as whether they are sacred or profane, accessible or off-limits, or oriented toward work or play.

- Different cultural identities and status categories influence the ways people experience and understand their environments, as well as how they are shaped by—and able to shape—them.

- Landscape serves as a kind of archive of society. It is a reflection of our culture and our experiences. Like a book, landscape is a text that is written by individuals and groups, and read by them as well.

- The emergence of the most recent phase of globalization has occurred in parallel with a transition from modernity to postmodernity.

2008, found that the average migrant construction worker was forced to work for 10 hours a day, 27 days a month, in blatant violation of Chinese law. Most who were injured because of dangerous building-site conditions were denied access to medical care and other social benefits.

In addition to tolerating widespread deplorable working conditions, the Chinese government has also been criticized for its callous appropriation of residences and whole neighborhoods in order to clear land for the Olympic venues as well as related public works and commercial and residential projects. It is estimated that hundreds of thousands of people were forcibly evicted with nowhere to go as they watched their homes demolished in order to make way for Beijing's Olympic makeover. Many tried to fight the evictions but their trial dates were constantly postponed, or never scheduled, which meant that their homes were lost anyway. And, because there is no legal right to assembly in China, residents of the city center were unable to come together to protest these actions collectively. In Beijing, more than one-half of the historic center, a good portion of it dating back to the Ming Dynasty, has been destroyed. And, it should be noted, the practices that enabled these employment abuses and land seizures occurred despite China's promise upon submitting the Olympic bid, to uphold international human rights standards.

The underside of Beijing's Olympic facelift serves to remind us that landscape is a material form produced by complex processes that are both hidden and apparent. In this chapter, we explore how to approach landscape as the manifestation of multifaceted relationships among people, places, and spaces in order to assess how individuals and groups experience their environments, create and struggle within places, and find meanings in the landscapes they create.

BEHAVIOR, KNOWLEDGE, AND HUMAN ENVIRONMENTS

In addition to attempting to understand how the environment shapes and is shaped by people, geographers also seek to identify how it is perceived and understood by people. Arguing that there is an interdependence between people and places, geographers explore how individuals and groups acquire knowledge of their environments and how this knowledge shapes their attitudes and behaviors. Some geographers have focused their research on natural hazards as a way of addressing environmental knowledge, while others have tried to understand how people ascribe meaning to landscape and places. In this chapter, we take the key geographical concepts of place, landscape, and space and explore the ways in which people understand them, create them, and operate within them.

In their attempts to understand environmental perception and knowledge, geographers share a great deal with other social scientists, but especially with psychologists. Human cognition and behavior are at the center of psychology. What makes environmental knowledge and behavior uniquely geographical is their relation to both the environmental context and the humans who struggle to understand and operate within it. Much of what we as humans know about the environment we live in is learned through direct and indirect experience. Our environmental knowledge is also acquired through a filter of personal and group characteristics, such as race, gender, stage of the life cycle, religious beliefs, and where we live (**Figure 6.1**).

For instance, children have interesting and distinct relationships to the physical and cultural environment. How do children acquire knowledge about their environments? How do boys differ from girls in the ways they learn about and negotiate their environments? What kind of knowledge do they acquire, and how do they use it? What role do cultural influences play in the process? What happens when larger social, economic, and environmental changes take place?

Geographer Cindi Katz conducted research in rural Sudan to find answers to these questions. Working with a group of 10-year-old children in a small village, she sought to discover how they acquired environmental knowledge. What she also learned through this work was how the transformation of agriculture in the region changed not only the children's relationships to their families and community but also their perceptions of nature.[1] In this Sudanese village, as in similar communities elsewhere in the periphery, children are important contributors to subsistence activities, especially planting, weeding, and harvesting. The villagers were strict Muslims and thus had stringent rules about what female members of the community were allowed to do and where they were allowed to

[1]C. Katz, *Growing Up Global: Economic Restructuring and Children's Everyday Lives.* Minneapolis: University of Minnesota Press, 2004.

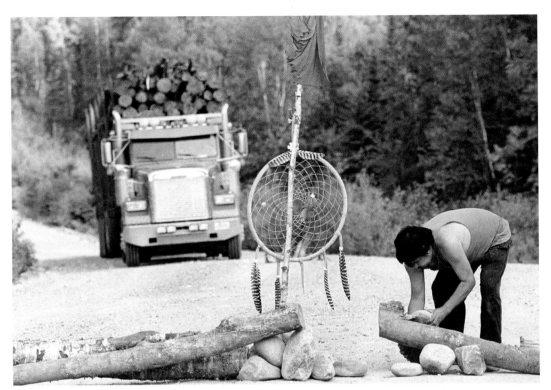

Figure 6.1 Conflicting environmental perceptions Pictured here is a member of the Algonquin Indian Tribe attempting to prevent a logging truck from passing along a forest road. In conflict here are opposing perceptions of the forest, where some members of the tribe see it as a place of spiritual renewal while lumber companies see it as the source of harvestable commodities.

go. Many of the subsistence activities that required leaving the family compound were customarily delegated to male children. In fact, within the traditional subsistence culture, boys predominated in all agricultural tasks except planting and harvesting and were responsible for herding livestock as well. Many boys (and occasionally girls) were also responsible for fetching water and helping to gather firewood. Both boys and girls collected seasonal foods from lands surrounding the village. Work and play were often mixed together, and play, as well as work, provided a creative means for acquiring and using environmental knowledge and for developing a finely textured sense of the home area (**Figure 6.2**).

What happens when globalization reshapes the agricultural production system, as it did in the village when, through an international development scheme, irrigated cash-crop cultivation was introduced? With the financial assistance of outside donors, a Sudanese government development project transformed the agriculture of the region from subsistence-level livestock raising and cultivation of sorghum and sesame to cultivation of irrigated cash crops such as cotton. The new cash-crop regime, which required management of irrigation works, application of fertilizers, herbicides, and pesticides, and more frequent weeding, required children as well as adults to work longer and harder. Parents were often forced to keep their children out of school because many of the tasks had to be done during the school term.

Loss of forests required children to range farther afield to procure fuelwood and to make gathering trips more frequently (**Figure 6.3**). Soon wealthier households began buying wood rather than increasing the demands on their children or other household members. For the children of more marginalized households, the selling of fuelwood, foods, and other items provided a new means for earning cash to support their families but also placed increasing demands on their energies and resourcefulness and changed their whole experience of their world.

For the children of this village, globalization (in the form of the transition to cash-crop agriculture) changed their relationships with their environments and with their futures. The kinds of skills the children had learned for subsistence production were no longer useful for cash-crop production. As they played less and worked longer hours in more specialized roles, their experience of their environment became narrowed. As their roles within the family changed, they attended school less and learned less about their world through formal education. As a result, their perceptions of their environment changed, along with their values and attitudes toward the landscape and the place they knew as home.

Today many of the children face a considerable challenge, for there are simply not enough agricultural jobs to go around. Many of the boys—especially those whose parents are tenant farmers and lacking status—will be left with few options but to become agricultural wage laborers or to seek nonagricultural work outside the village. The girls, trained largely to assume their mothers' household roles, face another challenge—how to take over the agricultural tasks previously performed by the boys, but without the advantage of knowing how to do those tasks or use the proper tools. Traditional relationships within village society have been undermined by globalization, and traditional opportunities to acquire important environmental knowledge are being lost to new generations of children.

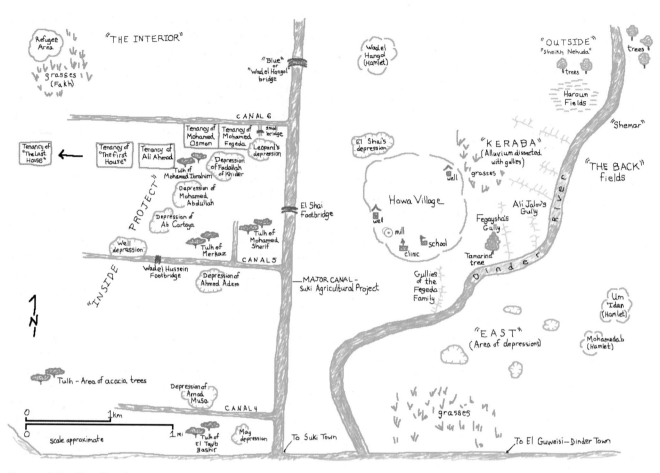

Figure 6.2 Shepherd's map This map, drawn by a 10-year-old Sudanese boy, shows the area over which the sheep are herded. It illustrates the detailed environmental knowledge Sudanese children possess of the landscape that surrounds their village. The village is an Islamic one, and norms determine the kinds of tasks in which boys and girls can participate. Only boys are allowed to tend sheep, which requires a particular environmental knowledge about grazing areas and water availability. (*Source:* Image courtesy of C. Katz)

PLACE-MAKING

As we saw in Chapter 1, places are socially constructed—given different meanings by different groups for different purposes. Most people identify with places as part of their personal identity, drawing on particular images and particular histories of places in order to lend distinctiveness to both their individuality and their sense of community. But identifying with place may also imply the exclusion of other people and the stereotyping of

Figure 6.3 Sudanese children working and playing The introduction of an agricultural development project brought the forces of globalization very directly to this small village in Sudan. The project dramatically changed both work and play for children and created a gap between the environmental knowledge they traditionally learned and the knowledge they would need to survive under different economic circumstance. In this photograph, two 10-year-old girls are shown gathering firewood for domestic use. The wood is of very poor quality, and the girls had to range far to procure it. These branches were gathered in and around the irrigation canal, which for all members of the village is a risky environmental site. The waters in the canals contain the parasitic blood flukes that cause schistosomiasis, a chronic disease that can affect the liver, urinary bladder, lungs, or central nervous system.

other places. People often reinforce their sense of place and of who they are by contrasting themselves with places and people they feel are very different from them. Seen in this context, place-making stands at the center of issues of culture and power relations, a key part of the systems of meaning through which we make sense of the world.

Territoriality

Some social scientists believe that wanting to have a place where you feel you belong is a natural human attribute, part of a strong territorial instinct. Humans, it is argued, have an innate sense of territoriality, just like many other species. The concept of **territoriality** refers to the persistent attachment of individuals or peoples to a specific location or territory. The concept is important to geographers because it can be related to fundamental place-making forces.

The specific study of people's sense of territoriality is part of the field of **ethology**, the scientific study of the formation and evolution of human customs and beliefs. The term is also used to refer to the study of the behavior of animals in their natural environments. According to ethologists, humans carry genetic traits produced by our species' need for territory. Territory provides a source of physical safety and security, a source of stimulation (through border disputes), and a physical expression of identity. These needs add up to a strong territorial urge that can be seen in the microgeography of people's behavior: claims to space in reading rooms or on beaches, for example, and claims made by gangs to neighborhood turf (**Figure 6.4**). Ethologists argue that the territorial urge also can be observed when people become frustrated because of overcrowding. They become stressed and, in some circumstances, begin to exhibit aggressive or deviant behavior. Ethologists and environmental psychologists link crowding to everything from vandalism and assault to promiscuity, listlessness, and clinical depression.

While such claims are difficult to substantiate, as is the whole notion that humans have an inborn sense of territoriality, the idea of territoriality as a product of *culturally* established meanings is supported by a large body of scientific evidence. Some of this evidence comes from the field of **proxemics**, the study of the social and cultural meanings that people give to personal space. These meanings make for unwritten territorial rules (rather than a biological urge) that can be seen in the microgeography of people's behavior. It has been established, for example, that people develop unwritten protocols about how to claim space. One common protocol is simply regular use (think of students' habits in classroom seating arrangements). Another is through the use of spatial markers such as a newspaper or a towel to fix a space in a reading room or on a beach. There are also bubbles, or areas, of personal space that we try not to

Figure 6.4 Graffiti as territorial markers Graffiti are used by neighborhood gangs to establish and proclaim their identity. Some graffiti, such as these from a Cambodian gang in Long Beach, California, also function as simple territorial markers that help to stake out turf in high-density environments where there exist few other clues about claims to territory.

invade (or allow to be invaded by others). Varying in size and shape according to location and circumstance, these bubbles tend to be smaller in public places and in busier and more crowded situations; they tend to be larger among strangers and in situations involving members of different social classes; and they tend to vary from one social class or cultural group to another.

At larger spatial scales, territoriality is mostly a product of forces that stem from political relations and cultural systems. This dimension of territoriality underpins a great deal of human geography. All social organizations and the individuals who belong to them are bound at some scale or another through formal or informal territorial limits. Many of them—nations, corporations, unions, clubs—actually claim a specific area of geographic space under their influence or control. In this context, territoriality can be defined as any attempt to assert control over a specific geographic area to achieve some control over other people, resources, or relationships. Territoriality also is defined as any attempt to fulfill socially produced needs for identity, defense, and stimulation. Territoriality covers many phenomena, including the property rights of individuals and private corporations; the neighborhood covenants of homeowners' associations; the market areas of commercial businesses; the heartlands of ethnic or cultural groups; the jurisdictions of local, state, and national governments; and the reach of transnational corporations and supranational organizations.

Territoriality thus provides a means of meeting three social and cultural needs:

- The regulation of social interaction
- The regulation of access to people and resources
- The provision of a focus and symbol of group membership and identity

Territoriality meets these needs because, among other things, it facilitates classification, communication, and enforcement. We can classify people and/or resources in terms of their location in space much more easily than we can classify them in relation to personal or social criteria. All that is necessary to communicate territory is a simple marker or sign that constitutes a boundary. This in turn makes territory an efficient device for determining whether people are subject to the enforcement of a particular set of rules and conditions.

Territoriality also gives tangible form to power and control, but does so in a way that directs attention away from the personal relationships between the controlled and the controllers. In other words, rules and laws become associated with particular spaces and territories rather than with particular individuals or groups. Finally, territoriality allows people to create and maintain a framework through which to experience the world and give it meaning. Bounded territories, for example, make it easier to differentiate "us" from "them."

People and Places, Insiders and Outsiders

As we saw in Chapter 1, places are constantly under social construction by people responding to the opportunities and constraints of their particular locality. As people live and work in places, they gradually impose themselves on their environment, modifying and adjusting it to suit their needs and express their values. At the same time, they gradually accommodate both to their physical environment and to the values, attitudes, and comportment of people around them. People are constantly modifying and reshaping places, and places are constantly coping with change and influencing their inhabitants.

Places are both centers of meaning for people and the external frameworks for their actions and behavior. It is important to remember that places are constructed by their inhabitants from their own subjective point of view and that they are simultaneously constructed and seen as an external "other" by outsiders. A neighborhood, for example, is both an area of special meanings to its residents as well as an area containing houses, streets, and people that others may view from an outsider's ("decentered") perspective.

A fundamental element in individuals' construction of place is the **existential imperative** for people to define themselves in relation to their material world—their basic need and capacity to achieve a form of spiritual or psychic unity between themselves and their material worlds. People's subjective "creation" of space and place provides them with roots and a sense of identity—a phenomenon known to philosophers in terms of "dwelling." People's innate capacity for "dwelling" allows them to give meanings to places, through repeated experience, that are deepened and qualified over time with multiple nuances.

Yet the construction of place by individual insiders cannot occur independently of local societal norms that codify and frame the social construction of spaces and places. People's territoriality and sense of "dwelling," in other words, are shaped by locally shared notions of social distance, rules of comportment, forms of social organization, and so on. There is, therefore, an important two-way relationship between social structures and the everyday practices of the "insiders" of subjectively constructed spaces and places. We live both *in* and *through* places.

As we saw in Chapter 1, a crucial concept here is that of the *lifeworld*, the taken-for-granted pattern and context for everyday living through which people conduct their day-to-day lives without having to make it an object of conscious attention. People's familiarity with one another's vocabulary, speech patterns, dress codes, gestures, and humor, and with shared experiences of their physical environment, often carries over into people's attitudes and feelings about themselves and their locality and to the symbolism they attach to that place. When this happens, the result is a collective and self-conscious "structure of feeling": a sociocultural frame of reference generated among people as a result of the experiences and memories that they associate with a particular place.

Over the past couple of decades, however, people and places have been confronted with change on an unprecedented scale and at an extraordinary rate. As we saw in Chapter 2, globalization has generated a world within which commonalities among places are intensifying. The more that places change, the more they seem to look alike, the less they are able to retain a distinctive sense of place, and the less they are able to sustain public social life. As a result, the experience of spectacular and distinctive places, physical settings, and landscapes has become an important element of consumer culture. Responding to this shift, developers have created theme parks, shopping malls, festival marketplaces, renovated historic districts, and neo-traditional villages and neighborhoods. But the more developers have competed to provide distinctive settings and the larger and more spectacular their projects, the more convergent the results. The inevitable outcome is that the "authenticity" of places is undermined. City spaces, in particular, become inauthentic and "placeless."

Experience and Meaning

The interactions between people and places raise some fundamental questions about the meanings that people attach to their experiences: How do people process information from external settings? What kind of information do they use? How do new experiences affect the way they understand their worlds? What meanings do particular environments have for individuals? How do these meanings influence behavior? How do people develop and modify their sense of a place, and what does it mean to them? The answers to these questions are by no means complete, but it is clear that people not only filter information from their environments through neurophysiological processes but also draw on personality and culture to produce cognitive images of their environment, pictures or representations of the world that can be called to mind through the imagination (**Figure 6.5**). Cognitive images are what people see in the mind's eye when they think of a particular place or setting.

Two of the most important attributes of cognitive images are that they both simplify and distort real-world environments. Research on the ways in which people simplify the world through such means has suggested, for example, that many people tend to organize their cognitive images of particular parts of their world in terms of several simple elements (**Figure 6.6**):

Paths: The channels along which they and others move; for example, streets, walkways, transit lines, canals.

Edges: Barriers that separate one area from another; for example, shorelines, walls, railroad tracks.

Districts: Areas with an identifiable character (physical and/or cultural) that people mentally "enter" and "leave"; for example, a business district or an ethnic neighborhood.

Nodes: Strategic points and foci for travel; for example, street corners, traffic junctions, city squares.

Landmarks: Physical reference points; for example, distinctive landforms, buildings, or monuments.

For many people, individual landscape features may function as more than one kind of cognitive element. A freeway, for instance, may be perceived as both an edge and a path in a person's cognitive image of a city. Similarly, a railroad terminal may be seen as both a landmark and a node.

Distortions in people's cognitive images are partly the result of incomplete information. Once we get beyond our immediate living area, we know few spaces in complete detail. Yet our worlds—especially those of us in the fast world who are directly tied to global networks of communication and knowledge—are increasingly large in geographic scope. As a result, these worlds must be conceived, or understood, without many direct stimuli. We have to rely on fragmentary and often biased information from

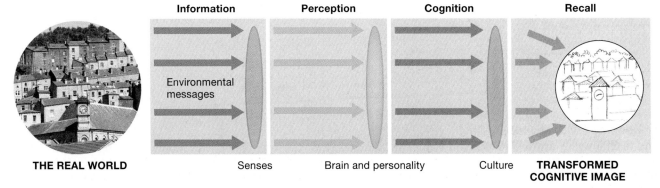

Figure 6.5   The formation of cognitive images People form cognitive images as a product of information about the real world, experienced directly and indirectly, and filtered through their senses, brain, personality, and the attitudes and values they have acquired from their cultural background. (*Source:* Adapted from R. G. Golledge and R. J. Stimpson, *Analytical Behavior Geography.* Beckenham, UK: Croom Helm, 1987, Fig. 3.2, p. 3.)

Figure 6.6   Cognitive image of Boston This map was compiled by Kevin Lynch, one of the pioneer researchers into cognitive images, from interviews with a sample of Boston residents. Lynch found that the residents of Boston tended to structure their cognitive images of the city with the same elements. He produced ingenious maps, such as this one, to demonstrate the collective "mental map" of the city, using symbols of different boldness or color to indicate the proportion of respondents who had mentioned each element. (*Source:* Adapted from K. Lynch, *The Image of the City.* Cambridge, MA: M.I.T. Press, 1960, p. 146.)

other people, from books, magazines, television, and the Internet. Distortions in cognitive images are also partly the result of our own biases. What we remember about places, what we like or dislike, what we think is significant, and what we impute to various aspects of our environments all are functions of our personalities, our experiences, and the cultural influences to which we have been exposed.

Images and Behavior

Cognitive images are compiled, in part, through behavioral patterns. Environments are "learned" through experience. Meanwhile, cognitive images, once generated, influence behavior. In these two-way relationships, cognitive images are constantly changing. Each of us also generates—and draws on—different kinds of cognitive images in different circumstances.

Elements such as districts, nodes, and landmarks are important in the kinds of cognitive images that people use to orient themselves and to navigate within a place or region. The more of these elements an environment contains—and the more distinctive they are—the more legible that environment is to people and the easier it is to get oriented and navigate. In addition, the more first-hand information people have about their environment and the more they are able to draw on secondary sources of information, the more detailed and comprehensive their images will be.

This phenomenon is strikingly illustrated in **Figure 6.7**, which shows the collective image of Los Angeles as seen by the residents of three different neighborhoods: Westwood, an affluent neighborhood; Avalon, a poor, inner-city neighborhood; and Boyle Heights, a poor, immigrant

neighborhood. The residents of Westwood have a well-formed, detailed, and comprehensive image of the entire Los Angeles basin. At the other end of the socioeconomic spectrum, residents of the black ghetto neighborhood of Avalon, near Watts, have a vaguer image of the city, structured only by the major east–west boulevards and freeways and dominated by the gridiron layout of streets between Watts and the city center. The Spanish-speaking residents of Boyle Heights—even less affluent, less mobile, and somewhat isolated by language—have an extremely restricted image of the city. Their world consists of a small area around Brooklyn Avenue and First Street, bounded by the landmarks of city hall, the bus depot, and Union Station.

The importance of these images goes beyond people's ability simply to navigate around their environments. The narrower and more localized people's images are, for example, the less they will tend to venture beyond their home area. Their behavior becomes circumscribed by their cognitive imagery. People's images of places also shape particular aspects of their behavior. Research on shopping behavior in cities, for example, has shown that customers do not necessarily go to the nearest store or to the one with the lowest prices; they are influenced by the configuration of traffic, parking, and pedestrian circulation within their imagery of the retail environment. The significance of this clearly has not been lost on the developers of shopping malls, who always provide extensive space for free parking and multiple entrances and exits.

In addition, shopping behavior, like many other aspects of behavior, is influenced by people's values and feelings. A district in a city, for example, may be regarded as attractive or repellent, exciting or relaxing, fearsome

reckless p
discount
tends to
overestima
adopt diff
tions, mal
models ba

Finally
importance
tal and sy
their daily
ences and
fective bor
different g
neighborh
dency for
Topophilia
use the te
meanings
reason or
The result
town, or h
ment or se

LANDS

As we saw
different th
brings to n
as in lands
fies a bucc
plantings
still others,
of scenery,

Geogra
landscapes)
of particula
of power,"
"landscapes
and **derelic**
rienced abar
Geographer
reflect certa
builders or
by other gr
are so pow
entire cities
Rome, and
(**Figure 6.10**
are most int

Ordinar

As we saw
symbolize e
landscapes
are underst

Figure 6.7 Images of Los Angeles These images, as seen by residents of (a) Westwood, (b) Avalon, and (c) Boyle Heights are drawn to the same scale. The intensity of particular elements in the urban images of residents of different areas is measured by the percentage of residents who recall a particular element as being part of their own mental map of the city. (*Source:* Adapted from P. Orleans, in R. Downs and D. Stea (eds.), *Image and Environment.* Chicago: Aldine, 1973, pp. 120–122.)

or reassuring, or, more likely, a combination of such feelings. As with all other cognitive imagery, such images are produced through a combination of direct experience and indirect information, all filtered through personal and cultural perspectives. Images such as these often exert a strong influence on behavior. Returning to the example of consumer behavior, one of the strongest influences on shopping patterns relates to the imagery evoked by retail environments—something else that has not escaped the developers of malls, who spend large sums of money to establish the right atmosphere and image for their projects.

Shopping behavior is one narrow example of the influence of place imagery on behavior; other examples can be drawn from every aspect of human geography and at every spatial scale. The settlement of North America, for example, was strongly influenced by the changing images of the Plains and the West. In the early 1800s, the Plains and the West were perceived as arid and unattractive, an image that was reinforced by early atlases. When the railroad companies wanted to encourage settlement in these regions, they changed people's image of the Plains and the West with advertising campaigns that portrayed them as fertile and hospitable regions. The images

Figure 6.8
of the Unite
isoline map il
preferences f
coterminous l
places in whic
work, as expre
ture students
in 1996. It is a
based on the s
dents gave to
cities in the cc
the score, the
preference for
working there

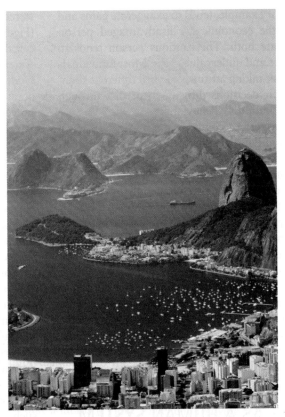

associated w
to shape settl
tive imagery,
migrating fro
composite im
Virginia Tech
ness of cities

Another
on people's b
environmenta
quakes, storm

Figure 6.9   In
and attitudes
sons tend to ove
a loss while und
of a gain. For re
true. Many peop
or reckless, their
where between
toward risk-taki
level of affluenc
gerate both larg
rich persons tend
(Source: Adapted fr
Analytical Behavior
Croom Helm, 1987,

Figure 6.10   Famous landmarks Some cities are immediately recognizable because of certain buildings, landmarks, or cityscapes that have come to symbolize them. These examples are known worldwide.

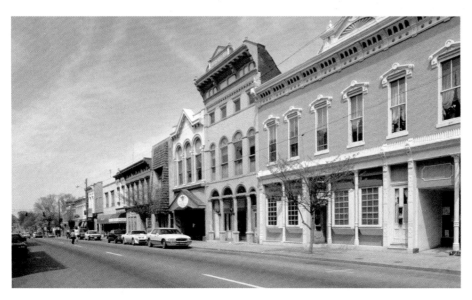

Figure 6.11 Ordinary landscapes
Some ordinary cityscapes are power-fully symbolic of particular kinds of places. The New England village and the Main Street of Middle America are in this category, so much so that they have been taken as symbolizing the United States itself—part of the "iconography of nationhood," the symbolic landscapes that give the country a sense of identity, both at home and abroad.

Figure 6.12 Vulgaria
The dominant theme in upscale residential development in the United States is size and ostentation.

landscape differently is central to the **humanistic approach** in geography, which places the individual—especially individual values, meaning systems, intentions, and conscious acts—at the center of analysis. As the Sudanese example given earlier suggests, children's perceptions of their worlds are different from those of their parents, and girls see their world differently from boys, even in the same family.

Environmental perception, and its close relative behavioral geography, are interdisciplinary, drawing together geographers, landscape architects, psychologists, architects, and others. Professionals in these disciplines investigate what preferences people have in landscapes, how they construct cognitive images of their worlds, and how they find (or fail to find) their way around in various settings. Recall from Chapter 1 that cognitive images are representations of the world that can be called to mind through the imagination. The humanistic approach's focus on the perceptions of individuals is an important counterweight to the tendency to talk about a social group or society more generally. Nevertheless, some critics argue that humanistic research has limited utility because individual attitudes and views do not necessarily add up to the views held by a group or a society.

One alternative to the humanistic approach explores both the role of larger forces, such as culture, gender, and the state, and the ways in which these forces enhance or constrain individuals' lives. Much recent cultural geographical work, therefore, conceptualizes the relationship of people and the environment as interactive, not one-way, and emphasizes the role that landscapes play in shaping and reinforcing human practices. This most recent conceptualization of landscape is more dynamic and complex than the one Carl Sauer advanced, and it encourages geographers to look outside their own discipline—to anthropology, psychology, sociology, and even history—to fully understand its complexity.

Landscape as Text

Such a dynamic and complex approach to understanding landscape is based on the conceptualization of **landscape as text**, by which we mean that, like a book, landscape can be read and written by groups and individuals. This approach departs from traditional attempts to systematize or categorize landscapes based on the different elements they contain. The landscape-as-text view holds that landscapes do not come ready-made with labels on them. Rather, there are "writers" who produce landscapes and give them meaning, and there are "readers" who consume the messages embedded in landscapes. Those messages can be read as signs about values, beliefs, and practices, though not every reader will take the same message from a particular landscape (just as people may differ in their interpretation of a passage from a book). In short, landscapes both produce and communicate meaning, and one of our tasks as geographers is to interpret those meanings.

CODED SPACES

Landscapes, then, are embedded with meaning, which can be interpreted differently by different people and groups. But in order to interpret or read our environment, we need to understand the language in which it is written. We must learn how to recognize the signs and symbols that go into the making of landscape. The practice of writing and reading signs is known as **semiotics**.

Semiotics in the Landscape

Semiotics assumes that innumerable signs are embedded or displayed in landscape, space, and place, sending messages about identity, values, beliefs, and practices. The signs that are constructed may have different meanings for those who produce them and those who read, or interpret, them. Some signs are so subtle as to be recognizable only when pointed out by a knowledgeable observer; others may be more readily available and more ubiquitous in their spatial range. For example, semiotics enables us to recognize that college students, by the way they dress, send messages to each other and the wider world about who they are and what they value. For some of us, certain types, such as the "skater" or the "emo," are readily identifiable by their clothes, hairstyle, or footwear, by the books they carry, or even by the food they eat.

Semiotics, however, is not only about the signs that people convey by their mode of dress. Messages are also deployed through the landscape and embedded in places and spaces. Consider the very familiar landscape of the shopping mall. Although there is certainly a science to the size, scale, and marketing of a mall based on demographic research as well as environmental and architectural analysis, there is more to the mall than these concrete features. The placement and mix of stores and their interior design, the arrangement of products within stores, the amenities offered to shoppers, and the ambient music all combine to send signals to the consumer about style, taste, and self-image (**Figure 6.13**). Called by some "palaces of consumption," malls are complex semiotic sites, directing important signals not only about what to buy but also about who should shop there and who should not.

To fully appreciate the importance of shopping to life in the United States, consider the following statistics. In 2007, according to the National Research Bureau, there were nearly 50,000 shopping centers in the United States with annual sales of around $1,500 billion (not including online shopping). In a typical month nearly 200 million individuals over age 18 shop at shopping centers. It is not clear how many in the 13–17 age group are also shopping, but we do know this number is growing rapidly.

As much as Americans seem to enjoy shopping, a great many express disdain for shopping and the commercialism and materialism that accompany it. Thus, shopping is a complicated activity about which people feel ambivalence.

Figure 6.13   The Forums Mall, Caesar's Palace, Las Vegas Casinos in Las Vegas, such as this one, have invested heavily in creating opulent malls that are situated close to the gaming rooms. Developers of shopping malls know that consumer behavior is heavily conditioned by the spatial organization and physical appearance of retail settings. As a result, they find it worthwhile to spend large sums creating what they consider to be the appropriate atmosphere for their projects.

It is not surprising, therefore, that developers have promoted shopping as a kind of tourism. The mall is a "pseudoplace" meant to encourage one sort of activity—shopping—by projecting the illusion that something else besides shopping (and spending money) is actually going on. Because of their important and complex function, malls are places with rich semiotic systems expressed through style, themes, and fantasy. The South Coast Plaza in Orange County, California, is a highly popular retail center in the United States, and with almost 3 million square feet of enclosed space covering 128 acres, it is one of the largest malls in the country. It is also probably one of, if not the most profitable, boasting over $20 billion annually in sales. The mall contains luxury goods stores like Gucci, Versace, Chanel, Tiffany, Jimmy Choo, and Cartier—among a wide range of other upscale venues—more popular, less pricey upscale stores like Bloomingdales and Nordstrom, as well as more middle-brow stores like Macy's and Sears. The latter stores anchor the mall at its outside corners, while the more expensive and luxurious shops and boutiques occupy interior locations. Thus the central stretches of the mall convey signs of affluence and luxury appealing to upper-class patrons while the periphery is more oriented to necessity and practicality for middle- and lower-middle-class patrons.

However complex the messages that malls send, one focus is consistent across class, race, gender, age, ethnicity, and other cultural boundaries: consumption, a predominant aspect of globalization. Indeed, malls are the early twenty-first century's spaces of consumption, where just about every aspect of our lives has become a commodity. Consumption—or shopping—defines who we are more than ever before, and what we consume sends signals about who we want to be. Advertising and the mass media tell us what to consume, equating ownership of products with happiness, a good sex life, and success in

general. Within the space of the mall these signals are collected and resent. The architecture and design of the mall are an important part of the semiotic system shaping our choices and molding our preferences. As architectural historian Margaret Crawford writes:

> All the familiar tricks of mall design—limited entrances, escalators placed only at the end of corridors, fountains and benches carefully positioned to entice shoppers into stores—control the flow of consumers through the numbingly repetitive corridors of shops. The orderly processions of goods along endless aisles continuously stimulate the desire to buy. At the same time, other architectural tricks seem to contradict commercial consideration. Dramatic atriums create huge floating spaces for contemplation, multiple levels provide infinite vistas from a variety of vantage points, and reflective surfaces bring near and far together. In the absence of sounds from the outside, these artful visual effects are complemented by the "white noise" of MUZAK and fountains echoing across enormous open courts.[3]

Malls, condominium developments, neighborhoods, university campuses, and any number of other possible geographic sites possess codes of meaning. By linking these sites with the forces behind globalization, it is possible to interpret them and understand the implicit messages they contain. And it is certainly not necessary to restrict our focus to sites in the core.

Consider Brasilia, the capital city of Brazil. As early as independence from Portugal in 1822, Brazilian politicians

---

[3]M. Crawford, "The World in a Shopping Mall," in M. Sorkin (ed.), *Variations on a Theme Park.* New York: Noonday Press, 1992, p. 14.

Figure 6.14 Costa's master plan for the new capital of Brasilia In 1957 Lucio Costa submitted his entry to the master plan competition for the new capital of Brazil on five medium-size cards, each containing a set of freehand sketches and some commentary. Three essential elements comprised his plan: a cross created by the intersection of two highway axes, an equilateral triangle superimposed on this cross that defined the geographical area of the city, and two terraced embankments and a platform. The more detailed rendition of the plan Costa drew for Brazil's new capital, illustrated here, reveals how he thought the various activities and material forms of city life should be arranged geographically within the city. (*Source:* Adapted from B. Marshall (ed.), *The Real World.* London: Marshall Editions/Houghton Mifflin, 1991, p. 171.)

began suggesting that a new capital be established on the central plateau in the undeveloped interior of the country, but the government was not officially transferred there until 1960. A symbol of the taming of the wild interior of the country and the conquest of nature through human ingenuity, Brasilia is also a many-layered system of signs conveying multiple and frequently contradictory messages. Interestingly, Brasilia, which was intended to symbolize a new age in Brazilian history, was also literally an attempt to construct one. That is why its plan and its architecture are so self-consciously rich with messages meant to transform Brazilian society through a new and radical form of architecture.

To launch both the idea and the reality of a "city in the wilderness," the Brazilian government sponsored a design contest in the 1950s hoping to encourage the development of a new vision for the new capital (and by extension a new society). The winner of the contest was engineer Lucio Costa. His original plan, submitted to an

international jury, was a simple sketch of a series of three crosses, each more elaborate than the previous one. In a semiotic reading of Costa's plan, as well as of the city, now over five decades old, the sign-of-the-cross ordering is an intentional use of a well-known mark. In its graphic form, the crossed axes represent the cross of Christianity. This aspect of the sign is important because it suggests that Brasilia was to be built on a sacred site, an important endorsement for the founding of a new capital.

Many observers of the plan and the completed city have said that it resembles an airplane. This observation seems especially apt when one looks at the master plan of the city (**Figure 6.14**). The plan shows that the residential districts were to be located along the north and south wings, and administrative government offices on the part corresponding with the fuselage. The commercial district was to be constructed at the intersection of the wings and the fuselage, with a cathedral and museum along the monumental axis. Like the sign of the cross, the significance of an airplane is obvious. Politicians and planners envisioned Brasilia as both the engine and the symbol of the rapid modernization of the country. The image of an airplane in flight was an exciting, soaring, and uplifting symbol of modernization.

All of Brasilia's major public buildings were designed by the internationally famous Brazilian architect Oscar Niemeyer (**Figure 6.15**). The residential axes were designed with clusters of apartment buildings, each cluster surrounding a set of recreational facilities, school buildings, and shopping areas. The University of Brasilia was also part of the early vision of the city, as was the creation of a lake and the official home of the president of the country, the Palacio de Alvorada, or the Palace of Dawn (**Figure 6.16**).

The examples of the American shopping mall and the federal district of Brasilia illustrate the way that landscapes can be read, or decoded, by interpreting the signs and symbols they project. Not all the signs are consistent, even when planners and designers have complete control over their projects, because readers do not always interpret signs in ways the creators intended. In both cases, social and political realities can disrupt the plan and send very different messages.

Sacred Spaces

Religious places can also be read and decoded. Indeed, most religions designate certain places as sacred, often because a special event occurred there. Sites are often designated as sacred in order to distinguish them from the rest of the landscape, which is considered ordinary or profane. **Sacred spaces** are areas of the globe recognized by individuals or groups as worthy of special attention because they are the sites of special religious experiences and events. They do not occur naturally; rather, they are assigned sanctity through the values and belief systems

Figure 6.15 The landscape of Brasilia The landscape of Brasilia, as designed and constructed by architect Oscar Niemeyer, is that of a dynamic, modern world city. Aesthetic beauty, a sense of permanence, the expression of power, and the idiosyncrasies of the individual creative process are among the primary elements of this designed landscape.

of particular groups or individuals. Geographer Yi-Fu Tuan insists that what defines the sacredness of a space goes beyond the obvious shrines and temples. Sacred spaces are those that rise above the commonplace and interrupt ordinary routine.

In almost all cases, sacred spaces are segregated, dedicated, and hallowed sites that are maintained as such generation after generation. Believers—including mystics, spiritualists, religious followers, and pilgrims—recognize sacred spaces as being endowed with divine meaning. The range of sacred spaces includes sites as different as a an elegant and elaborate pagoda in Burma, a

Figure 6.16 Performing Arts Center, Brasilia This structure combines the homogenized international style of "glass box" architecture with culturally distinctive artistic elements, such as the sweeping colonnades pictured here.

holy spot for Burmese Buddhists, and the Black Hills of South Dakota, the sacred mountains of the Lakota Sioux (**Figures 6.17 and 6.18**).

Often, members of a specific religion are expected to journey to especially important sacred spaces to renew their faith or to demonstrate devotion. A pilgrimage is a journey to a sacred space, and a pilgrim is a person who undertakes such a journey. In India many of the sacred pilgrimage sites for Hindus are concentrated along the seven sacred rivers: the Ganges, the Yamuna, the Saraswati, the Naramada, the Indus, the Cauvery, and the Godavari. The Ganges is India's holiest river, and many sacred sites are located along its banks (**Figure 6.19**). Hindus visit sacred pilgrimage sites for a variety of reasons, including seeking a cure for sickness, washing away sins, and fulfilling a promise to a deity. Perhaps the most well-known pilgrimage is the hajj, the obligatory once-in-a-lifetime journey of Muslims to Mecca. For one month every year, the city of Mecca in Saudi Arabia swells from its base population of 150,000 to over 1,000,000 as pilgrims from all over the world journey to fulfill their obligation to pray in the city and receive the grace of Allah. **Figure 6.20** shows the principal countries that send pilgrims to Mecca.

Pilgrimages to sacred sites are made all over the world, including Christian Europe. The most visited sacred site in Europe is Lourdes, at the base of the Pyrenees in southwest France, not far from the Spanish border (**Figure 6.21**). Another sacred site that attracts pilgrims throughout the world is the city of Jerusalem, and the Holy Land more generally, which is visited by Jews, Orthodox, Catholics, Protestants, Christian Zionists, and followers of many other religions (see Box 6.1, "Jerusalem, the Holy City"). As with most sacred spaces, the codes that are embedded in the landscape of the Holy Land may be read quite differently by different religious and even secular visitors. Two students of pilgrimage observed:

> Each group brings to Jerusalem their own entrenched understandings of the sacred; nothing unites them save their sequential—and sometimes simultaneous—presence at the same holy sites. For the Greek Orthodox pilgrims, indeed, the precise definition of the site itself is largely irrelevant; it is the icons on display which are the principal focus of attention. For the Roman Catholics, the site is important in that it is illustrative of a particular biblical text relating to the life of Jesus, but it is important only in a historical sense, as confirming the truth of past events. Only for the Christian Zionists does the Holy Land itself carry any present and future significance, and here they find a curious kinship with indigenous Jews.[4]

[4]J. Eade and M. Sallnow (eds.), *Contesting the Sacred*. London: Routledge, 1991, p. 14.

Figure 6.17   Myit Aiya pagoda in Bago, Burma This structure is an example of an elaborate and highly constructed sacred site. For Burmese as well as Buddhists around the world, temples, pagodas, stupas, and statues of Buddha are holy places.

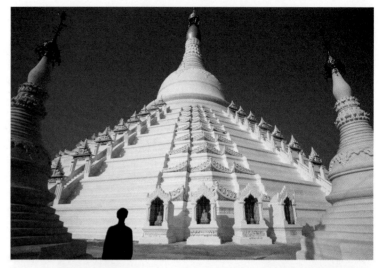

Figure 6.18   Black Hills of South Dakota The Black Hills are a sacred site of the Sioux Indians. In 1877 the Lakota Sioux lost control of the Black Hills after a series of bloody battles with the United States Cavalry, including the Battle of Little Bighorn. The Sioux are still fighting to reclaim their sacred site—the Paha Sapa—and have turned down $105 million in compensation.

Figure 6.19   Sacred sites of Hindu India India's many rivers are holy places within the Hindu religion, so sacred sites are located along the country's many riverbanks. Shrines closer to the rivers are regarded as holier than those farther away. (*Source:* Adapted from Ismail Ragi al Farugi and David E. Sopher, *Historical Atlas of the Religions of the World.* New York: Macmillan, 1974.)

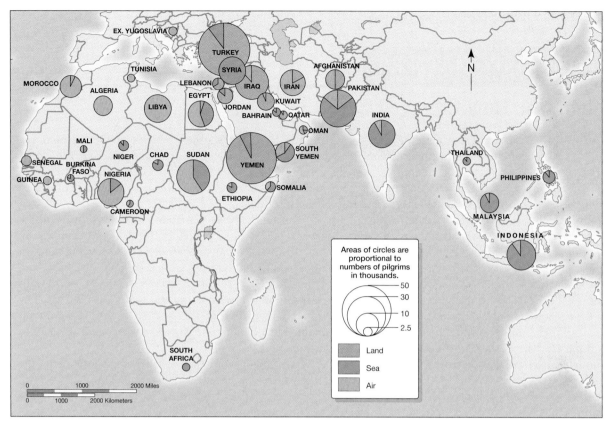

Figure 6.20 Source areas for pilgrims to Mecca Islam requires that every adult Muslim perform the pilgrimage to Mecca at least once in a lifetime. This obligation is deferred for four groups of people: those who cannot afford to make the pilgrimage; those who are constrained by physical disability, hazardous conditions, or political barriers; slaves and those of unsound mind; and women without a husband or male relative to accompany them. The pattern of actual pilgrimages to Mecca (which is located close to the Red Sea coast in Saudi Arabia) suggests a fairly strong distance-decay effect, with most traveling relatively short distances from Middle Eastern Arab countries. More distant source areas generally provide smaller numbers of pilgrims, though Indonesia and Malaysia are notable exceptions. (*Source:* After C. C. Park, *Sacred Worlds.* London: Routledge, 1994, p. 268.)

Figure 6.21 Source areas for pilgrims to Lourdes This map shows the points of origin of European group-organized pilgrims to Lourdes in 1978. These represent only about 30 percent of all pilgrims to Lourdes, most of whom travel to the shrine on their own. Improved transportation (mainly by train) and the availability of organized package trips have contributed to a marked increase in the number of pilgrims visiting. Many of the 5 million pilgrims who visit the town each year do so in the hope of a miraculous cure for medical ills at a grotto where the Virgin Mary is said to have appeared before 14-year-old Bernadette Soubirous in a series of 18 visions in 1858. (*Source:* After C. C. Park, *Sacred Worlds.* London: Routledge, 1994, p. 284.)

PLACE AND SPACE IN MODERN SOCIETY

While malls, planned cities, and sacred places illustrate the way that places are coded within specific settings or contexts, there are broad dimensions of social, economic, and political organization that are related to the ways in which people think about themselves and about the places they inhabit. For more than a century the philosophy of modernity has been a major influence on the interdependencies among culture, society, space, place, and landscape throughout much of the world. **Modernity** is a forward-looking view of the world that emphasizes reason, scientific rationality, creativity, novelty, and progress. Its origins can be traced to the European Renaissance and the emergence of the world-system of competitive capitalism in the sixteenth century, when scientific discovery and commerce began to displace traditional sociocultural views of the world that emphasized mysticism, romanticism, and fatalism. These origins were consolidated into a philosophical movement during the eighteenth century, when the so-called Enlightenment established the widespread belief in universal human progress and the sovereignty of scientific reasoning.

6.1 GEOGRAPHY MATTERS

Jerusalem, the Holy City

Although many cities in the world have been the object of struggle and conflict over the centuries, none seems as endlessly beset as Jerusalem. Visitors, writers, and residents believe Jerusalem to be the most beautiful city in the world. If there are other serious competitors to that coveted beauty title, Jerusalem certainly has few rivals for the title of the most sacred city in the world, possessing as it does an unmatched Christian, Jewish, and Islamic history. Jerusalem began as a small settlement on the slopes of Mount Moriah. In 997 B.C., it was captured by David, king of the Israelites, who made Jerusalem the capital of Israel. Solomon, David's son and successor, built the Great (First) Temple on Mount Moriah to commemorate the place where Abraham offered to sacrifice his son. Though the temple was destroyed centuries ago, the site is central to the Jewish faith.

The history of the city reflects the history of the various empires that dominated and were succeeded by new and yet more powerful empires (**Figure 6.A**). Nebuchadnezzar, king of Babylon, destroyed the Great Temple in 586 B.C. and banished the Jews. But the Babylonian control of Jerusalem eventually gave way to the Persians, under whose rule the Jews were allowed to return and rebuild their temple, known as the Second Temple. The Persian occupation of Jerusalem was swept aside by Alexander the Great (356–323 B.C.), the king of Macedonia and one of the world's greatest military leaders, who helped spread the Greek

Empire from the southern shores of the Caspian Sea into central Asia. The Romans entered the scene around 63 B.C., and Herod the Great was eventually installed to command the Roman Kingdom of Judea from Jerusalem.

During the early Christian period in Jerusalem, the Jews revolted openly against the Roman occupiers. In A.D. 132 the Romans responded by destroying the Second Temple and banishing all Jews from Jerusalem and Palestine. As a result, the Jews scattered north into Babylon and later into Europe and North Africa. The popular myth is that this ancient Jewish diaspora remained in exile until 1948, when the state of Israel was created.

The major Christian influence on Jerusalem, the "Holy City," began when Constantine I (A.D. 285?–337), the emperor of the Eastern Roman Empire, converted to Christianity in A.D. 313. This event led to the construction of churches and other buildings dedicated to celebrating the life of Jesus Christ. But Christian influence over the city ceased when Jerusalem eventually succumbed to Islam. In A.D. 638, Jerusalem was designated a holy city of Islam because it was believed that Muhammad had once visited heaven while in the city.

Although for several centuries Jews, Christians, and Muslims were all allowed access to the city of Jerusalem, by the tenth century, the persecution of non-Muslims became common. From the eleventh century

Figure 6.A Jerusalem, the Holy City This map of Jerusalem indicates the main sections of the city. Over the many years of the Israeli-Palestinian peace process, numerous proposals have been advanced about how to divide the city to satisfy the desire of both Palestinians and Israelis to possess it. The disposition of Jerusalem is one of the major issues in the ongoing peace process. (*Source:* Redrawn from *The Guardian*, October 14, 2000, p. 5.)

until the thirteenth, European Christians undertook military expeditions—called the Crusades—to the Holy Land in an attempt to wrest control of Jerusalem from the Muslims. In 1099, Crusaders captured the city; Christians lost it again to the Muslim military leader Saladin in 1187. In 1517, Jerusalem was absorbed into the Ottoman empire, and the city was ruled from Istanbul for more than 400 years. The Ottomans, however, had little interest in Jerusalem, and Jewish immigrants began returning to the city and Palestine starting in the mid-nineteenth century.

The contemporary history of the city derives from the political and geographical implications of the Balfour Declaration, which stipulated that Jerusalem should be an international city with no one state claiming it as entirely its own. Today, Jerusalem is a highly contested city as Palestinians, Christians, Muslims, and Israeli Jews fight for control of it. An example of this contest for control is the continuing dispute over the Dome of the Rock, which was constructed between A.D. 688 and 691. Muslims claim the Dome as their most sacred site (**Figure 6.B**). Yet the Dome sits on a site sacred to the Jews, the Temple Mount, the site where the Great Temple and Second Temple were built and later destroyed. Indeed, the Dome is believed to enclose the sacred rock upon which Abraham prepared to sacrifice his son, according to Jewish tradition, and, according to Islamic tradition, the same rock from

which the prophet Muhammad launched himself to visit heaven. Also located at the Temple Mount is the Al-Aqsa Mosque, a central sacred site to Muslims.

While nationalist Israelis maintain that Jerusalem will be the "eternal and undivided capital" of Israel, Palestinians believe that Jerusalem is the future capital of the Palestinian state. In fact, the 1992 Oslo peace accords that lead to a declaration of principles between Israel and Palestine hint at the possibility of negotiating the future of Jerusalem. Further peace negotiations, including those at Wye River, Maryland, in 1998; at Sharm al-Shaykh in the Egyptian Sinai Desert in 1999; in Camp David, Maryland, in 2000; in Washington, D.C., in 2001; and in Geneva in 2003 are cause for hope that a resolution to the future of Jerusalem that includes some control by the Palestinians can be achieved. At present, Jerusalem is entirely controlled by Israel. Yet in accordance with the Oslo peace agreements, in May 2000 Israel conceded three villages to the control of the Palestinian Authority. All three villages are part of greater Jerusalem in the West Bank, and all but one, Abu Dis, are actually a few meters beyond the Jerusalem municipal border. From Abu Dis, the Old City of Jerusalem is clearly visible, particularly the Dome of the Rock. In late 2008 it is still unclear how the contested control of Jerusalem might eventually be resolved. What is clear is that both sides are equally passionate about their claims.

Figure 6.B   Dome of the Rock Located in Haram Al-Shaif, or Temple Mount, the Dome of the Rock sits in the part of Jerusalem that is technically neither Jewish nor Muslim, but both. The Muslims want this entire holy site. The Israelis also want it because the Wailing Wall, sacred to Jews, also occupies part of the Temple Mount. Palestinian anger was unleashed—and the delicate peace process derailed—in the fall of 2000 when Ariel Sharon, a nationalist Israeli politician who several months later became Israel's president, visited the Temple Mount during the first days of the Muslim holy season of Ramadan.

At the beginning of the twentieth century, this philosophy developed into a more widespread intellectual movement. A series of sweeping technological and scientific developments not only triggered a new round of spatial reorganization (see Chapter 2) but also transformed the underpinnings of social and cultural life. These developments included the telegraph, the telephone, the X-ray, the cinema, the radio, the bicycle, the internal combustion engine, the airplane, the skyscraper, relativity theory, and psychoanalysis. Universal human progress suddenly seemed to be a realistic prospect.

Nevertheless, the pace of economic, social, cultural, and geographic change was unnerving and the outcomes uncertain. The intellectual response, developed among a cultural avant-garde of painters, architects, novelists, and photographers, was a resolve to promote modernity through radical changes in culture. These ideas were first set out in the "Futurist Manifesto," published in 1909 in the Paris newspaper *Le Figaro* in the form of a letter from the Italian poet Filippo Marinetti. Gradually the combination of new technologies and radical design contributed to the proliferation of landscapes of modernity. Among the most striking of these were modernist urban landscapes, from Helsinki to Hong Kong and from New York to Nairobi. Indeed, in a general sense almost all of the place-making and landscapes of the early and mid-twentieth century are the products of Modernity.

Throughout this period a confident and forward-looking Modernist philosophy remained virtually unquestioned, with the result that places and regions everywhere were heavily shaped by people acting out their notions of rational behavior and progress. Rural regions, for example, bore the imprint of agricultural modernization. The hedgerows of traditional European field patterns were torn up to make way for landscapes of large, featureless fields in which heavy machinery could operate more effectively. At a different scale, peripheral countries sought to remake traditional landscapes through economic modernization. Economic development and social progress were to be achieved through a modern infrastructure of highways, airports, dams, harbors, and industrial parks.

Globalization and Place-Making

The spread of the philosophy of Modernity to peripheral regions can be seen as part of the processes of globalization. These processes have not only brought about a generalization of forms of industrial production, market behavior, trade, and consumption but reinforced and extended the commonalities among places. Three factors are especially important in this context. First, mass communications media have created global culture markets in print, film, music, television, and the Internet. Indeed, the Internet has created an entirely new *kind* of space—cyberspace—with its own "landscape" (or technoscape) and its own embryonic cultures (see Box 6.2, "The Cultural Geography of Cyberspace"). The instantaneous character of contemporary communications has also made possible the creation of

a shared, global consciousness from the staging of global events such as the Olympic Games, and the World Cup. Second, mass communications media have diffused certain values and attitudes toward a wide spectrum of sociocultural issues, including citizenship, human rights, child rearing, social welfare, and self-expression. Third, international legal conventions have increased the degree of standardization and level of harmonization not only of trade and labor practices but of criminal justice, civil rights, and environmental regulations.

These commonalities have been accompanied by the growing importance of material consumption within many cultures. Increasingly, people around the world are eating the same foods, wearing the same clothes, and buying the same consumer products. Yet the more people's patterns of consumption converge, the more fertile the ground for countercultural movements. The more transnational corporations undercut the authority of national and local governments to regulate economic affairs, the greater the popular support for regionalism. The more universal the diffusion of material culture and lifestyles, the more local and ethnic identities are valued. The faster the information highway takes people into cyberspace, the more they feel the need for a subjective setting—a specific place or community—they can call their own. The faster the pace of life in search of profit and material consumption, the more people value leisure time. And the faster their neighborhoods and towns acquire the same generic supermarkets, gas stations, shopping malls, industrial estates, office parks, and suburban subdivisions, the more people feel the need for enclaves of familiarity, centeredness, and identity. The United Nations Center for Human Settlements (UNCHS) notes:

> In many localities, people are overwhelmed by changes in their traditional cultural, spiritual, and social values and norms and by the introduction of a cult of consumerism intrinsic to the process of globalization. In the rebound, many localities have rediscovered the "culture of place" by stressing their own identity, their own roots, their own culture and values and the importance of their own neighbourhood, area, vicinity, or town.[5]

One example of the impulse for people to recover a sense of place is provided by the slow city (CittaSlow) movement. The CittaSlow movement is a grassroots response to globalization and is closely related to the longer-established and better-known slow food movement. The aims of the two movements are different but complementary: In broad terms, both organizations are in favor of local, traditional cultures, a relaxed pace of life, and conviviality. Both are a response to the increasing pace of everyday life associated with the acceleration of money around local, national, and global circuits of capital. Both

[5]United Nations Center for Human Settlements, *Global Report on Human Settlements 2001*. London: Earthscan, 2001, p. 4.

6.2 GEOGRAPHY MATTERS

The Cultural Geography of Cyberspace

The rapid growth of the Internet is of great cultural significance, for it has created the basis for a massive shift in patterns of social interaction, a seedbed for new forms of human consciousness, and a new medium for cultural change. Culture is fundamentally based on communication, and in cyberspace we have an entirely new form of communication: uncensored, multidirectional, written, visual, and aural.

At face value, the Internet represents the leading edge of the globalization of culture. In broad terms the culture propagated by the Internet is very much core-oriented. The Internet portends a global culture based on English as the universal world language, with a heavy emphasis on core-area cultural values, such as novelty, spectacle, fashionability, material consumption, and leisure. It is unlikely, however, that the Internet will simply be a new medium through which core-area values and culture are spread.

To begin with, the impact of the Internet is likely to be highly uneven because of the digital divide. Moreover, there is resistance in some places and regions to the cultural globalization associated with cyberspace. The French and French-Canadian authorities, already sensitive about the influence of English-language popular culture, have actively sought ways to give Francophone cybernauts access to the Internet without submitting to English, the dominant language of Web sites. The French government has also subsidized an all-French alternative to the Internet: Minitel, an online videotext terminal that plugs into French telecommunication networks. A free Minitel terminal is available to anyone who stops by a France Telecom office.

In much of Asia, the Internet's basic function as an information-exchange medium clashes with local cultures in which information is a closely guarded commodity. Whereas many U.S. World Wide Web sites feature lengthy government reports and scientific studies, as well as lively debates about government policy, comparable Asian sites typically offer little beyond public relations materials from government agencies and corporations. In puritanical Singapore, political leaders, worried that the Internet will undermine morality, have taken to reading private e-mail as part of an all-out effort to beat back the menace of online pornography. Chinese authorities fear that the Internet will foment political rebellion, so officials have limited access to ensure that the Chinese portion of the Internet can easily be severed from the world in the event of political unrest. The reluctance of major Asian organizations to put important information on their Web sites—along with the need for Westerners to use special software to read any local-language documents that do exist—has resulted in a largely one-way flow of information, from America to Asia.

Nevertheless, the greatest potential of the Internet in terms of cultural change resides in the liberating and empowering potential of its vast resource of knowledge and information. By its very nature—a decentralized and complex web of computer hosts—the Internet empowers individuals (rather than social groups or institutions), allowing millions to say whatever they want to each other, free (for the first time in history) from state control. As such, it is an important vehicle for the spread of participatory democracy to much of the world.

---

are hostile to big business and globalization, though their driving motivation is not so much political as ecological and humanistic. Slow food is devoted to a less hurried pace of life and to the true tastes, aromas, and diversity of good food. The movement also serves as a rallying point against globalization, mass production, and the kind of generic fast food represented by U.S.-based franchises like McDonald's, Burger King, Pizza Hut, Taco Bell, and Kentucky Fried Chicken. By 2008, the slow food movement, based in Bra, near Turin in northern Italy, had established convivia (local branches) in more than 100 countries, with over 80,000 members worldwide. Its campaigns cover a range of specific causes, from protecting the integrity of chocolate to promoting the cultivation of traditional crop varieties and livestock breeds and opposing genetically engineered foods.

The slow city movement was formed in October 1999, when Paolo Saturnini, mayor of Greve-in-Chianti, a

Tuscan hill town, organized a meeting with the mayors of three other municipalities (Orvieto, Bra, and Positano) to define the attributes that might characterize a *città lente*— slow city. At their founding meeting in Orvieto, the four mayors committed themselves to a series of principles that included working toward calmer and less polluted physical environments, conserving local aesthetic traditions, and fostering local crafts, produce, and cuisine. They also pledged to use technology to create healthier environments, to make citizens aware of the value of more leisurely rhythms to life, and to share their experience in seeking administrative solutions for better living. The goal is to foster the development of places that enjoy a robust vitality based on good food, healthy environments, sustainable economies, and the seasonality and traditional rhythms of community life (see Box 6.3, "Slow Cities").

Many commentators on cultural change have noted this broad shift in cultural sensibilities within the world's

6.3 WINDOW ON THE WORLD

Slow Cities

In Western Europe a backlash is under way in reaction to the culture of speed imposed by the 24/7 global marketplace and the manic pace of life dictated by globalization. In France, a 35-hour work week was introduced in 2002, an innovation that is also being considered by Belgium and Sweden. In Italy, writer and activist Carlo Petrini was so offended by the temerity of McDonald's wanting to open a store in the Piazza di Spagna in 1989 that he started the "slow food" movement.

In 1999, the slow food movement spurred an offshoot called the CittaSlow (slow cities) movement, made up of municipalities whose leaders reject the North Americanization of European cities. The first 28 slow cities were certified in 2001. All 28 charter members were Italian, the majority of them located in northern and central Italy, particularly in Tuscany and Umbria. By mid-2008, the list had grown to more than 70 (**Figure 6.C**). The majority are located in Italy but towns in Australia (Goolwa, Kaoomba, and Willunga), Austria (Enns), Germany (Deidesheim, Hersbruck, Lüdinghausen, Marihn, Schwarzenbruck, Überlingen, Waldkirch, and Wirsberg), Norway (Eidskog, Levanger, and Sokndal), New Zealand (Matakana), Poland (Biscupiek, Bisztynek, Lidzbark, Reszel, and Warminski), South Korea (Cheongsan Island, Juengdo, Cheongpyeong, and Yuchi), Spain (Mungia), Sweden (Falköping), and the United Kingdom (Aylsham, Diss, Ludlow, Mold, and Perth) are now certified. More than 300 other towns from around the world have inquired about joining. The movement's charter specifies that a slow city is one with a population of fewer than

Figure 6.C   Even on a normal working day, Abbiategrasso, Italy, seems like the perfect place to get away from it all. Locals linger over coffee at sidewalk tables, gossiping with friends or watching the world drift by.

50,000. Member towns must pledge to promote organically produced, traditional foods; a clean environment; quiet neighborhoods; urban charm; and the idea that the good life is an unhurried sensual experience. The slow city charter contains over 50 pledges, such as cutting noise pollution and traffic, increasing green spaces and pedestrian zones, supporting farmers who produce local delicacies and the shops and restaurants that sell them, and preserving local aesthetic traditions.

Promoting local distinctiveness and a sense of place is almost as important to the movement as the enjoyment of good local food, wine, and beer. This means that

more affluent countries. Most pronounced in the more affluent core countries, this shift is broadly characterized as one from modernity to postmodernity and has involved both avant-garde and popular culture. **Postmodernity** is a view of the world that emphasizes openness to a range of perspectives in social inquiry, artistic expression, and political empowerment. Postmodernity abandons modernity's emphasis on economic and scientific progress, focusing instead on living for the moment. Above all, postmodernity is consumption-oriented, with an emphasis on the possession of particular *combinations* of things and on the style of consumption.

Postmodern society has been called a "society of the spectacle" in which the symbolic properties of places and material possessions have assumed unprecedented importance. Since the mid-1970s, postmodernity has permeated every sphere of creative activity, including art, architecture, advertising, philosophy, clothing design, interior design, music, cinema, novels, television, and urban design. The shift to postmodern cultural sensibilities is of particular

importance to cultural geography because of the ways in which changed attitudes and values have begun to influence place-making and the creation of landscapes. Some of the most striking postmodern landscapes are to be found in the redeveloped waterfronts, revitalized downtown shopping districts, and neotraditional suburbs of major cities (**Figure 6.22**).

Places as Objects of Consumption

In much of the world, people's enjoyment of material goods now depends not just on their physical consumption or use. It is also linked more than ever to the role of material culture as a social marker. A person's home, automobile, clothing, reading, viewing, eating and drinking preferences, and choice of vacations are all indicators of that person's social distinctiveness and sense of style. This pressures individuals to continuously search for new and distinctive styles. The wider the range of foods, products, and ideas from around the

the charter also covers many aspects of urban design and planning. Candidate cities must be committed not only to supporting traditional local arts and crafts but to supporting modern industries whose products lend distinctiveness and identity to the region. They must also be committed to the conservation of the distinctive character of their built environment and must pledge to plant trees; create more green space; increase cycle paths and pedestrianized streets; keep public squares and piazzas free of advertising billboards and neon; ban car alarms; reduce noise pollution, light pollution, and air pollution; foster the use of alternative sources of energy; improve public transport; and promote eco-friendly architecture in any new developments.

Bra, one of the founding municipalities of the slow cities movement (along with Greve, Orvieto, and Positano) has banned cars, as well as supermarkets and lurid neon signs, from parts of its historic core. In order to promote a more leisurely pace of life, every small food shop in Bra is obliged to close on Thursdays and Sundays. Small, family-run businesses selling locally produced crafts, fabrics, cheeses, roasted peppers, white truffles, fresh pasta, olive oil, and specialty meats are granted the best commercial real estate. The municipality subsidizes building renovations that use the honey-colored stucco typical of the Piedmont region. In schools, children are served organic fruit and vegetables grown by local producers.

One obvious critique of the CittaSlow movement is that it could all too easily produce enervated, backward-looking, isolationist communities: living mausoleums where the puritanical zealotry of slowness has displaced the fervent materialism of the fast world. Aware of the dangers of prescriptive slowness, the CittaSlow movement hopes to propagate vitality through farmers' markets, festivals, and the creation of inviting public spaces. It aims to deploy technology in air, noise and light pollution-control systems, modern waste-cycling plants, and composting facilities. It seeks to encourage business through ecologically sensitive, regionally authentic, and gastronomically oriented tourism. Here, though, is another danger: that, paradoxically, slow city designation becomes a form of brand recognition within the heritage industry. Because they are small, the charming attractions of slow cities could all too easily be overwhelmed by tourism. So the more they flaunt their gentle-paced life, the faster they may end up changing. In this scenario, shop prices will rise and cafés will lose their spilled-drink, smoky, messy authenticity. The better known that slow cities become, the more affluent outsiders will choose to make their second homes in them. House prices will go up, and the poor and the young will be pushed out.

Nevertheless, whatever the eventual outcomes of the slow city movement *per se*, its principles speak directly to the concepts of "dwelling" and intersubjectivity that are key to the social construction of place. Fostering respect for seasonality and traditional rhythms of community life propagates recurring and interlocking patterns of events that make for cultural transactions and public sociability in the public realm.

world—and from past worlds—the greater the possibilities for establishing such styles.

Given that material consumption is so central to the repertoire of symbols, beliefs, and practices of postmodern cultures, the "culture industries"—advertising, publishing, communications media, and popular entertainment—have become important shapers of spaces, places, and landscapes. Because the symbolic meanings of material culture must be advertised (in the broadest sense of the word) in order to be shared, advertising (in its narrower sense) has

Figure 6.22 Landscape of postmodernity This photograph shows the Piazza d'Italia in New Orleans, one of the first public projects in the United States to consciously celebrate an eclectic, over-the-top decorative approach to design.

become a key component of contemporary culture and place-making. In addition to stimulating consumer demand, advertising has always had a role in teaching people how to dress, how to furnish a home, and how to signify status through groupings of possessions.

In the 1970s and 1980s, however, the emphasis in advertising strategies shifted away from presenting products as newer, better, more efficient, and more economical (in keeping with modernist sensibilities) to identifying them as the means to self-awareness, self-actualization, and group stylishness (in keeping with postmodern sensibilities). Increasingly, products are advertised in terms of their association with a particular lifestyle rather than in terms of their intrinsic utility. Many of these advertisements deliberately draw on international or global themes, and some entire advertising campaigns (for Coca-Cola, Benetton, and American Express, for example) have been explicitly based on the theme of cultural globalization. Many others rely on stereotypes of particular places or kinds of places (especially exotic, spectacular, or "cool" places) in creating the appropriate context or setting for their product. Images of places therefore join with images of global food, architecture, pop culture, and consumer goods in the global media marketplace. Advertisements both instruct and influence consumers about products and also about spaces, places, and landscapes.

One result of these trends is that contemporary cultures rely much more than before not only on material consumption but also on *visual* and *experiential* consumption: the purchase of images and the experience of spectacular and distinctive places, physical settings, and landscapes. Visual consumption can take the form of magazines, television, movies, sites on the World Wide Web, tourism, window shopping, people-watching, or visits to galleries and museums. The images, signs, and experiences that are consumed may be originals, copies, or simulations.

The significance of the increased importance of visual consumption for place-making and the evolution of landscapes is that settings such as theme parks, shopping malls, festival marketplaces, renovated historic districts, museums, and galleries have all become prominent as centers of cultural practices and activities. The number of such settings has proliferated, making a discernible impact on metropolitan landscapes. The design of such settings, however, has had an even greater impact. Places of material and visual consumption have been in the vanguard of postmodern ideas and values, incorporating eclecticism, decoration, and a heavy use of historical and vernacular motifs and spectacular features in an attempt to create stylish settings that are appropriate to contemporary lifestyles.

One interesting aspect of the increasing trend toward the consumption of experiences is the emergence of restaurants as significant cultural sites. Restaurants often represent a synthesis of the global and the local, and they

Figure 6.23 Consumption in style Restaurant design contributes to a city's visual style as architects and interior designers, restaurant consultants, and restaurant-industry magazines diffuse global trends adapted to local styles. Shown here is the interior of the Buddakan restaurant, in Philadelphia.

can be powerful cultural symbols in their own right (**Figure 6.23**). The dining experience in a particular restaurant can be an important symbolic good. By the same token, the social standing or celebrity status of customers contributes to the value of the dining experience offered by a restaurant. Restaurants themselves can be both theater and performance—particularly in big cities like New York and Los Angeles, where underpaid or out-of-work actors, dancers, and other artists form an important part of the restaurant labor force. Restaurants bring together a global and a local labor force (immigrant owners, chefs, and waiters, as well as locals) and clientele (tourists and business travelers, as well as locals). Finally, restaurant design also contributes to a city's visual style as architects and interior designers, restaurant consultants, and restaurant-industry magazines adapt global trends to local styles.

The idea of the emergence of a single global culture is too simplistic, but the postmodern emphasis on material, visual, and experiential consumption means that many aspects of contemporary culture transcend local and national boundaries. Furthermore, many residents of the fast world are world travelers—either directly or via the TV in their living room—so that they are knowledgeable about many aspects of others' cultures. This contributes to **cosmopolitanism**, an intellectual and aesthetic openness toward divergent experiences, images, and products from different cultures.

Cosmopolitanism is an important geographic phenomenon because it fosters a curiosity about all places, peoples, and cultures, together with at least a rudimentary ability to map, or situate, such places and cultures geographically, historically, and anthropologically. It also suggests an ability to reflect upon, and judge aesthetically between, different places and societies. Furthermore, cosmopolitanism allows people to locate their own society and its culture in terms of a wide-ranging historical and geographical framework. For travelers and tourists, cosmopolitanism encourages both the willingness and the ability to take the risk of exploring off the beaten track of tourist locales. It also develops in people the skills needed to be able to interpret other cultures and to understand their visual symbolism.

Place Marketing

Economic and cultural globalization has meant that places and regions throughout the world are increasingly seeking to influence the ways in which they are perceived by tourists, businesses, media firms, and consumers. As a result, places are increasingly being reinterpreted, re-imagined, designed, packaged, and marketed. Through place marketing, sense of place has become a valuable commodity and culture has become an important economic activity. Furthermore, culture has become a significant factor in the ability of places to attract and retain other kinds of economic activity. Seeking to be competitive within the globalizing economy, many places have sponsored extensive makeovers of themselves, including the creation of pedestrian plazas, cosmopolitan cultural facilities, festivals, and sports and media events—what geographer David Harvey has described as the "carnival masks" and "businessmen's utopias" of global capitalism. An increasing number of places have also set up home pages on the Internet containing maps, information, photographs, guides, and virtual spaces in order to promote themselves in the global marketplace for tourism and commerce. Meanwhile, the question of who does the reimagining and cultural packaging, and on whose terms, can become an important issue for local politics.

Central to place marketing is the deliberate manipulation of material and visual culture in an effort to enhance the appeal of places to key groups. These groups include the upper-level management of large corporations, the higher-skilled and better-educated personnel sought by expanding high-technology industries, wealthy tourists, and the organizers of business and professional conferences and other income-generating events. In part, this manipulation of culture depends on promoting traditions, lifestyles, and arts that are locally rooted (see Box 6.4, "Place Marketing and Economic Development"); in part, it depends on being able to tap into globalizing culture through new cultural amenities and specially organized events and exhibitions. Some of the most widely adopted strategies for the manipulation and exploitation of culture include funding for facilities for the arts, investment in public spaces, the re-creation and refurbishment of distinctive settings like waterfronts and historic districts, the expansion and improvement of museums (especially with blockbuster exhibitions of spectacular cultural products that attract large crowds and can be marketed with commercial tie-ins), and the designation and conservation of historic landmarks.

Examples of the re-creation and refurbishment of distinctive settings like waterfronts and historic districts can be found in many cities. The popularity of waterfront redevelopments can be traced to the success of Harborplace in Baltimore (**Figure 6.24**). Harborplace is a waterfront complex that was redeveloped from a semiderelict wharf and dock area. Built in 1980, it has become one of the best-known examples of the kind of large, mixed-use developments that industrial cities have sponsored in an effort to bolster their image and attract tourists and shoppers. By 1995, Harborplace was attracting more than 30 million visitors a year. Not surprisingly, this kind of success has encouraged other cities to re-create themselves through major waterfront developments. Other examples include South Street Seaport in New York, the Faneuil Hall Marketplace and Harborwalk in Boston, and

Figure 6.24 Harborplace, Baltimore The developer of this reclaimed waterfront, James Rouse, sought to provide a setting that would be attractive to daytime office workers from the city's nearby downtown, to visitors to the adjacent convention center and hotel, and to tourists visiting the nearby aquarium. Rouse provided a new waterfront promenade with decorative paving, benches, and streetlights; a food court offering a variety of ethnic, prepared, and fresh foods from restaurants, fast-food counters, stalls, and pushcarts; and a total of 142,000 square feet of shopping space.

Place Marketing and Economic Development

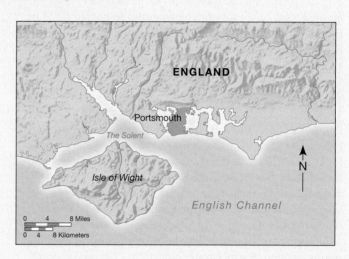

Portsmouth, England, with its superb natural harbor, has been England's chief naval port and dockyard since the reign of Henry VIII. At the height of the British Empire, the growth of the dockyard fueled the development of the city, helping to create a specialized economy based on marine engineering. With the decline of the empire after the 1960s, the British Navy was much reduced in size, and Portsmouth's dockyard functions shrank dramatically. After a period of economic adjustment, the city has begun to remake its image, exploiting its past to market itself as a distinctive conference center and tourist site.

The *Mary Rose*. (*Source:* Pepys Library, Magdalene College, Cambridge.)

Entrance to the *Mary Rose* exhibit in the Heritage Area of Portsmouth naval base.

HMS *Victory*.

HMS *Warrior*.

The legacy of Portsmouth's naval history has allowed the city to develop a heritage industry that has helped to make up for the loss of jobs in the naval dockyard. HMS *Victory*, Nelson's flagship at the Battle of Trafalgar in 1805, had been a long-standing attraction in the Royal Dockyard. In 1979, a volunteer organization recovered the

July 18, 1545: Henry VIII's fleet, in Portsmouth Harbor, fends off an attack by the French fleet.

Mary Rose, a Tudor warship that had been sunk just off Portsmouth in 1545 during an engagement with a French invasion fleet. It has since been restored and is a major attraction in the dockyard's heritage area. In 1988, HMS *Warrior,* built in 1860 as Britain's first ironclad battleship, was added to the dockyard's collection of major tourist attractions.

Throughout the second half of the nineteenth century and the first half of the twentieth, Portsmouth's economy included a lucrative resort and recreation function. When international travel became more affordable in the 1960s, many British vacationers chose the warmer beaches of the Mediterranean, and Portsmouth's tourist industry began to decline sharply. Since the 1980s, Portsmouth has repositioned itself as a different kind of resort, drawing on the city's naval history in order to attract overnight visitors traveling to and from France on the cross-Channel car ferries that now operate from an unused part of the naval dockyard. In 1996, the city formed a partnership with neighboring municipalities to redevelop Portsmouth

Southsea Castle.

Harbor. The £90 million ($135 million) redevelopment project is being substantially funded by proceeds from the National Lottery and will include a Baltimore-style wharf with a conference center, shops, cafés, hotels, and viewing terraces for maritime events.

Sixteenth- and seventeenth-century coastal fortifications, formerly boarded up and fenced off as eyesores, have been landscaped in order to take advantage of their historic interest. In one formerly abandoned fort, Southsea Castle, a museum of Tudor life has been established exploiting the city's link with Henry VIII, who first developed the British Navy into a significant military force.

June 5, 1944: D-Day.

Restored naval cutter, in the Mast Pond, part of the Heritage Area within Portsmouth naval base.

Royal Marines Museum, Portsmouth.

D-Day Museum, Portsmouth.

During World War II, Portsmouth was the main port of departure for Allied forces invading German-held France on D-Day. In 1982, the city opened the D-Day museum, reinforcing the city's emerging identity as a specialized heritage site. A few years later, the closure of a Royal Marine barracks provided the opportunity for the development of another specialized military museum.

Darling Harbour in Sydney, Australia. The largest and most ambitious of all such developments is in London's Docklands (**Figure 6.25**). The remaking of the Docklands in the 1980s was a deliberate attempt by Prime Minister Margaret Thatcher's government not simply to market this part of London to global tourists and investors but to sell the whole idea of the United Kingdom as a rejuvenated, postindustrial economy.

Examples of the re-creation and refurbishment of historic districts and settings are even more widespread—so widespread, in fact, that they have become a mainstay of the heritage industry. This industry, based on the commercial exploitation of the histories of peoples and places, is now worldwide, as evidenced by the involvement of the Economic, Social, and Cultural Organization of the United Nations (UNESCO) in identifying places for inclusion on World Heritage Lists. In countries such as the United Kingdom, with a high density of historic districts and settings, place marketing relies heavily on the heritage industry. In the United Kingdom, more than 90 million people pay to view about 650 historic properties each

year, and millions more visit free-of-charge heritage sites such as cathedrals, field monuments, remote ruins, and no longer useful industrial waterways. In 2007, more than 200 million tourists visited designated heritage sites in the United Kingdom, spending about $50 billion on entry fees, retail sales, travel, and hotel accommodations.

One important consequence of the heritage industry is that urban cultural environments are vulnerable to a debasing and trivializing "Disneyfication" process. The United Nations Center for Human Settlements (UNCHS) *Global Report on Human Settlements 2001* notes:

> The particular historic character of a city often gets submerged in the direct and overt quest for an international image and international business. Local identity becomes an ornament, a public relations artifact designed to aid marketing. Authenticity is paid for, encapsulated, mummified, located and displayed to attract tourists rather than to shelter continuities of tradition or the lives of its historic creators.[6]

[6]Ibid., p. 38.

Figure 6.25 London's Docklands (a) Once the commercial heart of Britain's empire, employing over 30,000 dockyard laborers, London's extensive Docklands fell into a sharp decline in the late 1960s because of competition from specialized ports using new container technologies. In 1981, the London Docklands Development Corp. was created by the central government and given extensive powers to redevelop the derelict dock areas. (b) The Docklands are now recognized as the largest urban redevelopment scheme in the world, with millions of square feet of office and retail space and substantial amounts of new housing. The photo shows Canary Wharf.

(a)

(b)

Figure 6.26 Las Vegas The boundaries between the heritage industry and the leisure and entertainment industries have become increasingly blurred, with the result that a great deal of investment has been channeled toward the creation of inauthentic "historic" settings whose characteristics owe as much to movies and popular stereotypes as to historic realities. Shown here is the Venetian Hotel in Las Vegas.

As a result, contemporary landscapes contain increasing numbers of inauthentic settings—what David Harvey has called the "degenerative utopias" of global capitalism. These are as much the product of contemporary material and visual culture as they are of any cultural heritage. Particularly influenced by images and symbols derived from movies, advertising, and popular culture, examples include the re-creation of the Wild West in the fake cowboy town of Old Tucson in Arizona; the Bavarian village created in Torrance, California; and the simulation of European tourist hot spots in Las Vegas, Nevada (**Figure 6.26**). The town of Helen, Georgia, boasts a full-scale reconstruction of a Swiss mountain village, complete with costumed staff and stores selling Swiss merchandise. In Japan, a reproduction village called British Hills, complete with church, pub, and school, has been constructed north of Tokyo.

CONCLUSION

Geographers study the interdependence between people and places and are especially interested in how individuals and groups acquire knowledge of their environments and how this knowledge shapes their attitudes and behaviors. People ascribe meanings to landscapes and places in many ways, and they also derive meanings from the places and landscapes they experience. Different groups of people experience landscape, place, and space differently. For instance, the experience that rural Sudanese children have of their landscapes and the ways in which they acquire knowledge of their surroundings differ from how middle-class children in an American suburb learn about and function in their landscapes. Furthermore, both landscapes elicit a distinctive sense of place that is different for those who live there and those who simply visit.

As indicated in previous chapters, the concepts of landscape and place are central to geographic inquiry. They are the result of intentional and unintentional human action, and every landscape is a complex reflection of the operations of the larger society. Geographers have developed categories of landscape to help distinguish the different types that exist. Ordinary landscapes, such as neighborhoods and drive-in movie theaters, are ones that people create in the course of their everyday lives. By contrast, symbolic landscapes represent the particular values and aspirations that their developers and financiers want to impart to a larger public. An example is Mount Rushmore in the Black Hills of South Dakota, designed and executed by sculptor Gutzon Borglum. Chiseling the heads of George Washington, Thomas Jefferson, Theodore Roosevelt, and Abraham Lincoln into the granite face of the mountain, Borglum intended to construct an enduring landscape of nationalism in the wilderness.

More recently, geographers have come to regard landscape as a text, something that can be written and read, rewritten, and reinterpreted. The concept of landscape as text suggests that a landscape can have more than one author, and different readers may derive different meanings from what is written there. The idea that landscape can be written and read is further supported by the understanding that the language in which the landscape is written is a code. To understand the significance of the code is to understand its semiotics, the language in which the code is written. The code may be meant to convey many things, including a language of power or of playfulness, a language that elevates one group above another, or a language that encourages imagination or religious devotion and spiritual awe.

The global transition from modernity to postmodernity has altered cultural landscapes, places, and spaces differently as individuals and groups have struggled to negotiate the local impacts of this widespread shift in cultural sensibilities. The shared meanings that insiders derive from their place or landscape have been disrupted by the intrusion of new sights, sounds, and smells as values, ideas, and practices from one part of the globe have been exported to another. The Internet and the emergence of cyberspace have meant that new spaces of interaction have emerged that have neither a distinct historical memory attached to them nor a well-established sense of place. Because of this, cyberspace carries with it some unique possibilities for cultural exchange. It remains to be seen, however, whether access to this new space will be truly open—or whether the Internet will become another landscape of power and exclusion.

MAIN POINTS REVISITED

- In addition to understanding how the environment shapes (and is shaped by) people, geographers seek to identify how it is perceived and understood by people.

  People not only filter information from their environments through neurophysiological processes but draw on personality and culture to produce cognitive images of their environment—pictures or representations of the world that can be called to mind through the imagination.

- The language in which a landscape is written is a kind of code. The code includes signs that direct our attention toward certain features and away from others.

  To interpret our environment, we must learn how to read the codes that are written into the landscape.

- The written code of landscape is also known as semiotics. Codes signify important information about landscapes, such as whether they are sacred or profane, accessible or off-limits, or oriented toward work or play.

  Landscapes as different from each other as shopping malls and national capitals can be understood in terms of their semiotics.

- Different cultural identities and status categories influence the ways in which people experience and understand their environments, as well as how they are shaped by—and are able to shape—them.

  Among the most important of these are the cultural identities of race, class, gender, ethnicity, and sexuality. Often these identities come together in a group, and their influence in combination becomes central to our understanding of how group identity shapes space and is shaped by it.

- Landscape serves as a kind of archive of society. It is a reflection of our culture and our experiences. Like a book, landscape is a text that is written by individuals and groups and read by them as well.

  It is therefore possible to have one landscape convey different meanings for different groups. Landscapes can be constructed to reflect the everyday words of social groups, as well as to represent power and the values of a particular society.

- The emergence of the most recent phase of globalization has occurred in parallel with a transition from modernity to postmodernity.

  This transition involves a shift in cultural sensibilities that affects every sphere of creative activity, from art and architecture to television and urban design. It seems to have originated in parts of the world's core countries and is currently spreading throughout the rest of the world.

KEY TERMS

cosmopolitanism (p. 240)
derelict landscapes (p. 225)
ethology (p. 219)
existential imperative (p. 220)

humanistic approach (p. 228)
landscape as text (p. 228)
modernity (p. 233)

postmodernity (p. 238)
proxemics (p. 219)
sacred spaces (p. 230)

semiotics (p. 228)
territoriality (p. 219)
topophilia (p. 225)

ADDITIONAL READING

Agnew, J. A., and J. M. Smith (eds.), *American Space/American Place: Geographies of the Contemporary United States.* Edinburgh: Edinburgh University Press, 2002.

Blesser, B., and L. R. Salter, *Spaces Speak, Are You Listening?: Experiencing Aural Architecture.* Cambridge, MA: MIT Press, 2007.

Burnett, R., *The Global Jukebox: The International Music Industry.* New York: Routledge, 1996.

Coleman, R., *Reclaiming the Streets: Surveillance, Social Control and the City.* Portland, OR: Willan Publishing, 2004.

Gold, J. R., and S. V. Ward (eds.), *Place Promotion: The Use of Publicity and Marketing to Sell Towns and Regions.* New York: Wiley and Sons, 1994.

Golledge, R. G., and R. J. Stimpson, *Spatial Behavior: A Geographic Perspective.* New York: Guilford Press, 1996.

Groth, P., and T. W. Bressi (eds.), *Understanding Ordinary Landscapes.* New Haven, CT: Yale University Press, 1997.

Hackworth, J., *The Neoliberal City: Governance, Ideology, and Development in American Urbanism.* Ithaca, NY: Cornell University Press, 2007.

Harvey, D. W., *Spaces of Hope.* Berkeley: University of California Press, 2000.

Honoré, C., *In Praise of Slowness: How a Worldwide Movement Is Challenging the Cult of Speed.* New York: HarperCollins, 2004.

Jackson, J. B., *A Sense of Place, A Sense of Time.* New Haven, CT: Yale University Press, 1994.

Jackson, J. B., *Discovering the Vernacular Landscape.* New Haven, CT: Yale University Press, 1984.

Judd, D. R., and S. S. Fainstein (eds.), *The Tourist City.* New Haven, CT: Yale University Press, 1999.

Katz, C., *Growing up Global: Economic Restructuring and Children's Everyday Lives.* Minneapolis; St. Paul: University of Minnesota Press, 2004.

Kearns, G., and C. Philo (eds.), *Selling Places.* Oxford, UK: Pergamon Press, 1993.

Knox, P. L., "Creating Ordinary Places: Slow Cities in a Fast World," *Journal of Urban Design,* 10 (1), 2005, pp. 3–13.

Kohn, M., *Brave New Neighborhoods: The Privatization of Public Space.* New York; London: Routledge, 2004.

Massey, D., and P. Jess (eds.), *A Place in the World: Places, Cultures, and Globalization.* New York: Oxford University Press, 1995.

Mitchell, D., *Cultural Geography: A Critical Introduction.* Malden, MA: Blackwell, 2000.

Relph, E., *Place and Placelessness.* London: Pion, 1976.

Ritzer, G., *The Globalization of Nothing.* Thousand Oaks, CA: Pine Forge Press, 2004.

Schama, S., *Landscape and Memory.* New York: Vintage Books, 1995.

Shurmer-Smith, P., and K. Hannam, *Worlds of Desire, Realms of Power.* London: Edward Arnold, 1994.

Sluyter, A., *Colonialism and Landscape: Postcolonial Theory and Applications.* Lanham, MD; Oxford: Rowman & Littlefield Publishers, 2002.

Urry, J., *Consuming Places.* London: Routledge, 1995.

Vale, L. J., and S. B. Warner, Jr. (eds.), *Imaging the City.* New Brunswick, NJ: CUPR Press, 2002.

Vaneigem, R., The Revolution of Everyday Life. Available online at http://library.nothingness.org/articles/SI/en/pub_contents/5.

Ward, S. V., *Selling Places: The Marketing and Promotion of Towns and Cities 1850–2000.* New York: Routledge, 1998.

Wylie, J., *Landscape.* Routledge, 2007.

Zukin, S., *The Cultures of Cities.* Cambridge, MA: Blackwell, 1995.

EXERCISES

On the Internet

The Companion Website at www.mygeoscienceplace.com contains a number of tools designed to reinforce the geographic concepts presented in this chapter. Structured within a four-step learning path, these online resources include: chapter Key Concepts; a fully-searchable myeBook version of the chapter, which allows you to add notes and highlights; RSS news feeds from various media agencies; links to additional web resources; Human Geography Videos; a digital Glossary; Concept Review questions; Thinking Spatially mapping exercises; and Critical Thinking essay questions.

Unplugged

1. Write a short essay (500 words or two double-spaced, typed pages) that describes, from your personal perspective, the sense of place that you associate with your hometown or county. Write about the places, buildings, sights, and sounds that are especially meaningful to you.

2. Draw up a list of the top 10 places in the United States in which you would like to live and work, then draw up a list of the bottom 10. How do these lists compare with the map of student preferences shown in Figure 6.8? Why might your preferences be different from theirs?

3. On a clean sheet of paper and without reference to maps or other materials, sketch a detailed map of the town or city in which you live. When you have finished, turn to page 228 and compare your sketch to Figure 6.6. Does your sketch contain nodes? Landmarks? Edges? Districts? Paths? How does "your" city compare to the "real" city?

4. Using telephone books that can be accessed at your university library and a map of your local area, map the sacred landscape of your locality by identifying the locations of different churches, temples, mosques, and other places of worship. What patterns do you see? How do they fit with your knowledge of particular areas? What other sorts of sacred sites might exist beyond the institutionalized religious sites?

5. Your campus provides an institutional landscape that has been "written" to convey certain important relationships. "Read" your campus landscape and compare the more "powerful" sites with the less powerful, and discuss which groups or academic disciplines occupy them. Map what you think are the most powerful and important places and sites and the least powerful and important. Why do different places and sites fit into each category? Are there significant differences in the architecture, in the location (central or on the edges of campus), in accessibility? How might your reading of the campus landscape differ from readings created by other members of the campus community (as well as outsiders)?

The Geography of Economic Development

7

In the past quarter-century, many of China's regions have gone through an extraordinary transformation as the Chinese economy has become increasingly integrated into the global economy. In Guangdong province, near Hong Kong, for example, industrial growth, real estate development, and investment in highway construction has increased by an average of 15 to 25 percent per year throughout the past decade. Once known mainly for its abundant and high-quality long-grain rice, Guangdong itself now imports fragrant jasmine rice from Thailand. Along Guangdong's Pearl River Delta, it is difficult to photograph one of the classic rice paddies without capturing a construction site as well. Incomes and consumer spending in Guangdong suddenly skyrocketed in the 1990s. In Shandong Province, 1,500 kilometers (930 miles) north, conditions were also changing. Traditionally a region of peasant agriculture and a modest prosperity based on winter wheat and summer corn, Shandong is now beginning to join in China's industrial surge. In the tiny village of Xishan, on the plain of the Yellow River, a slaughterhouse, a pharmaceuticals plant, and a furniture factory now employ workers who until recently were peasant farmers.

As individual regions develop new specializations, China itself is changing from a self-sufficient agricultural producer to a more diversified, affluent country that is beginning to export substantial amounts of manufactured items as well as import substantial amounts of food. Meanwhile, in the southwestern province of Sichuan, there remain more than 20,000 state-owned enterprises, most of which are not able to keep pace with China's economic transformation. Exports from the region's textile factories have declined. The region's economy had become increasingly depressed even before the severe disruption of the 2008 earthquake that killed at least 80,000 and made 4.8 million homeless. Although the official unemployment rate is just 3 percent, the actual rate is closer to 20 percent.

Economic development is always an uneven geographic phenomenon, and countries and regions can follow various pathways to development. Places and regions in China are beginning to experience many of the processes of capitalist economic development that have shaped the fortunes of places and regions in most of the rest of the world. Explaining how and why these processes occur is an important goal of human geography.

Main Points

- Geographically, the single most important feature of economic development is that it is highly uneven.

- Geographical divisions of labor have evolved with the growth of the world-system of trade and politics and with the changing locational logic of successive technology systems.

- Regional cores of economic development are created cumulatively, following some initial advantage, through the operation of several basic principles of spatial organization.

- Spirals of economic development can be arrested in various ways, including the onset of disinvestment and deindustrialization, which follow major shifts in technology systems and international geopolitics.

- The globalization of the economy has meant that patterns and processes of local and regional economic development are much more open to external influences than before.

Wal-Mart superstore, Jinan, Shandong Province, China

PATTERNS OF ECONOMIC DEVELOPMENT

Economic development is often discussed in terms of levels and rates of change in prosperity, as reflected in bottom-line statistical measures of productivity, incomes, purchasing power, and consumption. Increased prosperity is only one aspect of economic development, however. For human geographers and other social scientists, the term *economic development* refers to processes of change involving the nature and composition of the economy of a particular region as well as to increases in the overall prosperity of a region. These processes can involve three types of changes:

■ changes in the structure of the region's economy (for example, a shift from agriculture to manufacturing);

■ changes in forms of economic organization within the region (for example, a shift from socialism to free-market capitalism);

■ changes in the availability and use of technology within the region.

Economic development is also expected to bring with it some broader changes in the economic well-being of a region. The most important of these are changes in the capacity of the region to improve the basic conditions of life (through better housing, health care, and social welfare systems) and to improve the physical framework, or infrastructure, on which the economy rests.

The Unevenness of Economic Development

Geographically, the single most important feature of economic development is that it is *uneven*. At the global scale, this unevenness takes the form of core-periphery contrasts within the evolving world-system (see Chapter 2). These global core-periphery contrasts are the result of a competitive economic system that is heavily influenced by cultural and political factors. The core regions within the world-system—currently, the tripolar core of North America, Europe, and Japan—have the most diversified economies, the most advanced technologies, the highest levels of productivity, and the highest levels of prosperity. They are commonly referred to as *developed regions* (though processes of economic development are, of course, continuous, and no region can ever be regarded as fully developed).

Other countries and regions—the periphery and semiperiphery of the world-system—are often referred to as *developing* or *less developed*. Indeed, the nations of the periphery are often referred to as LDCs (less developed countries). Another popular term for the global periphery, developed as a political label but now synonymous with economic development, is the *Third World*. As we saw in Chapter 2, this term had its origins in the early Cold War era of the 1950s and 1960s, when the newly independent countries of the periphery positioned themselves as a distinctive political bloc, aligned with neither the First World of developed, capitalist countries nor the Second World of the Soviet Union and its satellite countries.

Global Core-Periphery Patterns At this global scale, levels of economic development are usually measured by economic indicators such as gross domestic product and gross national income. **Gross domestic product (GDP)** is an estimate of the total value of all materials, foodstuffs, goods, and services that are produced by a country in a particular year. To standardize for countries' varying sizes, the statistic is normally divided by total population, which gives an indicator, *per capita* GDP, that is a good yardstick of relative levels of economic development. **Gross national income (GNI)** is a measure of the income to a country from production wherever in the world it occurs. For example, if a U.S.-owned company operating in another country sends some of its income (profits) back to the United States, this adds to the U.S. GNI.

In making international comparisons, GDP and GNI can be problematic because they are based on each nation's currency. As a result, it is now common to compare national currencies based on *purchasing power parity* (PPP). In effect, PPP measures how much of a common "market basket" of goods and services each currency can purchase locally, including goods and services that are not traded internationally. Using PPP-based currency values to compare levels of economic prosperity usually produces lower GNI figures in wealthy countries and higher GNI figures in poorer nations, compared with market-based exchange rates. Nevertheless, even with this compression between rich and poor, economic prosperity is very unevenly distributed across nations.

As **Figure 7.1** shows, most of the highest levels of economic development are to be found in northern latitudes (very roughly, north of 30° N), which has given rise to another popular shorthand for the world's economic geography: the division between the "North" (the core) and the "South" (the periphery). Viewed in more detail, the global pattern of per capita GNI (measured in the "international dollars" of PPP) in 2005 is a direct reflection of the core-semiperiphery-periphery structure of the world-system. In many of the core countries of North America, northwestern Europe, and Japan, annual per capita GNI (in PPP) exceeds $30,000. The only other countries that match these levels are Australia, Hong Kong, and Singapore, where annual per capita GNI in 2005 was $31,794, $34,833, and $29,663, respectively. Semiperipheral countries such as Brazil, Russia, and Thailand have an annual per capita GNI ranging between $7,000 and $9,000. In the rest of the world—the periphery—annual per capita GNI (in PPP) is typically less than $5,000.

The gap between the highest per capita GNIs ($71,240 in Luxembourg and $68,440 in Norway,) and the lowest ($130 in Liberia, $100 in Burundi) is huge. The gap between the world's rich and poor is also getting

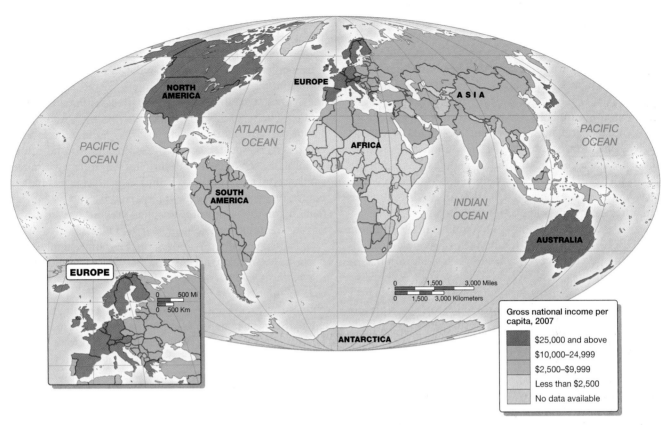

Figure 7.1   Gross national income (GNI) per capita GNI per capita is one of the best single measures of economic development. This map, based on 2005 data, shows the tremendous gulf in affluence between the core countries of the world economy—like the United States, Norway, and Switzerland, with annual per capita GNI (in PPP "international dollars") of more than $25,000—and peripheral countries like Angola, Haiti, and Mali, where annual per capita GNI was less than $2,500. In semiperipheral countries like South Korea, Brazil, and Mexico, per capita GNI ranged between $5,000 and $10,000. (Adapted from Map projection, Buckminster Fuller Institute and Dymaxion Map Design, Santa Barbara, CA. The word *Dymaxion* and the Fuller Projection Dymaxion™ Map design are trademarks of the Buckminster Fuller Institute, Santa Barbara, CA, © 1938, 1967 & 1992. All rights reserved.)

wider (**Figure 7.2**). In 1970, the average GNI per capita of the 10 poorest countries in the world was just one-fiftieth of the average GNI per capita of the 10 most prosperous countries. By 1990, the relative gap had doubled, and by 2007 the average GNI per capita of the 10 poorest countries was $190 compared to $52,105 for the 10 most prosperous countries. Overall, more than 80 percent of the world's population lives in countries where income differentials are widening.

Spatial Justice? Inequality in income is reflected—and reinforced—by many aspects of human well-being. Patterns of infant mortality, a reliable indicator of social well-being, show the same steep core-periphery gradient (refer back to Figure 3.16). For adults in the industrial core countries, life expectancy is high and continues to increase. Life expectancy at birth in Australia and Switzerland in 2007 was 81 years, and in the United States it was 78. In contrast, life expectancy in the poorest countries is dramatically shorter. In 2007 in Malawi, life expectancy at birth was 40 years, in Zimbabwe it was 34, and in Botswana it was only 34. In most African

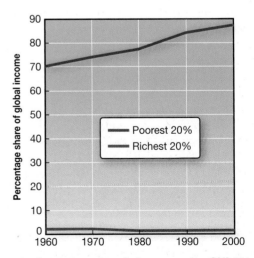

Figure 7.2   Long-term trends in per capita GNI This graph shows the steady divergence in international economic prosperity between the richest and poorest of the world's population. In 1960, the richest 20 percent of the world's population accounted for 70.2 percent of global income, while the poorest 20 percent accounted for 2.3 percent: a ratio of 30 to 1. By 1980, the ratio had increased to 45 to 1; by 1990, it was 64 to 1; and by 2000, it was 72 to 1.

countries, only 60 to 75 percent of the population can expect to survive to age 40.

Patterns of food consumption, one of the most basic of all human needs, reflect another key dimension of inequality in human well-being. Every day around 800 million people—12 percent of the world's population—go hungry (**Figure 7.3**). In South Asia one person in four goes hungry, and in sub-Saharan Africa the rate is as high as one in three. In terms of sheer numbers, however, there are more chronically hungry people in Asia and the Pacific. The fact that many of them are located in countries that are net exporters (by value) of foods is a telling indictment of the world economic system.

The United Nations Development Programme (UNDP) has devised an overall index of human development based on measures of life expectancy, educational attainment, and personal income. The index is calculated so that a country with the best combined scores among all countries in the world on all three indicators has a perfect

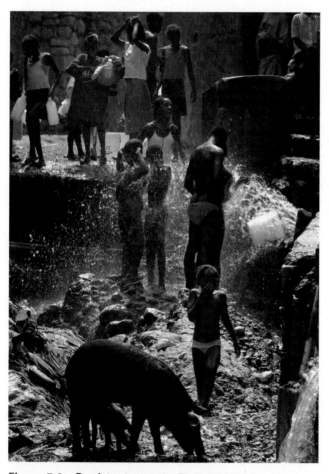

Figure 7.3 Persistent poverty The poorest country in the western hemisphere, Haiti has hundreds of thousands of very young children who have been handed over to "host" families by desperately poor parents lured by the promise of a better life. Instead, the children receive no education and are forced to do hard, menial jobs; oftentimes they are forbidden to join the host family for meals and sleep on concrete or dirt floors with little clothing. Host families often throw them out on the street when they reach 15 years of age (by law the age to receive a wage) and replace them with younger children.

index score of 1.0, while a country that ranked worst in the world on all three indicators has an index score of zero. **Figure 7.4** shows the international map of human development in 2005. Iceland and Norway (0.968), Australia (0.962), Canada (0.961), and Ireland (0.959) had the highest overall levels of human development, while Guinea-Bissau (0.374), Burkina Faso (0.370), and Sierra Leone (0.336) had the lowest levels. The same fundamental pattern is repeated across the entire array of indicators of human development: adult literacy, poverty, malnutrition, access to physicians, public expenditure on higher education, telephone lines, Internet users, and so on. Inequality on this scale poses the most pressing as well as the most intractable questions of spatial justice.

These questions are underscored by some simple comparisons between the needs of the periphery and the spending patterns in core countries. The UNDP has calculated that the annual cost of providing a basic education for all children in peripheral countries would be around $6 billion, which is less than the annual sales of cosmetics in the United States. Providing water and sanitation for everyone in peripheral countries is estimated at $9 billion per year, which is less than Europeans' annual expenditure on ice cream. Providing basic health and nutrition for everyone in the peripheral countries would cost an estimated $13 billion per year—less than the annual expenditure on pet foods in Europe and the United States. Reducing the military expenditures of core countries (in the region of $700 billion per year) by less than 10 percent each year would pay for the costs of basic education, water and sanitation, basic health and nutrition, and reproductive health programs for everyone in peripheral countries.

America's score on the human development index has increased steadily over the last 25 years. But other affluent countries have progressed more quickly, and as a result the United States dropped from number 2 in 1980 to number 12 in 2005, falling behind other countries that have been more efficient in transforming national income into positive outcomes in terms of social well-being (**Table 7.1**). The United States also provides a good example of the point that, even within countries that have a high level of income per capita, there are wide variations from place to place. **Figure 7.5** shows an index of human development for the congressional districts of the United States, based on similar variables to those used in the UNDP international index. There is a clear urban-rural dimension to the pattern of human development in the United States, along with a north-south regional dimension, with the rural South attaining the worst levels of overall human development.

Development and Gender Equality Core-periphery patterns are also reflected in indicators that measure economic development by *gender equality*. The UNDP has established a gender-sensitive development index that adjusts the overall human development index for gender inequality in life expectancy, educational attainment, and income. According to this index, in no country are women better off than men. Perhaps most revealing is the UNDP's

Patterns of Economic Development   **253**

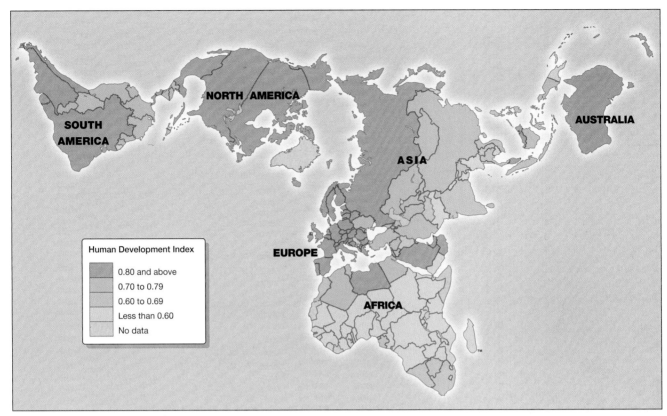

Figure 7.4   An index of human development, 2005 This index, calculated by the United Nations Development Programme, is based on measures of life expectancy, educational attainment, and personal income. A country with the best scores among all countries in the world on all three measures would have a perfect index score of 1.0; a country that ranked worst in the world on all three indicators would have an index score of zero. Most of the affluent core countries have index scores of 0.9 or more; the worst scores—those less than 0.4—are concentrated in Africa. (Adapted from Map projection, Buckminster Fuller Institute and Dymaxion Map Design, Santa Barbara, CA. The word *Dymaxion* and the Fuller Projection Dymaxion™ Map design are trademarks of the Buckminster Fuller Institute, Santa Barbara, CA, © 1938, 1967 & 1992. All rights reserved.)

TABLE 7.1 Top-Ranked Countries in Human Development, 1980–2005

1980	1985	1990	1995	2000	2005
Switzerland	Canada	Canada	Norway	Norway	Iceland
United States	United States	United States	Canada	Sweden	Norway
Iceland	Switzerland	Iceland	Sweden	Australia	Australia
Norway	Norway	Japan	Netherlands	Netherlands	Canada
Canada	Iceland	Switzerland	Australia	Iceland	Ireland
Japan	Japan	Netherlands	United States	Canada	Sweden
Netherlands	Netherlands	Norway	Belgium	Switzerland	Switzerland
Denmark	Sweden	France	Japan	Belgium	Netherlands
Sweden	Denmark	Finland	United Kingdom	United States	Japan
France	France	Sweden	Switzerland	Japan	Finland
Belgium	Finland	Belgium	France	Finland	France
Australia	Belgium	Austria	Iceland	France	United States

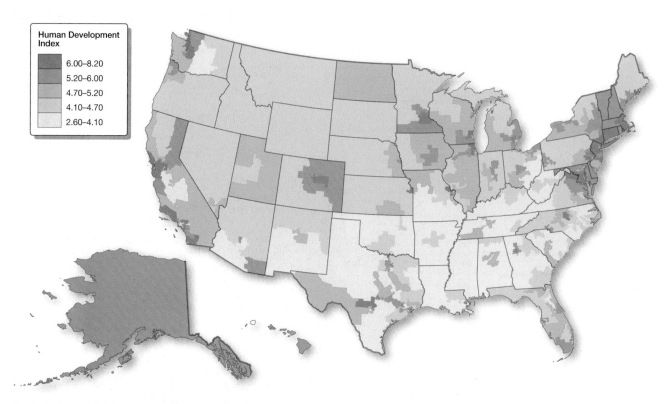

Figure 7.5 Human development in the United States, 2007–2008 Among the nation's 436 congressional districts, New York's Fourteenth District, in New York City, ranks first, and California's Twentieth District, around Fresno, ranks last. The average resident of New York's Fourteenth District earns over 3 times as much as the average resident of California's Twentieth District, lives 4½ years longer, and is 10 times as likely to have a college degree. (*Source*: Adapted from "The Measure of America: American Human Development Report 2008–2009," http://www.measureofamerica.org/.)

Gender Empowerment Index, which is based on measures of women's incomes, their participation in the labor force as administrators and managers, professional and technical workers, and the percentage of parliamentary seats held by women. As in the overall index of human development, a country with a perfect score would score 1.0, with zero representing the worst possible score. In 2007–2008, the top four countries were Iceland (0.962), Norway (0.957), and Australia (0.960). In many countries, data on women are insufficiently accurate to calculate the index, but the countries with the worst recorded index scores were Burkina Faso (0.364), Guinea-Bissau (0.355), and Sierra Leone (0.320).

Women play a central and increasing role in processes of development and change in the global economy. In many peripheral countries, women constitute the majority of workers in the formal sector that is dependent on the new international division of labor (**Figure 7.6**). In many poor countries, it is women who keep households afloat in a world economy that has resulted in localized recession and intensified poverty. In these countries, on average, women earn 40 to 50 percent less than men for the same work. They also tend to work longer hours than men: 12 to 13 hours a week more (counting both paid and unpaid work) in Africa and Asia.

Child labor also remains a serious problem around the world. According to International Labor Organization

(ILO) estimates, more than 250 million children between the ages of 5 and 14 are sent out to work, many of them in dangerous conditions (**Figure 7.7**). The ILO also estimates that some 1 million children work in small-scale mining and quarrying around the world. While most of these children live in peripheral countries, hundreds of thousands of underage workers are exploited in sweatshops, farm fields, and other workplaces in core countries, including up to 1.5 million children, as estimated by the United Nations International Children's Fund (UNICEF), in the United States.

Regional Patterns Inequality in economic development often has a regional dimension. Initial conditions are a crucial determinant of regional economic performance. Scarce resources, a history of neglect, lack of investment, and concentrations of low-skilled people all combine to explain the lagging performance of certain areas. In some regions, for example, initial disadvantages are so extreme as to constrain the opportunities of individuals born there. A child born in the Mexican state of Chiapas, for example, has much bleaker prospects than a child born in Mexico City. The child from Chiapas is twice as likely to die before age five, less than half as likely to complete primary school, and 10 times as likely to live in a house without access to running water. On reaching working age, he or she will earn 20 to 35 percent less than a comparable

Figure 7.6 Women in development The changing global economy has placed unprecedented demands on women. Much of the industrialization in peripheral countries depends on female labor because wage rates for women are significantly lower than those for men. In times of rapid economic change and adjustment, women are typically called upon to help sustain household incomes, in addition to their traditional household responsibilities. When employed, women are often more vulnerable than men, disproportionately concentrated in occupations with little or no job security and few opportunities for advancement. This photo shows women peeling and grading cashew nuts by hand at a large processing factory in Mtwara, Tanzania, which employs 2,400 women. The workers are paid by the weight of the nuts they process and this works out to approximately $1.80 per day.

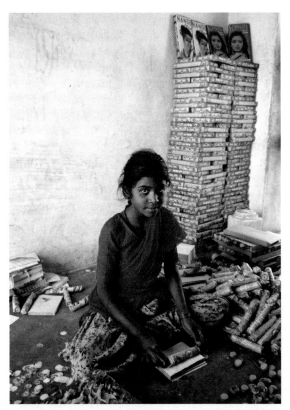

Figure 7.7 Child labor A 13-year-old girl makes fireworks at a factory at Sivakasi, India. One in twelve of the world's children (180 million young people aged 5 to 17) are involved in the worst forms of child labor—hazardous work, slavery, forced labor, the military, commercial sexual exploitation, and illicit activities. Of these children, 97 percent are in peripheral and semiperipheral countries. Globally, an estimated 114 million children of primary school age are not enrolled in school, depriving one in five children of an education. They become exposed to exploitation and abuse and miss out on developing the knowledge and employable skills that could lift them and their own children out of the poverty cycle.

worker living in Mexico City and 40 to 45 percent less than one living in northern Mexico.

Other examples of regional inequality can be found throughout the world. Gansu, with an income per capita 40 percent below the national average, is one of the poorest and most remote regions in China. With poor soils highly susceptible to erosion, low and erratic rainfall, and few off-farm employment opportunities, a high proportion of its inhabitants live in poverty. Chaco Province, in Argentina, has a GDP per capita that is only 38 percent of the national average. Low educational attainment and lack of infrastructure, especially roads, explain much of this deviation.

Even when growth in other parts of a country's economy is strong, relative and absolute regional poverty can persist for long periods. In the seventeenth century, for example, the decline of the sugar economy pushed northeastern Brazil into a decline from which it has never fully recovered. In the United States, the fortunes of coal-mining West Virginia waned with the collapse of the coal industry and the increased importance of oil and gas in energy production. It remains one of the poorest regions in the United States to this day. In Thailand, rapid development has failed to reach the northern hill people. Fewer than 30 percent of their villages have schools, and only 15 percent of the hill people can read and write. Their average annual income is less than a quarter of the country's overall GNI per capita.

Globalization has been associated with increasing regional inequality within many countries since the 1980s. In China, for example, disparities have widened dramatically between the export-oriented regions of the coast and the interior. The transition economies of the countries of the former Soviet Union and its Eastern European satellites have registered some of the largest increases in regional inequality, and some core countries—especially Sweden, the United Kingdom, and the United States—have also registered significant increases in regional inequality since the

1980s. At this regional scale, as at the global scale, levels of economic development often exhibit a fundamental core-periphery structure. Indeed, within-country core-periphery contrasts are evident throughout the world: in core countries such as France and the United States, in semiperipheral countries such as South Korea, and in peripheral countries such as Nigeria and Indonesia.

Resources and Development

These patterns of economic development are the result of many different factors. One of the most important is the availability of key resources such as cultivable land, energy sources, and valuable minerals. Unevenly distributed across the world, however, are both key resources and—just as important—the *combinations* of energy and minerals crucial to economic development. A lack of natural resources can, of course, be remedied through international trade (Japan's success is a prime example of this). For most countries, however, the resource base remains an important determinant of development.

Energy One particularly important resource in terms of the world's economic geography is energy. The major sources of commercial energy—oil, natural gas, and coal—are all very unevenly distributed across the globe. Most of the world's core economies are reasonably well off in terms of energy *production,* the major exceptions being Japan and parts of Europe. Most peripheral countries, on the other hand, are energy poor. The major exceptions are Algeria, Ecuador, Gabon, Indonesia, Libya, Nigeria, Venezuela, and the Gulf states—all major oil producers.

Because of this unevenness, energy has come to be an important component of world trade. As more of the world becomes industrialized and developed, demand for oil increases (**Figure 7.8**). Yet, as we saw in Chapter 4, it is generally agreed that the peak of oil *discovery* was

passed in the 1960s and that the world started using more oil than was contained in new fields in 1981. As a result, oil prices have risen sharply: more than 300 percent between 2003 and 2008. Oil is now the most important single commodity in world trade, making up more than 20 percent of the total by value in 2008.

For many peripheral countries the costs of importing energy represent a heavy burden. Consider, for example, the predicament of countries like India, Ghana, Paraguay, Egypt, and Armenia, where in 2007 the cost of energy imports amounted to more than one-quarter of the total value of exported merchandise. Few peripheral countries can afford to consume energy on the scale of the developed economies, so patterns of commercial energy *consumption* tend to mirror the fundamental core-periphery cleavage of the world economy. In 2005, energy consumption per capita in North America was 20 times that of India, 35 times that of Mozambique, and over 230 times that of Niger. The world's high-income countries, with 15 percent of the world's population, use half its commercial energy and 10 times as much per capita as low-income countries.

It should be noted that these figures do not reflect the use of firewood and other traditional fuels for cooking, lighting, heating, and sometimes industrial needs. In total, such forms of energy probably account for around 20 percent of total world energy consumption. In parts of Africa and Asia, they account for up to 80 percent of energy consumption. This points to yet another core-periphery contrast. Whereas massive investments in exploration and exploitation are enabling more of the developed, energy-consuming countries to become self-sufficient through various combinations of coal, oil, natural gas, hydroelectric power, and nuclear power, 1.5 billion people in peripheral countries depend on collecting fuelwood as their principal source of energy. The collection of wood fuel causes considerable deforestation, and the problem is most serious in densely populated locations, arid and

Figure 7.8 Offshore oil drilling
Offshore oil drilling escalating prices of crude oil in 2008 prompted the U.S. government to reconsider long-standing restrictions on offshore oil drilling. Restrictions were imposed in 1990 in response to environmental concerns—in particular, concerns about pollution. In this photo, a rare and endangered blue whale spouts near offshore oil rigs after a long dive near Long Beach, California in July, 2008.

semiarid regions, and cooler mountainous areas, where the regeneration of shrubs, woodlands, and forests is particularly slow. Nearly 100 million people in 22 countries (16 of them in Africa) cannot meet their minimum energy needs even by overcutting remaining forests.

Cultivable Land The distribution of cultivable land represents another important factor in international economic development. Much more than half of Earth's land surface is unsuitable for any productive form of arable farming, as suggested by **Figure 7.9**. Poor soils, short growing seasons, arid climates, mountainous terrain, forests, and conservation limit the extent of agricultural land across much of the globe. As a result, the distribution of the world's cultivable land is highly uneven, being concentrated in Europe, west-central Russia, eastern North America, the Australian littoral, Latin America, India, eastern China, and parts of sub-Saharan Africa. In detail, of course, some of these regions may be marginal for arable farming because of marshy soils or other adverse conditions, while irrigation, for example, sometimes extends the local frontier of productive agriculture. We also have to bear in mind that not all cultivable land is of the same quality. This leads to the concept of the **carrying capacity** of agricultural land: the maximum population that can be maintained in a place with rates of resource use and waste production that are sustainable in the long term without damaging the overall productivity of that or other places.

Industrial Resources A high proportion of the world's key industrial resources—basic raw materials—are concentrated in Russia, the United States, Canada, South Africa, and Australia. The United States, for example, in addition to having 42 percent of the world's known resources of hydrocarbons (oil, natural gas, and oil shales) and 38 percent of the lignite ("brown coal," used mainly in power stations), has 38 percent of the molybdenum (used in metal alloys), 21 percent of the lead (used chiefly for batteries, gasoline, and construction), 19 percent of the copper (used mainly for electrical wiring and components and for coinage), 18 percent of the bituminous coal (used mainly for fuel in power stations and in the chemical industry), and 15 percent of the zinc. Russia has 68 percent of the vanadium (used in metal alloys), 50 percent of the lignite, 38 percent of the bituminous coal, 35 percent of the manganese, 25 percent of the iron, and 19 percent of the hydrocarbons.

The concentration of known resources in just a few countries is largely a result of geology, but it is also partly a function of countries' political and economic development. Political instability in much of postcolonial Africa, Asia, and Latin America has seriously hindered the exploration and exploitation of resources. In contrast, the relative affluence and great political stability of the United States have led to a much more intensive exploration of its resources. We should also bear in mind that the significance of particular resources is often tied to particular technologies. As technologies change, so do resource requirements, and the geography of economic development is "rewritten." One important example of this was the switch in the manufacture of mass-produced textiles from natural fibers like wool and cotton to synthetic fibers in the 1950s and 1960s. When this happened, many farmers

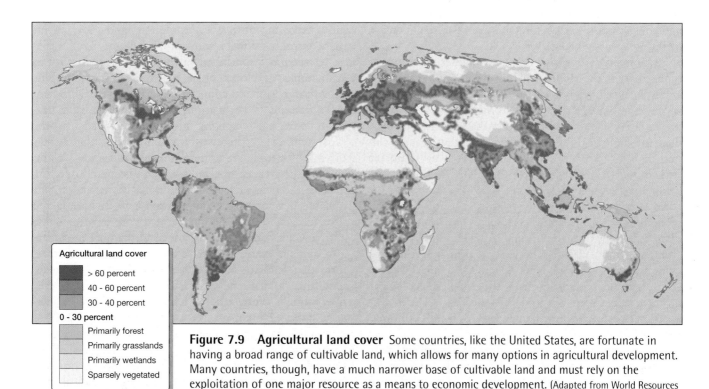

Agricultural land cover

> 60 percent
40 - 60 percent
30 - 40 percent
0 - 30 percent
Primarily forest
Primarily grasslands
Primarily wetlands
Sparsely vegetated

Figure 7.9 Agricultural land cover Some countries, like the United States, are fortunate in having a broad range of cultivable land, which allows for many options in agricultural development. Many countries, though, have a much narrower base of cultivable land and must rely on the exploitation of one major resource as a means to economic development. (Adapted from World Resources Institute, *World Resources 2000–2001: People and Ecosystems*. Washington, DC: World Resources Institute, 2000, p. 57.)

in the U.S. South, for example, had to switch from cotton to other crops.

Regions and countries that are heavily dependent on one particular resource are vulnerable to the consequences of technological change. They are also vulnerable to fluctuations in the price set for their product on the world market. These vulnerabilities are particularly important for countries whose economies are especially dependent on nonfuel minerals, such as the Democratic Republic of the Congo (copper), Mauritania (iron ore), Namibia (diamonds), Niger (uranium), Sierra Leone (diamonds), Togo (phosphates), and Zambia (copper).

Sustainable Economic Development

In 1987, the World Commission on Environment and Development, chaired by former Norwegian Prime Minister Gro Harlem Brundtland, issued a report, *Our Common Future,* that stressed the intensification and interdependence of ecological and economic crises and made a strong plea for the principle of sustainable development—economic development that seeks to meet the needs and aspirations of the present without compromising the ability to meet those of the future.

The importance of sustainability is cogently illustrated by the concept of an **ecological footprint,** which is a measure of the human pressures on the natural environment from the consumption of renewable resources and the production of pollution. It represents a quantitative assessment of the biologically productive area required to produce the resources (food, energy, and materials) and to absorb the wastes of an individual, city, region, or country. The ecological footprint of a country or region changes in proportion to population size, average consumption per person, and the resource intensity of the technology being used.

Humanity's footprint first grew larger than global biocapacity in the 1980s, and this overshoot has been increasing every year since. In 2003, demand exceeded supply by about 25 percent. This means that it took approximately a year and three months for Earth to produce the ecological resources we used in that year. By the mid-2000s, humanity's collective ecological footprint was about 30 percent greater than Earth's biologically productive capacity to sustain. At 9.57 hectares (23.6 acres) per person, the United States currently has the second-largest per capita ecological footprint on the planet, just behind the United Arab Emirates. Other countries with extremely large ecological footprints include Canada, Norway, New Zealand, Kuwait, Sweden, and Australia. Countries with the smallest ecological footprints, between 0.5 and 0.75 hectares (1.2 to 1.9 acres) per person, include Bangladesh, Burundi, Ethiopia, Haiti, Malawi, Nepal, and Pakistan.

Sustainable development means using renewable natural resources in a manner that does not eliminate or degrade them—by making greater use, for example, of solar and geothermal energy and recycled materials. It means

managing economic systems so that all resources—physical and human—are used optimally. It means regulating economic systems so that the benefits of development are distributed more equitably (if only to prevent poverty from causing environmental degradation). It also means organizing societies so that improved education, health care, and social welfare can contribute to environmental awareness and sensitivity and an improved quality of life. A final and more radical aspect of sustainable development is to move away from wholesale globalization toward increased "localization": a desire to return to more locally based economies where production, consumption, and decision making can be oriented to local needs and conditions.

Put this way, sustainable development sounds eminently sensible yet impossibly utopian. A widespread discussion of sustainability took place in the early 1990s and was the focus of the "Earth Summit" (the United Nations Conference on Environment and Development) meeting in Rio de Janeiro in 1992. Attended by 128 heads of state, the meeting attracted intense media attention. After the UN conference, however, many observers commented bitterly on the deep conflicts of interest between core countries and peripheral countries that were exposed by the summit.

One of the most serious obstacles to prospects for sustainable development is continued heavy reliance on fossil fuels as the fundamental source of energy for economic development. This not only perpetuates international inequalities but also leads to transnational problems such as acid rain, global warming, climatic changes, deforestation, health hazards and, many would argue, war. The sustainable alternative—renewable energy generated from the sun, tides, waves, winds, rivers, and geothermal features—has been pursued half-heartedly because of the commercial interests of the powerful corporations and governments that control fossil-fuel resources.

A second important challenge to the possibility of sustainable development is the rate of demographic growth in peripheral countries. Sustainable development is feasible only if population size and growth are in harmony with the changing productive capacity of the ecosystem. It is estimated that 1.2 billion of the world's 6.7 billion people are undernourished and underweight.

But the greatest single obstacle to sustainable development is the inadequacy of institutional frameworks. Sustainable development requires economic, financial, and fiscal decisions to be fully integrated with environmental and ecological decisions. In practice, national and local governments everywhere have evolved institutional structures that tend to separate decisions about what is economically rational and what is environmentally desirable. International organizations, while better placed to integrate policy across these sectors and better able to address economic and environmental "spillovers" from one country to another, have (with the notable exception of the European Union) not acquired sufficient power to promote integrated, harmonized policies. Without radical and widespread changes in value systems and unprecedented changes in political will, "sustainable development" is likely to remain an embarrassing contradiction in terms.

The Economic Structure of Countries and Regions

The relative share of primary, secondary, tertiary, and quaternary economic activities determines the *economic structure* of a country or region. **Primary activities** are those concerned directly with natural resources of any kind; they include agriculture, mining, fishing, and forestry. **Secondary activities** are those that process, transform, fabricate, or assemble the raw materials derived from primary activities or that reassemble, refinish, or package manufactured goods. Secondary activities include, for example, steelmaking, food processing, furniture making, textile manufacturing, automobile assembly, and garment manufacturing. **Tertiary activities** are those involving the sale and exchange of goods and services; they include warehousing, retail stores, personal services such as hairdressing, commercial services such as accounting and advertising, and entertainment. **Quaternary activities** are those dealing with the handling and processing of knowledge and information. Examples include data processing, information retrieval, education, and research and development (R&D).

Geographical Divisions of Labor Variations in economic structure—according to primary, secondary, tertiary, or quaternary activities—reflect *geographical divisions of labor*. Geographical divisions of labor are national, regional, and locally based economic specializations that have evolved with the growth of the world-system of trade and politics (see Chapter 2) and with the locational needs of successive technology systems. They represent one of the most important dimensions of economic development. For instance, countries whose economies are dominated by primary-sector activities tend to have a relatively low per capita GDP. The exceptions are oil-rich countries such as Saudi Arabia, Qatar, and Venezuela. Where the geographical division of labor has produced national economies with a large secondary sector, per capita GDP is much higher (as, for example, in Argentina and South Korea). The highest levels of per capita GDP, however, are associated with economies that are *postindustrial*: economies where the tertiary and quaternary sectors have grown to dominate the workforce, with smaller but highly productive secondary sectors.

As **Figure 7.10** shows, the economic structure of much of the world is dominated by the primary sector (that is, primary activities such as agriculture, mining, fishing, and forestry). In much of Africa and Asia, between 50 and

Figure 7.10 The geography of primary economic activities Primary economic activities are those that are concerned directly with natural resources of any kind. They include agriculture, mining, forestry, and fishing. The vast majority of the world's population, concentrated in China, India, Southeast Asia, and Africa, is engaged in primary economic activities. This map shows the percentage of the labor force in each country that was engaged in primary employment in 2002. In some countries, including China, primary activities account for more than 70 percent of the workforce. In contrast, primary activities always account for less than 10 percent of the labor force in the world's core countries, and often for less than 5 percent.

75 percent of the labor force is engaged in primary-sector activities. In contrast, the primary sector of the world's core regions is typically small, occupying only 5 to 10 percent of the labor force.

The secondary sector is much larger in the core countries and in semiperipheral countries, where the world's specialized manufacturing regions are located. In 2005, core countries accounted for almost three-quarters of world manufacturing value added (MVA). This share has been slowly decreasing, however. The core countries had an average annual growth rate for MVA of around 2 percent during 1990–2004, while the growth rate in the rest of the world was closer to 7 percent. This growth has been concentrated in semiperipheral, newly industrializing countries (NICs). **Newly industrializing countries** are countries, formerly peripheral within the world-system, that have acquired a significant industrial sector, usually through direct foreign investment. Of the 20 biggest manufacturing countries in 2005, seven were NICs: China, South Korea, Mexico, Brazil, India, Argentina, and Thailand (in order of importance). The vast majority of peripheral countries have a very small manufacturing output. The share of world MVA for Africa has changed little over the last two decades, remaining at about 1 percent.

In terms of individual countries, the United States remains by far the most important source of manufactured goods, accounting for just over 25 per cent of global MVA in 2005. Just five countries—the United States, Japan, Germany, China, and the United Kingdom—together produced over 60 percent of the world total. Another important aspect of secondary activities concerns *productivity*. In general, the highly capitalized manufacturing industries of the developed countries have been able to maintain high levels of worker productivity, with the result that the contribution of manufacturing to their GDP has remained relatively high even as the size of their manufacturing labor forces has decreased.

Within the framework of this continuing dominance of the advanced industrial economies there are several important trends. Although the United States has retained its leadership as the world's major producer of manufactured goods, its dominance has been significantly reduced. In the 1960s, its share of world manufacturing output was 40 percent, compared to its current share of around 25 percent. Meanwhile, Japan increased its share from less than 6 percent in the 1960s to about 18 percent by 2005. The emergence of a dozen or so NICs as settings for manufacturing is particularly striking. Several of these are in Latin America (Brazil, Mexico, and Argentina), but in terms of the rate of manufacturing growth, the Asian NICs are most impressive.

The growth of manufacturing in Pacific Asia has generated agglomerations of economic activity at a scale that sometimes crosses national boundaries, as with the Southern China–Hong Kong–Taiwan triangle and the Singapore–Batam–Johor triangle (**Figure 7.11**). The 1,500-kilometer (932-mile) urban belt in Northeast Asia that runs from Beijing to Tokyo via Pyongyang and Seoul

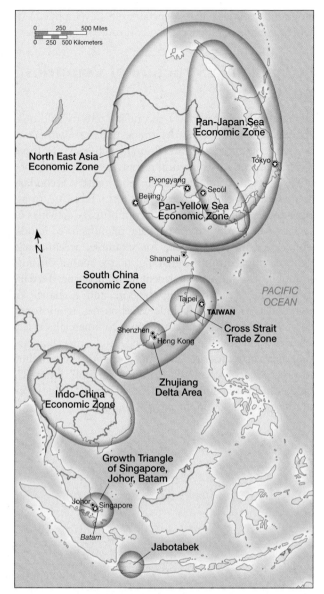

Figure 7.11 Emerging growth zones in Pacific Asia The globalization of manufacturing industries has spurred the growth of a series of extended metropolitan regions. (Adapted from P. Dicken, *Global Shift*, 4th ed. New York: Guilford, 2003, Figure 3.28, p. 78.)

connects some 80 cities of over 200,000 inhabitants each, encompassing nearly 100 million urban dwellers altogether. In terms of individual countries, China has experienced a dramatic increase in manufacturing production, achieving annual average growth rates during the 1970s, 1980s, and 1990s of about 8 percent, 11 percent, and 14 percent, respectively (see Box 7.1, "China's Economic Development"). Of the four Asian "Tigers"—South Korea, Hong Kong, Taiwan, and Singapore—South Korea enjoyed the most spectacular increase in manufacturing production (**Figure 7.12**), achieving annual average growth rates of almost 18 percent in the 1960s, 17 percent in the 1970s, 12 percent in the 1980s, and over 7 percent in the 1990s (despite a financial crisis that occurred in Asia in the late 1990s). In fact, the four Asian "Tigers" have made remarkable progress up the world league table of exporters.

Figure 7.12 Manufacturing in South Korea Part of an engine assembly line for Hyundai trucks and buses at the JeonJu plant in JeonJu, South Korea.

In 2007 Hong Kong ranked 4th in the world (up from 26 in 1980), Taiwan ranked 17th (up from 22), South Korea ranked 12th (up from 27), and Singapore ranked 14th (up from 30). More recently, other Pacific Rim NICs, such as Malaysia and Thailand, have experienced rapid growth in manufacturing production.

These shifts are part of a globalization of economic activity that has emerged as the overarching component of the world's economic geography. As we shall see, it has been corporate strategy, particularly the strategies of large transnational corporations (TNCs), that has created this globalization of economic activity.

The tertiary and quaternary sectors are significant only in the most affluent countries of the core. In the United States, for example, the primary sector in 2007 accounted for less than 4 percent of the labor force, the secondary sector for about 22 percent, the tertiary sector for just over 50 percent, and the quaternary sector for 22 percent. In every core country, the tertiary sector has grown significantly in the past several decades as consumption and marketing became the hallmarks of postindustrial economies. More recently, globalization has meant that knowledge-based activities have become a critical aspect of economic development, resulting in the rapid growth of quaternary industries.

For the world's core economies, knowledge has become more important than physical and human resources in determining levels of economic well-being. More than half of the GDP of major core countries is based on the production and distribution of knowledge. In the United States, more workers are engaged in producing and distributing knowledge than in making physical goods.

For the world's peripheral economies, lack of knowledge—along with a limited capacity to absorb and communicate knowledge—has become an increasingly important barrier to economic development. Poor countries have fewer resources to devote to research, development, and the acquisition of information technology. They also have fewer institutions that can provide high-quality education, fewer bodies that can enforce standards and performance, and only weakly developed organizations for gathering and disseminating the information needed for business transactions. As a result, economic productivity tends to fall relative to the performance of places and regions in core economies, where new knowledge is constantly generated and rapidly and effectively disseminated.

International Trade, Aid, and Debt

The geographical division of labor on a world scale means that the geography of international trade is very complex. One significant reflection of the increased economic integration of the world-system is that global trade has grown much more rapidly over the past few decades than global production. Between 1985 and 2005, the average annual growth rate of the value of world exports was twice that of the growth of world production and several times greater than that of world population growth.

The fundamental structure of international trade is based on a few **trading blocs**—groups of countries with formalized systems of trading agreements—with most of the world's trade taking place within four trading blocs:

- Western Europe, together with some former European colonies in Africa, South Asia, the Caribbean, and Australasia;

- North America, together with some Latin American states;

- the countries of the former Soviet world-empire; and

- Japan, together with other East Asian states and the oil-exporting states of Saudi Arabia and Bahrain.

Nevertheless, a significant number of countries exhibit a high degree of **autarky** from the world economy. That is, they do not contribute significantly to the flows of imports and exports that constitute the geography of trade. Typically, these are smaller peripheral countries, such as Bolivia, Burkina Faso, Ghana, Malawi, Samoa, and Tanzania.

China's Economic Development

Under the leadership of Deng Xiaoping (1978–1997), China embarked on a thorough reorientation of its economy, dismantling Communist-style central planning in favor of private entrepreneurship and market mechanisms, and began integrating itself into the world economy. Saying that he did not care whether the cat was black or white, as long as it caught mice, Deng Xiaoping established a program of "Four Modernizations" (industry, agriculture, science, and defense) and an "open-door policy" that allowed China to be plugged in to the interdependent circuits of the global economy. As a result, China has completely reorganized and revitalized its economy. Agriculture has been decollectivized, with Communist collective farms modified to allow a degree of private profit-taking. State-owned industries have been closed or privatized, and centralized state planning has been dismantled in order to foster private entrepreneurship.

In the 1980s and early 1990s, when the world economy was sluggish, China's manufacturing sector grew by almost 15 percent each year. Almost all of the shoes once made in South Korea or Taiwan are now made in China. More than 60 percent of the toys in the world, accounting for more than $10 billion in trade, are made in China. Since 1992, China has extended its open-door policy, permitted foreign investment aimed at Chinese domestic markets, and normalized trading relationships with the United States and the European Union. In 2001, China was admitted to the World Trade Organization, allowing China to trade more freely than ever before with the rest of the world.

China's increased participation in world trade has created an entirely new situation within the world economy. Chinese manufacturers, operating with low wages, have imposed a deflationary trend on world prices for manufacturers. The Chinese economy's size makes it a major producer, and its labor costs stay flat year after year because there is an endless supply of people happy to work for 60 cents an hour. Meanwhile, the rapid expansion of consumer demand in China has begun to drive up commodity prices in the world market. Overall, China's economy is already the third largest in the world after those of the United States and Japan.

Nowhere have China's "open-door" policies had more impact than in South China, where the Chinese government has deliberately built upon the prosperity of Hong Kong, the former British colony that was returned to China in 1997. The coastline of South China provides many protected bays suitable for harbors, and a series of ports has developed, including Quanzhou, Shantou, Xiamen, and, on either side of the mouth of the Zhu Jiang (Pearl River), Macão and Hong Kong. These ports were a precondition for South China's emergence as a core manufacturing region, providing an interface with the world economy. The established

trade and manufacturing of Macão—a Portuguese colony that was returned to China in 1999—and Hong Kong provided another precondition for success.

When Deng Xiaoping established his "open-door" policy, a third factor kicked in: capital investment from Hong Kong, Taiwan, and the Chinese diaspora. By 1993, more than 15,000 manufacturers from Hong Kong alone had set up businesses in neighboring Guangdong Province, and a similar number had established subcontracting relationships, contracting out processing work to Chinese companies. Today, the cities and special economic zones of South China's "Gold Coast" provide a thriving export-processing platform that has driven double-digit annual economic growth for much of the past two decades. The city of Shenzhen (**Figure 7.A**) has grown from a population of just 19,000 in 1975 to 8.5 million in 2007, with an additional 2 million in the surrounding municipalities. Such growth has generated a substantial middle class with significant spending power (**Figure 7.B**); it has also created significant inflation, especially in real estate values (**Figure 7.C**).

Much of China's manufacturing growth has been based on a strategy of import substitution (see page 266). In spite of China's membership in the World Trade Organization (which has strict rules about intellectual property), a

Figure 7.A Shenzhen The population of the city of Shenzhen, just across the border from the Special Administrative Region of Hong Kong, grew from just 5,000 in 1970 to more than 8.5 million in 2007.

significant amount of China's industry is based on counterfeiting and reverse engineering (making products that are copied and then sold under different or altered brand names) and piracy (making look-alike products passed off as the real thing). Copies of everything from DVDs, movies, designer clothes and footwear, drugs, motorcycles, and automobiles to high-speed magnetic levitation (maglev) cross-country trains save Chinese industry enormous sums in research and development and licensing fees, while saving the country even greater sums in imports.

Foreign investors, meanwhile, have been keen to develop a share of China's rapidly expanding and increasingly affluent market. The automobile market is particularly attractive to Western manufacturers. Volkswagen was the first to establish a presence in China, in 1985. By 2003, Volkswagen had claimed around 40 percent of China's annual production of almost 4 million cars and light trucks. General Motors, in partnership with Shanghai

Automotive Industry Corporation, has a 10 percent market share. Other foreign manufacturers operating in China include Honda, Toyota, Nissan, and, most recently, BMW and Mercedes.

Overall, most foreign investment in China comes from elsewhere within East Asia. Japan, Taiwan, and South Korea, having developed manufacturing industries that undercut those of the United States, now face deindustrialization themselves through the inexorable process of "creative destruction" (see p. 275). More than 10,000 Taiwanese firms have established operations in China, investing an estimated $150 billion. Pusan, the center of the South Korean footwear industry that in 1990 exported $4.3 billion worth of shoes, is full of deserted factories. South Korean footwear exports are down to less than $700 million, while China's footwear exports have increased from $2.1 billion in 1990 to $24 billion in 2007. Several Japanese electronics giants,

Figure 7.B   New affluence A saleswoman waits for customers at an outlet of the French fashion brand Hermes in Shanghai. China's economic boom has led to a rapid increase in the size of the country's middle class, up nearly 25 percent in the past two years from 65 to 80 million people.

Figure 7.C   Real estate boom In the rapidly growing regions of coastal China, house price inflation has risen as high as 9.5 percent a month. Shown here is a small fishing village near Sanya Harbor in Hainan province, where new luxury buildings are displacing the older homes and their inhabitants.

including Toshiba Corp., Sony Corp., Matsushita Electric Industrial Co., and Canon, Inc., have expanded operations in China even as they have shed tens of thousands of workers at home. Olympus manufactures its digital cameras in Shenzhen and Guangzhou. Pioneer has moved its manufacture of DVD recorders to Shanghai and Dongguan.

Toshiba's factory in Dalian illustrates the logic. Toshiba is one of about 40 Japanese companies that have built large-scale production facilities in a special export-processing zone established by Dalian in the early 1990s with generous financial support from the Japanese government and major Japanese firms. By shifting production of digital televisions here from its plant in Saitama, Japan, in 2001, Toshiba cut labor costs per worker by 90 percent.

Patterns of world trade have been shifting rapidly, however, in response to several factors. In general, the trend has been toward an intensification of the long-standing domination of trade within and between core regions at the expense of trade between core countries and peripheral countries—with the major exception of trade in oil. Innovations in transport, communications, and manufacturing technology have diminished the importance of the distance-based factors that have underpinned traditional trading blocs. Shifts in global politics have also affected the geography of trade. One important shift was the breakup of the former Soviet Union. Other significant geopolitical changes include the trend toward the political as well as economic integration of Europe and the increasing participation of China in the world economy. But perhaps the most important shift in global politics in relation to world trade has been the shift toward open markets and free trade through neoliberal policies propagated by core countries.

The globalization of economic activity has created new flows of materials, components, information, and finished products. As a global system of manufacturing has emerged, significant quantities of manufactured goods are now imported *and* exported across much of the world through complex commodity chains; no longer do developed economies export manufactures and peripheral countries import them. African countries are an important exception, with many of them barely participating in world trade in manufactures.

The most striking aspect of contemporary patterns of trade is the persistence of the dependence of peripheral countries on trade with core countries that are geographically or geopolitically close. Thus, for example, the United States is the central focus for the exports and the origin of the bulk of the imports of most Central American countries, while France is the focus for commodity flows to and from French ex-colonies such as Algeria, Cambodia, Benin, and the Ivory Coast. These flows, however, represent only part of the action for the core economies, whose trading patterns are dominated by flows to and from other core countries.

One implication of this situation is that the smaller, peripheral partners in these trading relationships are highly dependent on levels of demand and the overall economic climate in developed economies. Another aspect of dependency, in this context, is the degree to which a country's export base lacks diversity. **Dependency** involves a high level of reliance by a country on foreign enterprises, investment, or technology. External dependence for a country can mean that it is highly reliant on levels of demand and the overall economic climate of other countries. Dependency for a peripheral country, for example, can result in a narrow economic base in which the balancing of national accounts and the generation of foreign exchange depend on the export of one or two agricultural or mineral resources. **Figure 7.13** shows one reflection of dependency: the index of commodity concentration of exports. Countries with low scores on this index have diversified export bases. They include Argentina, Brazil, China, India, and North and South Korea, as well as most of the core countries. At the other extreme are peripheral countries where the manufacturing sector is poorly developed and the balancing of national accounts and the generation of foreign exchange depend on the export of one or two agricultural or mineral resources, as in Angola, Chad, the Dominican Republic, Iran, Iraq, Libya, and Nigeria, for example.

Patterns of International Debt In many peripheral countries, debt service—the annual interest on international debts—is a significant handicap to economic development. Overall, for every $1 that developing countries receive in aid, they end up having to return $5 in debt service payments to core countries. In many countries, 20 percent or more of all export earnings are swallowed up by debt service (**Figure 7.14**).

At the root of the international debt problem is the structured inequality of the world economy. The role inherited by most peripheral countries within the **international division of labor** (the specialization, by countries, in particular products for export) has been one of producing primary goods and commodities for which both the elasticity of demand and price elasticity are low. The **elasticity of demand** is the degree to which levels of demand for a product or service change in response to changes in price. Where a relatively small change in price induces a significant change in demand, elasticity is high; where levels of demand remain fairly stable in spite of price changes, demand is said to be inelastic. Demand for the products of peripheral countries in their principal markets

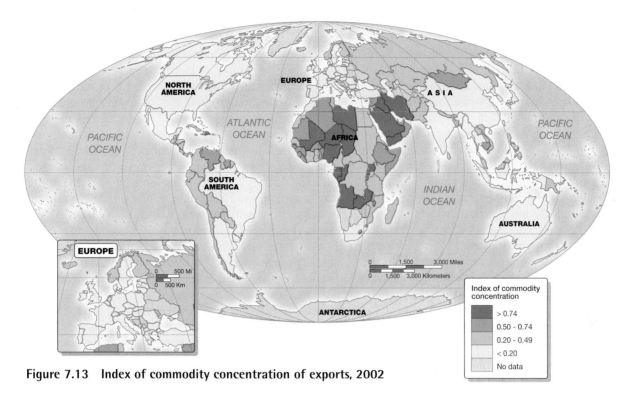

Figure 7.13   **Index of commodity concentration of exports, 2002**

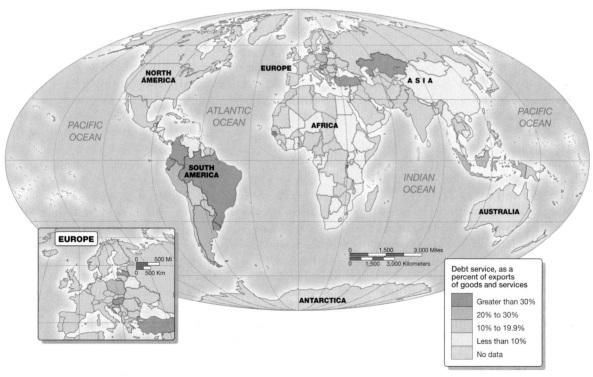

Figure 7.14   The debt crisis In some countries, the annual interest on international debts (their "debt service") accounts for more than 20 percent of the annual value of their exports of goods and services. Many countries first got into debt trouble in the mid-1970s, when Western banks, faced with recession at home, offered low-interest loans to the governments of peripheral countries rather than being stuck with idle capital. When the world economy heated up again, interest rates rose and many countries found themselves facing a debt crisis. The World Bank and the International Monetary Fund (IMF), in tandem with Western governments, worked to prevent a global financial crisis by organizing and guaranteeing programs that eased poor countries' debt burdens. Western banks were encouraged to swap debt for equity stakes in nationalized industries, while debtor governments were persuaded to impose austere economic policies. These policies have helped ease the debt crisis, but often at the expense of severe hardship for ordinary people. In dark humor, among radical development theorists IMF came to stand for "imposing misery and famine."

(the more developed countries) has a low elasticity: It tends to increase by relatively small amounts in response to significant increases in the incomes of their customers. Similarly, significant reductions in the price of their products tend to result in only a relatively small increase in demand.

Think, for example, of the cocoa-producing regions of West Africa. No matter how they improve productivity in order to keep prices low, and no matter how much more affluent their customers in core countries become, there is a limit to the demand for cocoa products. In contrast, the elasticity of demand and price elasticity of high-tech manufactured goods and high-order services (the specialties of core economies within the international division of labor) are both high. As a result, the terms of trade are stacked against the producers of primary goods.

The **terms of trade** are determined by the ratio of the prices at which exports and imports are exchanged. When the price of exports rises relative to the price of imports, the terms of trade reflect an improvement for the exporting country. No matter how efficient primary producers may become, or how affluent their customers, the balance of trade will be tilted against them. Quite simply, they must run in order to stand still.

An obvious counterstrategy for peripheral countries is to attempt to establish a new role in the international division of labor, moving away from a specialization in primary commodities toward a more diversified manufacturing base. This strategy is known as **import substitution.** It is a difficult strategy to pursue, however, because building up a diversified manufacturing base requires vast amounts of start-up capital. With the terms of trade running against them, it is extremely difficult for peripheral countries to accumulate this capital.

By 2005, the total debt owed by low- and middle-income countries to high-income core countries was just over $2.7 trillion. Practically none of this money is ever likely to be repaid, but so long as the North American, European, and Japanese banks continue to receive interest on their loans, they will be satisfied. Indeed, core countries are doing extremely well from this aspect of international finance. In 2005, the world's core countries took in about $514 billion in debt servicing, while paying out a total of less than $60 billion in new loans. There is always a danger to the future well-being of the better-off countries, however, in that debtor countries might act together to deliberately default on their debts, which would cause a major disturbance to the global financial system.

The debt problem has led to calls for affluent lending countries to provide debt relief to some of the poorest countries. In 2005, the world's richest countries—the G8 group—agreed to write off $40 billion in debts owed by 18 of the world's poorest countries, most of them in Africa. Addressing the full magnitude of LDC debt, however, will require continued substantial and sustained efforts if the poorest countries are to break free of the crushing financial obligations of their accumulated debts.

Patterns of International Aid For many countries, the only alternative to crippling debt, short of opting for self-sufficiency (attempted, for example, by Tanzania and Burma) or opting out of the capitalist world economy altogether (for example, Cuba), is to raise the capital as loans. International aid programs provide low interest rates for many of these loans. But if the borrowed capital is invested in economic development projects that do not yield sufficient returns, further loans have to be undertaken in order to service the original debt and/or to finance new development projects. In this way, international debt tends to pile up. The syndrome of having constantly to borrow in order to fund "development" has come to be known as the **debt trap.**

Large-scale movements of international development aid began shortly after World War II with the Marshall Plan, financed by the United States to bolster war-torn European allies whose economic weakness, it was believed, made them susceptible to communism. During the 1950s and 1960s, as more peripheral countries gained independence, aid became a useful weapon in Cold War offensives to establish and preserve political influence throughout the world. By the late 1960s, the list of donor countries had expanded beyond the superpowers to include smaller countries such as Denmark and Sweden, whose motivation in aid giving must be seen as more philanthropic than political. In addition, there was a greater geographic dispersal of aid, thanks largely to the activities of multilateral financial agencies such as the International Monetary Fund and the World Bank. Private aid organizations, meanwhile, have focused their efforts on emergency aid, on campaigning for debt-relief programs, and on fair trade practices (see Box 7.2, "Fair Trade").

Nevertheless, the geography of aid still has a strong political flavor, the United States in particular using foreign aid to support its international political agenda. In the past, this meant a disproportionately high level of support for a few strategically important countries, as the United States sought to limit the influence of the former Soviet bloc. More recently, the chief beneficiary of U.S. aid has been Israel, which is scheduled to receive $3 billion in military aid from the United States each year through 2018. Aid from several other countries reflects localized political aspirations and colonial ties. Much British and French aid, for example, is directed toward former African colonies, while Japanese aid is disbursed largely within Asia, and aid from the OPEC countries is directed mainly toward the "frontline" Arab countries.

The end of the Cold War resulted in diminished levels of international aid. Thus, whereas official development assistance from **OECD** countries (the Organization for Economic Cooperation and Development, an association of 30 industrialized countries) amounted to nearly 0.5 percent of their total GNI in 1965, it had fallen to 0.25 percent by 2007. The most striking decrease was in aid from the United States, whose overseas development assistance as a percentage of its GNI fell from 0.58 to 0.15. Japan, meanwhile, has never exceeded 0.33 percent of its GNI in aid. In contrast, Denmark, Luxembourg, the Netherlands, Sweden, and Norway have steadily increased their aid giving to become the only ones to have met or surpassed the United Nations target of 0.7 percent of GNI (**Table 7.2**). Perhaps more disturbing is the World Bank's assertion that a large proportion of developed countries' pledged aid is "phantom" aid because it ends up going to international consultants and donor countries' own firms. Altogether, around 60 percent of all international aid becomes phantom in this way; and in the case of aid from France and the United States, the figure is closer to 90 percent.

Nevertheless, the impact of aid on some countries is clearly significant. Liberia, for example, received aid amounting to 54 percent of its GDP in 2007; and 16 other countries—including Afghanistan, Burundi, Congo, Mozambique, Rwanda, and Sierra Leone—received aid equivalent to more than 20 percent of their GDP. These figures, however, say as much about these countries' extremely small GDP as anything else. They are also exceptions. In general, the poorest countries are by no means the biggest recipients of aid; the amount of aid received per capita is generally very low, and a large share of it is deployed as emergency food aid (**Figure 7.15**). In 2002, low-income countries like Burundi, Chad, and Uganda received an average of around US$17 per capita in overseas development assistance; middle-income developing economies like the Dominican Republic, Egypt, and Sri Lanka received around US$8 per capita. At these levels, aid cannot seriously be regarded as a catalyst for development or as an instrument for redressing core-periphery inequalities.

Interpretations of Patterns of Development

The overall relationship between economic structure and levels of prosperity makes it tempting to interpret economic development in terms of distinctive stages. Each developed region or country, in other words, might be thought of as progressing from the early stages of development, with a heavy reliance on primary activities (and relatively low

TABLE 7.2 International Economic Aid: Net Official Development Assistance Disbursed, 2007

	ODA, USD million current	As % of GNI
Norway	3,727	0.95
Sweden	4,334	0.93
Luxembourg	365	0.90
Denmark	2,563	0.81
Netherlands	6,215	0.81
Ireland	1,190	0.54
Austria	1,798	0.49
Belgium	1,953	0.43
Spain	5,744	0.41
Finland	973	0.40
France	9,940	0.39
Germany	12,267	0.37
Switzerland	1,680	0.37
United Kingdom	9,921	0.36
Australia	2,741	0.30
Canada	3,922	0.28
New Zealand	315	0.27
Italy	3,929	0.19
United States	21,753	0.16

Figure 7.15 Food aid Displaced Kikuyu people wait for food distribution January 2008 at a feeding center at a school in the village of Noigam, near Eldoret, Kenya. Thousands of displaced Kikuyu people were unable to return to their home after Kenyan elections for fear of attacks by other tribal groups.

7.2 GEOGRAPHY MATTERS

Fair Trade

The Fair Trade movement highlights the interdependencies involved in international trade. The movement is a result of increasing awareness within developed countries of the weak bargaining position of many small producers at the beginning of the commodity chains that underpin the global economy. Fair Trade has become part of the "mobilization against globalization" that we described in Chapter 2, an attempt to raise consumers' consciousness about the relationships embodied in their purchases.

The Fair Trade movement is a global network of producers, traders, marketers, advocates, and consumers focused on building equitable trading relationships between consumers and the world's most economically disadvantaged artisans and farmers. As such, it is fundamentally a strategy for poverty alleviation and sustainable development. The key principles of Fair Trade include (1) creating opportunities for economically disadvantaged producers, (2) capacity building, (3) ensuring that women's work is properly valued and rewarded, (4) ensuring a safe and healthy working environment for producers, and (5) payment of a fair price—one that covers not only the costs of production but enables production that is socially just and environmentally sound. The result is often very modest in relation to the retail price of Fair Trade goods in developed countries—only about 9 cents of the $3.10 cost of an average fairly traded 100-gram bar of chocolate goes to producers in developing countries, the rest being accounted for by core-country food processors, designers, packagers, photographers, marketing staff, advertisers, shopkeepers, and tax authorities. Nevertheless, Fair Trade demonstrably helps producers in developing countries (**Figure 7.D**) and typically involves democratic decision making over how the extra money earned is to be distributed.

There are several types of Fair Trade organizations that perform various roles along the commodity chains linking producers to consumers. At the beginning of the commodity chain are producer organizations—village or community groups or cooperatives, for example, often joined together under export marketing organizations. In early 2008 there were more than 630 Fair Trade–certified producer groups (including many umbrella bodies) in 58 countries. These organizations typically sell their products to a second kind of Fair Trade organization: registered importers and

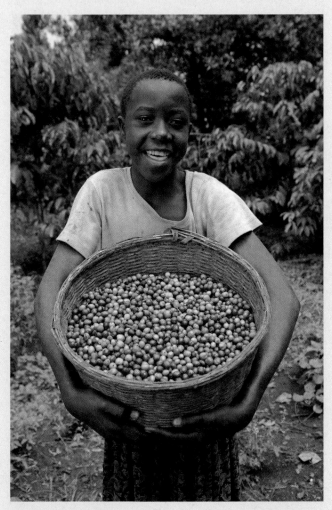

Figure 7.D Fair Trade coffee Susan Nangobi, aged 14, works with her father, a coffee farmer in the Kamuli region of Uganda. Their freshly picked coffee beans are ready to be dried before being taken to the Kulika Sustainable Organic Agricultural Training Program center and then sold to Ibero Coffee company under Fair Trade guidelines.

wholesalers in more developed countries. In 2008, these existed in about 60 countries. They in turn sell to Fair Trade retailers: "world shops" and catalog- or Internet-based retailers (**Figure 7.E**). In some countries, Fair Trade products are now mainstream, available in major supermarkets and independent shops and beginning to gain market share. In Switzerland, for example, Fair Trade bananas account for 20 percent of the retail market. In the United States, it is coffee that is

Figure 7.E   Fair Trade retailing
Transfair USA's founder and CEO, Paul Rice, inspects Fair Trade labels on foreign coffee products at company offices in Oakland, California. Transfair is a nonprofit group that audits the books of U.S. companies to ensure that the ingredients come from farmers who get fair prices for their goods.

the most important certified Fair Trade product, accounting for between $200 and $600 million in sales in 2007.

The fourth category of Fair Trade organizations are labeling organizations that certify the chain of supply of certain commodities in order to guarantee adherence to Fair Trade practices. Fair Trade labeling was created in the Netherlands in the late 1980s when Max Havelaar launched the first Fair Trade consumer guarantee label in 1986 on coffee sourced from Mexico. Today, there are 20 organizations that together run the international standard-setting and monitoring body Fairtrade Labeling Organizations International (FLO). Producers registered with FLO receive a minimum price that covers the cost of production and an extra premium that is invested in the local community.

Finally, there are organizations that are focused on labor practices, such as the Ethical Trading Initiative (ETI), which evolved from Fair Trade campaigns run by British aid organizations. ETI involves a multiagency grouping of companies, nongovernmental organizations (NGOs), and trades unions who have together devised a basic code of labor practices covering the right to collective bargaining, safe and hygienic working conditions,

living wages, and a standard working week of no more than 48 hours.

Fair Trade sales are growing at double-digit rates and the movement is expanding rapidly. There are approximately 30,000 outlets in the United States where Fair Trade–labeled products can be purchased. In May 2000, the town of Garstang in Lancashire, England, declared itself the world's first Fair Trade Town, and by 2008 there were more than 350 towns and cities in Britain that had committed to Fair Trade principles and products. There were also several universities committed to Fair Trade, including the University of Birmingham, the University of Bristol, the University of Edinburgh, the London School of Economics and Political Science, and Sheffield University. In 2007, a Fair Trade Towns campaign was launched in the United States, with Media, Pennsylvania, as the first member town. Nevertheless, the total value of Fair Trade is miniscule in relation to the overall flows of international trade. Britain is the largest Fair Trade market in the world, with over $1.1 billion in total sales in 2007. Total sales for the Fair Trade industry in North America in 2007 were just under $1.28 billion, with worldwide sales amounting to $3.6 billion, an increase of 46 percent from the previous year.

levels of prosperity), through a phase of industrialization and on to a "mature" stage of postindustrial development (with a diversified economic structure and relatively high levels of prosperity). This, in fact, has been a commonly held view of economic development, first conceptualized by a prominent economist W. W. Rostow (**Figure 7.16**).

Rostow's model, however, is too simplistic to be of much help in understanding human geography. The reality is that places and regions are interdependent. The fortunes of any given place are increasingly tied up with those of many others. Furthermore, Rostow's model perpetuates the myth of "developmentalism," the idea that every country and region will eventually make economic progress toward "high mass consumption" provided that they compete to the best of their ability within the world economy. The main weakness of developmentalism is that it is simply not reasonable to compare the prospects of late starters to the experience of those places, regions, and countries that were among the early starters. For these early starters the horizons were clear: free of effective competition, free of obstacles, and free of precedents. For the late starters the situation is entirely different. Today's

less developed regions must compete in a crowded field while facing numerous barriers that are a direct consequence of the success of some of the early starters.

Indeed, many writers and theorists of international development claim that the prosperity of the core countries in the world economy has been based on *under*development and squalor in peripheral countries. Peripheral countries, it is argued, could not "follow" the previous historical experience of developed countries in stages-of-development fashion because their underdevelopment (that is, exploitation) was a structural requirement for development elsewhere. The development of Europe and North America, in other words, required the systematic underdevelopment of peripheral countries. By means of unequal trade, exploitation of labor, and profit extraction, the underdeveloped countries became increasingly rather than decreasingly impoverished.

The writing of André Gunder Frank exemplifies the explanations of international economic change that arose from this critique. Frank rejected the idea that underdevelopment is an original condition, equivalent to "traditionalism" or "backwardness." To the contrary, he argues, it is

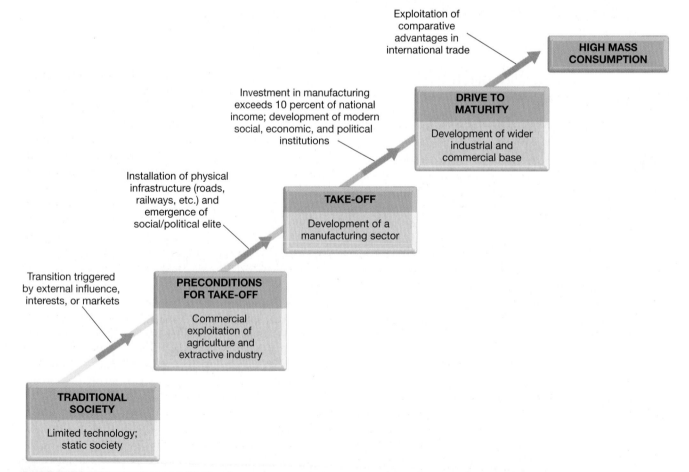

Figure 7.16 Stages of economic development This diagram illustrates a model of economic development based on the idea of successive stages of development. Each stage is seen as leading to the next, though some regions or countries may take longer than others to make the transition from one stage to the next. According to this view, now regarded as overly simplistic, places and regions can be seen as following parallel courses within a world that is steadily modernizing. Late starters will eventually make progress, but at speeds determined by their resource endowments, their productivity, and the wisdom of their people's policies and decisions.

a condition created by integration into the worldwide system of exchange that is the world capitalist system. The world economy, Frank argues, has been unequally structured since Europeans first ventured out into the world in the sixteenth century. Although the form of the dominance of core over periphery has changed from colonialism and imperialism to neocolonialism, an overall transfer of wealth from periphery to core has continued to fuel growth in some places at the expense of others.

Frank's approach is an example of what can be called "dependency theory." This has been a very influential approach in explaining global patterns of development and underdevelopment. It states, essentially, that development and underdevelopment are reverse sides of the same process: *Development somewhere requires underdevelopment somewhere else.* Immanuel Wallerstein's world-system theory (see Chapter 2) takes this kind of dependency into account. According to this perspective, the entire world economy is to be seen as an evolving market system with an economic hierarchy of states—a core, a semiperiphery, and a periphery. The composition of this hierarchy is variable: Individual countries can move from periphery to semiperiphery, core to periphery, and so on.

PATHWAYS TO REGIONAL DEVELOPMENT

Patterns of *regional* economic development are also historical in origin and cumulative in nature. Recognizing this, geographers are interested in **geographical path dependence**, the relationship between present-day activities in a place and the past experiences of that place. When spatial relationships and regional patterns emerge through the logic of fundamental principles of spatial organization, they do so in ways guided and influenced by preexisting patterns and relationships.

These observations lead to an important principle of regional economic development, the principle of initial advantage. **Initial advantage** highlights the importance of an early start in economic development. It represents a special case of external economies. Other things being equal, new phases of economic development will take hold first in settings that offer **external economies**: existing labor markets, existing consumer markets, existing frameworks of fixed social capital, and so on. External economies are cost savings resulting from advantages that are derived from circumstances beyond a firm's organization and methods of production. These initial advantages will be consolidated by **localization economies**—cost savings that accrue to particular industries as a result of clustering together at a specific location—and so form the basis for continuing economic growth. Examples of localization economies include sharing a pool of labor with special skills or experience, supporting specialized technical schools, joining to create a marketing organization or a research institute, and drawing on specialized subcontractors, maintenance firms, suppliers, distribution agents,

and lawyers. Where such advantages lead to a reputation for high-quality production, localization will be intensified, because more producers will want to cash in on the reputation. Among many examples are the electronics and software industries of Silicon Valley; recording companies in Los Angeles; Sheffield steel and steel products, especially cutlery; the U.S. auto industry in Detroit; Swiss watches made in the towns of Biel, Geneva, and Neuchâtel; Belgian lace from Bruges, Brussels, and Mechelen; English worsted in Bradford and Huddersfield; Stoke-on-Trent English china; Irish linen of Athlone; and French perfume made in Grasse.

For places and regions with a substantial initial advantage, therefore, the trajectory of geographical path dependence tends to be one of persistent growth—reinforcing the core-periphery patterns of economic development found in every part of the world and at every spatial scale. That said, geographers recognize there is no single pathway to development. The consequences of initial advantage for both core and peripheral regions can be—and often are—modified. Old core-periphery relationships can be blurred, and new ones can be initiated.

How Regional Economic Cores are Created

Regional cores of economic development are created cumulatively, following some initial advantage, through the operation of several of the basic principles of economic geography. Commercial and industrial location decisions all take place within complex webs of *functional interdependence*. These webs include the relationships and linkages between different kinds of industries, different kinds of stores, and different kinds of offices. Particularly important here are the **agglomeration effects** that are associated with various kinds of economic linkages and interdependencies—the cost advantages that accrue to individual firms because of their location among functionally related activities. The trigger for these agglomeration effects can be any kind of economic development—the establishment of a trading port or the growth of a local industry or any large-scale enterprise. The external economies and economic linkages generated by such developments represent the initial advantage that tends to stimulate a self-propelling process of local economic development.

Given the location of a new economic activity in an area, a number of interrelated effects come into play. *Backward linkages* develop as new firms arrive to provide the growing industry with components, supplies, specialized services, or facilities. *Forward linkages* develop as new firms arrive to take the finished products of the growing industry and use them as inputs to their own processing, assembly, finishing, packaging, or distribution operations. Together with the initial growth, the growth in these linked industries helps to create a threshold of activity large enough to attract **ancillary industries** and activities (maintenance and repair, recycling, security, and business services, for example).

The existence of these interrelated activities establishes a pool of specialized labor with the kinds of skills and experience that make the area attractive to still more firms. Meanwhile, the linkages among all these firms help to promote interaction between professional and technical personnel and allow the area to support R&D (research and development) facilities, research institutes, and so on, thus increasing the likelihood of local inventions and innovations that might further stimulate local economic development.

Another part of the spiral of local economic growth is a result of the increase in population represented by the families of employees. Their presence creates a demand for housing, utilities, physical infrastructure, retailing, personal services, and so on—all of which generate additional jobs. This expansion, in turn, helps to create populations large enough to attract an even wider variety and more sophisticated kinds of services and amenities. Last—but by no means least—the overall growth in local employment creates a larger local tax base. The local government can then provide improved public utilities, roads, schools, health services, recreational amenities, and so on—all of which serve to intensify agglomeration economies and so enhance the competitiveness of the area in being able to attract further rounds of investment.

Swedish economist Gunnar Myrdal was the 1974 Nobel Prize winner who first recognized that any significant initial local advantage tends to be reinforced through geographic principles of agglomeration and localization. He called the process **cumulative causation** (**Figure 7.17**), meaning the spiraling buildup of advantages that occurs in specific geographic settings as a result of the development of external economies, agglomeration effects, and localization economies. Myrdal also pointed out that this spiral of local growth tends to attract people—enterprising young people, usually—and investment funds from other areas. According to the basic principles of spatial interaction, these flows tend to be strongest from nearby regions and those with the lowest wages, fewest job opportunities, or least attractive investment opportunities.

In some cases this loss of entrepreneurial talent, labor, and investment capital is sufficient to trigger a cumulative negative spiral of economic disadvantage. With less capital, less innovative energy, and depleted pools of labor, industrial growth in peripheral regions tends to be significantly slower and less innovative than in regions with an initial advantage and an established process of cumulative causation. This in turn tends to limit the size of the local tax base, making it difficult for local governments to furnish a competitive **infrastructure** of roads, schools, and recreational amenities. Myrdal called these disadvantages **backwash effects,** the negative impacts on a region (or regions) of the economic growth of some other region. These negative impacts take the form, for example, of out-migration, outflows of investment

capital, and the shrinkage of local tax bases. Backwash effects are important because they help to explain why regional economic development is so uneven and why core-periphery contrasts in economic development are so common.

How Core-Periphery Patterns are Modified

Although very important, cumulative causation and backwash effects are not the only processes affecting the geography of economic development. If they were, the world's economic geography would be even more starkly polarized than it is now. There would be little chance for the emergence of new growth regions, like Guangdong in Southeast China, and there would be little likelihood of stagnation or decline in once-booming regions, like northern England.

Myrdal himself recognized that peripheral regions do sometimes emerge as new growth regions and partially explained them in what he called *spread* (or trickledown) effects. **Spread effects** are the positive impacts on a region (or regions) of the economic growth of some other region, usually a core region. This growth creates levels of demand for food, consumer products, and other manufactures that are so high that local producers cannot satisfy them. This demand gives investors in peripheral regions (or countries) the opportunity to establish a local capacity to meet the demand. Entrepreneurs who attempt this are also able to exploit the advantages of cheaper land and labor in peripheral regions. If strong enough, these spread effects can enable peripheral regions to develop their own spiral of cumulative causation, thus changing the interregional geography of economic patterns and flows. The economic growth of South Korea, for example, is partly attributable to the spread effects of Japanese economic prosperity.

Another way in which peripheral regions or countries can develop their own spiral of cumulative causation is through the process of *import substitution*. In this process, goods and services previously imported from core regions come to be replaced by locally made goods and locally provided services. Some things are hard to copy because of the limitations of climate or natural resources. However, many products and services *can* be copied by local entrepreneurs, thus capturing local capital, increasing local employment opportunities, intensifying the use of local resources, and generating profits for further local investment. The classic example is in Japan, where import substitution, especially for textiles and heavy engineering, played an important part in the transition from a peripheral economy to a major industrial power in the late nineteenth century. Import substitution also figured prominently in the Japanese "economic miracle" after World War II, this time featuring the automobile industry and consumer electronics. Today countries like Brazil, Peru, and Ghana are seeking to follow the same sort of strategy, subsidizing domestic

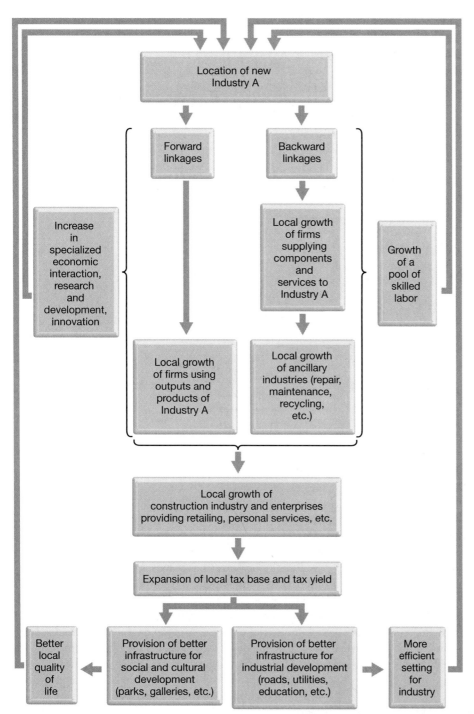

Figure 7.17 Processes of regional economic growth Once a significant amount of new industry becomes established in an area, it tends to create a self-propelling process of economic growth. As this diagram shows, the initial advantages of industrial growth are reinforced through geographic principles of agglomeration and localization. The overall process is sometimes known as *cumulative causation.*

industries and protecting them from outside competitors through tariffs and taxes.

Core-periphery patterns and relationships can also be modified by changes in the dynamics of core regions—internal changes that can slow or modify the spiral of cumulative causation. The main factor is the development of **agglomeration diseconomies,** the negative economic effects of urbanization and the local concentration of industry. Such effects include the higher prices that must be paid by firms competing for land and labor; the costs of delays resulting from traffic congestion and crowded port and railroad facilities; the increasing unit costs of solid waste disposal; and the burden of higher taxes that eventually

have to be levied by local governments in order to support services and amenities previously considered unnecessary—traffic police, city planning, and transit systems, for example.

Diseconomies that are imposed through taxes can often be passed on by firms to consumers in other regions and other countries in the form of higher prices. Charging higher prices, however, decreases the competitiveness of a firm in relation to firms operating elsewhere. Agglomeration diseconomies that cannot be "exported"—noise, air pollution, increased commuting costs, and increased housing costs, for example—require local governments to tax even more of the region's wealth in attempts to compensate for a deteriorating quality of life. In California this issue

Figure 7.18 Regional economic decline When the locational advantages of manufacturing regions are undermined for one reason or another, profitability declines and manufacturing employment falls. This can lead to a downward spiral of economic decline, as experienced by many of the traditional manufacturing regions of Europe and North America during the 1960s, 1970s, and 1980s. (Reprinted with permission of Prentice Hall, from P. L. Knox, *Urbanization,* © 1994, p. 55.)

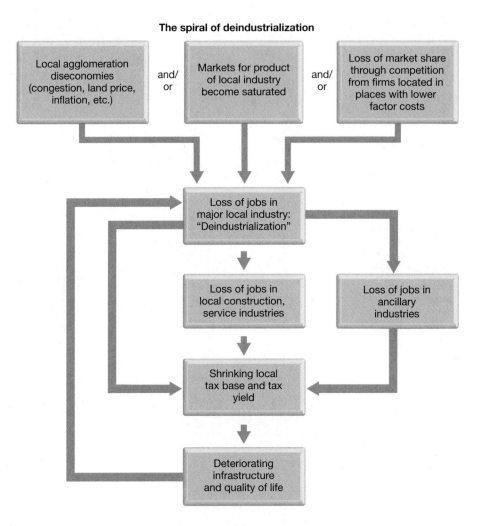

The spiral of deindustrialization

became so pressing in the 1990s that a report prepared jointly by the state, the Bank of America, the Greenbelt Alliance, and the Low Income Housing Fund concluded that "California businesses cannot compete globally when they are burdened with the costs of sprawl. An attractive business climate cannot be sustained if the quality of life continues to decline and the cost of financing real estate development escalates."[1] California's response has been to enact some of the nation's toughest traffic and environmental regulations.

Deindustrialization and Creative Destruction

The most fundamental cause of change in the relationship between initial advantage and cumulative causation is to be found in the longer-term shifts in technology systems and in the competition between states within the world-system. The innovations associated with successive technology systems generate new industries that are not yet tied down by enormous investments in factories or allied to existing industrial agglomerations. Combined with innovations in transport and communications, this creates *windows of locational opportunity* that can result in new industrial districts, with small

[1]*Beyond Sprawl,* Sacramento, CA: State of California-Resources Agency, 1995.

towns or cities growing into dominant metropolitan areas through new rounds of cumulative causation.

Equally important as a factor in how core-periphery patterns are modified are the consequent shifts in the profitability of old, established industries in core regions compared to the profitability of new industries in fast-growing new industrial districts. As soon as the differential is large enough, some disinvestment will take place within core regions. This disinvestment can take place in several ways. Manufacturers can reduce their wage bill by cutting back on production; they can reduce their fixed costs by closing down and selling off some of their factory space and equipment; or they can reduce their spending on research and development for new products. This disinvestment, in turn, leads to *deindustrialization* in formerly prosperous industrial core regions.

Deindustrialization involves a relative decline (and in extreme cases an absolute decline) in industrial employment in core regions as firms scale back their activities in response to lower levels of profitability (**Figure 7.18**). This is what happened to the Manufacturing Belt (sometimes called the "Rustbelt") of the United States in the 1960s and 1970s (**Figure 7.19**). It also occurred in many of the traditional industrial regions of Europe during the 1960s, 1970s, and 1980s. In France, Belgium, the Netherlands, Norway, Sweden, and the United Kingdom,

**Figure 7.19
Deindustrialization** This derelict steel mill in New Jersey is testament to the downward economic spiral in what was once one of the world's most important heavy manufacturing regions.

manufacturing employment decreased by between one-third and one-half from 1960 to 1990. The most pronounced example of this deindustrialization has been in the United Kingdom, where a sharp decline in manufacturing employment has been accompanied by an equally sharp rise in service employment.

Meanwhile, the capital made available from disinvestment in these core regions becomes available for investment by entrepreneurs in new ventures based on innovative products and production technologies. Old industries—and sometimes entire old industrial regions—have to be "dismantled" (or at least neglected) in order to help fund the creation of new centers of profitability and employment. This process is often referred to as creative destruction, something that is inherent to the dynamics of capitalism. **Creative destruction** involves the withdrawal of investments from activities (and regions) that yield low rates of profit in order to reinvest in new activities (and new regions). The concept of creative destruction provides a powerful image, helping us to understand the entrepreneur's need to withdraw investments from activities (and regions) yielding low rates of profit in order to reinvest in new activities (and, often, in new regions). In the United States, for example, the deindustrialization of the Manufacturing Belt provided the capital and the locational flexibility for firms to invest in the Sunbelt of the United States and in semiperipheral countries like Mexico and South Korea.

The process does not stop there, however. If the deindustrialization of the old core regions is severe enough, the relative cost of their land, labor, and infrastructure may decline to the point where they once again become attractive to investors. As a result, a see-saw movement of investment capital occurs, which over the long term tends to move from developed to less developed regions—then back again, once the formerly developed region has expe-

rienced a sufficient relative decline. "Has-been" regions can become redeveloped and revitalized, given a new lease on life by the infusion of new capital for new industries. This is what happened, for example, to the Pittsburgh region in the 1980s, resulting in the creation of a postindustrial economy out of a depressed industrial setting. USX, the U.S. Steel group of companies, reduced its workforce in the Pittsburgh region from more than 20,000 to less than 5,000 between 1975 and 1995. These losses have been more than made up, however, by new jobs generated in high-tech electronics, specialized engineering, and finance and business services.

Government Intervention In addition to the processes of deindustrialization and creative destruction, core-periphery patterns can also be modified by government intervention. National governments realize that regional planning and policy can be an important component of broad economic strategies designed to stabilize and reorganize their economies, as well as to maximize their overall competitiveness. Without regional planning and policy, the resources of peripheral regions can remain underutilized, while core regions can become vulnerable to agglomeration diseconomies. For political reasons, too, national governments are often willing to help particular regions adjust to changing economic circumstances. At the same time, most local governments take responsibility for stimulating economic development within their jurisdiction, if only in order to increase the local tax base.

The nature and extent of government intervention has varied over time and by country. In some countries, special government agencies have been established to promote regional economic development and reduce core-periphery contrasts. Among the best-known examples are the Japanese MITI (Ministry of International Trade and Industry), the Italian Cassa del Mezzogiorno

(Southern Development Agency, replaced in 1987 by several smaller agencies), and the U.S. Economic Development Administration. Some governments have sought to help industries in declining regions by undertaking government investment in infrastructure and providing subsidies for private investment; others have sought to devise tax breaks that reduce the cost of labor in peripheral regions. Still others have sought to deal with agglomeration diseconomies in core regions through increased taxes and restrictions on land use.

While each approach has its followers, one of the most widespread governmental approaches to core-periphery patterns involves the exploitation of the principle of cumulative causation through the creation of growth poles. **Growth poles** are places of economic activity deliberately organized around one or more high-growth industries. Economists have noted, however, that not all industries are equal in the extent to which they stimulate economic growth and cumulative causation. The ones that generate the most pronounced effects are known as "propulsive industries," and they have received a great deal of attention from geographers and economists who are interested in helping to shape strategic policies that might promote regional economic development. In the 1920s, shipbuilding was a propulsive industry. In the 1950s and 1960s, automobile manufacturing was a propulsive industry, and today biotechnology and digital technologies are propulsive industries. The basic idea is for governments to promote regional economic growth by fostering propulsive industries in favorable locations. These locations are intended to become growth poles—places that, given an artificial start, develop a self-sustaining spiral of economic prosperity.

Many countries have used the growth-poles approach as a basis for regional development policies. For example, French governments have designated certain locations as *technopoles*—sites for the establishment of high-tech industries (such as computers and biotechnology)—under the assumption that these leading-edge activities will stimulate further development. In southern Italy various heavy industries were located in a number of remote areas after World War II in order to stimulate ancillary development. In the United States a whole series of growth centers were designated within Appalachia in an attempt to establish a manufacturing base in the region. The Appalachian Regional Commission, established by Congress in 1965 to promote the economic development of the region, initiated a massive program of road building in order to open up the region for modern economic development. The commission also designated over 200 growth centers that were intended to attract industrialists by furnishing a new physical infrastructure and offering economic incentives in the form of tax breaks. The new industries, it was hoped, would trigger the geographic process of cumulative causation.

The results of such policies have been mixed, however. The French technopoles have been fairly successful because the French government invested large sums of money in establishing propulsive industries in favorable locations. But the Italian and U.S. growth-pole efforts, like many others, have been disappointing. In practice, governments often fail to invest in the right industries, and they nearly always fail to invest heavily enough to kickstart the process of cumulative causation. In Appalachia, for example, the growth centers were too small and too numerous. As a consequence, none of them has been able to generate self-sustaining processes of local economic growth.

GLOBALIZATION AND ECONOMIC DEVELOPMENT

As we saw in Chapter 2, the globalization of the world economy involves a new international division of labor in association with the internationalization of finance; the deployment of a new technology system (using robotics, information technology, biotechnology, and other new technologies); and the homogenization of consumer markets. The first wave of corporate globalization, in the 1970s, was led by manufacturing giants like General Motors and General Electric, whose global reach had the triple objectives of reducing labor costs, outflanking national labor unions, and increasing overseas market penetration. In the 1980s, as the globalization of manufacturing spread and the information economy began to grow, the leading firms in advanced business services—accountancy, advertising, banking, and law—established global networks of their own. The globalization of advanced financial and business service firms was initially a reaction to the global practices of their most important clients, the global manufacturing corporations. If, say, a law firm or advertising agency wished to keep the business of a major corporation, it had to be able to provide its services in places where that corporation needed them. Thus advanced business service firms followed their clients along the globalization path in the late 1970s and especially in the 1980s. This meant creating an office network to match clients' needs. After a while, some advanced business service firms used their global office network to win more clients in new markets. By the 1990s, the leading business service firms themselves became global corporations that could offer a seamless service with offices in key cities around the world.

The dynamics of economic globalization rest on the flows of capital, knowledge, goods, and services among countries. In 2006, companies around the world invested more than $1.3 trillion in business ventures beyond their own shores. This level of **foreign direct investment** is 20 times that of the mid-1970s. Approximately 40 percent of foreign direct investment is targeted at peripheral and semiperipheral countries, the rest going to core countries. These investments are reflected in increased levels of world trade in goods and services. Between 1975 and 2006, world exports of goods and services increased tenfold.

At a very general level, foreign direct investment can be expected to be good for the places and regions that are targeted for investment. Foreign direct investment increases competition among local producers, forcing them to improve their performance. At the same time, knowledge of new business practices and production technology spreads through the regional economy as regional manufacturers become suppliers to the enterprises funded through foreign investment and as personnel move from one firm to another. The overall effect is for higher levels of productivity all around.

Now that the world economy is much more globalized, the ability to acquire, absorb, and communicate knowledge is more important than ever before in determining the fortunes of places and regions. Patterns of local and regional economic development are much more open to external influences and much more interdependent with economic development processes elsewhere. Shrinking space, shrinking time, and disappearing borders are linking people's lives more deeply, more intensely, and more immediately than ever before. This new framework for economic geography has already left its mark on the world's economic landscapes and has also meant that the lives of people in different parts of the world have become increasingly intertwined. Not everyone has benefited, however, and the new international division of labor has come under attack by some in industrial countries, where rising unemployment and wage inequality are making people feel less secure about the future. Some workers in core countries are fearful of losing their jobs because of cheap exports from lower-cost producers. Others worry about companies relocating abroad in search of low wages and lax labor laws.

Most of the world's population now lives in countries that are either integrated into world markets for goods and finance or rapidly becoming so. As recently as the late 1970s, only a few peripheral countries had opened their borders to flows of trade and investment capital. About a third of the world's labor force lived in countries like the Soviet Union and China, with centrally planned economies, and at least another third lived in countries insulated from international markets by prohibitive trade barriers and currency controls. Today, with nearly half the world's labor force among them, three giant population blocs—China, the republics of the former Soviet Union, and India—are entering the global market. Many other countries, from Mexico to Thailand, have already become involved in deep linkages. According to World Bank estimates, fewer than 10 percent of the world's labor force remain isolated from the global economy. (The World Bank, properly called the International Bank for Reconstruction and Development, is a UN affiliate established in 1948 to finance productive projects that further the economic development of member nations.)

In this section, we examine some specific impacts of three of the principal components of the global economy: the global assembly line, resulting from the operations of transnational manufacturing corporations; the global office, resulting from the internationalization of banking, finance, and business services; and the pleasure periphery, resulting from the proliferation of international tourism.

The Global Assembly Line

The globalization of the world economy represents the most recent stage in a long process of internationalization. At the heart of this process has been the emergence of private companies that participate not only in international trade but also in production, manufacturing, and/or sales operations in several countries. Almost 80 percent of Ford's workforce, for example, is employed overseas, and foreign sales account for 55 percent of its total revenues. Over 50 percent of IBM's workforce is employed overseas, and 61 percent of its revenues are derived from foreign sales; 82 percent of the workforce of Philips, the Dutch electronics firm, is employed overseas, and 95 percent of its revenues are derived from foreign sales. Many of these transnational corporations have grown so large through a series of mergers and acquisitions that their activities now span a diverse range of economic activities. **Transnational corporations** are companies that participate not only in international trade but also in production, manufacturing, and/or sales operations in several countries.

Corporations that consist of several divisions engaged in quite different activities are known as **conglomerate corporations,** those having diversified into various economic activities, usually through a process of mergers and acquisitions. Altria (formerly known as Philip Morris), for example, primarily known for its tobacco products (such as Marlboro cigarettes), also controls a large group of assets in the beverage industry (including Miller Brewing) and has extensive interests in real estate, import-export, publishing, and foods (including General Foods, Tobler, Terry's, and Suchard chocolate). Nestlé, the world's largest packaged-food manufacturer, is the largest company in Switzerland but derives less than 2 percent of its revenue from its home country. Its major U.S. product lines and brand names include beverages (Calistoga, Nescafé, Nestea, Perrier, Taster's Choice), chocolate and candy (Butterfinger, Crunch, KitKat), culinary products (Buitoni, Carnation, Libby, Maggi), frozen foods (Lean Cuisine), pet foods (Purina, Fancy Feast, Friskies, Mighty Dog), and drugs and cosmetics (L'Oreal). In addition to its 480 factories in 63 countries around the world, Nestlé operates more than 40 Stouffer's hotels.

Transnationals and Globalization Transnational corporations first began to appear in the nineteenth century, but until the mid-twentieth century there were only a few, most of them U.S.- or European-based transnationals that were concerned with obtaining raw materials, such as oil or minerals, for their domestic manufacturing operations. After World War II, an increasing number of large corporations began to invest in overseas production and manufacturing operations as a means of establishing a foothold in foreign consumer markets. Between 1957 and 1967, 20 percent of all new U.S. machinery plants, 25 percent of all new chemical plants, and over 30 percent of all new transport equipment plants were located abroad. By 1970, almost 75 percent of U.S. imports were transactions between the domestic and foreign subsidiaries of transnational conglomerates. By the end of the 1970s, overseas profits accounted for a third or more of the overall profits of the 100 largest transnational corporations. By the early 1980s, 40 percent of all world trade was in the form of intrafirm trade (that is, between different branches and companies of the same transnational conglomerate).

Beginning in the 1970s, a sharp increase occurred in the growth of transnational conglomerates, not only in the United States but also in Europe, Japan, and even some semiperipheral countries. By 2006, there were about 78,000 transnational corporations in the world. Of these, the top 300 controlled approximately one-quarter of the world's productive assets. Many of the largest transnational corporations are now more powerful, in economic terms, than most sovereign nations. General Motors' economy is larger than Portugal's; Toyota's is larger than Ireland's; and Wal-Mart's annual sales exceed Norway's gross domestic product.

The reason for such growth in the number and scale of transnational conglomerate corporations has been that international economic conditions have changed. A recession, triggered by a massive increase in the price of crude oil in 1973, meant that companies everywhere had to reexamine their strategies. At around the same time, technological developments in transport and communications provided larger companies with the flexibility and global reach to exploit the steep differentials in labor costs that exist between core countries and peripheral countries. Meanwhile, these same developments in transport and communications made for intensified international competition, which forced firms to search more intensely for more efficient and profitable global production and marketing strategies. Concurrently, a homogenization of consumer tastes (also facilitated by new developments in communications technologies) has made it possible for companies to cater to global markets.

It was the consequent burst of transnational corporate activity that formed the basis of the recent globalization of the world economy. In effect, the playing field for large-scale businesses of all kinds had been marked out anew. Companies have had to reorganize their operations in a variety of ways, restructuring their activities and redeploying their resources *among different countries, regions, and places* (see Box 7.3, "The Changing Geography of the Clothing Industry"). Local patterns of economic development have been recast again and again as these processes of restructuring, reorganization, and redeployment have been played out.

The advantages to manufacturers of a global assembly line are several. First, a standardized global product for a global market allows them to maximize economies of scale. Second, a global assembly line allows production and assembly to take greater advantage of the full range of geographical variations in costs. Basic wages in manufacturing industries, for example, are between 25 and 75 times higher in core countries than in some peripheral countries. With a global assembly line, labor-intensive work can be done where labor is cheap, raw materials can be processed near their source of supply, final assembly can be done close to major markets, and so on. Third, a global assembly line means that a company is no longer dependent on a single source of supply for a specific component, thus reducing its vulnerability to industrial troubles and other disturbances. Fourth, global sourcing allows transnational conglomerates better access to local markets. For example, Boeing has pursued a strategy of buying a significant number of aircraft components in China and has therefore succeeded in opening the Chinese market to its products.

The automobile industry was among the first to develop a global assembly line. In 1976, Ford introduced the Fiesta, a vehicle designed to sell in Europe, South America, and Asia as well as North America. The Fiesta was assembled in several locations from components manufactured in an even greater number of locations. It became the first of a series of Ford "world cars" that now includes the Focus, the Mondeo, and the Contour. The components of the Ford Escort, for example, are made and assembled in 15 countries across three continents. Ford's international subsidiaries, which used to operate independently of the parent company, are now being functionally integrated, using supercomputers and video teleconferences.

The other automobile companies have organized their own global assembly systems (**Figure 7.20**). The global production networks of these companies allow them to process raw materials near sources of supply, to undertake labor-intensive work where labor is cheap, to complete final assembly close to major markets, and to establish multiple sources for components (thus reducing their vulnerability to work stoppages arising from local labor disputes). They employ modular manufacturing for their world cars based on a common underbody platform yet have the flexibility to adapt the interior, trim, body, and ride characteristics to conditions in different countries. Ford, for example, was able to offer the Focus with different engines, transmissions, and other features. Honda has produced three distinct versions of the same car from its Accord world car platform—the bigger, family-oriented Accord for American drivers, the smaller, sportier Accord aimed at young Japanese professionals, and the shorter and narrower Accord, offering the stiff and sporty ride preferred by European drivers.

More than two-thirds of the 60 to 70 million motor vehicles that roll off the production lines each year are made by just 10 supergroups of global corporations.

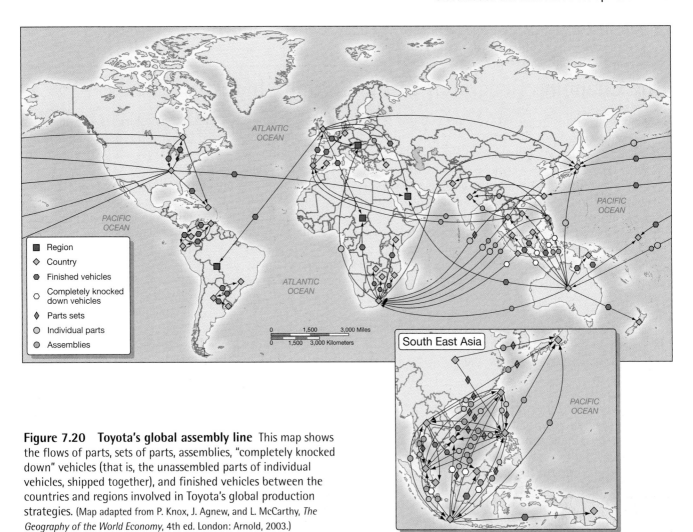

Figure 7.20 Toyota's global assembly line This map shows the flows of parts, sets of parts, assemblies, "completely knocked down" vehicles (that is, the unassembled parts of individual vehicles, shipped together), and finished vehicles between the countries and regions involved in Toyota's global production strategies. (Map adapted from P. Knox, J. Agnew, and L. McCarthy, *The Geography of the World Economy*, 4th ed. London: Arnold, 2003.)

In order of size, they are Toyota (including Daihatsu, Hino, Lexus, Scion, Toyota), General Motors (which includes Buick, Cadillac, Chevrolet, Daewoo, GMC, Holden, Opel, Saab, Saturn, and Vauxhall), Ford (incorporating Lincoln, Mercury, Troller, and Volvo Cars), Volkswagen (including Audi, Bentley, Bugatti, Lamborghini, Scania, SEAT, and Škoda), Honda (including Acura), PSA Peugeot Citroën, Nissan (including Infiniti), Renault (including Dacia, and Renault Samsung Motors), and Hyundai.

The global assembly line is constantly being reorganized as transnational corporations seek to take advantage of geographical differences between places and regions and as workers and consumers in specific places and regions react to the consequences of globalization. Nike, the athletic footwear and clothing marketer, provides a good illustration. Nike once relied on its own manufacturing facilities in the United States and the United Kingdom. Today, however, most of its production is subcontracted to suppliers in East, South, and Southeast Asia. The geography of this subcontracting evolved over time in response to the changing pattern of labor costs in Asia. The first Asian production of Nike shoes took place in Japan. The company then switched most of its subcontracting to South Korea and Taiwan. As labor costs rose there, Nike's

subcontracting was spread across more and more peripheral countries—China, Indonesia, Malaysia, and Vietnam—in search of low labor costs. By 2006, Nike subcontractors employed more than 800,000 people in more than 680 different factories. Nike was the largest foreign employer in Vietnam, where its factories accounted for 5 percent of Vietnam's total exports. China, Indonesia, and Thailand were also major components in Nike's expanded global assembly line because of their low wage costs—around $60 per month.

Flexible Production Systems The strategies of transnational corporations are an important element in the transition from Fordism to Neo-Fordism in much of the world. **Fordism** is named after Henry Ford, the automobile manufacturer who pioneered the principles involved: mass production, based on assembly-line techniques and "scientific" management, together with mass consumption, based on higher wages and sophisticated advertising techniques. In **neo-Fordism** the logic of mass production coupled with mass consumption has been modified by the addition of more flexible production, distribution, and marketing systems. This flexibility is rooted in forms of production that enable manufacturers to shift quickly and

7.3 GEOGRAPHY MATTERS

The Changing Geography of the Clothing Industry

The clothing industry provides a good example of the way that local economic geographies are affected by an industry's response to globalization. In the nineteenth century the clothing industry developed in the metropolitan areas of core countries, with many small firms using cheap, migrant, or immigrant labor. In the first half of the twentieth century, the industry, like many others, began to modernize. Larger firms emerged, their success based on the exploitation of mass-production techniques for mass markets and on the exploitation of principles of spatial organization within national markets. In the United States, for example, the clothing industry went through a major locational shift as a great deal of production moved out of the workshops of New York to big, new factories in smaller towns in the South, where labor was not only much cheaper but less unionized.

The global textile and apparel industry has undergone several shifts in production since the 1950s. The first was from North America and Western Europe to Japan in the 1950s and early 1960s, when Western textile and clothing production was displaced by a sharp rise in imports from Japan. The second shift was from Japan to Hong Kong, Taiwan, and the Republic of Korea, which dominated global textile and clothing exports in the 1970s and early 1980s. In the late 1980s and the 1990s, there was a third migration, from these countries to other developing economies. In the 1980s, production moved principally to mainland China, but also to Indonesia, Malaysia, the Philippines, Sri Lanka, and Thailand. In the 1990s, new suppliers included India and Mexico but the largest newcomer in the 1990s was Turkey, whose total of $3.4 billion in clothing exports placed it fifth in world rankings. By 2005, important new clothing-producing countries (with a billion dollars or more of exports) included Bangladesh, the Czech Republic, Hungary, Mauritius, Morocco, Poland, Romania, and Vietnam.

Much of this shifting globalization of production was "buyer-driven" by retailers like Wal-Mart, Sears, and JC Penney, and fashion-oriented apparel companies like Liz Claiborne, Gap, and The Limited. Their cost-reduction strategies led them to retain design and marketing functions but to contract out the production of their apparel to firms in low-wage countries. In effect, they became "manufacturers without factories." In an attempt to protect domestic manufacturers, the United States, Canada, and 13 countries in Europe entered into a trade pact called the Multifiber Arrangement (MFA) in 1974. The MFA used quotas to regulate access to domestic clothing and textile markets, restricting the amount of imports from any one country. Designed to protect MFA signatories from competition from Japan, Hong Kong, Taiwan, and the Republic of Korea, the import quotas ended up working as a kind of affirmative action program for countries that had large workforces and low wage rates—hence the spread of production around the world.

The result was that by 1980 more than half of all apparel purchased in the United States was imported (compared to less than 7 percent in 1960). Leisure wear—jeans, shorts, T-shirts, polo shirts, and so on—was an important component of the homogenization of consumer tastes around the world at that time, and such clothing could be produced most profitably by the cheap labor of young women in the peripheral metropolitan areas of the world. The same holds true today for fashion wear as well as leisure wear. The hourly compensation (including benefits) of apparel workers in the United States is $10–$12 for a 37-hour week, whereas for a 50- or 60-hour week, their counterparts in Hong Kong are paid around US$5 an hour, and in many parts of China the pay rate is US$0.20 an hour for 11–12-hour shift patterns for six days a week. Not surprisingly, the retail margin on domestically made clothing sold in Europe and the United States is 70 percent or so; the retail margin on clothing made in workshops in countries like Indonesia and Thailand is 100 to 250 percent.

This globalization of production has resulted in a complex set of commodity chains. Many of the largest clothing companies, such as Liz Claiborne, have most of their products manufactured through arrangements

efficiently from one level of output to another and, more importantly, from one product configuration to another.

Flexible production systems involve flexibility both within firms and between them. *Within* firms, new technologies now allow a great deal of flexibility. Computerized machine tools, for example, are capable of producing a variety of new products simply by being reprogrammed, often with very little downtime between production runs

for different products. Different stages of the production process (sometimes located in different places) can be integrated and coordinated through computer-aided design (CAD) and computer-aided manufacturing (CAM) systems. Computer-based information systems can be used to monitor retail sales and track wholesale orders, thus allowing producers to reduce the costs of raw materials stockpiles, parts inventories, and warehousing through with many different independent suppliers, with no one supplier producing more than a fraction of the company's total output. These manufacturers are scattered throughout the world, making the clothing industry one of the most globalized of all manufacturing activities (**Figure 7.F**). In some countries, clothing manufacture is now the primary driver of their economy. Fifty percent of Sri Lanka's export earnings, for example, are derived from clothing, and for Bangladesh, El Salvador, and Mauritius the figure is 63 percent and for Cambodia it is 76 percent. As we have seen, the actual geography of commodity chains in the clothing industry is rather volatile, with frequent shifts in production and assembly sites as companies and their suppliers continuously seek out new locations with lower costs.

Although cheap leisure wear can be produced most effectively through arrangements with multiple suppliers in peripheral low-wage regions, higher-end apparel for

Figure 7.F   The changing distribution of clothing manufacturing Most of the world's clothing exports come from just a few countries. In detail, however, the geography of clothing manufacturing changes rapidly in response to the changing patterns of costs and opportunities within the world economy. This photograph shows Cambodian garment workers in a factory in Phnom Penh, Cambodia.

the global marketplace requires a different geography of production. These products—women's fashion, outerwear, and lingerie, infants' wear, and men's suits—are based on frequent style changes and high-quality finish. This requires short production runs and greater contact between producers and buyers. The most profitable settings for these products are in the core countries' metropolitan areas—London, Paris, Stuttgart, Milan, New York, and Los Angeles—where, once again, migrant and immigrant labor provides a workforce for "designer" clothing that can be shipped in small batches to upscale stores and shopping malls around the world.

The result is that commodity chains in the clothing industry are quite distinctive in the origins of products destined for different segments of the market. Fashion-oriented retailers in the United States who sell "designer" products to up-market customers obtain most of their goods from manufacturers in a small group of high-value-added countries, including France, Italy, Japan, the United Kingdom, and the United States. Department stores that emphasize "private label" products (that is, store brands, such as Nordstrom) and premium national brands obtain most of their goods from established manufacturers in semiperipheral East Asian countries. Mass merchandisers who sell lower-priced brands buy primarily from a third tier of lower-cost, mid-quality manufacturers, and large-volume discount stores like Wal-Mart import most of their goods from low-cost suppliers in peripheral countries like China, Bangladesh, and the Dominican Republic.

Currently, a further realignment of global production is taking place as a result of the phasing out of the MFA's network of quotas, which began in the mid-1990s. The last of the quotas expired on January 1, 2005, prompting significant shifts in production to China and India—the two countries with the greatest comparative advantage in terms of labor costs— from countries that had prospered under the quota system, such as Bangladesh, Cambodia, El Salvador, Lesotho, and Mauritius.

sophisticated small-batch, just-in-time production and distribution systems. **Just-in-time production** employs vertical disintegration within large, formerly functionally integrated firms, such as automobile manufacturers, in which daily and even hourly deliveries of parts and other supplies from smaller (often nonunion) subcontractors and suppliers now arrive "just in time" to maintain "last-minute" and "zero" inventories. The combination of

computer-based information systems, CAD/CAM systems, and computerized machine tools has also given firms the flexibility to exploit specialized niches of consumer demand so that economies of scale in production can be applied to upscale but geographically scattered markets.

The Benetton clothing company provides an excellent case-study example of the exploitation of flexible production systems within a single firm. In 1965, the

Benetton company began with a single factory near Venice. In 1968. it acquired a single retail store in the Alpine town of Belluno, marking the beginning of a remarkable sequence of corporate expansion. Benetton is now a global organization with over 5,000 retail outlets in more than 120 countries and with its own investment bank and financial services organizations. It achieved this growth by exploiting computers, new communications and transportation systems, flexible outsourcing strategies, and new production-process technologies (such as robotics and CAD/CAM systems) to the fullest possible extent.

Only about 400 of Benetton's employees are located in the company's home base of Treviso, Italy (**Figure 7.21**). From Treviso, Benetton managers coordinate the activities of more than 250 outside suppliers in order to stock its worldwide network of retail outlet franchises. In Treviso, the firm's designers create new shirts and sweaters on CAD terminals, but their designs are produced only for orders in hand, allowing for the coordination of production with the purchase of raw materials. In factories, rollers linked to a central computer spread and cut layers of cloth in small batches according to the numbers and colors ordered by Benetton stores around the world. Sweaters, gloves, and scarves, knitted in volume in white yarn, are dyed in small batches by machines similarly programmed to respond to sales orders. Completed garments are warehoused briefly (by robots) and shipped out directly (via private package delivery firms) to individual stores to arrive on their shelves within 10 days of manufacture.

Sensitivity to demand, however, is the foundation of Benetton's success. Niche marketing and product differentiation have been central to this sensitivity, which requires a high degree of flexibility in exploiting new product lines. Key stores patronized by trendsetting consumers (such as the store on the Rue Faubourg St. Honoré in Paris and the megastore on Omotesando Street in Tokyo) are monitored closely, and many Benetton stores' cash registers operate as point-of-sale terminals so that immediate marketing data are available to company headquarters daily. Another notable feature of the company's operations is the way that different market niches are exploited with the same basic products. In Italy, Benetton products are sold through several different retail chains, each with an image and decor calculated to bring in a different sort of customer.

Between firms, the flexibility inherent in neo-Fordism has been achieved through the externalization of certain functions. One way of doing this has been to reorganize administrative, managerial, and technical functions into flatter, leaner, and more flexible forms of organization that can make increased use of outside consultants, specialists, and subcontractors. This has led to a degree of vertical disintegration among firms. **Vertical disintegration** involves the evolution from large, functionally integrated firms within a given industry toward networks of specialized firms, subcontractors, and suppliers. Another route to externalization has been for firms to participate in joint ventures, in the licensing or contracting of technology, and in strategic alliances involving design partnerships, collaborative R&D projects, and the like. **Strategic alliances** involve commercial agreements between transnational corporations, usually involving shared technologies, marketing networks, market research, or product development. They are an important contributor to the intensification of economic globalization.

For example, the Nestlé food company has a number of strategic alliances, including a joint venture with General Mills called Cereal Partners Worldwide (CPW); an alliance since 1990 with the Walt Disney Company in Europe, the United States, Latin America, and other markets that includes making Nestlé the main food company for Disneyland Paris; and a joint marketing venture with Pillsbury Häagen-Dazs since 1999 called Ice Cream Partners USA. Nestlé also has had a joint-venture partnership with the Coca-Cola Company since 1991 in which the Swiss company cooperates with Coca-Cola in exchanging technologies and in marketing. Nestlé, for example, uses Coca-Cola's distribution network for products such as Nescafé instant coffee.

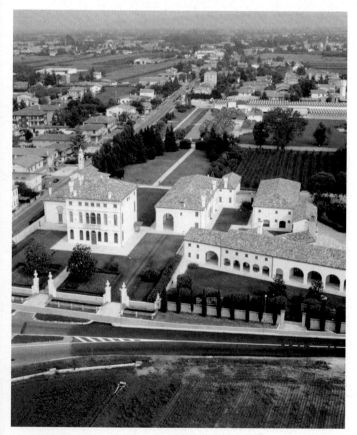

Figure 7.21 Headquarters of the Benetton Group, Villa Minelli, Treviso, Italy

Maquiladoras and Export-Processing Zones The type of subcontracting carried out by Nike is encouraged by the governments of many peripheral and semiperipheral countries, who see participation in global assembly lines as

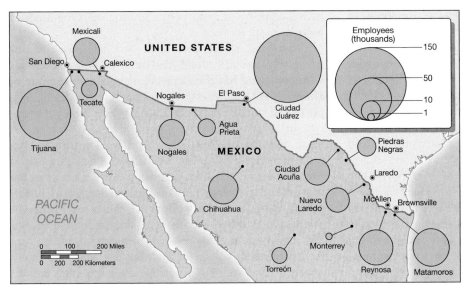

Figure 7.22 Principal *maquiladora* centers on the United States–Mexico border
Cheap labor and tax breaks for firms manufacturing and assembling goods for re-export have made many Mexican border towns attractive to U.S. companies. Around half a million workers are employed in these *maquiladora* plants, producing electronic products, textiles, furniture, leather goods, toys, and automotive parts. (Adapted from P. Dicken, *Global Shift*, 4th ed. New York: Guilford, 2003.)

a pathway to export-led industrialization. They offer incentives such as tax "holidays" (not having to pay taxes for a specified period) to transnational corporations. In the 1960s Mexico enacted legislation permitting foreign companies to establish "sister factories"—*maquiladoras*—within 19 kilometers (12 miles) of the border with the United States for the duty-free assembly of products destined for re-export (see **Figures 7.22** and **7.23**). By 2005, more than 3,500 such manufacturing and assembly plants had been established, employing around a million Mexican workers, most of them women, and accounting for more than 30 percent of Mexico's exports. But since 2000, over *500 maquiladoras* have closed, resulting in the loss of tens of thousands of jobs. The tax breaks that favored the establishment of the *maquiladoras* are due to expire under the terms of the North American Free Trade Agreement

(NAFTA), while lower wage rates and better incentives in other countries—particularly China—have proved more attractive to manufacturers.

Export-processing zones (EPZs) are small areas within which especially favorable investment and trading conditions are created by governments in order to attract export-oriented industries. These conditions include minimum levels of bureaucracy surrounding importing and exporting, the absence of foreign exchange controls, the availability of factory space and warehousing at subsidized rents, low tax rates, and exemption from tariffs and export duties. In 1985, it was estimated that there were a total of 173 EPZs around the world, which together employed 1.8 million workers. In 1998, the International Labor Organization (ILO) estimated that there were 850 EPZs, employing about 27 million people, 90 percent of whom were women.

Figure 7.23 Hyundai *maquiladora* factory, Tijuana, Mexico

China alone had 124 EPZs, which together employed 18 million workers. The ILO's report on EPZs[2] criticized these "vehicles of globalization" because very few of them have any meaningful links with the domestic economies around them, and most trap large numbers of people in low-wage, low-skill jobs.

In addition to tax incentives and EPZs, many governments also establish policies that ensure cheap and controllable labor. Sometimes countries are pressured to participate in global assembly lines by core countries and by the transnational institutions they support. The United States and the World Bank, for example, have backed regimes that support globalized production and have pushed for austerity programs that help to make labor cheap in peripheral countries. Countries pursuing export-led industrialization as an economic development strategy do not plan to remain the providers of cheap labor for foreign-based transnational corporations, however. They hope to shift from labor-intensive manufactures to capital-intensive, high-technology goods, following the path of semiperipheral Asian countries like Singapore and South Korea.

Big-Box Retailing, Chains, and Global Sourcing

At the other end of the commodity chain from the farms, mines, *maquiladoras*, and factories of the world are the retail outlets and restaurants where the products are sold and consumed. Traditionally, retailing and food services in developed countries have been dominated by small, specialized, independent stores and local cafes, restaurants, pubs, and bars. But the logic of economic rationalization and economies of scale has displaced the traditional pattern of downtown department stores, Main Street shops, corner stores, and local bars with big-box superstores and national and international chains of retail outlets and restaurants. Town centers once filled with a thriving mix of independent butchers, cafes, newsagents, tobacconists, bars, book shops, greengrocers, and family-owned general stores now have "cloned" settings consisting of standardized supermarket retailers, fast-food chains, mobile phone shops, and the down-market fashion outlets of global conglomerates. With their cost advantages, these chains have become the economic equivalent of invasive species: voracious, indiscriminate, and often antisocial. Their big, centralized logistical operations have not only put small independent stores out of business but are driving the homogenization of consumption, eating, farming, the landscape, the environment, and people's daily lives. The retail sector of most towns and cities is now characterized by external control, the decisions about hiring, labor policies, wages, stock, and

menus being made in corporate headquarters hundreds of miles or more away.

Fast-food restaurants have become icons of this trend. McDonald's alone has some 28,000 restaurants worldwide and has been opening new ones at the rate of almost 2,000 each year. It is the largest purchaser of beef, pork, and potatoes and the largest owner of retail property in the world. In the United States, 40 percent of meals are eaten outside of the home, most of them at fast-food restaurants. One in four adults visits a fast-food restaurant every day. Not surprisingly, the majority of the population is overweight and the frequency of health problems associated with obesity—such as early-onset diabetes and high cholesterol—is rising rapidly. The cost of these problems to personal well-being and to health care systems is already daunting. Meanwhile, fast food's low-paying service sector has become an increasingly significant component of the economy.

Equally significant in terms of local economies, local development patterns, and global supply chains has been the success of "big-box" retail outlets such as Circuit City, Best Buy, Office Max, Home Depot, Target, and Wal-Mart. To many observers, Wal-Mart has come to symbolize the worst characteristics of globalization, including corporate greed, low wages, the decline of small-town mom-and-pop stores, and the proliferation of sweatshops in less-developed countries (see Box 7.4, "Wal-Mart's Economic Landscape").

Supermarket chains have also become particularly influential. In the United Kingdom, for example, the top four supermarket chains—Tesco, Asda (owned by Wal-Mart), Sainsbury, and Safeway—have come to dominate the retail food market through a combination of out-of-town superstores and convenience supermarkets such as Tesco Express, Tesco Metro, and Sainsbury's Local along high streets. As a consequence, they have killed off small general stores in the United Kingdom at the rate of one a day and specialist shops—such as butchers, bakers, and fishmongers—at the rate of 50 per week.

The centralized supply chains of supermarket chains have not only killed off small local businesses but also impacted local farmers. Supermarket chains rely on big suppliers in agribusiness—national and transnational firms that are often highly subsidized and whose global reach depends heavily on monoculture and extensive husbandry that, in turn, require the extensive use of antibiotics in animals and pesticides, fertilizer, and genetic engineering for crops. As a result, small farmers and fishermen have been squeezed from the market. And with them many traditional local foods have disappeared or are in danger of disappearing. Meanwhile, supermarket shelves are lined with highly processed foods, out-of-season fruit and vegetables, and produce that has traveled a long way and, often, been stored for a while. When the average North American or European family sits down to eat, most of the ingredients have typically traveled at least 1,500 miles from farm, processing, packing, distribution, to consumption.

[2]International Labor Organization, *Labour and Social Issues Relating to Export Processing Zones*. Geneva: International Labor Office, 1998.

The external control inherent to centralized chains also brings the prospect of local closures through rationalization. In Britain, traditional community pubs are increasingly under threat because of this. According to the Campaign for Real Ale, an average of 56 pubs close in Britain every month as the ownership of pubs becomes concentrated in fewer and fewer hands. Corporate chains such as Greene King, Punch Taverns, and Enterprise Inns each own thousands of outlets and constantly acquire more in the cause of market penetration and economies of scale. Smaller pubs that do not fit into their business model are sold to a developer to be converted into apartments or restaurants. Following a similar pattern, the number of local bistros in France has fallen to 45,000 from around 225,000 in the 1970s. Many of the surviving local bistros are in danger of having to transform themselves into theme bars, of being swallowed up by large chains, or of going out of business altogether.

The Global Office

The globalization of production and the growth of transnational corporations have brought about another important change in patterns of local economic development. Banking, finance, and business services are no longer locally oriented ancillary activities, but important global industries in their own right. They have developed some specific spatial tendencies of their own—tendencies that have become important shapers of local economic development processes.

The new importance of banking, finance, and business services was initially a result of the globalization of manufacturing, an increase in the volume of world trade, and the emergence of transnational corporate empires. It was helped along by advances in telecommunications and data processing. Satellite communications systems and fiber-optic networks made it possible for firms to operate key financial and business services 24 hours a day around the globe, handling an enormous volume of transactions. Linked to these communications systems, computers permit the recording and coordination of the data. The world's fourth-largest stock market, the National Associated Automated Dealers Quotation System (NASDAQ), has no trading floor at all: Telephone and fiber-optic lines connect its half-million traders worldwide.

As banking, finance, and business services grew into important global activities, however, they were themselves transformed into something quite different from the old, locally oriented ancillary services. The global banking and financial network now handles trillions of dollars every day (about $6 trillion in 2005)—no more than 10 percent of which has anything to do with the traditional world economy of trade in goods and services. International movements of money, bonds, securities, and other financial instruments have now become an end in themselves because they are a potential source of high profits from speculation and manipulation. Several factors have supported this development:

- In core countries the institutionalization of savings (through pension funds and so on) established a large pool of capital managed by professional investors with few local or regional allegiances or ties.

- The quadrupling of crude oil prices in 1973 (organized by OPEC, the Organization of Petroleum Exporting Countries) generated so much capital for oil-rich countries that their banks opened overseas branches in order to find enough borrowers. In many cases, the borrowers were companies and governments in underdeveloped, peripheral countries that had previously been considered poor investment prospects. The internationalization of financial services soon paid off for the big banks. By the mid-1970s, about 70 percent of Citibank's overall earnings came from its international operations, with Brazil alone accounting for 13 percent of the bank's earnings in 1976. For the borrowers, however, the result was a huge increase in international debt, and many peripheral countries became net exporters of capital as they started to make the interest payments on their loans.

- The initial response of many governments (including the U.S. government) to balance-of-payments problems in the 1960s and 1970s was to print more money—a short-term solution that eventually contributed to a significant surge in **inflation** in the world economy. This inflation, because it promoted rapid change and international differentials in financial markets, gave a further boost to speculative international financial transactions of all kinds.

- In many countries, governments have lifted restrictions and regulations relating to banking, finance, and business services in the hope of capturing more of their growth.

- The persistent trade deficit of the United States vis-à-vis the rest of the world (a result of the postwar recovery of Europe and Japan) created a growing pool of dollars outside the United States, known as Eurodollars. This supply in turn created a pool of capital that was beyond the direct control of U.S. authorities.

- "Hot" money (undeclared business income, proceeds of securities fraud, trade in illegal drugs, and syndicated crime), easily laundered through international electronic transactions, also found its way into the growing pool of Eurodollars. It is estimated that 100 billion U.S. dollars are laundered each year through the global financial system.

Together these factors have amounted to a change so important that a deep-seated restructuring of the world economy has resulted. The effects of this restructuring have altered economic geography at every scale, affecting the lives of people everywhere. Banks and financial

Wal-Mart's Economic Landscape

Wal-Mart is the largest company in the world in terms of revenue, with profits of $11 billion in 2007. It is also the most highly rationalized, centralized retail chain in the world, designed to run like an assembly line. To some, Wal-Mart represents an opportunity to consume high-quality products at low cost from a global market economy. To others, Wal-Mart has displaced McDonald's as the company most closely associated with U.S. political-economic and cultural power, the negative cultural consequences of globalization, and the negative consequences of corporate power.

Wal-Mart was founded on a distinctive—and unusual—geographic marketing strategy, based on a network of stores in small, isolated towns in the American South. From the firm's headquarters in Bentonville, Arkansas, Wal-Mart's founder, Sam Walton, built a retail empire that reflects the values embedded in the small-town rural South: conservatism, idealized views of family and community, and the principles of hard work, frugality, and competitiveness. Until 1962, the focus was on general merchandise stores, but then it switched to discount stores. Wholesale clubs (Sam's Clubs) were added in 1982, and supercenters (a general merchandise/supercenter combination) in 1988. The company has moved out of the small-town South and into metropolitan markets and international ventures but is still characterized by peripheral store locations, usually a short distance from interstate highways and often not even within the boundaries of an incorporated place (**Figure 7.G**)—partly because of land costs and tax benefits and partly because of the NIMBY-ism (NIMBY stands for "Not In My Backyard") associated with established neighborhoods (**Figure 7.H**).

From the start, the emphasis was on a business model that rests on high-volume turnover through low prices. This required a rapid expansion of stores in order to strengthen the firm's position with suppliers and this, in turn, meant that Wal-Mart quickly developed expertise in **logistics**—the movement and storage of goods and the management of the entire supply chain, from purchase of raw materials through sale of the final product and back again for the replenishment of goods as they are sold. Wal-Mart put together its own distribution facilities, fleet of trucks, and satellite communications network in order to maximize the supplier discounts required by the business model. Another key feature of Wal-Mart's economic landscape is cost control. Pressure is continuously exerted on local governments, employees, and suppliers to get the best deal possible for Wal-Mart. Wal-Mart's labor policies are virulently antiunion; the stores are deliberately understaffed; there is no grievance policy for employees; vendor agreements are tough; suppliers do not give up ownership of the goods until they are sold to the customer; and suppliers are required to drop prices by as much as 5 percent annually.

Figure 7.G   Wal-Mart stores in the Atlanta metropolitan region. (Adapted from Figure 2.6 in S. D. Brunn (ed.), *Wal-Mart World*. New York: Routledge, 2006, p. 23.)

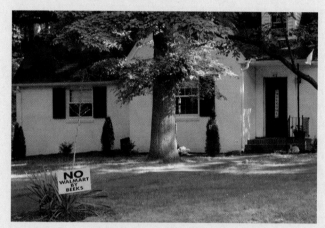

Figure 7.H   Not In My Backyard Local resistance to a proposed new Wal-Mart store in Blacksburg, Virginia.

The company's rapid expansion and tremendous profitability "allowed it a market power unequalled by any of its large corporate competitors, a power that is reshaping the nature of America's and the world's retail industry."[1] Wal-Mart's low prices have reduced consumer inflation and brought many products within the range of previously underserved consumers. But there has been a great deal of concern over the impact of Wal-Mart stores on smaller stores and towns. These concerns center on the impact on local businesses and employment because of the many examples of Wal-Mart's predatory pricing having put independent local stores out of business. The result is external control of the local economy. With fewer locally owned businesses, money spent on goods no longer stays within the community but rather is funneled back to corporate headquarters. At the same time, Wal-Mart wages tend to be significantly lower—in the region of 30 percent lower—than in independent, locally owned businesses, so that wages in local labor markets tend to be depressed.

[1] E. Rosen, "Wal-Mart: The New Retail Colossus." In S. Brunn (ed.), *Wal-Mart World*. New York: Routledge, 2006, p. 92.

Other concerns include the aesthetic and environmental impact of the big cinderblock boxes and their enormous adjoining parking lots (**Figure 7.1**). Also, although large retailers such as Wal-Mart carry an extensive array of goods, they often stock only a limited variety of any specific product, thereby leaving the consumer with less choice overall. For example, whereas independent local newsagents typically carry a broad range of magazine titles, big-box chain stores typically concentrate on only the titles with the biggest turnover, in order to maximize profit. The same is true for the sale of CDs and DVDs. So, not only do corporate chains reduce the range of shops available in small towns, they also reduce the choice of goods readily available. This is compounded, in Wal-Mart's case, by the company's censorship, "protecting" consumers from products that the management deems offensive. This includes refusing to sell CDs with parental warning stickers and obscuring the covers of popular magazines like *Redbook, Marie Claire, Cosmopolitan,* and *Rolling Stone.* Because of the enormous size of the Wal-Mart network, some magazines willingly send advance copies to corporate headquarters in Bentonville for approval before publication and will even alter cover artwork to avoid losing sales. "Thus, far from being simply a store, Wal-Mart is also a moral universe external to the community."[2] Wal-Mart advertising campaigns have always emphasized happy staff, customers from cozy families, and support for local communities. Patriotism was a strong theme in the 1990s, the firm's Buy-American campaign coinciding with Wal-Mart becoming China's sixth-largest trading partner. There are, however, no references to fair trade, fair employment, ethical trading, or environmental consciousness.

The most recent development in Wal-Mart's economic landscape has been internationalization. The company entered markets in Mexico, Puerto Rico, and Canada in the early 1990s, later expanding into Hong Kong, Indonesia, Brazil, Argentina, China, Germany, and South Korea. There were mixed results. More recently, the approach has been to acquire "Wal-Mart ready" chains of stores and then convert their operations to the Wal-Mart model, as with the company's takeover of Asda (the United Kingdom's second-largest supermarket chain) in 1999.

Wal-Mart's greatest global impact, though, is through its supply chains. Overall, 50 to 60 percent of the goods Wal-Mart imports into the United States each year by ship come from China, and the home office of Wal-Mart Global Procurement is in Shenzhen, China. The company has a network of over two dozen field offices in other countries and oversees the sourcing of products from more than 5,000 factories in 65 countries, including Bangladesh, Brazil, Guatemala, India, Indonesia, Malaysia, the Philippines, Sri Lanka, and Thailand. The reason, of course, is labor costs. But the "natural" cheapness of labor in these countries is not enough for Wal-Mart; the company systematically imposes pressure on suppliers to reduce costs still further. This, of course, has led to abuse and violations of labor laws. An example that received widespread public attention came to light in 1999 as a result of a report by the National Labor Committee, which revealed that workers in the Liang Shi Handbag Factory in Dongguan City, Guangdong, China, earned as little as 12 cents an hour for 60- to 70-hour weeks producing Kathie Lee Gifford handbags.[3] This was not an isolated example. Recent evidence from *China Labor Watch,* for example, found that four apparel factories in China that supplied Adidas and Wal-Mart were paying their workers just 48 cents an hour—which was not enough to sustain themselves unless they did overtime. A 2007 report by the National Labor Committee, *A Wal-Mart Christmas,*[4] featured Guangzhou Huanya Gift Ltd. Company of Guangdong Province, China, where the 8,000 workers must put in 10- to 12- to 15-hour shifts, seven days a week during the long, eight-month busy season manufacturing Christmas ornaments for Wal-Mart. At least half the workers, some 4,000 people, were working 95 hours, including 55 hours of (mandatory) overtime. The legal minimum wage in Guangzhou, is 55 cents an hour, but workers were paid by a piece rate, with some workers earning just 26 cents an hour, half the legal minimum wage.

Figure 7.1   Wal-Mart superstore

[2] B. Warf and T. Chapman, "Cathedrals of Consumption: A Political Phenomenology of Wal-Mart." In S. Brunn (ed.), *Wal-Mart World.* New York: Routledge, 2006, p. 165.

[3] China Labor Watch, 2007, "Textile Sweatshops: Adidas, Bali Intimates, Hanesbrands Inc., Piege Co. (Felina Lingerie), Quiksilver, Regina Miracle Speedo, Walcoal America Inc., and Wal-Mart Made In China," http://www.chinalaborwatch.org/200711204textile.htm.

[4] National Labor Committee, 2007, "A Wal-Mart Christmas. Brought to You by a Sweatshop in China," http://www.nlcnet.org/article.php?id=498.

corporations with the size and international reach of Citigroup, Crédit Suisse, or Deutsche Bank are able to influence local patterns and processes of economic development throughout the world, just like the major transnational conglomerates involved in the global assembly line.

Electronic Offices, Decentralization, and Outsourcing

It is clear that an important shift has occurred in the economic structure of the world's core economies, with the rapid growth of banking, financial, and business services contributing to the expansion of the quaternary sector. What is particularly important from a *geographical* perspective is that this growth has been localized—that is, concentrated in relatively small and distinctive settings within major metropolitan centers. This phenomenon is surprising to some observers, who had expected that new communications technologies would allow for the dispersion of "electronic offices" and, with it, the decentralization of an important catalyst for local economic development. A good deal of geographic decentralization of offices has occurred, in fact, but it has mainly involved "back-office" functions that have been relocated from metropolitan and business-district locations to small-town and suburban locations.

Back-office functions are record-keeping and analytical functions that do not require frequent personal contact with clients or business associates. The accountants and financial technicians of main street banks, for example, are back-office workers. Developments in computing technologies, database access, electronic data interchanges, and telephone call routing technologies are enabling a larger share of back-office work to be relocated to specialized office space in cheaper settings, freeing space in the high-rent locations occupied by the bank's front office. For example, the U.S. Postal Service uses optical character readers (OCRs) to read addresses on mail, which is then bar-coded and automatically sorted to its appropriate substation. Addresses that the OCRs cannot read are digitally photographed and transmitted to a computer screen, where a person manually types the address into a terminal. In Washington, DC., OCR sorting takes place at the central mail facility, but the manual address entry is done in Greensboro, NC., where wage rates are lower. Workers in Greensboro view images of letters as they are sorted in Washington and enter correct addresses, which are electronically transmitted back to be bar-coded on the piece of mail. Similarly, when residents of London, England, call to inquire about city parking-ticket fines, the calls are processed in a small city in northern England.

Some prominent examples of back-office decentralization from U.S. metropolitan areas have included the relocation of back-office jobs at American Express from New York to Salt Lake City, Fort Lauderdale, and Phoenix; the relocation of Metropolitan Life's back offices to Greenville, SC., Scranton, PA., and Wichita, KS; the relocation of Hertz's data-entry division to Oklahoma City, Dean

Witter's to Dallas, and Avis's to Tulsa; and the relocation of Citibank's MasterCard and Visa divisions to Tampa and Sioux Falls. Some places have actually become specialized back-office locations as a result of such decentralization. Omaha and San Antonio, for example, are centers for a large number of telemarketing firms, and Roanoke, VA, has become something of a mail-order center.

Internationally, this trend has taken the form of offshore back offices. By decentralizing back-office functions to offshore locations, companies save even more in labor costs. Several New York–based life-insurance companies, for example, have established back-office facilities in Ireland. Situated conveniently near Ireland's main international airport, Shannon, they ship insurance claim documents from New York via Federal Express, process them, and beam the results back to New York via satellite or the transatlantic fiber-optics line.

The logical next step from decentralized back-office functions is outsourcing. The outsourcing of services is one of the most dynamic sectors of the world economy. Global outsourcing expenditures are expected to grow significantly as small and medium-sized enterprises follow the example of large transnational companies in taking advantage of low wages in semiperipheral and peripheral countries (**Figure 7.24**). Typically, international outsourcing in service industries involves the work of "routine producers" (who process data by following instructions, perform repetitive tasks, and respond to explicit procedures) rather than "symbolic analysts" (who work with abstract images; are involved with independent problem-identifying, problem-solving, and strategic-brokering activities; and make decisions based on critical judgment sharpened by experience). Outsourced services range from simple business-process activities (for example, data entry, word processing, transcription) to more sophisticated, high-value-added activities (for example, architectural drawing, product support, financial analysis, software programming, and human resource services). India has become one of the most successful exporters of outsourced

Figure 7.24 Outsourcing Indonesian workers in a Samsung factory in Cikarang, Indonesia, manufacturing consumer electronics.

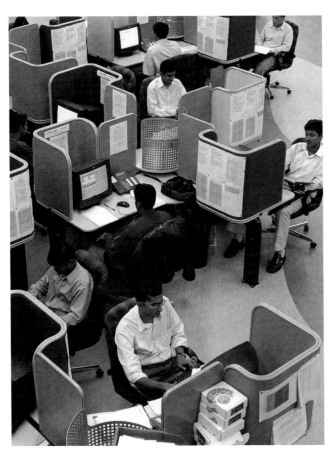

Figure 7.25 Globalized office work Employees of India's largest tobacco-maker, ITC Limited, and U.S.-based Client Logic Corporation work in a call center in Bangalore, India.

service activities, ranging from call centers and business-process activities to advanced IT (information technology) services (**Figure 7.25**). More than 150 of the *Fortune* 500 companies, for example, now outsource software development to India. In the Philippines, special electronic "enterprise zones" have been set up with competitive international telephone rates for companies specializing in telemarketing and electronic commerce. Mexico, South Africa, and Malaysia have also become important locations for call centers and business-process activities.

Clusters of Specialized Offices Decentralization
is outweighed, however, by the tendency for a disproportionate share of the new jobs created in banking, finance, and business services to cluster in highly specialized financial districts within major metropolitan areas. The reasons for this localization are to be found in another special case of the geographical agglomeration effects that we discussed earlier in this chapter.

Metropolitan areas such as New York City, London, Paris, Tokyo, and Frankfurt have acquired the kind of infrastructure—specialized office space, financial exchanges, teleports (office parks equipped with satellite Earth stations and linked to local fiber-optic lines), and communications networks—that is essential for delivering services to clients with a national or international

scope of activity. These metropolitan areas have also established a comparative advantage both in the mix of specialized firms and expert professionals on hand and in the high-order cultural amenities available (both to high-paid workers and to their out-of-town business visitors). Above all, these metro areas have established themselves as centers of authority, with a critical mass of people in the know about market conditions, trends, and innovations—people who can gain one another's trust through frequent face-to-face contact, not just in business settings but also in the informal settings of clubs and office bars. They have become **world cities**—places that, in the globalized world economy, are able not only to generate powerful spirals of local economic development but also to act as pivotal points in the reorganization of global space: control centers for the flows of information, cultural products, and finance that collectively sustain the economic and cultural globalization of the world (see Chapter 10 for more details).

A good example of the clustering of business services in major world cities is provided by advertising services in Europe. In the early 1980s, these services were distributed among major European cities, with Paris, London, Amsterdam, and Stockholm accounting for the headquarters of most firms, but with smaller concentrations in Brussels, Düsseldorf, Frankfurt, and Zürich. By 1990, the headquarters of most advertising agencies in Europe had moved to London (**Figure 7.26**), which, along with New York and Tokyo, has become one of the three most dominant world cities in the contemporary world-system.

Offshore Financial Centers The combination of metropolitan concentration and back-office decentralization fulfills most of the locational needs of the global financial network. There are, however, some needs—secrecy and shelter from taxation and regulation, in particular—that call for a different locational strategy. The result has been the emergence of a series of **offshore financial centers**: islands and microstates such as the Bahamas, Bahrain, the Cayman Islands, the Cook Islands, Luxembourg, Liechtenstein, and Vanuatu that have become specialized nodes in the geography of worldwide financial flows.

The chief attraction of offshore financial centers is simply that they are less regulated than financial centers elsewhere. They provide low-tax or no-tax settings for savings, havens for undeclared income and for hot money. They also provide discreet markets in which to deal currencies, bonds, loans, and other financial instruments without coming to the attention of regulating authorities or competitors. The U.S. Internal Revenue Service estimates that about $400 billion ends up in offshore financial centers each year as a result of tax-evasion schemes. Overall, about 60 percent of all the world's money now resides offshore.

The Cayman Islands provide the classic example of an offshore financial center. This small island state in the Caribbean has transformed itself from a poor, underdeveloped colony to a relatively affluent and modern setting for upscale tourism and offshore finance. With a

Figure 7.26 The clustering of advertising agencies in European cities This map shows the changing distribution of the headquarters offices of the top 100 advertising agencies in Europe between 1982 and 1990. Note the striking concentration of agency headquarters in London. (Adapted from P. Daniels, "Services in a Shrinking World," *Geography*, 80, 1995, Fig. 4, p. 102.)

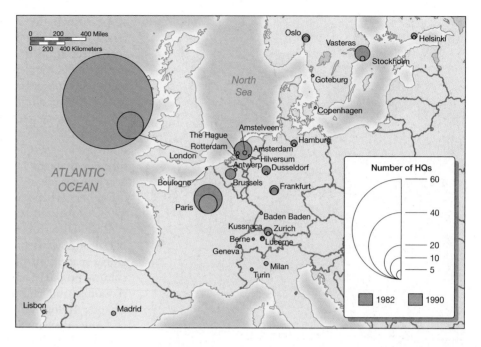

population of just over 40,000, it has more than 30,000 registered companies, including 350 insurance companies and over 580 banks from all around the world. Yet fewer than 75 of the 580 or so registered banks in the Cayman Islands maintain a physical presence on the islands, and only half a dozen offer the kind of local services that allow local residents to maintain checking accounts or tourists to cash traveler's checks. Most exist as post office boxes, as nameplates in anonymous office buildings, as fax numbers, or as entries in a computer system. In 2004, over a trillion dollars passed through the Caymans, making it the most successful of all offshore financial centers.

The Pleasure Periphery: Tourism and Economic Development

Many areas of the world, including parts of the world's core regions, do not have much of a primary base (that is, in agriculture, fishing, or mineral extraction), are not currently an important part of the global assembly line, and are not closely tied into the global financial network. For these areas, tourism can offer the otherwise unlikely prospect of economic development. Tourism has, in fact, become enormously important. In 2003, international tourism earnings amounted to $514 billion. Tourism is already the world's largest nonagricultural employer, with 1 in every 12 workers worldwide involved in transporting, feeding, housing, guiding, or amusing tourists. The global stock of lodging, restaurant, and transportation facilities is estimated to be worth about $3 trillion.

The globalization of the world economy has been paralleled by a globalization of the tourist industry. In aggregate, there were almost 700 million international tourist trips in 2003, compared with just 147 million in 1970. The majority of these trips were made by tourists

from the more affluent countries of the world. Spending by U.S., German, U.K., and Japanese tourists accounted for almost 40 percent of total international tourist dollars in 2003. Tourists from France, Italy, Canada, and the Netherlands accounted for another 12 percent. What is most striking, though, is not so much the growth in the number of international tourists as the increased range of international tourism. Thanks largely to cheaper long-distance flights, a significant proportion of tourism is now transcontinental and transoceanic. While Europe (58 percent) and the Americas (18 percent) continue to be the main tourist destinations, visits to countries in Africa, Asia, and the Pacific have grown to account for almost one-quarter of the industry. This, of course, has made tourism a central component of economic development in countries with sufficiently exotic wildlife (Kenya, for example), scenery (Nepal, Vietnam), beaches (the Seychelle Islands), shopping (Singapore and Hong Kong), culture (China, India, Japan, and Indonesia), or sex (Thailand).

In addition, "alternative" tourism has been widely advanced as a more sustainable strategy for economic development in peripheral regions (**Figure 7.27**). Alternative tourism emphasizes self-determination, authenticity, social harmony, preservation of the existing environment, small-scale development, and greater use of local techniques, materials, and architectural styles. Alternative tourism includes ecotourism (birdwatching in Costa Rica; helping with programs for endangered sea turtles in Bali; "working" in elephant camps in Thailand), cultural tourism (visiting Machu Picchu, the lost city of the Incas), adventure tourism ("exploring" the Amazon), and industrial tourism (touring the Potteries district of northern England by canal boat).

Costa Rica, despite being a poor country, has won high praise from environmentalists for protecting

Figure 7.27 Ecotourism
Tourists walk toward the Perito
Moreno glacier in the Park and
National Reservation Los Glaciares,
an ecotourism destination in
Patagonia, Argentina, declared by
the UNESCO as a Natural World
Heritage Site.

30 percent of its territory in biosphere and wildlife pre-
serves. Costa Rica has more bird species (850) than are
found in the United States and Canada combined and
more varieties of butterflies than in all of Africa. It has
12 distinct ecosystems that among them contain more
than 6,000 kinds of flowering plants, over 200 species of
mammals, 200 species of reptiles, and more than 35,000
species of insects. The payoff for Costa Rica is the escalating number of tourists who come to visit its active vol-
canoes, palm-lined beaches, cloud forests, and tropical
parks. In 2007, Costa Rica received over 1.9 million
tourists. Tourism is now the country's largest source of
foreign exchange, followed closely by bananas, coffee,
sugar, the textile industry, and more recently the production of microchips.

Ecuador is another country that has fostered alternative tourism. With 6 national parks, 7 nature reserves, and
20 privately protected areas, Ecuador offers great variety
for tourists. In a small area straddling the equator, it has
some of the world's oldest rain forests and the world's
second-highest active volcano—Cotopaxi, at 5,897 meters
(19,347 feet). In Ecuador, tourists can see Amazon tribes,
spectacular wildlife in the Galapagos Islands, a thriving Andean culture, and a well-preserved legacy of Spanish colonialism. Two-thirds of all organized travel to Ecuador is
handled by members of the Ecuadorean Ecotourism
Association, an organization sponsored by the private sector as well as the government in an effort to ensure sustainable development through environmental awareness. Other
examples of alternative tourism include guest-house developments in Papua New Guinea, bungalows in French Poly-
nesia, packaged ecotourism in Belize, and "integrated rural
tourism" in Senegal. Such developments must be aimed,
however, at tourists who are both wealthy and environmentally conscious—not, perhaps, a large-enough market
on which to pin hopes of significant increases in levels of
economic development.

Tourism is by no means confined to the less-developed
("unspoiled") peripheral regions. Only 10 percent of
Americans, for instance, have passports. Most U.S. tourist
dollars are spent in safe and predictable settings where
English is spoken—in national parks, specialized resorts,

theme parks, big cities, rural idylls, health spas, and renovated historic towns and districts. The growth of tourism
and the economic success of places like Baltimore's renovated Harborplace, which attracts about 30 million visitors a year, has meant that few localities exist in
America—or anywhere else in the developed world, for
that matter—that do not encourage tourism as one of the
central planks of their economic development strategy. It
has been estimated that because tourism requires only a
basic infrastructure, no heavy plant, and little high-tech
equipment, the cost of creating one new job in tourism is
less than one-fifth the cost of creating a job in manufacturing and less than one-fiftieth the cost of creating a high-
tech engineering job. Consequently, as we saw in Chapter
6, place marketing has become an extremely important aspect of local efforts to promote economic development.

Tourism can provide a basis for economic development, but it is often a mixed blessing. It certainly creates
jobs, but they are often seasonal. Dependence on tourism
also makes for a high degree of economic vulnerability.
Tourism, like other high-end aspects of consumption,
depends very much on matters of style and fashion. As a
result, once-thriving tourist destinations can suddenly
find themselves struggling for customers. Some places are
sought out by tourists because of their remoteness and
their "natural," undeveloped qualities, which are most
vulnerable to shifts of style and fashion. Nepal and New
Zealand are recent examples of this phenomenon; they
are now too "obvious" as destinations and are consequently having to work hard to continue to attract sufficient numbers of tourists. Bhutan, Bolivia, Estonia,
Patagonia, and Vietnam have been "discovered" and are
coping with their first significant growth in tourism.
China, meanwhile, is likely to become one of the most
popular destinations of all by 2010.

It is not only exotic tourist locations that are vulnerable to changing tastes and fashions, however. The
Mediterranean beach resort of Rimini, Italy, for example,
has been all but deserted by its former northwest European, middle-class clientele, who now prefer more distant,
exotic, and distinctive locations. Consequently, Rimini has
had to look to the middle classes of Eastern Europe and

the former Soviet Union in order to fill its hotel rooms. Unfortunately for Rimini, this is a much less affluent population that generates far fewer tourist dollars.

The vulnerability of tourism to political events and economic trends was demonstrated very emphatically in the wake of the 2001 terrorist attack on New York's World Trade Center. The immediate result was a significant drop in total worldwide air traffic. The immediate cause of the decline in travel was a loss of faith in airport security, especially within the United States. The second-order effect of the terrorist attack, however, was to intensify an economic downturn. As in any economic downturn, the tourist industry soon felt the adverse effects. The net result was not only localized economic recession in tourist-dependent places and regions but also a shift in the geography of tourism. Tourist destinations that were more distant from their principal markets, such as Australia and East Africa, tended to lose market share, while closer-to-home and less expensive destinations tended to gain.

Tourism in more exotic tourist destinations, meanwhile, is vulnerable in other ways: to political disturbances, natural disasters, outbreaks of disease or food poisoning, and atypical weather. For ski resorts, warm weather represents the equivalent of a harvest failure for an agricultural region. Moreover, although tourism is a multibillion-dollar industry, the financial returns for tourist areas are often not as high as might be expected. The greater part of the price of a package vacation, for example, stays with the organizing company and the airline. Typically, only 40 percent is captured by the tourist region itself. If the package involves a foreign-owned hotel, this number may fall below 25 percent.

The costs and benefits of tourism are not only economic, of course. On the positive side, tourism can help sustain indigenous lifestyles, regional cultures, arts, and crafts, and it can provide incentives for wildlife preservation, environmental protection, and the conservation of historic buildings and sites. On the negative side, tourism can adulterate and debase indigenous cultures and bring unsightly development, pollution, and environmental degradation. In the Caribbean, sewage has poisoned mangrove trees and polluted coastal waters, and boats and divers have damaged coral reefs. In Kenya, the Maasai Mara National Reserve has been severely degraded by unregulated off-road driving by tour operators. In the United States, congestion in Yosemite National Park (with over 4 million visitors per year) has become so acute that Park Service rangers frequently have to turn away as many as 1,000 vehicles per day, and park administrators are considering the establishment of a strict advance reservation system. In the European Alps, where an incredible 40,000 ski runs attract a winter tourist population 10 times greater than the resident population, forests have been ripped up, pastures obliterated, rivers diverted, and scenic valleys and mountainsides covered with chalets, cabins, and hotels.

Tourism can also involve exploitative relations that debase traditional lifestyles and regional cultural heritages as they become packaged for outsider consumption. In the process, the behaviors and artifacts that are made available to an international market of outsiders can lose much of their original meaning. Traditional ceremonies that formerly had cultural significance for the performers are now enacted only to be watched and photographed. Artifacts like masks and weapons are manufactured not for their original use but as curios, souvenirs, and ornaments. In the process, indigenous cultures are edited, beautified, and altered to suit outsiders' tastes and expectations.

Toward the Experience Economy

Tourism is just part of what some observers claim to be the fourth in a sequence of phases of economic organization. Over the last 200 years, they argue, there has been a shift from an agrarian economy based on extracting commodities, to an industrial economy based on manufacturing goods, to a service economy based on delivering services and, now, to an experience economy based on staging experiences. As profitability becomes more challenging in agrarian, industrial, and service enterprises, so businesses must orchestrate memorable events for their customers, with memory itself becoming the "product"—the experience. James Gilmore, one of the originators of the concept of the experience economy, notes that "Vans has launched a network of Skateparks, where customers pay a $40 annual membership fee, plus $4 to $11 per two-hour skateboarding session. Atlantis, a resort in the Bahamas, charges people who aren't guests just to tour the place! In New Zealand, the country that introduced bungee jumping to the world, people pay to go zorbing, an adventure that involves being strapped into a huge ball and then pushed down a hill."[3]

The experience economy is by no means new. Think, for example, of sports entertainment, arena rock concerts, fancy megamalls with themed entertainment, and visits to museums and galleries, as well as a great deal of traditional tourism. But as more affluent consumers satisfy more and more of their needs and wants in terms of commodities, products, and services, businesses are increasingly combining these things with compelling experiences. There are now dental practices and general medical practices, for example, that double as day spas. In addition, some businesses are based on selling new kinds of experiences. Ciudad de los ninos (Kids City), for example, charges an admission fee to an entire "shopping mall" where children playshop and play-work. Similarly, American Girl Place offers a combination of attractions specifically targeted at young girls: a doll hair salon, special café, a theater featuring musicals, and shops with dolls and outfits. Their Web site advertises "Café. Theater. Shops. *Memories.*"[4]

---

[3]www.darden.edu/batten/pdf/bf_gilmore.pdf.
[4]http://www.americangirl.com/stores/brand_agplace.php.

CONCLUSION

The growth of alternative tourism in Costa Rica, like the growth of the Cayman Islands as an offshore financial center, the emergence of Ireland as a center for back-office activities, and the decline of northern England as a manufacturing region, shows that economic development is not simply a sequential process of modernization and increasing affluence. Various pathways to development exist, each involving different ways of achieving increased economic productivity and higher incomes, together with an increased capacity to improve the basic conditions of life. Economic development means not just using the latest technology to generate higher incomes but also improving the quality of life through better housing, health care, and social welfare systems and enhancing the physical framework, or infrastructure, on which the economy rests.

Local, regional, and international patterns and processes of economic development are of particular importance to geographers. Levels of economic development and local processes of economic change affect many aspects of local well-being and so contribute to many aspects of human geography. Economic development is an important place-making process that underpins much of the diversity among regions and nations. At the same time, it is a reflection and a product of variations from place to place in natural resources, demographic characteristics, political systems, and social customs.

Economic development is always an uneven geographic phenomenon. Regional patterns of economic development are tied to the geographic distribution of resources and to the legacy of the past specializations of places and regions within national economies. A general tendency exists toward the creation of regional cores with dependent peripheries. Nevertheless, such patterns are not fixed or static. Changing economic conditions can lead to the modification or reversal of core-periphery patterns, as in the stagnation of once-booming regions like northern England and the spectacular growth of Guangdong Province in Southeast China. Over the long term, core-periphery patterns have most often been modified as a result of the changing locational needs and opportunities of successive technology systems. Today, economic globalization has exposed more places and regions than ever to the ups and downs of episodes of creative destruction—episodes played out ever faster, thanks to the way that telematics have shrunk time and space.

At the global scale, the unevenness of economic development takes the form of core-periphery contrasts within the world-system framework. Most striking about these contrasts today are the dynamism and pace of change involved in economic development. The global assembly line, the global office, and global tourism are all making places much more interdependent and much faster changing. Parts of Brazil, China, India, Mexico, and South Korea, for example, have developed quickly from rural backwaters into significant industrial regions. The Cayman Islands have been transformed from an insignificant Caribbean colony to an upscale tourist resort and a major offshore financial center. Countries like Ecuador and Costa Rica, with few comparative advantages, suddenly find themselves able to earn significant amounts of foreign exchange through the development of ecotourism. This dynamism has, however, brought with it an expanding gap between rich and poor at every spatial scale: international, regional, and local.

MAIN POINTS REVISITED

- **Geographically, the single most important feature of economic development is that it is highly uneven.**

  At the global scale, this unevenness takes the form of core-periphery contrasts. These contrasts raise important issues of spatial justice that are closely bound up with gender inequality and social justice. Similar core-periphery contrasts—and equally important issues of spatial justice—exist at the regional scale.

- **Geographical divisions of labor have evolved with the growth of the world-system of trade and politics and with the changing locational logic of successive technology systems.**

  Geographical divisions of labor are national, regional, and locally based economic specializations in primary, secondary, tertiary, or quaternary activities. The relationship between changing regional economic specialization and changing levels of prosperity has prompted the interpretation of economic development in distinctive stages. In reality, however, various pathways exist to development, as well as various processes and outcomes of development.

- **Regional cores of economic development are created cumulatively, following some initial advantage, through the operation of several basic principles of spatial organization.**

  Any significant initial local economic advantage—existing labor markets, consumer markets, frameworks of fixed social capital, and so on—tends to be reinforced through a process of cumulative causation, a spiral buildup of advantages that occurs in specific geographic settings as a result of the development of external economies, agglomeration effects, and localization economies.

- **Spirals of economic development can be arrested in various ways, including the onset of disinvestment and deindustrialization, which follow major shifts in technology systems and in international geopolitics.**

  The capital made available from disinvestment in core regions becomes available for investment by entrepreneurs in new ventures based on innovative products and innovative production technologies. Old industries—and sometimes entire old industrial regions—have to be "dismantled" (or at least neglected) in order to help fund the creation of new centers of profitability and employment. This process is often referred to as *creative destruction*.

- **The globalization of the economy has meant that patterns and processes of local and regional economic development are much more open to external influences than before.**

  The globalization of the world economy involves a new international division of labor in association with the internationalization of finance, the deployment of a new technology system, and the homogenization of consumer markets. This new framework for economic geography has meant that the lives of people in different parts of the world have become increasingly intertwined.

KEY TERMS

agglomeration diseconomies (p. 273)
agglomeration effects (p. 271)
ancillary industries (p. 271)
autarky (p. 261)
backwash effects (p. 272)
carrying capacity (p. 257)
conglomerate corporations (p. 277)
creative destruction (p. 275)
cumulative causation (p. 272)
debt trap (p. 266)
deindustrialization (p. 274)
dependency (p. 264)
ecological footprint (p. 258)

elasticity of demand (p. 264)
export-processing zones (EPZs) (p. 283)
external economies (p. 271)
flexible production systems (p. 280)
Fordism (p. 279)
foreign direct investment (p. 276)
geographical path dependence (p. 271)
gross domestic product (GDP) (p. 250)
gross national income (GNI) (p. 250)

growth poles (p. 276)
import substitution (p. 266)
inflation (p. 285)
infrastructure (fixed social capital) (p. 272)
initial advantage (p. 271)
international division of labor (p. 264)
just-in-time production (p. 281)
localization economies (p. 271)
logistics (p. 286)
neo-Fordism (p. 279)
newly industrializing countries (p. 260)
offshore financial centers (p. 289)

OECD (p. 266)
primary activities (p. 259)
quaternary activities (p. 259)
secondary activities (p. 259)
spread effects (p. 272)
strategic alliances (p. 282)
sustainable development (p. 258)
terms of trade (p. 266)
tertiary activities (p. 259)
trading blocs (p. 261)
transnational corporations (p. 277)
vertical disintegration (p. 282)
world cities (p. 289)

ADDITIONAL READING

Beder, S., *Suiting Themselves: How Corporations Drive the Global Agenda*. London: Earthscan, 2006.

Brunn, S. D., *Wal-Mart World: The World's Biggest Corporation in the Global Economy*. London: Routledge, 2006.

Bryson, J., N. Henry, D. Keeble, and R. Martin (eds.), *The Economic Geography Reader*. New York: Wiley, 1999.

Castells, M., *The Information Age*. 3 vols. Oxford: Blackwell, 1996–1998.

Clark, G., M. Feldman, and M. S. Gertler (eds.), *The Oxford Handbook of Economic Geography*. New York: Oxford University Press, 2000.

Dicken, P., *Global Shift*, 4th ed. New York: Guilford Press, 2003.

Gereffi, G., and O. Memedovic, *The Global Apparel Value Chain*. Vienna: United Nations Industrial Development Organization, 2003.

Harvey, D., *Spaces of Global Capitalism*. London: Verso, 2006.

Hughes, A., and S. Reimer (eds.), *Geographies of Commodity Chains*. New York: Routledge, 2004.

Hugill, P., *Global Communications Since 1844: Geopolitics and Technology*. Baltimore: Johns Hopkins University Press, 1999.

Knox, P. L., J. Agnew, and L. McCarthy, *The Geography of the World Economy*, 5th ed. London: Edward Arnold, 2008.

Lechner, F. J., and J. Boli (eds.), *The Globalization Reader*. Malden, MA: Blackwell, 2000.

Mead, W. R., *God and Gold: Britain, America, and the Making of the Modern World*. New York: Knopf, 2007.

O'Loughlin, J., L. Staeheli, and E. Greenburg (eds.), *Globalization and Its Outcomes*. New York: Guilford Press, 2004.

Pine, J., and J. Gilmore, *The Experience Economy*. Boston: Harvard Business School Press, 1999.

Smith, N., *The Endgame of Globalization*. New York: Routledge, 2004.

Stiglitz, J. E., *Globalization and its Discontents*. New York: W. W. Norton, 2003.

UNDP (United Nations Development Programme), *Human Development Report 2007/8: Fighting Climate Change: Human Solidarity in a Divided World*. New York and Oxford: Oxford University Press, 2004.

United Nations, *Trade and Development Report 2007. Regional Cooperation for Development*. New York: UN Department of Economic and Social Affairs, 2007.

United Nations, *World Economic Situation and Prospects, 2007*. New York: UN Department of Economic and Social Affairs, 2007.

United Nations, *World Economic and Social Survey 2007: International Migration*. New York: UN Department of Economic and Social Affairs, 2007.

Wallach, B. *Understanding the Cultural Landscape*. New York: Guilford Press, 2005. See especially Chapters 10 ("Globalization") and 12 ("Manufacturing").

World Bank, *World Development Report 2007: Development and the Next Generation*. Washington, DC: The World Bank, 2004.

World Bank, *Global Economic Prospects: Managing the Next Wave of Globalization, 2005*. Washington, DC: The World Bank, 2007.

EXERCISES

On the Internet

The Companion Website at www.mygeoscienceplace.com contains a number of tools designed to reinforce the geographic concepts presented in this chapter. Structured within a four-step learning path, these online resources include: chapter Key Concepts; a fully-searchable myeBook version of the chapter, which allows you to add notes and highlights; RSS news feeds from various media agencies; links to additional web resources; Human Geography Videos; a digital Glossary; Concept Review questions; Thinking Spatially mapping exercises; and Critical Thinking essay questions.

Unplugged

1. While India's per capita income is well below that of the United States (Figure 7.1), India has more people who earn the equivalent of $70,000 a year than the United States does. How can you explain this, and what might be some of the consequences from the point of view of economic geography?

2. Write a short essay (500 words, or two double-spaced typed pages) on any specialized manufacturing region or office district with which you are familiar. Describe the different kinds of firms that are found there, and suggest the kinds of linkages between them that might be considered examples of agglomeration effects.

Agriculture and Food Production

8

While the soaring cost of food, and the increasing numbers of people unable to feed themselves adequately because of it, has been a persistent news item since early 2006, a back story about the ramping up of global food prices has been less prominent. This story, which is about finance, investment, stock markets, and profit, is as crucial to understanding the geography of agriculture and the roots of the current food crisis as the farm worker, the crop, and farm land. Market forces—largely in the form of stock market speculation—have been identified by economists and market analysts as significantly implicated in the growing price of food. And yet commodities speculation has received far less media attention than other factors—drought, biofuels popularity, rising new demand, the price of oil, the sinking U.S. dollar—in explaining reduced food and nutrition security for many of the world's poorest people.

The modern stock market is the site where the globe's future food prices are determined. Bids by investors on "commodities futures"—for example, the likely price a crop will yield at the time of harvest—help to shape the price of any commodity when it is finally offered for sale. Market speculation occurs around food when investors buy, hold, and sell agricultural commodities in order to profit from fluctuations in their prices as opposed to buying them for use or income. Food experts and stock market observers see the recent increase in commodities market speculation as a response to the collapse of the sub-prime mortgage market. No longer able to channel investment into the once very profitable real estate sector, investors have turned to commodities—an article or raw material, such as grains or metals, which is interchangeable with another product of the same type and which investors buy or sell, usually through futures contracts.

The International Food Policy Research Institute, in a 2008 publication, points out that investors typically turn to commodities speculation for two reasons. The first is the fact that, historically, commodity investments tend to increase in value when other classes of assets decline. In the second case, many investors believe that the commodity markets are in the midst of a "super cycle," which means that commodity prices will continue to increase for years to come. Commodities, especially food, are seen as relatively safe investments. Whereas many commodities can be foregone in a time of crisis, it would be difficult to stop buying food. Because of pressure from the World Bank and the International Monetary Fund, national governments have deregulated trade in agriculture over the last several years, and it is now much easier for the private sector to invest in a global food market.

But how does speculative investment drive up price? Just as with the real estate bubble of the last 15 years, commodity price inflation feeds on itself. The more prices rise, and big profits are made, the more others invest, hoping for big returns. And as the U.S. dollar and other currencies have been sinking, commodities, including gold, oil, and grains, are seen as "hedges"—investment

Main Points

- Agriculture has been transformed into a globally integrated system; the changes producing this result have occurred at many scales and have had many sources.

- Agriculture has proceeded through three revolutionary phases, from the domestication of plants and animals to the latest developments in biotechnology and industrial innovation.

- The introduction of new technologies, political concerns about food security and self-sufficiency, and changing opportunities for investment and employment are among the many forces that have dramatically shaped agriculture as we know it today.

- The industrialized agricultural system of today's world has developed from—and in much of the world has displaced—older agricultural practices, including shifting cultivation, subsistence agriculture, and pastoralism.

- The contemporary agro-commodity system is organized around a chain of agribusiness components that begin at the farm and end at the retail outlet. Different economic sectors, as well as different corporate forms, have been involved in the globalization process.

- Transformations in agriculture have had dramatic impacts on the environment, including soil erosion, desertification, deforestation, and soil and water pollution, as well as the elimination of some plant and animal species.

Searching for food in a Haitian garbage dump, 2008

297

■ The biggest issues food-policy experts, national governments, consumers, and agriculturalists face revolve around the availability and quality of food in a world where access to safe, healthy, and nutritious foodstuffs is unevenly distributed.

made to reduce risk—against these falling values. In short, as a response to the increasing economic malaise that is emanating from the United States but also moving across the global landscape, investors are looking for new ways to improve their chances of turning a profit when interest rates are below the rate of inflation. In the 1990s, investors did this by moving into Internet stocks until that bubble burst. Then they moved into the property market until that option soured. Funds began to be moved into commodities in search of higher returns in the mid 2000s.

The September 2008 collapse of global financial markets had stock analysts continuing to carefully assess the impact on investment trends to see whether agricultural markets will remain viable. Third-quarter profit statements for the same year for the biggest investors indicate that they are still strong. Still, the most crucial question right now about the effects of speculation on commodities markets is whether an agricultural price bubble may yet be underway. If it is, the consequences of it bursting could be severe for farmers and consumers—and not just poor ones—around the world.

In this chapter, we examine the history and geography of agriculture from the global to the household level. We begin by looking at traditional agricultural practices and proceed through the three major revolutions of agricultural change. Much of the chapter is devoted to exploring the ways geographers have investigated the dramatic transformations in agriculture over the last half century as it has become increasingly industrialized through technological, political, social, and economic forces and the effects the globalization of production has had on producing, marketing, delivering, and consuming food. As the above example makes clear, in order to understand the geography of agriculture, one needs also to take into account a whole host of related factors besides crops and farming.

TRADITIONAL AGRICULTURAL GEOGRAPHY

The study of agriculture has a long tradition in geography. Because of geographers' interest in the relationships between people and land, it is hardly surprising that agriculture has been of primary concern. Research on agriculture is strongly influenced by geography's commitment to viewing the physical and human systems as interactively linked. Such an approach combines an understanding of spatial differentiation, the importance of place, and the fact that practices such as agriculture affect and are affected by local, regional, national, and globally extensive processes. It also provides geographers with a powerful perspective for understanding the dynamics of contemporary agriculture.

One of the most widely recognized and appreciated contributions that geographers have made to the study of agriculture is the mapping of the factors that shape it. They have mapped soil, temperature, and terrain, as well as the areal distribution of different types of agriculture and the relationships among and between agriculture and other practices or variables.

Major changes in agriculture worldwide have occurred in the last four decades. Of these, the decline in the number of people employed in farming in both the core and the periphery is perhaps the most dramatic. Meanwhile, the use of chemical, mechanical, and

biotechnological innovations and applications has significantly intensified farming practices (**Figure 8.1**). Agriculture has also become increasingly integrated into wider regional, national, and global economic systems at the same time that it has become more directly linked to other economic sectors, such as manufacturing and finance (**Figure 8.2**). The repercussions from these profound changes range from the structure of global finance to the social relations of individual households.

By examining agricultural practices, geographers have sought to understand the myriad ways humans have learned to modify the natural world around them to sustain themselves, their kin, and ultimately the global community. In addition to understanding agricultural systems, geographers are also interested in investigating the lifestyles and cultures of different agricultural communities. They and other social scientists often use the adjective **agrarian** to describe the way of life that is deeply embedded in the demands of agricultural production. *Agrarian* not only defines the culture of distinctive agricultural communities but also refers to the type of tenure (or landholding) system that determines who has access to land and what kind of cultivation practices will be employed there.

Agriculture is a science, an art, and a business directed at the cultivation of crops and the raising of livestock for sustenance and profit. The unique and ingenious methods by which humans have learned to

Figure 8.1   Pesticide spraying, Nicaragua Pictured here is a plantation in Nicaragua where this group of workers is getting ready to apply pesticides to the crops.

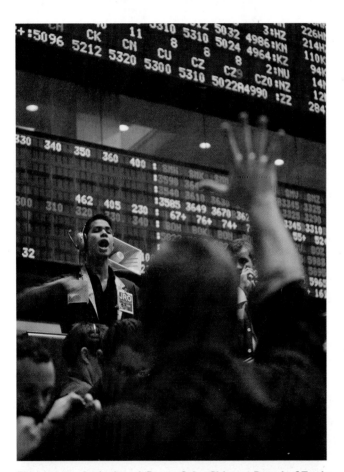

Figure 8.2   Agricultural floor of the Chicago Board of Trade Farming has always been tied up with trading. Pictured here is the main site for the global trading of agricultural commodities. While trade remains an important aspect of agricultural production, it is finance that has had one of the most significant impacts on farming over the last several decades. Agricultural finance provides mortgages and credit for farm expansion, land acquisition, and re-financing of debt and is made available through everything from transnational commercial banks and local credit institutions to federal and international agencies and pension funds.

transform the land through agriculture are an important reflection of the two-way relationship between people and their environments (**Figure 8.3**). Just as geography shapes our choices and behaviors, so we are able to shape the physical landscape. Most introductory textbooks give considerable attention to tracing the origins of agriculture and the distribution of different agricultural practices across the globe. Although agricultural origins are important, the impact of twentieth-century political and economic changes in agriculture are so transformative that in this textbook we focus on the state of global agriculture at the beginning of the new millennium.

While there is no definitive answer as to where agriculture originated, we know that before humans discovered the advantages of agriculture, they procured their food through hunting (including fishing) and gathering. **Hunting and gathering** characterizes activities whereby people feed themselves by killing wild animals and gathering fruits, roots, nuts, and other edible plants. Hunting and gathering are considered subsistence activities, in that people who practice them procure only what they need to consume. Subsistence agriculture replaced hunting and gathering activities in many parts of the globe when people came to understand that the domestication of animals and plants could enable them to settle in one place over time rather than having to go off frequently in search of edible animals and plants (**Figure 8.4**). **Subsistence agriculture** is a system in which agriculturalists consume all they produce. The practice of subsistence agriculture is declining, but it is still practiced in many areas of the globe.

During the twentieth century, the dominant agricultural system in the core countries became **commercial agriculture**, a system in which farmers produce crops and animals primarily for sale rather than for direct consumption by themselves and their families. Worldwide, subsistence agriculture is diminishing as increasing numbers of

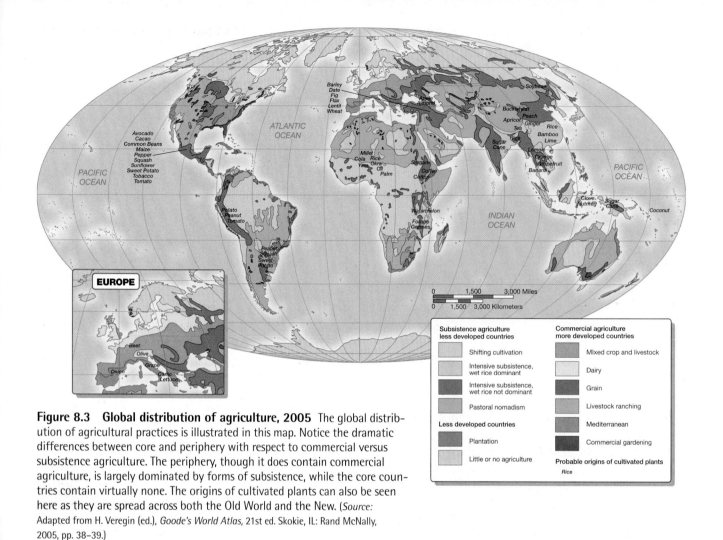

Figure 8.3 Global distribution of agriculture, 2005 The global distribution of agricultural practices is illustrated in this map. Notice the dramatic differences between core and periphery with respect to commercial versus subsistence agriculture. The periphery, though it does contain commercial agriculture, is largely dominated by forms of subsistence, while the core countries contain virtually none. The origins of cultivated plants can also be seen here as they are spread across both the Old World and the New. (*Source:* Adapted from H. Veregin (ed.), *Goode's World Atlas,* 21st ed. Skokie, IL: Rand McNally, 2005, pp. 38–39.)

Figure 8.4 Areas of plant and animal domestication Plant and animal domestication did not predominate in any one continent but was spread out across the globe. The origins of plant and animal domestication, however, are not definitively known, and much of what is represented on this map is speculative. Archaeological evidence to date supports the distribution shown here and developed in the mid-twentieth century by Carl Sauer. (*Source:* Adapted from J. M. Rubenstein, *The Cultural Landscape: An Introduction to Human Geography,* 7th ed., Prentice Hall © 2003, p. 319.)

places are irresistibly incorporated into a globalized economy with a substantial commercial agricultural sector. Still widely practiced in the periphery, however, subsistence activities usually follow one of three dominant forms: shifting cultivation, intensive subsistence agriculture, and pastoralism. Although many people in the periphery rely on these traditional practices to feed themselves, traditional practices are increasingly being abandoned or modified as peasant farmers convert from a subsistence and barter economy to a cash economy.

Shifting Cultivation

In **shifting cultivation**, a form of agriculture usually found in tropical forests, farmers aim to maintain soil fertility by rotating the fields they cultivate. Shifting cultivation contrasts with another method of maintaining soil fertility, **crop rotation**, in which the fields under cultivation remain the same but the crops planted are changed to balance the types of nutrients withdrawn from and delivered to the soil.

Shifting cultivation is globally distributed in the tropics—especially in the rain forests of Central and West Africa; the Amazon in South America; and much of Southeast Asia, including Thailand, Burma, Malaysia, and Indonesia—where climate, rainfall, and vegetation combine to produce soils lacking nutrients. The practices involved in shifting cultivation have changed very little over thousands of years (**Figure 8.5**). As a land rotation system, shifting cultivation requires less energy than modern forms of farming, though it can successfully support only low population densities.

The typical agrarian system that supports shifting cultivation is one in which small groups of villagers hold land in common tenure. Through collective agreement

Figure 8.5 Shifting cultivation Shifting cultivation is usually practiced in tropical forests. It is a system of agriculture that maintains soil fertility by rotating the fields within which cultivation occurs. This photograph shows a plot under cultivation as well as the burned stumps of the trees that used to occupy the site. Corn and bean shoots are scattered throughout the plot.

or a ruling council, sites are distributed among village families and then cleared for planting by family members. As villages grow, tillable sites must be located farther and farther away from their center. When population growth reaches a critical stage, several families within the village normally split off to establish another village in one of the more remote sites.

Because tropical soils are poor in nutrients, the problem of the rapid depletion of soil fertility through cultivation means that fields are actively planted for less than 5 years. The biggest culprits in soil depletion are the cultivated plants and the heavy tropical rains that draw off and wash out the few nutrients that are present in the soil. Once the soil nears exhaustion, a new site is identified and the process of clearing and planting, described in the next paragraph, begins again. It may take over two decades for a once-cleared and cultivated site to become tillable again, after decomposition returns sufficient organic material to the soil. When this occurs, the site is reintegrated into cultivation.

The typical method for preparing a new site is through **slash-and-burn** agriculture, in which existing plants are cropped close to the ground, left to dry for a period, and then ignited. The burning process adds valuable nutrients to the soil, such as potash, which is about the only readily available fertilizer for this form of agricultural practice. Once the land is cleared and ready for cultivation, it is known as **swidden** (see Chapter 2). While generally an agricultural practice that is workable when undertaken by small populations on limited portions of land, slash-and-burn is also seen to be ecologically destructive when large numbers of farmers are involved, especially in forests where vulnerable and endangered species occur, as in rain forests (**Figure 8.6**). For example, the use of slash-and-burn techniques to plant marijuana, coca, and opium poppies in Columbia since the late 1990s has led to significant devastation in the country's portion of the Amazon rain forest.

The practice of shifting cultivation usually occurs without the aid of livestock or plow to turn the soil. Thus, this type of agriculture relies largely on human labor, as well as extensive acreage for new plantings because old sites are abandoned frequently when soil fertility is diminished. Although a great deal of human labor is involved in cutting and clearing vegetation, once the site is planted, there is little tending of crops until harvest time.

From region to region the kinds of crops grown and their arrangement in the swidden vary depending upon local taste and plant domestication histories. In the warm, humid tropics, tubers—sweet potatoes and yams—predominate, while grains such as corn or rice are more widely planted in the subtropics. The practice of mixing different seeds and seedlings in the same swidden is known as **intertillage** (**Figure 8.7**). Not only are different plants cultivated but their planting is usually staggered so that harvesting can continue throughout the year. Such staggered planting and harvesting reduces the risk of disasters from crop failure and increases the nutritional balance of the diet.

Figure 8.6 Slash-and-burn, Xishuangbanna, China Slash-and-burn is a process of preparing low-fertility soils for planting. In the mid-twentieth century, the Xishuangbanna region of China pictured here was a remote tropical rain forest, China's only one. Since then, the government and people of Xishuangbanna have begun to convert the area to rubber tree and sugar cane plantations. While the transformations have lifted much of the region out of abject poverty, soil erosion is occurring rapidly. Some estimates indicate that Xishuangbanna will be a dust bowl in 30 years.

Shifting cultivation also involves a gender division of labor that may vary from region to region (**Figure 8.8**). For the most part, men are largely responsible for the initial tasks of clearing away vegetation, cutting down trees, and burning the stumps. Women are usually involved with sowing seeds and harvesting crops. Research on shifting cultivation indicates that the actual division of tasks between men and women (and sometimes children) results from traditional cultural practices, as well as the new demands placed upon households by globalization (recall the discussion of Sudanese children in Chapter 6). For instance, many women have found it necessary to complement their subsistence agricultural activities with craft production for local tourist markets. Their absence from routine agricultural activities means those tasks must be taken up by other household members.

Although heralded by many as an ingenious, well-balanced response to the environmental constraints of the tropics and subtropics, shifting cultivation is not without limitations. Its most obvious limitation is that it can be effective only with small populations. Increasing populations cause cultivation sites to be located farther from villages, with the result that cultivators expend as much energy traveling to sites as they garner from the crops they produce. Indeed, at any one time, it is not unusual for land closest to the village to be entirely fallow or unseeded because the soil is exhausted from previous plantings.

Increasingly, population pressures and ill-considered government policies are undermining the practicality of

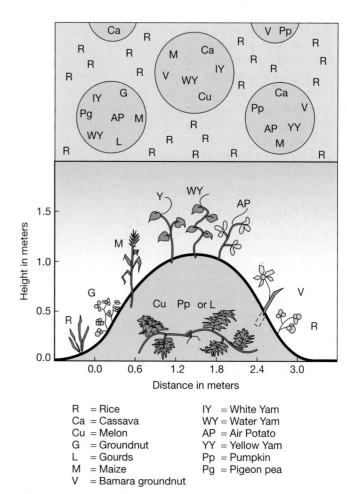

R	= Rice	IY	= White Yam
Ca	= Cassava	WY	= Water Yam
Cu	= Melon	AP	= Air Potato
G	= Groundnut	YY	= Yellow Yam
L	= Gourds	Pp	= Pumpkin
M	= Maize	Pg	= Pigeon pea
V	= Bamara groundnut		

Figure 8.7 Intertillage Planting different crops together in the same field has many benefits, not the least of which are the spreading out of food production over the farming season, reduction of disease and pest loss, greater protection from loss of soil moisture, and control of soil erosion. This diagram provides an illustration of what an intertilled site might contain and how the planting is arranged. Hill-planted seeds have tall stalks and a deeper root system; those planted on flat earth tend to be spreading plants that produce large leaves for shading.

shifting cultivation, resulting in irreparable damage to the environment in many parts of the world. In Central and South America, for example, national governments have used rural resettlement programs to address urban population pressures. In some cases, individuals not familiar with shifting cultivation techniques have employed them improperly. In others, individuals have been relocated to areas unsuitable for such cultivation practices. In parts of the Brazilian Amazon, for example, shifting cultivators have acted in concert with cattle grazers, resulting in accelerated environmental degradation.

Despite its negative impacts on the environment, shifting cultivation can be an elegant response to a fragile landscape. The fallow period, which is an essential part of the process, is a passive and perfectly effective way of restoring plant nutrients to the soil. The burning of stumps and other debris makes the soil more workable, and seeding can proceed with a minimum of effort. Intertillage mimics the natural pattern of differing plant heights and types

Figure 8.8   Gender division of labor Pictured here are women thinning rice plants by hand while a machine plow operated by a man is used to till the earth along the Mekong River, Vietnam. It is not unusual to find women having different access to technology—whether it be agricultural or information—than men in both the core and the periphery. Differential access to technology, training, and jobs helps to constitute a gender division of labor in workplaces as well as at home.

Need: More intensive Agriculture.

characteristic of the rain forest. It also helps protect the soil from leaching and erosion. Shifting cultivation requires no expensive inputs (except possibly where native seeds are not available) because no manufactured fertilizers, pesticides, herbicides, or heavy equipment—mechanical or otherwise—are necessary. Finally, the characteristically staggered sowing allows for food production throughout the year. Although shifting cultivation was likely once practiced throughout the world, population growth and the greater need for increased outputs per acre have led to its replacement by more intensive forms of agriculture.

Intensive Subsistence Agriculture

The second dominant form of subsistence activity is **intensive subsistence agriculture**, a practice involving the effective and efficient use of a small parcel of land in order to maximize crop yield; a considerable expenditure of human labor and application of fertilizer are also usually involved. Unlike shifting cultivation, intensive subsistence cultivation often can support large rural populations. While shifting cultivation is more characteristic of low agricultural densities, intensive subsistence normally reflects high agricultural density. Consequently, intensive subsistence usually occurs in the region of the world where agricultural densities are especially high: Asia, and especially India, China, and Southeast Asia.

Whereas shifting cultivation involves the application of a relatively limited amount of labor and other resources to cultivation, intensive subsistence agriculture involves fairly constant human labor in order to achieve high productivity from a small amount of land. With population pressures fierce and the amount of arable land limited, intensive subsistence agriculture also reflects the inventive ways in which humans confront environmental constraints and reshape the landscape in the process. In fact, the landscape of intensive subsistence agriculture is often a distinctive one, including raised fields and hillside farming through terracing (**Figure 8.9**).

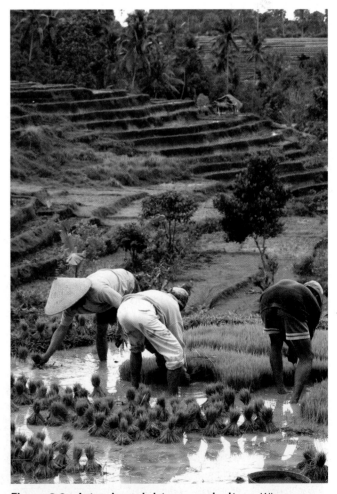

Figure 8.9   Intensive subsistence agriculture Where usable agricultural land is at a premium, agriculturalists have developed ingenious methods for taking advantage of every square inch of usable terrain. Landscapes like this one—of a terraced rice field in Bali, Indonesia—can be extremely productive when carefully tended and can feed relatively large rural populations.

Intensive subsistence agriculture is able to support large rural populations. Unlike shifting cultivation, fields are planted year after year as fertilizers and other soil enhancers are applied to maintain soil nutrients. For the most part, the limitations on the size of plots have more to do with demography than geography. In Bangladesh and southern China, for example, where a significant proportion of the population is engaged in intensive subsistence agriculture, land is passed down from generation to generation—usually from fathers to sons—so that each successive generation, where there are multiple male offspring, receives a smaller and smaller share of the family holdings. Yet each family must produce enough to sustain itself.

Under conditions of a growing population and a decreasing amount of arable land, it is critical to plant subsistence crops that produce a high yield per hectare. Different crops fulfill this need, depending on the regional climate. Generally speaking, the crops that dominate intensive subsistence agriculture are rice and other grains.

Rice production predominates in those areas of Asia—South China, Southeast Asia, Bangladesh, and parts of India—where summer rainfall is abundant. In drier climates and where the winters are too cold for rice production, other sorts of grains—among them wheat, barley, millet, sorghum, corn, and oats—are grown for subsistence. In both situations, the land is intensively used. In fact, it is not uncommon in milder climates for intensive subsistence fields to be planted and harvested more than once a year, a practice known as **double cropping**.

Pastoralism

Although not obviously a form of agricultural production, pastoralism is a third, dominant form of subsistence activity associated with a traditional way of life and agricultural practice. **Pastoralism** involves the breeding and herding of animals to satisfy the human needs for food, shelter, and clothing. Usually practiced in the cold and/or dry climates of savannas (grasslands), deserts, and steppes (lightly wooded, grassy plains), where subsistence agriculture is impracticable, pastoralism can be either sedentary (pastoralists live in settlements and herd animals in nearby pastures) or nomadic (they travel with their herds over long distances, never settling in any one place for very long). Although forms of commercial pastoralism exist—the regularized herding of animals for profitable meat production, as among Basque Americans in the Basin and Range regions of Utah and Nevada and among the gauchos of the Argentinean grasslands—we are concerned here with pastoralism as a subsistence activity.

Pastoralism is largely confined to parts of North Africa and the savannas of central and southern Africa, the Middle East, and central Asia. Pastoralists generally graze cattle, sheep, goats, and camels, although reindeer are herded in parts of Eurasia. The type of animal herded is related to the culture of the pastoralists, as well as the animals' adaptability to the regional topography and foraging conditions (**Figure 8.10**).

As a subsistence activity, nomadism involves the systematic and continuous movement of groups of herders,

Figure 8.10 Pastoralism In this image, sheep forage near the summer settlement of yurts—circular tents of felt or skins on a collapsible framework—at the base of Tsaast Uul mountain in Mongolia, where pastoralism is the main livelihood. Note the dryness of the landscape. Pastoralism usually occurs where agriculture is not feasible.

their families, and the herds in search of forage. Most pastoralists practice **transhumance**, the movement of herds according to seasonal rhythms: warmer, lowland areas in the winter, and cooler, highland areas in the summer. Although the herds are occasionally slaughtered and used directly for food, shelter, and clothing, often they are bartered with sedentary farmers for grain and other commodities. Women and children in pastoralist groups may also be involved with cultivation. They usually split off from the larger group and plant crops at fixed locations in the spring. The women and children may tend the crops throughout the growing season, or they may rejoin the group and return to the fields when the crops are ready for harvesting. The distinguishing characteristic of pastoralists is that they depend on animals, not crops, for their livelihood.

Like the two other traditional forms of agriculture previously mentioned, pastoralism is not simply a subsistence activity but part of a social system as well. Pastoralists consist of groups of families who are governed by a leader or chieftain. Groups are divided into units that follow different routes with the herds. The routes are well known, with members of the group intimately conversant with the landscape, watering places, and opportunities for contact with sedentary groups. Not surprisingly, pastoralism as a subsistence activity is on the decline as more and more pastoralists have become integrated into a global economy that requires more efficient and regularized forms of production. They have also been forced off the land by competition from other land uses and the state's need to track citizens for taxation and military reasons.

AGRICULTURAL REVOLUTION AND INDUSTRIALIZATION

For a long time human geography textbooks treated the differences in agricultural practices worldwide as systems to be described and cataloged, as we have just done. In the last three decades, however, new conceptual approaches to the agricultural sector have transformed the ways we view it. Agriculture has become less a human activity to be described through classification and more a complex component of the global economic system to be explained. Indeed, while the importance and persistence of traditional agricultural forms are acknowledged, such description must be balanced with an understanding of the ways in which new commercial practices are undermining and otherwise changing the older forms.

Increasingly, geographers and others have come to see world agricultural practices as having proceeded through "revolutionary" phases, just as manufacturing did. As in manufacturing, practices have not been transformed everywhere at the same time; consequently, some parts of the world are still largely unaffected by certain aspects of agricultural change. By seeing agriculture in this new light, we can recognize that, as in manufacturing, the changes that have occurred in agricultural practices have transformed geography and society as the global community has moved from predominantly subsistence to predominantly capital-intensive, market-oriented practices.

To understand the new agricultural geography, it is necessary to review the history of world agriculture. This history has proceeded in alternating cycles: long periods of very gradual change punctuated by short, explosive periods of radical change. Geographers and others have divided the history of world agriculture into three distinct revolutionary periods.

The First Agricultural Revolution

The first agricultural revolution is commonly recognized as having been founded on the development of seed agriculture and the use of the plow and draft animals (**Figure 8.11**). Aspects of this transformation have been discussed in Chapters 2 and 4. The emergence of seed agriculture through the domestication of crops such as wheat and rice,

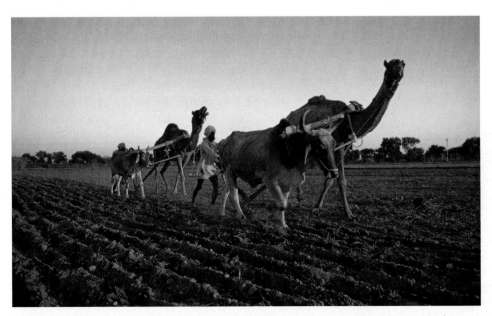

Figure 8.11 Plowing with yoked oxen and camels In many parts of the world, agriculturalists rely on draft animals to prepare land for cultivation. Using animals to assist in agricultural production was an important element in the first agricultural revolution. By expanding the amount of energy applied to production, draft animals enabled humans to increase food supplies. Many contemporary farmers view draft animals as their most valuable possessions. Pictured here are camels and oxen pulling the plows of Sikh farmers in Punjab State, India.

and animals such as sheep and goats, replaced hunting and gathering as a way of living and sustaining life. In fact, seed agriculture occurred during roughly the same period in several regions around the world. The result was a broad belt of cultivated lands across Southwest Asia, from Greece in the west into present-day Turkey and part of Iran in the east, as well as in parts of Central and South America, northern China, northeast India, and East Africa.

The domestication of plants and animals allowed for the rise of settled ways of life. Villages were built, creating different types of social, cultural, economic, and political relationships than those that dominated hunter-gatherer societies. Especially important were floodplains along the Tigris, Euphrates, and Nile rivers, where complex civilizations were built upon the fruits of the first agricultural revolution (**Figure 8.12**). Over time, the knowledge and skill underlying seed agriculture and the domestication of plants diffused outward from these original areas, having a revolutionary impact throughout the globe.

The Second Agricultural Revolution

A great deal of debate exists among historians as to the timing and location of the second agricultural revolution. Though most historians agree that it did not occur

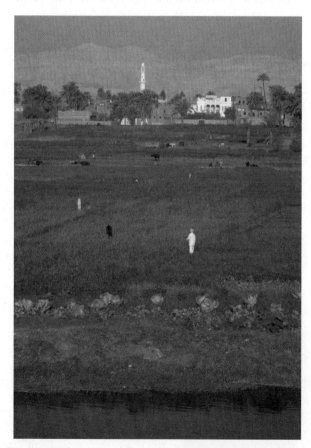

Figure 8.12   Present-day agriculture along the Nile
Agriculture along the Nile River dates far back into prehistory. The Nile floodplain was one of the important cultural hearths for sedentary agriculture, providing the foundation for the growth of complex civilizations in Egypt. The Nile floodplain remains a remarkably productive area.

everywhere at the same time, they disagree over which elements were essential to the fundamental transformation of subsistence agriculture. Important elements included:

- dramatic improvements in outputs, such as crop and livestock yields;

- such innovations as the improved yoke for oxen and the replacement of the ox with the horse; and

- new inputs to agricultural production, such as the application of fertilizers and field drainage systems.

The apex of the second agricultural revolution coincided historically and geographically with the Industrial Revolution in England and Western Europe. Although many important changes in agriculture preceded the Industrial Revolution, none had more of an impact on everyday life than the rise of an industrialized manufacturing sector, the effects of which spread rapidly to agriculture.

On the eve of the Industrial Revolution—in the middle of the eighteenth century—in Western Europe and England, subsistence peasant agriculture was predominant, though partial integration into a market economy was underway. Many peasants were utilizing a crop-rotation system that, in addition to the application of natural and semiprocessed fertilizers, improved soil productivity and led to increased crop and livestock yields. Additionally, the feudal landholding system was breaking down and yielding to a new agrarian system based not on service to a lord, but on an emerging system of private-property relations. Communal farming practices and common lands were being replaced by enclosed, individually owned land or land worked independently by tenants or renters.

Such a situation was logical in response to the demands for food production that emerged from the dramatic social and economic changes accompanying the Industrial Revolution. Perhaps most important of all these changes was the development—through the creation of an urban industrial workforce—of a commercial market for food. Many innovations of the Industrial Revolution, such as improvements in transportation technology, had substantial impacts on agriculture. Innovations applied directly to agricultural practices, such as the new types of horse-drawn farm machinery, improved control over—as well as the quantity of—yields.

By helping to usher in the second agricultural revolution, the Industrial Revolution changed rural life as profoundly as the sedentary requirements of seed agriculture had transformed a hunting and gathering society. As geographer Ian Bowler writes, this revolution moved rapidly from Europe to other parts of the world:

From its origins in Western Europe, the new commercialized system of farming was diffused by European colonization during the nineteenth and twentieth centuries to other parts of the world. A dominant agrarian model of commercial capitalist farming was established, based on a structure of numerous, relatively small family farms. From this period can be traced both the dependence of agriculture on the manufacturing industry for many farm inputs, and the increasing productivity of farm labor, which released large numbers of workers from the land to swell the ranks of factory workers and city dwellers. Moreover, the production of food surplus to domestic demand enabled international patterns of agricultural trade to be established.[1]

The Third Agricultural Revolution

The third agricultural revolution is a fairly recent one that, unlike the previous two, emanates mostly from the New World. Scholars identify the third agricultural revolution as beginning in the late nineteenth century and gaining momentum throughout the twentieth century. Each of its three important developmental phases originated in North America. Indeed, the globalization trends framing all of our discussions in this text are the very same ones that have shaped the third agricultural revolution. The difference between the second and third agricultural revolutions is mostly a matter of degree, so that by the late twentieth century, technological innovations have virtually industrialized agricultural practices. Additionally, a great deal of farming is now undertaken through contract.

Contract farming is becoming an increasingly important aspect of the contemporary agro-food system, whether the products are purchased by multinationals, smaller companies, government agencies, farmer cooperatives, or individual entrepreneurs. In its most recent manifestation, it is an agreement between farmers and processing and/or marketing firms for the production, supply, and purchase of agricultural products by way of a legal agreement. The arrangement requires the firm to provide specified support through, for example, the supply of fertilizer or seeds and the provision of technical advice. The farmer is in turn obliged to produce a specific commodity in quantities and at quality standards determined by the firm. A great deal of agricultural production in the contemporary global system proceeds according to contracts. One example of contract farming is shrimp farming, which occurs mostly in South and Southeast Asia as well as in Mexico, as discussed in Box 8.1, "The Blue Revolution and Global Shrimp." But contract farming occurs with many other commodities, from beef and cotton to flowers, poultry, milk, and vegetables.

Contract farming has been seen by some economic development agencies like the Food and Agricultural Organization and the World Bank as a strategy for maintaining the livelihoods of small farmers marginalized by the growth of more corporate forms of farming. For farmers, contractual arrangements can provide access to services and credit as well as knowledge of new technology. And the price offered for the product can help to reduce risk and uncertainty. On the other hand, there are risks associated with the contract. The most worrisome is the indebtedness that often occurs when farmers are unable to meet the conditions of the contract or when firms fail to honor their agreements.

The three phases of the third agricultural revolution are mechanization, chemical farming with synthetic fertilizers (**Figure 8.13**), and globally widespread food manufacturing. **Mechanization** is the replacement of human farm labor with machines. Tractors, combines, reapers, pickers, and other forms of motorized machines have, since the 1880s and 1890s, progressively replaced human and animal labor inputs to the agricultural production process in the United States (**Figure 8.14**). In Europe, mechanization did not become widespread until after World War II. **Figure 8.15** shows the global distribution of tractors as a measure of the mechanization of worldwide agriculture.

Chemical farming is the application of synthetic fertilizers to the soil—and herbicides, fungicides, and pesticides to crops—to enhance yields. Becoming widespread in the 1950s in the United States, chemical farming diffused to Europe in the 1960s and to peripheral regions of the world in the 1970s. The widespread application of synthetic fertilizers and their impact on the environment is what Rachel Carson wrote about in her pathbreaking book *Silent Spring*, which we discuss in Chapter 4.

Food manufacturing also had its origins in late-nineteenth-century North America. **Food manufacturing** involves adding economic value to agricultural products through a range of treatments—processing, canning, refining, packing, packaging, and so on—occurring off the farm and before the products reach the market (**Figure 8.16**). The first two phases of the third revolution affected inputs to the agricultural production process, whereas the final phase affects agricultural outputs. While the first two are related to the modernization of farming as an economic practice, the third involves a complication of the relationship of farms to firms in the manufacturing sector, which had increasingly expanded into the area of food early in the 1960s. Considered together, these three developmental phases of the third agricultural revolution constitute the industrialization of agriculture.

The Industrialization of Agriculture

Advances in science and technology—including mechanical as well as chemical and biological innovations—have determined the industrialization of agriculture over time. As with industrialization more generally, the industrialization of agriculture has unfolded as the capitalist economic system became more advanced and widespread. We regard **agricultural industrialization** as the process whereby the farm has moved from being the centerpiece of agricultural production to being one part of an integrated multilevel (or vertically organized) industrial process that includes production, storage, processing, distribution, marketing, and retailing. Experts in the study of agriculture

[1]I. Bowler (ed.), *The Geography of Agriculture in Developed Market Economies.* Harlow, England: Longman Scientific and Technical, 1992, pp. 10–11.

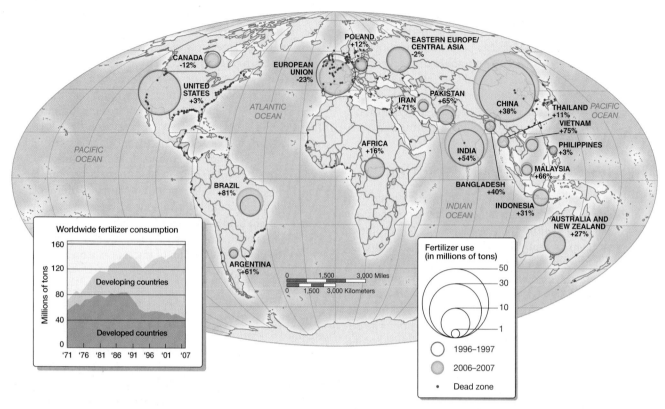

Worldwide fertilizer consumption

Figure 8.13   Worldwide Growth in Fertilizer Use, 2005–2007   As the map indicates, fertilizer use has been growing in peripheral countries faster than in the core, though core countries are still the largest users of fertilizer. One of the biggest problems, and one that is only expected to grow, is increased runoff from fertilizer and resultant dead zones along ocean shores. (*Source:* Adapted from *New York Times*, http://www.nytimes.com/imagepages/2008/04/30/business/20080430_FERTILIZER_GRAPHIC.html, accessed August 17, 2008.)

(b)

Figure 8.14   Old and new farm machines   In the late nineteenth and early twentieth centuries, agriculture in the United States became increasingly linked to mechanization. (a) This photograph shows a young Vassar College student driving a tractor to plow a field in 1917. It helps to illustrate how transformations in personal transportation diffused to agriculture. (b) This photograph shows contemporary harvesting equipment on a farm in the midwestern United States. While both machines are operated by humans, the contemporary machinery relies on computer chips to send and receive information about the multiple operations of the vehicle.

(a)

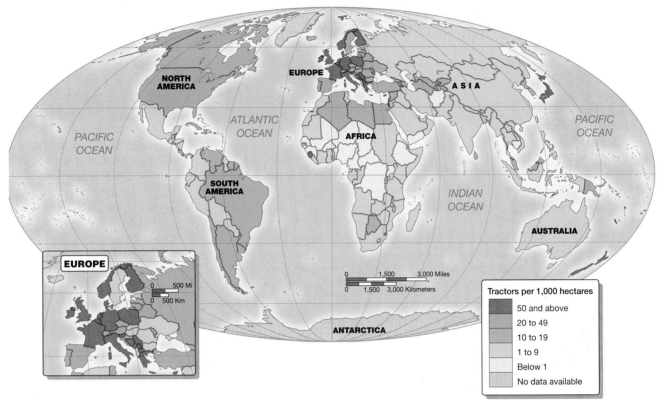

Figure 8.15   Tractors per 1,000 hectares Tractor use, a measure of the mechanization of agriculture, is highest in the core countries. Mechanized farming is an expensive undertaking requiring not only machines but the ability to afford fuels and repairs. Mechanization, however, also allows for more extensive areas of land to be brought into cultivation. (*Source:* Reprinted with permission of Prentice Hall, from J. M. Rubenstein, *The Cultural Landscape: An Introduction to Human Geography,* 6th ed., © 1999, p. 341.)

have come to see it as clearly linked to industry and the service sector, thus constituting a complex agro-commodity production system.

Geographers have helped demonstrate the changes leading to the transformation of an agricultural product into an industrial food product. This transformation has been accomplished not only through the indirect and/or direct altering of agricultural outputs, such as tomatoes or wheat, but also through changes in rural economic activities. Agricultural industrialization involves three important developments:

■ changes in rural labor activities as machines replace and/or enhance human labor;

■ the introduction of innovative inputs—fertilizers and other agrochemicals, hybrid seeds, and biotechnologies—to supplement, alter, or replace biological outputs; and

Figure 8.16   Food manufacturing Pictured here are tomatoes being processed through an assembly-line operation. Food processing is one of the ways that economic value is added to agricultural products before they reach the market.

The Blue Revolution and Global Shrimp

Written by
Brian J. Marks

The first decade of the twenty-first century has been a disastrous one for shrimp fishers across the U.S. Southeast, even before the devastating hurricanes of 2005. The price of shrimp dropped precipitously: Between 2000 and 2003, average Gulf of Mexico dockside prices fell between 38 and 57 percent, depending on shrimp size. Much of this decline occurred in just a few months in 2001. As a result, almost half the Gulf shrimp fleet stopped fishing; ice houses, processing plants, and docks closed; and coastal communities dependent on shrimp for revenues were economically depressed. Fishers responded by intensifying their fishing effort, diversifying their labor into other means of earning income, and replacing hired deckhands on their boats with the unpaid labor of family members (**Figure 8.A**). Some people left the fishery altogether, and others held on, though at a much reduced standard of living.

The cause of this calamity was not because of declining shrimp populations in the Gulf or reduced consumer demand: Shrimp landings have been stable and people in the United States are eating more shrimp than ever before. From 1999 to 2003 alone, per capita consumption increased from 3 to 4 pounds per person, and has almost tripled since 1980.[1] Why, then, the price

[1]Statistics for this piece are drawn largely from the following: National Marine Fisheries Service, *Draft Shrimp Business Options: Proposals to Develop a Sustainable Shrimp Fishery in the Gulf of Mexico and South Atlantic.* Silver Spring, MD: National Oceanic and Atmospheric Administration, 2004; and National Marine Fisheries Service, *Current Fisheries Statistics No. 2004-2: Imports and Exports of Fishery Products—Annual Summary, 2004.* Silver Spring, MD: National Oceanic and Atmospheric Administration.

collapse? To find the answer, one has to look at everything from the workings of the international economy to the microbiology of shrimp. The reasons for the price collapse speak to the global nature of the contemporary food system and how it is complexly affected by economics, politics, ecology, and technology.

Shrimp have been both fished from the ocean and farmed in ponds for hundreds of years. Commercial shrimp fisheries in the United States first developed in Louisiana as an export industry of dried shrimp to Asia and to Asian immigrants in the United States, then into canned products. Around World War II, frozen shrimp began to predominate, tapping into the growing affluence of the U.S. domestic market as more people went out to eat more often and infrastructures for frozen foods (home freezers, refrigerated transport, cold storage warehouses, etc.) both for restaurant service and in homes became commonplace. This paralleled the overall shift in core countries' food systems at that time in terms of household labor (particularly of women), food processing, and production, a process researchers David Goodman and Michael Redclift refer to as "food into freezers: women into factories."[2]

Also after World War II, many countries in Latin America, Asia, and Africa received considerable investment from the United States, Western Europe, and Japan to modernize their economies. Much of this assistance took the form of loans and grants directed at modernizing agriculture (the so-called "Green Revolution" discussed in Box 8.2, pp. 314–316), which brought chemical fertilizers, pesticides, agricultural machinery, and new hybrid strains of crops to farmers.

[2]D. Goodman and M. Redclift, *Refashioning Nature: Food, Ecology, and Culture.* London: Routledge, 1991.

Figure 8.A Louisiana shrimpers
Pictured here are fishers harvesting and sorting wild-caught shrimp near Bayou Lafourche, Louisiana.

Alongside the Green Revolution, came a **"Blue Revolution,"** which was the introduction of motorized and larger boats, processing technology and infrastructure, and new production techniques into peripheral country fisheries. A prominent component of the Blue Revolution is aquaculture, which is the growing of aquatic creatures in ponds on shore or in pens suspended in water (**Figure 8.B**). Both the Green and Blue Revolutions moved primary-sector activities toward a greater dependence on capitalized inputs—like tractors, boats, diesel engines, and petroleum—instead of basic human labor. On the one hand, the Blue and Green revolutions increased food production in many places; on the other, they engendered conflict over how the new practices redistributed power and wealth among producers.

Aquaculture claimed to be an answer to feeding the periphery a cheap form of protein. Advances in farming fish like carp and tilapia in countries like China has increased the availability of fish for millions of people in producing countries. Yet to date, aquaculture has found its biggest economic successes in catering to the demand of affluent consumers in the core for products like shrimp and salmon. Modern shrimp aquaculture both created new shrimp farming industries where none had existed before and revolutionized wild shrimp fisheries and old forms of aquaculture already in place.

The growth of the shrimp trade and aquaculture were rapid. From one-twentieth of the global shrimp supply in 1980, aquaculture amounted to over one-third by 2001. The majority of world shrimp production of all types was exported in 2001, whereas approximately only one-quarter was exported in 1980. The major consumers of shrimp are the United States, Japan, and the European Union (**Figure 8.C**). In 2001, Thailand, India, Indonesia, Vietnam, and Mexico were the top five exporting countries by volume. Shrimp is big business, the most consumed seafood in the United States, and a major source of foreign exchange for many countries.

The global shrimp industry is also a major source of controversy. Fisherpeople and other coastal residents of exporting countries have long denounced shrimp exports, both caught at sea and raised in ponds, for damaging their catches of fish for local consumption, for destroying coastal wetlands and agricultural land to build shrimp ponds, and by using wild fish stocks to make fish meal to feed shrimp. There have been many incidents of violence, even killings, of people opposed to shrimp farming in their communities by shrimp farming interests. The so-called "Pink Gold Rush" of shrimp exports has come with a high social and ecological cost.

The industry also faces another problem: that of its own success. Aquaculture means that the limits on biological production from the ocean have been partially overcome so that production can increase much faster than demand, depressing prices. According to a National Marine Fisheries Service report, by volume world shrimp trade increased by 240 percent between 1980 and 2001, but only by 70 percent in terms of value. Global oversupply relative to demand contributed to the fragility of the shrimp industry in the early 2000s that led to the price collapse felt on the Gulf Coast.

The immediate triggers of the decline in prices were many. In the mid- to late 1990s, the U.S. economy was strong, fueling high demand for shrimp. That demand drove global increases in production of shrimp at the same time that there were stagnant markets in Japan and the European Union. New exporting countries, like Vietnam and Brazil, dramatically accelerated their production to get their share of the profits. An international economic order that stressed earning foreign exchange

Figure 8.B Shrimp farm Phang-Nqa Province, Thailand Shrimp aquaculture is based on the use of ponds where shrimp larvae can be introduced and tended. Many of the ponds are built on coastal farmlands, mangrove swamps, or other wetlands.

Figure 8.C   Leading importers of shrimp, 2004 As this graph makes clear, imported shrimp is a luxury enjoyed almost exclusively by core countries.

in order to service debt and fund development projects was an added incentive to exporting countries to invest heavily in shrimp farming. Shrimp exports to the United States brought in dollars, which were very valuable at

that time relative to other major world currencies, allowing these countries to buy more with the proceeds from each pound of shrimp they sold.

This growth in exports might have continued at stable prices if it were not for the faltering of the U.S. economy beginning in 2001 while demand remained flat in other consuming countries. As U.S. demand declined, shrimp exports had nowhere else to go, and as a result inventories started to pile up and prices began to fall. Other markets were not as desirable to shrimp exporters for a variety of reasons. For example, the U.S. market had no significant tariffs on imported shrimp, unlike Japan and Europe, which imposed between 6 and 12 percent tariffs on many countries.

Global economics does not explain everything that occurred in 2001 in the shrimp industry, however. Part of the answer can be found in changing attitudes about health and diet in the consuming countries. The "low-fat" and "low-carb" diet crazes fit well with increased shrimp consumption. The promotion of seafood as a health food in the reconfiguration of the U.S. diet was a big part of the increase in demand for the product, and as prices fell, shrimp became even more attractive as a part of restaurant menus and frozen-food aisles.

The politics of food safety also played a part. The European Union (EU) had much more stringent food inspection requirements than the United States on imported seafood and detected and rejected numerous shipments of imported shrimp in 2001 and 2002 containing chloramphenicol, a powerful antibiotic used to control outbreaks of disease in shrimp ponds that is potentially dangerous to human health and illegal for use in food. In the wake of the chloramphenicol scare in Europe, much shrimp destined for the EU was rerouted

■ the development of industrial substitutes for agricultural products (Nutrasweet instead of sugar, and artificial thickeners instead of cornstarch or flour, for example).

Recall, however, that the industrialization of agriculture has not occurred simultaneously everywhere throughout the globe. Changes in the global economic system affect different places in different ways as different states and social groups respond to and shape these changes. For example, the use of fertilizers and high-yielding seeds occurred much earlier in core-region agriculture than in the periphery, where many people still farm without them. Beginning in the late 1960s, however, core countries exported a technological combination of fertilizers and high-yielding seeds to regions of the periphery (largely in Asia and Mexico) in an attempt to boost agricultural production. In a development known as the **Green Revolution**, this combination also included

new machines and institutions, all designed to increase global agricultural productivity, as described in Box 8.2, "A Look at the Green Revolution."

Economic crises, a reduction in government programs, and reduced trade barriers have slowed the progress of the Green Revolution in many countries. For example, fertilizer use in countries such as Brazil and Mexico has declined with high prices, fewer subsidies, and increased competition from imported corn and wheat, especially from the United States. Many governments have shifted from giving top priority to self-sufficiency in basic grains to encouraging crops that are apparently more competitive in international trade, such as fruit, vegetables, and flowers (**Figure 8.17**). These **nontraditional agricultural exports (NTAEs)**—new export crops that contrast with traditional exports, such as sugar and coffee—have become increasingly important in areas of Mexico, Central America, Colombia, to the United States, where standards were not as strict and the chances of detection were smaller.

The ecology of shrimp also contributed to the price collapse. Shrimp raised in ponds are susceptible to many forms of disease, which is why antibiotics are sometimes used on farmed shrimp, causing the food safety concerns mentioned above. This is especially true for farmers using "intensive" techniques, which means they stock high densities of shrimp, feed them processed feeds, and use other technological methods to increase production and profitability. Diseases enter ponds from wild-caught or hatchery-raised larval shrimp used to stock ponds and from the effluent from nearby already infected ponds. In an effort to reduce outbreaks caused by local wild-caught larvae, farms increased their use of shrimp from hatcheries. This attempt backfired as hatcheries, that sell larvae internationally, helped spread diseases rapidly by unknowingly exporting infected shrimp to previously unaffected countries. The unpredictable increases and decreases in supply caused by expansion of shrimp farming and disease led to considerable price volatility in the late 1990s. High prices caused by temporary disruptions in supply and strong demand led to waves of new investment in the industry, that flooded the market and caused price declines, leading to a boom–bust cycle that accelerated global overproduction. The year 2000 was one such boom year. Declining production from some exporters due to disease drove prices to very high levels. Many U.S. fishers responded by making new investments in boats, sometimes taking on new debts in order to do so. Shrimp exporters also saw an opportunity to invest and increased production. In 2001, when the U.S. economy began to slow down, disease outbreaks in shrimp ponds had declined, and new exporters entered the market, the result was a crash in the price of shrimp that caught many fishers off guard, leading to the economic hardships that have been experienced since then.

Average nominal Gulf of Mexico shrimp prices were $1.63 in 1980 and $1.43 in 2003. But if we adjust the real 1980 price of shrimp using the CPI, then the price of 1980 shrimp in 2003 dollars was actually $4.40 on average. What this means is that the real buying power of the earnings from shrimp fell by over 60 percent; shrimp fishers were in effect working for less than half the money they were almost a quarter century earlier as cheap imported shrimp flooded the market each year and especially recently. While U.S. imports of shrimp increased by over 50 percent from 345,000 metric tons to 555,299 between 2000 and 2007, the value of that shrimp actually *declined*. Per pound, the price of imports decreased from $5.50 to $3.25 over the same period.[3]

As domestic U.S. shrimp production accounts for barely 15 percent of U.S. shrimp supply, imports will continue to make up the majority of consumption. Stopping imports is not an option. Domestic shrimp fishers need better prices in order to maintain their livelihoods; fishers and farmers abroad likewise need the earnings from exports. Controlling supplies by means of quotas, floor prices, or other means could achieve higher prices for all producers, more stability in the international market, and an ability to address some of the negative consequences to people and the environment from shrimp production.

[3]National Marine Fisheries Service Office of Science and Technology, Fisheries Statistics Division, "Fisheries of the United States 2007," Silver Spring, MD, July 2008, http://www.st.nmfs.noaa.gov/st1/fus/fus07/.pdf, accessed July 19, 2008.

Source: Brian Marks is a Ph.D. candidate in geography at the University of Arizona.

and Chile, replacing grain production and traditional exports, such as coffee and cotton. These new crops obtain high prices but also require heavy applications of pesticides and water to meet export-quality standards and fast refrigerated transport to market. They are vulnerable to climatic variation and to the vagaries of the international market, including changing tastes for foods and health scares about pesticide or biological contamination.

Biotechnology and Agriculture

The final and most dramatic "mini-revolution" we discuss here—along with the Blue and the Green revolutions—is the **Biorevolution,** the genetic engineering of plants and animals with the potential to greatly exceed the productivity improvements of the Green Revolution. Ever since the nineteenth century, when Austrian botanist Gregor Mendel identified hereditary traits in plants and French chemist Louis Pasteur explained fermentation, the manipulation and management of biological organisms has been of central importance to the development of agriculture. The central feature of the Biorevolution is **biotechnology,** any technique that uses living organisms (or parts of organisms) to improve, make, or modify plants and animals or to develop microorganisms for specific uses. Recombinant DNA techniques, tissue culture, cell fusion, enzyme and fermentation technology, and embryo transfer are some of the most-talked-about aspects of the use of biotechnology in agriculture (**Figure 8.18**).

A common argument for applying biotechnology to agriculture is the belief that it helps reduce agricultural production costs, as well as acts as a kind of resource-management technique (where certain natural resources are replaced by manufactured ones). Biotechnology has been hailed as a way to address growing concern for the rising costs of cash-crop production;

Figure 8.17   Commercial flower production, Argentina Pictured here is a field of allium, flowers that produce a blue, lilac, or purple-colored bloom, on a flower plantation near Mendoza, Argentina. In the background are vineyards. Once grown only in home gardens, alliums are becoming commercialized all over the world, from Argentina to New Zealand.

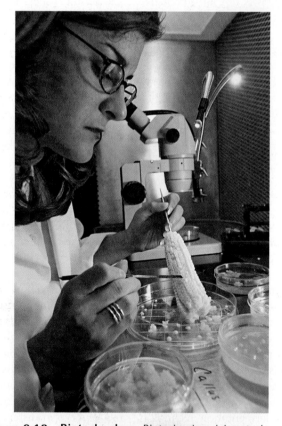

Figure 8.18   Biotechnology Biotechnology laboratories are typically high-technology greenhouses where plants are raised under carefully monitored conditions, from the implanting of special seeds to the applications of fertilizers and water and light. Biotechnology offers both benefits and costs. The benefits include increased yields and more pest-resistant strains, but too often the costs of such technology are too high for the world's neediest populations. In this photograph a Monsanto Corporation worker is extracting corn embryo for the development of a genetically modified crop.

surpluses and spoilage; environmental degradation from chemical fertilizers and overuse; soil depletion; and related challenges now facing profitable agricultural production.

Indeed, biotechnology has provided impressive responses to these and other challenges. One particularly spectacular aspect of biotechnology is **biopharming,** an application of biotechnology in which genes from other life forms (plant, animal, fungal, bacterial, or human) are inserted into the host plant. In short, this means bioengineering plants so that they produce pharmaceuticals. The resulting "pharma crops"—currently in the experimental stage—are expected to include therapeutic proteins; medical and veterinary drugs for treating diarrhea, heart disease, cancer, AIDS, and other illnesses; and vaccines that would be administered by consuming the modified grain or fruit. The engineering of pharma crops is still in the experimental stage and the research is highly confidential. It is estimated that there are over 400 pharma crops currently at the experimental stage, with many of them being anonymously field tested in open-air settings. One of the benefits of pharma crops is the expected lowered cost of medication, though it is too soon to tell how this decrease might work in practice. The major disadvantage seen by farmers is that only a small part of the pharma crops' value is expected to come to them.

While the pharma sector of biotechnology is still in an experimental stage, there are other aspects of biotechnological research that are already fully developed. For example, "super plants" that produce their own fertilizers and pesticides can be grown on nutrient-lacking soils, are high-yielding varieties, and are resistant to disease or the development of microorganisms are now being grown. Additionally, biotechnologists have been able to clone plants—that is, take cells of tissues from one plant and use them to form new plants. The tissue culture may be no more than 1 cubic centimeter in size, but it has the potential to produce millions of identical plants. Such a procedure has decreased the time needed to grow mature plants ready for reproduction.

While such technological innovations can seem miraculous, there is a downside to biotechnological solutions to agricultural problems. For example, cloned plants are more susceptible to disease than natural ones, probably because they have not developed tolerances. This susceptibility leads to an increasing need for chemical treatment. And while industry may reap economic benefits from the development and wide use of tissue cultures, farmers may suffer because they lack the capital or the knowledge to participate in biotechnological applications.

Biotechnology has truly revolutionized traditional agriculture. Its proponents argue that it provides a new pathway to the sustainable production of agricultural commodities. By streamlining the growth process with such innovations as tissue cultures, disease- and pest-resistant and fertilizer-independent plants, optimists believe that the Biorevolution can maximize global agricultural production to keep up with global requirements of population and demand. Moreover, the intensification of agriculture, which the Biorevolution (along with the Green and Blue revolutions) enables, has meant that forest resources are at a reduced risk of being converted into agricultural land. According to this view, known as the **Borlaug hypothesis** (after Norman Borlaug, the force behind the Green Revolution), because global food demand is on the rise, restricting crop usage to traditional low-yield methods (such as organic farming) would require either the world population to decrease or the further conversion of forest land into cropland. There is no sign that world population is on the decline, but the hypothesis—which is controversial—indicates that high-yield biotechnologicial techniques would ultimately save forest ecosystems from destruction.

Just as with the Green Revolution, however, biotechnology may have deleterious effects on peripheral countries (and on poor laborers and small farmers in core countries). For example, biotechnology has enabled the development of plants that can be grown outside of their natural or currently most suitable environment. Yet cash crops are critical to economic stability for many peripheral nations—such as bananas in Central America and the Caribbean, sugar in Cuba, and coffee in Colombia and Ethiopia (**Figure 8.19**). These and other export crops are threatened by the development of alternative sites of production. Transformations in agriculture have ripple effects throughout the world system. As an illustration, **Table 8.1** compares the impacts of the Biorevolution and the Green Revolution on various aspects of global agricultural production.

GLOBAL CHANGE IN FOOD PRODUCTION AND CONSUMPTION

When geographers talk about the globalization of agriculture, they are referring to the incorporation of agriculture into the world economic system of capitalism.

Figure 8.19 Workers in an Ethiopian coffee plantation harvesting beans For many peripheral countries, the production of cash crops is a way to boost exports and bring in needed income for the national economy. In Ethiopia, coffee has for decades been a cash crop grown for export. Luxury exports such as coffee generate some of the capital needed to import staple foods such as wheat.

A useful way to think about the term **globalized agriculture** is to recognize that, as both an economic sector and a geographically distributed activity, modern agriculture is increasingly dependent on an economy and set of regulatory practices that are global in scope and organization.

Forces of Globalization

Several forces, institutions, and organizational forms play a role in the globalization of agriculture. As mentioned, technology, the economy, and politics have played a central role in propelling national and regional agricultural systems into becoming global in scope. One important way these forces of change have been harnessed for global transactions is through new institutions, especially trade and financial organizations. The result has been a virtually unprecedented form of production that is built upon an integrated, globally organized, agro-production system.

As a result of these forces, the globalization of agriculture has dramatically changed relationships among and within different agricultural production systems. Important outcomes of these changed relationships have been the elimination of some forms of agriculture and the erosion or alteration of some systems as they are integrated into the global economy. Two examples include the current decline of traditional agricultural practices, such as shifting cultivation, and the erosion of a national agricultural system based on family farms (**Figure 8.20**).

Agriculture is one part of a complex and interrelated worldwide economic system. Consequently, important changes in the wider economy—whether technological, social, political, or otherwise—affect all sectors, including agriculture. National problems in agriculture, such as production surpluses, soil erosion, and food price stability, affect agriculture as well as other economic sectors globally, nationally, and locally in different ways. The same is true of more global factors—such as the

TABLE 8.1 Biorevolution Compared with Green Revolution

Characteristics	Green Revolution	Biorevolution
Crops affected	Wheat, rice, maize	Potentially all crops, including vegetables, fruits, agro-export crops, and specialty crops
Other sectors affected	None	Pesticides, animal products, pharmaceuticals, processed food products, energy, mining, warfare
Territories affected	Some developing countries	All areas, all nations, all locations, including marginal lands
Development of technology and dissemination	Largely public or quasi-public sector, international agricultural research centers (IARCs); R&D millions of dollars	Largely private sector, especially corporations; R&D billions of dollars
Proprietary considerations	Plant breeders' rights and patents generally not relevant	Genes, cells, plants, and animals patentable as well as the techniques used to produce them
Capital costs of research	Relatively low	Relatively high for some techniques; relatively low for others
Access to information	Restricted due to privatization and proprietary considerations	Relatively easy, due to public policy of IARCs
Research skills required	Conventional plant breeding and parallel agricultural sciences	Molecular and cell biology expertise as well as conventional plant-breeding skills
Crop vulnerability	High-yielding varieties relatively uniform; high vulnerability	Tissue culture crop propagation produces exact genetic copies; even more vulnerability
Side effects	Increased monoculture and use of farm chemicals, marginalization of small farmer, ecological degradation. Increased foreign debt due to decrease in biomass fuels and the increasing reliance on costly, usually imported, petroleum	Crop substitution replacing Third World exports; herbicide tolerance; increasing use of chemicals; engineered organisms might affect environment; further marginalization of small farmer

Source: Adapted from M. Kenney and F. Buttel, "Biotechnology: Prospects and Dilemmas for Third-World Development," *Development and Change 16,* 1995, p. 70; and H. Hobbelink, *Biotechnology and the Future of World Agriculture: The Fourth Resource.* London: Red Books, 1991.

Figure 8.20 Family farm Pictured is an idyllic scene of a prosperous U.S. family farm. Since the third agricultural revolution, the number of family farms in the United States and other core countries has declined dramatically as more corporate forms of farming have emerged. In 1920, about one in every three U.S. citizens lived on farms. By 1978, that number had dropped dramatically, to about 1 in 28. This drastic change in the U.S. farm population has caused some commentators to observe that the family farm in the United States is now just a myth, because many of them have been bought out by larger corporations or have become incorporated themselves. The local food movement, however, has begun to reinvigorate small-scale farming in the United States and elsewhere.

price and availability of oil and other petroleum products critical to commercialized agriculture, the stability of the dollar in the world currency market, and recessions or inflationary tendencies in the economy.

Because of the systemic impact of many problems, integration and coordination of the global economy is needed to anticipate or respond to them. In the last several decades, global and international coordination efforts among states have occurred. These include policies advanced by the World Trade Organization (WTO), as well as the formation of supranational economic organizations such as the European Union (EU) and the Association of SouthEast Asian Nations (ASEAN). However, these new forms of cooperation are not without their opponents, as the protest against the WTO in Seattle in 1999 made very clear (see also Chapter 12).

At the same time that supranational organizations and coordination efforts have been addressing global problems, states continue to be essential in mediating crises at national levels. By changing public policy, the state attempts to regulate agro-industries in order to maintain production, consumption, and corporate profits. One way that states try to maintain the profitability of the agricultural sector while keeping food prices affordable is through direct and indirect subsidies to agricultural producers. For example, the U.S. government subsidizes agriculture in a number of ways. One is by paying farmers not to grow certain crops that are expected to be in excess supply. Another is by buying up surplus supplies and guaranteeing a fixed price for them.

Subsidies, however, while perhaps stabilizing agricultural production in the short term, can lead to problems within the larger national and international agricultural system. For instance, guaranteeing a fixed price for surplus food can act as a disincentive for producers to lower their production, so the problem of overproduction continues. Once in possession of the surplus, governments must find ways to redistribute it. The U.S. government often sells or donates its surplus to foreign governments, where the "dumping" of cheap foodstuffs may undermine the local price structure for food, as well as reduce economic incentives for farmers to farm. Billions of dollars are paid out each year in agricultural subsidies, the effects of which are complex and global in impact.

Many reasons exist for state intervention in agriculture. Economic interests can be both internally and externally driven. At the internal level, governments routinely intervene in one economic sector or another in order, for instance, to correct wider problems of inflation or depression. In the 1930s, the U.S. government attempted to address the problem of economic depression through policies intended to reduce overproduction—policies that had been initiated over a century earlier.

States can also intervene in the agricultural sector with respect to consumers' interests. Because of subsidies to farm income, the real cost of food may be quite high, but many states in both the developed and less developed world also subsidize the price of food in the marketplace. Such policies are meant to keep the workforce well-fed and healthy, as well as to avoid problems of civil unrest should food prices exceed the general population's ability to pay. Nineteenth-century bread riots—a response to the high cost of flour and bread—were a common occurrence in Europe, and similar forms of civil unrest in response to food prices are not uncommon today in many peripheral countries, where states lack the capital to provide adequate subsidies or where subsidies have been eliminated.

In addition to the internal regulation of and assistance to agricultural practices, many countries—especially in the core—are also involved directly and indirectly in the agricultural sectors of other countries, primarily those in the periphery. Food as well as agricultural development aid are widespread and popularly accepted ways in which core states intervene in the agricultural sector of peripheral states. Such intervention is one way that peripheral states are incorporated into the global economy. In addition to straightforward food aid, core states also attempt to improve the capacity of the agricultural sector of peripheral states (**Figure 8.21**). Unsuccessful agricultural development projects—whether because of poor design, implementation, or some other reason—as well as successful ones illustrate the many ways in which global forces produce different local consequences and reactions from local people. International development organizations and institutions like the World Bank and the Food and Agriculture Organization (FAO) have been involved in agricultural development projects in the periphery for five decades.

The Organization of the Agro-Food System

Although the changes that have occurred in agriculture worldwide are complex, examining certain elements

Figure 8.21 Ostrich-rearing project, Kenya
The men pictured here are Masai, who once were nomadic but are now involved in an international development project focused on ostrich rearing and ecosystem management. The ostriches are a rare species that is under threat because of habitat destruction and the bush meat trade.

A Look at the Green Revolution

The Green Revolution was an attempt by agricultural scientists to find ways to feed the world's burgeoning population. The effort began in 1943, when the Rockefeller Foundation funded a group of U.S. agricultural scientists to set up a research project in Mexico aimed at increasing that country's wheat production. Only 7 years later scientists distributed the first Green Revolution wheat seeds. The project was eventually expanded to include research on maize as well. By 1967, Green Revolution scientists were exporting their work to other parts of the world and had added rice to their research agenda (**Figure 8.D**). Norman Borlaug, one of the founders of the Green Revolution, went on to win the Nobel Peace Prize in 1970 for an important component of the project: promoting world peace through the elimination of hunger.

The initial focus of the Green Revolution was on the development of seed varieties that would produce higher yields than those traditionally used in the target areas. However, in developing new, higher-yielding varieties, agricultural scientists soon discovered that plants were limited in the amount of nitrogen they could absorb and use. The scientists' solution was to increase the nitrogen absorption capacity of plants by delivering nitrogen-based fertilizers in water (this led to the need to build major water and irrigation development projects). Then the scientists discovered that the increased nitrogen and water caused the plants to develop tall stalks. The tall stalks, with heavy heads of seed on top, fell over easily, thus reducing the amount of seed that could be harvested. The scientists went back to the drawing board and came up with dwarf varieties of grains that would support the heavy heads of seeds without falling over. Then another problem arose: The short plants were growing in very moist conditions, which encouraged the growth of diseases and pests. The scientists responded by developing a range of pesticides.

Thus, the Green Revolution came to constitute a package of inputs: new "miracle seeds," water, fertilizers, and pesticides. Farmers had to use all of the inputs—and use them properly—to achieve the yields the scientists produced in their experimental plots. Green Revolution crops, if properly watered, fertilized, and treated for pests, can generate yields two to five times larger than those of traditional crops. In some countries, yields are high enough to engage in export trade, thus generating important sources of foreign exchange. Furthermore, the creation of varieties that produce faster-maturing crops has allowed some farmers to plant two or more crops per year on the same land, thus increasing their individual production—and wealth—considerably.

Thanks to Green Revolution innovations, rice production in Asia grew 66 percent between 1965 and 1985. India, for example, became largely self-sufficient in rice and wheat by the 1980s. Worldwide, Green Revolution seeds and agricultural techniques accounted for almost 90 percent of the increase in world grain output in the 1960s and about 70 percent in the 1970s. In the late 1980s and 1990s, at least 80 percent of the additional production of grains could be attributed to the use of Green Revolution techniques. **Figure 8.E** shows the distribution of maize production worldwide. Thus, although hunger and famine persist, many argue that they would be much worse if the Green Revolution had never occurred (**Figure 8.F**).

The Green Revolution has not been an unqualified success, however, everywhere in the world. One important reason is that wheat, rice, and maize are unsuitable as crops in many areas, and research on more suitable crops, such as sorghum and millet, has lagged far behind. In Africa, poor soils and lack of water make progress even more difficult to achieve. Another important factor is the vulnerability of the new seed strains to pest and disease infestation, often after only a couple of years of planting. Whereas traditional varieties often have a built-in resistance to the pests and diseases characteristic of an area, the genetically engineered varieties often lack such resistance.

Another problem is that Green Revolution technology has decreased the need for human labor. In southeastern

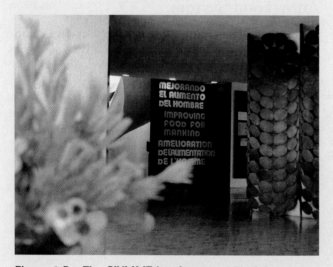

Figure 8.D   The CIYMMT headquarters The Centro International de Mejorimiento de Maiz y Trigo (CIMMYT) (International Center for the Improvement of Maize and Wheat) in Texcoco, Mexico, is involved in plant breeding and research. High-yield-variety seeds were developed here for the Green Revolution. The center holds the world's premiere collection of corn and wheat germplasm. Modern, refrigerated storage vaults store many thousands of varieties. Today, in addition to the static, cold storage, scientists are working with farmers to preserve seeds dynamically.

Figure 8.E Global distribution of maize production The widespread production of grains throughout the globe, particularly maize, has been one of the successes of the Green Revolution. (*Source:* Adapted from J. P. Goode, J. C. Hudson, and E. P. Espanshade, Jr., *Rand McNally's Goode's World Atlas,* 20th ed., Rand McNally, 2000, p. 41.)

Figure 8.F Effects of the Green Revolution This map illustrates the increased yields of protein crops, root crops, other cereals, maize, rice, and wheat brought about by the Green Revolution in selected countries in Latin America, Asia, sub-Saharan Africa, and the Middle East and North Africa. (Data from: R. E. Evenson and D. Gollin, "Assessing the Impact of the Green Revolution, 1960–2000," *Science, 300,* May 2, 2005, p. 759.)

Brazil, machines replaced workers, creating significant unemployment. Green Revolution technology and training have also tended to exclude women, who play important roles in food production. In addition, the new agricultural chemicals, especially pesticides, have contributed to ecosystem pollution and worker poisonings, and the more intensive use of irrigation has created salt buildup in soils (*salinization*) and water scarcity.

Yet another criticism is that the Green Revolution has magnified social inequities by allowing more wealth and power to accrue to a small number of agriculturalists while causing greater poverty and landlessness among poorer segments of the population. In Mexico, a black market developed in Green Revolution seeds, fertilizers, and pesticides when poorer farmers, who were coerced into using them, accrued high debts that they could not begin to repay. Many ended up losing their lands and becoming migrant laborers or moved to the cities and joined the urban poor. Some critics who have monitored the effects of the Green Revolution suggest that political and economic conditions may, in fact, be more important than levels of production with regard to a country's food security.

Even regarding quality, the Green Revolution crops often fall short. The new seed varieties may produce grains that are less nutritious, less palatable, or less flavorful. The chemical fertilizers and pesticides that must be used are derived from fossil fuels—mainly oil—and are thus subject to the vagaries of world oil prices. Furthermore, the use of these chemicals, as well as monocropping practices, has produced worrisome levels of environmental contamination and soil erosion. In many countries these practices have posed substantial threats to public health, especially among farm workers who are frequently exposed to poisonous (if not lethal) chemicals. Water developments have benefited some regions, but less-well-endowed areas have experienced a deterioration of already existing regional inequities. Worse, pressures to build water projects and to acquire foreign exchange to pay for importation of Green Revolution inputs have increased pressure on countries to grow even more crops for export, often at the expense of production for local consumption.

In recent years scientists have endeavored to develop seeds with greater pest and disease resistance and more drought tolerance. The new focus is best revealed in Africa. The International Institute of Tropical Agriculture

in Ibadan, Nigeria, focuses on foods for the humid and subhumid tropics of Africa, including cassava (imported to Africa from South America by the Portuguese in the sixteenth century), yams, sweet potatoes, maize, soybeans, and cowpeas. The International Crops Research Institute for the Semi-Arid Tropics (located in Hyderabad, India, but with a major research center near Niamey, Niger) focuses on researching staples of the Sahel region, such as sorghum, millet, pigeonpea, and groundnut. Research in Africa on new varieties emphasizes testing under very adverse conditions (such as no plowing or fertilizing). New varieties are chosen not just for good yield but because they will provide stable yields over good and bad years. A focus also exists on developing plants that will increase production of fodder and fuel residues, as well as of food, and that give optimal yields when intertilled—a very common practice in Africa. In the Sahel, scientists are working on crops that mature more quickly to compensate for the serious drop in the average length of the rainy season recently experienced in the region.

There are two final criticisms that have raised concern about the overall benefits of the Green Revolution. The first is that it has decreased the production of biomass fuels—wood, crop residues, and dung—traditionally used in many peripheral areas of the world. For example, in India, as tractors have replaced draft animals, less dung is produced and thus less is available as fuel. Instead, a greater reliance is being placed upon oil to fuel both tractors and other energy needs; this means that if farmers are to be successful, they increasingly must depend upon the most costly of energy resources. The second is that the Green Revolution has contributed to a worldwide loss of genetic diversity by replacing a wide range of local crops and varieties with a narrow range of high-yielding varieties of a few crops. Planting single varieties over large areas (monocultures) has made agriculture more vulnerable to disease and pests.

Although the Green Revolution has come under much justified attack over the years, it has focused attention on finding innovative new ways to feed the world's peoples. In the process, the world system has been expanded into hitherto very remote regions, and important knowledge has been gained about how to conduct science and how to understand the role that agriculture plays at all geographical scales of resolution, from the global to the local.

that serve as important indicators of change helps us understand the complexity. Geographers and other scholars interested in contemporary agriculture have noted three prominent and nested forces that signal a dramatic departure from previous forms of agricultural

practice: agribusiness, food chains, and integration of agriculture with the manufacturing, service, finance, and trade sectors. Box 8.3, "The New Geography of Food and Agriculture in New Zealand" illustrates the deployment of these forces in one country.

The concept of agribusiness has received a good deal of attention in the last two decades, and in the popular mind it has come to be associated with large corporations, such as ConAgra or DelMonte. Our definition of agribusiness departs from this popular conceptualization. Although multi- and transnational corporations (TNCs) are certainly involved in agribusiness, the concept conveys more than a corporate form. **Agribusiness** is a system rather than a kind of corporate entity. It is a set of economic and political relationships that organizes food production from the development of seeds to the retailing and consumption of the agricultural product. Defining agribusiness as a system, however, does not mean that corporations are not critically important to the food production process. In the core economies, the transnational corporation is the dominant player, operating at numerous strategically important stages of the food production process. TNCs have become dominant for a number of reasons, but mostly because of their ability to negotiate the complexities of production and distribution in many geographical locations. That capability requires special knowledge of national, regional, and local regulations and pricing factors.

The concept of a food chain (a special type of commodity chain) is a way to understand the organizational structure of agribusiness as a complex political and economic system of inputs, processing and manufacturing, and outputs. A **food chain** is composed of five central and connected sectors (inputs, production, processing, distribution, and consumption) with four contextual elements acting as external mediating forces (the state, international trade, the physical environment, and credit and finance). **Figure 8.22** illustrates these linkages and relationships, including how state farm policies shape inputs, product prices, the structure of the farm, and even the physical environment.

The food chain concept illustrates the complex connections among producers and consumers and regions and places. For example, important linkages connect cattle production in the Amazon and Mexico, the processing of canned beef along the United States–Mexico border, the availability of frozen hamburger patties in core grocery stores, and the construction of McDonald's restaurants in Moscow (**Figure 8.23**). Because of complex food chains such as this, it is now common to find that traditional agricultural practices in peripheral regions have been displaced by expensive, capital-intensive practices.

That agriculture is not an independent or unique economic activity is not a particularly new realization. Beginning with the second agricultural revolution, agriculture began slowly but inexorably to be transformed by industrial practices. What is different about the current state of the food system is the way in which farming has become just one stage of a complex and multidimensional economic process. This process is as much about distribution and marketing—key elements of the service sector—as it is about growing and processing agricultural products.

Food Regimes and Alternative Food Movements

A **food regime** is a specific set of links that exists among food production and consumption and capital investment and accumulation opportunities. Like the agricultural revolutions already described, food regimes have developed out of different historical periods, during which different political and economic forces are in operation. In addition to describing the complex ways in which specific food items are produced, manufactured, and marketed, a food chain indicates the ways in which a particular type of food item is dominant during a specific temporal period. Although hundreds of food chains may be in operation at any one time, agricultural researchers believe that only one food regime dominates a particular period.

During the decades surrounding the turn of the nineteenth century, an independent system of nation-states emerged and colonization expanded (see Chapters 2 and 9). At the same time, the industrialization of agriculture began. These two forces of political and economic change were critical to the fostering of the first food regime, in which colonies became important sources of exportable foodstuffs by supplying the industrializing European states with cheap food in the form of wheat and meat. The expansion of the colonial agriculture sectors, however, created a crisis in production. The crisis was the result of the higher cost-efficiency of colonial food production, which undercut the prices of domestically produced food, put domestic agricultural workers out of work, and forced members of the agricultural sector in Europe to look for new ways to increase cost-efficiency. The response was to industrialize agriculture, which helped to drive down operating costs and restabilize the sector (reducing even more the need for farm workers) while moving toward the integration of agriculture and industry (also known as agro-industrialization).

A wheat and livestock food regime characterized global agriculture until the 1960s; researchers now believe that a fresh fruit and vegetable regime has emerged. This new pattern of food consumption and production has been called the "postmodern diet" because it represents an important addition to the predominance of grains and meats, as the more perishable agro-commodities of fresh fruits and vegetables become central to diets in the core. Integrated networks of food chains, using integrated networks of refrigeration systems, deliver fresh fruits and vegetables from all over the world to the core regions of Western Europe, North America, and Japan. Echoing the former food networks that characterized nineteenth-century imperialism, peripheral production systems supply core consumers with fresh, often exotic and off-season, produce. Indeed, consumers in the core regions have come to expect the full range of fruits and vegetables to be available year-round in the produce sections of supermarkets, and unusual and exotic produce has become increasingly popular.

The food supply chain

Agricultural inputs
Fertilizers
Biotechnology
Plant and Machinery
Agrichemicals
Advisory services
Energy

Farm production
Farm size
Enterprise type
Farmland
Farm labor
Land tenure

Product processing
Washing and grading
Freezing and packing
Transforming
Slaughtering and jointing

Food distribution
Wholesalers
Retailers
Catering trade

Food consumption
Population and growth
Dietary preferences
Purchasing power
Household structure
Employment

Physical environment
Space
Time
Soil
Topography
Climate

Credit/Financial markets
Agricultural market
Mortgages
Clearing banks
Pension funds
Credit companies

State's farm policies
Influence on inputs
Influence on product prices
Influence on farm structures
Fiscal policy
Land inheritance
Environment

International food trade (exports and imports)
Competitive products
Noncompetitive products
Export subsidies
Food aid

Dominant finance and material flows
Dominant power relations

Figure 8.22 The food–supply chain The production of food has been transformed by industrialization into a complex system that comprises distinctly separate and hierarchically organized sectors. Mediating forces (the state, the structure and processes of international trade, credit and finance arrangements, and the physical environment) influence how the system operates at all scales of social and geographical resolution. (*Source:* After I. Bowler (ed.), *The Geography of Agriculture in Developed Market Economies.* New York: J. Wiley & Sons, 1992, p. 12.)

This emergence of a new food regime based on fresh fruit and vegetables has been helped by retailers, who provide symbolic cues and incentives to shoppers to consume the more exotic products. Store managers have introduced them by providing associations between the fruit or vegetable and prevailing ideas about health, class attachment, and epicurean eating. Thus, the transformation of agricultural practices at the global level has enabled

the emergence of a new food regime, accompanied by new cultural messages that promote and persuade at the local level. Furthermore, just as traditional agricultural practices worldwide have been affected by globalization trends, so too are the mainstream eating habits of consumers in core as well as peripheral regions.

Alongside the emergence of a global food chain of fresh fruits and vegetables, it is important to note an

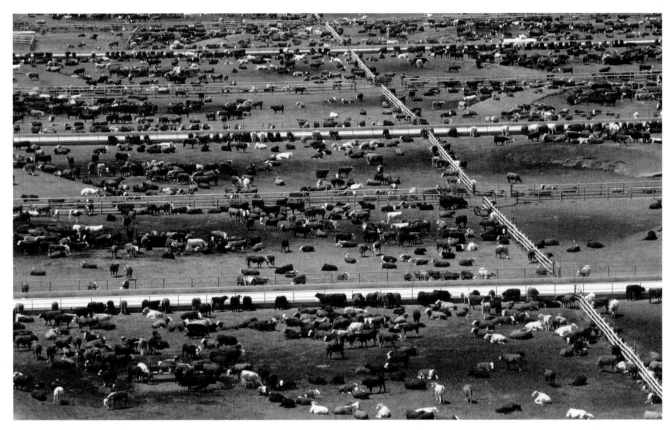

Figure 8.23   World's largest cattle feedlot With over 120,000 head, this feedlot near Greeley, Colorado is a subsidiary of the food giant, ConAgra.

additional aspect of food production practices that has been taking hold in core regions over the last 25 years and accelerating especially over the last ten. One goal of sustainability in agriculture has been increased commitment to organic crops, in terms of both protein (such as beef, eggs, soy) as well as fruits and vegetables. **Organic farming** describes any farming or animal husbandry that occurs without commercial fertilizers, synthetic pesticides or growth hormones. It is important to point out that organic food production is not the primary mode of practice in the global agro-food system, but rather that it has become a growing force alongside the dominant **conventional farming** (an approach that uses chemicals in the form of plant protectants and fertilizers, or intensive, hormone-based practices in breeding and raising animals) (**Table 8.2**). We discuss issues of food safety in more detail later in the chapter, but safety issues as well as overall health concerns and an evolving commitment to food as a gastronomic experience (the idea of relating food to science, society, and culture) among mostly white, middle-class households is at the center of a number of alternative movements that are driving organically produced food, local food, and slow food.

Local food is usually also organically grown, but its designation as local means that it is produced within a fairly limited distance from where it is consumed. Most understandings of local food set a 100-mile radius as the border of what it means for it to be truly local. Thus, if

one is a "locavore," the food one consumes should be produced on farms that are no further than 100 miles from the point of distribution. Local food movements have resulted in the proliferation of communities of individuals who have joined together to support the growth of new farms within the local "food shed" (the area within the 100-mile radius). These individuals have been behind a movement known as Community Supported Agriculture, or CSA. The CSA operates by collecting subscriptions and then paying a farmer up-front to grow food locally for them for a season or a specified time frame. The farmer is then able to purchase seeds, hire workers, cultivate produce and livestock, and deliver the harvest without relying on interest-bearing loans. The CSA members, in return for their investment in the farm, receive weekly shares of produce (and sometimes meats, eggs and cheese, flowers, and milk) that reflect the season and the constraints and possibilities of local growing conditions. Thus, while a CSA in Tucson, Arizona, might receive verdolagas (a green that is native to the region) in additon to squash, snow peas, beets, and carrots, they are unlikely to ever receive a pineapple or celery because neither can be grown naturally in the desert. CSA farmers are expected to grow their food organically and members accept what the farmers grow at the same time that they assume the financial risks that are part of the vagaries of any agricultural enterprise. CSA farms are usually small, independent, and labor-intensive, and the

TABLE 8.2 Conventional vs. Alternative Food

Conventional	Alternative
Cheap food to most of the Western world, where the amount of money spent on food in the household budget has steadily decreased.	Logic of sustainable development, not mass production.
Quantity over quality where agriculture is part of consumption; food is a commodity like any other.	The idea of (real or idealized) "quality" predominates—expressed in such parameters as farmers' markets, CSAs, free-range, organic, GMO-free, integrated pest-management schemes, authenticity, traceability.
Fast, convenient foods: for nutritionists, the year 2000 was significant since it was the first time the number of overweight people matched the number of undernourished people—at 1.1 billion.	Slow, pleasurable food and consumers with an ethical consciousness; a new moral economy surfaces where issues of provenance take a central role and consumers demand knowledge of where their food comes from, whether it contains additives, and whether pesticides were used in production.
"Hard Power"—retailer-led supply chains, emasculation of primary producers as supermarkets drive down costs to below that of even production.	"Soft Power"—food sells itself through ethical persuasion and the new "moral economy."
Corporate capital, combined with science and technology enable large-scale food processing and a reliance on large retailers—at the expense of small, independent farms.	Social, as opposed to corporate capital. Direct farmer-to-consumer interaction, relational, trust-based, sense of "community," and attempts to restore small farm enterprises.
Food is "liberated" from nature; technological fixes predominate.	Food is relocated in specific ecological niches and the externalized costs of the conventional system—such as soil degradation, water pollution, animal welfare issues, and health care costs—are addressed.
Logic of DETERRITORIALIZATION—where geographies are a result of the intensification and stretching of the links and networks between production and consumption.	Logic of RETERRITORIALIZATION—where geographies are shaped by the resurgence of local and regional practices as central forces in new articulations of production and consumption.
Geographic specialization, monoculture, and spatial homogenization in North American and Europe produce mass, standardized products.	The "local" as a set of social networks and cultural contexts that are harnessed in place—which provide the foundation for development, innovation, and economic relations.
Placeless landscape of food production (U.K., U.S.) versus place-based landscapes (France, Italy)—where region-specific foods are much more prevalent and provenance is understood and as a selling point.	Transparent food chain, traceable, where producers attempt to regain power against the conventional price-driven "race to the bottom" and consumers' knowledge of provenance helps to produce local networks and promotion of social capital.

CSA movement is seen to be helping to restore the family farm to the national landscapes (where they have been disappearing for 25 years as more corporate forms of farming have taken hold). The CSA movement originated in Japan but it has spread from there to other core regions, including the North America and Europe.

Europe has contributed its own focus to alternative food production and consumption through the slow food movement. **Slow food** (see also Chapter 6), as the name suggests, is an attempt to resist fast food by preserving the cultural cuisine and the associated food and farming of an ecoregion. Started in Italy in the mid-1980s, the Web site of the Slow Food headquarters—there are over 85,000 chapters (known as convivia)—declares the following philosophy:

We believe that everyone has a fundamental right to pleasure and consequently the responsibility to protect the heritage of food, tradition and culture that make this pleasure possible. Our movement is founded upon this concept of eco-gastronomy—a recognition of the strong connections between plate and planet.

Slow Food is good, clean and fair food. We believe that the food we eat should taste good; that it should be produced in a clean way that does not harm the environment, animal welfare or our health; and that food producers should receive fair compensation for their work.

We consider ourselves co-producers, not consumers, because by being informed about how our food is produced and actively supporting those who produce it, we become a part of and a partner in the production process. (Slow Food, http://www.slowfood.com/welcome_eng.lasso, accessed August 11, 2008.)

Although slow food, local food and CSAs, and organic and ecologically sustainable agricultural practices are movements that signal a shift in food production and consumption, they in no way challenge the dominance of more conventionally produced, distributed, marketed, and consumed food. Moreover, as a number of critics of the movements have pointed out, these alternative practices are largely organized and promoted by white, middle-class members of core regions and exclude, often simply through cost and associated accessibility, poor people of color everywhere. The latter are denied these purportedly healthier eating opportunities because they lack the knowledge, income, and proximity to access them. In core countries, the result is that more than any other economic class, poor people turn to cheap, easily accessible food, also known as fast food.

Fast Food

Fast food was born in the United States as a product of the post–World War II economic boom and the social, political, and cultural transformations that occurred in its wake. The concept of **fast food**—edibles that can be prepared and served very quickly, sold in a restaurant and served to customers in packaged form—actually preceded the word for it. What made fast food *fast*, was the adoption of industrial organizational principles applied to food preparation in the form of the Speedee Service System. The system was invented by Richard and Maurice McDonald in their San Bernardino, California, McDonald Brothers Burger Bar Drive-In in 1948, and it revolutionized the restaurant business. Looking to shed the labor and material costs of standard restaurant food preparation, each Burger Bar worker was assigned one task in an assembly-line operation—cooking the burger, placing it on the bun, or packaging it for take-away, for instance—and the product was standardized so that the same condiments—ketchup, onions, mustard, and two pickles—were added to each patty. As Eric Schlosser writes in his *New York Times* best-selling book *Fast Food Nation*, once Raymond Kroc came along in 1954 and convinced the McDonald brothers to agree to franchise (licensing of trademarks—the golden arches—and methods of doing business), the fast-food concept began to expand all over the United States, so that by 1951, *Merriam-Webster's Dictionary* included it as a new word; and by 1964, there were 657 McDonald franchises. At the same time, other enterprising individuals seized the fast-food idea and opened hundreds of Wendy's, Carl's Jr., Kentucky Fried Chicken, and Taco Bell restaurants.

Today, fast food is so ubiquitous in the United States—and is becoming increasingly so in core areas throughout the world—that it is taken for granted by many consumers. But health practitioners, alternative food activists, journalists, and others have begun to turn a critical eye on fast food in order to expose some of its dietary, labor-related, and ecological shortcomings. Along with Schlosser's book and other cultural products like the independent film about the link between obesity and fast food, *Super Size Me* (2004), the underbelly of fast food has become a widely popularized issue. To appreciate just how ubiquitous fast food is, consider the following statistics published by the National Restaurant Association. In the United States in 2007 there were 945,000 fast-food locations nationwide. Serving over 70 billion meals or snacks, their annual sales totaled $558,000 billion. Over 13 million workers are employed in the fast-food industry, making it one of the largest private-sector employers in the country. Every day, 1 in 4 Americans visits a fast-food establishment. French fries are the most eaten vegetable in the country and McDonald's feeds more than 46 million people a day—more than the entire population of Spain.

The concerns about the effects of fast food on the nation's health have risen as our consumption of (or, as some health experts would argue, addiction to) fast food increases and our levels of routine exercise decrease (**Figure 8.24**). Specific worries are focused on obesity (an amount of fatty body mass that is a danger to health) and diabetes (a disorder that affects the human body's use of food for energy), particularly among children. Since the late 1970s, the rate of obesity and diabetes among U.S. children has doubled. In a 2005 study published in the prestigious medical journal *Lancet*, that followed children's eating habits for 15 years, researchers found that study subjects who consumed fast food two or more times a week gained approximately 10 more pounds over the 15-year period and had twice as great an increase in insulin resistance than subjects who consumed fast food less than once per week. These numbers, though certainly alarming, are not surprising given how heavily the fast-food industry advertises to children, a large percentage of whom can recognize the golden arches of McDonald's before they know their own name.

It is important to understand that diseases related to the increased consumption of fast food and many prepared, processed foods—energy-dense, nutrient-poor foods with high levels of sugar and saturated fats—as well as reduced physical activity, have led to adult obesity rates that have risen threefold or more since 1980 across North America, the United Kingdom, Eastern Europe, the Middle East, the Pacific Islands, Australasia, and China. Moreover, the obesity epidemic—as it has come to be called—is not restricted to industrialized societies; this increase is often faster in rapidly developing peripheral countries than in the core. In offering an explanation for the problem, the World Health Organization writes:

> The rising epidemic reflects the profound changes in society and in behavioral patterns of communities over recent decades. While genes are important in determining a person's susceptibility to weight gain, energy balance is determined by calorie intake and physical activity. Thus societal changes and worldwide nutrition transition are driving the obesity epidemic. Economic growth, modernization, urbanization and globalization of food markets are just some of the forces thought to underlie the epidemic.
>
> As incomes rise and populations become more urban, diets high in complex carbohydrates give way to more varied diets with a higher proportion of fats, saturated fats and sugars. At the same time, large shifts towards less physically demanding work have been observed worldwide. Moves towards less

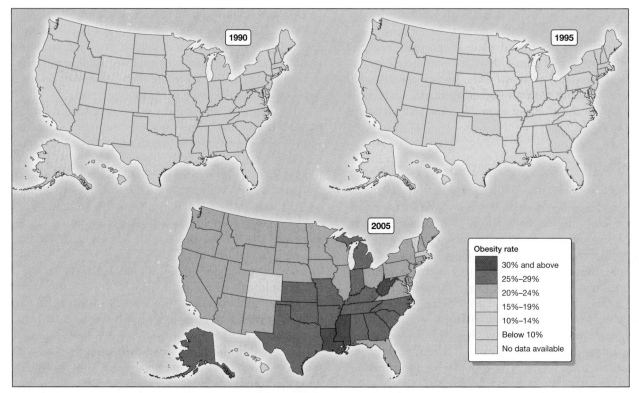

Figure 8.24   U.S. obesity epidemic, 1990, 1995, and 2005 These maps show change over time as well as the geographic variation in obesity among Americans based on a calculation called the Body Mass Index (BMI) that calculates weight with respect to height into a percentage. Individuals with a BMI of 30 per cent or higher are said to be obese.

physical activity are also found in the increasing use of automated transport, technology in the home, and more passive leisure pursuits. (World Health Organization, http://www.who.int/dietphysicalactivity/publications/facts/obesity/en/, accessed August 11, 2008.)

Obesity is of concern to the international health community because it poses a major risk for serious diet-related chronic diseases, including type 2 diabetes, cardiovascular problems, hypertension and stroke, and certain forms of cancer. The consequences range from increased risk of premature death, to grave chronic conditions that reduce the overall quality of life.

Besides the impact on human health, the rapacious growth of fast-food production and distribution processes is having dramatic effects on Earth's resources, especially forests and farm land. The Beyond Beef Campaign, an international coalition of environment, food safety, and health activist organizations, lists the following facts about beef production and its environmental impacts.

■ Beef raised for fast-food hamburgers is one of the largest components of the fast-food sector.

■ One quarter-pound of hamburger beef imported from Latin America requires the clearing of 6 square yards of rain forest and the destruction of 165 pounds of living matter including 20 to 30 different plant species, 100 insect species, and dozens of bird, mammal, and reptile species.

■ Cattle degrade the land by stripping vegetation and compacting the earth. Each pound of feedlot steak costs about 35 pounds of eroded topsoil.

■ Nearly half of the total amount of water used annually in the United States goes to grow feed and provide drinking water for cattle and other livestock. Producing a pound of grain-fed steak requires almost 1200 gallons of water.

■ Cattle produce nearly 1 billion tons of organic waste each year. The average feedlot steer produces more than 47 pounds of manure every twenty-four hours.

■ Much of the carbon dioxide released into the atmosphere is directly attributable to beef production: burning forests to make way for cattle pasture and burning massive tracts of agricultural waste from cattle feed crops.

■ U.S. cattle production has caused a significant loss of biodiversity on both public and private lands. More plant species in the United States have been eliminated or threatened by livestock grazing than by any other cause. (Beyond Beef, http://www.mcspotlight.org/media/reports/beyond.html#9, accessed August 11, 2008.)

The impacts of fast-food production and consumption on people and places are complex, from the convenience for busy consumers and the provision of play spaces for children to the elimination of forests and the debilitating health effects of overconsumption.

How we manage these effects and address the causes will have wide-ranging implications for people and the planet.

The availability of technology to peripheral nations is limited because most advances in biotechnology are the property of private companies. For example, patents protect both the process and the end products of biotechnological techniques. Utilizing biotechnological techniques requires paying fees for permission to use them, and the small farmers of both the core and the periphery are unlikely to be able to purchase or utilize the patented processes. The result of private ownership of biotechnological processes is that control over food production is removed from the farmer and put into the hands of biotechnology firms. Under such circumstances it becomes possible for world food security to be controlled not by publicly accountable governments, but by privately held biotechnology firms. Finally, with the refinement and specialization of plant and animal species, laborers who are currently employed in ancillary activities could face the loss of their jobs. For example, if a grower chooses to plant a bioengineered type of wheat that does not require winnowing (the removal of the chaff, a normally labor-intensive process), laborers who once were involved in that activity are no longer needed.

The biorevolution in agriculture is so recent that we are just beginning to understand both its negative and positive impacts. At this point, it seems quite clear that these impacts will be distributed unevenly across countries, regions, and places and certainly across class, race, and gender lines.

THE ENVIRONMENT AND AGRICULTURAL INDUSTRIALIZATION

Agriculture always involves the interaction of biophysical as well as human systems. In fact, this relationship makes agriculture distinct from forms of economic activity that do not depend so directly on the environment. This relationship also requires determining how best to manage the environment in order to facilitate the continued production of food. Because the relationships between the human system of agriculture and the biophysical system of the environment are highly interactive, it is important to look at the ways each shapes the other.

The Impact of the Environment on Agriculture

Farmers have increasingly managed the environment over the course of the three agricultural revolutions. In fact, the widespread use of fertilizers, irrigation systems, pesticides, herbicides, and industrial greenhouses suggests that agriculture has become an economic practice that can ignore the limitations of the physical environment (**Figure 8.25**). Yet it is exactly because agriculture

Figure 8.25   Modern irrigation system A self-propelling irrigation system can be electronically programmed to deliver different amounts of water at different times of the day or days of the week. Irrigation is just one way that humans can alter the environment to serve their agricultural needs. In many parts of the core, water prices are heavily subsidized for agricultural users in order to ensure food supplies. For many parts of the periphery, however, access to water is limited to the amount of rain that falls and can be stored behind small dams and in impoundments. (*Source:* After Agricultural Research Service, USDA.)

is an economic activity that management of the environment in which it occurs becomes critical. As geographer Martin Parry writes:

> Soil, terrain, water, weather and pests can be modified and many of the activities through the farming year, such as tillage and spraying, are directed toward this. But these activities must be cost-effective; the benefits of growing a particular crop, or increasing its yield by fertilizing, must exceed the costs of doing so. Often such practices are simply not economic, with the result that factors such as soil quality, terrain and climate continue to affect agriculture by limiting the range of crops and animals that can profitably be farmed. In this way the physical environment still effectively limits the range of agricultural activities open to the farmer at each location.[2]

Though the impact of the environment on industrialized agricultural practices may not at first seem obvious, the reverse is more readily observable. In fact, there are many contemporary and historical examples of the ways that agriculture destroys, depletes, or degrades the environmental resources on which its existence and profitability depend.

[2]M. Parry, "Agriculture as a Resource System," in I. Bowler (ed.), *The Geography of Agriculture in Development Market Economics*. Harlow, England: Longman Scientific and Technical, 1992, p. 208.

8.3 WINDOW ON THE WORLD

The New Geography of Food and Agriculture in New Zealand

Geographers who study agriculture and food are using several new approaches to understand the ways the restructuring of international trade, the activities of transnational corporations, and rapid shifts in government policies are affecting agriculture and the rural landscape in different countries. For example, agriculture has been transformed through horizontal integration in which smaller enterprises are merged to create larger units (for example, when adjacent farms are consolidated into one large landholding, resulting in the disappearance of small, family-run farms) and through vertical integration, in which a single firm takes control of several stages in the production process (when a company owns the fertilizer and seed companies as well as the food-processing plant and supermarkets). The international corporation ConAgra, which owns grain companies, feedlots, meat processing, and wholesale distribution facilities is an agribusiness that organizes food production from the manufacturing of chemical inputs and the genetic manipulation of animal breeds or crop varieties to the processing, retailing, and consumption of the agricultural product.

Geographers such as Richard Le Heron and Guy Robinson have written extensively about how New Zealand agriculture has changed in response to the restructuring of the global food system. They document how New Zealand's agricultural system evolved during the nineteenth century with an orientation to exports of wool and lamb based on a pastoral landscape and a guaranteed market in the core economy of the United Kingdom. After World War II, a second regime developed that included dairy cows on small farms and processing of products such as butter for export using refrigerated shipping. By the mid-twentieth century, the New Zealand government was heavily involved in the agricultural system through "marketing boards" that mediated farmers' relationships with international markets through quality controls, price supports, and marketing.

In the 1970s, the shock of the oil crisis (increasing the cost of agricultural inputs) and the loss of the imperial-preference market when Britain joined the European Community resulted in further state support for producers, with price supports, incentives, and subsidies for inputs such as fertilizers providing more than a third of farm revenues. However, even these institutional supports could not fully buffer farmers against the increasing cost of inputs and loss of markets for the staples of wool, meat, and dairy, and some farmers began to diversify into nontraditional exports,

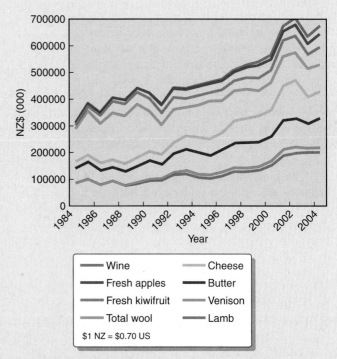

Figure 8.G   New Zealand agricultural production The structure of New Zealand agricultural exports has changed in response to the restructuring of global agricultural trade and demand. Although the traditional exports of lamb meat and butter have remained high, wool exports have fallen. Exports of nontraditional crops such as kiwi, berries, avocados, wine, and venison have increased considerably.

The Impact of Agriculture on the Environment

As discussed in Chapter 4, one of the earliest treatises on the impact of chemical pesticides on the environment was Rachel Carson's *Silent Spring*, which identified the detrimental impacts of synthetic chemical pesticides—especially DDT—on the health of human and animal populations (**Figure 8.26**). Although the publication of the book and the environmental awareness that it generated led to a ban on the use of many pesticides in most industrialized nations, chemical companies continued to produce and market them in peripheral countries. Some of these pesticides were effective in combating malaria and other insect-borne diseases, but many were applied to crops that were later sold in the markets of developed countries. Thus, a kind of "circle of poison" was set into motion, encompassing the entire global agricultural system.

One of the most pressing issues facing agricultural producers today is soil degradation and denudation, which

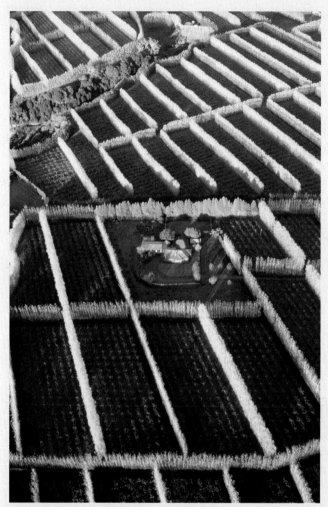

Figure 8.H   Kiwi production These kiwi orchards on the North Island of New Zealand are surrounded by lines of trees that protect the delicate fruit from strong winds.

such as venison, produced on deer farms, and fruit such as kiwi and Asian pears, responding to a new global food regime of specialty foods and the export of fruit and vegetables (**Figure 8.G and Figure 8.H**).

A dramatic change in agricultural policies in 1984 abruptly removed most price supports, trade protections, and farm subsidies and required farms to pay for extension services, water, and quality inspections. Farm income fell by up to half, debt increased, 10 percent of farms were sold, herds were significantly reduced, and 10,000 farmers protested in front of Parliament. New Zealand agriculture was thrown into a global free market and the full impact of what has been called the "international farm crisis," while most other developed countries, including the United States, Canada, and those in Europe, maintained considerable state regulation and support for their agricultural systems. Although New Zealand farmers coped by adjusting herd sizes and changing crop mixes, some went out of business and their properties were horizontally integrated into larger farms. But New Zealand was also one of the first countries to adopt certification for organic agricultural products, and there is a thriving domestic market for sustainably grown foods. Transnational agribusiness firms, such as H.J. Heinz, purchased New Zealand agricultural processing enterprises with the goal of supplying growing Asian markets. Michael Moore, the New Zealander who had spearheaded his country's plunge into the free market as minister of trade, became the head of the World Trade Organization, charged with reducing barriers to trade worldwide.

Many geographers are reexamining agriculture in the context of global restructuring and changing government policies and contributing important insights into how the new international geography of food and agriculture is changing the economy and environments of countries such as New Zealand. The geographic perspective allows us to link international trade, regulation, and corporations to the decisions of national and local governments and to the impacts on and responses of agricultural regions, communities, and farm families.

are occurring at rates more than a thousand times natural rates. Although we in the United States tend to dismiss soil problems such as erosion as an artifact of the 1930s Dust Bowl, the effects of agriculture on worldwide soil resources are dramatic, as **Table 8.3** illustrates. Unfortunately, most forms of agriculture tend to increase soil degradation. Although severe problems of soil degradation persist in the United States—which has a federal agency devoted exclusively to managing soil conservation—more severe problems are occurring in peripheral countries.

The problem of soil degradation and loss is particularly critical in the humid tropical areas of the globe—especially in South America and Asia. Arguably the most critical hotspot with respect to soil and related environmental impacts is the moist Brazilian *cerrado*, or grassland, which is being converted for soybean production in order to feed the growing biofuels industry (which we discuss in more detail in the next section). In addition to soil erosion and degradation, the conversion of the *cerrado*, premised increasingly on industrial farming techniques, is also threatening the

Figure 8.26 Poisoned crane, Hungary In spring 2005, carcasses of more than a hundred cranes were found in an area 150 kilometers southeast of Budapest. Experts say the birds of passage on their way to the northern regions of Europe were apparently poisoned by eating grain containing toxic pesticides.

species richness of birds, fishes, reptiles, amphibians, and insects, with some animal and plant species facing extinction.

The loss of topsoil worldwide is a critical problem because it is a fixed resource that cannot be readily replaced. It takes, on average, 100 to 500 years to generate 10 millimeters (one-half inch) of topsoil, and it is estimated that nearly 50,000 million metric tons (55,000 million tons) of topsoil are lost each year to erosion. The quantity and quality of soil worldwide are thus important determining factors in the quantity and quality of food that can be produced.

Soil erosion due to mismanagement in the semiarid regions of the world has led to desertification, in which topsoil and vegetation losses have been extensive and largely permanent. As explained in Chapter 4, desertification is the spread of desertlike conditions in arid or semiarid lands resulting from climatic change or human influences. Desertification means not only the loss of topsoil but possibly the deterioration of farm and grazing lands and the decimation of forests (**Figure 8.27**). In addition to causing soil degradation and denudation problems, agriculture affects water quality and quantity through the overwithdrawal of groundwater and the pollution of the same water through agricultural runoff contaminated with herbicides, pesticides, and fertilizers. Deforestation can also result from poor agricultural practices.

The nature-society relationship discussed in Chapter 4 is very much at the heart of agricultural practices. Yet as agriculture has industrialized, its impacts on the environment have multiplied and in some parts of the globe have reached crisis stage. In some regions the agricultural system leads to overproduction of foodstuffs, but in others the quantity and quality of water and soil severely limit the ability of a region's people to feed themselves.

PROBLEMS AND PROSPECTS IN THE GLOBAL FOOD SYSTEM

The future of the global food system is being shaped at this very moment in food science laboratories, in corporate boardrooms, on the street in organized protests, and in homely settlements throughout the world. The biggest issues food policy experts, national governments, consumers, and agriculturalists face revolve around the availability and quality of food in a world where access to safe, healthy, and nutritious foodstuffs is unevenly distributed. For the periphery, the most pressing concern is adequate food supplies to feed growing populations. For the core, there are concerns about food quality in a system that is increasingly industrialized and biologically engineered. Sometimes the solutions to one problem for one population become a new problem for another population.

TABLE 8.3 Global Soil Degradation

Region	Degrading area (km²)	mi²	% Territory	% Global degrading area	% Total population	Affected people
Asia	9,128,498	3,524,533	25.39	0.890	23.58	1,070,737,071
Africa	6,596,641	2,546,977	26.65	0.441	28.31	230,604,253
Europe	656,007	253,286	12.88	0.089	9.75	48,457,913
South America	4,719,162	1,822,079	24.86	0.457	22.90	149,905,245
Oceania	2,364,959	913,116	34.68	1.224	26.29	5,494,554
North America	3,968,971	1,532,428	20.25	9.755	14.24	36,654,152
World Total (land, excluding inland water body)	35,058,104	1,35,360,010	23.54	100.000	23.89	1,537,679,148

Source: Adapted from Z. G. Bai, D. L. Dent, L. Olsson, and M. E. Schaepman, *Global Assessment of Land Degradation and Improvement 1: Identification by Remote Sensing.* Report 2008/01, FAO/ISRIC, Rome and Wageningen, 2008, p. 24, Table 1.

Figure 8.27 Desertification, Gansu, China A woman walks along the leading edge of the Kumtag desert as it threatens to engulf her onion farm. The sand is advancing at the rate of up to 4 meters (13 feet) a year.

In this final section, we examine two problematic issues in the world food system, as well as an encouraging prospect. These cases certainly do not illuminate the myriad challenges and possibilities facing food producers and policymakers today, but they do provide a broad sense of the range.

Food and Health

We have spent most of this chapter describing the ways food is cultivated, processed, engineered, marketed, financed, and consumed throughout the world. What we have yet to do is talk about access to this most essential of resources. Although there is more than enough food to feed all the people who inhabit Earth, access to food is uneven, and many millions of individuals in both the core and the periphery have had their lives shortened or harmed because war, poverty, or natural disaster has prevented them from securing adequate nutrition. In fact, hunger is very likely the most pressing problem facing the world today.

Hunger occurs in two basic ways: chronic or acute. Chronic hunger is nutritional deprivation that occurs over a sustained period of time: months or even years. Acute hunger is short-term and is often related to catastrophic events—personal or systemic. Chronic hunger, also known as **undernutrition,** is the inadequate intake of one or more nutrients and/or of calories. Undernutrition can occur in individuals of all ages, but its effects on children are dramatic, leading to stunted growth, inadequate brain development, and a host of other serious physical ailments.

Perhaps the most widely publicized examples of acute hunger are famines, especially those that have occurred in parts of the periphery over the last few decades. **Famine** is acute starvation associated with a sharp increase in mortality. The most widely publicized famines of the late twentieth century occurred in Bangladesh in 1974 and Ethiopia in 1984–1985. The causes of these two famines (and the other twentieth-century famines that preceded them) were complex. The crisis of starving people so often publicized by the news media is usually just the final stage of a process that has been unfolding for a far longer period, sometimes years or even decades. Experts who study famine argue that there are at least two critical factors behind long-standing vulnerability to famine. The first has to do with a population's command over food resources in terms of their livelihood (for instance, rural laborer or farmer). The second has to do with a trigger mechanism, which may be a natural phenomenon like drought or a human-made situation such as civil war.

The famine in Bangladesh resulted from a sharp rise in food prices combined with a drop in employment opportunities that resulted in part from a very serious flood that had wiped out crops in lowland areas (but left highland agricultural production intact). Adequate food supplies were still available, but individuals who labored in the countryside to produce, distribute, or process the lowland crops could not afford them. Exacerbating the problem was the fact that negotiations between the Bangladeshi government and foreign-aid donors had broken down to such an extent that the distribution of food aid to needy individuals was seriously impaired. The lesson to be drawn from the Bangladesh famine is that when access to livelihood collapses, famine can result, even when overall levels of food availability are adequate.

People who study famine and other forms of hunger have come to conceptualize vulnerability in terms of the notion of food security. **Food security** means that a person, a household, or even a country has assured access to enough food at all times to ensure active and healthy lives. And while famine is a dramatic reminder of the precarious nature of food security, it is important to appreciate that chronic hunger resulting from food insecurity is a far more widespread and devastating problem than famine, which tends to be shorter in duration and more contained geographically. Related to food security is the concept of **food sovereignty,** the right of peoples, communities, and countries to define their own agricultural, labor, fishing, food, and land policies that are ecologically, socially, economically, and culturally appropriate to their unique circumstances. It is being argued by international governance organizations as well as activist groups that only through attention to food sovereignty will the problems of food security for vulnerable populations be addressed.

Globally, 24,000 people a day die from the complications brought about by undernutrition. Today in the United States, where food is abundant and overeating is a national problem, 10 percent of the population is undernourished or experiences food-security problems at one time or another each year. In the periphery, where food availability is more limited than in core countries like the United States, undernutrition is far more pervasive. Food-security experts agree that undernutrition is the result of an overarching set of factors ultimately caused by poverty. Wealthy people, whether they live in the core or the periphery, rarely if ever go hungry. But impoverished people everywhere are likely to have experienced short-term or long-term undernutrition, if not both. Thus, in a world where there is more than enough food to adequately and consistently nourish every man, woman, and child, many go hungry because they simply do not possess the livelihoods adequate to gain sufficient access to food resources. Furthermore, in some parts of the world, among some social classes, there are higher levels of undernutrition among women and girls than among men and boys. This is largely because different cultural and social norms favor men and boys, who eat first, leaving the leftovers for females, or who eat certain high-status foods, such as proteins like meat or fish, that women are not allowed to eat.

One of the factors thought to be adding to the emergence of the most recent food emergency is the increasing amount of cropland around the globe once dedicated to food production now being redirected to raising **biofuels**, renewable fuels derived from biological materials that can be regenerated. (Biofuels include ethanol, biodiesel, and methanol). Replacing even a fraction of fossil fuels with biofuels requires the acquisiton and conversion of vast areas of land. National governments around the world have passed laws and are providing inducements for the conversion of tens of millions of hectares to biofuel monocultures like corn, soy, sugar, cotton, and grains. Land conversion on this scale will likely result in evictions of small farmers and poor communities who depend on land sometimes classified as "wasteland" for growing food and/or for grazing because the plots are small and often disconnected. For example, the Tanzanian government has handed over 9,000 hectares of land on which over 11,000 people depend for their livelihoods to the U.K. firm Sun Biofuels PLC for a jatropha (a shrubby plant that produces seeds that are 40 percent oil) plantation. Land grabs for biofuels are happening across Asia, Latin America, and Africa and often involve violence. Some 150,000 families in Argentina and 90,000 families in Paraguay have already been displaced by soy production. An unpublished 2008 report by the World Bank states that biofuels production has forced global food prices up by 75 per cent; it is our view that the price of food is a complex variable affected by a host of causes that include the emergence and growth of biofuels production but is also related, among other causes, to rising demand among increasingly more affluent populations, commodities speculation, rapidly escalating energy prices, and poor harvests.

While it is important to understand the complex causes of the food crisis, perhaps the most relevant point to take away from this discussion of hunger and its multiple causes and complications is that it is a problem that can be solved. Because neither short-term nor long-term hunger involves inadequate supply, the solution must lie in improving access to those supplies. This could occur, under a radical scenario, through a massive redistribution that would give all the world's people access to the same amount of food resources. Barring such a dramatic restructuring of the current economic system, the solution to the problem of hunger lies in improving access to livelihoods that pay well enough that adequate nutrition becomes a human right and not dependent upon the vagaries of economic or natural systems.

At the same time hunger is a serious problem for peripheral populations worldwide, for both core and peripheral populations, issues of food safety have also grown in significance. Over the last several years, "food scares" caused by tainted foodstuffs have been routinely in the headlines. In 2006, U.S. newspapers carried the story of spinach grown in California and tainted with *E. coli* bacteria causing severe illness in over 200 people in 26 states. The kind of *E. coli* that make humans sick are produced in the guts of ruminants—cows, elk, sheep, goats, and deer. The bacteria leaves the animals' system through feces. The tainted spinach in the 2006 case was ultimately linked to a nearby cattle ranch (**Figure 8.28**). In summer 2008, another food scare occurred in the United States, this time from *salmonella*, a food-borne

Figure 8.28   Spinach crop, Salinas, California While consumers worried about tainted spinach, farmers were laying off workers and plowing under their crops as government inspectors examined their fields. The economic loss was estimated to be nearly $100 million.

microorganism that causes poisoning in humans. The U.S. Department of Agriculture is still not certain how the *salmonella* entered the food system. Again that summer, beef being sold by Whole Foods Market, a grocery retailer of "natural" and organic products, was discovered to be tainted by *E. coli*. The list could go on and on, especially if we added food scares from around the world such as toxic chemicals routinely added to pickled vegetables in a factory in China; *salmonella* contamination in hummus in the United Kingdom; cancer-causing dioxins in buffalo milk used in the production of mozzarella produced in Italy, which entered grazing pastures through illegally dumped toxic waste—just to name a few of the hundreds that could be included.

The impact of food scares is widespread, affecting not only those who have become ill or died but also farmers and grocers as well as the credibility of government agencies meant to protect consumers from tainted and toxic foods. Mostly, the recent apparent escalation in food scares reminds consumers of the vulnerability of the global food system.

Genetically Modified Organisms and the Global Food System

In 1999, protesters in London, Chicago, and Seattle dressed up as monarch butterflies and informed passersby that genetically modified corn might be posing a threat to that insect. Shouting chants like "Hey, hey, ho, ho, Frankenfoods have got to go!" these protesters represented the emergence of a now rapidly growing global movement that opposes the production and sale of genetically modified food products—or demands, at the very least, that these foods be labeled as such (**Figure 8.29**). In Brazil, a federal judge banned the sale of Monsanto

Corporation's Ready Roundup genetically engineered soybean seeds. Japan has required that all genetically engineered food be labeled. In the United States, however, the Food and Drug Administration has made regulation of genetically engineered food voluntary: Companies can decide for themselves whether they want the agency to review their product. Since these early protests, popular opposition to GMOs has continued to grow.

A **genetically modified organism**, or **GMO** as it is more commonly known, is any organism that has had its DNA modified in a laboratory rather than through cross-pollination or other forms of evolution. Examples of GMOs are a bell pepper with DNA from a fish added to make it more drought-tolerant, a potato that releases its own pesticide, and a soybean that has been engineered to resist fungus.

Genetic modification has both critics and supporters. Proponents argue that it allows great advances in agriculture (for instance, making plants more resistant to certain diseases or water shortages), as well as other beneficial creations, such as the petroleum-eating bacteria that can help clean up oil spills. Opponents worry that genetically modified organisms may have unexpected and irreversible effects on human health and the environment, causing maturation problems in children or mutant plant and animal species.

In the United States, genetic modification is permitted, on the principle that there is no evidence yet that it is dangerous. GMO foods are fairly common in the United States, and estimates of their market saturation vary widely. It is not, however, easy to recognize GMOs in the grocery store, as there are no labeling requirements. While the food-safety establishment in the United States maintains that GMOs are safe until proved otherwise, countries in Europe have taken the opposite position: that genetic modification has not been proved safe, so they will not accept genetically modified food from the United States or

Figure 8.29 Protest over GMOs, Paris, France Food activist and leader of the Confederation paysanne, a French agricultural union that supports sustainable agriculture, Jose Bove, (center) leads thousands of GMO activists in protest in the streets of Paris close to the Assemblee Nationale where ministers were voting on GMO laws in spring 2008.

any other country. The World Trade Organization has determined that not allowing GMO food into a country creates an unnecessary obstacle to international trade.

GMOs represent perhaps the most highly technological way to date in which nature and society come together in the global food system. At present, so little is known about the wider impacts of GMOs on human health, the environment, or even the wider global economic system that it is difficult to sort the costs from the benefits of their increasing incorporation into global food production. What is clear is that genetic modification is no passing fad and the debate cannot be reduced to a simple "good" or "bad." GMOs are neither entirely evil nor entirely good; certain applications may be widely beneficial, while others may not. Regulatory structures are crucial to protecting human health and the environment, as well as spreading whatever benefits may accrue from GMOs beyond the core and into the periphery, where food-security issues are at their most critical. But regulatory structures are not so easily accomplished.

Protests against GMO regulatory structures have been most effective in Europe and parts of Asia, Africa, and Latin America as well as in Canada and Mexico. In those regions, national governments have begun to devise regulations to control or publicize the entry of GMO commodities into the food system or are requiring more research in order to better understand the long-term effects of consuming GMOs on humans as well as the impact of GMOs on the food chain. Widespread protests against GMO foods have also been organized in the United States, but the government has not responded with strong support for these movements. It has passed domestic legislation making the labeling of GMOs voluntary, and at the international level it has threatened action against countries who are arguing for the mandatory labeling of GMO foods traded in the global marketplace. It is important to be aware that the position taken by the United States is a protrade one, rather than a strictly pro-GMO food one. In short, the U.S. government is mostly concerned with the economic impact GMO labeling is likely to have on trade and is reluctant to put the trade of U.S. agricultural products at risk. This strategy has been followed because the United States is the largest exporter of agricultural goods in the world as well as the largest producer of GMO foods. If GMO labeling were mandated by the World Trade Organization, it is likely that many U.S. trading partners—both large and small—would be legally bound either to refuse GMO commodities or the commodities would be seen as less valuable to potential buyers. At present, there is no sign that the WTO is likely to take such a step. In fact, its current regulations specify that if a country imposes additional restrictions and safety concerns on products that the country thinks are unsafe, the WTO can overrule it. Issues such as this one threaten the independence of individual nations and create governance problems for the WTO.

Global debate and activism—both popular and governmental—over GMOs is still in its early stages. Some of the most vocal opponents of GMOs are concerned that engineered food is destined for consumption by poor people in the periphery, while "real" food, produced by artisans and organic growers, will only be available to rich people in the core with no attention paid to the importance of food to reducing world hunger or to ensuring a safe diet as a global human right.

Urban Agriculture

Although most people think of agriculture as a rural activity, urban agriculture made possible the emergence of the world's first cities. Until recently, however, urban agriculture was largely ignored in the development of urban economic policies, apparently because produce generated from it was seen as belonging to the informal sector of the local economy and not significant in terms of income-generating potential. Most definitions of **urban agriculture** take it to mean the establishment or performance of agricultural practices in or near an urban or citylike setting. In countries like China, official policies have long recognized and even fostered urban agricultural practices. In many core countries, however, particularly since the Industrial Revolution, urban agriculture has been officially discouraged or made difficult as arable land has been taken for real estate development or seriously degraded through industrial processes.

Whether encouraged or discouraged by official policy, urban residents across the globe are increasing their participation in growing crops and raising livestock, for reasons ranging from food security to income production to taste and health concerns. Up to 30 percent of agricultural production occurs within metropolitan areas in the United States, for example, and up to 15 percent globally. Throughout the core, urban agriculture is largely a leisure activity that helps to supplement the routine purchase of commercial foodstuffs. In the periphery, however, it can often be the sole means of economic and personal survival. As wage cuts, inflation, job loss, civil strife, and natural disasters become more frequent, urban agriculture in the periphery has become a way to address greater food insecurity.

As we discuss in Chapter 10, urban populations throughout the world are growing more than twice as fast as rural populations. By the year 2015, according to the UN Center for Human Settlements, more than half of humanity will live in cities. And as development experts look to sustainable development as a way of maintaining economic growth without destroying the environment, urban agriculture has increasingly drawn their attention as a way of making cities sustainable. Proponents of urban agriculture contend that it should not be understood as an alternative to conventional agriculture, but rather as a supplementary branch of modern agricultural systems. For most development experts, an ideal urban agricultural system would incorporate various elements of modern sustainable agriculture based on reusable, self-contained waste and nutrient cycles through resource conservation and management based on nonchemical fertilizers and pest-management techniques.

It is important to recognize that urban agriculture cannot solve the world's food-security problems. For example, small urban gardens will not replace agribusiness as the primary players in the global food system. Moreover, there are legitimate health concerns surrounding urban agriculture, particularly in terms of recycling urban wastes into agricultural inputs. In arid parts of the world, where water is scarce, the use of wastewater from domestic or commercial uses seems an obvious solution to the irrigation needs of urban agriculture. Yet while in some parts of the world wastewater is effectively treated and used for secondary applications, in others it is not treated and can easily carry disease, which can then be spread into the food system when applied to crops. Clearly, policies and practices need to be developed that very carefully address the health implications of urban agriculture in the very different settings in which it is being practiced.

Currently, urban agriculture is practiced in a variety of ways, including rooftop, hydroponic, and community gardens; roadside urban fringe agriculture; field-to-direct-sale farmers' markets; and livestock grazing in parks and feedlots (**Figure 8.30**). As a growing practice worldwide, urban agriculture may help establish sustainable food systems in predominately urban areas. While health concerns about urban agricultural practices should not be taken lightly, there is evidence that they are far outweighed by the current and potential benefits of urban agriculture. Particularly in developing countries and poorer inner-city neighborhoods throughout the world, urban agriculture can be crucial to a family's survival. This ability of urban agriculture to enhance household food security seems to be currently drawing the most attention. Urban agriculture can also play a very important role in the absorption of labor—particularly women and youth—so that urban

Figure 8.30 Urban garden, Hong Kong One of the most densely populated urban areas in the world, Hong Kong is also home to hundreds of urban gardens that residents of apartment blocks use to supplement market purchases. Throughout the world, in both the core and the periphery, private as well as communal gardens are becoming a more familiar feature of the urban landscape.

households are better able to take full advantage of their own human resources. It may even be a way of turning urban waste into a resource when incorporated safely into the food system.

CONCLUSION

Agriculture has become a highly complex, globally integrated system. While traditional forms of agricultural practices, such as subsistence farming, continue to exist, they have been overshadowed by the global industrialization of agriculture. This industrialization has included not only mechanization and chemical applications but also the linking of the agricultural sector to the manufacturing, service, and finance sectors of the economy. In addition, states have become important players in the regulation and support of agriculture at all levels, from the local to the global.

The dramatic changes that have occurred in agriculture have affected different places and different social groups. Households in both the core and the periphery have strained to adjust to these changes, often disrupting existing patterns of authority and access

to resources. Just as people have been affected by the transformations in global agriculture, so have the land, air, and water.

The geography of agriculture at the turn of the twentieth century is a far cry from what it was 100 or even 50 years ago. As the globalization of the economy has accelerated in the last few decades, so has the globalization of agriculture. The changes in global agriculture do not necessarily mean increased prosperity in the core, nor are the implications of these changes simple. For example, the production of oranges in Florida is directly influenced by the newer Brazilian orange industry. Both industries, in turn, affect the prices of oranges in the marketplaces of Europe and Asia. Additionally, the European unease with genetically engineered foods will affect U.S. producers, agricultural research, trade, and a host of other factors that have repercussions throughout the world system.

MAIN POINTS REVISITED

- Agriculture has been transformed into a globally integrated system; the changes producing this result have occurred at many and have had many sources.

  In addition to the restructuring of entire national farming systems, farming households have been transformed as well in core, periphery, and semiperiphery regions.

- Agriculture has proceeded through three revolutionary phases, from the domestication of plants and animals to the latest developments in biotechnology and industrial innovation.

  These three revolutionary phases have not occurred simultaneously throughout the globe but have been adopted and adapted to differing degrees, based on levels of development, culture, and physical geography.

- The introduction of new technologies, political concerns about food security and self-sufficiency, and changing opportunities for investment and employment are among the many forces that have dramatically shaped agriculture as we know it today.

  Two of the most important forces behind these transformations in agriculture have been multinational and transnational corporations and states. The World Trade Organization is another important influence.

- The industrialized agricultural system of today's world has developed from—and largely displaced—older agricultural practices, including shifting cultivation, subsistence agriculture, and pastoralism.

  Although these systems no longer dominate agricultural practices on a global scale, they are still practiced in many areas of the world, in some cases alongside more mechanized forms.

- The contemporary agro-commodity system is organized around a chain of agribusiness components that begins at the farm and ends at the retail outlet. Different economic sectors, as well as different corporate forms, have been involved in the globalization process.

  The farm is no longer the central piece in this chain of agricultural organization, but one of several important components that include seed and fertilizer manufacturers, food processors, food distributors, and consumers.

- Transformations in agriculture have had dramatic impacts on the environment, including soil erosion, desertification, deforestation, and soil and water pollution, as well as the elimination of some plant and animal species.

  While most of the core countries have instituted legislation to address some of the problems associated with environmental degradation, these problems exist throughout the global agricultural system to greater and lesser degrees. In peripheral countries, where governments are often too poor to monitor and enforce such legislation, they are being encouraged by international agencies and environmental organizations to limit their degradational practices through relief of part of their national debt.

- The biggest issues food-policy experts, national governments, consumers, and agriculturalists face revolve around the availability and quality of food in a world where access to safe, healthy, and nutritious foodstuffs is unevenly distributed.

  Genetic modification is one way of improving productivity, though it does not address issues of access to food. Increasing opportunities for the world's poor—who are increasingly residing in urban settings—to grow their own food is another way.

KEY TERMS

agrarian (p. 298)
agribusiness (p. 320)
agricultural industrialization (p. 308)
agriculture (p. 298)
biofuels (p. 331)
biopharming (p. 317)
Biorevolution (p. 313)
biotechnology (p. 313)
Blue Revolution (p. 311)
Borlaug hypothesis (p. 317)
chemical farming (p. 308)

commercial agriculture (p. 299)
contract farming (p. 307)
conventional farming (p. 323)
crop rotation (p. 301)
double cropping (p. 304)
famine (p. 331)
fast food (p. 326)
food chain (p. 321)
food manufacturing (p. 308)
food regime (p. 322)
food security (p. 331)
food sovereignty (p. 331)

globalized agriculture (p. 318)
GMO (genetically modified organism) (p. 332)
Green Revolution (p. 312)
hunting and gathering (p. 299)
intensive subsistence agriculture (p. 303)
intertillage (p. 301)
local food (p. 324)
mechanization (p. 308)
nontraditional agricultural exports (NTAEs) (p. 312)

organic farming (p. 323)
pastoralism (p. 304)
shifting cultivation (p. 301)
slash-and-burn (p. 301)
slow food (p. 326)
subsistence agriculture (p. 299)
swidden (p. 301)
transhumance (p. 304)
undernutrition (p. 331)
urban agriculture (p. 334)

ADDITIONAL READING

Brar, K. K., *Green Revolution: Ecological Implications.* Delhi: Dominant Publishers, 1999.

Brown, A. D., *Feed or Feedback: Agriculture, Population Dynamics and the State of the Planet.* Utrecht: International Books, 2003.

Conway, G., *The Doubly Green Revolution: Food for All in the Twenty-First Century.* Ithaca, NY: Comstock Publishers, 1998.

Corum, V., M. Rosenzweig, and E. Gibson, *The New Farmers' Market :Farm-fresh Ideas for Producers, Managers & Communities.* Auburn, CA: New World Publishing, 2001.

Devereux, S., *The New Famines: Why Famines Persist in an Era of Globalization.* London: Routledge, 2005.

Ezcurra, E. (ed.), *Global Deserts Outlook: Global Environment Outlook Series.* United National Environment Program, available online at http://www.unep.org/geo/gdoutlook/.

Fold, N., and B. Pritchard, *Cross-Continental Agro-Food Chains: Structures, Actors and Dynamics in the Global Food System.* London: Routledge, 2005.

Food and Agriculture Organization of the United Nations, *The State of Food Insecurity in the World 2006: Eradicating*

World Hunger—Taking Stock 10 Years after the World Food Summit. Rome, Italy, 2006.

Gruhn, P., F. Goletti, and M. Yudelman, *Integrated Nutrient Management, Soil Fertility, and Sustainable Agriculture: Current Issues and Future Challenges.* 2002 Vision Discussion Paper 32. Washington, DC: IFPRI.

Halweil, B., *Eat Here: Reclaiming Homegrown Pleasures in a Global Supermarket.* New York: W.W. Norton, 2004.

Henderson, E., and R. Van En, *Sharing the Harvest: A Citizen's Guide to Community Supported Agriculture.* White River Junction, VT: Chelsea Green Publishing, 2007.

Lang, T., and M. Heasman, *Food Wars: The Global Battle for Minds, Mouths, and Markets.* London: Earthscan, 2004.

Lappe, M., and B. Bailey, *Engineering the Farm: The Social and Ethical Aspects of Agricultural Biotechnology.* Washington, DC: Island Press.

Lebel, L., N. H. Tri, A. Saengnoree, S. Pasong, U. Buatama, and L. K. Thoa, "Industrial Transformation and Shrimp Aquaculture in Thailand and Vietnam: Pathways to Economic, Social and Ecological Sustainability?" *Ambio*, 31(4), 2002, pp. 311–323.

Manning, R., *Against the Grain: How Agriculture Has Hijacked Civilization.* New York: North Point Press, 2004.

Morgan, K., T. Marsden, and J. Murdoch, *Worlds of Food: Place, Power, and Provenance in the Food Chain.* Oxford; New York: Oxford University Press, 2006.

Mougeot, L. J. A. (ed.), *Agropolis: The Social, Political and Environmental Dimensions of Urban Agriculture.* London; Sterling, VA: Earthscan, 2005.

Public Citizen, *The Shrimp Stockpile: America's Favorite Imported Seafood.* Washington, DC: Public Citizen, 2005.

Raworth, K., *Trading Away Our Rights: Women Working in Global Supply Chains.* Oxford, England: Oxfam GB, 2004.

Scoones, I., *Science, Agriculture and the Politics of Policy: The Case of Biotechnology in India.* New Delhi: Orient Longman, 2006.

Shiva, V., *Stolen Harvest: the Hijacking of the Global Food Supply.* Cambridge, MA: South End Press, 2000.

Shiva, V., and G. Bedi, *Sustainable Agriculture and Food Security: The Impact of Globalisation.* London: Sage Publications, 2002.

Viljoen, A., K. Bohn, and J. Howe, *Continuous Productive Urban Landscapes: Designing Urban Agriculture for Sustainable Cities.* Oxford; Boston: Architectural Press, 2005.

EXERCISES

On the Internet

The Companion Website at www.mygeoscienceplace.com contains a number of tools designed to reinforce the geographic concepts presented in this chapter. Structured within a four-step learning path, these online resources include: chapter Key Concepts; a fully-searchable myeBook version of the chapter, which allows you to add notes and highlights; RSS news feeds from various media agencies; links to additional web resources; Human Geography Videos; a digital Glossary; Concept Review questions; Thinking Spatially mapping exercises; and Critical Thinking essay questions.

Unplugged

1. Your neighborhood grocery store is a perfect location to begin to identify the "global" in the globalization of agriculture. Go to the produce section there, and document the source of at least ten fruits and vegetables you find. You may need to ask the produce manager where they come from; once you have established that, locate those sources on a world map. You might also determine if the items are conventionally grown, organically grown or GMO crops and learn more about the firm that produces them

2. The Food and Agriculture Organization (FAO) has been publishing a range of yearbooks containing statistical data on many aspects of global food production since the mid-1950s. Using the *State of Food and Agricultural Production* yearbooks, compare the changes that have occurred in agricultural production between the core and the periphery since midcentury. You can use just two yearbooks for this exercise, or you may want to use several to get a better sense of when and where the most significant changes have occurred. Once you have identified where the changes have been most significant, try to explain why these changes may have occurred.

3. The U.S. Department of Agriculture (USDA) also provides statistics on food and agricultural production, though, of course, limited to the United States. Contained in volumes simply called *Agricultural Statistics,* a range of important variables are included, from what is being grown where, to who is working on farms, and what kinds of subsidies the government is providing. Using the USDA's annual publication *Agricultural Statistics,* examine the changing patterns of federal subsidies to agriculture over time. Using a map of the United States, show which states since the 1940s (just following the Great Depression) have received subsidies for the decades 1945, 1965, 1985, and the present. Have subsidies increased for some parts of the country and not others? If so or if not, why? Have subsidies increased or decreased overall for the entire country? Which farm sectors and, therefore, which regions have most heavily benefited from federal agricultural subsidies? Why?

4. Your breakfast is the result of the activities of a whole chain of producers, processors, distributors, and retailers whose interactions provide insights into both the globalization of food production and the industrialization of agriculture. Consider the various foods you consume in a typical breakfast and describe not only where (and by whom) they were produced—grown and processed—how they were transported (by whom) from the processing site—but also where and by whom they were retailed. Summarize how the various components of your breakfast illustrate the two concepts of globalization and the industrialization of agriculture.

The Politics of Territory and Space

9

Cambodia, Guatemala, East Timor, Bosnia, Rwanda, Sudan—all have experienced genocide, the deliberate and systematic killing of a people based on their ethnicity, nationality, race, religion, or politics. The deaths of over 40 million people in the past 100 years have been attributed to genocide and other acts of state violence. The term itself is a mid-twentieth-century formulation coined by Jewish scholar Raphael Lemkin to describe the Nazi Holocaust. Genocide appears to have been a widespread feature in the history of civilization, but it wasn't until after the Holocaust that steps were taken by the international community to create a set of laws forbidding it. In December 1948, the UN General Assembly acted to prevent genocide through the Convention on the Prevention and Punishment of the Crime of Genocide. When the convention first became effective in 1951, only two of the five permanent members of the UN Security Council were parties to the treaty, France and the Republic of China. Eventually, the Soviet Union ratified it in 1954, the United Kingdom in 1970, the People's Republic of China in 1983, and the United States in 1988. But only in the 1990s did the international law on the crime of genocide begin to be enforced, this with respect to the atrocities in Rwanda.

According to Gregory Stanton, president of Genocide Watch in Washington D.C., genocide develops in eight stages:

1. Classification: Dividing people into "us and them" occurs.

2. Symbolization: Hateful symbols are imposed on "them," including hate speech.

3. Dehumanization: The pariah group is made to seem inhuman so that the normal human revulsion against murder can be overcome.

4. Organization: Special military units or militias are often trained and armed.

5. Polarization: Hate groups broadcast polarizing propaganda.

6. Identification: Victims are separated out because of their different identities.

7. Extermination: Murder, rape, torture, and kidnapping are perpetrated.

8. Denial: The perpetrators deny they committed any crimes.

Although eight stages characterize the unfolding of genocide, it is important to be aware that at any stage preventive measures are possible and can be undertaken by a state or combination of states, though issues of sovereignty make some steps exceedingly difficult. With the arrival of the new century, it does not appear that either the international community or individual states have been able to use the convention to prevent genocide, nor has the process of enacting punishment for those guilty of genocide been swift. Today, in Darfur, Sudan, populations are still very much at risk from state-sponsored

Main Points

- Political geography, a subfield of the discipline of geography, examines complex relationships between politics and geography (both human and physical).

- Political geographers recognize that the relationship between politics and geography is two-way. Political geography can be seen both as the geography of politics and the politics of geography.

- The relations between politics and geography are often driven by particular theories and practices of the world's states. Understanding imperialism, colonialism, heartland theory, domino theory, the end of the Cold War, and the emergence of the new world order is key to comprehending how, within the context of the world system, geography has influenced politics and how politics has influenced geography.

- Political geography deals with the phenomena occurring at all scales of resolution, from the global to the individual. Important East/West and North/South divisions dominate international politics, whereas regionalism, sectionalism, and similar divisions dominate intrastate politics.

Mass graves, Choeung Ek killing fields, Cambodia

violence. In Cambodia, a court was finally convened in 2006 to try the members of the Khmer Rouge, who between 1975 and 1979 caused the deaths of nearly 2 million people (one-third of the total population) through starvation, forced labor, disease, and execution.

A report by the International Rescue Committee has shown that for every death caused by violence in a genocidal war zone, another 62 nonviolent deaths occur due to malnutrition, respiratory disease and diarrhea, anemia, tuberculosis, accidents, and fever, as well as other causes. And while institutions exist to prevent and punish, the world is still a very long way from making those practices a reality. Globalization has certainly created the conditions for economic cooperation across state territories, as international organizations and trade agreements like the World Trade Organization and North American Free Trade Agreement make clear. What is less clear is how cooperation among states will bring about the end of political violence through the enforcement of a global regime of fundamental, protected human rights.

This chapter explores how globalization continues to create new maps, politics, and geopolitical arrangements at the same time that established boundaries persist. Exploration, imperialism, colonization, decolonization, and the Cold War between East and West are powerful forces that have created and re-created national boundaries. Much of the political strife that currently grips the globe involves local or regional responses to the impacts of globalization of the economy, aided by the practices of the state. The complex relationships between politics and geography—both human and physical—are two-way. In addition, political geography is not just about global or international relationships. It is also about the many other geographic and political divisions, from the globe to the neighborhood, from large, far-reaching processes to the familiar sites of our everyday lives.

THE DEVELOPMENT OF POLITICAL GEOGRAPHY

Political geography is a long-established subfield in the wider discipline of geography. The ancient Greek philosopher Aristotle is often considered the first political geographer because his model of the state is based upon factors such as climate, terrain, and the relationship between population and territory. Other important political geographers, from Strabo to Montesquieu, have promoted theories of the state that incorporated elements of the landscape and the physical environment, as well as the population characteristics of regions. From about the fourteenth through the nineteenth century, scholars interested in political geography theorized that the state operated cyclically and organically. They believed that states consolidated and fragmented based on complex relationships among and between factors such as population size and composition, agricultural productivity, land area, and the role of the city.

As these factors indicate, political geography at the turn of the nineteenth century was influenced by two important traditions within the wider discipline of geography: the people-land tradition and environmental determinism. Different theorists placed more or less emphasis on each of these traditions in their own political geographic formulations and the traditions' effects are evident in the factors deemed important to state growth and change. Why these factors were identified as central undoubtedly had much to do with the widespread influence of Charles Darwin on intellectual and social life during this period. His theory of competition inspired political geographers to conceptualize the state as a kind of biological organism that grew and contracted in response to external factors and forces. It was also during the late nineteenth century that foreign policy as a focus of state activity began to be studied. This new field came to be called *geopolitics*.

The Geopolitical Model of the State

Geopolitics is the state's power to control space or territory and shape the foreign policy of individual states and international political relations. Within the discipline of geography, geopolitical theory originated with Friedrich Ratzel (1844–1904), a German trained in biology and chemistry. Ratzel was greatly influenced by the theories embodied in *social Darwinism* that emerged in the mid-nineteenth century. Ratzel employed biological metaphors adopted from Charles Darwin to describe his seven laws of state growth:

1. The space of the state grows with the expansion of the population having the same culture.

2. Territorial growth follows other aspects of development.

3. A state grows by absorbing smaller units.

4. The frontier is the peripheral organ of the state that reflects the strength and growth of the state; hence, it is not permanent.

5. States in the course of their growth seek to absorb politically valuable territory.

6. The impetus for growth comes to a primitive state from a more highly developed civilization.

7. The trend toward territorial growth is contagious and increases in the process of transmission.[1]

Ratzel's model portrays the state as behaving like a biological organism, with its growth and change seen as "natural" and inevitable. Although Ratzel advanced his model of the state at the turn of the nineteenth century, his views continue to influence theorizing about the state today. What has been most enduring about Ratzel's

[1]Adapted from Martin I. Glassner and Harm deBlij, *Systematic Political Geography*, 3rd ed. New York: J. Wiley & Sons, 1980, p. 164.

conceptualization is the conviction that geopolitics stems from the interactions of power and territory.

Although it has evolved since Ratzel first introduced the concept, geopolitical theory has become one of the cornerstones of contemporary political geography and state foreign policy more generally. And, though the organic view of the state has been abandoned, the twin features of power and territory still lie at the heart of political geography. In fact, the changes that occurred in Europe and the former Soviet Union during the late twentieth century suggest that Ratzel's most important insights about geopolitics are still valid.

Figure 9.1 demonstrates Ratzel's conceptualization of the interaction of power and territory through the

Figure 9.1   The changing map of Europe: 1924, 1989, 2008 The boundaries of the European states have undergone dramatic changes since World War I. The changing map of Europe illustrates the instability of international politics and the resultant dynamism in the geography of political boundaries.

changing face of Europe from the end of World War I to the present. The fluidity of the maps reflects the unstable relationship between power and territory, especially some states' failure to achieve stability. The most recent map of Europe portrays the precarious nation-state boundaries in the post-Cold War period. Estonia, Latvia, and Lithuania have regained their sovereign status. Czechoslovakia has dissolved into the Czech Republic and Slovakia. Yugoslavia has dissolved into four states, but not without much civil strife and loss of life. The former Soviet Union is now the Commonwealth of Independent States, with Russia the largest and most powerful. And still the map is not fully settled.

The tensions within the former Soviet Union, for instance, are especially high as new republics such as Georgia and Ukraine move politically closer to Europe and the United States and further away from Russia. These emerging Western allegiances signal a seismic geopolitical shift that is causing retaliation by Russia, particularly against Georgia, as evidenced by Russia's attack on the country in the summer of 2008.

Tensions in the region began to mount most recently when Mikhail Saakashvili was elected Georgian president in 2004. He promised to reunite the country, reduce corruption, seek membership in NATO and the European Union, and improve relations with Russia. Georgia had already witnessed breakaway movements by South Ossetia and Abkhazia, two regions containing large ethnic minorities near the Caucasus Mountains (**Figure 9.2**). South Ossetia has a population of around 70,000 and Abkhazia of approximately 250,000. The South Ossetians have close ethnic ties with North Ossetia, a republic that is nominally autonomous but part of the Russian Federation, where it is a "federal subject."

Because these two regions remain officially part of Georgia—at the same time that Georgia is working toward autonomy for them—tensions are understandably high. After a series of long-simmering conflicts, on August 8, 2008 fighting escalated between Georgia and South Ossetia and the former began launching aerial and ground attacks against the latter. South Ossetia and the Russian army responded with artillery fire and bombings, as well as a full-scale Russian incursion throughout Georgia. Georgia's problems were compounded when pro-Russian separatists in Abkhazia attacked Georgian forces. Civilian causalities are thought to be around 2,000 killed; the UNHCR estimates that approximately 22,000 people in the region have been displaced. There have also been reports of war crimes on both sides.

It seems Moscow was determined to dismantle Georgia's military capability and infrastructure. A week after the conflict began, a peace settlement was brokered by French President and current head of the European Union Nicolas Sarkozy. The result has been an escalation of tensions between Russia and the West, with the threat of EU sanctions looming. Russia's military remains in a "security zone" around South Ossetia that dates back to agreements in 1999 its and illegal (vis-à-vis UN Security Council resolutions) recognition of South Ossetia and Abkhazia's independence has led to condemnation from around the world, with increasingly heated and Cold War-like rhetoric from both sides.

The situation between Russia and Georgia and the continuing shifting map of Europe also illustrate more

Figure 9.2 Disputed territories in Georgia South Osettia and Abkhazia are both pushing for independence from Georgia. While Georgia has allowed the regions a large measure of autonomy, they are still part of its national territory. Russia has recognized their independence, though no other country has.

control specific sets of people and resources within them. Encompassed within a clearly defined territory, all sorts of activity can be controlled and regulated—everything, in fact, from birth to death. The delimited area over which a state exercises control and which is recognized by other states is **territory**. Such an area may include both land and water.

Boundaries can also be exclusionary. They are designed to control people and resources outside them. National boundaries, for example, can control the flow of immigrants or imported goods (**Figure 9.3**). Municipal boundaries and land-use zoning boundaries can regulate access to upscale residential neighborhoods, field boundaries can regulate access to pasture, and so on (**Figure 9.4**).

One aspect of boundaries that is of increasing interest to geographers is how they are organized and policed. Since 9/11, for example, the relevance of policing borders with respect to international travel has grown dramatically. States across the world, led by front-runners like Israel, the United Kingdom, and the United States, have been instituting a wide range of practices to secure borders by electronically screening and assessing those who can cross them. By way of passports that contain RFID (radio frequency identification) chips, as well as retinal scanning and related biometrics (electronic technologies for recognizing an individual based on one or more physical or behavioral traits), state agencies are increasingly able to know more and more personal information about air travelers. These practices have the effect of creating virtual (technologically generated) boundaries that fortify the territorial ones.

Once established, boundaries tend to reinforce spatial exclusion as well as differentiation. The outcomes result partly from different sets of rules, both formal and informal that apply within different territories, and partly because boundaries often restrict contact between

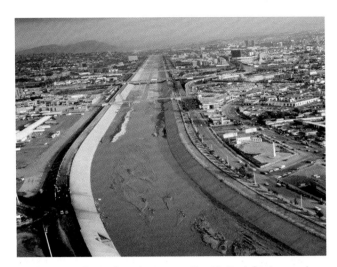

Figure 9.3   Boundary between the United States and Mexico The U.S.–Mexico border is exclusionary. It is heavily patrolled and lined with barbed-wire chain-link fences along the highly urbanized parts. Aerial surveillance is also extensive along the border. This photo shows the U.S.–Mexico border along the Tijuana River estuary, with southern California on the left and Mexico on the right.

generally the centrality of territorial boundaries to the operations of the state.

Boundaries and Frontiers

Boundaries are important phenomena because they enable territoriality to be defined and enforced and allow conflict and competition to be managed and channeled. The creation of boundaries is, therefore, an important element in place-making. It follows from the concept of territoriality that boundaries are normally inclusionary. That is, they are constructed in order to regulate and

Figure 9.4   Boundary between rural and urban places Some boundaries signal differences in settlement activities that may actually be governed by land-use regulations. The difference between agricultural activities and suburban living is clearly shown in this image.

people and so foster the development of stereotypes about "others." This restricted contact, in turn, reinforces the role of boundaries in regulating and controlling conflict and competition between territorial groups.

Boundaries can be established in many ways, however, and with differing degrees of permeability. At one extreme are informal, implied boundaries that are set by markers and symbols but never delineated on maps or in legal documents. Good examples are the "turf" of a city gang, the "territory" of an organized crime "family," and the range of a pastoral tribe. At the other extreme are formal boundaries established in international law, delimited on maps, demarcated on the ground, fortified, and aggressively defended against the movement not only of people but also of goods, money, and even ideas. An extreme example of this sort of boundary is the one between India and Pakistan (**Figure 9.5**). In between are formal boundaries that have some degree of permeability. The boundaries between the states of the European Union, for example, have become quite permeable, and people and goods from member states can now move freely between them with no customs or passport controls.

Figure 9.6 Berlin Wall The boundary between East and West Germany was virtually impermeable for more than 40 years. The photograph here shows the scene on November 12, 1989, when Berliners tore the wall down in celebration of the reunification of Germany.

Impermeability does not necessarily mean immutability, however. The boundary between East and West Germany, part of the Iron Curtain for more than 40 years, was as aggressively defended as the present boundary between India and Pakistan, yet it was removed in 1989 when Germany was reunified (**Figure 9.6**). Similarly, the boundaries of the former Soviet Union have been dramatically redrawn since 1989, allowing states like Lithuania and Estonia to reappear. Boundaries are an important element of geopolitics and of the geography of domestic politics.

Frontier Regions Frontier regions occur where boundaries are very weakly developed. They involve zones of underdeveloped territoriality, areas that are distinctive for their marginality rather than for their belonging. In the nineteenth century, for example, some vast frontier regions still existed—major geographic realms that had not yet been conquered, explored, and settled (such as Australia, the American West, the Canadian North, and sub-Saharan Africa). All of these are now largely settled, with boundaries set at a range of jurisdictional levels from individual land ownership to

Figure 9.5 Border between India and Pakistan Indian Border Security Force soldiers patrol near the fenced border with Pakistan. Both countries exchanged hostilities regularly for much of the latter half of the twentieth century until a ceasefire that has remained in place since 2004.

local and national governmental borders. Only Antarctica, virtually unsettled, exists today as a frontier region in this strict sense of the term.

There remain, nevertheless, many regions that are still somewhat marginal in that they have not been fully settled or do not have a recognized economic potential, even though their national political boundaries and sovereignty are clear-cut. Examples are the Rondonia region of the Amazon rain forest and the Sahelian region of Africa. Such regions often span national boundaries simply because they are inhospitable, inaccessible, and (at the moment) economically unimportant. Political boundaries are drawn through them because they represent the line of least territorial resistance.

At the local level many examples of frontier, or marginal, regions exist. Although the residents of most towns and cities recognize a series of distinctive districts and neighborhoods, these are often separated by zones or spaces that are marginal. Not fully integrated into the territorial realm of any one sociocultural group, these spaces are often transitional in nature, with a relatively rapidly changing pattern of land use and an equally rapidly changing profile of residents.

Boundary Formation   Generally speaking, formal boundaries tend first to follow natural barriers, such as rivers, mountain ranges, and oceans. Good examples of countries with important mountain-range boundaries include France with Spain (the Pyrenees); Italy with France, Switzerland, and Austria (the Alps); and India and Nepal with China (the Himalayas). Chile, though, provides the ultimate example: a cartographic freak, restricted by the Andes to a very long and relatively thin strip along the Pacific coast. Examples of countries with boundaries formed by rivers include China and North Korea (the Yalu Tumen), Laos and Thailand (the Mekong), and Zambia and Zimbabwe (the Zambezi). Similarly, major lakes divide Canada and the United States (along the Great Lakes), France and Switzerland (Lake Geneva), and Kenya and Uganda (Lake Victoria).

Where no natural features occur, formal boundaries tend to be fixed along the easiest and most practical cartographic device: a straight line. Examples include the boundaries between Egypt, Sudan, and Libya, between Syria and Iraq, and between the western United States and Canada. Straight-line boundaries are also characteristic of formal boundaries that are established through colonization, which is the outcome of a particular form of territoriality. The reason, once again, is practicality. Straight lines are easy to survey and even easier to delimit on maps of territory that remain to be fully charted, claimed, and settled. Straight-line boundaries were established, for example, in many parts of Africa during European colonization in the nineteenth century.

In detail, however, formal boundaries often detour from straight lines and natural barriers in order to accommodate special needs and claims. Colombia's border, for instance, was established to contain the source of the Orinoco River, the Democratic Republic of Congo's border was established to provide a corridor of access to the Atlantic Ocean, and Sudan's border detoured to include the settlement Wadi Halfa (**Figure 9.7**).

After primary divisions have been established, internal boundaries tend to evolve as smaller, secondary territories are demarcated. In general, the higher the population density, the smaller these secondary units tend to be. Their configuration tends to follow the same generalizations as for

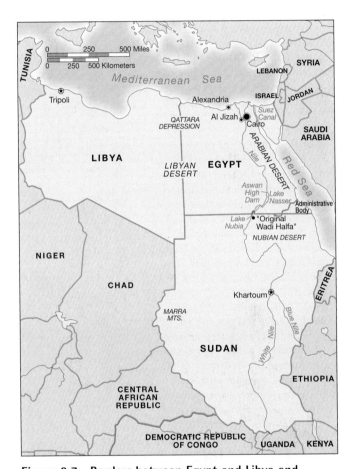

Figure 9.7   Borders between Egypt and Libya and Sudan Where no natural barriers to human activity exist, boundaries are often drawn as straight lines, easily described in a treaty or convention, as in the border between Egypt and Libya. Small natural barriers and isolated settlements sometimes require special detours, however. In the border between Sudan and Egypt, the small half-circle that straddles Lake Nubia in Sudan (Lake Nasser in Egypt) is the original location of the settlement of Wadi Halfa. When the Aswan Dam was under construction, the residents of the settlement were moved eastward as the rising waters of the Nile River flooded their land. That small area marks the location of the original settlement. The triangular area along the Red Sea on the eastern end of the border of Sudan and Egypt, called "Administrative Body," once belonged to Egypt but is now in Sudan. That area was given to Sudan by Egypt in an attempt to induce it to join the now disbanded United Arab Republics (UAR). Sudan never joined the UAR but retained the land, which is a point of recurring controversy and sometimes outright conflict between the two countries.

larger units, following physical features, accommodating special needs, and following straight lines where there are no appropriate natural features or where colonization has made straight lines expedient. It is this last reason, for example, that explains the rectilinear pattern of administrative boundaries in the United States to the west of the Mississippi (**Figure 9.8**).

Territories delimited by formal boundaries—national states, states, counties, municipalities, special districts, and so on—are known as *de jure* spaces or regions. *De jure* simply means "legally recognized." Historically, the world has evolved from a loose patchwork of territories (with few formally defined or delimited boundaries) to nested hierarchies (**Figure 9.9**) and overlapping systems of *de jure* territories.

These *de jure* territories are often used as the basic units of analysis in human geography, largely because they are both convenient and significant units of analysis. They are often, in fact, the only areal units for which reliable data are available. They are also important units of analysis in their own right because of their status as units of governance or administration. A lot of regional analysis and nearly all attempts at regionalization, therefore, are based on a framework of *de jure* spaces.

Figure 9.8 Township-and-range system The checkerboard landscapes of the midwestern and western United States can be traced to the U.S. Land Ordinance of 1785 and the Northwest Territories Act of 1803, which divided much of the country into "townships" in order to facilitate the surveying of federal land. Most townships were laid out as squares, oriented north-south and east-west, with sides 8.9 kilometers (6 miles) long. Some were given irregular shapes that followed natural features such as rivers and mountains. The square-shaped townships were each divided into 36 "sections," each one being 1 square mile, or 640 acres (259 hectares). Sections, in turn, were divided into four "quarter-sections." The Homestead Act of 1863, designed to encourage the settlement of the Great Plains, stipulated that the quarter-section would be the basic unit of land ownership, giving each settler 160 acres of land. With this rectilinear pattern of land ownership, it also made sense for the boundaries of administrative areas—counties and states—to be rectilinear.

Source: U.S.G.S. Capital Cities CD-ROM, 1996.
Original map was: *Phoenix Quadrangle* 1:24,000, U.S.G.S., 1982.

Figure 9.9 Nested hierarchy of *de jure* territories *De jure* territories are constructed at various spatial scales depending on their origin and function. Administrative and governmental territories are often "nested," with one set of territories fitting within the larger framework of another, as in this example of region, department, arrondissement, canton, and commune in France.

GEOPOLITICS AND THE WORLD ORDER

There is, arguably, no other concept to which political geographers devote more of their attention than the state. The state is one of the most powerful institutions—if not the most powerful—implicated in the process of globalization. The state effectively regulates, supports, and legitimates the globalization of the economy.

States and Nations

As described in Chapter 2, the state is an independent political unit with recognized boundaries, even if some of these boundaries may be in dispute. In contrast to a state, a **nation** is a group of people sharing certain elements of culture, such as religion, language, history, or political identity. Members of a nation recognize a common identity, but they need not reside within a common geographical area. For example, the Jewish nation refers to members of the Jewish culture and faith throughout the world, regardless of their places of origin. The term **nation-state** refers to an ideal form consisting of a homogeneous group of people governed by their own state. In a true nation-state, no significant group exists that is not part of the nation. **Sovereignty** is the exercise of state power over people and territory, recognized by other states and codified by international law. **Citizenship** is a category of belonging to a nation-state that includes civil, political, and social rights.

Following the overthrow or decline of monarchies in Europe in the late eighteenth to mid-nineteenth centuries, a number of new republics were created. Republican government, as distinct from monarchy, requires the democratic participation and support of its population. Monarchical political power is derived from force and subjugation; republican political power derives from the support of the governed. By creating a sense of nationhood, the newly emerging states of Europe were attempting to homogenize their multiple and sometimes conflicting constituencies so that they could govern with the population's active cooperation according to a sense of common purpose. Modern citizenship as a political category was a product of the popular revolutions—from the English Civil War to the American War of Independence—that transformed monarchies into republics. In the process, these revolutions produced the need to reimagine the socially and culturally diverse population occupying the territory of the state. Where once they were considered subjects with no need for a unified identity, nationhood required of these same people a sense of an "imagined community," one that rose above divisions of class, culture, and ethnicity. This new identity was called *citizen* and citizenship came to be based on a framework of civil, political, and social rights and responsibilities.

Given that nations were created out of very diverse populations, it is not surprising that no entirely pure nation-states exist today. Rather, multinational states are the norm—states composed of more than one regional or ethnic group. Spain is such a multinational state (composed of Catalans, Basques, Gallegos, and Castilians), as are France, Kenya, the United States, and Bolivia, to name a few. Since World War I, it has become increasingly common for groups of people sharing an identity different from the majority, yet living within the same political unit with the majority, to agitate to form their own state. This has been the case with the Québecois in Canada and the Basques in Spain. It is out of this desire for autonomy that the term nationalism emerges. **Nationalism** is the feeling of belonging to a nation, as well as the belief that a nation has a natural right to determine its own affairs. Nationalism can accommodate itself to very different social and cultural movements, from the white supremacy of the Aryan Nation movements in the United States and Europe to the movements for independence in Estonia, Latvia, and Lithuania. The impact of minority nationalism on the world map has been pronounced during the twentieth century.

The history and the present status of the former Soviet Union also clearly illustrates the tensions among and between states, nations, and nationalism. Demonstrating how enduring are both nationalism and the desire for sovereignty is the history of the Russian Empire: the overthrow of the czar, the subsequent establishment of the Union of Soviet Socialist Republics, and the recent events that have resulted in the establishment of the Commonwealth of Independent States.

The Russian Empire, like other colonial empires in Europe such as Spain and Britain, had a long history. Although it had a medieval origin, most relevant to issues of nationalism was the territorial expansion of the Muscovite state beginning in the fifteenth century. In 1462, Muscovy was a principality of approximately 5,790 square kilometers (2235.5 square miles) centered on the present-day city of Moscow. Over a 400-year period, the Muscovite state expanded at a rate of about 20 square kilometers (7.7 square miles) per day, so that by 1914, on the eve of the Russian Revolution, the empire occupied more than 3.25 million square kilometers (roughly 1.25 million square miles), or one-seventh of the land surface of Earth. The expansion of the Muscovite state was mostly eastward but also to the west and south. This remarkable expansion occurred in discernible stages (**Figure 9.10**).

By the end of the fifteenth century, the Muscovite state had annexed most of the other principalities of present-day European Russia. During the sixteenth century, Moscow conquered the non-Russian Tartar states. Desirous of more forest resources—especially furs—Moscow expanded into Siberia. By the mid-seventeenth century, Russia had wrested the eastern and central parts of Ukraine from Poland. In the eighteenth century, under Catherine the Great, Russia secured the territory of what would eventually become southern Latvia, Lithuania, Belarus, and western Ukraine. The majority of the remaining expansion of the Russian state occurred in the late eighteenth and nineteenth centuries with the acquisition of territory in the Transcaucasus, what is now Kazakhstan,

Figure 9.10 Territorial growth of the Muscovite/Russian state The vast Russian empire grew over the period that corresponds with the globalization of the world economy (fifteenth century to the present). What makes the Russian empire different from many others is that the lands conquered were adjacent, not overseas. When the Bolsheviks came to power early in the twentieth century, some of the territory was lost. Eventually, however, the Bolsheviks were able to control most of the territories formerly ruled by the czars, and it was upon this that they built the Soviet state. (Adapted from D. J. B. Shaw (ed.), *The Post-Soviet Republics: A Systematic Geography.* New York: John Wiley & Sons, 1995, p. 164.)

and desert lands to the south bordering on China and Afghanistan.

Russia's imperial expansion was driven by the same forces that drove expansion of other European empires—the need for more territorial resources (especially a warm-water port) and additional subjects. Russia, however, annexed vast stretches of adjacent land on the Eurasian continent, whereas other empires established new territories overseas. What was strikingly similar about Russian (as well as other European) expansion was that different nations had to be incorporated into one state—and *not* as colonies. Numerous problems attendant on that fact came to challenge Russia's consolidation of its extensive empire, and have reemerged in new ways today.

To meet the challenge of incorporating different nationalities within one state, Russia needed to apply centripetal forces to counter centrifugal forces. **Centripetal forces** strengthen and unify something, whereas **centrifugal forces** divide or tend to pull something apart. Russia's centripetal strategies to bind the 100-plus nationalities (non-Russian ethnic peoples) into a unified Russian state were often punitive and not at all successful. Non-Russian

nations were simply expected to conform to Russian cultural norms. Those that did not were more or less persecuted. The result was opposition and, among many if not most of the nationalities, sometimes rebellion and refusal to bow to Russian cultural dominance.

Such was the legacy that Lenin and the Bolsheviks inherited from the Russian Empire following the overthrow of the aristocracy in 1917. The solution to the "national problem" orchestrated by Lenin was *recognition* of the many nationalities through the newly formed Union of Soviet Socialist Republics (USSR). Lenin believed that a *federal system,* with *federal units* delimited according to the geographic extent of ethnonational communities, would ensure political equality among at least the major nations in the new state. This political arrangement recognized the different nationalities and provided them a measure of independence. Federation was also a way of bringing reluctant areas of the former Russian Empire into the Soviet fold. A **federal state** allocates power to units of local government within the country. The United States is a federal state with its system of state, county, and city/town government. A federal state can be contrasted with a **unitary state**, in which power is concentrated in the central government. The Russian state under the czar was a unitary state.

Lenin was optimistic that once international inequalities were diminished and the many nationalities became one Soviet people, the federated state would no longer be needed. Nationalism would be replaced by communism.

Lenin's vision was short-lived, and following his death in 1924, a true federal system in the USSR also declined. Stalin came to power and enforced a new nationality policy. Although the federal system remained in place, nations increasingly lost their independence and by the 1930s were punished for displays of nationalism. **Figure 9.11** shows the administrative units and nationalities that were part of

Figure 9.11   Soviet state expansionism, 1940s and 1950s World War II gave the Soviet state the opportunity to move westward in order to gain additional territories. Insisting that these countries would never again be used as a base for aggression against the USSR, Stalin retained control over Poland, East Germany, Czechoslovakia, Hungary, Romania, Bulgaria, Albania, and western Austria as well as Yugoslavia for a short time. (Mongolia gained its independence from China in 1921 with the help of the Soviets.) In 1945, Stalin promised democratic elections in these territories. After 1946, however, Soviet control over Eastern and Central Europe became complete as noncommunist parties were dissolved and Stalinist governments installed. (Adapted from *Atlas of Twentieth Century World History.* New York: HarperCollins Cartographic, 1991, pp. 86–87.)

the USSR during Stalin's tenure as premier. It also shows how after World War II Stalin expanded the power of the Soviet state westward to include domination over Albania, Austria, Bulgaria, Czechoslovakia, the German Democratic Republic, Hungary, Poland, Romania, and Yugoslavia.

While a federal system theoretically remained in place through the administration of Mikhail Gorbachev, the USSR actually operated as a unitary state with power concentrated in Moscow. When Gorbachev came to lead the USSR in 1985, he assumed that the diverse nationalities had been so "Sovietized" that an international worker orientation had replaced nationalism. This was not the case, however, and his economic and political policies created conditions that encouraged the reemergence of nationalism in the USSR (**Figure 9.12**).

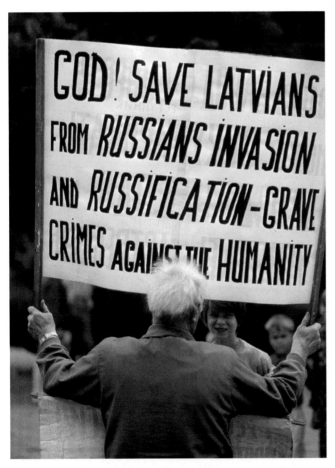

Figure 9.12 Nationalism in Riga, Latvia The Baltic republics (Latvia, Estonia, and Lithuania) were the first to push for reestablishing their sovereignty in the wake of *perestroika* introduced by Gorbachev. All three had a history of independence movements that began with the early expansion of the Russian empire. One obvious reason for their resistance to Russian and later Soviet state incorporation was that all were largely populated by non-Russian people until their forced incorporation into the Soviet Union in 1940. These three republics had been independent nation-states from 1918 to 1940. With a well-developed economic base and higher standards of living than most of the rest of the USSR, Latvia, Estonia, and Lithuania demanded separate nation-state status when the former Soviet Union crumbled in 1991, as illustrated by the Latvian pictured here.

Gorbachev's goal was a massive restructuring of the Soviet economy through radical economic and governmental reforms (*perestroika*) and the direct democratic participation of the republics in shaping those reforms through open discussions, freer dissemination of information, and independent elections (*glasnost*). Effectively, Gorbachev lifted the restrictions that had been placed on the legal formation of national identity. By 1988, grassroots national movements were emerging, first in the Baltic republics and later in Transcaucasia, Ukraine, and Central Asia. By 1991, the relatively peaceful breakup of the Soviet Union was under way, and new states had emerged to claim their independence, facilitated by the federated structure that had existed under the USSR. **Figure 9.13** is a recent map of the former USSR, including all the newly independent states now in the Commonwealth of Independent States (CIS) and the Baltic states of Estonia, Latvia, and Lithuania who chose not to be members of the CIS. The CIS is a **confederation**, a group of states united for a common purpose. The newly independent states that chose confederation did so mostly for economic (and to a lesser extent for military) purposes. A similar case is that of the Confederate States of America, the 11 southern states that seceded from the United States between 1860 and 1861 to achieve economic and political solidarity. This secession led ultimately to the Civil War, a bloody conflict that caused a massive loss of American lives for both the Union and the Confederacy.

With the fall of communism in Eastern Europe in 1991, regions such as the Balkans (the mountainous isthmus of land between the Danube River and the plains of northern Greece that includes Albania, Bulgaria, continental Greece, southeast Romania, European Turkey, and most of the territories formerly organized as Yugoslavia) have also experienced national movements resulting in the redrawing of political boundaries. And while the breakup of the former USSR has been mostly peaceful (with significant exceptions such as Georgia and Chechnya), the recent redrawing of national boundaries in the Balkans resulted in bitter and widespread ethnic conflict. The region is situated at a geopolitical crossroads where East meets West, Islam meets Christendom, the Ottoman Empire met the Austro-Hungarian Empire, and communism once confronted capitalist democracy. In fact, conflict has been characteristic of the region for centuries. In the twentieth century, for instance, ethnic tensions spilled over the boundaries of the Balkan region, eventually pulling all of Europe (and later the United States as well as other, more far-flung countries) into World War I in 1914–1915.

The recent conflicts have been concentrated in the region organized in 1918 as the Kingdom of Serbs, Croats, and Slovenes following the collapse of the Austro-Hungarian Empire. In 1929, this region was renamed Yugoslavia. After 1946, the country included six republics: Slovenia, Croatia, Bosnia-Herzegovina, Macedonia, Serbia, and Montenegro. The first four declared independence in 1991, while Serbia and Montenegro formed a new Yugoslavian state in 1992.

Figure 9.13   Independent states of the former Soviet Union For the most part, the administrative structure of the USSR has remained in place. The differences are that autonomous regions and republics now have
more than nominal local control, and the former federal republics have become independent states (pictured here
in color), as have the popular democracies of Eastern and Central Europe. Despite, or perhaps because of, recent
democratic reforms, nationalist movements continue to plague the consolidation of the state in autonomous
republics like Chechnya, located on the map in the beige area north of Georgia. (*Source:* Reprinted with permission
from Prentice Hall, from J. M. Rubenstein, *The Cultural Landscape: An Introduction to Human Geography,* © 1996, p. 318.)

The most horrifying conflict of the late 1990s was
the war in Kosovo, an autonomous region within Serbia
that lost its autonomy in 1989 when Serbian nationalist
leader Slobodan Milosevic placed it under military occupation. A year later Kosovo's parliament was abolished
and its political leaders fled. Milosevic orchestrated numerous attempts to rid the region of ethnic Albanians,
and in 1995 Serb refugees from Croatia began flooding
into Kosovo. Fueled by the racist rhetoric and military
support of Milosevic, a civil war between Serbs and ethnic Albanians eventually erupted. By late 1999, 800,000
ethnic Albanians had fled Kosovo and tens of thousands
had been massacred. An 11-week air war by NATO
helped to bring the conflict to a halt, though atrocities
continued as ethnic Albanians retaliated against Serbs.
An international war tribunal was set up in The Hague
to try Milosevic and others on human rights violations,
including the possible charge of genocide. Milosevic
acted in his own defense at the trial; he was found in his
cell in 2006, dead of apparent heart failure, before a
verdict was reached.

Theories and Practices of States

The definition of the state provided in the previous section
is a static one. In fact, the state, through its institutions—
such as the military or the educational system—can act to
protect national territory and harmonize the interests of
its people. Therefore, the state is also a *set of institutions*
for the protection and maintenance of society. Thus, it is
not only a place, a bounded territory, it is also an active
entity that operates through the rules and regulations of
its various institutions, from social-service agencies to
governing bodies to the courts, in order to shape national
populations.

An especially influential state theorist whose work
has inspired a large number of political geographers is
Louis Althusser. Althusser developed an approach to the
state that identifies its operations in two ways. The first
is to see the state as an ideological force operating
through the institutions of the schools, the media, the
family, and religion to produce citizens who conform to
state expectations. The second, and complementary, way

is to see the state as repressive: It uses force, through different institutions like the courts, army, and the police, to compel citizens to comply with its rules. These two related aspects of a model for understanding the state have been employed by political geographers in a number of ways, but perhaps the most common is to explore how the spaces of these various institutions, such as the school or the family, are produced and operate to do the work of shaping citizens.

A similar theorist who has also been a significant influence on political geography is Michel Foucault. Foucault's theoretical contribution has been to push Althusser's ideas further by formulating a way for understanding just *how* the various institutions of the state operate to do the work of shaping citizens. Foucault explored the way that power, knowledge, and discourse operate in concert to produce us as particular kinds of subjects: citizens, women, homosexuals, terrorists. For Foucault, **discourse** is an institutionalized way of thinking. For example, an army turns civilians into soldiers through the conjunction of power (in the form of both force and cooperation) and knowledge (accumulated insights about training, discipline, warfare) that come together as discourse, the institutionalized way of thinking that produces rules, identities, practices, exclusions, and a range of other elements that make large numbers of individuals function as an effective collective.

Geographers have incorporated these and other theories and models of the state to explain the unfolding and impact of different kinds of political spaces, from the organization and influence of the classroom to the development and effects of international laws. For political geographers, theories and models of geopolitics have been their most prominent contributions to understanding the role and behavior of the state.

As mentioned previously, an important influence on the field of geopolitics up until the middle of the twentieth century was the nineteenth-century German geographer Friedrich Ratzel. Ratzel's organic theory of the state used biological constructs to describe its actions. He believed that states, like living organisms, progress through stages of youth, maturity, old age, and, unlike living organisms, even a possible return to youth. He also believed that one could determine the general well-being of the state by regarding its size as measured according to its geographic expansion or contraction over time. Ratzel, though certainly an environmental determinist, did not believe that the state is an organism, only that it *acts* like one and grows—increases its territory—as its population grows.

Ratzel wrote during a period of tremendous change in Europe, when the system of states that largely persists to this day was being solidified. It was also a period when European states, Japan, and the United States were maintaining or initiating imperialist practices. The states' efforts to expand were seen as a response to population pressures for more territory and resources. These pressures also led to the need for new markets for manufactured goods and increased colonization of less economically developed places. Eventually, all-out war within Europe occurred as states fought each other over territories.

Imperialism, Colonialism, and the North/South Divide

Geopolitics may involve the extension of power by one group over another. Two ways in which it may occur are through the related processes of imperialism and colonialism. As discussed in Chapter 1, imperialism is the extension of state authority over the political and economic life of other territories. As Chapter 2 describes, over the last 500 years imperialism has resulted in the political or economic domination of strong core states over the weaker states of the periphery. Imperialism does not necessarily imply formal governmental control over the dominated area. It can also involve a process by which some countries pressure the independent governments of other countries to behave in certain ways. This pressure may take many forms, such as military threat, economic sanctions, or cultural domination. Imperialism always involves some form of *authoritative control* of one state by another.

The process of imperialism begins with exploration (**Figure 9.14**), often prompted by the state's perception that there is a scarcity or lack of a critical natural resource. It culminates in development via colonization, the exploitation of indigenous people and resources, or both.

In the first phases of imperialism, the core exploits the periphery for raw materials. As the periphery becomes developed, colonization may occur and cash economies may be introduced where none previously existed. The periphery may also become a market for the manufactured goods of the core. Eventually, though not always, the periphery—because of the availability of cheap labor, land, and other inputs to production—can become a new arena for large-scale capital investment. Some peripheral countries improve their status to become semiperipheral or even core countries. **Figure 9.15** illustrates the colonies created by European imperialism in Africa.

Colonialism is a form of imperialism in that it involves the formal establishment and maintenance of rule by a sovereign power over a foreign population through the establishment of settlements. The colony does not have any independent standing within the world system, but it is considered an adjunct of the colonizing power. From the fifteenth to the early twentieth centuries, colonialism constituted an important component of core expansion. Between 1500 and 1900 the primary colonizing states were Britain, Portugal, Spain, the Netherlands, and France. **Figure 9.16** illustrates the colonization of South America, largely by Spain and Portugal.

Other important states more recently involved in both colonization and imperialist wars include the United States and Japan. Although colonial penetration often results in political dominance by the colonizer, such is not always the case. For example, Britain may have succeeded in setting up British colonial communities in China, but it never succeeded in imposing British administrative or legal structures in any widespread way.

Figure 9.14   Principal steps in the process of exploration
This diagram illustrates the main elements in the process of exploration, beginning with a need in the home country that prompts a desire to look outward to satisfy that need. Geographers have figured prominently in the process of exploration by identifying areas to be explored as well as actually traveling to those places and cataloging resources and people. Nineteenth-century geography textbooks are records of these explorations and the ways geographers conceptualized the worlds they encountered. Exploration is one step in the process of imperialism; colonization is another. (Adapted from J. D. Overton, "A Theory of Exploration," *Journal of Historical Geography* 7, 1981, p. 57.)

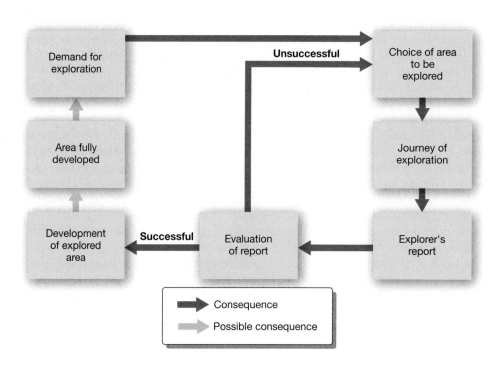

Figure 9.15   European colonies in Africa, 1496–1912 The partitioning of the African continent by the imperial powers created a crazy quilt that cross-cut preexisting affiliations and alliances among the African peoples. Lying directly within easy reach of Europe, Africa was the most likely continent for early European expansion. The Belgian, Italian, French, German, and Portuguese states all laid claim to various parts of Africa and in some cases went to war to protect those claims. (Adapted from *Harper Atlas of World History.* New York: HarperCollins, 1992, p. 139.)

Figure 9.16 Colonization in South America and the Caribbean, 1496–1667 The Spanish and Portuguese dominated the colonization and settlement of South America, with the Dutch, French, and English maintaining only a minor and largely tentative presence. African colonization focused upon gaining new subjects and the simple acquisition of additional territories; South American colonization yielded rich commodity and mineral returns. (Adapted from *Rand McNally Atlas of World History*. Skokie, IL: Rand McNally, 1992, p. 85.)

And at the end of the colonial era a few former colonies, such as the United States, Canada, and Australia, became core states themselves. Others, such as Rwanda, Bolivia, and Cambodia, remain firmly within the periphery. Some former colonies, such as Mexico and Brazil, have come close to the core but have not fully attained core status and therefore are categorized as being within the semiperiphery.

Two examples of colonialism are the extension of British rule to India and French rule to Algeria. The substantial British presence in India began with the establishment of the East India Trading Company in the mid-eighteenth century. The British government gave the company the power to establish forts and settlements, as well as to maintain an army. The company soon established settlements—including factories—in Mumbai (formerly Bombay), Chennai (formerly Madras), and Calcutta. What began as a small trading and manufacturing operation burgeoned over time into a major military, administrative,

and economic presence by the British government, which did not end until Indian independence in 1947. During that 200-year period, Indians were brutalized and killed and their society transformed by British influence. That influence permeated nearly every institution and practice of daily life—from language and judicial procedure to railroad construction and cultural identity. The reasons Britain was able to be so callous in its colonial practices are complex. Theorist Edward Said has proposed the concept of Orientalism to explain them, at least in part. For Said, **Orientalism** is a discourse (as described above) that positions the West as culturally superior to the East. Said developed the concept to describe the way the West has both historically and contemporarily treated Arabs. This same argument can be applied to the British in India or other Western powers with respect to their colonies: The colonized is seen as inferior and in need of disciplining in the eyes of the superior and enlightened colonizer. The colonizer has a moral obligation to colonize, according to this logic.

The postcolonial history of the Indian subcontinent has included partition and repartition, as well as the eruption of regional and ethnic conflicts. In 1947, Pakistan split off from India and became a separate Muslim state. In 1971, Bangladesh, previously part of Pakistan, declared its independence. Regional conflicts include radical Sikh movements for independence in the states of Kashmir and Punjab. Ethnic conflicts include decades of physical violence between Muslims and Hindus over religious beliefs and the privileging of Hindus over Muslims in the national culture and economy. It would be misleading, however, to attribute all of India's current strife to colonialism. The preexisting caste system also plays a significant role in political conflict. The Hindu caste system, which distinguishes social classes based on heredity, preceded British colonization and persists to this day.

The French presence in Algeria is a story of 132 years of colonialism, with accompanying physical violence as well as cultural, social, political, and economic dislocation. Over the course of more than a century, the French appropriated many of the best agricultural lands and completely transformed Algiers, the capital, into a Westernized city (**Figure 9.17**). They also imposed a veneer of Western religious and secular practice over the native and deeply rooted Islamic culture.

In the cases of both India and Algeria, achieving independence has been a painful and sometimes bloody process for both colonizer and colonized. In Algeria, for example, in the aftermath of the protracted war of independence, there occurred a major exodus of French settlers, many of whose families had lived in Algeria for several generations. With few Westerners in positions of power, important aspects of Islamic society and culture were restored. More recently, Western-oriented politicians, bureaucrats, and citizens have been threatened and killed by radical Islamists who seek to return Algeria to a state governed by strict religious tenets (see Chapter 5 on cultural nationalism).

Since the turn of the nineteenth century, the effects of colonialism have continued to be felt as peoples all over the globe struggle for political and economic independence. The 1994 civil war in Rwanda is a sobering

Figure 9.17   British colonialism in India The British presence in India affected culture, politics, the economy, and the layout of cities, as well as numerous other aspects of everyday life. This painting shows how Indian and British cultural practices intermingled, changing both in the process. (Note that the British child is being held by the Indian woman.) Indian society absorbed and remolded many British political and cultural practices so that contemporary Indian government, for example, embodies both British and Indian ideals and practices. British society continues to be shaped by its colonial history in India, most obviously through the large numbers of Indians who have migrated to Britain, affecting all aspects of society and culture.

example of the ill effects of colonialism. As occurred in India, where an estimated 1 million Hindus and Muslims died in a civil war when the British pulled out, the exit of Belgium from Rwanda left colonially created tribal rivalries unresolved and seething. Although the Germans were the first to colonize Rwanda, the Belgians, who arrived after World War I, established political dominance among the Tutsi by allowing them special access to education and the bureaucracy.

Previously a complementary relationship had existed between the Tutsi, who were cattle herders, and the more numerous Hutu, who were agriculturists. In effect, colonialism introduced difference into an existing political and social structure that had operated more or less peacefully for centuries. In 1959, the Hutus rebelled and the Belgians abandoned their Tutsi favorites to side with the Hutus. In 1962, the Belgians ceded independence to Rwanda, leaving behind a volatile political situation that has erupted periodically ever since, most tragically in 1994's civil war. After a year of violence in which over half a million Tutsis were killed, the Hutus were driven across the border to the Democratic Republic of Congo (DRC) and a new Tutsi-led Rwandan government was formed. The Hutu refugees gathered in UN camps that gradually came to be controlled by armed extremists, who transformed them into virtual military bases and used them to attack the Tutsis in Rwanda (**Figure 9.18**). When Rwanda's Tutsi-led military, with Uganda's support, invaded the DRC to break up the camps, over a million refugees were released. Many have since fanned out across Central Africa.

To help to address the culture of impunity that angered so many Rwandans, the UN established a truth commission, known as "Gacaca Court System" based on Rwandan traditional justice. Although its work has had some important moral impacts, the aim of enabling Tutsis and Hutus to work together to rebuild their society and economy has not been particularly successful. Hundreds of thousands of children have become orphans as their parents either died in the atrocities or their mothers were infected with HIV/AIDS through rape and have since died. The country spends more on debt repayments to international banks than it does on education and health. Seventy percent of all households in Rwanda live below the national poverty line. Half the adult population can neither read nor write and one in three children does not attend school. In fact, the economic situation in Rwanda is worse than it was before the 1994 ethnic cleansing, which doesn't bode well for the country's long-term political stability.

The colonization of Africa, South America, parts of the Pacific, Asia, and smaller territories scattered throughout the Southern Hemisphere resulted in a political geographic division of the world into North and South, known as the **North/South divide**. In the North are the imperialist states of Europe, the United States, Russia, and Japan. In the South are the colonized states. Though the equator has been used as a dividing line, some so-called Southern territories, such as Australia and New Zealand, actually are part of the North in an economic sense.

The crucial point is that a relation of dependence was set up between countries in the South, or periphery, and those in the North, or the core, that began with colonialism and persists even today. Very few peripheral countries have become prosperous and economically competitive since achieving political autonomy. Political independence is markedly different from economic independence, and the South remains very much oriented to the economic demands of the North.

Twentieth-Century Decolonization The reacquisition by colonized peoples of control over their own territory is known as **decolonization**. In many cases, sovereign statehood has been achievable only through armed conflict. From the Revolutionary War in the United States to the twentieth-century decolonization of Africa, the world map created by the colonizing powers has repeatedly been redrawn. Today this map comprises an almost universal mosaic of sovereign states.

Many former colonies achieved independence after World War I. Deeply desirous of averting wars like the one that had just ended, the victors (excluding the

Figure 9.18 Refugees returning to Rwanda Fleeing civil unrest in their own country, Rwandans from the Hutu tribe increasingly sought refuge in the Democratic Republic of Congo (formerly Zaire) when the Tutsi-led government assumed power in 1994. Two-and-a-half years later, over half a million Rwandan refugees in the Democratic Republic of Congo occupied some of the largest refugee camps in the world. In late 1996, they began streaming back into Rwanda when the Tutsi-led government urged them to come help rebuild the country. Faced with two difficult alternatives—dire conditions in the camps or possible violence in Rwanda—many refugees chose to go home. Tens of thousands of Rwandans jammed the road between eastern Democratic Republic of Congo and Rwanda for over three days.

United States, which entered a period of isolationism following the war) established the League of Nations. One of the first international organizations ever formed, the League of Nations had a goal of international peace and security. **Figure 9.19** shows the member countries of the League. An **international organization** is one that includes two or more states seeking political and/or economic cooperation with each other.

Within the League, a system was designed to assess the possibilities for independence of colonies and to ensure that the process occurred in an orderly fashion. Known as the *colonial mandate system,* it had some success in overseeing the dismantling of numerous colonial administrations. **Figures 9.20, 9.21,** and **9.22** illustrate decolonization during the twentieth century of Africa, Asia, and the South Pacific and during the nineteenth century in South and Central America. (Although the League of Nations proved effective in settling minor international disputes, it was unable to prevent aggression by major powers and dissolved itself in 1946. It did, however, serve as the model for the more enduring United Nations.)

Decolonization does not necessarily mean an end to domination within the world system, however. Even though a former colony may exhibit all the manifestations of independence, including its own national flag, governmental structure, currency, educational system, and so on, its economy and social structures may continue to be dramatically shaped in a variety of ways by core states. Participation in foreign aid, trade, and investment arrangements originating from core countries subjects the periphery to relations that are little different from those they experienced as colonial subjects.

For example, core countries' provision of foreign aid monies, development expertise, and educational opportunities to selected individuals in Kenya has created a class of native civil servants that is in many ways more strongly connected to core processes and networks than those operating within Kenya. This relatively small group of men and women, often foreign educated, now comprises the first capitalist middle class in Kenyan history.

Commercial relations also enable core countries to exert important influence over peripheral, formerly colonized, countries. For example, as mentioned in Chapter 8, *contract farming* has become a central mechanism around which agricultural production in the periphery is organized for core consumption. Dictated by core countries to

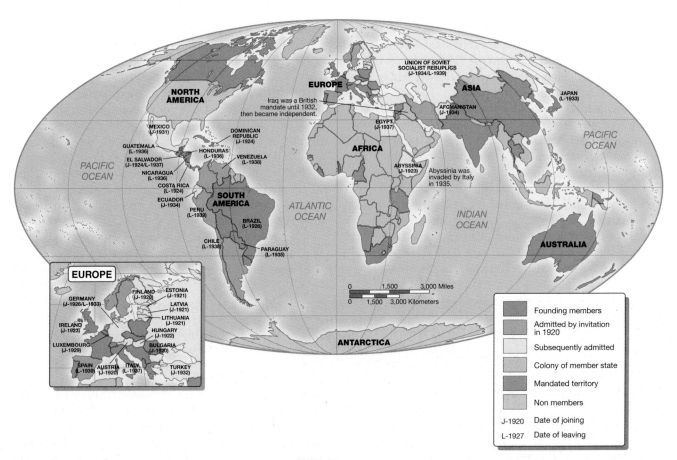

Figure 9.19   Countries participating in the League of Nations Although Woodrow Wilson was a central figure in the creation of the League of Nations, his inability to convince the U.S. Senate to join was a major blow to the effectiveness of this, the first international organization of the twentieth century. Britain and France played important roles in the League but were never able to secure arms limitations and security agreements among the membership. Perhaps its greatest success before it was dissolved in 1946 was pressing for the decolonization of Africa.

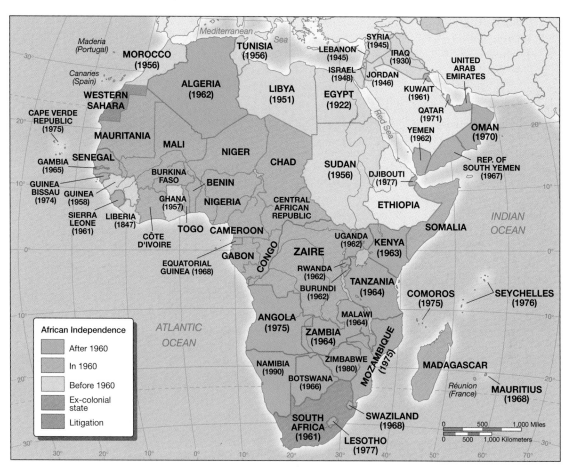

Figure 9.20   Decolonization of Africa, before and after 1960 Britain, France, and Belgium—the dominant European presences in African colonization—were also the first to divest themselves of their colonies. Britain was the first colonial power to grant independence. France granted independence to its African colonies soon after Britain made the first move. In the French-speaking former colonies, the transition to independence occurred largely without civil strife. Belgium's withdrawal as well as the withdrawal of Britain from the remainder of its colonial holdings did not go at all smoothly, with civil wars breaking out. Portugal did not relinquish its possession of Guinea Bissau, Mozambique, or Angola until 1974. (Adapted from *The Harper Atlas of World History, Revised Edition,* Librairie Hachette, p. 285. Copyright © 1992 by HarperCollins Publishers, Inc. Reprinted by permission of HarperCollins Publishers, Inc.)

growers in the periphery are the conditions of production of specified agricultural commodities. Whether the commodity is tea, processed vegetables, fresh flowers, rice, or something else, a contract sets the standards that are involved in producing the commodity.

For example, Japanese firms issue contracts that set the conditions of production for the Thai broiler (chicken) industry; the United Fruit Company, a U.S. firm, issues contracts for Honduran banana production. Thus, a core country can invoke a new form of colonialism in places it never formally colonized. As explained in Chapter 2, this new form, known as *neocolonialism,* is the domination of peripheral states by core states not by direct political intervention (as in colonialism), but by economic and cultural influence and control.

Exploration and colonization did not cease at the midpoint of the twentieth century, however. In fact, exploration, and to a lesser extent, colonization are still occurring in Antarctica (**Figure 9.23**). This ice-covered landmass is an unusual example of twentieth-century

imperialism in which strong states exerted power in an area where no people and, therefore, no indigenous state power existed. At present, while no one country exclusively "owns" the continent, 15 countries lay claim to territory and/or have established research stations there: Argentina, Australia, Belgium, Brazil, Chile, China, France, Germany, India, Japan, New Zealand, Norway, the United Kingdom, the United States, and Uruguay.

Heartland Theory   Because imperialism and colonialism shaped the existing world political map, it is helpful to understand one of the theories that drove them. By the end of the nineteenth century, numerous formal empires were well established, and imperialist ideologies were dominant. To justify the strategic value of colonialism and explain the dynamic processes and possibilities behind the new world map created by imperialism, Halford Mackinder (1861–1947) developed a theory. Mackinder was a professor of geography at Oxford University and director of the London School of Economics.

Figure 9.21 Independent South America, nineteenth century In comparison to Africa, independence came much earlier to South America (it had also been colonized much earlier than Africa). What was most influential in the independence movements there was the presence of local Spanish and Portuguese elites. This colonial ruling class became frustrated with edicts and tax demands from the home country and eventually waged wars—not unlike the U.S. Revolutionary War—against Spain and Portugal for independence. (Adapted from *Rand McNally Atlas of World History*. Skokie, IL: Rand McNally, 1992, p. 113.)

Annotations in map legend:

Annexations
1 Brazil 1905
2 Brazil 1904
3 Ecuador 1880-1922
4 Ecuador 1880-1942
5 Ecuador 1880-1942
6 Peru 1942
7 Brazil 1904-1905
8 Brit.Guiana 1899
9 Brazil 1900
10 Bolivia 1867-1903
11 Bolivia 1867-1903
12 Bolivia 1867
13 Bolivia 1867-1909
14 Bolivia 1867-1902
15 Brazil 1927
16 Paraguay 1935
17 Bolivia 1880-1935
18 Chile 1883
19 Brazil 1870
20 Brazil 1870
21 Argentina 1874
22 Brazil 1895
23 Chile 1884
24 Argentina 1874
25 Argentina 1881
26 Chile 1902

Borders c. 1830
Subsequent border changes
Border of Greater Colombia 1819–1830
Border of Federation of Bolivia and Peru 1835–1839
European immigration (showing country of origin)

He later went on to serve as a member of Parliament from 1910 to 1922 and as chairman of the Imperial Shipping Committee from 1920 to 1945. With his background in geography, economics, and government, it is not surprising that his theory highlighted the importance of geography to world political and economic stability and conflict.

Mackinder believed that Eurasia was the most likely base from which a successful campaign for world conquest could be launched. He considered its closed heartland to be the "geographical pivot," the location central to establishing global control. Mackinder premised his model on the conviction that the age of maritime exploration, beginning with Columbus, was drawing to a close. He theorized that land transportation technology, especially railways, would reinstate land-based power rather than sea prowess as essential to political dominance. Eurasia, which had been politically powerful in earlier centuries, would rise again because it was adjacent to the borders of so many important countries, it was not accessible to sea power, and it was strategically buttressed by an inner and outer crescent of landmasses (**Figure 9.24**).

When Mackinder presented his geostrategic theory in 1904, Russia controlled a large portion of the Eurasian landmass protected from British sea power. In an address to the British Royal Geographical Society, Mackinder suggested that the "empire of the world" would be in sight if one power, or combination of powers, came to control the heartland. He believed that Germany allied with Russia and China organized by Japan were alliances to be feared. Mackinder's theory was a product of the age of imperialism. To understand why

Figure 9.22   Independence in Asia and the South Pacific, before and after 1960
Decolonization and resultant independence are not uniform phenomena. Different factors influence the shape that independence takes. The form of colonial domination that was imposed is as much a factor as the composition and level of political organization that existed in an area before colonization occurred. Some former colonies gained independence without wars of liberation; in Asia these include India and Australia. In other places the former colonizers were prepared to surrender their colonies only after wars of liberation were waged. This was the case in Indochina, where the domino theory influenced the French colonizers—and later the United States—to go to war against the Vietcong. The war for independence in Vietnam lasted from 1954 until 1976, exacting huge costs from all involved. Mostly, decolonization and political independence forced societies into a nation-state mold for which they had little, if any, preparation. It is little wonder, then, that few former colonies have succeeded in competing effectively in a world economy. (Adapted from *The Harper Atlas of World History, Revised Edition,* Librairie Hachette, p. 283. Copyright © 1992 by HarperCollins Publishers, Inc. Reprinted by permission of HarperCollins Publishers, Inc.)

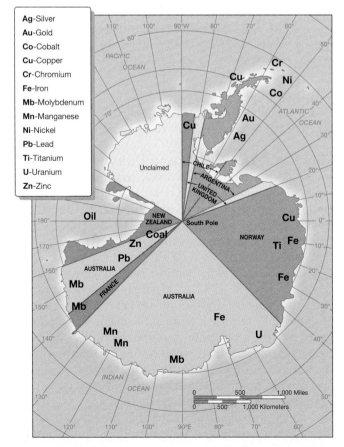

Figure 9.23   Territorial divisions of Antarctica Even the uninhabitable terrain of Antarctica has become a site for competition among states. The radial lines delineating the various claims bear no relationship to the physical geography of Antarctica; rather, they are cartographic devices designed to formalize and legitimate colonial designs on the region. The overlay provides an indication of the mineral wealth of the continent and the drive behind much of the territorial claims. (*Source:* Reprinted with permission from Prentice Hall, from J. M. Rubenstein, *The Cultural Landscape: An Introduction to Human Geography,* 5th ed., © 1996, p. 294; and http://www.coolantarctica.com/Antarctica%20fact%20file/science/threats_mining_oil.htm, accessed August 30, 2008.)

361

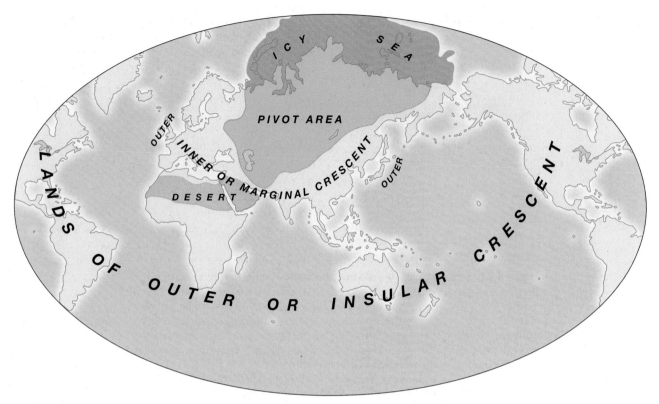

Figure 9.24   The heartland A quintessential geographical conceptualization of world politics, Mackinder's heartland theory has formed the basis for important geopolitical strategies throughout the decades since the theory's inception. While the pivot area of Eurasia is wholly continental, the outer crescent is mostly oceanic and the inner crescent part continental and part oceanic. It is interesting to compare the Mercator map projection, which Mackinder used to promote his geostrategic theory, with the Dymaxion projection used in several maps in this text. This is a classic example of how maps can be used for ideological purposes. The Mercator projection decreases the size of the northern and southern oceans, which are vast and significant natural barriers. The spatial distortions inherent in the Mercator projection overemphasize the importance of Asia, and the splitting of North and South America so that they appear on both sides of the map exaggerates the centrality of Asia. The Dymaxion projection, as a northern polar representation, deemphasizes the centrality of any one landmass but exaggerates distances between continents (see Chapter 2). Mackinder's world-view map provides a good example of how cartographic representations can be employed to support ideological arguments. (Adapted from M. I. Glassner and H. DeBrij, *Systematic Political Geography*, 3rd ed. New York: J. Wiley & Sons, 1980, p. 291.)

Britain adopted this theory, remember that antagonism was increasing among the core European states, leading to World War I a decade later.

Mackinder's geopolitical theory has certain resonance today with respect to the aggressive nationalism of Vladimir Putin, the current prime minister of Russia. As mentioned earlier in this chapter, the breakup of the Soviet Union created extensive regional realignments in the former confederation. Putin's response to the Western leanings of Estonia and Ukraine and invasion of Chechnya and more recently Georgia have some observers suggesting that a new round of Russian empire-building is under way. Although Putin is not likely to attempt a further expansion into Central Europe, he could undermine those countries by capturing their next-door neighbor, Ukraine. And expanding into Georgia would give Russia access to the Mediterranean, the transit route between the Caspian Sea and the Black Sea. Many political geographers today believe that Mackinder's theories are outdated, but Prime Minister Putin's recent aggressions and animosity toward what he regards as the "traitorous"

former republics and breakaway regions underscore the current relevance of his geostrategic formulations.

The East/West Divide and Domino Theory
In addition to a North/South divide based on imperialism and colonialism, the world order of states could also be seen as dividing along an East/West split. The **East/West divide** refers to the gulf between communist and noncommunist countries, respectively. Though the Cold War appears to have ended, the East/West divide has played a significant role in global politics since at least the end of World War II in 1945 and, perhaps more accurately, since the Russian Revolution in 1917. By the second decade of the twentieth century, the major world powers were backing away from colonization. Still, many were reluctant to accelerate decolonization for fear that independent countries in Africa and elsewhere would choose communist political and economic systems instead of some form of Western-style capitalism.

Cuba provides an interesting illustration of an East/West tension that persists despite the official end of the Cold War. Although Cuba did not become independent from Spain until 1902, U.S. interest in the island dates back to the establishment of trade relations in the late eighteenth century. U.S. commercial expansion to Cuba in the first half of the twentieth century wrought major changes in the island's global linkages as U.S. economic imperialism replaced Spanish colonialism. Cuba, at that time, was experiencing a series of reform and revolutionary movements that culminated in the rise of Fidel Castro in the 1950s.

Following the end of World War II, anticommunist sentiment was at its peak in the United States, and U.S. fears about Cuba's "going communist" intensified. With Castro in power by 1959, the United States began training Cuban exiles in Central America for an attack on Cuba. This training proceeded despite the fact that in a 1959 visit to the United States, Castro had publicly declared his alliance with the West in the Cold War. In 1961, an unsuccessful attempt by U.S.-trained Cuban exiles to invade Cuba and overthrow Castro further undermined Cuba's relations with the United States and paved the way for improved Cuban-Soviet relations.

Since that time, the U.S. method of dealing with communism in Cuba has been economic, not military. Having failed to overthrow Castro militarily, the United States fell back upon a 1960 embargo designed to destabilize the Cuban economy. The intent of the embargo, which is still in place, was to create such popular dissatisfaction with economic conditions that Cubans themselves would eventually overthrow Castro.

The nearly 50-year-old embargo forbids economic trade between the United States and Cuba across a wide range of activities and products and it prevents U.S. citizens from traveling there. But a thaw in relations between the two countries began about a decade ago when, in 2000, the United States began selling food and agricultural products to the Cuban government. Today, the United States is Cuba's main supplier of food. And in 2003, President Bush established a Cabinet-level Commission for Assistance to a Free Cuba, whose mission is to explore ways the United States can help hasten and ease a democratic transition there. In February 2008, Fidel Castro announced that he would step down as Cuba's president and his brother Raul has since been elected by the Cuban parliament to that position. Raul Castro has indicated that he is open to talks with the United States but only after the 2008 presidential election.

Though the United States was not able to stop Castro's rise to power in Cuba, the end of World War II marked the rise of the United States to a dominant position among countries of the core. Following the war, the tension that arose between East and West translated into an American foreign policy that pitched it against the former Soviet Union. Domino theory underlay that foreign policy, which included economic, political, and military objectives directed at preventing Soviet world domination. **Domino theory** held that if one country in a region chose or was forced to accept a communist political and economic system,

neighboring countries would fall to communism as well, just as one falling domino in a line of dominos causes all the others to fall. The means of preventing the domino-like spread of communism was often military aggression.

Domino theory first took hold in 1947, when the postwar United States feared communism would spread from Greece to Turkey to Western Europe. It culminated in U.S. wars in Korea, Vietnam, Nicaragua, and El Salvador. Yet preventing the domino effect was based not just on military aggression. Cooperation was also emphasized, as in the establishment of international organizations like NATO (North Atlantic Treaty Organization) in 1949, which had the stated purpose of *safeguarding* the West against Soviet aggression. After World War II, the core countries set up a variety of foreign aid, trade, and banking organizations. All were intended to open foreign markets and bring peripheral countries into the global capitalist economic system. The strategy not only improved productivity in the core countries but also was seen as a way of strengthening the position of the West in its confrontation with the East.

The Vietnam War and its aftermath were probably the most serious global manifestation of Cold War competition and wrought terrible social and environmental damage to Southeast Asia as well as serious impacts on U.S. domestic and international politics. More than a million Vietnamese died, together with 58,000 Americans. U.S. forces sprayed 2 million hectares (5 million acres) of Vietnam with defoliants, such as Agent Orange, that poisoned ecosystems and caused irreparable damage to human health (**Figure 9.25**). Cambodia and Laos were also bombed with napalm and defoliated to disrupt communist supply lines and camps.

The media images of destruction, the loss of U.S. lives, and the financial cost of the war created considerable opposition in the United States, including protests on college campuses and marches on Washington. The United States began to withdraw its forces in the early 1970s and left South Vietnam altogether as the Vietcong approached Saigon in 1973; in 1975, Vietnam was unified under communist rule.

Two million people left South Vietnam fearing repression after unification, many (the so-called Vietnamese boat people) sailing away in small, fragile boats. The communist government confiscated farms and factories from owners to create state- and worker-owned enterprises, resettled hill tribes into intensive agricultural zones, and moved 1 million people into new economic development regions. But U.S.-led economic sanctions from 1973 to 1993 limited the potential for exports and restricted some critical imports, such as medicines. At present, sanctions continue to be lifted and trade and diplomatic relations with the Socialist Republic of Vietnam have improved dramatically since the mid-1990s.

In Cambodia, the Khmer Rouge communist revolutionaries overcame the U.S.-backed military government in 1975 and instituted a cruel regime under the leadership of Pol Pot. The Khmer Rouge suspended formal education,

Figure 9.25 Agent Orange in the Vietnam War Approximately 20 million gallons of herbicides, given the code name Agent Orange, were used in Vietnam between 1962 and 1971 to defoliate the jungle in order to deny cover to enemy forces. The dense jungle created significant challenges for the U.S. military in fighting the Viet Cong. The toxins contained in Agent Orange resulted in serious health problems for service personnel exposed to the spray as well as for Vietnamese civilian and military personnel on the ground. These include diabetes, chloracne, Hodgkin's disease, multiple myeloma, non-Hodgkin's lymphoma, prostate cancer, respiratory cancer, soft tissue sarcoma, and neuropathic problems as well as birth defects in the children of those exposed.

emptied the cities, and set out to eliminate the rich and educated and to isolate the country from the world, renaming it Kampuchea. It is estimated that nearly 2 million people—approximately 30 percent of Cambodia's population—died in the brutal death march out of the capital, Phnom Penh, in 1975–1976 and in the so-called killing fields of the countryside where civilians were tortured and executed. Conflict did not end even with the 1991 UN peace settlement and the return of the monarch, King Sihanouk, because the Khmer Rouge retook power in a coup in 1997. Today the Khmer Rouge have been defeated, and the Cambodian people, with the help of the international community, are building a democratic society as they recover from the emotional, physical, and economic brutality of the recent past.

The New World Order and the Increase in Terrorism With the fall of the Berlin Wall in 1989 and the opening up of former socialist and communist countries, such as China and Russia, to Western-style capitalist economic development, the Cold War is widely regarded as over. In March 1991, President George H. W. Bush made a speech referring to a "new world order" following the collapse of the Soviet Union. Francis Fukayama, author of *The End of History and the Last Man,* subjected the term to a more thorough analysis and put it into circulation in the popular imagination. The notion of the **new world order** assumes that with the triumph of capitalism over communism, the United States becomes the world's only superpower and therefore its policing force. With the political, economic, and cultural dominance of the United States comes the worldwide promotion of liberal democracy and of a global economy predicated on transnational corporate growth through organizations like the World Bank and the World Trade Organization.

However, the move toward liberal, Western-style democracies and the capitalist consumption practices necessary to the success of the new global economy have created instability in some parts of the world. This instability is especially problematic where the Cold War struggle between the United States and the Soviet Union was once waged, in countries that appeared ripe for succumbing to communism. The recent history of Afghanistan provides a telling illustration of this instability and its geopolitical implications (see Box 9.1, "Afghanistan: From the Cold War to the New World Order").

The emergence of a new world order has also meant that radical forms of warfare and political practices have replaced more conventional ones. The attacks on the World Trade Center and the Pentagon on September 11, 2001 and the resulting war on terrorism are prime examples of the political and economic restructuring impulses of the new world order. The terrorist attacks and the subsequent response by the United States and other governments make clear that terrorism is becoming an increasingly important factor in global geopolitics.

Terrorism is a complicated concept whose definition very much depends upon social and historical context. A very simple definition is that **terrorism** is the threat or use of force to bring about political change. It is most commonly understood as actions by individuals or groups of individuals against civilian populations to undermine state practices or institutional organizations. But the state can also be an agent of terrorism, as the original use of the term makes clear. Because terrorism involves violent acts directed against society—whether by antigovernment actors, governments themselves, angry mobs or militants, or even psychotic individuals—it will always mean different things to different people.

The term *terrorism* was first used during the French Revolution (1789–1795) to describe the new revolutionary government's repression of its people during the "Reign of Terror." Fifty years later, the term began to be used to describe revolutionaries who violently opposed existing governments. As the nineteenth century came to a close, "terrorism" was often expanded to apply to militant labor and nationalist political organizations. By the mid-twentieth century, the term was used to describe many left-wing groups, as well as subnationalist (minority groups within the nation-state) or radical ethnic groups. In the 1980s, the violent activity of hate movements in the United States was defined as terrorist, while terrorism internationally was seen as a brand of ethnic or subnational warfare sponsored by rogue regimes.

Ethnic and subnational terrorism affects many countries today, including Russia, Uzbekistan, India, China, Colombia, the Philippines, Israel, and Palestine. In Sri Lanka the Tamil minority rebelled in the early 1980s against the oppressive Sinhalese majority. Although they started as a political movement, eventually the Liberation Tigers of Tamil Eelam launched a brutal campaign of terror that has left over 65,000 people dead. Another ethnic or subnational violent struggle involves the Russian region of Chechnya, where the population has demanded independence. For the Muslim Chechens, Russia is acting as a terrorist state by employing military force to keep them from gaining independence. Chechens believe they have little in common politically, historically, or culturally with Russia. The Russian state has identified the defiant Chechens as extremists and terrorists (see Box 9.2, "State Terrorism in Chechnya").

While subnational resistance organizations using terrorist tactics continue to operate throughout the world, the most widely recognized terrorism of the new century has religious roots. The September 11 attacks have helped to bring the realities of religious terrorism sharply into public focus. The connection between religion and terrorism is nothing new, as terrorism has been perpetrated by religious fanatics for more than 2000 years. Indeed, words like *zealot*, *assassin*, and *thug* all stem from fundamentalist religious movements of previous eras. And while the links between Muslims and terrorism in the world today are especially strong and geographically widespread, it is critical to understand that Muslim terrorism is not the only form of terrorism and that domestic terrorism sparked by fundamentalist Christian organizations has taken the lives of hundreds of innocent people and continues to be a real threat in the United States.

The most recent and widely known example of domestic, Christian, white supremacist–based terrorism in the United States is the bombing of the Alfred P. Murrah Federal Office Building in Oklahoma City in April 1995, when 168 people were killed (**Figure 9.26**). In June 1997, Timothy McVeigh, a U.S. Army veteran, was convicted in federal court of perpetrating the attack. McVeigh was executed by lethal injection in June 2001, and his accomplice, Terry L. Nichols, was sentenced to

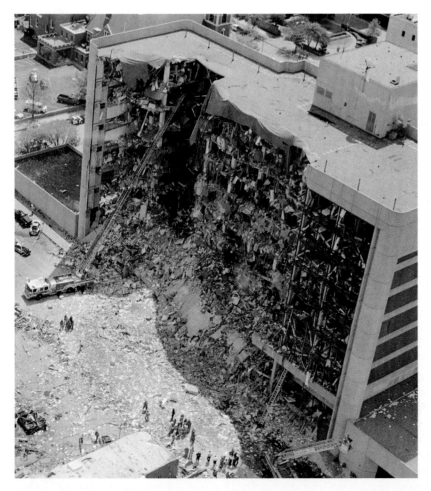

Figure 9.26   The destroyed Alfred P. Murrah Federal Office Building in Oklahoma City, 1995 Until the attacks of September 11, 2001, the Oklahoma City bombing was the worst terrorist incident—domestic or foreign—to be perpetrated on U.S. soil. The blast was caused by a deadly combination of fuel oil and fertilizer placed in a van parked outside the building and detonated remotely. In remembrance of the bombing, the Oklahoma City National Memorial was recently completed and includes a shimmering reflecting pool bordered by 149 chairs for the adult victims and 19 smaller chairs for the children who were killed in the bombing.

Afghanistan: From the Cold War to the New World Order

Afghanistan, known in ancient times as Gandhar, was once famous for its wealth, art, and culture (**Figure 9.A**). Its trading centers were important links on ancient trading routes between Central Asia and South Asia, and their wealth soon attracted invaders. Alexander the Great swept into Afghanistan—then a part of the Persian Empire—in 329 B.C. This invasion paved the way for a cultural awakening and the emergence of the Gandhar school of art, known for its amalgamation of Indian and Greek styles.

But invading the mountain passes of Afghanistan is easier than maintaining control of them (**Figure 9.B**). The problem, put simply, is physical geography. The country is dominated by the rugged Hindu Kush Mountains, which sweep from the east to the west, petering out near the northwestern city of Herat, where they sink into the desert. Tens of thousands of square kilometers of the Hindu Kush form an intricate and seemingly endless maze of valleys and ravines. Jagged scree-strewn mountains and rugged valleys and caves provide ideal territory in which to fight a guerrilla war against invaders or occupying forces. The problems of topography are compounded by the weather. By late October swirling snow descends on the mountains, sealing off many of the passes, valleys, and high plateaus and making the movement of troops almost impossible until late spring.

Figure 9.B   The Khyber Pass Over the centuries, the Khyber Pass has been a gateway to Afghanistan's pivotal geopolitical position between Central Asia and South Asia. Persians, Greeks, Mongols, Afghans, and the British have all taken this route, and the surrounding mountains and gulches became the graveyard for many of them.

Figure 9.A   Afghanistan As this map illustrates, Afghanistan is a land-locked, mountainous country sharing borders with six other countries. When war occurred here in the twentieth century, many of the people fled to these neighboring countries for safety.

Nevertheless, Afghanistan's geopolitical significance has attracted one invader after another. In A.D. 642, Arabs invaded the region and introduced Islam. Arabs quickly gave way to Persians, who controlled the region until they were conquered by Turkic Ghaznavids in A.D. 998. The Turks turned Ghazni into a great cultural center and a base for frequent forays into India. In A.D. 1219, a Mongol invasion, led by Genghis Khan, resulted in the destruction of many cities and the despoliation of fertile agricultural areas. Following Genghis Khan's death in 1227, a number of petty chieftains and princes struggled for supremacy until late in the fourteenth century, when one of his descendants, Timur, incorporated Afghanistan into his vast Asian empire. Babur, a descendant of Timur and founder of the Mughal dynasty, made

Kabul the capital of his principality at the beginning of the sixteenth century. In 1747, Ahmad Shah Durrani consolidated the many chieftainships, petty principalities, and fragmented provinces of the region into one country. His rule extended from Mashhad in the west to Delhi in the east, and from the Amu Darya river in the north to the Arabian Sea in the south.

Late in the eighteenth century, Afghanistan's geopolitical significance increased still more. For the eastward-expanding Russian empire, Afghanistan represented the last barrier to a thrust toward the rich plains of India. For the British, who were establishing a hold on India, Afghanistan represented a bastion against Russian expansion. Both the Russians and the British desperately wanted to control Afghanistan, and so began the Great Game—the struggle between the two imperial powers for control of Afghanistan. The British were able to block the Russians in the Great Game but were not able to establish territorial control. After three wars with the stubborn Afghans, the British finally granted Afghanistan independence in 1921. The first Anglo-Afghan war (1839–1842) resulted in the destruction of the British Army and illustrated the ferocity of Afghan resistance to foreign rule and the difficulty of moving troops and supplies through the Hindu Kush. The second war (1878–1880) was sparked by the Afghans' refusal to accept a British mission in Kabul and resulted in the British and Russians together establishing the official boundaries of modern Afghanistan. The third war began in 1919 with an attack by Afghans on India and ended later that year when the war-weary British relinquished control over Afghan foreign affairs by signing the Treaty of Rawalpindi.

A brief period of Afghan independence followed, during which Mohammad Zahir Shah, who reigned from 1933 to 1973, established a relatively liberal constitution. Although Zahir's experiment in democracy produced few lasting reforms, it permitted the growth of extremist political movements, including the communist People's Democratic Party of Afghanistan (PDPA), which had close ideological ties to the Soviet Union, Afghanistan's northern neighbor.

In 1973, following charges of corruption against the royal family and poor economic conditions caused by a severe drought the previous year, a military coup, led by former prime minister Sardar Mohammad Daoud, abolished the monarchy and declared Afghanistan a republic with Daoud as its first president. His attempts to carry out badly needed economic and social reforms met with little success and were followed in April 1978 by a bloody coup led by the PDPA. During its first 18 months of rule, the PDPA brutally imposed a Marxist-style reform program that ran counter to deeply rooted Islamic traditions. As a result, opposition to the Marxist government emerged almost immediately.

The Soviet Union moved quickly to take advantage of the 1978 coup, signing a new bilateral treaty of friendship and cooperation with Afghanistan that included a military assistance program. Before long, as the opposition insurgency intensified, the PDPA regime's survival was wholly dependent upon Soviet military equipment and advisers. In December 1979, faced with a rapidly deteriorating security situation, the Soviet Union sent a large airborne force and thousands of ground troops to Kabul under the pretext of a field exercise. More than 120,000 Soviet troops were eventually sent to Afghanistan, but they were unable to establish authority outside Kabul. An overwhelming majority of Afghans opposed the communist regime, and Islamic freedom fighters (*mujahideen*) made it almost impossible for the regime to maintain control outside major urban centers. Poorly armed at first, the *mujahideen* began receiving substantial assistance in the form of weapons and training from the United States, Pakistan, and Saudi Arabia in 1984. The *mujahideen* exploited Afghanistan's terrain expertly, firing from surrounding ridges to disable the first and last vehicles in the Soviet columns and then slowly picking off the soldiers trapped in the center. The guerrilla war ended in 1989 with the withdrawal of the Soviet Union, but as the victorious *mujahideen* entered Kabul to assume control over the city and the central government, a new round of fighting began between the various militia groups that had coexisted uneasily during the Soviet occupation.

With the demise of their common enemy, the militias' ethnic, clan, religious, and personality differences surfaced, and civil war ensued. Large-scale fighting in Kabul and in northern provinces caused thousands of civilian deaths and created new waves of displaced persons and refugees, hundreds of thousands of whom trekked across the mountains to Pakistan for sanctuary. Eventually, the hard-line Islamist faction of the *mujahideen*, the Taliban (**Figure 9.C**), gained control of Kabul and most of Afghanistan. By 2001, the only remaining resistance to the Taliban regime was the Northern Alliance, a loose coalition of minority ethnic groups, including the Tajiks, who controlled between 5 and 10 percent of Afghan territory, including the beautiful Panjshir valley 65 miles (105 kilometers) to the northeast of Kabul, and a small enclave around the far northern provinces of Badakhshan and Takhar. The Taliban regime not only imposed harsh religious laws and barbaric social practices on the Afghan population but also harbored an entirely new geopolitical force with worldwide implications: Osama bin Laden and his al-Qaeda terrorist

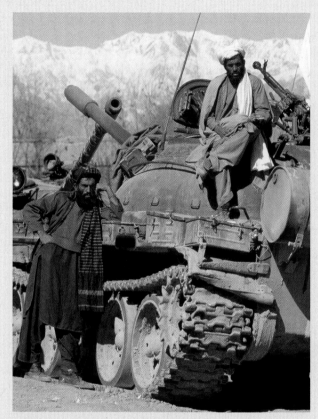

Figure 9.C   Taliban militiamen The militia that dubbed itself the Taliban (which translates as "Islamic students") emerged in 1994 from the rural southern hinterlands of Afghanistan under the guidance of a reclusive former village preacher, Mullah Mohammed Omar. Fed by recruits from conservative religious schools across the border in Pakistan (most of whom were destitute refugees from the 1979–1989 war against the Soviet invasion), the Taliban won military and political support from Pakistan. It rose to power by promising peace and order to a country ravaged by corruption and civil war and the prospect of reestablishing the traditional dominance of the majority ethnic group, the Pashtun.

network, which was responsible for numerous attacks on the West, including the September 11, 2001 attacks on the Pentagon and the World Trade Center.

The United States, with support from the United Kingdom, Australia, Canada, and the Northern Alliance, invaded Afghanistan in October 2001 as part of its "War on Terrorism" campaign. The military campaign is called Operation Enduring Freedom and its aim has been to capture Osama bin Laden, who has been resident there since 1996, destroy the al-Qaeda network, and overthrow the Taliban regime. Bombing raids over Afghanistan officially ceased in 2002 but coalition forces are still operating there despite the democratic election of a president, Hamid Karzai. Although optimism surrounded the democratic elections, the Taliban units and other insurgent fighters have since reconstituted in the country's south and east, the al-Qaeda network persists, and Osama bin Laden has yet to be captured.

For U.S. forces in Afghanistan, the situation is escalating and casualties are on the rise. Increasingly, troops are being killed by roadside bombs, small-arms fire and rocket attacks, and in unspecified combat operations, similar to the situation in Iraq. The biggest concern for defense officials is that the sheer levels of violence are higher than they ever have been. Experts, including those at the Pentagon, say that the war in Afghanistan will probably become more violent, meaning the next U.S. president could inherit an increasingly bloody conflict.

The stability of the country for the Afghan civilian population continues to deteriorate as there is rampant inflation (for example, in 2006 a bag of flour cost 900 Afghanis (US $19); in 2008 it costs 1400 Afghanis (US$30) and Taliban violence is spreading to include Afghan government officials. The UN Security Mission declared in 2008 that Afghanistan is in danger of becoming a "failed state" unless more resources are invested in shoring up the fledgling democracy.

life in prison. Both McVeigh and Nichols were connected to the Christian Identity movement through the Michigan Militia, a 12,000-strong paramilitary survivalist organization. The Christian Identity movement is just one of several religious extremist groups in the United States, including other forms of white supremacy movements, apocalyptic cults, and Black Hebrew Israelism. The Christian Identity movement is based on a belief in the superiority of whiteness as ordained by God. As geographer Carolyn Gallaher has shown, militant tax resistance and a form of regressive populism ("the people" are superior to government) are also characteristic of the U.S. militia groups. An estimated 800 militia groups are believed to exist in almost every state of the United States today. The orientations and objectives of these groups vary, but all are antigovernment and have a

strong belief that God has ordained the superiority of white people. Conversely, members of the Black Hebrew Israelism movement believe that whites are evil incarnate and that blacks are the true Jews of old.

The most feared form of terrorism in the twenty-first century is bioterrorism. **Bioterrorism** is the deliberate use of microorganisms or toxins from living organisms to induce death or disease. Biological and chemical agents that could be used for bioterrorist purposes range from anthrax to West Nile virus. In 2002, the president of the United States signed into law the Bioterrorism Act, intended to protect the nation's food and drug supply from bioterrorist threats. Since 9/11, research about and public and commercial responses to possible bioterrorist threats have grown significantly. Legislation, such as the Bioterrorism Act of 2002, is the driving force behind this new growth industry.

The emergence and persistence of a wide range of extremist religious groups in the United States and throughout the world represents a potentially far more violent threat than terrorist groups of the early and mid-twentieth century. The rise in faith-based terrorism since the new millennium suggests that the potential for even greater acts of violence should not be ignored.

The War on Terror in Iraq

The United States responded to the terrorist attacks of September 11, 2001 by declaring a global war against terrorism and identifying first Afghanistan and then Iraq as the greatest threats to U.S. security. Although the evidence of involvement in the 9/11 attacks by Iraq and its leader, Saddam Hussein, was highly questionable, on March 19, 2003, after amassing over 200,000 U.S. troops in the Persian Gulf region, President George W. Bush ordered the bombing of the city of Baghdad, Iraq. The declaration of war and invasion occurred without the explicit authorization of the UN Security Council, and some legal authorities take the view that the action violated the UN Charter. Some of the staunchest U.S. allies (Germany, France, and Canada) as well as Russia opposed the attack and hundreds of thousands of antiwar protestors repeatedly took to the streets throughout the world for the weeks and months preceding and following the onset of war launched by coalition forces of the United Kingdom and the United States. The

motivation for the war, as expressed by British Prime Minister Tony Blair and George W. Bush, was that Iraq had stockpiled "weapons of mass destruction"—chemical and biological weapons capable of massive human destruction. In the days leading up to the war, the UN weapons inspector Hans Blix and his team were unable to locate any weapons despite an intensive search of the country. President Bush, however, proceeded to justify a dramatically stepped-up "war on terrorism" (following the war in Afghanistan) on the grounds that "neutralizing" Iraq's leader, Saddam Hussein, was necessary to global security.

On May 1, 2003, President Bush gave a speech after having landed a Lockheed S-3 Viking on the aircraft carrier USS *Abraham Lincoln*, where he announced the end of major combat operations in the Iraq War. The fact that "major combat" has ended, however, does not mean that peace has returned to Iraq. Iraq continues to experience violent conflict between U.S. and Iraqi soldiers and forces described by the occupiers as insurgents. The tactics in use include mortars, suicide bombers, roadside bombs, small-arms fire, and rocket-propelled grenades, as well as sabotage against the oil infrastructure of the country. As of summer 2008, the total number of deaths of U.S. soldiers as a direct result of the Iraq invasion reached 4,155, with 4,016 of these occurring since President Bush first declared the mission accomplished. According to official estimates, 30,568 military personnel have been wounded (**Figure 9.27**), but nongovernmental estimates put the number at

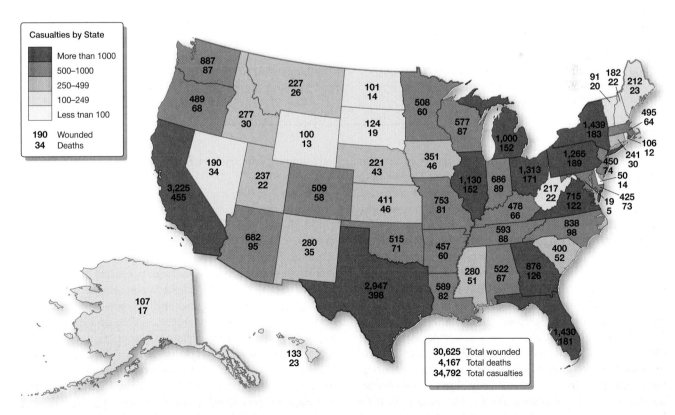

Figure 9.27 Casualties from the war in Iraq, 2008 This map includes both the wounded and dead among military and civilian personnel participating in the U.S. war effort in Iraq. More than a third of those who died have come from six states: California, Florida, Pennsylvania, Ohio, New York, and Texas; more than half were between the ages of 18 and 24.

9.2 GEOGRAPHY MATTERS

State Terrorism in Chechnya

In Chechnya, a region of the northern Caucasus mountain range (**Figure 9.D**), clan, not territory, is the traditional form of political organization. Ever since imperial Russia began expanding into the region in the late 1700s, the Sunni Muslim Chechens put up strong resistance, periodically waging holy wars against Christian Russia. When the Russian revolution occurred in 1917, they did not view the Bolsheviks as an improvement, not least because the newly created Soviet Union formally adopted scientific atheism as its state religion. Following a brief, failed attempt by the peoples of the North Caucasus and Transcaucasus (including the countries of Armenia, Azerbaijan, and Georgia) to resist Soviet domination, the Soviets decided to divide and conquer by creating administrative regions that encompassed a mixture of clans and ethnic groups. The anti-Soviet Chechens were put in the same region as the Ingush peoples to the south.

The Chechens remained defiant but paid a terrible price for doing so. In the late 1930s, tens of thousands were liquidated by Stalin's purges aimed at suspected anti-Soviet elements. In 1944, after invading German troops were forced to retreat from the North Caucasus, Stalin accused the Chechens of having collaborated with the Nazis and ordered the entire Chechen population—then numbering about 700,000—exiled to Kazakhstan and Siberia. Brutal treatment during this mass deportation led to the deaths of more than 200,000 Chechens.

In 1957, Nikita Khrushchev embarked on a program of de-Stalinization that included the rehabilitation of Chechens. But when the Chechens returned, they found newcomers had taken over many of their homes and possessions. Over the next 30 years many of these newcomers withdrew, while the Chechen population consolidated and grew to almost 1 million. When Mikhail Gorbachev initiated his policy of glasnost in 1985, Chechens finally saw a chance for self-determination, and with the breakup of the Soviet Union in 1989, they wasted no time unilaterally declaring their complete independence.

Ingushetia decided to separate from Chechnya in 1992 and signed the Treaty of Federation with Russia. The Russian Federation chose at first to ignore

Figure 9.D The Northern Caucasus This reference map shows the principal features, political boundaries, and major cities of the northern Caucasus.

over 100,000. An especially distressing aspect of the war is its impact on veterans' mental health; the Department of Veteran Affairs director has admitted that there are about 1,000 suicide attempts per month among the nation's 25 million veterans. Besides the mental health impacts, there are, of course, also very significant physical ones, including lost limbs and serious brain and organ injuries.

There have been over 1,225,000 Iraqi deaths—both civilian and military—from all causes as of summer 2008, according to the well-respected British polling agency Opinion Research Business (a detailed discussion of these statistics is available at http://www.antiwar.com/casualties/; for a comparison of the different assessments of Iraqi casualties, see http://en.wikipedia.org/wiki/Casualties_of_the_conflict_in_Iraq_since_2003, accessed September 7, 2008). No reliable figures exist for the number of Iraqis injured since the 2003 onset of the war. UNHCR has indicated that a total of 4.7 million Iraqis have fled the country and have been displaced internally and to surrounding countries such as Syria and Jordan since the war broke out.

In early 2004, the 9/11 Commission (more formally known as the National Commission on Terrorist Attacks Upon the United States) concluded that there was no credible evidence that Saddam Hussein, the captured and executed former president of the country, had assisted the terrorist organization al-Qaeda in preparing for or carrying out the 9/11 attacks. There is also general agreement among U.S. intelligence and military personnel that Iraq had most likely completely destroyed its programs for biological and chemical weapons production before the UN team began its inspection.

In 2007, President Bush ordered, over congressional opposition, a surge of 140,000 troops to Iraq with the intention of providing security to the city of Baghdad and the Al Anbar Province. The surge has met with some success, however patchy and unstable, particularly in Al Anbar, where Sunni tribes are cooperating with the U.S. Army to fight al-Qaeda terrorism. And Sunni tribes

Chechnya's declaration of independence but could not tolerate the possibility of losing the region, particularly since the area around Grozny is one of the Russian Federation's major oil-refining centers and has significant natural-gas reserves. In December 1994, Russian troops invaded Chechnya. The ensuing conflict brought terrible suffering to the Chechen population and resulted in mass migrations away from the fighting. Chechen resistance stiffened, with increased popular support because of the invasion. Russian forces were eventually worn down, and in 1996 the Russian Federation settled for peace, leaving Chechnya with *de facto* independence. For three years there were protracted negotiations over the nature of the peace settlement. In the summer of 1999, however, after Chechen rebels took the fight to the neighboring republic of Dagestan and to the Russian heartland with a series of terrorist bombings of apartment blocks, the Russian military effort was renewed. After bitter and intense fighting, during which the Russian army suffered over 400 deaths and nearly 1,500 wounded while hundreds of thousands of Chechens were made homeless and several thousand were dead or missing, Russian troops took the capital, Grozny, in February 2000 (**Figure 9.E**). By that time, Grozny was virtually uninhabitable. Since early 2000, Russian Federation troops have maintained control of Grozny, though they continue to be harassed by Chechen rebel guerrillas. And in 2008, although the war has ceased, in Chechnya, everything and everyone has been brutalized. Particularly in

Figure 9.E   Grozny refugees In December 1999, Chechens were advised by Russian troops to evacuate Grozny, the capital of Chechnya, before it was destroyed.

Grozny, murders and kidnappings—initiated by both Chechens and the Russian intelligence service—are still commonplace, and organized crime flourishes through the Chechen and Russian mafias throughout the country.

Chechnya provides a useful example of state terrorism. It also provides an illustration of the complexity of terrorism as a concept by showing that it can be practiced by both individuals as well as institutions, by rogue forces as well as legitimate ones.

around the country have formed "Awakening Councils" paid for by and cooperating with U.S. troops and leading to an overall decrease in sectarian violence. Unfortunately, sectarian violence is still occurring on a daily basis throughout the rest of the country, especially in Baghdad. Moreover, the civilian humanitarian situation is among the most critical in the world, as millions of Iraqis have insufficient access to clean water, sanitation, and health care.

For the United States, the costs of the war are still mounting. U.S. intelligence agencies have suggested that the global jihadist movement has been fueled by the coalition forces' occupation of Iraq to carry out a cycle of revenge attacks known as "blowback," which could threaten the United States and other coalition countries for decades. The U.S. Treasury estimates the cost of the war at $845 billion directly. Nobel-prize winning economist Joseph Stiglitz and coauthor Linda Bilmes argue that the true cost of the war is closer to $3 trillion when the indirect costs are also assessed, including interest on the debt raised to fund the war, the rising cost of oil, health care costs for returning veterans, and

replacing the destroyed military hardware and degraded operational capacity caused by the war.[2] Most politically disturbing is that global perceptions of the U.S. role in world affairs have significantly deteriorated over the last year according to a BBC World Service poll of more than 26,000 people across 25 different countries.

INTERNATIONAL AND SUPRANATIONAL ORGANIZATIONS AND NEW REGIMES OF GLOBAL GOVERNANCE

Just as states are key players in political geography, so too have international and supranational organizations become important participants in the world system in the last

[2]J. Stiglitz and L. Bilmes, *The Three Trillion Dollar War*. New York: W.W. Norton, 2008.

century. These organizations have become increasingly important ways of achieving goals that could otherwise be blocked by international boundaries. These goals include, among other things, the freer flow of goods and information and more cooperative management of shared resources, such as water.

Transnational Political Integration

Perhaps the best-known international organization operating today is the United Nations (**Figure 9.28**). The postwar period has seen the rise and growth not only of large international organizations but also of new regional arrangements. These arrangements vary from highly specific, such as the Swiss-French cooperative management of Basel-Mulhouse airport, to the more general, such as the North American Free Trade Agreement (NAFTA),

which joins Canada, the United States, and Mexico into a single trade region. Regional organizations and arrangements now address a wide array of issues, including the management of international watersheds and river basins, such as the Great Lakes of North America and the Danube and Rhine Rivers in Europe. They also oversee the maintenance of health and sanitation standards, coordinated regional planning, and tourism management. Such regional arrangements seek to overcome the barriers to the rational solution of shared problems posed by international boundaries. They also provide larger arenas for the pursuit of political, economic, social, and cultural objectives.

Unlike international organizations, supranational organizations reduce the independence of individual states. A **supranational organization** is a collection of individual states with a common goal that may be economic and/or

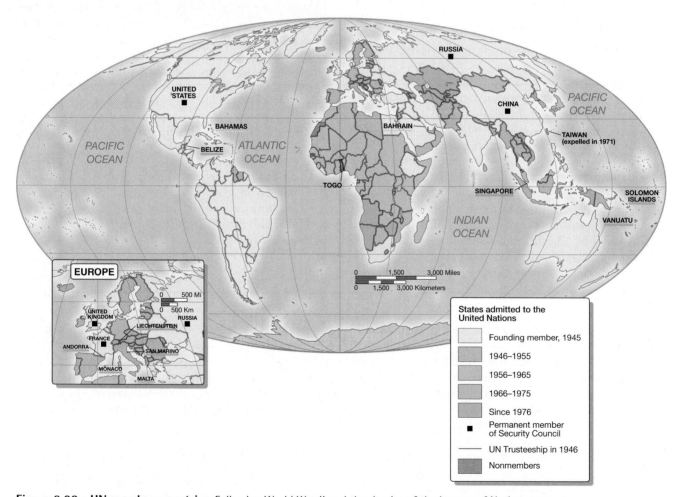

Figure 9.28   UN member countries Following World War II and the demise of the League of Nations, a renewed effort was made to establish an international organization aimed at instituting a system of international peace and security. The UN Charter was approved by the U.S. Senate in July 1945, raising hopes for a more long-lived organization than the ineffective League of Nations. Located in New York City, the United Nations is composed of a Security Council, which includes permanent members the United States, Britain, China, France, and Russia, and a General Assembly, which includes all those countries identified on the map. At the same time that the United Nations was set up, the United States lobbied for the creation of the International Monetary Fund (IMF) and the World Bank. The U.S. government had believed that World War II resulted from the collapse of world trade and financial dislocation caused by the Great Depression. The task of the IMF and the World Bank is to provide loans to stabilize currencies and enhance economic growth and trade.

political in nature. By organizing and regulating designated operations of the individual member states, these organizations diminish, to some extent, individual state sovereignty in favor of the collective interests of the entire membership. The European Union (EU) is perhaps the best example of a supranational organization.

At the end of World War II, European leaders realized that Europe's fragmented state system was insufficient to the demands and levels of competition coalescing within the world political and economic system. They endeavored to create an entity that would preserve important features of state sovereignty and identity. They also intended to create a more efficient intra-European marketing system and a more competitive entity in global transactions. **Figure 9.29** shows the original member countries of the European Economic Community, which evolved into the EU in 1992, the existing members of the EU, and those countries that are currently applying for admission.

The EU holds elections, has its own parliament and court system, and decides whether and when to allow new members to join. Generally speaking, the EU aims to create a common geographical space within Europe in which goods, services, people, and information move freely and in which a single monetary currency prevails. Whether an EU foreign policy will ever be accomplished remains to be seen, but a common European currency, the euro, has been in circulation since 2002. Despite the members' concession to a common currency, indicators of nationalism within the individual member countries

remain strong. For example, just as the European system of states is on the threshold of dissolving into the larger EU organizational form, national and regional movements (described in Chapter 5) have become potent forces in resisting full integration, including the rejection of a European Union Constitution that would supersede already existing national ones.

Globalization, Transnational Governance, and the State

As we have already noted, globalization has been as much about restructuring geoeconomics as it has been about reshaping geopolitics. In fact, some globalization scholars believe that the impact of globalization on politics has been so profound that it is leading to the diminution of the powers of the modern state, if not its ultimate disappearance. These scholars, known as hyperglobalists, believe that because the modern state is organized around a bounded territory and because globalization is creating a new economic space that is transnational, the state is increasingly incapable of responding to the needs of the new transnational economy. Although we do not subscribe to this position, we do recognize that the state is undergoing dramatic changes that are restructuring its role with respect to both local, domestic concerns as well as global, transnational ones. Moreover, these changes are very much part of recent history.

Figure 9.29 Membership in the European Union The goal of the European Union is to increase economic integration and cooperation among the member states. The EU was established in 1992, when the Maastricht Treaty was ratified by the 12 members of the European Economic Community, or EEC (Belgium, Denmark, France, Germany, Greece, Ireland, Italy, Luxembourg, Netherlands, Portugal, Spain, and the United Kingdom), which was created in 1967. Upon ratification of the treaty, the countries of the EEC became members of the EU, and the EEC became the policymaking body of the EU. The Maastricht Treaty established European citizenship for citizens of each member state, enhanced EEC customs and immigration agreements, and allowed for the establishment of a common currency, the euro, which is currently in circulation among all of the original 12 members except Denmark, the United Kingdom, and Greece. Newer members, including Finland, Austria, Slovakia, Latvia, Lithuania, and Malta, also use the euro in addition to their own currencies. The EU is governed through both supranational European institutions (the European Commission and the European Parliament, both administered by the EU) and the governments of the member states, which send representatives to the Council of Ministers (the main lawmaking body of the EU). Membership in the EU is much sought after, and numerous European countries, such as Turkey, have applied and are on the waiting list for admission.

In the twentieth century, from the end of World War II until 1989, when the Berlin Wall was dismantled, world politics were organized around two superpowers. The capitalist West rallied around the United States, and the communist East around the Soviet Union. But with the fall of the Berlin Wall signaling the "end of communism" as a world force, the bipolar world order came to an end, and the new world order, which was organized around global capitalism, emerged and has increasingly solidified around a new set of political powers and institutions that have recast the role of the state.

We have referred throughout the text, and especially in Chapters 2 and 7, to the importance for the contemporary global economy of such regional and supranational organizations as the EU, NAFTA, the Association of Southeast Asian Nations (ASEAN), the Organization of Petroleum Exporting Countries (OPEC), and the World Trade Organization (WTO). We have also noted that these organizations are unique in modern history, as they aim to treat the world and different regional clusters as seamless trading areas unhindered by the rules that ordinarily regulate national economies. The increasing importance of these trade-facilitating organizations is the most telling indicator that the world, besides being transformed into one global economic space, is also experiencing global geopolitical transformations. But rather than disappearing altogether, the powers and roles of the modern state are changing as it is forced to interact with these sorts of organizations, as well as with a whole range of other political institutions, associations, and networks (**Figure 9.30**).

The point is not that the state is disappearing, but rather that it must now contend with a whole new set of processes and other important political actors on the international stage as well as within its own territory. For instance, geographer Andrew Leyshon has shown how in the 1980s a transnational financial network was established that is far beyond the control of any one state, even a very powerful state like the United States, to regulate effectively. In fact, what the increasing importance of transnational flows and connections—from flows of capital to flows of migrants—indicates is that the state is

less a container of political or economic power and more a site of flows and connections.

The increasing importance of flows and connections means that contemporary globalization has made possible a steadily shrinking world. In addition to allowing people and goods to travel farther faster and to receive and send information more quickly, time-space convergence—or the smaller world that globalization has created—means that politics and political action have also become global. In short, politics can move beyond the confines of the state into the global political arena, where rapid communications enable complex supporting networks to be developed and deployed, thereby facilitating interaction and decision making. One indication of the increasingly global nature of politics outside of formal political institutions is the increase in environmental organizations whose purview and membership are global, as discussed in Chapter 4.

What has been most interesting about the increasing institutionalization of global politics is that it has been less involved with the traditional preoccupations of relations between states and military security issues and more involved with issues of economic, ecological, and social security. The massive increases in flows of trade, foreign direct investment, financial commodities, tourism, migration, crime, drugs, cultural products, and ideas have been accompanied by the emergence and growth of global and regional institutions whose role is to manage and regulate these flows. The impact of these twin forces on the modern state has been to draw it increasingly into this complex of global, regional, and multilateral systems of governance. And as the state has been drawn into these new activities, it has shed or deemphasized some of its previous responsibilities, such as maintenance of social welfare.

The involvement of the state in these new global activities, the growth of supranational and regional institutions and organizations, the critical significance of transnational corporations to global capital, and the proliferation of transnational social movements and professional organizations are captured by the term **international regime**. The term reflects the fact that the arena of contemporary politics is now international, so much so that even city

Figure 9.30 Growth of states and intergovernmental organizations and nongovernmental organizations in the twentieth century The number of states has grown steadily over the twentieth century, but intergovernmental organizations (IGOs) and international nongovernmental organizations (INGOs) have experienced dramatic growth, particularly since the 1960s. While states remain the main forms of national government, they have turned over many of their governing responsibilities to international governing organizations and nongovernmental organizations.

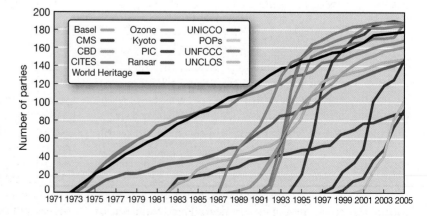

governments and local interests groups—from sister-city organizations to car clubs—are making connections and conducting their activities both beyond and within the boundaries of their own states. An example of this is the human rights movement that has gained ascendancy over the last four to five decades or so. **Human rights**, including the rights to justice, freedom, and equality, are considered by most societies to belong automatically to all people.

Until World War II, safeguarding human rights was the provenance of states whose rules and regulations legislated the proper treatment of its citizens, from prisoners to schoolchildren. Since the late 1940s and 1950s, nearly all states have come to accept the importance of a comprehensive political and legal framework that focuses on human rights and that allows an international organization to intervene in the operations of a sovereign state that is in violation of the International Bill of Human Rights adopted by the United Nations in 1948. In 1998, the United Nations realized another step in the protection of human rights by adopting a treaty to establish a permanent International Criminal Court (ICC). In establishing the court, Kofi Anan, secretary-general of the United Nations, stated: "Our hope is that, by punishing the guilty, the ICC will bring some comfort to the surviving victims and to the communities that have been targeted. More important, we hope it will deter future war criminals, and bring nearer the day when no ruler, no State, no junta and no army anywhere will be able to abuse human rights with impunity."

In forming the treaty, the United Nations aimed to create a permanent mechanism to bring to justice the perpetrators of such crimes as genocide, ethnic cleansing, sexual slavery, and maiming, in order to put an end to the impunity so often enjoyed by those in positions of power. The court has the mandate to try individuals, not states, for crimes committed in the present (after July 2002). What makes the ICC unique is that its basic premise is the principle of complementarity, which means that the court can only exercise its jurisdiction when a national court is unable or unwilling to genuinely do so itself. Thus, the importance of the international scale has increased over the last 50 years of the twentieth century and appears likely to continue to do so into the twenty-first century. It is important to note, however, that not all states support the ICC. Seven UN members voted against the treaty to establish the ICC. These were the United States, China, Iraq, Israel, Libya, Qatar, and Yemen.

An additional and often overlooked aspect of human rights is **children's rights**. In 1989, the United Nations adopted the Convention on the Rights of the Child, and despite the convention's nearly universal ratification (only the United States and Somalia have not ratified it), many of the most basic rights of children are still not being met. The convention promises children around the world the fundamental right to life, liberty, education, and health care. Among the many other fundamental safeguards the convention provides are protection of children in armed conflict; protection from discrimination; protection from torture or cruel, inhuman, or degrading treatment or punishment; protection within the justice system; and protection from economic exploitation. And yet, while these promises are in place and every one but two of the world's nations have ratified the convention, around the world street children are killed or tortured by police, they are recruited or kidnapped to serve as soldiers in military forces or labor under extremely difficult conditions, they are forced into prostitution, or they are imprisoned. Refugee children, often separated from their families, are vulnerable to exploitation, sexual abuse, or domestic violence. Keeping the promises made in the Convention on the Rights of the Child is one of the biggest challenges of the twenty-first century.

A final example is the UN Declaration on Rights of Indigenous People, approved in 2007 when 143 countries voted for it; 11 countries abstained; and 4—Australia, Canada, New Zealand, and the United States, all countries with very significant numbers of indigenous people—voted against it. The declaration emphasizes the rights of indigenous peoples to maintain and strengthen their own institutions, cultures, and traditions and to pursue their development in keeping with their own needs and aspirations. It also prohibits discrimination against them and promotes their full and effective participation in all matters that concern them. The four "no" voting countries expressed concern about the vague language of the declaration and the impact this might have on existing, already settled treaties.

The emergence of human rights as a globally relevant issue has occurred as groups and organizations, both governmental and nongovernmental, have been able to debate and discuss issues that concern all people everywhere and can do so at the international level through conferences, e-mail, listservs, and direct action at international events. The phenomenon of different people and groups across the world in common cause is known as global civil society. **Global civil society** is composed of the broad range of institutions that operate between the private market and the state.

In the next section, we move from the international level to national, regional, and local levels. Our aim here is to show that political geography occurs at all levels of political organization and that each scale enables significant insights about the politics of geography and the geography of politics.

THE TWO-WAY STREET OF POLITICS AND GEOGRAPHY

Political geography can be viewed according to two contrasting orientations. The first orientation sees it as the *politics of geography*. This perspective emphasizes that *geography*—or the areal distribution/differentiation of people and objects in space—has a very real and measurable impact on politics. Regionalism and sectionalism, discussed later in this section, illustrate how geography shapes politics. The politics-of-geography orientation is also a reminder that politics occurs at all levels of the human experience, from the international order down to the neighborhood, household, and body.

The second orientation sees political geography as the *geography of politics*. This approach analyzes how *politics*—the tactics or operations of the state—shapes geography. Mackinder's heartland theory and the domino theory attempt to explain how the geography of politics works at the international level. In the heartland theory, the state expands into new territory in order to relieve population pressures. In the domino theory, as communism seeks new members, it expands geographically to incorporate new territories. An examination of a series of maps of Palestine/Israel since 1923 reveals how the changing geography of this area is a response to changing international, national, regional, and local politics (see Box 9.3, "The Palestinian-Israeli Conflict").

The Politics of Geography

Territory is often regarded as a space to which a particular group attaches its identity. Related to this concept of territory is the notion of **self-determination**, which refers to the right of a group with a distinctive politico-territorial identity to determine its own destiny, at least in part, through the control of its own territory.

Regionalism and Sectionalism Different groups with different identities—religious or ethnic—sometimes coexist within the same state boundaries. At times, discordance between legal and political boundaries and the distribution of populations with distinct identities leads to movements to claim or reclaim particular territories. These movements, whether conflictual or peaceful, are known as *regional movements*. **Regionalism** is a feeling of collective

identity based on a population's politico-territorial identification within a state or across state boundaries.

Regionalism often involves ethnic groups whose aims include autonomy from an interventionist state and the development of political power. For example, in spring 1993 several leading Basque guerrillas were arrested in France, raising hopes for an end to Basque terrorism in Europe. The Basque people are one of the oldest European peoples, with a distinctive culture and language. Despite over a century of nationalism, they are still governed by both France and Spain, with only limited autonomy in selected parts of their territory.

The Basque provinces of northern Spain and southern France have sought autonomy from those states for most of the twentieth century. Since the 1950s, agitation for political independence has included—especially for the Basques in Spain—terrorist acts. Not even the Spanish move to parliamentary democracy and the granting of autonomy to the Basque provinces could squelch the Basque thirst for self-determination. On the French side, the Basque separatist movement is neither as violent nor as active as the movement in Spain.

Another separatist movement is the EZLN (Ejército Zapatista de Liberación Nacional), the Zapatista Army of Liberation, which practices a different type of revolutionary autonomy. Composed largely of indigenous peasants from Chiapas, one of the poorest states in Mexico, the EZLN opposes the Mexican federal state and its embrace of corporate globalization, arguing that it oppresses them by denying the peasant way of life. A key element of the Zapatista ideology is their aspiration to practice politics in a truly participatory way, from the "bottom-up," by guaranteeing the right of indigenous peoples to form and govern traditionally their own municipalities. In doing so the Zapatistas reject official authorities and elect their own at the same time that they refuse federal government

Figure 9.31 Kurdish independence protest At a demonstration in Germany, Kurdish immigrants hold up signs on behalf of Abdullah Öcalan, the Kurdish rebel leader who founded the Kurdistan Workers Party (the PKK) and undertook armed attacks on the Turkish government in order to secure an independent Kurdish state. Since his capture in Greece in 1999, öcalan has been held in solitary confinement as the only prisoner on the imrali Island in the Turkish Sea of Marmara. Despite the fact that all other prisoners formerly at imrali were transferred, there are still over 1,000 Turkish military personnel stationed there guarding him. Since his imprisonment, Ocalan has advocated more peaceful means for securing Kurdish independence. The PKK has also become an important force for protecting the Kurds as a minority population within Turkey.

involvement and control. The Zapatistas have been building this autonomous form of governing since 1994 with dramatically beneficial results in community health, education, housing and general welfare.

We need only look at the long list of territorially based conflicts that have emerged in the post–Cold War world to realize the extent to which territorially based ethnicity remains a potent force in the politics of geography. For example, the Kurds continue to fight for their own state separate from Turkey and Iraq insert (**Figure 9.31**). A significant proportion of Québec's French-speaking population, already accorded substantial autonomy, continues to advocate complete independence from Canada. In the 1996 plebiscite on the issue, the separatists were only very narrowly defeated. Consider also the former Yugoslavia, whose geography has fractured along ethnic and religious lines (**Figure 9.32**). Regionalism also underlies efforts to sever Scotland from the United Kingdom.

Regionalism may also be based on economics. For example, in a nonbinding referendum put before the California voters in 1993, drawing on sentiment over a century old, California's northern and mostly rural counties voted to separate from the southern, more urban counties. The desire for separation was based on the belief that political representation in the state legislature economically and politically advantaged the south over the north. Many citizens felt that the state had grown unmanageably large and that the government had become unresponsive to the people. Although many additional steps would have to be taken—such as state senate

approval, approval by the governor and by the voters, and finally by the U.S. Congress—before separation could occur, the California separatist movement represents an interesting example of *economic* regionalism.

Although they sometimes coexist, **sectionalism,** an extreme devotion to local interests and customs, should not be confused with regionalism. Sectionalism has been identified as an overarching explanation for the U.S. Civil War. It was an attachment to the institution of slavery and to the political and economic way of life slavery made possible that prompted the southern states to secede from the Union. The Union went to war to ensure that sectional interests would not take priority over the unity of the whole, that is, that states' rights would not undermine the power of the federal government. Although the Civil War was waged around the real issue of permitting or prohibiting slavery, it also involved disagreement over the power of the states. As **Figure 9.33** shows, the election of Abraham Lincoln to the presidency in 1860 reflected the sectionalism that dominated the country: He received no support from slave states.

Suburbs vs. Cities, and Rural vs. Urban

Sectionalism persists today, but in different forms. In the United States one of the most apparent manifestations of sectionalism may be found in the politics of differentiation between suburbs and cities. Two examples are the "taxpayer revolts," expressed in the passage of Proposition 13 in California and Proposition 2½ in Massachusetts, and

Figure 9.32 Map of the former Yugoslavia The former Yugoslavia now consists of five nations: Slovenia, Croatia, Bosnia and Herzegovina, the new Yugoslavia (made up of Serbia and Montenegro), and Macedonia. For the most part, the boundaries of the Yugoslav states were laid out only in the twentieth century, across segments of the Austro-Hungarian and Ottoman empires that had acquired a complex mixture of ethnic groups. The history of these boundaries has also been the history of ethnic conflict revolving around claims to territory, as well as religious intolerance. As this map shows, with the exception of Slovenia, the new states are home to a mix of nationalities. (Redrawn with permission from Prentice Hall, from J. M. Rubenstein, *The Cultural Landscape: An Introduction to Human Geography*, 6th ed., © 1999, p. 260.)

Figure 9.33 The presidential election The U.S. presidential election of 1860 graphically illustrates the role of sectionalism in determining who gets votes from which geographical regions. In a four-way race, Abraham Lincoln failed to win the support of any of the slave states. (Adapted from *Presidential Elections Since 1789*, 4th ed. Washington, DC: Congressional Quarterly Inc., 1987.)

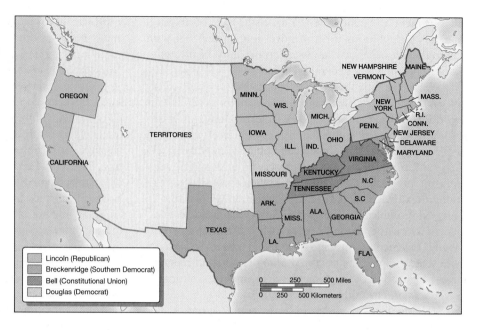

growth-control movements. The former were voter-approved mandates to cut property taxes and limit government spending, while the latter were government-enforced caps on population growth approved by residents to limit the density and extent of growth. Both types of actions have been dominated by suburban U.S. voters.

Arising out of the taxpayer revolts, the growth-control movements of the 1980s also drew their support from suburban home-owning constituencies. Arguing for the maintenance of open space, low-density, large lots, and community control over land-use planning and development, the growth-control movements have attempted to keep the city out of the suburb by retaining a "country feel." Critics of growth control complain that it discriminates against newcomers to the community. Related to growth-control movements is *NIMBYism*, which stands for "Not In My Backyard" actions. NIMBYism is action by neighborhood residents against the introduction of unwanted land uses, that can range from group homes for developmentally disabled adults or battered women's shelters to low-income housing and other residential forms. NIMBY supporters see these land uses as a threat to the composition and quality of the neighborhood and the market value of single-family residences.

The politics of geography, in terms of regionalism, also finds strong focus today in rural versus urban politics. In France, for example, attitudes about birth control (and birth rates themselves) are significantly different between the urbanized north of the country and the more rural south. Throughout the EU, farmers have fought the removal of farm subsidies and tariff arrangements advocated by urban-based policymakers because the existing arrangements have long protected agricultural productivity. Here the dispute pits the politics of local farmers against an international organization.

In England and Wales, a rural versus urban conflict over foxhunting has been ongoing. In the countryside, foxhunting is largely seen as a ritual of rural life. In the early 2000s, a largely urban constituency of animal welfare proponents opposed foxhunting and pressured the government to act to protect the foxes. The result was the Hunting Act of 2004 that made it illegal to hunt a mammal using a dog. The act has pitted a group called the Countryside Alliance, which continues to seek its repeal, against animal rights activists, mostly headquartered in large British cities, and the London-based Parliament.

Competition also exists among and between cities, as well as among and between states. The most ubiquitous form of this competition revolves around the desire by local and state authorities to attract corporate investment (as discussed in Chapter 7). Often corporations play the jurisdictions against each other in attempts to obtain the most attractive investment packages. At other times, cities and states compete to induce the government to locate government facilities within their jurisdictions. The location of military bases during and after World War II illustrates this point. Cities and states were especially keen to attract the military because it meant increased employment opportunities and all the economic growth attached to employment generation.

The Geography of Politics and Geographical Systems of Representation

An obvious way to show how politics shapes geography is to demonstrate how systems of political representation are geographically anchored. For instance, the United States has a political system in which democratic rule

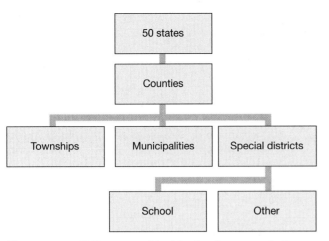

Figure 9.34   U.S. geographical basis of representation
This diagram gives a breakdown of the types of voter districts at each level of political representation. Each type of district is territorially defined, creating a complicated and overlapping pattern of political units.

and territorial organization are linked by the concept of territorial representation.

Democratic rule is a system in which public policies and officials are directly chosen by popular vote. **Territorial organization** is a system of government formally structured by area, not by social groups. Thus, voters vote for officials and policies that will represent them and affect them *where they live*. The territorial bases of the U.S. system of representation are illustrated in **Figure 9.34**. The United States is a federation of 50 states, themselves subdivided into over 3,000 counties or parishes. Counties and parishes are further broken down into municipalities, townships, and special districts, which include school districts, water districts, library districts, and others.

The electoral divisions established for choosing elected officials in the United States range from precincts and wards to congressional districts and states. State power is applied within geographical units, and state representatives are chosen from geographical units. The bottom line is that in the United States—as in many other representative democracies—politics is geography. People and their interests gain representation in government through the location of their interests in particular places and through their relative ability to capture political control of *geographically based* political units.

For example, election of the president involves a popular vote carried out at the precinct level but totaled at the state level. Thus, even though particular precincts, cities, or counties may give a majority to one candidate, if the majority of votes at the state level supports the opposing candidate, then that person is declared the winner in that state. This arrangement is particularly important because the president is not elected by the popular vote, but by the electoral college. The electoral college is composed of a specified number of delegates allocated to each state based on that state's population

as of the most recent official census. Thus, it is the state-level voting tally that drives the process.

A candidate may win the countrywide popular vote but lose in the electoral college if that candidate fails to win enough states to secure the required majority of the electoral votes. This is what happened to candidate Al Gore in the 2000 presidential elections. Furthermore, as in the 1992 election of Bill Clinton, a candidate may win with considerably less than 50 percent of the popular vote if a third-party candidate (in this case, H. Ross Perot) siphons off enough of the popular vote to prevent the opposing candidate (in this case, George H. W. Bush) from winning a sufficient number of electoral votes. The geographical implications of the U.S. presidential voting arrangement are crucial to candidates' campaign strategies. To be a winner requires concentrating time and energy on capturing a majority of votes in some of the nation's most populous states as Barack Obama did in 2008.

The U.S. tradition of territorial representation promotes acceptance of the legitimacy of local interests. Such an arrangement also makes it difficult for third parties to succeed. Unless third parties are geographically concentrated, they can never win enough states to win a presidential election. The failed presidential campaign of H. Ross Perot in 1992 is an excellent example of this (**Figure 9.35**). Although Perot received 19 percent of the vote, he did not win a single state.

Other systems of representation exist throughout the world. For example, many electoral systems are based on representing special constituencies in the legislative branch of government. In Pakistan, for example, there are four seats for Christians, four seats for Hindus and people belonging to the scheduled castes, and one seat each for the Sikh, Buddhist, and Parsi communities. The point is that systems of representation are very much tied to the history of a country, with some very sensitive to the way that history and geography (who lives where) come together. These systems are both a product of and an important influence on the political culture of a country.

Reapportionment and Redistricting   The U.S. Constitution determines the allocation of legislators among states, guaranteeing that each state will have a representative system of government. For U.S. presidential and senatorial elections, candidates are elected at large within each state, not on the basis of electoral districts. U.S. representatives, however, are elected from congressional districts of roughly equal population size. This is also the case for state senators, representatives, and, often, other elected officials, from city council members to school board members. It is the responsibility of each state's legislature to create the districts that will elect most federal and state representatives. Other levels of government—from counties to special districts—also establish their own electoral districts. The result is

The Palestinian–Israeli Conflict

The Palestinian-Israeli conflict and the Palestinians's passionate desire for self-rule continues to be complex and highly volatile despite persistent local and international efforts to bring peace to the region. The violence that erupted in fall of 2000 (and has persisted), just as the peace process seemed to be most promising, underscores the complexity of the problem and the difficulty of resolution. As with the Iran/Iraq/Kuwait case, the chief factors that have created this seemingly intractable political problem were exacerbated by the British partitioning of the region.

The official Jewish state of Israel is a mid-twentieth-century construction that has its roots in the emergence of **zionism**, a late-nineteenth-century European movement. Zionism's chief objective has been the establishment of a legally recognized home in Palestine for the Jewish people. Thousands of European Jews, inspired by the early zionist movement, began migrating to Palestine at the turn of the nineteenth century. When the Ottoman Empire was defeated in 1917, the British gained control over Palestine and the Transjordan area and issued the Balfour Declaration. The Balfour Declaration was highly problematic, however, because indigenous peoples, the Palestinians, already occupied the area and viewed the arrival of increasing numbers of Jews and European sympathy for the establishment of a Jewish homeland as an incursion into the sacred lands of Islam. In response to increasing Arab-Jewish tensions in the area, the British decided to limit Jewish immigration to Palestine in the late 1930s through the end of World War II. In 1947, with conflict continuing between the two groups, Britain announced that it had despaired of ever resolving the problems and would withdraw from Palestine in 1948, turning it over to the United Nations when that happened. The United Nations, under heavy pressure from the United States, responded by voting to partition Palestine into Arab and Jewish states and designated Jerusalem as an international city, preventing either group from having exclusive control. The Jewish state was to have 56 percent of the mandate of Palestine; an Arab state was to have 43 percent; and Jerusalem, a city sacred to Jews, Muslims, and Christians, was to be administered by the United Nations. The proposed UN plan was accepted by the Jews and angrily rejected by the Arabs, who argued that a mandated territory could not legally be taken from an indigenous population.

When Britain withdrew in 1948, war broke out. In an attempt to aid the militarily weaker Palestinians, combined forces from Egypt, Jordan, and Lebanon, as well as smaller units from Syria, Iraq, and Saudi Arabia, confronted the Israelis. Their goal was not only to prevent the Jewish forces from gaining control over additional Palestinian territory but also to wipe out the newly formed Jewish state altogether. This war, however, which came to be known as the first Arab-Israeli war, resulted in the defeat of the Arab forces in 1949, and later armistice agreements enabled Israel to expand beyond the UN plan by gaining the western sector of Jerusalem, including the Old City. In 1950, Israel declared Jerusalem its national capital, though very few countries have recognized this.

Israel maintained the new borders gained during the first Arab-Israeli war for another 18 years until the Six-Day War in 1967, which resulted in further gains for Israel, including the Sinai Peninsula, the Golan Heights, and the southwestern corner of Syria. The eastern sector of Jerusalem, previously held by Jordan, was also annexed during the Six-Day War. As **Figure 9.F** shows, relatively little territorial change occurred until the 1970s and 1980s, when Israel moved toward reconciliation with Egypt through a series of withdrawals that eventually returned all of Sinai to Egyptian control by 1988.

The territorial expansion of Israel has meant that hundreds of thousands of Palestinians have been driven from their homelands, and the landscape of Palestine has been dramatically transformed. Today, Palestinians live as refugees either in other Arab countries in the region, abroad, or under Israeli occupation in the West Bank, the Golan Heights, and the Gaza Strip (also known as the "Occupied Territories"). The Arabs of the Middle East and North Africa and many other international observers are convinced that Israel has no intention of allowing the diasporic Palestinian population to return to their homelands. By the late 1980s, in fact, Palestinians who had remained in their homeland had become so angered by Israeli territorial aggression that they rose up in rebellion. This rebellion, known as the **intifada** ("uprising"), has involved frequent clashes between fully armed Israeli soldiers and rock-throwing Palestinian young men. The intifada, more than anything, is a reaction against over three decades of Israeli occupation of the Palestinian homeland and increasing Israeli settlement, particularly in the West Bank and the Gaza Strip. In addition to the intifada, other Palestinian groups have coalesced in opposition to the Israeli occupation. The Palestinian Liberation Organization (PLO) was formed in 1964 as an organization devoted to returning Palestine to the Palestinians. Since its official recognition, the PLO has become the Palestinian Authority and is seen as the only legitimate representative

Figure 9.F Changing geography of Israel/Palestine, 1923–2005 Since the creation of Israel out of much of what had been Palestine in 1947, the geography of the region has undergone significant modifications. A series of wars between Israelis and Arabs and a number of political decisions regarding how to cope with both resident Palestinians and large volumes of Jewish people immigrating to Israel from around the world have produced the changing geographies we see here. (*Source:* Reprinted with permission from Prentice Hall, from J. M. Rubenstein, *The Cultural Landscape: An Introduction to Human Geography,* 5th ed., © 1996, p. 233.)

of the Palestinian people. Besides the Palestinian Authority, however, other, more extreme groups exist claiming to represent the Palestinian cause. One of the most well known is Hamas (Harakat al-Muqawama al-Islmiyya, or the Islamic Resistance Movement), whose activities are largely centered in the West Bank and Gaza Strip.

Since the mid-1990s, hopes for peace in the region have risen, fallen, risen, fallen, and most recently risen once again. In October 2000, after weeks of very difficult but promising U.S.-sponsored peace negotiations between Yasir Arafat, then chairman of the Palestinian Authority, and Ehud Barak, then Israeli prime minister, violence broke out again in the West Bank. This new violence left little hope in Israel, the Occupied Territories, or elsewhere that the Palestinian-Israeli conflict would be resolved anytime in the near future. Renewed hope emerged, however, following the death of Yasir Arafat in November 2004. Arafat was a controversial figure throughout his lengthy political career; his supporters viewed him as a heroic freedom fighter who symbolized the national aspirations of the Palestinian people, but his opponents often described him as a terrorist who promoted violence. His death made room for a new leader to step forward, one who might negotiate a peace where Arafat could not.

On January 9, 2005, Mahmoud Abbas was elected president of the Palestinian Authority by voters in the West Bank and Gaza, in the first Palestinian election held since 1996. Although the most militant Islamist organizations, Hamas and the Islamic Jihad, boycotted the elections, it is estimated that about 66 percent of eligible voters went to the polls (compared to 42.45 percent in the 2005 U.S. presidential elections). Israel has been withdrawing Israeli settlers (many of whom are extremely resistant to leaving) from Palestinian territory, but infighting between Fatah and Hamas in Gaza, following the 2005 election, has spilled into the West Bank, with mass arrests, abductions, and summary executions deepening the discord.

As Israel has begun to cede territory to Palestine, it is continuing to construct a physical barrier between Israelis and Palestinians. The Gaza Strip barrier was constructed in 1994. It consists of 52 kilometers (30 miles) of mainly wire fence with posts, sensors, and buffer zones. Israel argues that the barrier is essential to protect the security of its citizens from Palestinian terrorism. Palestinians and other opponents of the barrier contend that its purpose is geographical containment of the Palestinians in order to pave the way for an expansion of Israeli sovereignty and to preclude any negotiated border agreements in the future. But Israel argues that the fence is purely a security obstacle, not a part of a future border. In 2002, the West Bank wall was begun and, when completed, will seal off that portion of the Palestinian territories from Israel (**Figure 9.G**). The barrier continues to uproot and destroy Palestinian settlements and separate them from their livelihood. In

Figure 9.G Israeli security fence, 2005 Shown are the planned and completed portions of the security fence, called the "the wall" or the "apartheid wall" by Palestinians and other opponents. The physical barrier consists of a network of fences, walls, and trenches. Israel's stated purpose in constructing it is to create a zone of security between itself and the West Bank. (*Source:* Adapted from BBC News, Web edition; http://news.bbc.co.uk/1/hi/world/middle_east/3111159.stm, accessed July 4, 2005.)

October 2003, the UN General Assembly voted 144–4 that the wall was "in contradiction to international law" and therefore illegal. Israel called the resolution a "farce." The barrier construction had already destroyed an estimated 102,320 Palestinian olive and citrus trees, demolished 75 acres (304,000 m^2) of greenhouses and 23 miles (37 km) of irrigation pipes. And the network of road blocks and checkpoints has been a major factor in stifling the West Bank's economy because the Palestinians are being cut off from jobs and markets in Israel. As a result, unemployment continues to climb at the same time that GDP per capita has plummeted. The World Bank deemed the West Bank "a shattered economic space" and more than half the population lives below the poverty line.

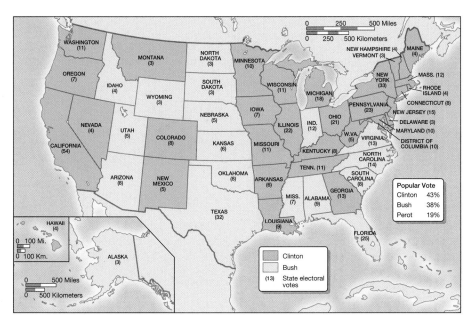

Figure 9.35 The vote for president, 1992 The total popular vote, aggregated at the state level, determines the number of electoral college votes for each candidate. Voter preferences show a clear geographical pattern, with George Bush running particularly strong in the West and South. H. Ross Perot won 19 percent of the popular vote but was unable to translate that into any electoral votes.

that representatives are elected at any number of levels of government in a collection of districts that is complicated, extensive, and by no means systematic.

Problems of the proper "fit" between political representation and territory emerge when population changes. Because most forms of representation are based on population, it often becomes necessary to change electoral district boundaries to distribute the total population more evenly among districts. For example, the number of congressional representatives in the United States as a whole is fixed at 435. These 435 seats must be reapportioned in accordance with population change every 10 years. (Recall from Chapter 3 that the federal government is required to count the U.S. population every 10 years. One of the chief reasons for this is to maintain the proper match between population and representation.) **Reapportionment** is the process of allocating electoral seats to geographical areas. **Redistricting** is the defining and redefining of territorial district boundaries. Both are political, geographical, and statistical exercises. As geographer Richard Morrill writes:

The process is *political* in that the design and approval of systems of districts is usually done by bodies of elected representatives; the balance of power between groups and areas is often involved; identification of citizens with a traditional electoral territory is altered; and the incumbency of individuals is usually at stake. . . . The process is *geographic* in that areas must be allocated to districts and boundaries drawn (or territories partitioned into districts); communities of interest which may have arisen in part from pre-existing systems of districts, may well be affected; restructuring of basic electoral geography is altered; and accessibility of voters to

their representatives or centers of decision-making may be changed. . . . Redistricting is also *statistical* or mathematical in that there is a requirement for reasonably current and accurate data on population and its characteristics and, sometimes, of property and its valuation.[3]

Gerrymandering The purpose of redistricting is to ensure the equal probability of representation among all groups. In the practice of redistricting for partisan purposes, known as **gerrymandering**, boundaries of districts are redrawn to advantage a particular political party or candidate or to prevent or ensure a loss of power of a particular subpopulation (like African Americans). The term immortalized Governor Elbridge Gerry of Massachusetts, who signed into law a bill designed to maximize the election of Republican-Democrats over Federalists in the election of 1812. Although Federalists won the most votes overall, Republican-Democrats took 29 of 40 seats because they won the most districts. An electoral district north of Boston was redrawn so that most Federalist votes were contained there. With the majority of Federalist votes restricted to a small number of districts, their voting power was reduced.

Gerrymandering persists today. Although the Federal Voting Rights Act intended that redistricting enhance minority representation in Congress, a fine line exists between "enhancement" and creating a district solely to ensure that a minority person be elected. **Figure 9.36** shows North Carolina's reengineered Twelfth District, the constitutionality of which was upheld by a Supreme Court decision in 2001.

[3]R. Morrill, *Political Redistricting and Geographic Theory.* Washington, DC: Association of American Geographers, 1981, p. 1.

Figure 9.36   North Carolina's Twelfth Congressional District The drawing of electoral district boundaries remains a politically volatile exercise. A 1996 case before the Supreme Court concerned the redrawn Twelfth Congressional District of North Carolina, whose shape (in red on the map) was as contorted as the original salamander. Accusations of gerrymandering circulated around the drawing of the district, which was intended to consolidate African American voting strength. Although this redistricting was immediately challenged legally, the Supreme Court upheld its constitutionality in a 2001 decision.

CONCLUSION

The globalization of the economy has been largely facilitated by the actions of states extending their spheres of influence and paving the way for the smooth functioning of markets and industries. Political geography is as much about what happens at the global level as it is about what happens at other levels of spatial resolution, from the region to the neighborhood to the household and the individual.

Theories of the state have been one of geography's most important contributions to understanding politics. Ratzel's emphasis on the relationship between power and territory and Mackinder's model of the geographical pivot remind us that space and territory shape the actions of states in both dramatic and mundane ways. Time and space shape politics, and events distant in time and space—such as colonialism—continue to have impacts long afterwards. For example, the civil war in Northern Ireland, instigated by English colonial practices now centuries old, has only recently shown to cease. The impacts of English colonization have also been felt in countries throughout the Northern Hemisphere, as well as by neighbors living unhappily for several generations side by side in cities like Belfast and Boston.

Continuing strife also characterizes the enduring North/South divide that pits core countries against peripheral, mostly formerly colonial, countries. Perhaps the most surprising political geographical transformation of this century has been the near dissolution of the East/West divide. Although it is too soon to tell whether communism has truly been superseded by capitalism, the distinctions between them are certainly more blurred than they once were.

Perhaps the most significant aspect of contemporary globalization for political geographical understandings is the emergence of a new world order and transnational institutions of governance. Both are reshaping not only governing structures and economic processes (such as creating new layers of rules, regulations, and policies, as well as new ways of political interaction among and between nation-states) but also the practices of everyday life (such as increased personal security measures and the transformation of human rights).

The pairing of the terms *politics* and *geography* serves to remind us that politics is clearly geographical at the same time that geography is unavoidably political. The simple divisions of area into states, counties, cities, towns, and special districts mean that where we live shapes our politics, and vice versa. Geography is politics, just as politics is geography. And geographical systems of representation, as well as identity politics based on regional histories, confirm these interactive relationships.

MAIN POINTS REVISITED

- Political geography, a subfield of the discipline of geography, examines complex relationships between politics and geography (both human and physical).

    As societies are organized around territorial units, geography and access to it are often at the center of political conflicts but can also make possible the resolution of conflicts.

- Political geographers recognize that the relationship between politics and geography is two-way. Political geography can be seen both as the geography of politics and the politics of geography.

    The politics of geography is best illustrated through regional and sectional movements and conflicts, such as the India-Pakistan conflict or the American Civil War. The geography-of-politics approach recognizes that systems of political representation are geographically anchored and shape the opportunities of the people who live within them.

- The relations between politics and geography are often driven by particular theories and practices of the world's states. Understanding imperialism, colonialism, heartland theory, domino theory, the end of the Cold War, and the emergence of the new world order is key to comprehending how, within the context of the world system, geography has influenced politics and politics has influenced geography.

    At present, theories about globalization and the interconnectedness of places are particularly important, whereas the intellectual and popular appeal of the domino theory and heartland theory have waned.

- Political geography deals with the phenomena occurring at all scales of resolution, from the global to the body. Important East/West and North/South divisions dominate international politics, whereas regionalism, sectionalism, and similar divisions dominate intrastate politics.

    No one scale necessarily predominates, and changes emanating from a locality may have international impacts and vice versa.

KEY TERMS

bioterrorism (p. 368)
centrifugal forces (p. 349)
centripetal forces (p. 349)
children's rights (p. 375)
citizenship (p. 348)
confederation (p. 351)
decolonization (p. 357)
discourse (p. 353)
democratic rule (p. 379)
domino theory (p. 363)

East/West divide (p. 362)
federal state (p. 349)
geopolitics (p. 340)
gerrymandering (p. 383)
global civil society (p. 375)
human rights (p. 375)
international organization (p. 358)
international regime (p. 374)
intifada (p. 380)

nation (p. 348)
nation-state (p. 348)
nationalism (p. 348)
new world order (p. 364)
North/South divide (p. 357)
Orientalism (p. 355)
reapportionment (p. 383)
redistricting (p. 383)
regionalism (p. 376)
sectionalism (p. 377)

self-determination (p. 376)
sovereignty (p. 348)
supranational organization (p. 372)
territorial organization (p. 379)
territory (p. 343)
terrorism (p. 364)
unitary state (p. 350)
zionism (p. 380)

ADDITIONAL READING

Agnew, J. A., *Hegemony: The New Shape of Global Power.* Philadelphia: Temple University Press, 2005.

Agnew, J., K. Mitchell, and G. Toal, *A Companion to Political Geography.* Oxford: Blackwell, 2002.

Appelbaum, R. P., and W. I. Robinson (eds.), *Critical Globalization Studies.* New York: Routledge, 2005.

Blomley, N., D. Delaney, and R. T. Ford (eds.), *The Legal Geographies Reader: Law, Power, and Space.* Oxford; Malden, MA: Blackwell, 2001.

Cox, K. R., *Political Geography: Territory, State, Society.* Oxford: Blackwell, 2002.

Enloe, Cynthia H., *Bananas, Beaches and Bases: Making Feminist Sense of International Politics.* Berkeley: University of California Press, 2000.

Ferguson, Y. H., and R. J. B. Jones, *Political Space: Frontiers of Change and Governance in a Globalizing World.* Albany, NY: State University of New York Press, 2002.

Epic Project Web site, *Comparative and Country-by-Country Data on Election Systems, Laws, Management and Administration.* http://www.epicproject.org/en/, accessed on September 24, 2005.

Gallaher, C., *On the Fault Line: Race, Class, and the American Patriot Movement.* Lanham, MD: Rowman and Littlefield, 2003.

Gettleman, M. E., and S. Schaar (eds.), *The Middle East and Islamic World Reader.* New York: Grove Press, 2003.

Gregory, D., *The Colonial Present: Afghanistan, Palestine, and Iraq.* Malden, MA: Blackwell, 2004.

Harff, B., *Early Warning of Communal Conflict and Genocide: Linking Empirical Research to International Responses.* Boulder, CO: Westview Press, 2003.

Harvey, D., *Spaces of Global Capitalism: Towards a Theory of Uneven Geographical Development.* London; New York: Verso, 2006.

Harvey, D., *The New Imperialism.* Oxford; New York: Oxford University Press, 2005.

Heiberg, M., *The Making of the Basque Nation.* New York; Cambridge: Cambridge University Press, 2007.

Held, D., and Kaya, A. (eds.), *Global Inequality: Patterns and Explanations.* Cambridge, UK; Malden, MA: Polity, 2007.

Held, D., A. McGrew, D. Goldblatt, and J. Perraton, *Global Transformations.* Malden, CT: Blackwell, 1999.

Hirst, P., and G. Thompson, *Globalization in Question.* Cambridge, U.K.: Polity Press, 1999.

Hoffman, B., *Inside Terrorism.* New York: Columbia University Press, 2006.

Riesebrodt, M., *Pious Passion: The Emergence of Modern Fundamentalism in the United States and Iran;* translated from the German *Fundamentalismus als patriarchalische Protestbewegung* by Don Reneau. Berkeley: University of California Press, 1993.

Ruthven, M., *Fundamentalism: The Search for Meaning.* Oxford: Oxford University Press, 2005.

Salamon, L. M., S. Sokolowski, S. Wojciech, and R. List, *Global Civil Society: An Overview.* Baltimore, MD: Johns Hopkins University Institute for Policy Studies, 2003.

Smith, N., *The Endgame of Globalization.* New York: Routledge, 2005.

Staeheli, L. A., E. Kofman, and L. Peake, *Mapping Women, Making Politics: Feminist Perspectives on Political Geography.* New York: Routledge, 2004.

Taylor, P., and C. Flint, *Political Geography: World-Economy, Nation-State, and Locality,* 4th ed. Upper Saddle River, NJ: Prentice Hall, 2000.

Totten, S., W. S. Parsons, and I. W. Charny (eds.), *Century of Genocide: Critical Essays and Eyewitness Accounts,* 2nd ed. New York: Routledge, 2004.

Tuathail, G., *Critical Geopolitics: The Politics of Writing Global Space.* Minneapolis, MN: University of Minnesota Press, 1996.

UN Office for Coordination of Humanitarian Affairs and UN Relief and Works Agency for Palestinian Refugees, *The Humanitarian Impact of the West Bank Barrier on Palestinian Communities,* 2005.

van den Anker, C., *The Political Economy of New Slavery.* New York: Palgrave Macmillan, 2004.

White, J., *Terrorism: An Introduction,* 3rd ed. Toronto: Wadsworth, 2002.

EXERCISES

On the Internet

The Companion Website at www.mygeoscienceplace.com contains a number of tools designed to reinforce the geographic concepts presented in this chapter. Structured within a four-step learning path, these online resources include: chapter Key Concepts; a fully-searchable myeBook version of the chapter, which allows you to add notes and highlights; RSS news feeds from various media agencies; links to additional web resources; Human Geography Videos; a digital Glossary; Concept Review questions; Thinking Spatially mapping exercises; and Critical Thinking essay questions.

Unplugged

1. International boundaries are a prominent feature of the political geography of the contemporary world. In this exercise you are asked to explore the impact of a boundary on nationalist attitudes and behaviors. You will need to use *The New York Times Index* to complete this assignment. Using the United States–Mexico border as your keyword, describe the range of issues that derive from this juxtaposition of two very different nations. You should concentrate on a five-year period and show which issues grew in importance, which issues declined, and which issues continued to have a consistent news profile throughout the period.

2. National elections usually tell a story about the ways in which regional ideas and attitudes shape the national political agenda. In the 2000 and 2004 U.S. presidential election, pollsters considered religion an important issue, with the religious right playing a crucial role in reelecting George W. Bush. Using national election result data available through the Government Documents Division of your college or university library, describe the political geography of fundamentalist or evangelical Christians. Did these Christians vote for Bush in all regions of the country? If not, which ones did not, and what might explain the regional distribution of this powerful voting block?

3. Using two maps of Europe (up to, but not including, Russia and the former Soviet Union), one from 1930 and one from the present, compare the differences. How do issues of ethnicity, religion, and political system help to explain these changes? Identify any areas on the map that you feel may be the sites of future border changes, and explain why.

4. U.S. presidential elections provide a snapshot of the changing political geography of the country. Using maps of the 1992 and 2008 presidential elections, compare the 1992 results with those for the year 2008. What are the most significant differences between the two maps? What are some reasons for these differences? If you are able to get maps that are disaggregated by race, ethnicity, or gender, what further explanations can you offer for the differences in the maps based on these additional variables?

Urbanization

10

Urbanization is one of the most important geographic phenomena in today's world. The United Nations Center for Human Settlements (UNCHS)[1] notes that the growth of cities and the urbanization of rural areas are now irreversible because of the global shift to technological-, industrial-, and service-based economies. The proportion of the world's population living in urban settlements is growing at a rapid rate, and the world's economic, social, cultural, and political processes are increasingly being played out within and between the world's systems of towns and cities.

The UNCHS has concluded that few countries are able to handle the urban population crush, which is causing problems on an unprecedented scale with everything from clean water to disease prevention. Already 10 million people are dying annually in densely populated urban areas from conditions produced by substandard housing and poor sanitation. In 2005, there were 1.6 billion inadequately housed people across the world and an estimated 100 million who were completely homeless. According to the World Bank, 2.6 billion people worldwide have one thing in common—they do not have access to sanitation. About 1.7 million deaths a year—90 percent of which are children—are attributed to unsafe water, poor sanitation and hygiene, mainly through infectious diarrhea.[2]

The United Nations International Children's Fund (UNICEF)[3] has blamed "uncontrollable urbanization" in less developed countries for the widespread creation of "danger zones" in which increasing numbers of children are forced to become beggars, prostitutes, and laborers before reaching their teens. Pointing out that urban populations are growing at twice the general population rate, UNICEF has concluded that too many people are being squeezed into cities that do not have enough jobs, shelter, or schools to accommodate them. As a consequence, the family and community structures that support children are being destroyed, with the result that increasing numbers of children have to work. For hundreds of thousands of street kids in less developed countries, "work" means anything that contributes to survival: shining shoes, guiding cars into parking spaces, chasing other street kids away from patrons at an outdoor café, working as domestic help, making fireworks, selling drugs. In Abidjan, in the Ivory Coast, 15-year-old Jean-Pierre Godia, who cannot read or write, spends about six hours every day trying to sell 10-roll packets of toilet paper to motorists at a busy intersection. He buys the packets for about $1.20 and sells them for $2. Some days he doesn't sell any. In the same city, 7-year-old Giulio guides cars into

Main Points

- The urban areas of the world are the linchpins of human geographies at the local, regional, and global scales.

- The earliest towns and cities developed independently in the various hearth areas of the first agricultural revolution.

- The expansion of trade around the world, associated with colonialism and imperialism, established numerous gateway cities.

- The Industrial Revolution generated new kinds of cities—and many more of them.

- Today the single most important aspect of world urbanization, from a geographical perspective, is the striking difference in trends and projections between the core regions and the peripheral regions.

- A small number of "world cities," most of them located within the core regions of the world-system, occupy key roles in the organization of global economics and culture.

- Many megacities of the periphery are primate and exhibit a high degree of centrality within their urban systems.

- There is a close relationship between globalization and urbanization, mediated by networked infrastructures of transportation, information, and communications technologies.

[1]United Nations Center for Human Settlements (Habitat), *Cities in a Globalizing World: Global Report on Human Settlements, 2001.* London: Earthscan Publications, 2001.

[2]The World Bank, 2008, "The Economic Impacts of Poor Sanitation," http://web.worldbank.org/WBSITE/EXTERNAL/NEWS/0,,contentMDK:21693343~pagePK:64257043~piPK:437376~theSitePK:4607,00.html, accessed July 27, 2008.

[3]*The Progress of Nations.* New York: United Nations International Children's Fund (UNICEF), 1995.

parking spaces outside a chic pastry shop. He has been doing this since he was 5, to help his mother and four siblings, who beg on a nearby corner.

In this chapter we describe the extent and pattern of urbanization across the world, explaining its causes and the resultant changes wrought in people and places.

URBAN GEOGRAPHY AND URBANIZATION

From small market towns and fishing ports to megacities of millions of people, the urban areas of the world are the linchpins of human geographies. They have always been a crucial element in spatial organization and the evolution of societies, but today they are more important than ever. Between 1980 and 2005, the number of city dwellers worldwide rose by 1.35 billion. Cities now account for almost half the world's population. Much of the developed world has become almost completely urbanized (**Figure 10.1**), and in many peripheral and semiperipheral regions the current rate of urbanization is without precedent (**Figure 10.2**). The United Nations Human Settlements Program (UN-Habitat) estimates

that 60 percent of the world's population will live in cities by 2030, a trend that equals the addition of a city of 1 million residents every week. Around 80 percent of all city dwellers in 2030 will be in peripheral and semi-peripheral countries. Urbanization on this scale is a re-markable geographical phenomenon—one of the most important processes shaping the world's landscapes.

Towns and cities are centers of cultural innovation, social transformation, and political change. They can also be engines of economic development. The gross product of large cities like London, Los Angeles, Mexico City, and Paris is roughly equivalent to that of entire countries like Australia and Sweden; New York's gross product is almost as large as that of China. Although they often pose social and environmental problems, towns and cities are essential elements in human economic and

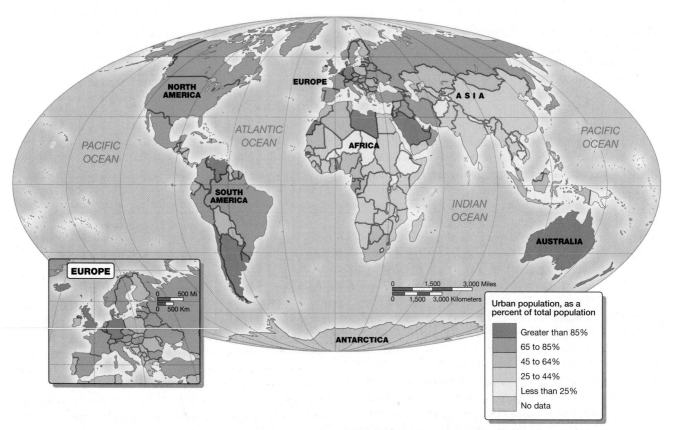

Figure 10.1 Percentage of each country's population living in urban settlements, 2003 The lowest levels of urbanization—less than 25 percent—are found in Central Africa and South and Southeast Asia. Most of the core countries are highly urbanized, with between 65 and 95 percent of their populations living in urban settlements. (*Source:* Data from United Nations Department of Economic and Social Affairs, Population Division, *World Urbanization Prospects: The 2003 Revision,* 2004, pp. 25–29.)

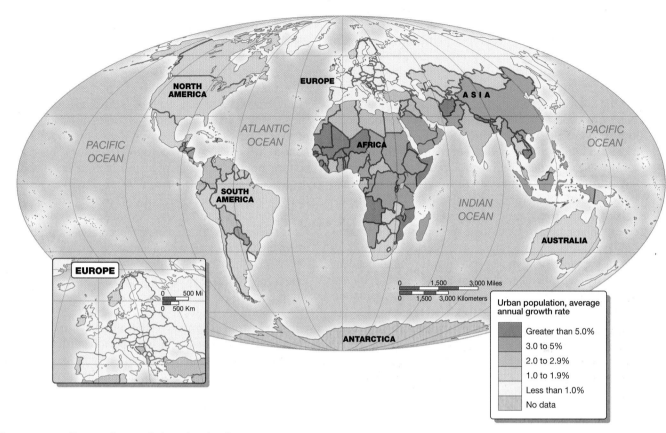

Figure 10.2   Rates of growth in urbanization, 2000–2005 This map shows the annual average growth rate between 2000 and 2005 in the proportion of people in each country living in urban settlements. Core countries, already highly urbanized, grew quite slowly. The urban populations of peripheral countries such as Angola, Afghanistan, Burundi, Liberia, Rwanda, and Somalia, on the other hand, grew by more than 5 percent each year, creating tremendous pressure on cities' capacity to provide jobs, housing, and public services. (*Source:* Data from United Nations Department of Economic and Social Affairs, Population Division, *World Urbanization Prospects: The 2003 Revision,* 2004, pp. 70–78.)

social organization. Experts on urbanization point to four fundamental aspects of the role of towns and cities in human economic and social organization:

- *The mobilizing function of urban settlement.* Urban settings, with their physical infrastructure and their large and diverse populations, are places where entrepreneurs can get things done. Cities, in other words, provide efficient and effective environments for organizing labor, capital, and raw materials and for distributing finished products. In developing countries, urban areas produce as much as 60 percent of total gross domestic product with just one-third of the population.

- *The decision-making capacity of urban settlement.* Because urban settings bring together the decision-making machinery of public and private institutions and organizations, they come to be concentrations of political and economic power.

- *The generative functions of urban settlement.* The concentration of people in urban settings makes for much greater interaction and competition, which

facilitates the generation of innovation, knowledge, and information.

- *The transformative capacity of urban settlement.* The size, density, and variety of urban populations tend to have a liberating effect on people, allowing them to escape the rigidities of traditional, rural society and to participate in a variety of lifestyles and behaviors.

The study of urban geography is concerned with the development of towns and cities around the world, with particular reference to the similarities and differences both *among* and *within* urban places. For urban geographers, some of the most important questions include: What attributes make towns and cities distinctive? How did these distinctive identities evolve? What are the relationships and interdependencies between particular sets of towns and cities? What are the relationships between cities and their surrounding territories? Do significant regularities exist in the spatial organization of land use within cities, in the patterning of neighborhood populations, or in the layout and landscapes of particular kinds of cities?

Urban geographers also want to know about the causes of the patterns and regularities they find. How, for example, do specialized urban subdistricts evolve? Why did urban growth occur in a particular region at a particular time? And why did urban growth exhibit a distinctive physical form during a certain period? In pursuing such questions, urban geographers have learned that the answers are ultimately to be found in the wider context of economic, social, cultural, and political life. In other words, towns and cities must be viewed as part of the economies and societies that maintain them.

Urbanization, therefore, is not simply the demographic growth of towns and cities. It also involves many other changes, both quantitative and qualitative. From the geographer's perspective, these changes can be conceptualized in several different ways. One of the most important of these is by examining the attributes and dynamics of urban systems. An **urban system**, or city system, is any interdependent set of urban settlements within a given region. Thus, for example, we can speak of the French urban system, the African urban system, or even the global urban system. As urbanization takes place, the attributes of urban systems will, of course, reflect the fact that increasing numbers of people are living in ever-larger towns and cities. They will also reflect other important changes, such as changes in the relative size of cities, changes in their functional relationships with one another, and changes in their employment base and population composition.

Another important aspect of change associated with urbanization processes concerns urban form. **Urban form** refers to the physical structure and organization of cities in their land use, layout, and built environment. As urbanization takes place, not only do towns and cities grow bigger physically, extending upward and outward, but they also become reorganized, redeveloped, and redesigned in response to changing circumstances.

These changes, in turn, are closely related to a third aspect of change: transformations in patterns of urban ecology. **Urban ecology** is the social and demographic composition of city districts and neighborhoods. Urbanization not only brings more people to cities, it also

brings a greater variety of people. As different social, economic, demographic, and racial subgroups become sorted into different territories, distinctive urban ecologies emerge. As new subgroups arrive or old ones leave, these ecologies change.

A fourth aspect of change associated with urbanization concerns people's attitudes and behavior. New forms of social interaction and new ways of life are brought about by the liberating and transformative effects of urban environments. These changes have given rise to the concept of urbanism, which refers to the distinctive nature of social and cultural organization in particular urban settings. **Urbanism** describes the way of life fostered by urban settings, in which the number, physical density, and variety of people often result in distinctive attitudes, values, and patterns of behavior. Geographers are interested in urbanism because of the ways in which it varies both within and between cities.

URBAN ORIGINS

It is important to put the geographic study of towns and cities in historical context. After all, many of the world's cities are the product of a long period of development. We can only understand a city, old or young, if we know something about the reasons behind its growth, about the rate at which it has grown, and about the processes that have contributed to this growth.

In broad terms, the earliest urbanization developed independently in the various hearth areas of the first agricultural revolution. The very first region of independent urbanism was in the Middle East, in the valleys of the Tigris and Euphrates (in Mesopotamia) and in the Nile Valley from around 3500 B.C. (see Chapter 4). Together, these intensively cultivated river valleys formed the so-called Fertile Crescent. In Mesopotamia, the growth in size of some of the agricultural villages located on the rich alluvial soils of the river floodplains formed the basis for the large rival city-states of the Sumerian empire. They included Ur (in present-day Iraq), the capital from about 2300 to 2180 B.C., as well as Eridu, Uruk, and Erbil (Ancient Arbela—**Figure 10.3**). These fortified city-states

Figure 10.3 Erbil Erbil (Ancient Arbela) in northeast Iraq is located atop a *tell*, a mound representing the remains of generations of sun-dried mud-brick buildings, visible as a hill rising high above the surrounding plain. The 100-foot-high Erbil tell is believed to represent perhaps 6,000 years of continuous occupation.

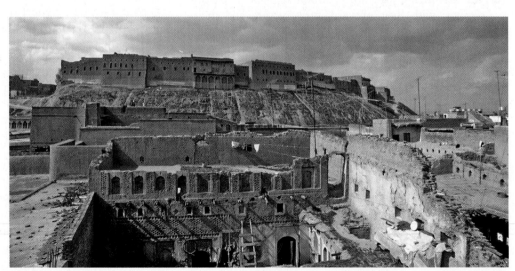

contained tens of thousands of inhabitants; social stratification, with religious, political, and military classes; innovative technologies, including massive irrigation projects; and extensive trade connections. By 1885 B.C., the Sumerian city-states had been taken over by the Babylonians and then the Neo-Babylonians, who governed the region from their capital city, Babylon. In Egypt, which became a unified state as early as 3100 B.C., large irrigation projects controlled the Nile's waters for agricultural and other uses, supporting a series of capital cities that included Thebes, Akhetaten (Tell el-Amarna), and Tanis. Internal peace in Egypt meant that there was no need for massive investments in these cities' defensive fortifications. Also, each pharaoh was free to locate a new capital at any site he selected for his tomb, and after his death the city was usually abandoned to the priests.

By 2500 B.C., cities had appeared in the Indus Valley, and by 1800 B.C., they were established in northern China. Other areas of independent urbanism include Mesoamerica (from around 100 B.C.) and Andean America (from around A.D. 800). Meanwhile, the original Middle Eastern urban hearth continued to produce successive generations of urbanized world-empires, including those of Greece, Rome, and Byzantium.

Experts differ in their explanations of these first transitions from subsistence minisystems to city-based world-empires. The classical archaeological interpretation emphasizes the availability of an agricultural surplus large enough to allow the emergence of specialized, nonagricultural workers. Some urbanization, however, may have resulted from the pressure of population growth. This pressure, it is thought, disturbed the balance between population and resources, causing some people to move to marginal areas. Finding themselves in a region where agricultural conditions were unfavorable, these people either had to devise new techniques of food production and storage or establish a new form of economy based on services such as trade, religion, or defense. Any such economy would have required concentrations of people in urban settlements.

Most experts agree that changes in social organization were an important precondition for urbanization. Specifically, urbanization required the emergence of groups who were able to exact tributes, impose taxes, and control labor power, usually through some form of religious persuasion or military coercion. Once established, this elite group provided the stimulus for urban development by using its wealth to build palaces, arenas, and monuments in order to show off its power and status. This activity not only created the basis for the physical core of ancient cities but also required an increased degree of specialization in nonagricultural activities—construction, crafts, administration, the priesthood, soldiery, and so on—which could be organized effectively only in an urban setting. By A.D. 1000, city-based world-empires had emerged in Europe, the Middle East, and China, including a dozen major cities with populations of 100,000 or more (**Figure 10.4**).

The urbanized economies of world-empires were a precarious phenomenon, however, and many of them lapsed into ruralism before being revived or recolonized. In a number of cases, the decline of world-empires was a result of demographic setbacks associated with wars or epidemics. Such disasters left too few people to maintain the social and economic infrastructure necessary for urbanization. This lack of labor power seems to have been a major contributing factor to the eventual collapse of the Mesopotamian empire, and it may also have contributed to the abandonment of much of the Mayan empire more than 500 years before the arrival of the Spanish. Similarly, the population of the Roman empire began to decline in the second century A.D., giving rise to labor shortages, abandoned fields, and depopulated towns and allowing the infiltration of "barbarian" settlers and tribes from the German lands of East-Central Europe.

The Roots of European Urban Expansion

In Europe the urban system introduced by the Greeks and reestablished by the Romans almost collapsed during the Dark Ages of the early medieval period (A.D. 476–1000). During this time, feudalism gave rise to a fragmented landscape of inflexible and inward-looking world-empires. Feudalism was a rigid, rurally oriented form of economic and social organization based on the communal chiefdoms of Germanic tribes that had invaded the disintegrating Roman empire. From this unlikely beginning, however, an elaborate urban system developed, its largest centers eventually growing into what would become the nodal centers of a global world-system.

Early medieval Europe, divided into a patchwork of feudal kingdoms and estates, was mostly rural. Each feudal estate was more or less self-sufficient regarding foodstuffs, and each kingdom or principality was more or less self-sufficient regarding the raw materials needed to craft simple products. Most regions, however, did support at least a few small towns. The existence of these towns depended mainly on their role:

- *Ecclesiastical or university centers*—Examples include St. Andrews in Scotland (**Figure 10.5**); Canterbury, Cambridge, and Coventry in England; Rheims and Chartres in France; Liège in Belgium; Bremen in Germany; Trondheim in Norway; and Lund in Sweden.

- *Defensive strongholds*—Examples include the hilltop towns of central Italy such as Foligno, Montecompatri, and Urbino; the bastide, or fortress, towns of southwestern France, such as Aigues-Mortes and Montauban; and gateway towns such as Bellinzona, Switzerland (**Figure 10.6**).

- *Administrative centers (for the upper tiers of the feudal hierarchy)*— Examples include Cologne (**Figure 10.7**), Mainz, and Magdeburg in Germany; Falkland in Scotland; Winchester in England; and Toulouse in France.

Córdoba (population 450,000)
The largest and most prosperous city of the time, Córdoba was at the cultural forefront in A.D. 1000, renowned for its architecture, craftwork, and dedication to learning.

Rayy (population 100,000)
Known for its superior silks and ceramics, the city was described at the time as stunningly beautiful.

Kaifeng (population 400,000)
Situated near the Yellow River, this Song dynasty capital benefited from its proximity to the empire's industrial center and canal network.

Seville (population 90,000)
One of the wealthiest and most cultured cities in the Muslim state of Andalusia, Seville excelled in science and the arts.

Constantinople (population 300,000)
Located at a strategic crossroads between Europe and Asia, Constantinople was the center of the Byzantine Empire and a major trading hub.

Isfahan (population 100,000)
Located high atop a fertile plain, Isfahan was a producer of grains and silk and was well known for its metalwork and rugs.

Neyshabur (population 125,000)
One of Persia's most progressive cities, Neyshabur also served as a major source of turquoise.

Cairo (population 135,000)
Capital of the Fatimid dynasty, Cairo was known for its many libraries and colleges.

Baghdad (population 125,000)
The capital of the Abassid caliphate, Baghdad was known in A.D. 1000 as the intellectual center of the world. Persian influence pervaded the city's architecture, literature, and court life.

Al Hasa (population 110,000)
Al Hasa was the center of the Qarmatian movement, a radical arm of the Shiite Muslim sect that advocated widespread social equality.

Anhilvada (population 100,000)
The size and location of Anhilvada, like many Indian cities, were subject to changes in the path and flow of nearby rivers.

Angkor (population 200,000)
This Khmer capital was the political center of Southeast Asia and the main market for rice produced by the Khmer empire's high-yield irrigation system.

Kyoto (population 175,000)
Japan's capital since the late eighth century, Kyoto was a religious and cultural center. It was also renowned for its silk works.

Thanjavur (population 90,000)
Thanjavur was the capital of India's Chola dynasty. There King Rajaraja built a massive stone temple dedicated to the god Shiva.

Dali (population 90,000)
Dali peaked in A.D. 986, but the fine marble that was widely sought for buildings and sculptures is still quarried there today.

Figure 10.4   Major cities in A.D. 1000 The most important cities in A.D. 1000 were the seats of world-empires—the Islamic caliphates, the Byzantine Empire, the Chinese Empire, and Indian kingdoms—that had developed well-established civilizations with urban systems based on regional trade and protected by strong military rule. (*Source:* Data from T. Chandler, *Four Thousand Years of Urban Growth: A Historical Census.* Washington, DC: Worldwatch Institute, 1987; "The Year 1000," *U.S. News & World Report,* August 16, 1999, pp. 66–70.)

Figure 10.5 Chartres, France Chartres was an important ecclesiastical center. The cathedral, built in the thirteenth century, is widely considered to be the finest gothic cathedral in France.

Figure 10.6 (a) Aigues-Mortes, France The walled medieval town of Aigues-Mortes in southern France is one of the best-preserved examples of thirteenth-century military architecture. The town of rectilinear streets is surrounded by a wall with five towers and ten fortified gates. Aigues-Mortes was originally built as a port, with access to the sea via a canal dug through the ponds and marshes of the Camargue.

(b) Urbino, Italy An important strategic center in the thirteenth century with a classic hilltop defensive site, Urbino became an important artistic center during the Renaissance.

(a)

(b)

Figure 10.7   Cologne In the late 1400s, when this woodcut was made, Cologne had a population of less than 25,000 but was already an important commercial and manufacturing center, with an important cathedral and a university that was already more than 100 years old.

From the eleventh century onward, however, the feudal system faltered and disintegrated in the face of successive demographic, economic, and political crises that were caused by steady population growth in conjunction with only modest technological improvements and limited amounts of cultivable land. To bolster their incomes and raise armies against one another, the feudal nobility began to levy increasingly higher taxes. Peasants were consequently obliged to sell more of their produce for cash on the market. As a result, a more extensive money economy developed, along with the beginnings of a pattern of trade in basic agricultural produce and craft manufactures. Some long-distance trade even began in luxury goods, such as spices, furs, silks, fruit, and wine. Towns began to increase in size and vitality on the basis of this trade.

The regional specializations and trading patterns that emerged provided the foundations for a new phase of urbanization based on merchant capitalism (**Figure 10.8**). Beginning with networks established by the merchants of Venice, Pisa, Genoa, and Florence (in northern Italy) and the trading partners of the Hanseatic League (a federation of city-states around the North Sea and Baltic coasts), a trading system of immense complexity soon came to span Europe from Bergen to Athens and from Lisbon to Vienna. By 1400, long-distance trading was well established, based not on the luxury goods of the pioneer merchants, but on bulky staples such as grains, wine, salt, wool, cloth, and metals. Milan, Genoa, Venice, and Bruges had all grown to populations of 100,000 or more. Paris was the dominant European city, with a population of about 275,000. This was the Europe that stood poised to extend its grasp to a global scale.

Between the fifteenth and seventeenth centuries, a series of changes occurred that transformed not only the cities and city systems of Europe but the entire world economy. Merchant capitalism increased in scale and sophistication; economic and social reorganization was stimulated by the Protestant Reformation and the scientific revolution. Meanwhile, aggressive overseas colonization made Europeans the leaders, persuaders, and shapers of the rest of the world's economies and societies.

Spanish and Portuguese colonists were the first to extend the European urban system into the world's peripheral regions. They established the basis of a Latin American urban system in just 60 years, between 1520 and 1580. Spanish colonists founded their cities on the sites of Indian cities (in Oaxaca and Mexico City, Mexico; Cajamarca and Cuzco, Peru; and Quito, Ecuador) or in regions of dense indigenous populations (in Puebla and Guadalajara, Mexico; and Arequipa and Lima, Peru). These colonial towns were established mainly as administrative and military centers from which the Spanish Crown could occupy and exploit the New World. Portuguese colonists, in contrast, situated their cities—Recife, Salvador, São Paulo, and Rio de Janeiro—with commercial rather than administrative considerations in mind. They, too, were motivated by exploitation, but their strategy was to establish colonial towns in locations best suited to organizing the collection and export of the products of their mines and plantations.

In Europe, reorganization during the Renaissance saw the centralization of political power and the formation of national states, the beginnings of industrialization, and the funneling of plunder and produce from distant colonies. In this new context, the port cities of the North Sea and Atlantic coasts enjoyed a decisive locational advantage. By 1700, London had grown to 500,000 people, while Lisbon and Amsterdam had each grown to about 175,000. The cities of continental and Mediterranean Europe expanded at a more modest rate. By 1700, Venice had added only 30,000 to its 1400 population of 110,000, and Milan's population had not grown at all between 1400 and 1700.

The most important aspect of urbanization during this period, however, was the establishment of gateway cities around the rest of the world (**Figure 10.9**). **Gateway cities** are those that serve as a link between one country or region

Figure 10.8   The towns and cities of Europe, ca. 1350
Cities with more than 10,000 residents were uncommon in
medieval Europe except in northern Italy and Flanders, such
as the cities of Florence and Delft, where the spread of cloth
production and the growth of trade permitted relatively
intense urbanization. Elsewhere, large size was associated with
a complex of administrative, religious, educational, and
economic functions. By 1350, many of the bigger towns (for
example, Barcelona, Cologne, Prague) supported universities as
well as a variety of religious institutions. Most urban systems,
reflecting the economic and political realities of the time, were
relatively small. (*Source:* Adapted from P. M. Hohenberg and L. H. Lees,
The Making of Urban Europe 1000–1950.)

Ghent, Belgium

Venice, Italy

Florence, Italy

Prague, Czech Republic

New York, at first a modest Dutch fur-trading port, became the gateway for millions of European immigrants and for a large volume of U.S. agricultural and manufacturing exports.

Boston first flourished as the principal colony of the Massachusetts Bay Company, exporting furs and fish and importing slaves from West Africa, hardwoods from central America, molasses from the Caribbean, manufactured goods from Europe, and tea (via Europe) from South Asia.

Salvador, Brazil, was the landfall of the Portuguese in 1500. They established plantations that were worked by slave labor from West Africa. Salvador became the gateway for most of the 3.5 million slaves who were shipped to Brazil between 1526 and 1870.

Guangzhou was the first Chinese port to be in regular contact with European traders—first Portuguese in the sixteenth century and then British in the seventeenth century.

Nagasaki was the only port that feudal Japanese leaders allowed open to European traders, and for more than 200 years Dutch merchants held a monopoly of the import-export business through the city.

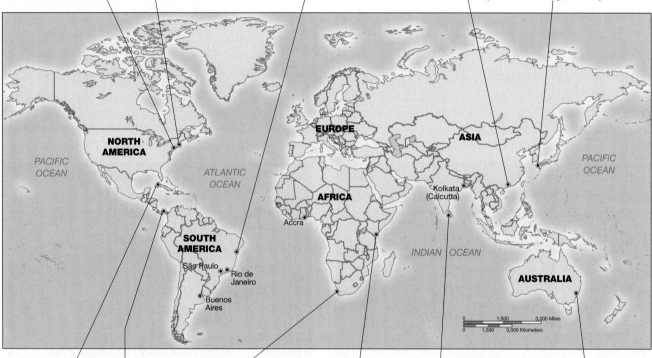

Havana was founded and developed by the Spanish in 1515 because of its excellent harbor. It was used as the assembly point for annual convoys returning to Spain.

Panama City, founded by the Spanish in 1519, became the gateway for gold and silver on its way by galleon to Spain.

Cape Town was founded in 1652 as a provisioning station for ships of the Dutch East India Company. Later, under British rule, it developed into an import-export gateway for South Africa.

Mombasa (in present-day Kenya) was already a significant Arab trading port when Vasco da Gama visited it in 1498 on his first voyage to India. The Portuguese used it as a trading station until it was recaptured by the Arabs in 1698. It did not become an important gateway port until it fell under British Imperial rule in the nineteenth century, when railroad development opened up the interior of Kenya, along with Rwanda, Uganda, and northern Tanzania.

Colombo's strategic situation on trade routes saw it occupied successively by the Portuguese, the Dutch, and the British. It became an important gateway after the British constructed an artificial harbor to handle the exports from tea plantations in Ceylon (now Sri Lanka).

Sydney, Australia, was not settled until the late eighteenth century, and even then many of the settlers were convicts who had been forcibly transported from Britain. It soon became the gateway for agricultural and mineral exports (mostly to Britain) and for imports of manufactured goods and European immigrants.

Figure 10.9 Gateway cities in the world–system periphery Many of the world's most important cities grew to prominence as gateway cities because they commanded routeways into and out of developing colonies. Gateway cities are control centers that command entrance to and exit from their particular country or region.

and others because of their physical situation. They are control centers that command entrance to, and exit from, their particular country or region. European powers founded or developed literally thousands of towns in other parts of the world as they extended their trading networks and established their colonies. The great majority of them were ports. Protected by fortifications and European naval power, they began as trading posts and colonial administrative centers. Before long they developed manufacturing of their own to supply the pioneers' needs, along with more extensive commercial and financial services.

As colonies were developed and trading networks expanded, some of these ports grew rapidly, acting as gateways for colonial expansion into continental interiors. Into their harbors came waves of European settlers; through their docks were funneled the produce of continental interiors. Rio de Janeiro (Brazil) grew on the basis of gold mining; Accra (Ghana) on cocoa; Buenos Aires (Argentina) on mutton, wool, and cereals; Kolkata (India, formerly Calcutta) on jute, cotton, and textiles; São Paulo (Brazil) on coffee; and so on. As they grew into major population centers, they became important markets for imported European manufactures, adding even more to their functions as gateways for international transport and trade.

Industrialization and Urbanization

It was not until the late eighteenth century, however, that urbanization began to become an important dimension of the world-system in its own right. In 1800, less than 5 percent of the world's 980 million people lived in towns and cities. By 1950, however, 16 percent of the world's population was urban, and more than 900 cities of 100,000 or more existed around the world. The Industrial Revolution

and European imperialism had created unprecedented concentrations of humanity that were intimately linked in networks and hierarchies of interdependence.

Cities were synonymous with industrialization. Industrial economies could be organized only through the large pools of labor; the transportation networks; the physical infrastructure of factories, warehouses, stores, and offices; and the consumer markets provided by cities. As industrialization spread throughout Europe in the first half of the nineteenth century and then to other parts of the world, urbanization increased at a faster pace. The higher wages and greater variety of opportunities in urban labor markets attracted migrants from surrounding areas. The countryside began to empty. In Europe the *demographic transition* caused a rapid growth in population as death rates dropped dramatically (see Chapter 3). This growth in population provided a massive increase in the labor supply throughout the nineteenth century, further boosting the rate of urbanization, not only within Europe itself but also in Australia, Canada, New Zealand, South Africa, and the United States as emigration spread industrialization and urbanization to the frontiers of the world-system.

The shock city of nineteenth-century European industrialization was Manchester, England, which grew from a small town of 15,000 in 1750 to a city of 70,000 in 1801; a metropolis of 500,000 in 1861; and a world city of 2.3 million by 1911 (see Box 10.1, "Shock City: Manchester"). A **shock city** is one that is seen at the time as the embodiment of surprising and disturbing changes in economic, social, and cultural life. As industrialization took hold in North America, the shock city was Chicago, which grew from under 30,000 in 1850 to 500,000 in 1880, 1.7 million in 1900, and 3.3 million in 1930 (**Figure 10.10**). When

Figure 10.10 Growth of Chicago In 1870, when Manchester was already a thriving metropolis, Chicago was at the beginning of a period of explosive growth. A year later 9 square kilometers (4 sq. mi.) of the city, including the business district, were destroyed by fire. It was rebuilt rapidly, with prosperous industrialists taking the opportunity to build impressive new structures in the downtown area. The city's economic and social elite colonized the Lake Michigan shore, while heavy industry, warehouses, and railyards crowded the banks of the Chicago River, stretching northwestward from the city center. To the south of the city center were the Union Stockyards and a pocket of heavy industry where the Calumet River met Lake Michigan. All around were the homes of working families in neighborhoods that spread rapidly outward as wave after wave of immigrants arrived in the city.

Chicago was first incorporated as a city in 1837, its population was only 4,200. Its growth followed the arrival of the railroads, which made the city a major transportation hub. By the 1860s, lake vessels were carrying iron ore from the Upper Michigan ranges to the city's blast furnaces, and railroads were hauling cattle, hogs, and sheep to the city for slaughtering and packing. The city's prime geographic situation also made it the nation's major lumber-distributing center by the 1880s.

Both Manchester and Chicago were archetypal forms of an entirely new kind of city—the *industrial city*—whose fundamental reason for existence was not, as in earlier generations of cities, to fulfill military, political, ecclesiastical, or trading functions. Rather, it existed simply to assemble raw materials and to fabricate, assemble, and distribute manufactured goods. Both Manchester and Chicago had to cope, however, with unprecedented rates of growth and the unprecedented economic, social, and political problems that were a consequence of their growth. Both were also world cities, cities in which a disproportionate part of the world's most important business—economic, political, and cultural—is conducted. At the top of a global urban system, such cities experience growth largely as a result of their role as key nodes in the world economy.

During the Industrial Revolution and for much of the twentieth century, a close and positive relationship existed between rural and urban development in the core regions of the world (**Figure 10.11**). The appropriation of new land for agriculture, together with mechanization and the innovative techniques that urbanization allowed, resulted in increased agricultural productivity. This extra productivity released rural labor to work in the growing manufacturing sector in towns and cities. At the same time, it provided the additional produce needed to feed growing urban populations. The whole process was further reinforced by the capacity of urban labor forces to produce agricultural tools, machinery, fertilizer, and other products that made for still greater increases in agricultural productivity. This kind of urbanization is a special case of cumulative causation (Chapter 7), in which a spiral buildup of advantages is enjoyed by particular places as a result of the development of external economies, agglomeration effects, and localization economies.

Imperialism and Peripheral Urbanization

As we pointed out in Chapter 2, the industrialization of the core economies was highly dependent on the exploitation of peripheral regions. Inevitably, the new international division of labor that resulted from this relationship also had a significant impact on patterns and processes of urbanization in the periphery. European imperialism led to the creation of new gateway cities in peripheral countries and, as Europeans raced to establish economic and political control over continental interiors, colonial cities were established as centers of administration, political control, and commerce.

Colonial cities are those that were deliberately established or developed as administrative or commercial centers by colonial or imperial powers. In fact, geographers often distinguish between two types of colonial city. The pure colonial city was usually established, or "planted," by colonial administrations in a location where no significant

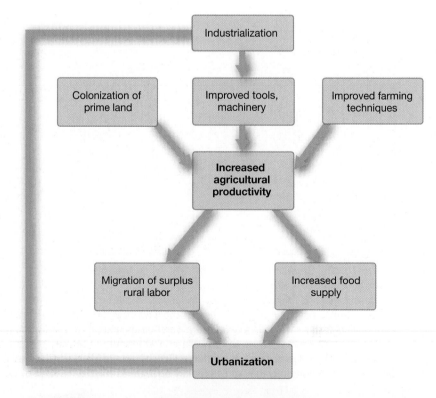

Figure 10.11 The urbanization process in the world's core regions Urbanization was stimulated by advances in farm productivity that (1) provided the extra food to support the increased numbers of townspeople and (2) made many farmers and farm laborers redundant, prompting them to migrate to cities. Labor displaced in this way ended up consuming food rather than producing it, but this was more than compensated for by the increases in agricultural productivity and by the increased capacity of enlarged urban labor forces to produce agricultural tools, machinery, fertilizers, and so on that contributed further to agricultural productivity.

urban settlement had previously existed. Such cities were laid out expressly to fulfill colonial functions, with ceremonial spaces, offices, and depots for colonial traders, plantation representatives, and government officials; barracks for a garrison of soldiers; and housing for colonists. Subsequently, as these cities grew, they added housing and commercial land uses for local peoples drawn by the opportunity to obtain jobs as servants, clerks, or porters. Examples of pure colonial cities were the original settlements of Mumbai (Bombay), Kolkata (Calcutta), Ho Chi Minh City (Saigon), Hong Kong, Jakarta, Manila, and Nairobi.

In the other type of colonial city, colonial functions were grafted onto an existing settlement, taking advantage of a good site and a ready supply of labor. Examples include Delhi, Mexico City, Shanghai, and Tunis. In these cities the colonial imprint is most visible in and around the city center in the formal squares and public spaces, the layout of avenues, and the presence of colonial architecture and monuments. This architecture includes churches, city halls, and railway stations (**Figure 10.12**); the palaces of governors and archbishops; and the houses of wealthy traders, colonial administrators, and landowners.

The colonial legacy can also be read in the building and planning regulations of many colonial cities. Often, colonial planning regulations were copied from those that had been established in the colonizing country. Because these regulations were based on Western concepts, many turned out to be inappropriate to colonial settings. Most colonial building codes, for example, are based on Western models of family and work, with a small family living in a residential area that is some distance from the adults'

places of work. This is at odds with the needs of large, extended families whose members are involved with a busy domestic economy and with family businesses that are traditionally integrated with the residential setting. Colonial planning, with its gridiron street layouts, zoning regulations that do not allow for a mixture of land uses, and building codes designed for European climates, ignored the specific needs of local communities and misunderstood their cultural preferences.

URBAN SYSTEMS

Every town and city is part of one of the interlocking urban systems that link regional-, national-, and international-scale human geographies in a complex web of interdependence. These urban systems organize space through hierarchies of cities of different sizes and functions. Many of these hierarchical urban systems exhibit common attributes and features, particularly in the relative size and spacing of individual towns and cities.

Geographers have long recognized the tendency for the functions of towns and cities as market centers to result in a hierarchical system of central places. A **central place** is a settlement in which certain types of products and services are available to consumers. **Central place theory** seeks to explain the tendency for central places to be organized in hierarchical systems, analyzing the relative size and geographic spacing of towns and cities as a function of consumer behavior. A fundamental tenet of central place theory is that the smallest settlements in an urban system will provide only those goods and services

Figure 10.12 Colonial architecture and urban design Cities in the periphery of the world-system have grown very rapidly since the colonial era, but the legacy of the colonial period can still be seen in the architecture, monuments, and urban design of the period. This painting of the Indian Peninsular Terminus in Mumbai (Bombay) shows the influence of Victorian British architecture.

10.1 VISUALIZING GEOGRAPHY

Shock City: Manchester

One day I walked with one of these middle-class gentlemen into Manchester. I spoke to him about the disgraceful unhealthy slums and drew his attention to the disgusting condition of that part of the town in which the factory worker lived. I declared that I had never seen so badly built a town in my life. He listened patiently and at the corner of the street he remarked: "And yet there is a great deal of money to be made here. Good morning, Sir!"

Friedrich Engels,
The Condition of the Working
Class in England in 1844

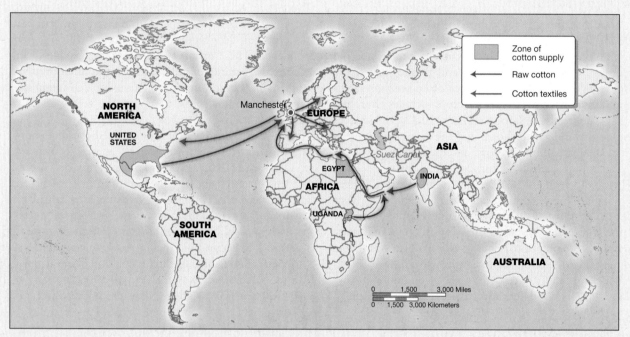

The opening of the Suez Canal in 1869 halved the traveling time between Britain and India. It ruined the Indian domestic cotton textile industry, but it allowed India to export its raw cotton to Manchester. Around the same time, British colonialists established cotton plantations in Egypt and Uganda, providing another source of supply.

Migrants from Ireland and northern England contributed to Manchester's rapid growth from the mid-nineteenth century.

Manchester's first cotton mill was built in the early 1780s, and by 1830 there were 99 cotton-spinning mills. As the city grew, it spilled out into the surrounding countryside, bringing its characteristic landscape of red-brick terrace housing and "Dark Satanic Mills" with their tall brick chimneys.

Railway viaducts, like this one in Stockport, just outside Manchester, brought rail transportation to Manchester early in the nineteenth century and helped to make the city a major transportation hub.

Manchester City Hall, a classic example of Victorian Gothic architecture, was built to show the world that the city had arrived. Manchester in the nineteenth century was a city of enormous vitality not only in its economic life but also in its political, cultural, and intellectual life.

In 1894 the Manchester Ship Canal was opened, allowing ocean-going vessels to reach docks close to the centre of Manchester. The development of container ports and motorway links outweighed the convenience of ships being able to reach Manchester and in the 1980s most of the docks were closed. The Salford docks have now been redeveloped as Salford Quays, with upscale condominiums and stores.

Canals were at the heart of the development of industrial Manchester. They enabled coal and raw materials to be carried right to the heart of the city and finished goods to be transported away easily. The Bridgewater Canal, one of the earliest modern canals, ran into Manchester from the Duke of Bridgewater's coal mine at Worsley, about 10 miles away. Later, other canals radiated outwards from the area and the Ship Canal brought sea-going vessels close to the city center.

In the late nineteenth century, working-class housing was built to conform with local building codes—but only just. Much of it has now been replaced, but a good deal still remains.

Part of the redeveloped Salford Quays, the Lowry Centre (named after L. S. Lowry, the Manchester artist famous for painting scenes of life in the industrial districts of northern England during the early 20th century) is a combined theatre and gallery complex.

Figure 10.13 The Spanish urban system Note how the smaller cities tend to be linked to middle-order cities, while these in turn are linked to regional metropolises, which are linked to the national metropolises, Madrid and Barcelona. These linkages represent the major flows of capital, information, and goods within the Spanish urban system. (*Source:* Adapted from L. Bourne, R. Sinclair, M. Ferrer, and A. d'Entremont (eds.), *The Changing Geography of Urban Systems.* Navarra, Spain: Department of Human Geography, Universidad de Navarra, 1989, fig. 2, p. 46.)

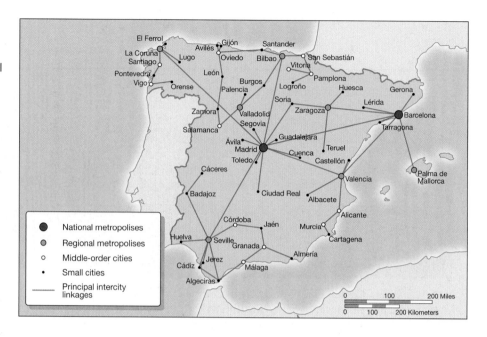

that meet everyday needs (bakery and dairy products and groceries, for example) and that these small settlements will be situated relatively close to one another because consumers, assumed to be spread throughout the countryside, will not be prepared to travel far for such items. On the other hand, people will be willing to travel farther for more expensive, less frequently purchased items, so that the larger the settlement, with a broader variety of more specialized goods and services, the farther apart it will be from others of a similar size.

Although consumer behavior certainly helps to explain some aspects of urban systems, there are relatively few regions today where the functions of towns and cities are dominated by local markets and shopping. Nevertheless, the urban systems of most regions do exhibit a clear hierarchical structure. This is partly a legacy of past eras, when towns and cities did function mainly as market centers for surrounding agricultural areas. **Figure 10.13** shows a typical example: the Spanish urban system, with smaller towns and cities functioning interdependently with successively larger ones, the whole system dominated by one or two metropolitan areas whose linkages are national in scope.

Urban systems also exhibit clear *functional* differences within such hierarchies, yet another reflection of the interdependence of places. The geographical division of labor resulting from such processes of economic development (Chapter 7) means that many medium- and larger-size cities perform quite specialized economic functions and so acquire quite distinctive characters. Thus, there are steel towns (for example, Pittsburgh, Pennsylvania; Sheffield, England), textile towns (for example, Lowell, Massachusetts; Manchester, England), and auto-manufacturing towns (for example, Detroit; Turin, Italy; Toyota City, Japan). Some towns and cities, of course, do evolve as general-purpose urban centers, providing an evenly balanced range of functions for their own particular sphere

of influence. **Figure 10.14** shows the urban system in the United States, where the top tier of cities consists of centers of global importance (including Chicago, New York, and Los Angeles) that provide high-order functions to an international marketplace. The second tier consists of general-purpose cities with diverse functions but only regional importance (including Atlanta, Miami, and Boston), and the third and fourth tiers consist of more specialized centers of subregional and local importance.

City–Size Distributions, Primacy, and Centrality

The functional interdependency between places within urban systems tends to result in a distinctive relationship between the population size of cities and their rank within the overall hierarchy. This relationship is known as the **rank-size rule**, which describes a certain statistical regularity in the city-size distributions of countries and regions. The relationship is such that the nth largest city in a country or region is $1/n$ the size of the largest city in that country or region. Thus, if the largest city in a particular system has a population of 1 million, the fifth-largest city should have a population one-fifth as big (that is, 200,000); the hundredth-ranked city should have a population one-hundredth as big (that is, 10,000), and so on. Plotting this relationship on a graph with a logarithmic scale for population sizes would produce a perfectly straight line. The actual rank-size relationship for the U.S. urban system has always come close to this (**Figure 10.15**). Over time the slope has moved to the right on the graph, reflecting the growth of towns and cities at every level in the urban hierarchy.

In some urban systems, the top of the rank-size distribution is distorted as a result of the disproportionate size of the largest (and sometimes also the second-largest)

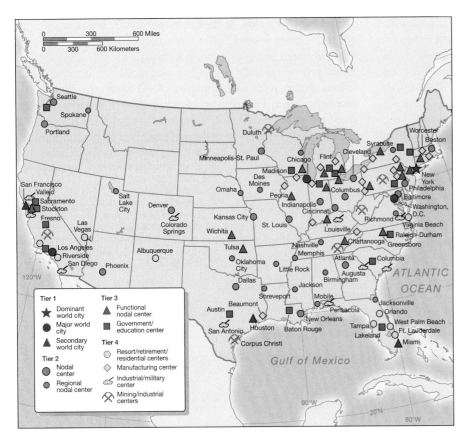

Figure 10.14 Functional specialization within an urban system Different cities tend to specialize in particular kinds of economic activities, and some provide a much greater range of functions than others. This map shows a functional classification of the U.S. urban system. The top tier of the system consists of "world cities" that provide high-order functions to a global marketplace. Among these, New York stands alone as the dominant metropolis. The next tier consists of cities with diverse functions but only regional importance (for example, Atlanta, Minneapolis). The third tier consists of more specialized centers of business, government, and producer services (for example Austin, Texas; Albany, New York; and Hartford, Connecticut). The fourth tier consists of still more specialized cities of various kinds: manufacturing centers (for example, Buffalo, New York; Chattanooga, Tennessee), mining/industrial centers (for example, Charleston, West Virginia; Duluth, Minnesota), industrial/military centers (for example, Newport News, Virginia; San Diego, California), and resort/retirement centers (for example, Las Vegas, Nevada; Orlando, Florida). (*Source:* Reprinted with permission of Prentice Hall from P. L. Knox, *Urbanization,* © 1994, p. 64.)

city. In Argentina, for example, Buenos Aires is more than 10 times the size of Rosario, the second-largest city. (According to the rank-size rule, the largest city should be just twice the size of the second-largest city.) In the United Kingdom, London is more than 9 times the size of Birmingham, the second-largest city. In France, Paris is more than 8 times the size of Marseilles, France's second-largest city. In Brazil, both Rio de Janeiro and São Paulo are 5 times the size of Belo Horizonte, the third-largest city. Geographers call this condition **primacy**, occurring when the population of the largest city in an urban system is disproportionately large in relation to the second- and third-largest cities in that system. Cities like London and Buenos Aires are termed *primate* cities.

Thus, primacy is not simply a matter of size. Some of the largest metropolitan areas in the world—Karachi, New York, and Mumbai (Bombay), for example—are not primate. Further, primacy is a condition that is found in both the core and the periphery of the world-system. This suggests that primacy is a result of the roles played by particular cities within their own national urban systems. A relationship does exist to the world economy, however.

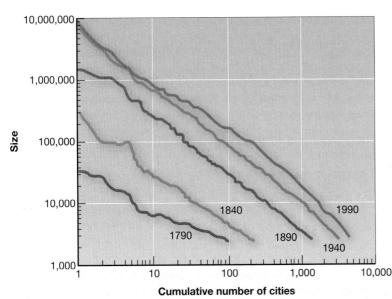

Figure 10.15 The rank-size distribution of cities in the U.S. urban system, 1790–1990 This graph shows that the U.S. urban system has conformed fairly consistently to the rank-size rule. As urbanization brought increased populations to cities at every level in the urban hierarchy, the rank-size graph has moved to the right. Meanwhile, the growth of some cities (for example, San Diego) has sent them from the lower end of the hierarchy to the very top, while other cities (for example, Savannah, Georgia) have declined, at least in relative terms, so that they have fallen down the hierarchy. (*Source:* Reprinted with permission of Prentice Hall from P. L. Knox, *Urbanization,* © 1994, p. 32.)

Primacy in peripheral countries is usually a consequence of primate cities' early roles as gateway cities. In core countries, it is usually a consequence of primate cities' roles as imperial capitals and centers of administration, politics, and trade for a much wider urban system than their own domestic system.

When cities' economic, political, and cultural functions are disproportionate to their population, the condition is known as **centrality**, or the functional dominance of cities within an urban system. Cities that account for a disproportionately high share of economic, political, and cultural activity have a high degree of centrality within their urban system. Very often primate cities exhibit this characteristic, but cities do not necessarily have to be primate in order to be functionally dominant within their urban system. **Figure 10.16** shows some examples of centrality,

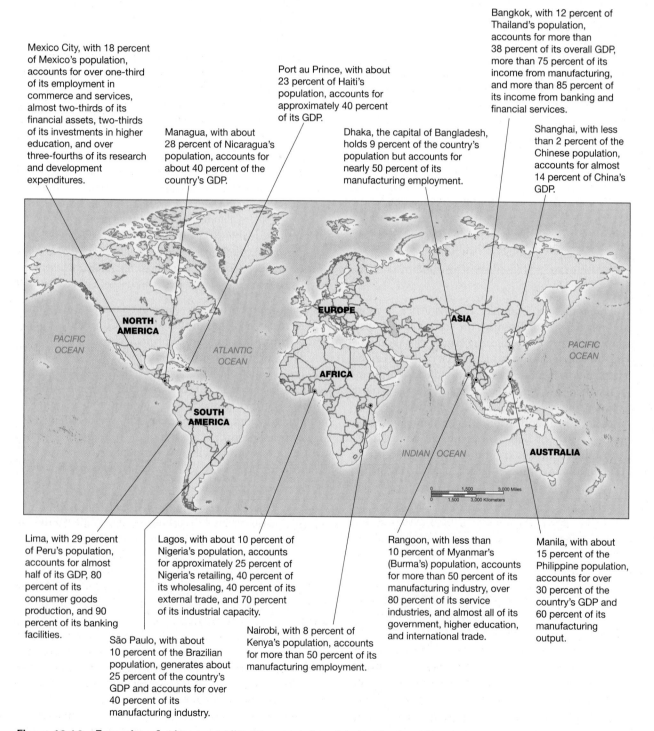

Mexico City, with 18 percent of Mexico's population, accounts for over one-third of its employment in commerce and services, almost two-thirds of its financial assets, two-thirds of its investments in higher education, and over three-fourths of its research and development expenditures.

Managua, with about 28 percent of Nicaragua's population, accounts for about 40 percent of the country's GDP.

Port au Prince, with about 23 percent of Haiti's population, accounts for approximately 40 percent of its GDP.

Dhaka, the capital of Bangladesh, holds 9 percent of the country's population but accounts for nearly 50 percent of its manufacturing employment.

Bangkok, with 12 percent of Thailand's population, accounts for more than 38 percent of its overall GDP, more than 75 percent of its income from manufacturing, and more than 85 percent of its income from banking and financial services.

Shanghai, with less than 2 percent of the Chinese population, accounts for almost 14 percent of China's GDP.

Lima, with 29 percent of Peru's population, accounts for almost half of its GDP, 80 percent of its consumer goods production, and 90 percent of its banking facilities.

Lagos, with about 10 percent of Nigeria's population, accounts for approximately 25 percent of Nigeria's retailing, 40 percent of its wholesaling, 40 percent of its external trade, and 70 percent of its industrial capacity.

São Paulo, with about 10 percent of the Brazilian population, generates about 25 percent of the country's GDP and accounts for over 40 percent of its manufacturing industry.

Nairobi, with 8 percent of Kenya's population, accounts for more than 50 percent of its manufacturing employment.

Rangoon, with less than 10 percent of Myanmar's (Burma's) population, accounts for more than 50 percent of its manufacturing industry, over 80 percent of its service industries, and almost all of its government, higher education, and international trade.

Manila, with about 15 percent of the Philippine population, accounts for over 30 percent of the country's GDP and 60 percent of its manufacturing output.

Figure 10.16 Examples of urban centrality The economic, political, and cultural importance of some cities is disproportionate to their population size, making them more central to their economies. This is a reflection of core-periphery differentials within countries and often becomes a political issue because of the economic disparities. The centrality of these cities also leads to localized problems of congestion, land price inflation, and pollution.

revealing the overwhelming dominance of some cities within the world-system periphery. Bangkok, for instance, with around 12 percent of the Thai population, accounts for approximately 38 percent of the country's overall gross domestic product (GDP), over 85 percent of the country's GDP in banking, insurance, and real estate, and 75 percent of its manufacturing.

World Cities and the Global Urban System

As explained in Chapter 7, ever since the evolution of a world-system in the sixteenth century, certain cities known as **world cities** (sometimes referred to as *global cities*) have played key roles in organizing space beyond their own national boundaries. In the first stages of world-system growth, these key roles involved the organization of trade and the execution of colonial, imperial, and geopolitical strategies. The world cities of the seventeenth century were London, Amsterdam, Antwerp, Genoa, Lisbon, and Venice. In the eighteenth century, Paris, Rome, and Vienna also became world cities, while Antwerp and Genoa became less influential. In the nineteenth century, Berlin, Chicago, Manchester, New York, and St. Petersburg became world cities, while Venice became less influential.

Today the globalization of the economy has resulted in the creation of a global urban system in which the key roles of world cities are concerned less with the deployment of imperial power and the orchestration of trade and more with transnational corporate organization, international banking and finance, supranational government, and the work of international agencies. World cities have become the control centers for the flows of information, cultural products, and finance that collectively sustain the economic and cultural globalization of the world.

World cities also provide an interface between the global and the local. They contain the economic, cultural, and institutional apparatus that channels national and provincial resources into the global economy and that transmits the impulses of globalization back to national and provincial centers. As such, world cities possess several functional characteristics:

- They are the sites of most of the leading global markets for commodities, commodity futures, investment capital, foreign exchange, equities, and bonds.
- They are the sites of clusters of specialized, advanced business services, especially those that are international in scope and that are attached to finance, accounting, advertising, property development, and law.
- They are the sites of concentrations of corporate headquarters—not just of transnational corporations but also of major national firms and large foreign firms.

- They are the sites of concentrations of national and international headquarters of trade and professional associations.
- They are the sites of most of the leading nongovernmental organizations (NGOs) and intergovernmental organizations (IGOs) that are international in scope (for example, the World Health Organization; United Nations Educational, Scientific, and Cultural Organization (UNESCO); the International Labor Organization, and the International Federation of Agricultural Producers).
- They are the sites of the most powerful and internationally influential media organizations (including newspapers, magazines, book publishing, and satellite television); news and information services (including news wires and online information services); and culture industries (including art and design, fashion, film, and television).
- Because of their importance and visibility, world cities are also the setting for many terrorist acts—see Box 10.2, "Urban Terrorism."

A great deal of synergy exists among the various functional dimensions of world cities. A city like New York, for example, attracts transnational corporations because it is a center of culture and communications. It attracts specialized business services because it is a center of corporate headquarters and of global markets, and so on. These interdependencies represent a special case of the geographical *agglomeration effects* that we discussed in Chapter 7. Agglomeration is the clustering of functionally related activities. In the case of New York City, corporate headquarters and specialized legal, financial, and business services cluster together because of the mutual cost savings and advantages of being close to one another. At the same time, different world cities fulfill different roles within the world-system, making for different emphases and combinations (that is, differences in the nature of their world-city functions) as well as for differences in the absolute and relative localization of particular world-city functions (that is, differences in their degree of importance as world cities). For example, Brussels is relatively unimportant as a corporate headquarters location but qualifies as a world city because it is the administrative center of the European Union and has attracted a large number of nongovernmental organizations and advanced business services that are transnational in scope. To take another example, Milan is relatively dependent in terms of corporate control and advanced business services but has global status in terms of cultural influence (especially fashion and design) and is an important regional financial center.

Today the global urban system is dominated by London and New York, whose influence is "pan regional"—that is, their world-city functions extend across all three main circuits of the global economy—the Americas,

10.2 WINDOW ON THE WORLD

Urban Terrorism

Not only has terrorism taken on a global cast, with international linkages, it is also apparent that cities have become the stage on which this tragic drama is played. Even before the stunning terrorist attacks on the World Trade Center and the Pentagon in 2001, on commuter trains in Madrid in 2004, and on London buses and underground trains in 2005 (**Figure 10.A**), cities had become the central venues of terror. Between 1993 and 2000 there were more than 500 terrorist incidents in cities around the world.

There are several reasons for this. First, cities—especially world cities—have considerable symbolic value. They are not only dense agglomerations of people and buildings but symbols of national prestige and military, political, and financial power. A blast in a mountain town or in the countryside may arouse local concern, but it is generally of little or no consequence for the rest of the world. But an attack on Wall Street (New York), a massacre in Piccadilly Circus (London), the bombing of the Eiffel Tower (Paris), or poison gas in a Tokyo metro arouses international alarm. Any such event will be instantly telegraphed to a larger world and alarm a much larger audience. Second, the assets of cities—densely packed and with a great mix of industrial and commercial infrastructure—make them rich targets. Third, cities have become nodes for a vast international network of communications. This is a reflection of their power, but it is also a vulnerability. A well-placed explosion can produce enormous reverberations, paralyze a city, and spread fear and economic dislocation. Finally, word gets around more quickly and socialization proceeds more rapidly in densely packed environments. This kind of environment provides an abundant source of recruitment for potential terrorists.

Heterogeneous urban settings, while providing rich synergies, can be a nesting ground for terrorist organizations under certain conditions. A sense of relative deprivation often sharpens as those struggling to get by come into closer proximity to other struggling groups and to the more affluent. Beirut provides a ready example of how different groups living under conditions of hopelessness and in proximity to one another turn on each other. Similar ecologies of terror pervade Belfast (Northern Ireland), Sarajevo (Bosnia-Herzegovina), Hyderabad (India), Karachi (Pakistan), and Baghdad (Iraq). Rather than being directed from lower classes upward toward elites, conflict occurs between groups operating at the same level—Hindus fighting Muslims in Mumbai (India) or rival narco-gangs in Bogotá (Colombia).

Statistically, the cities with the greatest incidence of terrorist events in the eight-year period between 1993 and 2000 were Srinagar (India), Athens (Greece), Sanaa (Yemen), Paris (France), Istanbul (Turkey) and Lima (Peru), followed closely by Jerusalem (Israel), Algiers (Algeria), and Dushanbe (Tajikistan). The largest number of fatalities occurred in Nairobi, Kenya (291 killed in 1 incident); Colombo, Sri Lanka (108 killed in 3 incidents); and Jerusalem (77 killed in 9 incidents). Altogether, more than 250 cities around the world experienced one or more terrorist acts in that period.

Source: Based on extracts from H. V. Savitch and G. Ardashev, "Does Terror Have an Urban Future?" *Urban Studies*, pp. 2525–2533.

Figure 10.A   Terrorist attack The wreckage of a public train near Atocha train station in Madrid, Spain, March 11, 2004. Thirteen bombs on four packed commuter trains killed 191 people and wounded more than 1,500. The attack was attributed to the Islamic militant group al-Qaeda.

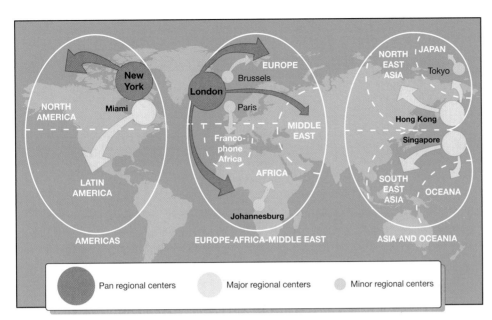

Figure 10.17 World cities in the global urban system This map shows the sphere of influence of world cities, based on an analysis of the regional headquarter functions of the world's largest advanced business services firms (accountancy, advertising, banking and finance, and law). (*Source:* Adapted from P. J. Taylor, "World Cities and Territorial States Under Conditions of Contemporary Globalization," *Political Geography, 19*, 2000, pp. 5–32.)

Europe/Africa/Middle East, and Asia/Oceania (**Figure 10.17**). A second tier of the global urban system consists of world cities with influence over large regions within the world-system. These are Miami (the regional center for Latin America), Hong Kong (the regional center for Northeast Asia), and Singapore (the regional center for Southeast Asia). A third tier consists of important international cities with more limited or more specialized international functions: Brussels, Paris, Johannesburg, and Tokyo.

As globalization continues, competition among world cities will intensify, and more functional and regional specializations will emerge. Already, world-city attributes are beginning to be exhibited by a few key cities in developing regions of the world. As the United Nations Human Settlements Programme observes:

> Cities such as Istanbul and Mumbai are already establishing the cultural trends in their countries and regions, and this influence could cross international borders through films, literature, satellite television networks and entertainment. Large cities in the developing world, including Nairobi, Addis Ababa and Bangkok, among others, are increasingly bringing together major national and international partners by hosting international agencies and development partners, and offering avenues for constructive peer exchange, mediation, and diplomacy.[4]

World Urbanization Today

It is difficult to say just how urbanized the world has become. In many areas, urban growth is taking place at such a pace and under such chaotic conditions that experts can provide only informed estimates. The most comprehensive source of statistics is the United Nations, whose data

suggest that almost half of the world's population is now urban. These data incorporate the very different definitions of *urban* used by different countries. Some countries (Australia and Canada, for example) count any settlement of 1,000 people or more as urban; others (including Italy and Jordan) use 10,000 as the minimum for an urban settlement, and Japan uses 50,000 as the cutoff. This, by the way, tells us something about the nature of urbanization itself: It is a relative phenomenon. In countries like Peru, where the population is thin and scattered, a settlement of 2,000 represents a significant center. In countries like Japan, however, with greater numbers, higher densities, and a tradition of centralized agricultural settlement, a much larger concentration of people is required to count as "urban."

Taking the definitions used in individual countries, almost one-half of the world's population is now urbanized. As **Table 10.1** shows, North America is the most urbanized continent in the world, with 80 percent of its population living in urban areas. In contrast, Africa and Asia are less than 40 percent urban.

To put these figures in perspective, only 29.7 percent of the world's population was urbanized in 1950, using the same definitions of urban settlements. In that year, there were only 83 metropolitan areas of a million or more, and only eight of 5 million or more existed; in 2007, there were approximately 412 metropolitan areas of a million or more people, and 30 with over 5 million. Looking ahead, population projections for 2025 suggest that more than 65 percent of the world's population will be living in urban areas, and there will be around 572 cities with a population of a million or more, including about 48 cities of 5 million or more.

Regional Trends and Projections

The single most important aspect of world urbanization, from a geographical perspective, is the striking difference in trends and projections between the core regions

[4]United Nations Human Settlements Programme, *The State of the World's Cities 2006/2007*. Sterling, VA: Earthscan, 2006, p. 8.

TABLE 10.1 Urbanization by Major World Regions, 2007

World Region	Population in Urban Areas, in thousands (2007)	Urban Percentage of Total Population (2007)	Percentage of Urban Population Living in Cities of 500,000 or Less (2010)	Percentage of Urban Population Living in Cities of 10 million or More (2010)
Africa	373,372	38.7	56.1	5.6
Asia	1,645,225	40.8	49.1	10.0
Europe	527,686	72.2	67.6	2.0
North America	275,494	81.3	36.5	11.3
Latin America and the Caribbean	448,006	78.3	47.9	13.7
Oceania	24,160	70.5	36.5	*no data*
World	3,293,944	49.4	51.5	8.8

Source: United Nations Department of Economic and Social Affairs/Population Division, *World Urbanization Prospects: The 2007 Revision,* New York: UN Department of Economic and Social Affairs, 2008.

and the semiperipheral and peripheral regions. In 1950, two-thirds of the world's urban population was concentrated in the more developed countries of the core economies. Since then, the world's urban population has increased threefold, the bulk of the growth having taken place in the less developed countries of the periphery (**Figure 10.18**). In 1950, 21 of the world's largest 30 metropolitan areas were located in core countries—11 of them in Europe and 6 in North America. By 1980, the situation was completely reversed, with 19 of the largest

Figure 10.18 Urban population growth, 1950–2030 While the metropolitan areas of the world's core countries have continued to grow, most of them have been overtaken by the startling growth of the "unintended" metropolises of peripheral and semiperipheral countries. (*Source:* Data from United Nations, *World Urbanization Prospects.* New York: UN Department of Economic and Social Affairs, 1998.)

30 located in peripheral and semiperipheral regions. By 2010, all but 7 of the 30 largest metropolitan areas are expected to be located in peripheral and semiperipheral regions (**Table 10.2**).

Asia provides some of the most dramatic examples of this trend. From a region of villages, Asia is fast becoming a region of cities and towns. Between 1950 and 2005, for example, its urban population rose more than tenfold, to over 1.5 billion people. By 2020, about two-thirds of Asia's population will be living in urban areas. Nowhere is the trend toward rapid urbanization more pronounced than in China, where for decades the communist government imposed strict controls on where people were allowed to live, fearing the transformative and liberating effects of cities. By tying people's jobs, school admission, and even the right to buy food to the places where people were registered to live, the government made it almost impossible for rural residents to migrate to towns or cities. As a result, more than 70 percent of China's 1 billion people still lived in the countryside in 1985. Now, however, China is rapidly making up for lost time (see Box 10.3, "The Pearl River Delta: An Extended Metropolis"). The Chinese government, having decided that towns and cities can be engines of economic growth within a communist system, has not only relaxed residency laws but also drawn up plans to establish over 430 new cities. Between 1981 and 2003, the number of people living in cities in China more than tripled, from 162 million to 504 million, and the number of cities with a population of half a million or more increased from 16 to 97.

High levels of urbanization have existed in the world's core countries for some time. According to their own national definitions, the populations of Australia, Belgium, and Canada are more than 90 percent urbanized, and those of France, Germany, Norway, Spain, Sweden, and the United Kingdom are all more than 75 percent urbanized. In these core countries, however, rates of urbanization are relatively low, just as the overall rates of population growth are slow (see Chapter 3).

TABLE 10.2 The World's 30 Largest Metropolitan Areas, Ranked by Population Size, 1950, 1980, and 2010 (in millions)

1950	Population	1980	Population	2010	Population
New York	12.3	Tokyo	28.5	Tokyo	36.0
London	8.7	New York	15.6	Mumbai	24.0
Tokyo	6.9	Mexico City	13.0	São Paulo	21.1
Paris	5.4	São Paulo	12.1	Mexico City	20.6
Moscow	5.4	Shanghai	11.7	New York-Newark	19.4
Shanghai	5.3	Osaka	10.0	Delhi	17.0
Essen	5.3	Buenos Aires	9.9	Shanghai	15.7
Buenos Aires	5.0	Los Angeles	9.5	Kolkata	15.6
Chicago	4.9	Kolkata	9.0	Dhaka	14.7
Kolkata	4.4	Beijing	9.0	Buenos Aires	13.0
Osaka	4.1	Paris	8.8	Karachi	13.0
Los Angeles	4.0	Mumbai	8.7	Los Angeles	12.7
Beijing	3.9	Rio de Janeiro	8.6	Cairo	12.5
Milan	3.6	Seoul	8.3	Rio de Janeiro	12.1
Berlin	3.3	Moscow	8.1	Beijing	11.7
Mexico City	3.1	London	7.7	Manila	11.6
Philadelphia	2.9	Cairo	7.3	Osaka-Kobe	11.3
St. Petersburg	2.9	Tianjin	7.3	Lagos	10.6
Mumbai	2.9	Chicago	7.2	Istanbul	10.5
Rio de Janeiro	2.9	Rhein-Ruhr	6.3	Moscow	10.5
Detroit	2.8	Jakarta	6.0	Paris	10.0
Naples	2.8	Metro Manila	6.0	Seoul	9.8
Manchester	2.5	Delhi	5.6	Jakarta	9.7
São Paulo	2.4	Milan	5.3	Guangzhou	9.4
Cairo	2.4	Tehran	5.0	Chicago	9.2
Tianjin	2.4	Karachi	5.0	Kinshasa	9.0
Birmingham	2.3	Bangkok	4.7	London	8.6
Frankfurt	2.3	St. Petersburg	4.6	Lima	8.3
Boston	2.2	Hong Kong	4.6	Bogotá	8.3
Hamburg	2.2	Philadelphia	4.5	Tehran	8.2

Source: Data from United Nations Department of Economic and Social Affairs/Population Division, *World Urbanization Prospects: The 2007 Revision.* New York: UN Department of Economic and Social Affairs, 2008, p. 167.

Levels of urbanization are also very high in many of the world's semiperipheral countries. Brazil, Hong Kong, Mexico, Taiwan, Singapore, and South Korea, for example, are all at least 75 percent urbanized. Unlike the core countries, however, rates of growth in semiperipheral countries are high. In peripheral countries, rates are even higher.

Whatever the current level of urbanization in peripheral countries, almost all are experiencing high rates of urbanization, with growth forecasts of unprecedented speed and unmatched size. Karachi, Pakistan, a metropolis of 1.03 million in 1950, had reached 8.5 million in 1995 and is expected to reach 16.2 million by 2015. Likewise, Cairo, Egypt, grew from 2.44 million to 9.7 million between 1950 and 1995 and is expected to reach 13.1 million by 2015. Mumbai (India; formerly Bombay), Delhi (India), Mexico City (Mexico), Dhaka (Bangladesh), Jakarta (Indonesia), Lagos (Nigeria), São Paulo (Brazil), and Shanghai (China) are all projected to have populations in excess of 17 million by 2015. The reasons for this urban growth vary. Wars in Liberia and Sierra Leone have pushed hundreds of thousands of people into their capitals, Monrovia and Freetown. In Mauritania, Niger, and other countries bordering the Sahara, deforestation and overgrazing have allowed the desert to expand and swallow up villages, forcing people toward cities. For the most part, though, urban growth in peripheral countries is a consequence of the onset of the demographic transition (see Chapter 3), which has produced fast-growing rural populations in regions that face increasing problems with agricultural development (see Chapter 8). As a response, many people in these regions migrate to urban areas seeking a better life.

Many of the largest cities in the periphery are growing at annual rates of between 4 and 7 percent; at the

10.3 WINDOW ON THE WORLD

The Pearl River Delta: An Extended Metropolis

The Pearl River Delta (**Figure 10.B**) is one of the fastest-growing urban regions in the world. Anchored by the major metropolitan centers of Guangzhou, Hong Kong, Macao, Shenzhen, and Zhuhai, the Pearl River Delta is an extended metropolitan region of nearly 50 million people. It is one of three extended metropolitan regions—Beijing-Tianjin and Shanghai are the others—that have been fostered by the Chinese government as engines of capitalist growth since liberal economic reforms were introduced in the late 1970s.

Hong Kong (**Figure 10.C**), a British colony until 1997, is a metropolis of 6.9 million with a thriving industrial and commercial base that is recognized as a capitalist economic dynamo by the Chinese government, which has created a Special Administrative District for the metropolis. As a result, Hong Kong's citizens have retained their British-based legal system and its guaranteed rights of property ownership and democracy. Hong Kong is the world's largest container port, the third-largest center for foreign-exchange trade, the seventh-largest stock market, and the tenth-largest trading economy.

Hong Kong's success encouraged the Chinese government to establish two of its first Special Economic Zones (SEZs) in nearby Shenzhen and Zhuhai. Designed to attract foreign capital, technology, and management practices, these SEZs were established as export-processing zones that offered cheap labor and land, along with tax breaks, to transnational corporations. Investors from Hong Kong and Taiwan responded quickly and enthusiastically. By 1993, more than 15,000 manufacturers from Hong Kong alone had set up businesses in Guangdong Province, and a similar number had established subcontracting relationships, contracting out assembly-line work to Chinese companies in the Pearl River Delta. Meanwhile, the entire delta region was subsequently designated an Open Economic Region, where local governments, individual enterprises, and farm households could enjoy a high degree of autonomy in economic decision making.

The relaxation of state control over the regional economy allowed the region's dense and growing rural population to migrate to urban areas in search of assembly-line jobs or to stay in rural areas and diversify agricultural production from paddy-rice cultivation to more profitable activities such as market-farming activities, livestock husbandry, and fishery. Economic freedom also facilitated rural industrialization—mostly low-tech, small-scale, labor-intensive, and widely scattered across the countryside—though the triangular area between Guangzhou, Hong Kong, and Macao has

Figure 10.B   Pearl River Delta One of the fastest-growing regions of the world, the Pearl River Delta is an extended metropolitan region of more than 50 million people.

including a metro system and an elevated railway network to link the city's new international airport, railway stations, and port. Throughout the region, enormous investments have been made in showpiece infrastructure projects geared to the needs of local and international capital. These include major airports, high-speed tolled highways, satellite ground stations, port installations, metro and light-rail networks, and new water-management systems. These, in turn, have attracted business and technology parks, financial centers, and resort complexes in a loose-knit sprawl of urban development.

Today the Pearl River Delta provides a thriving export-processing platform that has driven double-digit annual economic growth for much of the past two decades. The region's GDP grew from just over US$8 billion in 1980 to nearly US$221.2 billion in 2005. During that period, the average real rate of GDP growth in the Pearl River Delta Economic Zone exceeded 16 percent, well above the People's Republic of China national figure of 9.8 percent. By 2005, and with only 3.5 percent of the country's population, the region was contributing 9.9 percent of the country's GDP and 28.9 percent of its total trade.

Guangzhou is a megacity with a population of 10.5 million in 2005 (**Figure 10.D**). Shenzhen has grown from a population of just 19,000 in 1975 to 4.6 million in 2005, with an additional 2 million in the surrounding municipalities. The southern border of the Shenzhen Special Economic Zone adjoins Hong Kong, but the northern border is walled off from the rest of China by an electrified fence to prevent smuggling and to keep back the mass of people trying to migrate illegally into Shenzhen and Hong Kong.

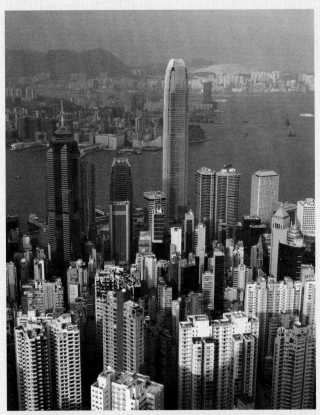

Figure 10.C City of Hong Kong Although most of its manufacturing has been transferred to neighboring Guangdong Province, where wages are much lower, thousands of companies are located in Hong Kong simply for the purpose of doing business with China. As a result, Hong Kong remains a major world city—a major financial hub with a thriving commercial sector and a population of 6.9 million.

quickly emerged as an especially important zone because of its relatively cheap land and labor and because of significant levels of investment by regional and local governments in the transport and communications infrastructure. The result is a distinctive "extended metropolis" in which numerous small towns play an increasingly important role in fostering the process of urbanization, with an intense mixture of agricultural and nonagricultural activities and an intimate interaction between urban and rural areas.

The metropolitan cores of the region, aiming to increase their competitiveness and prominence in the globalizing world economy, have meanwhile invested heavily in infrastructure improvements. The Guangzhou municipal government, for example, invested more than $10 billion between 1998 and 2004 in infrastructure construction—

Figure 10.D Guangzhou, China

higher rate their populations will double in 10 years, while at the lower rate they will double in 17 years. The *doubling time* of a city's population is the time needed for it to double in size, at current growth rates. To put the situation in numerical terms, metropolitan areas like Mexico City and São Paulo are adding half a million persons to their population each year, nearly 10,000 every week, even after making up for deaths and out-migrants. It took London 190 years to grow from half a million to 10 million. It took New York 140 years. By contrast, Mexico City, São Paulo, Buenos Aires, Kolkata (Calcutta), Rio de Janeiro, Seoul, and Mumbai (Bombay) all took less than 75 years to grow from half a million to 10 million inhabitants.

The Periphery and Semiperiphery: Overurbanization and Megacities

These statistics reflect the fact that urban growth processes in the world's peripheral regions differ widely from those in core regions. In contrast to the world's core regions, where urbanization has largely resulted from economic growth, the urbanization of peripheral regions has been a consequence of demographic growth that preceded economic development. Although the demographic transition is a fairly recent phenomenon in the peripheral regions of the world (see Chapter 3), it has generated large increases in population well in advance of any significant levels of industrialization or rural economic development.

The result, for the mainly rural populations of peripheral countries, has been more and more of worse and worse. Problems with agricultural development (see Chapter 8) have meant that fast-growing rural populations face an apparently hopeless future of drudgery and poverty. Emigration has provided one potential safety valve, but as the frontiers of the world-system closed, the more affluent core countries have put up barriers to immigration. The only option for the growing numbers of impoverished rural residents was—and still is—to move to the larger towns and cities, where at least there is the hope of employment and the prospect of access to schools, health clinics, piped water, and the kinds of public facilities and services that are often unavailable in rural regions. Cities also have the lure of modernization and the appeal of consumer goods—attractions that are now directly beamed into rural areas through satellite TV. Overall, the metropolises of the periphery have absorbed 4 out of 5 of the 1.2 billion city dwellers added to the world's population since 1970.

When natural disasters, environmental degradation, or civil war impacts highly populated rural regions in the periphery, rates of rural-urban migration increase dramatically:

> For several years now, relentless cycles of drought and flooding have wreaked havoc on the tiny country of Malawi, in the heart of southern Africa. In 2002 and 2003, torrential rains caused massive mudslides,

washing away bridges and homes and devastating harvests of maize, the main food staple. Unable to eke a living from the ravaged countryside, rural residents have flocked in droves to the country's bourgeoning cities—giving Malawi the dubious distinction of being the world's fastest-urbanizing nation today.[5]

Rural migrants have poured into cities out of desperation and hope rather than being drawn by jobs and opportunities. Because these migration streams have been composed disproportionately of teenagers and young adults, an important additional component of urban growth has followed—exceptionally high rates of natural population increase. In most peripheral countries the rate of natural increase of the population in cities exceeds that of net in-migration. On average, about 60 percent of urban population growth in peripheral countries is attributable to natural increase.

The consequence of all this urban population growth has been described as **overurbanization**, which occurs when cities grow more rapidly than they can sustain jobs and housing. In such circumstances, urban growth produces instant slums—shacks set on unpaved streets, often with open sewers and no basic utilities. The shacks are constructed out of any material that comes to hand, such as planks, cardboard, tarpaper, thatch, mud, and corrugated iron. Such is the pressure of in-migration that many of these instant slums are squatter settlements, built illegally by families who are desperate for shelter. **Squatter settlements** are residential developments on land that is neither owned nor rented by its occupants. Squatter settlements are often, but not always, slums. In Chile squatter settlements are called *callampas,* meaning "mushroom cities"; in Turkey they are called *gecekondu,* meaning that they were built after dusk and before dawn. In India they are called *bustees;* in Tunisia, *gourbevilles;* in Brazil, *favelas;* and in Argentina, simply *villas miserias.* They typically account for well over one-third and sometimes up to three-quarters of the population of major cities (**Figure 10.19**). The most extensive slums, according to United Nations statistics, are in the cities of sub-Saharan Africa, where over 70 percent of the population live in unfit accommodations.

Megacities are very large cities characterized by both primacy and a high degree of centrality within their national economy. Their most important common denominator is their sheer size—most of them number 10 million or more in population. This, together with their functional centrality, means that in many ways they have more in common with one another than with the smaller metropolitan areas and cities within their own countries. Examples of such megacities include Bangkok, Mexico City (**Figure 10.20**), Shanghai (**Figure 10.21**), Dhaka, Jakarta, Lagos, Manila, Mexico City, New Delhi, São Paulo, Shanghai, and Teheran. Each has more inhabitants

[5]*State of the World 2005.* New York: W. W. Norton & Company, 2005, p. 29.

Figure 10.19   Slum housing in peripheral cities Throughout much of the world, the scale and speed of urbanization, combined with the scarcity of formal employment, have resulted in very high proportions of slum housing, much of it erected by squatters. In Nairobi, Kenya, 40 percent of the city's population lives in unauthorized settlements. The largest—the Mathere Valley squatter area—grew from 4,000 inhabitants in 1964 to 90,000 in 1979 and to over 300,000 in 2004.

than 100 of the member countries of the United Nations. While most of them do not function as world cities, they do serve important intermediate roles between the upper tiers of the system of world cities and the provincial towns and villages of large regions of the world. They not only link local and provincial economies with the global economy but also provide a point of contact between the traditional and the modern, and between formal and informal economic sectors.

The **informal sector** of an economy involves a wide variety of economic activities whose common feature is that they take place beyond official record and are not subject to formalized systems of regulation or remuneration. As we shall see in the next chapter, the slums and squatter settlements in megacities are often associated with severe problems of social disorganization and environmental degradation. Nevertheless, many neighborhoods are able to develop self-help networks and organizations that form the basis of community amid dauntingly poor and crowded cities.

The Core: Mature Metropolises

The high levels of urbanization and relatively slow rates of urban growth within the world's core regions are reflected in relatively stable urban systems. There is constant change, nevertheless, in patterns and processes of urbanization as the metropolises, cities, and towns adjust to the opportunities of new technologies and new industries and to the constraints of obsolescent urban

Figure 10.20   Mexico City Every year another half million or more people are added to the city. In 2008, the population of the agglomeration had reached almost 20 million.

Figure 10.21 Shanghai
Shanghai is China's most populous city, with 15 million people in the greater metropolitan area.

infrastructure and land-use conflicts. New rounds of urbanization are initiated in the places most suited to new technologies and new industries, while those least suited are likely to suffer a spiral of deindustrialization and urban decline.

Deindustrialization and Agglomeration Diseconomies
Deindustrialization involves a decline in industrial employment in core regions as firms scale back their activities in response to lower levels of profitability (see Chapter 7). Such adversity has particularly affected cities such as Pittsburgh and Cleveland, Sheffield and Liverpool (United Kingdom), Lille (France), and Liège (Belgium)—places where heavy manufacturing constituted a key economic sector. Cities like these have suffered substantial reductions in employment since the 1970s and 1980s. During that period, better and more flexible transport and communications networks allowed many industries to choose from a broader range of potential locations. In many instances deindustrialization has been intensified by the dampening effects of *agglomeration diseconomies* (Chapter 7) on the growth of larger metropolitan areas. Agglomeration diseconomies, the negative effects of urban size and density, include noise, air pollution, increased crime, commuting costs, the costs of inflated land and housing prices, traffic congestion, and crowded port and railroad facilities. They also include higher taxes levied to rebuild decaying infrastructure and to support services and amenities previously considered unnecessary—traffic police, city planners, and homeless shelters, for example.

The result has been a *decentralization* of jobs and people from larger to smaller cities within the urban systems of core countries, and from metropolitan cores to suburban and ex-urban fringes. In some cases, routine production activities relocated to smaller metropolitan areas or to rural areas with low labor costs and more hospitable business climates. In other cases, these activities moved overseas—as part of the new international division of labor (see Chapter 2)—or were eliminated entirely.

Counterurbanization and Reurbanization
The combination of deindustrialization in core manufacturing regions, agglomeration diseconomies in major metropolitan areas, and the improved accessibility of smaller towns and rural areas can give rise to the phenomenon of counterurbanization. **Counterurbanization** occurs when cities experience a net loss of population to smaller towns and rural areas. This process results in the deconcentration of population within an urban system. This is, in fact, what happened in the United States, Britain, Japan, and many other developed countries as early as the 1970s and 1980s. Metropolitan growth slowed dramatically, while the growth rates of small and medium-size towns and of some rural areas increased. Counties that for decades had recorded stable populations grew by 15 or 20 percent. Some of the strongest gains were registered in counties that were within commuting range of metropolitan areas, but some remote counties also registered big population increases.

Counterurbanization was a major reversal of long-standing trends, but it seems to have been a temporary adjustment rather than a permanent change. The globalization of the economy and the growth of postindustrial activities in revamped and expanded metropolitan settings have restored the trend toward the concentration of population within urban systems. Most of the cities that were declining fast in the 1970s and 1980s are now either recovering (New York, London) or bottoming out (Paris, Chicago), while most of those that were growing only slowly (Tokyo,

Barcelona) are now expanding more quickly. This trend is one of **reurbanization**, involving the growth of population in metropolitan central cores following a period of absolute or relative decline in population. In the United States, two principal streams are driving reurbanization. One consists of new migrants, mainly from Latin America and Asia, who have moved into the central districts of major metropolitan areas, especially New York and Los Angeles, and some medium-size metropolitan areas of the West, Southwest, and South. A second, very different migration stream to the central districts of metropolitan areas consists of "baby boomers" (see Chapter 3) electing to pursue urban rather than suburban lifestyles. This stream has been most pronounced in the central districts of metropolitan areas with expanding high-tech and defense-oriented economies, such as Boston and Seattle.

GLOBALIZATION AND SPLINTERING URBANISM

The major cities and metropolitan regions of the world are pivotal settings for the processes involved in economic and cultural globalization. As a result, they also reflect these processes in their own patterns of growth and change. Central to this close relationship between urbanization and globalization are networked infrastructures of transportation, information, and communications technologies such as telephone systems, satellite television, computer networks, electronic commerce, and business-to-business Internet services.

According to the United Nations Center for Human Settlements (UNCHS),[6] information and communications technologies are intensifying global urbanization in three main ways. First, they allow specialist urban centers, with their high value-added services and manufacturing, to extend their powers, markets, and control to ever more distant regional, national, international, and global spheres of influence. Second, the growing speed, complexity, and riskiness of innovation in a global economy require a concentration of technological infrastructure and an associated, knowledgeable, technology-oriented culture in order to sustain competitiveness. Third, demand for information and communications technologies is overwhelmingly driven by the growth of metropolitan markets. World cities especially are of disproportionate importance in driving innovation and investment in networked infrastructures of information and communications technologies. This is because of their cultures of modernization, concentrations of capital, relatively high average disposable personal incomes, and concentrations of internationally oriented firms and institutions.

In contrast to the infrastructure networks of earlier technology systems that underpinned previous phases of urbanization, these information and communications technologies are not locally owned, operated, and regulated. Rather, they are designed, financed, and operated by transnational corporations to global market standards. Detached from local processes of urban development, these critical networked infrastructures are highly uneven in their impact, selectively serving only certain neighborhoods, certain cities, and certain kinds of metropolitan settings. This important new tendency has been called *splintering urbanism* by geographers Stephen Graham and Simon Marvin.

Splintering urbanism is characterized by an intense geographical differentiation, with individual cities and parts of cities engaged in different—and rapidly changing—ways in ever broadening and increasingly complex circuits of economic and technological exchange. The uneven evolution of networks of information and communications technologies is forging new landscapes of innovation, economic development, and cultural transformation, and at the same time intensifying social and economic inequalities between the fast world and the slow world. The conclusion of the 2001 UNCHS report *Cities in a Globalizing World* is that traditional patterns of urbanization are rapidly giving way across the globe to a very new dynamic, one that is dominated by "enclaves of superconnected people, firms, and institutions, with their increasingly broadband connections to elsewhere via the Internet, mobile phones and satellite TVs and their easy access to information services, often exist[ing] cheek-by-jowl with much larger numbers of people with at most rudimentary access to modern communications technologies and electronic information."[7]

The chief beneficiaries of splintering urbanism are world cities and major regional metropolitan centers, particularly in core countries where large corporations are able to undertake the necessary capital investments in new networked infrastructures and where national governments and entrepreneurial urban agencies are able to subsidize and facilitate them. We can identify several specific kinds of urban settings that are most directly involved in this dimension of globalization:

- *Enclaves of Internet and digital multimedia technology development, mostly in core-country world cities.* Examples include "Multimedia Gulch" in the SOMA (South of Market Street) district of downtown San Francisco, and New York's "Silicon Alley" (just south of Forty-first Street in Manhattan).

- *Technopoles and clusters of high-tech industrial innovation.* These have emerged in campus-like suburban and ex-urban settings around core-country world cities (as in London, Paris, and Berlin); in new and renewed industrial regions within core countries (as in southern California, Baden-Württemburg in Germany, and Rhône-Alps in France); and in emerging high-tech production and innovation spaces in semiperipheral countries (as in Bangalore, India, and the Multimedia Super Corridor south of Kuala Lumpur, Malaysia).

[6]United Nations Center for Human Settlements (Habitat), *Cities in a Globalizing World: Global Report on Human Settlements, 2001.* London: Earthscan Publications, 2001, p. 6.

[7]Ibid.

Figure 10.22 London's financial core London's status as a world city rests heavily on the "square mile" of the City of London, with its concentration of banking, insurance, and other advanced business services.

■ *Places configured for foreign direct investment in manufacturing, with customized infrastructure, expedited development approval processes, tax concessions, and in some cases exceptions to labor and environmental regulations.* Such places have emerged in economically depressed regions of core countries (including Northern England and parts of the U.S. Manufacturing Belt), but are mostly found in or near major cities in peripheral and semiperipheral countries (as in the Brazilian cities of Porto Alegre and Paranà, which have attracted foreign-owned automobile plants).

■ *Enclaves of international banking, finance, and business services in world cities and major regional*

centers. Examples include the business districts in Lower Manhattan, the City of London (**Figure 10.22**), Frankfurt, Hong Kong, and Kuala Lumpur.

■ *Enclaves of modernization in the megacities and major regional centers of peripheral countries.* This modernization typically includes advanced telecommunications and satellite complexes, world trade centers, retail and commercial centers, and new university precincts. Examples include the Pudong development zone in Shanghai (**Figure 10.23**) and "growth corridors" and "new towns in town" in Bangkok, Thailand.

■ *Enclaves of back-office spaces, data-processing, e-commerce, and call centers.* These have emerged

Figure 10.23 Pudong, Shanghai One of the most spectacular development zones in the world, Pudong (foreground) was a 556-square-kilometer (215-square-mile) area of farmland until the 1990s. It has been developed around a new deep-water port, a new airport, and a new infoport, with massive investments in the rail network, highway network, and cross-river transportation network, and has become symbolic of China's economic reform and growth.

in older industrial cities within core countries (for example, Roanoke, Virginia, and Sunderland, England) and in many cities within semiperipheral countries, most notably in the Caribbean, the Philippines, and India.

■ *Spaces customized as "logistics zones."* Airports, ports, export-processing zones—enclaves in major cities around the world within which the precise and rapid movement of goods, freight, and people are coordinated, managed, and synchronized among various transport modes.

Connected to one another through a complex dynamic of flows, these urban spaces and settings represent the spatial framework for the fast world. They are embedded within regions and metropolitan areas whose economic foundations derive from earlier technology systems and whose social and cultural fabric derives from more traditional bases. The result, as we shall see in Chapter 11, is that traditional patterns of land use and spatial organization in many parts of the world are being transformed by the local effects of splintering urbanism.

CONCLUSION

Urbanization is one of the most important geographic phenomena. Cities can be seedbeds of economic development and cultural innovation. Cities and groups of cities also organize space—not just the territory immediately around them, but in some cases national and even international space. The causes and consequences of urbanization, however, are very different in different parts of the world. The urban experience of the world's peripheral regions stands in sharp contrast to that of the developed core regions, for example. This contrast is a reflection of some of the demographic, economic, and political factors that we have explored in previous chapters.

Much of the developed world has become almost completely urbanized, with highly organized systems of cities. Today, levels of urbanization are high throughout the world's core countries, while rates of urbanization are relatively low. At the top of the urban hierarchies of the world's core regions are world cities such as London, New York, Tokyo, Paris, and Zürich, which have become control centers for the flows of information, cultural products, and finance that collectively sustain the economic and cultural globalization of the world. In doing so, they help to consolidate the hegemony of the world's core regions.

Few of the metropolises of the periphery, on the other hand, are world cities occupying key roles in the organization of global economics and culture. Rather, they operate as connecting links between provincial towns and villages and the world economy. They have innumerable economic, social, and cultural linkages to their provinces on one side and to major world cities on the other. Almost all peripheral countries, meanwhile, are experiencing high rates of urbanization, with forecasted growth of unprecedented speed and unmatched size. In many peripheral and semiperipheral regions, current rates of urbanization have given rise to unintended metropolises and fears of "uncontrollable urbanization," with urban "danger zones" where "work" means anything that contributes to survival. The result, as we shall see in Chapter 11, is that these unintended metropolises are quite different from the cities of the core as places in which to live and work.

MAIN POINTS REVISITED

■ **The urban areas of the world are the linchpins of human geographies at the local, regional, and global scales.**

Towns and cities are engines of economic development and centers of cultural innovation, social transformation, and political change. They now account for almost half the world's population.

■ **The earliest towns and cities developed independently in the various hearth areas of the first agricultural revolution.**

The very first region of independent urbanism, in the Middle East, produced successive generations of urbanized world-empires, including those of Greece, Rome, and Byzantium.

■ **The expansion of trade around the world, associated with colonialism and imperialism, established numerous gateway cities.**

European powers founded or developed literally thousands of towns as they extended their trading networks and established their colonies. The great majority of the towns were ports that served as control centers commanding entrance to, and exit from, their particular country or region.

■ **The Industrial Revolution generated new kinds of cities—and many more of them.**

Industrial economies could be organized only with the large pools of labor, the transportation networks, the physical infrastructure of factories, warehouses, stores, and offices, and the consumer markets provided by cities. As industrialization spread throughout Europe in the first half of the nineteenth century and then to other parts of the world, so urbanization increased at a faster pace.

■ **Today the single most important aspect of world urbanization, from a geographical perspective, is the striking difference in trends and projections between the core regions and the peripheral regions.**

In 1950, two-thirds of the world's urban population was concentrated in the more developed countries of the core economies. Since then, the world's urban population has increased threefold, the bulk of the growth having taken place in the less developed countries of the periphery.

■ **A small number of "world cities," most of them located within the core regions of the world-system, occupy key roles in the organization of global economics and culture.**

At the top of a global urban system, these cities experience growth largely as a result of their role as key nodes in the world economy. World cities have become the control centers for the flows of information, cultural products, and finance that collectively sustain the economic and cultural globalization of the world.

- **Many of the megacities of the periphery are primate and exhibit a high degree of centrality within their urban systems.**

  Primacy occurs when the population of the largest city in an urban system is disproportionately large in relation to the second- and third-largest cities in that system. Centrality refers to the functional dominance of cities within an urban system. Cities that account for a disproportionately high share of economic, political, and cultural activity have a high degree of centrality within their urban system.

- **There is a close relationship between globalization and urbanization, mediated by networked infrastructures of transportation, information, and communications technologies.**

  Because these critical networked infrastructures are highly uneven in their impact, selectively serving only certain kinds of metropolitan settings, globalization is exerting a "splintering" effect on patterns of urbanization.

KEY TERMS

central place (p. 401)
central place theory (p. 401)
centrality (p. 406)
colonial city (p. 400)
counterurbanization (p. 416)

gateway city (p. 396)
informal sector (p. 415)
megacity (p. 414)
overurbanization (p. 414)
primacy (p. 405)

rank-size rule (p. 404)
reurbanization (p. 417)
shock city (p. 399)
splintering urbanism (p. 417)
squatter settlements (p. 414)

urban ecology (p. 392)
urban form (p. 392)
urban system (p. 392)
urbanism (p. 392)
world city (p. 407)

ADDITIONAL READING

Abrahamson, M., *Global Cities*. New York: Oxford University Press, 2004.

Brenner, N., and R. Keil (eds.), *The Global Cities Reader*. New York: Routledge, 2006.

Brunn, S., and J. F. Williams (eds.), *Cities of the World*, 3rd ed. New York: Harper Collins, 2003.

Fyfe, N. R., and J. T. Kenny (eds.), *The Urban Geography Reader*. New York: Routledge, 2005.

Gugler, J. (ed.), *Cities in the Developing World: Issues, Theory, and Policy*. Oxford: Oxford University Press, 1997.

Hohenberg, P. M., and L. H. Lees, *The Making of Urban Europe 1000–1994*. Cambridge: Harvard University Press.

Knox, P. L., and L. McCarthy, *Urbanization: An Introduction to Urban Geography*. Englewood Cliffs, NJ: Prentice Hall, 2005.

Kotkin, J., *The City: A Global History*. New York: Random House.

Marcuse, P., and R. van Kampen (eds.), *Globalizing Cities*. Oxford: Blackwell, 2000.

Neuwirth, R., *Shadow Cities: A Billion Squatters in a New Urban World*. New York: Routledge, 2005.

Sassen, S., *Global Networks, Linked Cities*. New York: Routledge, 2002.

Scott, A. J. (ed.), *Global City-Regions: Trends, Theory, Policy*. Oxford: Oxford University Press, 2001.

Short, J. R., *Global Metropolitan: Globalizing Cities in a Capitalist World*. New York: Routledge, 2004.

Simmonds, R., and G. Hack (eds.), *Global City Regions: Their Emerging Forms*. New York: Spon Press, 2000.

Taylor, P., *World City Network: A Global Urban Analysis*. New York: Routledge, 2004.

Taylor, P., et al. (eds.), *Cities in Globalization: Practices, Policies and Theories*. New York: Routledge, 2007.

United Nations Center for Human Settlements (HABITAT), *Cities in a Globalizing World: Global Report on Human Settlements, 2001*. London: Earthscan Publications, 2001.

United Nations Human Settlements Programme, *The State of the World's Cities 2006/2007*. Sterling, VA: Earthscan, 2006.

Van Susteren, A., *Metropolitan World Atlas*. Rotterdam: 010 Publishers, 2005.

Verhulst, A., *The Rise of Cities in North-West Europe*. Cambridge: Cambridge University Press, 1999.

EXERCISES

On the Internet

The Companion Website at www.mygeoscienceplace.com contains a number of tools designed to reinforce the geographic concepts presented in this chapter. Structured within a four-step learning path, these online resources include: chapter Key Concepts; a fully-searchable myeBook version of the chapter, which allows you to add notes and highlights; RSS news feeds from various media agencies; links to additional web resources; Human Geography Videos; a digital Glossary; Concept Review questions; Thinking Spatially mapping exercises; and Critical Thinking essay questions.

Unplugged

1. Figures 10.1 and 10.2 show that the United States, like most core countries, is already highly urbanized and has a relatively low rate of urbanization. Nevertheless, some U.S. cities have been growing much faster than others. Which have been the fastest-growing U.S cities in recent times, and what reasons can you suggest for their relatively rapid growth? (*Hint:* The U.S. Bureau of the Census publishes data on population change by urban area, as does the Population Division of the UN Department of Economic and Social Affairs, in a volume entitled *World Urbanization Prospects.*)

2. From census volumes in your library, find out the population of the town or city you know best. Do the same for every census year, going back from 2000 to 1990, 1980, and so on, all the way back to 1860. Then plot these populations on a simple graph. What explanations can you offer for the pattern that the graph reveals? Now draw a larger version of the same graph, annotating it to show the landmark events that might have influenced the city's growth (or decline).

3. The following cities all have populations in excess of 2 million. How many of them could you locate on a world map? Their size reflects a certain degree of importance, at least within their regional economy. What can you find out about each? Compile for each a 50-word description that explains its chief industries and a little of its history.

   | Poona | Ibadan | Recife |
   | Bangalore | Turin | Ankara |

4. Figure 10.9 features two colonial gateway cities on the eastern seaboard of the United States—Boston and New York. Two other colonial gateway cities were Charleston, South Carolina, and Savannah, Georgia. What can you find out about the commodities and manufactures that each imported and exported in prerevolutionary times and about the origin and destination of these imports and exports? Which geographic concepts do you consider to be useful in explaining these facts?

City Spaces: Urban Structure

11

China seems to have caught onto sprawl. Yet, what is perhaps most distinctive about many of China's affluent new suburbs is not the sprawl itself, but the way it looks. Many of them are directly modeled on the tract homes that have defined American suburban growth in the past 30 years. Most of them carry few, if any, Asian influences in their design, layout, and ornamentation. The first ones were built to house mostly expatriates, such as diplomats or those heading the local offices of multinational companies; now, an increasing number of local Chinese with money are eager to get out of the crowded and dirty milieu of the contemporary Chinese city. Many are also buying them for investment purposes. "As China's economy continues its breakneck upward trajectory, with high economic growth and suburbanization occurring in major cities such as Beijing and Shanghai, the wealthy inhabitants and white-collar workers of such cities are showing a preference for the purchase of villas or houses located on the suburban fringes of these cities," says Andrew Ness, Asia executive director for research for CB Richard Ellis.

Wish to live in Australia? Beijing residents can buy a home at Sydney Coast, a subdivision that offers its residents a "seven-day Australian-style villa life." For those who would prefer to live in California, Beijingers have been able to set up shop in the Yosemite subdivision. Now there is the additional option of moving to Napa Valley, a new development under construction about 30 miles outside Beijing. Napa Valley attempts to capture a Californian Mediterranean lifestyle of laid-back, al fresco leisure. "Rustic stone is widely used, with rich stucco colors, along with wood shutters and wrought-iron accents, to create an intimate scale and village-like feel," according to Napa Valley's architects and planners, who are based in Palm Springs and Newport Beach (Orange County).

For those craving to re-create life in seventeenth-century France, there is Chateau Regalia, located on Beijing's northern outskirts. Here, potential buyers can choose from several different models of homes: the Duke I, the Duke II, the Marquis, the Earl, and the Viscount. In both form and decoration, Chateau Regalia's homes are an eccentric amalgam of French Baroque and neoclassical architecture. The craze has also caught on in Shanghai, where there are many more tract homes built in foreign styles. Thames Town, just outside Shanghai, is one of seven satellite towns built by the municipal government to house 500,000 people (the other six are themed with architectural styles adopted from Italy, Spain, Canada, Sweden, Holland, and Germany). Thames Town includes a gothic church, village green, and mock-Tudor pub selling real ale. Built from scratch in little over three years, the suburb encompasses five centuries of British architecture. At its center are half-timbered Tudor-style buildings. By the waterfront, Victorian redbrick warehouses have been preemptively "converted" into shops. The residential area includes gabled Edwardian houses bordered by privet hedges, manicured lawns, and leafy roads.[1]

[1]Based on D. Elsea, "China's Chichi Suburbs: American-style Sprawl All the Rage in Beijing," *San Francisco Chronicle*, April 24, 2005.

Thames Town development, Shanghai.

Main Points

- The internal structure of cities is shaped heavily by competition for territory and location. In general, all categories of land users—commercial and industrial, as well as residential—compete for the most convenient and accessible locations within the city.

- Social patterns in cities are heavily influenced by territoriality, which provides a means of establishing and preserving group membership and identity.

- The typical North American city is structured around a central business district (CBD); a transitional zone; suburbs; secondary business districts and commercial strips; industrial districts; and, in larger metropolitan areas, edge cities.

- Traditionally, North American cities have experienced high rates of in-migration, which has resulted in their becoming structured into a series of concentric zones of neighborhoods of different ethnicity, demographic composition, and social status through processes of invasion and succession.

- The most acute problems of the postindustrial cities of the world's core regions are localized in the central city areas that have borne the brunt of restructuring from an industrial to a postindustrial economy.

- The problems of the cities of the periphery stem from the way in which their demographic growth has outstripped their economic growth.

URBAN LAND USE AND SPATIAL ORGANIZATION

The internal organization of cities reflects the way they function, both to bring people and activities together and to sort them out into functional subareas and neighborhoods. While there are many complex processes at work in cities, there are some broad tendencies that go a long way toward shaping patterns of urban land use and spatial organization. We consider two of them here: people's need for accessibility and their sense of territoriality.

Accessibility and Land Use

Most urban land users want to maximize the *utility* they derive from a particular location. The utility of a specific place or location refers to its usefulness to particular persons or groups. The price they are prepared to pay for different locations—the bid-rent—is a reflection of this utility. In general, utility is a function of *accessibility*. Commercial land users want to be accessible to one another, to markets, and to workers; private residents want to be accessible to jobs, amenities, and friends; public institutions want to be accessible to clients. In an idealized city built on an isotropic surface, the point of maximum accessibility is the city center. An **isotropic surface** is a hypothetical, uniform plane: flat, and with no variations in its physical attributes. Under these conditions, accessibility decreases steadily with distance from the city center. Likewise, utility decreases, *but at different rates for different land users*. The result is a tendency toward concentric zones of different mixes of land use (**Figure 11.1**).

One counterintuitive implication of this model is that the poorest households will end up occupying the periphery of the city. Although this is true in some parts of the world, we know that in the core countries the farthest suburbs are generally the territory of wealthier households, while the poor usually occupy more accessible locations nearer city centers. Some modification of the assumptions is clearly required. In this case, we must assume that wealthier households trade off the convenience

of accessibility for the greater utility of being able to consume larger amounts of (relatively cheap) suburban space. Poorer households, unable to afford the recurrent costs of transportation, must trade off living space for accessibility to jobs so that they end up in high-density areas, at expensive locations, near their low-wage jobs. Because of the presumed trade-off between accessibility and living space, this urban land-use model is often referred to as a *trade-off model*.

Territoriality, Congregation, and Segregation

In cities, as at other geographic scales, territoriality provides a means of establishing and preserving group membership and identity. The first step in forming group identity is to define "others" in an exclusionary and stereotypical way. **Congregation**—the territorial and residential clustering of specific groups or subgroups of people—enables group identity to be consolidated in relation to people and places outside the group. Congregation is thus a place-making activity and an important basis for urban structure and land use. It is particularly important in situations where there are one or more distinctive minority groups. Defined in relation to a general population or host community, **minority groups** are population subgroups that are seen—or that see themselves—as somehow different from the general population. Their defining characteristics can be based on race, language, religion, nationality, caste, sexual orientation, or lifestyle.

Several specific advantages of congregation exist for minority groups:

- *Congregation provides a means of cultural preservation. It allows religious and cultural practices to be maintained and strengthens group identity through daily involvement in particular routines and ways of life.* Particularly important in this regard is the way that clustering fosters within-group marriage and kinship networks.

- *Congregation helps minimize conflict and provides defense against "outsiders."*

- *Congregation provides a place where mutual support can be established through minority institutions, businesses, social networks, and welfare organizations.*

- *Congregation helps establish a power base in relation to the host society.* This power base can be democratic, organized through local elections, or it can take the form of a territorial heartland for insurrectionary groups.

Congregation is not always voluntary, of course. Host populations are also impelled by territoriality, and they may respond to social and cultural differences by *discrimination* against minority groups. Discrimination can also have a strong territorial basis, the objective

Figure 11.1   Accessibility, bid-rent, and urban structure
Competition for accessible sites near the city center is an important determinant of land-use patterns. Different land users are prepared to pay different amounts—the bid-rents—for locations at various distances from the city center. The result is a tendency for a concentric pattern of land uses. (*Source:* Reprinted with permission of Prentice Hall, from P. L. Knox, *Urbanization* © 1994, p. 99.)

being to restrict the territory of minority groups and to resist their assimilation into the host society. This resistance can take a variety of forms. Social hostility and the voicing of "keep out" attitudes are probably the most widespread, although other forms of discrimination can have pronounced spatial effects. These forms include exclusion and prejudice in local labor markets, the manipulation of private land and housing markets, the steering of capital investment away from minority areas, and the institutionalization of discrimination through the practices and spatial policies of public agencies.

The combined result of congregation and discrimination is **segregation**, the spatial separation of specific subgroups within a wider population (**Figure 11.2**). Segregation varies a great deal in both intensity and form, depending on the relative degree and combination of congregation and discrimination. Geographers have identified three principal situations in terms of the spatial form of segregation:

- *Enclaves,* in which tendencies toward congregation and discrimination are long-standing but dominated by internal cohesion and identity. The Jewish districts of many of today's cities in Europe and the eastern United States are examples of enclaves.

- *Ghettos,* which are also long-standing, but are more the product of discrimination than congregation. Examples are the segregation of African Americans and Hispanics in American cities.

- *Colonies,* which may result from congregation, discrimination, or both, but in relatively weak and short-lasting ways. Their persistence over time depends on the continuing arrival of new minority-group members. For example, U.S. cities in the early twentieth century contained distinctive colonies of German, Scandinavian, Irish, and Italian immigrants that have now all but disappeared. Today there are colonies of Greek and Yugoslav neighborhoods in Australian cities, and colonies of Koreans in Japanese cities.

TRADITIONAL PATTERNS OF URBAN STRUCTURE

Economic competition for space and accessibility, along with the tendency toward social and ethnic discrimination, congregation, and segregation, can be traced in many of the world's cities, particularly in affluent core regions where economic, social, and cultural forces are broadly similar. Nevertheless, urban structure varies considerably because of the influence of history, culture, and the different roles that cities have played within the world-system. In this section, we examine the typical characteristics of North American, European, and Islamic cities, and the unintended metropolises of the periphery.

North American Cities

Traditionally, the very center of North American cities has been the principal hub of shops and offices, together with some of the major institutional land uses such as the city hall, libraries, and museums. This center, known as the **central business district**, or **CBD**, is a city's nucleus of commercial land uses. It traditionally contains the densest concentration of shops, offices, and warehouses and the tallest nonresidential buildings. It usually developed at the nodal point of transportation routes so that it also contains bus stations, railway terminals, and hotels. The CBD typically is surrounded by a zone of mixed land uses: warehouses, small factories and workshops, specialized stores, apartment buildings, public housing projects, and older residential neighborhoods (**Figure 11.3**). This zone is often referred to as the **zone in transition** because of its mixture of growth, change, and decline.

Figure 11.2 Segregation Congregation is often voluntary, but in many cases it is driven to some degree by discrimination. The combined result of congregation and discrimination is segregation, the spatial separation of specific subgroups within a wider population. This photograph is of a political mural in the Bogside district of Derry (Londonderry), Northern Ireland.

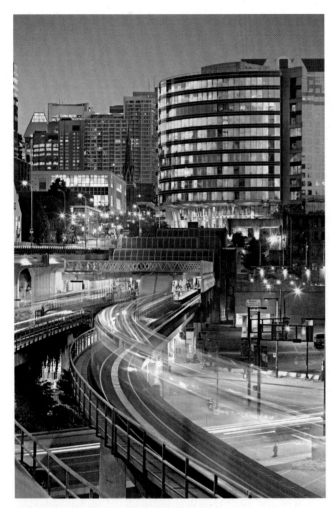

Figure 11.3   The central city This photograph of Vancouver, Canada, shows the concentration of high-rise office buildings on the skyline that is typical of central business districts (CBDs) in North American metropolitan areas. In Vancouver, as in other major cities, the modern CBD grew up around the point of maximum accessibility: near railway stations and at the intersection of the city's principal transit lines. It is surrounded by mixed land uses—older neighborhoods, apartment complexes, and housing projects—all pierced by freeway corridors that have taken the place of railway and transit lines.

Beyond this zone are residential neighborhoods, suburbs of various ages and different social and ethnic composition. Just as different categories of land use attract and repel one another, so do different social and ethnic groups. In North America, where urban population growth has been fueled by streams of migrants and immigrants with very different backgrounds, sociologists have developed an ecological perspective to describe neighborhoods as being structured by the "invasion" of successive waves of migrants and immigrants.

When immigrants first arrive in the city looking for work and a place to live, they have little choice but to cluster in the cheapest accommodations, typically found in the zone in transition around the CBD. The classic example was provided by Chicago in the 1920s and 1930s. Immigrants from Scandinavia, Germany, Italy, Ireland, Poland,

Bohemia (now part of the Czech Republic), and Lithuania established themselves in Chicago's low-rent areas, the only places they could afford. By congregating in these areas, immigrants accomplished several things—they were able to establish a sense of security; continue speaking their native language; have familiar churches or synagogues, restaurants, bakeries, butcher shops, and taverns; and support their own community newspapers and clubs. These immigrants were joined in the city's zone in transition by African American migrants from the South, who established their own neighborhoods and communities. In Chicago, as in other U.S. cities of the period, the various ethnic groups formed a patchwork or mosaic of communities encircling the CBD.

These ethnic communities lasted from one to three generations, after which they started to break up. Many of the younger, city-born individuals did not feel the need for the security and familiarity of ethnic neighborhoods. Gradually, increasing numbers of them were able to establish themselves in better jobs and move out into newer, better housing. As the original immigrants and their families moved out, their place in the transitional zone was taken by a new wave of migrants and immigrants. In this way, Chicago became structured into a series of *concentric zones* of neighborhoods of different ethnicity and social status (**Figure 11.4**).

Throughout this process of invasion and succession, people of the same background tend to stick together—partly because of the advantages of residential clustering and partly because of discrimination. **Invasion and succession** is a process of neighborhood change whereby one social or ethnic group succeeds another in a residential area. The displaced group, in turn, invades other areas, creating over time a rippling process of change throughout the city. The result is that within each concentric zone there exists a mosaic of distinctive neighborhoods. Classic examples include the Chinatowns, Little Italys, Koreatowns, and African American ghettos of big North American cities (**Figure 11.5**). Such neighborhoods can be thought of as *ecological niches* within the overall metropolis—settings where a particular mix of people have come to dominate a particular territory and a particular physical environment, or habitat.

For decades, the process of invasion and succession was a predominantly outward movement that gave North American cities their fundamental concentric structure. In the last quarter of the twentieth century, however, the flow of migrants and immigrants to many cities diminished significantly, prompting the gentrification of some central parts of many cities. **Gentrification** involves the invasion of older, centrally located, working-class neighborhoods by higher-income households seeking the character and convenience of less expensive and centrally located residences. These invasions result in the physical renovation and upgrading of housing (**Figure 11.6**), but they also displace many existing occupants.

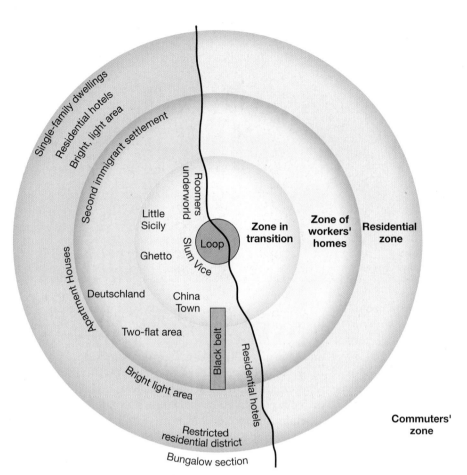

Figure 11.4 The ecological model of urban land use: Chicago in the 1920s Competition between members of different migrant and immigrant groups for residential space in the city often results in distinctive neighborhoods that have their own social "ecology." The classic example is Chicago of the 1920s, which had developed a series of concentric zones of distinctive neighborhoods as successive waves of immigrants established themselves. Over time, most immigrant groups made their way from low-rent, inner-city districts surrounding the CBD (known in Chicago as the Loop) to more attractive and expensive districts farther out. (*Source:* After R. E. Park, E. W. Burgess, and R. D. McKenzie, *The City.* Chicago: University of Chicago Press, 1925, p. 53.)

Problems of North American Cities

For all their relative prosperity, the cities of the world's core regions have their share of problems. The most acute are localized in central city areas and are interrelated: fiscal problems, infrastructure problems, and localized cycles of poverty and spirals of neighborhood decay. **Central cities** are the original, core jurisdictions of metropolitan areas.

Fiscal Problems The term *fiscal* refers to taxes. Economic restructuring and metropolitan decentralization over the past several decades have left central cities with a

Figure 11.5 Chinatown In most larger cities there is a patchwork of distinctive neighborhoods that results from processes of congregation and segregation. Most distinctive of all are neighborhoods of ethnic minorities, such as the Chinatowns, Little Italys, and Little Koreas of major American cities. Shown here is part of Chinatown in Los Angeles.

Figure 11.6 Gentrification In Old Town Alexandria, Virginia, a lengthy process of gentrification has established an elite area among neighborhoods of older townhouses, most of which had previously been occupied by lower- and middle-income households.

chronic "fiscal squeeze." A **fiscal squeeze** occurs when increasing limitations on tax revenues combine with increasing demands for expenditures on urban infrastructure and city services. The revenue-generating potential of most central cities has steadily fallen as metropolitan areas have lost both residential and commercial taxpayers to suburban jurisdictions. Growth industries, white-collar jobs, retailing, and more affluent households have moved out to suburban and exurban jurisdictions, taking their local tax dollars with them.

At the same time, growth in property-tax revenues from older, decaying neighborhoods has slowed as the growth of property values has slowed. Yet these older, decaying neighborhoods cost more to maintain and service. Older streets, water and sewage lines, schools, and transit systems have high maintenance costs. The residual populations of these neighborhoods, with high proportions of elderly and indigent households, are increasingly in need of municipal welfare services. Large numbers of low-income migrants and immigrants also bring increased demands for municipal services. Added to all this, central city governments are still responsible for services and amenities used by the entire metropolitan population: municipal galleries and museums, sports facilities, parks, traffic police, and public transport, for example. The net result is that central city jurisdictions tend to be locked permanently into a precarious financial position. This in turn results in a constant drive to develop revenue-generating projects, with cities competing fiercely with one another to finance and attract tourist developments, museums, sports franchises, and business and conference centers.

Infrastructure Problems As the rate of urban growth in North America has slowed and fiscal problems have intensified, public spending on the urban infrastructure of roads, bridges, parking spaces, transit systems, communications systems, power lines, gas supplies, street lighting, water mains, sewers, and drains has declined. Meanwhile, much of the original infrastructure, put in place 75 or 100 years ago with a design life of 50 or

75 years, is obsolete, worn out, and in some cases perilously near the point of collapse. In 2007, the I-35W bridge in Minneapolis collapsed (**Figure 11.7**), killing 13 and injuring 144; a year later, the American Association of State Highway and Transportation Officials released a report saying at least $140 billion was needed to shore up or replace 152,000 of the nation's 590,000 bridges.[2]

Infrastructure problems are easily overlooked because they build up slowly. Only when a bridge collapses or a water main bursts are infrastructure problems newsworthy. Yet cities' infrastructure is crucial not only to their economic efficiency but also to public health, safety, and the quality of life. One review of infrastructure problems in American cities estimated that between half and two-thirds of the country's cities are unable to support modernized investment until major new investments can be made in their basic infrastructure. This investment, of course, is unlikely, given the chronic fiscal problems they face. In Boston, three-quarters of the sewer system was built a century or more ago and has now deteriorated to the point where about 20 percent of the system's overall flow is lost to leaks. In Cleveland, the District of Columbia, and Philadelphia, the losses approach 25 percent. Hundreds of kilometers of water mains in New York City have been identified as being in need of replacement, at an estimated cost in 2004 of over $5 billion. Almost 50 percent of all wastewater treatment systems in America are operating at 80 percent or more of capacity, the level at which the federal government prohibits further industrial hookups in a community.

Freshwater supplies are also at risk: Old water systems are unable to cope with the leaching of pollutants into city water. Common pollutants include chlorides, oil, phosphates, and nondegradable toxic chemicals from industrial wastewater; dissolved salts and chemicals from highway deicing; nitrates and ammonia from fertilizers and sewage; and coliform bacteria from septic tanks and sewage. Many

---

[2]American Association of State Highway and Transportation Officials, *Bridging the Gap: Restoring and Rebuilding the Nation's Bridges*, http://www.transportation1.org/BridgeReport/front-page.html, accessed August 4, 2008.

Figure 11.7 Decaying Infrastructure The collapsed freeway bridge on Interstate 35W just outside of downtown Minneapolis, Minnesota, 2007. A report issued in 2005 by the American Society of Civil Engineers (ASCE)[3] had estimated that it would cost $1.6 trillion to bring the country's infrastructure up to acceptable standards. Most of the capital investment needed to repair and replace America's infrastructure of roads, bridges, transit systems, schools, water- and waste-treatment systems, and hazardous-waste disposal is needed in cities and metropolitan areas. In central cities, a fiscal squeeze has continuously postponed attention to obsolescent infrastructure. In suburban areas, a political ideology dominated by low taxation and small government has consistently diverted attention from underdeveloped infrastructure.

cities still use water-cleaning technology dating to World War I. About one-third of all towns and cities in the United States have contaminated water supplies, and about 8 million people are using water that is potentially dangerous.

Poverty and Neighborhood Decay

Inner-city poverty and neighborhood decay have become increasingly pronounced in the past several decades as manufacturing, warehousing, and retailing jobs have moved out to suburban and edge-city locations and as many of the more prosperous households have moved out to be near these jobs. The spiral of neighborhood decay typically begins with substandard housing occupied by low-income households who can afford to rent only a minimal amount of space. The consequent overcrowding not only causes greater wear and tear on the housing itself but also puts pressure on the neighborhood infrastructure of streets, parks, schools, and playgrounds. The need for maintenance and repair increases quickly but is rarely met. Individual households cannot afford it, and landlords have no incentive to do so, because they have a captive market. Public authorities face a fiscal squeeze and are in any case often indifferent to the needs of such neighborhoods because of their relative lack of political power.

Shops and privately run services such as restaurants and hair salons are afflicted with the same syndrome of decay. With a low-income clientele, profit margins must be kept low, leaving little to spare for upkeep or improvement. Many small businesses fail, or relocate to more favorable settings, leaving commercial property vacant for long periods. In extreme cases such property becomes abandoned, the owners unable to find either renters or buyers. Residential buildings may also be left derelict. Faced with escalating maintenance costs and rising property taxes and unable to increase revenues because of rent controls and the depressed state of inner-city housing markets, some landlords simply write off their property and abandon it.

There is, meanwhile, a dismal cycle of poverty that intersects with these localized spirals of decay. The **cycle of poverty** involves the transmission of poverty and

deprivation from one generation to another through a combination of domestic circumstances and local neighborhood conditions. This cycle begins with a localized absence of employment opportunities and, therefore, a concentration of low incomes, poor housing, and overcrowded conditions. Such conditions are unhealthy. Overcrowding makes people vulnerable to poor health, which is compounded by poor diets. This therefore contributes to absenteeism from work, which results in decreased income. Similarly, absenteeism from school through illness contributes to the cycle of poverty by constraining educational achievement, limiting occupational skills, and so leading to low wages. Crowding also produces psychological stress, which contributes to social disorganization and a variety of pathological behaviors, including crime and vandalism. Such conditions not only affect people's educational achievement and employment opportunities but also lead to the *labeling* of the neighborhood, whereby all residents may find their employment opportunities affected by the poor image of their neighborhood.

One of the most important elements in the cycle of neighborhood poverty is the educational setting. Schools, obsolete and physically deteriorated like their surroundings, are unattractive to teachers—partly because of the physical environment and partly because of the social and disciplinary environment. Because of fiscal squeeze, the schools are often resource-poor, with relatively small budgets for staff, equipment, and materials. Over the long term, poor educational resources translate into poor education, however positive the values of students and their parents. Poor education limits occupational choice and, ultimately, results in lower incomes. Students, faced with evidence all around them of unemployment or low-wage jobs at the end of school careers, find it difficult to be positive about school. The result becomes a self-fulfilling prophecy of failure, and people become trapped in areas of concentrated poverty (**Figure 11.8**).

[3]*Renewing America's Infrastructure.* American Society of Civil Engineers, 2005. http://www.asce.org/reportcard/

Figure 11.8   Poverty areas
Concentrations of poverty are found not only in decaying inner-city areas but also in newer public housing projects and in first- and second-tier suburbs that have filtered down the housing scale, as in this example in the District of Columbia, a short distance from the Capitol.

Many poverty areas are also racial ghettos, although not all ghettos are poverty areas. As we have seen, ethnic and racial congregation can mitigate the effects of poverty. Nevertheless, discrimination is usually the main cause of ghettoization. In the United States, discrimination in housing markets is illegal, but it nevertheless takes place in a variety of ways. One example of housing-market discrimination by banks and other lending institutions is the practice of redlining.

Redlining involves marking off bad-risk neighborhoods on a city map and then using the map to determine loans. This practice results in a bias against minorities, female-headed households, and other vulnerable groups, because they tend to be localized in low-income neighborhoods. Redlining tends to become another self-fulfilling prophecy, since neighborhoods starved of property loans become progressively run down and therefore increasingly unattractive to lenders. Discrimination affects education and labor markets as well as housing markets. In the case of ghetto poverty, all three types of discrimination come together, reinforcing the cycle of poverty and intensifying the disadvantages of the minority poor.

Social trends have compounded the problems of these areas in many instances. Increased divorce and teen pregnancy rates have led to more single-parent families and a feminization of poverty. These families have been portrayed as the core of a geographically, socially, and economically isolated underclass. The idea of an **underclass** refers to individuals who experience a form of poverty from which it is very difficult to escape because of their isolation from mainstream values and the formal labor market. Isolated from the formal labor force and from the social values and behavioral patterns of the rest of society, the underclass has been seen as subject to increased social disorganization and deviant behavior. In

cities in the United States, localized inner-city poverty is now characterized by senseless and unprovoked violence, premeditated and predatory violence, domestic violence, the organized violence of street gangs, and epidemic levels of HIV/AIDS and other communicable diseases—all closely associated with drug taking and drug dealing.

One consequence of extreme poverty is *homelessness*. Chronic, long-term homelessness means not having customary and regular access to a conventional dwelling. This includes people who have to sleep in shelters, flophouse cubicles, and emergency dormitories and missions, as well as those sleeping in doorways, bus stations, cars, tents, temporary shacks, and cardboard boxes, and on park benches and steam grates.

The number of homeless persons in North American cities rose sharply in the mid-1980s. This was mainly a consequence of the increased poverty and the economic and social dislocation caused by economic restructuring and the transition to a globalized, postindustrial economy. It was intensified by the fiscal squeeze confronting central cities and the tendency for governments to cut back on welfare programs of all kinds. It was also intensified by the adoption of revolving-door policies of mental health hospitals, which released large numbers of institutionalized patients. The homeless are now very visible throughout the major cities of North America (**Figure 11.9**): around the heating vents of shopping malls in Canadian cities, in the subterranean world of the New York subway system, under Miami's bridges and in Los Angeles's parks, in cardboard boxes by glass office towers in downtown Chicago, and near abandoned factories in Boston and St. Louis.

Estimates of the number of homeless people in U.S. cities vary a good deal. The estimate of the National Coalition for the Homeless is between 1.5 and 3 million. The National Law Center on Homelessness and Poverty

Figure 11.9   The new homeless In the 1980s and 1990s, changing conditions in U.S. cities created a huge increase in the number of homeless persons. The new homeless include significant numbers of women, children, and young adults, groups that had not previously been vulnerable to homelessness.

Figure 11.10   Foreclosures In the year 2008, through June, a total of more than 2.3 million properties in the United States had some sort of foreclosure action against them (default notice, auction notice, or bank repossession).

and the Urban Institute estimate that about 3.5 million people, 1.35 million of them children, are likely to experience homelessness in a given year. The United States Interagency Council on Homelessness reports that, in 2008, single men comprised 51 percent of the homeless population; families with children, 30 percent; single women, 17 percent; and unaccompanied youth, 2 percent. The homeless population is estimated to be 42 percent African American, 39 percent white, 13 percent Hispanic, 4 percent Native American, and 2 percent Asian. An average of 16 percent of homeless people are considered mentally ill; 26 percent are substance abusers; thirteen percent are employed. Estimates for Canadian cities suggest a total of more than 250,000 homeless persons. In Toronto alone more than 30,000 people—equivalent to 1.3 percent of the city's entire population—slept in shelters in 2000, a 40 percent increase from 1988.

What makes contemporary homelessness so striking a problem is not just the scale of the problem but also its nature. Whereas homelessness had previously involved white adult males, relatively few of whom actually had to sleep outdoors, the new homeless are of all races and include significant numbers of women, children, and the elderly. In Canada, for example, one quarter of the country's homeless are children. The new homeless are also much less likely to find shelter indoors and have become a very visible feature of the public spaces of many cities.

Mortgage Foreclosures   After a long housing boom, American cities were suddenly hit in 2007 by a wave of mortgage foreclosures and repossessions that has left many neighborhoods blighted by empty homes (**Figure 11.10**). The crisis was prompted by high default rates on mortgages that had been made by financial institutions during the boom years. The housing boom was driven by the affluence of the one-fifth of U.S. households at the top of the economy, who had been racing ahead of the other four-fifths since the mid-1980s. The number of households with

incomes of more than $100,000 grew 29 percent more than the overall growth in the number of U.S. households between 1986 and 1995, and 26 percent more than the growth in the number of U.S. households between 1996 and 2005. This was at a time when mortgage interest rates hit their lowest levels in more than four decades. The result was a tremendous increase in new home building at the top end of the market. After the stock market bubble of the late 1990s burst, the trend intensified still further as more people decided that the best bet for their investments was real estate. Inevitably, house prices soared; but mortgage financiers found increasingly novel ways to package loans to help people afford houses that otherwise would be beyond their reach: interest-only mortgages, graduated-payment mortgages, growing-equity mortgages, jumbo and super-jumbo mortgages, shared-appreciation mortgages, and step-rate mortgages that supercharged the market and contributed to the housing market bubble of rapidly inflating house prices in the first half of the 2000s.

In order to meet the soaring demand for loans, mortgage lenders began to sell on their mortgages on bond markets, raising cash to fund more mortgages. But, as noted in Chapter 1, this also led to abuses, as the relaxed controls on financial institutions resulting from the neoliberal policies of the Bush administration, meant that mortgage lenders no longer had an incentive to check carefully on the borrowers. Many lenders began to offer mortgages to the so-called subprime market—also pejoratively referred to as the NINJAs—those with no income, no jobs, and no assets. By 2005, one in five mortgages were subprime, and they were particularly popular among recent immigrants trying to buy a home for the first time in the expensive housing markets of big metropolitan areas.

The problem for lower-income borrowers was that subprime mortgages were typically "balloon" mortgages, whereby payments were fixed for two years with low-interest "teaser" deals that then became variable and much higher. Inevitably, this led many households to default on their monthly payments and, as the bad loans added up, mortgage lenders found themselves, in turn, in financial trouble. The internationalization of finance meant that the banking industry was thrown into crisis worldwide, prompting an economic recession and the first national decline in house prices since the Depression of the 1930s. This created a problem for many affluent homeowners, who had overreached themselves to buy big, flashy homes and now, in an economic recession, found themselves with homes worth less than their mortgage debt.

Some cities and some neighborhoods were hit harder by the resulting foreclosures and repossessions than others. Cleveland, for example, was particularly affected by mortgage defaults, repossessions, housing abandonment, and neighborhood decline because of the large number of subprime loans that had been issued to households there. Elsewhere, the problem has shown up more in the fast-growth fringes of America's metro areas. Most of Miami-Dade County's 29,923 foreclosures in the first three quarters of 2007, for example, occurred on its outer fringes. In Clark County, Nevada, where nearly 1 in 20 homes was in foreclosure at the beginning of 2008, foreclosures were concentrated in a ring around central Las Vegas.

European Cities

European cities, like North American cities, reflect the operation of competitive land markets and social congregation along ethnic lines. They also suffer from similar problems of urban management, infrastructure provision, and poverty. What makes most European cities distinctive in comparison with North American cities is that they are the product of several major epochs of urban development. As we saw in Chapter 2, because many of today's most important cities were founded in the Roman period, it is not uncommon for the outlines of Roman and medieval urban development to be preserved in their street plans. Many distinctive features of European cities derive from their long history. In the historic cores of

some older cities, the layout of streets reflects ancient patterns of rural settlement and field boundaries. Beyond these historic cores, narrow, complex streets are the product of the long, slow growth of European cities in the pre-automobile era, when hand-pushed and horse-drawn carts were the principal means of transportation and urban development was piecemeal and small-scale.

Plazas and squares are another important historical legacy in many European cities. Greek, Roman, and medieval cities were all characterized by plazas, central squares, and marketplaces, and those elements are still important nodes of urban activity (**Figure 11.11**). European history also means that its cities bear the accessories and scars of war. The legacy of defensive hilltop and clifftop sites and city walls has limited and shaped the growth of modern cities, while in more recent times the bombings and shellings of World War II destroyed many buildings in a large number of cities (**Figure 11.12**).

The legacy of a long and varied history includes a rich variety of symbolism in the built environment. Europeans are reminded of their past not only by large numbers of statues and memorials but also by cathedrals, churches, and monasteries; by guildhalls and city walls; by the palaces of royalty and the mansions of aristocracy; and by city halls and the libraries, museums, sports stadiums, and galleries that are monuments to civic achievement. European cities are also typically compact in form, resulting in high densities of population. A long history of pre-automobile urban development and the constraints of peripheral defensive walls all made urban land expensive and encouraged a tradition of high-density living in tenements and apartment houses.

Several other distinctive features of European cities should be noted:

■ *Low skylines*—Although the larger European cities have a fair number of high-rise apartment buildings and a sprinkling of office skyscrapers, they all offer a predominantly low skyline. This is partly because much of their growth came before the invention of the elevator

Figure 11.11 Vigevano, Italy Widely considered to be one of the finest piazzas in Italy, the Piazza Ducale in Vigevano is a product of early Renaissance town planning, designed by Bramante for Ludovico Maria Sforza in 1492–93 as a noble forecourt to the Castello Sforzesco. Unified by the arcades that completely surround the square, the piazza now provides an important social space for the citizens of the town.

Figure 11.12 Cologne, Germany, 1945 About 90 percent of Cologne's central area was destroyed or severely damaged by Allied bombing during World War II. One of the few structures to survive intact was the solid, stone–built cathedral, built between the thirteenth and nineteenth centuries.

and the development of steel-reinforced, concrete building techniques and partly because of master plans and building codes (some written as long ago as the sixteenth century) seeking to preserve the dominance of monumental buildings like palaces and cathedrals.

- *Lively downtowns*—The CBDs of European cities have retained their focal position in residents' shopping and social lives because of the relatively late arrival of the suburbanizing influence of the automobile and strong planning controls directed against urban sprawl.
- *Neighborhood stability*—Europeans change residence, on average, about half as often as Americans. In addition, the physical life cycle of city neighborhoods tends to be longer because of the past use of durable construction materials, such as brick and stone. As a result, European cities provide relatively stable socioeconomic environments.
- *Municipal socialism*—For decades European welfare states have provided a broad range of municipal services and amenities, from clinics to public transit systems. Perhaps the most important to urban structure is social housing (public housing), which accounts for 20 to 40 percent of all housing in most larger English, French, and German cities. In recent years neoliberal policies have resulted in a reduction in public service provision, especially in France and Great Britain.

We should note, however, that the richness of European history and the diversity of its geography mean that there are important regional variations: The industrial cities of northern England, northeastern France, and the Ruhr district of Germany, for example, are quite different in character from the cities of Mediterranean Europe. One of the most interesting regional variations is the Eastern European city, in which the legacy of an interlude of 44 years of socialism (1945–1989) was grafted onto cities that had already developed mature patterns of land use and social differentiation. Major examples include Belgrade, Budapest, Katowice, Kraków, Leipzig, Prague, and Warsaw. State control of land and housing meant

that huge public housing estates and industrial zones were created in outlying districts. The structure of the older city was little altered, however, apart from the addition of socialist monuments and the renaming of streets.

Urban Design and Planning
European city planning and design have a long history. As we have seen, most Greek and Roman settlements were deliberately laid out on grid systems, within which the siting of key buildings and the relationship of neighborhoods to one another were carefully thought out.

The roots of modern Western urban planning and design can be traced to the Renaissance and Baroque periods (between the fifteenth and seventeenth centuries) in Europe, when artists and intellectuals dreamed of ideal cities and rich and powerful regimes used urban design to produce extravagant symbolizations of wealth, power, and destiny. Inspired by the classical art forms of ancient Greece and Rome, Renaissance urban design sought to recast cities in a deliberate attempt to show off the power and the glory of the state and the Church. Spreading slowly from its origins in Italy at the beginning of the fifteenth century, Renaissance design had diffused to most of the larger cities of Europe by the end of the eighteenth century. Dramatic advances in military ordnance (cannon and artillery) brought a surge of planned redevelopment that featured impressive fortifications; geometric-shaped redoubts, or strongholds; and an extensive *glacis militaire*—a sloping, clear zone of fire. Inside new walls, cities were recast according to a new aesthetic of grand design (**Figure 11.13**)—fancy palaces and geometrical plans, streetscapes, and gardens that emphasized views of dramatic perspectives. These developments were often of such a scale that they effectively fixed the layout of cities well into the eighteenth and even into the nineteenth century, when walls and/or glacis eventually made way for urban redevelopment in the form of parks, railway lines, or beltways.

As societies and economies became more complex with the transition to industrial capitalism, national rulers and city leaders looked to urban design to impose order,

Figure 11.13 Sabbioneta, Italy Sabbioneta was built by Vespasiano Gonzaga in the mid-sixteenth century as an ideal town, with a central piazza, a ducal palace, churches, garden palace, theater, and residences all encompassed within a star-shaped plan, bounded by thick walls bearing the Gonzaga family crest.

safety, and efficiency, as well as to symbolize the new seats of power and authority. One of the most important early precedents was set in Paris by Napoleon III, who presided over a comprehensive program of urban redevelopment and monumental urban design. The work was carried out by Baron Georges Haussmann between 1853 and 1870. Haussmann demolished large sections of old Paris to make way for broad, new, tree-lined avenues (**Figure 11.14**), with numerous public open spaces and monuments. In doing so, he made the city not only more efficient (wide boulevards meant better flows of traffic) and a better place to live (parks and gardens allowed more fresh air and sunlight in a crowded city and were held to be a "civilizing" influence) but also safer from revolutionary politics (wide boulevards were hard to barricade; monuments and statues helped to instill a sense of pride and identity).

The preferred architectural style for these new designs was the **Beaux Arts** style, which takes its name from L'École des Beaux Arts in Paris. In this school, architects were trained to draw on Classical, Renaissance, and Baroque styles, synthesizing them in designs for new buildings for the Industrial Age. The idea was that the new buildings would blend artfully with the older palaces, cathedrals, and civic buildings that dominated European city centers. Haussmann's ideas were widely influential and extensively copied.

Early in the twentieth century there emerged a different intellectual and artistic reaction to the pressures of industrialization and urbanization. This was the **Modern movement**, which was based on the idea that buildings and cities should be designed and run like machines. Equally important to the Modernists was that urban design should not simply reflect dominant social and cultural values, but, rather, help to create a new moral and social order. The movement's best-known advocate was Le Corbusier, a Paris-based Swiss who provided the inspiration for technocratic urban design. Modernist buildings sought to dramatize technology, exploit industrial production techniques, and use modern materials and unembellished, functional

Figure 11.14 Boulevard Montmartre, Paris Central Paris owes much of its character to the *grandes boulevards* that were key to the urban renewal schemes of Baron Georges-Eugène Haussmann. (*Source:* From *The Boulevard Montmartre, Paris,* by Camille Pisarro, 1830–1903.)

Figure 11.15 La Ville Radieuse The modern era and the advent of new transportation and construction technologies encouraged the utopian idea that cities could be built as efficient and equitable "machines" for industrial production and progressive lifestyles. One of the most famous and influential examples was *La Ville Radieuse* (1933), a visionary design by Swiss architect Le Corbusier. His vision was for the creation of open spaces through collectivized, high-density residential areas, strictly segregated from industrial areas and highways through a geometric physical plan. (*Source:* After Le Corbusier, *La Ville Radieuse.* Paris: Editions de L'Architecture D'Aujourd'hui.)

design. Le Corbusier's ideal city (*La Ville Radieuse*—**Figure 11.15**) featured linear clusters of high-density, medium-rise apartment blocks, elevated on stilts and segregated from industrial districts; high-rise tower office blocks; and transportation routes—all separated by broad expanses of public open space.

After World War II, this concept of urban design became pervasive, part of what became known as the International Style: boxlike steel-frame buildings with concrete and glass façades. The International Style was avant-garde yet respectable and, above all, comparatively inexpensive to build. This tradition of urban design, more than anything else, has imposed a measure of uniformity on cities around the world. Globalization has brought the appearance of International Style buildings in big cities in every part of the world. Furthermore, the International Style has often been the preferred basis for large-scale urban design projects around the world. One of the best examples is Brasilia, the capital of Brazil, founded in 1956 in an attempt to shift the country's political, economic, and psychological focus away from the past; differentiate it from the former colonial cities on the coast; and orient the country toward the future and the interior (see Chapter 6).

Modern urban design has had many critics, mainly on the grounds that it tends to take away the natural life and vitality of cities, replacing varied and human-scale environments with monotonous and austere settings. In response to this, historic preservation has become an important element of urban planning in every city that can afford it.

Islamic Cities

Islamic cities provide good examples of how social and cultural values and people's responses to their environment are translated into spatial terms through urban form and the design of the built environment. Indeed, it is because of similarities in cityscapes, layout, and design that geographers are able to talk about the Islamic city as a meaningful category. It is a category that includes thousands of towns and cities, not only in the Arabian Peninsula and Middle East—the heart of the Islamic Empire

under the prophet Muhammad (A.D. 570–632)—but also in regions into which Islam spread later: North Africa, coastal East Africa, South-Central Asia, and Indonesia. Most cities in North Africa and South-Central Asia are Islamic, and many elements of the classic Islamic city can be found in towns and cities as far away as Seville, Granada, and Córdoba in southern Spain (the western extent of Islam), Kano in northern Nigeria and Dares-Salaam in Tanzania (the southern extent), and Davao in the Philippines (the eastern extent).

The fundamentals of the layout and design of the traditional Islamic city are so closely attached to Islamic cultural values that they are to be found in the Qur'an, the holy book of Islam. Although urban growth does not have to conform to any overall master plan or layout, certain basic regulations and principles are intended to ensure Islam's emphasis on personal privacy and virtue, on communal well-being, and on the inner essence of things rather than on their outward appearance.

The most dominant feature of the traditional Islamic city is the *Jami*—the city's principal mosque (**Figure 11.16**). Located centrally, the mosque complex is a center not only of worship but also of education and the hub of a broad range of welfare functions. As cities grow, new, smaller mosques are built toward the edge of the city, each out of earshot from the call to prayer from the Jami and from one another. The traditional Islamic city was walled for defense, with several lookout towers and a *Kasbah*, or citadel (fortress), containing palace buildings, baths, barracks, and its own small mosque and shops.

Traditionally, gates controlled access to the city, allowing careful scrutiny of strangers and permitting the imposition of taxes on merchants. The major streets led from these gates to the main covered bazaars or street markets (*suqs*) (**Figure 11.17**). The suqs nearest the Jami typically specialize in the cleanest and most prestigious goods, such as books, perfumes, prayer mats, and modern consumer goods. Those nearer the gates typically specialize in bulkier and less valuable goods such as basic foodstuffs, building materials, textiles, leather goods, and pots and pans. Within the suqs, every profession and line of business had its own alley, and

Figure 11.16 Al Hussein mosque, Baghdad, Iraq The dominant feature of traditional Islamic cities is the *Jami,* or main mosque.

the residential districts around the suqs were organized into distinctive quarters, or *ahya',* according to occupation (or sometimes ethnicity, tribal affiliation, or religious sect).

Privacy is central to the construction of the Islamic city. Above all, women must be protected, according to Islamic values, from the gaze of unrelated men. Traditionally, doors must not face each other across a minor street, and windows must be small, narrow, and above normal eye level (**Figure 11.18**). Cul-de-sacs (dead-end streets) are used where possible to restrict the number of persons needing to approach the home, and angled entrances are used to prevent intrusive glances. Larger homes are built around courtyards, which provide an interior and private focus for domestic life.

The rights of others are also given strong emphasis, the Qur'an specifying an obligation to neighborly cooperation and consideration—traditionally interpreted as applying to a minimum radius of 40 houses. Roofs, in traditional designs, are surrounded by parapets to preclude views of neighbors' homes, and drainage channels are steered away from neighbors' houses. Refuse and wastewater are carefully recycled. Public thoroughfares were originally de-

Figure 11.17 The suq The *suq,* a covered bazaar or open street market, is one of the most important distinguishing features of a traditional Islamic city. Typically, a suq consists of small stalls located in numerous passageways. Many important suqs are covered with vaults or domes. Within them is typically a marked spatial organization, with stalls that sell similar products clustered tightly together, as in these copper shops in Marrakech, Morocco.

Figure 11.18 Islamic architecture In Islamic societies, the privacy of individual residences is paramount, and elaborate precautions are taken through architecture and urban design to ensure the privacy of women. Entrances are L-shaped and staggered across the street from one another. Windows are screened and often placed high above pedestrian access, as in this example in Jeddah, Saudi Arabia. Architectural details also reflect climatic influences: Window screens and narrow, twisting streets help to maximize shade, while air ducts and roof funnels help create dust-free drafts.

signed to be wide enough to allow two fully laden camels to pass each other and high enough to allow a camel and rider to pass through. The overall effect is to produce a compact, cellular urban structure within which it is possible to maintain a high degree of privacy (**Figure 11.19**).

Because most Islamic cities are located in hot, dry climates, these basic principles of urban design have evolved in conjunction with certain practical solutions to intense heat and sunlight. Twisting streets, as narrow as

permissible, help to maximize shade, as does latticework on windows and the cellular, courtyard design of residential areas. In some regions, local architectural styles include air ducts and roof funnels with adjustable shutters that can be used to create dust-free drafts.

All these features are still characteristic of Islamic cities, though they are especially clear in their old cores, or *medinas*. Like cities everywhere, however, Islamic cities also bear the imprint of globalization. Although Is-

Figure 11.19 Housing in Kalaa Sghira, Tunisia Seen from above, the traditional Islamic city is a compact mass of residences with walled courtyards—a cellular urban structure within which it is possible to maintain a high degree of privacy.

lamic culture is self-consciously resistant to many aspects of globalization, it has been unable to resist altogether the penetration of the world economy and the infusion of the Western-based culture of global metropolitanism. The result can be seen in international hotels, skyscrapers and office blocks, modern factories, highways, airports, and stores. Indeed, the leading cities of some oil-rich states have become the "shock cities" of the early twenty-first century, with phenomenal rates of growth characterized by breathtakingly ambitious architectural and urban design projects (see Box 11.1, "Shock City: Dubai, United Arab Emirates"). Meanwhile, Islamic culture and urban design principles have not always been able to cope with the pressures of contemporary rates of urbanization, so the larger Islamic cities in less-affluent states—cities such as Algiers, Cairo, Karachi, and Teheran—now share with other peripheral cities the common denominators of unmanageable size, shanty and squatter development, and low-income mass housing. In the next section, we examine the problems of large and rapidly growing metropolises throughout the periphery.

Cities of the Periphery: Unintended Metropolises

The cities of the world-system periphery, often still referred to as Third World cities, are numerous and varied. What they have in common is the experience of unprecedented rates of growth driven by rural "push"—overpopulation and the lack of employment opportunities in rural areas—rather than the "pull" of prospective jobs in towns and cities. Faced with poverty in overpopulated rural areas, many people regard moving to a city much like playing a lottery: You buy a ticket (in other words, go to the city) in the hope of hitting the jackpot (in other words, landing a good job). As with all lotteries, most people lose, and the net result is widespread underemployment. **Underemployment** occurs when people work less than full-time even though they would prefer to work more hours. Underem-

ployment is difficult to measure with any degree of accuracy, but estimates commonly range from 30 to 50 percent of the employed workforce in peripheral cities.

Because of their rapid growth and high underemployment, the peripheral metropolises of the world, the shock cities of the late twentieth century—Mexico City (Mexico), São Paulo (Brazil), Lagos (Nigeria, **Figure 11.20**) Mumbai (India; formerly Bombay), Dhaka (Pakistan), Jakarta (Indonesia), Karachi (Pakistan), and Manila (the Philippines)—embodied the most remarkable and unprecedented changes in economic, social, and cultural life. Socioeconomic conditions in these cities are still shocking, in the sense of being deplorably bad, but they are no longer shocking because they are unprecedented. Rather, they have become all too familiar.

As we have seen, the typical peripheral metropolis plays a key role in international economic flows, linking provincial regions with the hierarchy of world cities and, thus, with the global economy. Within peripheral metropolises, this role results in a pronounced **dualism**, or juxtaposition in geographic space of the formal and informal sectors of the economy. This dualism is very evident in the built environment, with high-rise modern office and apartment towers and luxurious homes contrasting acutely with slums and shantytowns (**Figure 11.21**).

The Informal Economy In many peripheral cities, more than one-third of the population is engaged in the informal sector, and in some—for example Chennai, India; Colombo, Sri Lanka; Delhi, India; Guyaquil, Ecuador; and Lahore, Pakistan—more than one-half. People who cannot find regularly paid work must resort to various ways of gleaning a living. Some of these ways are imaginative, some desperate, some pathetic. Examples range from street vending, shoe shining, craft work, and street-corner repairs to scavenging in garbage dumps (**Figure 11.22**). The informal sector consists of a broad range of activities that represent an important coping mechanism (**Figure 11.23**). For too many, however, coping

Figure 11.20 Lagos, Nigeria Lagos, like most metropolises in the world's periphery, grew relatively slowly until quite recently. The combination of the demographic transition, political independence, and an economic boom stimulated by the discovery of oil reserves in southeastern Nigeria triggered an explosive growth in population. Because of its difficult site on sand spits and lagoons, this growth has resulted in an irregular sprawl and, in the central area, a density of population higher than that of Manhattan Island in New York.

Figure 11.21   Rio de Janeiro This photograph, looking toward Rio de Janeiro's famous Ipanema Beach, shows very clearly the dualism of peripheral metropolises, with shanty housing (*favelas*) in the foreground and luxury apartments near the beach.

Figure 11.22   Garbage picking Registered scavengers picking out recyclable garbage at the Rio de Janeiro city dump.

Shock City: Dubai, United Arab Emirates

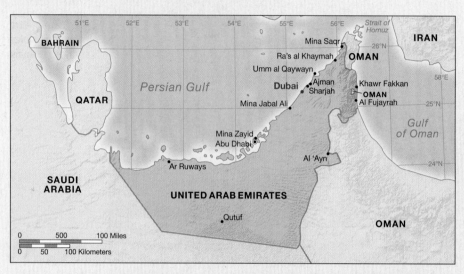

Dubai is one of seven emirates that comprise the United Arab Emirates (UAE). Since the discovery of oil in the UAE more than 30 years ago, the country has undergone a profound transformation, from an impoverished region of small desert principalities to a modern state with a high standard of living. Before the discovery of oil in the 1950s, the UAE's economy was mainly built around fishing and pearling. Oil is expected to last in the region only between 20 and 30 years, so there has been a major diversification program with investment in national infrastructure, health care, and education. Dubai has become a major financial and trading center and an important tourist destination for the wealthy. Current rates of growth exceed 10 percent a year, fueling a phenomenal construction boom that depends on immigrant labor drawn from Iran, Bangladesh, Pakistan, India, the Philippines, and other Asian nations.

Dubai's population of 1.3 million is dominated by immigrants from Asia and the Middle East, who account for over 70 percent of the city's residents. The construction boom is sustained by more than 500,000 low-skilled, poorly paid South Asian migrant workers who live in substandard conditions, with few rights. The UAE has a minimum legal wage, but it is rarely enforced, so that migrant construction workers receive, on average, the equivalent of US$175 a month. The average per capita income in the UAE is $2,106 a month.

Dubai's "shock city" status derives as much from its spectacular affluence as its phenomenal rate of growth from what had hitherto been an impoverished setting. The Jumeirah Palm is one of two artificial islands extending from the Dubai City waterfront. Built in the shape of date palm trees with a surrounding breakwater, the islands support more than 60 luxury hotels, 4,000 exclusive residential villas, 1,000 water homes, 5,000 shoreline apartments, marinas, water theme parks, restaurants, shopping malls, sports facilities, health spas, cinemas, and various dive sites.

Awash—for a while—with petrodollars, Dubai has been able to offer generous tax breaks to companies willing to relocate their activities to the city, creating a major business hub that will provide an economic base when the oil and gas run out. One notorious result was the move of Halliburton's headquarters from Houston, Texas, to Dubai. The city's skyline reflects its emerging role as a globally important business hub, the Burj Dubai tower, at 2,313 ft (818 meters), claiming the title of the tallest building in the world on completion in 2009—nearly 40 percent taller than the second-highest building in the world (Taipei 101, in Taipei, Taiwan).

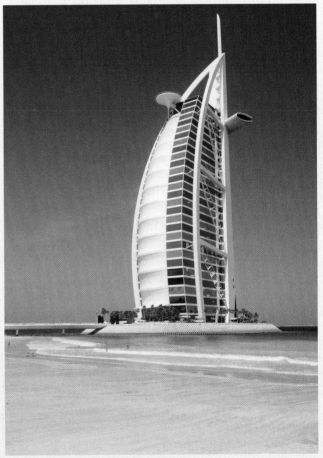

The Burj Al Arab hotel overlooking the Jumeira beach is the only hotel in the world rated as 7 stars.

Dubai's desert environment has been converted to a lush 18-hole championship golf course at the Dubai Creek Golf and Yacht Club, which also offers a 9-hole par-3 course, a golf academy, six restaurants and bars, a gymnasium, a tropical swimming pool, the 225-room Park Hyatt Dubai, 92 residential executive villas and a 121-berth marina.

Dubai has established a reputation for luxury shopping, with the duty-free stores at Dubai airport exploiting the city's role as a stopping-off point for travelers between Europe and Asia.

Ski Dubai, part of an enormous complex at the Mall of the Emirates, provides five different runs with fresh snow for skiing, snowboarding, and tobogganing.

means resorting to begging, crime, or prostitution. Occupations such as selling souvenirs, driving pedicabs, making home-brewed beer, writing letters for others, and dressmaking may seem marginal from the point of view of the global economy, but more than a billion people around the world must feed, clothe, and house themselves entirely from such occupations. In many peripheral cities more than half the population subsists in this way. Across Africa, the International Labor Office estimates, informal-sector employment is growing 10 times faster than formal-sector employment.

In most peripheral countries, the informal labor force includes children. In environments of extreme poverty, every family member must contribute something, and so children are expected to do their share. Industries in the formal sector often take advantage of this situation. Many firms farm out their production under subcontracting schemes that are based not in factories but in home settings that use child workers. In these settings, labor standards are nearly impossible to enforce. In the Philippines, for example, batches of rural children are ferried by syndicates to work in garment-manufacturing sweatshops in urban areas.

Despite this side of the picture, the informal sector has a few positive aspects. Pedicabs, for example, provide an affordable, nonpolluting means of transportation in crowded metropolitan settings. Garbage picking, while it may seem desperate and degrading in Western eyes, provides an important means of recycling paper, steel, glass, and plastic products. One study of Mexico City estimated that as much as 25 percent of the municipal waste ends up being recycled by the 15,000 or so scavengers who work over the city's official dump sites. This positive contribution to the economy, though, scarcely balances the lives of poverty and degradation experienced by the scavengers.

Urban geographers also recognize that the informal sector represents an important resource to the formal sector of peripheral economies. The informal sector provides a vast range of cheap goods and services that reduce the cost of living for employees in the formal sector, thus enabling employers to keep wages low. Although this network does not contribute to urban economic growth or help alleviate poverty, it does keep companies competitive within the context of the global economic system. For export-oriented companies, in particular, the informal sector provides a considerable indirect subsidy to production. We must recognize, too, that this subsidy is often passed on to consumers in the core regions in the form of lower prices for goods and consumer products made in the periphery.

Consider, for example, the paper industry in Cali, Colombia. This industry is dominated by one company, Cartón de Colombia, which was established in 1944 with North American capital and subsequently acquired by the Mobil Oil Company. Most of the company's lower-quality paper products are made from recycled waste paper, and 60 percent of this waste paper is gath-

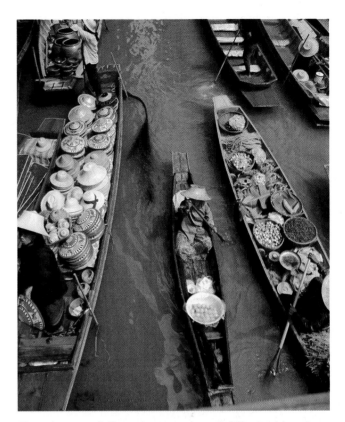

Figure 11.23 Informal economic activities In cities where jobs are scarce, people have to cope through the informal sector of the economy, which includes a very broad variety of activities, including agriculture (backyard hens, for example), manufacturing (craft work), and retailing (street vending).

ered by local garbage pickers. There are 1,200 to 1,500 garbage pickers in Cali. Some work the city's municipal waste dump, some work the alleys and yards of shopping and industrial areas, and some work the routes of municipal garbage trucks, intercepting trash cans before the truck arrives. They are part of Cali's informal economy, for they are not employed by Cartón de Colombia nor do they have any sort of contract with the company or its representatives. They simply show up each day to sell their pickings. This way the company can avoid paying both wages and benefits, while dictating the price it will pay for various grades of waste paper. The company can operate profitably while keeping the price of its products down—the arrangement is a microcosm of core-periphery relationships.

Slums of Hope, Slums of Despair   The informal labor market is directly paralleled in informal shantytowns and squatter housing: Because so few jobs with regular wages exist, few families can afford rent or house payments for sound housing. Unemployment, underemployment, and poverty mean overcrowding. In situations where urban growth has swamped the available stock of cheap housing and outstripped the capacity of builders to create affordable new housing, the inevitable outcome is makeshift shanty housing that offers, at best, precarious shelter. Such housing has to be constructed on the cheapest and least desirable sites. Often this means building on bare rock, over ravines, on derelict land, on swamps, or on steep slopes. Nearly always it means building without any basic infrastructure of streets or utilities. Sometimes it means adapting to the most extreme ecological niches, as in Lima, where garbage pickers actually live on the waste dumps, or in Cairo, where for generations the poor have adapted catacombs and cemeteries into living spaces. In many cities more than half of the housing is substandard. The United Nations estimated in 2005 that more than 1.6 billion people worldwide live in inadequate housing in urban areas.

Globalization and the accompanying trend toward neoliberal economic policies have intensified problems of poverty and slum housing in many cities. The United Nations Human Settlement Programme reported in 2004 that

The reduction of fiscal deficits has partly entailed reduction of public expenditure through downsizing of the civil service and privatization of state enterprises, resulting in the laying-off of large numbers of public-sector employees in many countries. Trade liberalization has often resulted in the closure of some industries that have been unable to compete against cheap imports, again leading to massive retrenchment and higher unemployment levels. Rising urban unemployment and increasing poverty have forced large numbers of the urban poor into the informal sector. Underpaid formal-sector employees have also entered the informal sector as a survival strategy. This, in turn, leads to the erosion of the tax base and decreasing ability of national and local governments to assist the poor through social and basic services. The removal of price controls on subsistence goods, and increased utility charges through privatization . . . have resulted in rising inequalities and increasing poverty. . . . As a result, cities end up with their prime resources largely appropriated by the affluent.[4]

Faced with the growth of slums, the first response of many governments has been to eradicate them. Encouraged by Western development economists and housing experts, many cities sought to stamp out unintended urbanization through large-scale eviction and clearance programs. In Caracas (Venezuela), Lagos (Nigeria), Bangkok (Thailand), Kolkata (India; formerly Calcutta), Manila (the Philippines), and scores of other cities in the periphery, hundreds of thousands of shanty dwellers were ordered out on short notice and their homes bulldozed to make way for public works, land speculation, luxury housing, urban renewal, and, on occasion, to improve the appearance of cities for special visitors. Seoul, South Korea, has probably had the most forced evictions of any city in the world. Since 1966, millions of people have been forced out of accommodations that they owned or rented as part of a sustained government clean-up campaign. The Beijing Olympics in 2008 displaced 1.5 million people since 2000, according to the Geneva-based Centre on Housing Rights and Evictions (COHRE). Their report[5] holds the Chinese government guilty of widespread forced evictions along with other human rights violations during preparations for the Beijing Olympics. Demolitions and evictions were often undertaken without due process, without the provision of adequate compensation sufficient to attain alternative accommodation, and without access to legal recourse. In some cases, tenants were given little or no notice of their eviction and did not receive the promised compensation. The city also subjected alleged unlicensed taxis operators, street vendors, vagrants, and beggars to "Re-education through Labour" (a form of imprisonment without charge).

Yet Seoul and Beijing, more than most other cities, could afford to build new low-income housing to replace the demolished neighborhoods. Most peripheral cities cannot do so, which means that displaced slum dwellers have no option but to create new squatter and shanty settlements elsewhere in the city. Most cities, in fact, cannot evict and demolish fast enough to keep pace with the growth of slums caused by in-migration. The futility of slum clearance has led to a widespread reevaluation of the wisdom of such policies. The thinking now is that informal-sector housing should be seen as a rational response to poverty. Shanty and squatter neighborhoods can not only provide affordable shelter but also function as important reception areas for migrants to the city, with supportive communal organizations and informal employment opportunities that help them to adjust to city life. They can, in other words, be "slums of hope." City authorities, recognizing the positive functions of informal housing and self-help improvements, are now increasingly disposed to be tolerant and even helpful toward squatters rather than sending in police and municipal workers with bulldozers.

In fact, many informal settlements are the product of careful planning. In parts of Latin America, for example, it is common for community activists to draw up plans for invading unused land, then quickly build shanty housing before land owners can react. Part of the activists' strategy is usually to organize a critical mass of people large enough to be able to negotiate with the authorities to resist eviction. It is also common for activists to plan their invasions for public holidays so that the risk of early detection is minimized. As the risk of eviction diminishes over time, some residents of informal housing are able to gradually improve their dwellings through self-help (**Figure 11.24**).

---

[4]United Nations Human Settlements Programme, *The State of the World's Cities 2004/2005. Globalization and Urban Culture.* Sterling, VA: Earthscan, 2004, p. 102.

[5]Centre on Housing Rights and Evictions (COHRE), 2008, "One World, Whose Dream?: Housing Rights Violations and the Beijing Olympic Games," http://www.cohre.org/store/attachments/ One_World_Whose_Dream_July08.pdf, accessed July 31, 2008.

Figure 11.24   Self-help as a solution to housing problems Self-help is often the only solution to housing problems because wages are so low and so scarce that builders cannot construct even the most inexpensive new housing and make a profit, and because municipalities cannot afford to build sufficient quantities of subsidized housing. One of the most successful ways of encouraging self-help housing is for municipal authorities to create the preconditions by clearing sites, putting in the footings for small dwellings, and installing a basic framework of water and sewage utilities. This "sites-and-services" approach has become the mainstay of urban housing policies in many peripheral countries. This photograph shows self-help housing in Ndola, Zambia.

Nevertheless, there are many shanty and squatter neighborhoods where self-help and community organization do not emerge. Instead, grim and desperately miserable conditions prevail. These are "slums of despair," where overcrowding, lack of adequate sanitation, and lack of maintenance lead to shockingly high levels of ill health and infant mortality and where social pathologies are at their worst. Consider, for example, the squatter settlement of Dharavi, a "mega-slum" in Mumbai, India.[6] Dharavi is about 1 square mile in size and home to over a million people. This means that as many as 18,000 people crowd into a single acre of land. In a city where house rents are among the highest in the world, Dharavi provides a cheap and affordable option to those who move to Mumbai to earn their living. Rents can be as low as 185 rupees ($4) per month. Recycling is one of the slum's biggest industries. In Dharavi nothing is considered garbage. Thousands of tons of scrap plastic, metals, paper, cotton, soap, and glass revolve through Dharavi each day. Ruined plastic toys are tossed into massive grinders, chopped into tiny pieces, melted down into multicolored pellets, ready to be refashioned into knockoff Barbie dolls. Dharavi also houses about 15,000 hutment factories, each typically stuffed with children as well as adults sewing cotton, melting plastic, hammering iron, and molding clay, as well as producing embroidered garments, export-quality leather goods, pottery, and plastic.

The aggregate annual turnover of these businesses is estimated to be more than $650 million a year, yet conditions for its residents are miserable (see Box 11.2, "Life in

a Mega-Slum"). In Dharavi, there is only one toilet per 1,440 people—and during the monsoon rains, flooded lanes run with human excrement. Most people have food intakes of less than the recommended minimum of 1,500 calories a day; 90 percent of all infants and children under 4 have less than the minimum calories needed for a healthy diet. More than half of the children and almost half the adults have intestinal worm infections. Infant and child mortality is high—though nobody knows just how high—with malaria, tetanus, diarrhea, dysentery, and cholera as the principal causes of death among under-fives.

Transport and Infrastructure Problems   The informal labor market is also reflected in cities' inability to provide a basic infrastructure of highways, transportation, schools, utilities, and emergency services. Because the informal sector yields no tax revenues, municipal funds are insufficient to provide an adequate infrastructure or to maintain a safe and sanitary environment.

Even though the governments of peripheral cities typically spend nearly all of their budgets on transport and infrastructure in their race to keep up with population growth, conditions are bad and getting rapidly worse. Peripheral cities have always been congested, but in recent years the modernizing influence of formal-sector activities has turned the congestion into near gridlock. In many of the world's peripheral and semiperipheral metropolises, sharp increases have occurred in the availability and use of automobiles. One of the most dramatic examples is Taipei, Taiwan, where the number of automobiles increased from about 11,000 in 1960 to over 1.35 million in 2000. India has become one of the world's fastest-growing automobile markets, with more than 2.3 million sold in 2007. Yet India's transportation infrastructure has not caught up with this sudden growth, nor have its drivers. The accident rate among cars is the highest in the world. India has about 1 percent of the world's cars yet still manages to kill over 100,000 people in traffic accidents each year. This amounts to 10 percent of the entire world's traffic fatalities

---

[6]See *The Economist*, 2007, "A Flourishing Slum," http://www.economist.com/world/asia/displaystory.cfm?story_id=10311293, accessed August 5, 2008; M. Jacobson, "Mumbai Slum," *National Geographic, 2007*, http://ngm.nationalgeographic.com/ngm/0705/feature3/index.html, accessed August 5, 2008; C. W. Dugger, "Toilets Underused to Fight Disease, U.N. Study Finds," *New York Times*, 2006, http://www.nytimes.com/2006/11/10/world/10toilet.html?_r=1&ex=1189828800&en=905358c57769b677&ei=5070&oref=slogin, accessed August 5, 2008.

11.2 WINDOW ON THE WORLD

Life in a Mega-Slum

"I came to stay with relatives in Dharavi when I was six, got married when I was 16 and gave birth to my first daughter soon after. She is now 11 years old. I have two more daughters, aged nine and five.

Both my husband and I work and we try to share our responsibilities equally. I work as a domestic maid and he works in an aluminium vessel manufacturing company.

Everyday I wake up at 6 am and finish off the household work such as getting my children ready for school, cooking food and preparing a lunchbox for my husband.

I work for about three-and-a-half hours in a home nearby, where I clean and dust and wash vessels and clothes.

Name: Vimla

Age: 27 years old

Occupation: Domestic maid and flower band maker

Earns: About $22 (£12) per month – husband earns $54 (£30)

Home: Hut by railway tracks, with no toilet or water

Time in slum: 16 years – hopes for relocation in Dharavi

I return home by about 1.30 pm. Then I make flower bands to earn some extra money. I cook dinner and supervise my children's studies.

My husband returns home around 8 pm and then we walk for about 10 minutes to get water for our use the next day.

I don't have a direct water supply in my house—none of the houses in my area have a tap with running water. All of us must cross the railway tracks to get it.

We don't earn enough money for our expenses and often end up borrowing. Right now, I owe a lender 15,000 rupees ($327/£177). I pay 1,500 rupees ($33/£18) as interest and he hasn't even started deducting anything from my loan amount yet.

Most of the money is spent on medicines for my girls because I want to give them the best possible treatment. That sometimes means I have to take to them to a private doctor.

I am constantly concerned for the safety of my children and I practically lock them in the house after 6 pm. Not only is it unsafe to play near the tracks but also the possibility of them being kidnapped or assaulted petrifies me."

Source: BBC News, "Life in a Slum: Domestic Maid," http://news .bbc.co.uk/1/shared/spl/hi/world/06/dharavi_slum/html/dharavi_ slum_1.stm, accessed August 5, 2008.

(the United States, with more than 40 perecent of the world's cars, creates just 43,000 fatalities). Not only are there more people and more traffic but the changing spatial organization of peripheral cities has increased the need for transportation. Traditional patterns of land use have been superseded by the agglomerating tendencies inherent in modern industry and the segregating tendencies inherent in modernizing societies. The greatest single change has been the separation of home from work, however, which has meant a significant increase in commuting.

In spite of many innovative responses to urban transportation needs, road transportation in many cities is breaking down, with poorly maintained roads, traffic jams, long delays at intersections, and frequent accidents. Many governments have invested in expensive new freeways and street-widening schemes, but because they tend to focus on city centers (which are still the settings for most jobs and most services and amenities), they ultimately fail, emptying vehicles into a congested and chaotic mixture of motorized traffic, bicycles, animal-drawn vehicles, and

hand-drawn carts. Some of the worst traffic tales come from Mexico City—where traffic backups total more than 90 kilometers (60 miles) each day, on average—and Bangkok, where the 15-mile trip into town from Don Muang Airport can take 3 hours. In São Paulo, Brazil, gridlock can span 160 kilometers (100 miles), rush-hour traffic jams average 85 kilometers (53 miles) in length, and 15-hour traffic jams are not unusual. The costs of these traffic backups are enormous. The annual costs of traffic delays in Singapore have been estimated at $305 million; in Bangkok, Thailand, they have been estimated at $272 million—the equivalent of around 1 percent of Thailand's gross national product.

Water supplies and sewerage also present acute problems for many cities (**Figure 11.25**). Definitions of what constitutes an adequate amount of safe drinking water and sanitation vary from country to country. Although many governments classify the existence of a water tap within 100 meters (328 feet) of a house as "adequate," such a tap does not guarantee that the household will be to one-third of all garbage and solid waste is collected and removed—the rest is partially recycled informally, tipped into gullies, canals, or rivers, or simply left to rot. Sewage services are just as bad. In Latin America, for example, only about 2 percent of collected sewage receives any treatment. In Mexico, more than 90 percent of wastewater treatment plants are nonfunctional, and in cities like Bogotá, Buenos Aires, Mexico City, and Santiago, some 50 to 60 million cubic meters of mostly untreated sewage is discharged every day into nearby bodies of water. São Paulo has over 1,600 kilometers

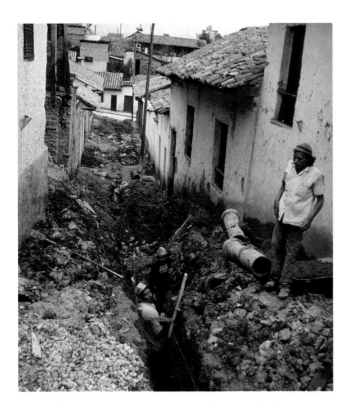

Figure 11.25 Infrastructure problems Low incomes mean that city governments are unable to raise sufficient revenues through local taxes, which in turn means that infrastructure is neglected. Putting water and sewage lines into neighborhoods that were built without these basic utilities is arduous and expensive, yet without basic utilities, public health is seriously threatened. This photograph shows water lines being installed in a low-income neighborhood of Cartagena, Colombia.

able to secure enough water for good health. Communal taps often function only a few hours each day, so residents must wait in long lines to fill even one bucket. In Rajkot, India, a city of 600,000 people, piped water routinely runs for only 20 minutes each day.

The World Bank estimates that around 65 percent of urban residents worldwide in less developed countries have access to a satisfactory water source, and only about 40 percent are connected to sewers (90 percent of which discharge their waste untreated into a river, a lake, or the sea). Hundreds of millions of urban dwellers have no alternative but to use contaminated water—or at least water whose quality is not guaranteed. A small minority, usually the residents of the most affluent neighborhoods, have water piped into their homes, while the majority have piped water nearby, which has to be collected. Those not served are obliged to carry water in small quantities over long distances or to use water from streams or other surface sources (**Figure 11.26**). In Colombo, Sri Lanka, about one-third of all houses have indoor piped water, and another one-fourth have piped water outside. In Dar es Salaam (Tanzania), Kinshasa (Democratic Republic of the Congo), and many other peripheral cities, almost half the population has no access to piped water, either indoors or outdoors.

In many cities, including Bangkok, Bogotá, Dar es Salaam, Jakarta, Karachi, and São Paulo, only one-fourth

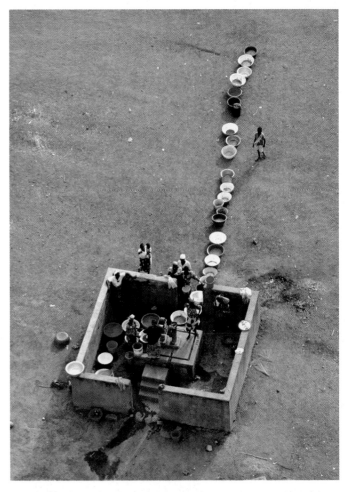

Figure 11.26 Water-supply problems Many peripheral cities have grown so quickly and under such difficult conditions that large sections of the population do not have access to supplies of clean water. Where there is a public supply—a well or an outdoor standpipe—water consumption is limited by the time and energy required to collect water and carry it home. It is not rare for 500 or more people to have to share a single pump. Because low-income people work very long hours, the time spent waiting in line for water and then transporting buckets home is time that could have been used in earning an income. Limited quantities of water mean inadequate supplies for personal hygiene and for washing food, cooking utensils, and clothes. Where public agencies provide no water supply—as is common in squatter settlements—the poor often obtain water from private vendors and can pay 20 to 30 times the cost per liter paid by households with piped supplies. Water vendors probably supply about one-fourth of the population of peripheral metropolises.

(1,000 miles) of open sewers, and raw sewage from the city's slums drains into the Billings reservoir, a major source of the city's drinking water. In Bangkok, less than 5 percent of the population is connected to a sewer system; human wastes are generally disposed of through septic tanks and cesspools with their effluents, as well as wastewater from sinks, laundries, baths, and kitchens, discharged into storm water drains or canals. Jakarta has no waterborne sewage system at all. Septic tanks serve about one-quarter of the city's population; others must use pit latrines, cesspools, and ditches along the roadside. A survey of over 3,000 towns and cities in India found that only 8 had full sewage-treatment facilities, and another 209 had partial treatment facilities. Along one river, the Ganga (better known to English speakers as the Ganges), 114 towns and cities dump untreated sewage into the river every day, along with waste from DDT factories, tanneries, paper and pulp mills, petrochemical and fertilizer complexes, and other industrial pollutants. Each day the Yamuna River picks up 200 million liters of untreated sewage and 20 million liters of industrial effluents as it passes through Delhi. In China, inadequate sewerage and the meager 4.5 percent coverage of municipal wastewater plants have resulted in widespread water-quality deterioration. Shanghai has had to move its water supply intake 40 kilometers (25 miles) upstream at a cost of US$300 million because of degradation of river water quality around the city.

These problems provide opportunities for the informal sector, however. Street vendors, who get their water from private tanker and borehole operators, sell water from 2- or 4-gallon cans. They typically charge 5 to 10 times the local rate set by public water utilities; in some cities, they charge 60 to 100 times as much. Similarly, many cities have evolved informal-sector mechanisms for sewage disposal. In many Asian cities, for example, human waste is removed overnight by handcart operators. Unfortunately, it is rarely disposed of properly and often ends up polluting the rivers or lakes from which the urban poor draw their water.

Environmental Degradation With pressing problems of poverty, slum housing, and inadequate infrastructure, it is not surprising that peripheral cities are unable to devote many resources to environmental problems. Because of the speed of population growth, these problems are escalating rapidly. Industrial and human wastes pile up in lakes and lagoons and pollute long stretches of rivers, estuaries, and coastal zones. Groundwater is polluted through the leaching of chemicals from uncontrolled dump sites, and the forests around many cities are being denuded by the demand of cities for timber and domestic fuels. This environmental degradation is, of course, directly linked to human health. People living in such environments have much higher rates of respiratory infections, tuberculosis, and diarrhea, and much shorter life expectancies, than people living in surrounding rural communities. Children in squatter settlements

may be 50 times as likely to die before the age of 5 as those born in affluent core countries.

In addition, air pollution has escalated to very harmful levels in many cities. With the development of a modern industrial sector and the growth of automobile ownership, but without enforceable regulations on pollution and vehicle emissions, tons of lead, sulfur oxides, fluorides, carbon monoxide, nitrogen oxides, petrochemical oxidants, and other toxic chemicals are pumped into the atmosphere every day in large cities. The burning of charcoal, wood, and kerosene for fuel and cooking in low-income neighborhoods also contributes significantly to dirty air. In cities where sewerage systems are deficient, the problem is compounded by the presence of airborne dried fecal matter. Worldwide, according to UN data, more than 1.1 billion people live in urban areas where air pollution exceeds healthful levels.

A UN study of 20 megacities found that every one of them had at least one major pollutant at levels exceeding World Health Organization (WHO) guidelines. Fourteen of the 20 had *two* major pollutants exceeding WHO guidelines, and seven had *three*. Such pollution is not only unpleasant but dangerous. The World Bank has cited China as having 16 of the 20 most air-polluted cities on Earth. China's spectacular economic growth has brought with it air pollution levels that are blamed for more than 400,000 premature deaths a year. The European Space Agency, utilizing satellite imagery, has found that Beijing and its surrounding areas have the world's highest levels of nitrogen dioxide, a substance poisonous to the lungs. In Manila, the Philippines, the Asian Development Bank found levels of suspended particulate matter in the air to be 200–400 percent above guideline levels. In Mexico City, where sulfur dioxide and lead concentrations are two to four times higher than the WHO guidelines, and where national ozone levels are exceeded on more than half of the days throughout the year, 7 in 10 newborns have dangerously high levels of lead in their bloodstream. WHO studies demonstrate that it is unhealthy for human beings to breathe air with more than 100 to 120 parts per billion (ppb) of ozone contaminants for more than one day a year. Yet Mexico City residents breathe this level, or more, for over 300 days a year. In Bangkok, Thailand, where air pollution is almost as severe as in Mexico City, research has shown that lead-bearing air pollutants reduce children's IQ by an average of 3.5 points per year until they are 7 years old. It has also been estimated that Bangkok's pall of dust and smoke causes more than 1,400 deaths each year and $3.1 billion each year in lost productivity resulting from traffic- and pollution-linked illnesses.

Proximity to industrial facilities, often the result of the need and desire of the poor to live near places of employment, poses another set of risks. A notorious accident at the Union Carbide factory in Bhopal, India, in 1984 caused 2,988 deaths and more than 100,000 injuries, mostly among residents of the shantytowns near the chemical factory.

NEW PATTERNS: THE POLYCENTRIC METROPOLIS

Traditional patterns of land use and spatial organization in many parts of the world are being transformed by the local effects of splintering urbanism. Economic and cultural globalization, together with the uneven evolution of networked infrastructures of information and communications technologies, is forging new landscapes of innovation, economic development, and cultural transformation, while intensifying social and economic inequalities between the fast world and the slow world. As we noted in Chapter 10, these trends are most pronounced in world cities and major regional metropolitan centers, particularly in core countries. Nevertheless, fragments of this splintering urbanism are increasingly evident throughout the world as new technologies, new forms of economic organization, and new sociocultural norms spread through the global urban system. The largest of the world's polycentric metropolises have become "100-mile cities"—metropolitan regions that are literally 100 miles or so across, consisting of a loose coalition of urban realms, or economic subregions, bound together through urban freeways.

The sheer scale and extent of contemporary development in America has challenged our lexicon of urbanization. It also severely challenges the models of urban form and structure that for so long have been the staples of urban studies. The suburbs now not only contain the largest fraction of America's households and population but also a significant fraction of America's industry, commercial office space, retailing, recreational facilities, and tourist attractions. Traditional "laws" of urbanization are being repealed by massive changes in real estate investment, in tandem with equally significant changes in the structure and functional organization of metropolitan regions. New rounds of economic restructuring, advances in digital telecommunications technologies, demographic shifts, and neoliberal policies are giving rise to new urban landscapes. Urban regions are being stretched and reshaped to accommodate increasingly complex and extensive patterns of interdependency, while the political economy of metropolitan America is being reshaped in response to socioeconomic realignments and cultural shifts.

Until the middle of the twentieth century, urban and metropolitan form could safely be conceptualized in terms of the outcomes of processes of competition for land and of ecological processes of congregation and segregation—all pivoting tightly around a dominant central business district and transportation hub (Figure 11.1). During the middle decades of the twentieth century, however, American metropolises were unbound by the combination of increased automobility and the blossoming of egalitarian liberalism in the form of massive federal outlays on highway construction and mortgage insurance. The resulting massive spurt of city building produced a dispersed spatial structure and the emergence of "urban realms"—semiautonomous sub-regions that resulted from a polycentric

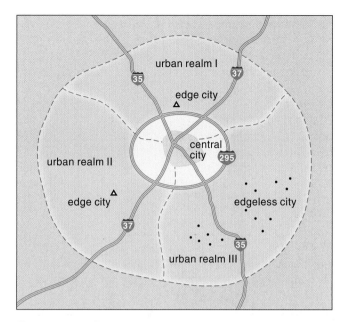

Figure 11.27 The twentieth-century metropolis Consisting of a central city, a ring of suburbs, and a series of far-flung urban realms, studded with edge cities (nodal concentrations of office and retail space) and "edgeless cities" of suburban and exurban office parks and shopping malls.

metropolitan structure (**Figure 11.27**) that displaced the traditional core-periphery relationship between central cities and their suburbs. Initially, the shift to an expanded polycentric metropolis was most pronounced in the northeastern United States, and geographer Jean Gottmann captured the moment in 1961 with his conceptualization of "megalopolis"—his term for the highly urbanized region that stretched between Boston and Washington, D.C.

Toward the end of the twentieth century the *nature* of metropolitan economic development began to change—as we saw in Chapter 7—with the rapid decline of the old base of manufacturing industries and the onset of a "new economy" based on digital technologies and featuring economic and cultural globalization, the international division of labor, international finance, and supranationalism. Capital investment switched increasingly into finance, insurance, and real estate and the structural transformation of the economy resulted in socioeconomic polarization, with increasing numbers of both upper-middle-class and impoverished households. The internationalization of economic geography meanwhile weakened the leverage of both big government and big labor, allowing a fundamental reassertion and intensification of the economic and spatial logic of capital—especially big capital.

The New Metropolis

Today, the traditional form of the metropolis has slipped into history. But the challenge of characterizing the evolving outcomes of urbanization have prompted a great variety of neologisms, including *postsuburbia, exurbia, exopolis, boomburbs, cosmoburbs, nerdistans, technoburbs, generica, satellite sprawl,* and *mallcondoville*. The *New*

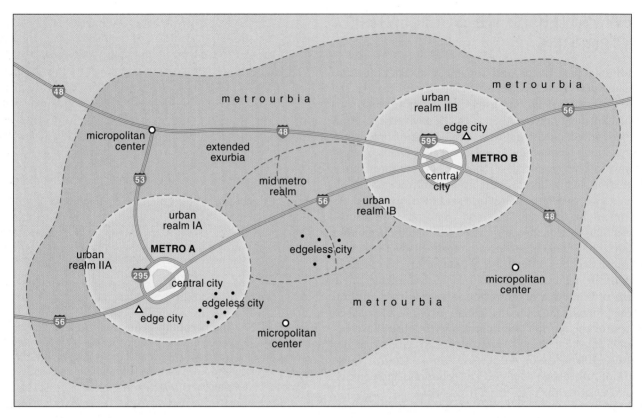

Figure 11.28   The new metropolis The largest metropolitan regions are now "megapolitan," with coalescing metropolitan areas merging into disjointed and decentralized urban landscapes with varying-sized urban centers, subcenters, and satellites and unexpected juxtapositions of form and function, the residential settings in suburban and exurban areas being thoroughly interspersed with office employment and high-end retailing, creating "metroburban" landscapes.

Metropolis, meanwhile, is an encompassing term for the stereotypical urbanized region that has been extended and reshaped to accommodate increasingly complex and extensive patterns of interdependency in polycentric networks of half a dozen or more urban realms and as many as 50 nodal centers of different types and sizes, physically separate but functionally networked and drawing enormous economic strength from a new functional division of labor (**Figure 11.28**). Bound together through urban freeways, arterial highways, beltways, and interstates, new metropolises are themselves now beginning to coalesce functionally into "megapolitan" regions that dominate national economies. More recently, the term *metroburbia* has emerged to capture the way that residential settings in suburban and exurban areas are thoroughly interspersed with office employment and high-end retailing.

The polycentric form of metroburbia is framed around a system of nodes and realms, bound together with ever-expanding four-, six- and eight-lane highways. It is interspersed with smaller clusters of decentralized employment, studded with micropolitan centers and filled out with booming stand-alone suburbs. Traditional nodal anchors—downtown commercial centers—remain very important, especially as settings for the advanced business services—advertising, banking, insurance, investment management, and logistics services. But in addition there

are other nodes within the polycentric form of metroburbia. These vary in character and include:

- "edge cities," decentralized clusters of retailing and office development, often located on an axis with a major airport, sometimes adjacent to a high-speed rail station, always linked to an urban freeway system;
- newer business centers, often developing in a prestigious residential quarter and serving as a setting for newer services such as corporate headquarters, the media, advertising, public relations, and design;
- outermost complexes of back-office and R&D operations, typically near major transport hubs 20 to 30 miles from the main core; and
- specialized subcenters, usually for education, entertainment, and sporting complexes and exhibition and convention centers.[7]

In many ways, the New Metropolis is a product of the extension and evolution of long-standing elements: demand for low-density single-family housing; access to mortgage finance; economies of scale and scope; fragmentation of land-use planning among multiple municipalities; aversion

[7]P. Hall, "Global City-Regions in the Twenty-first Century," in A. J. Scott (ed.), *Global City-Regions: Trends, Theory Policy.* New York: Oxford University Press, 2001, pp. 59–77.

to public transport and the reliance on private automobiles; federally subsidized road building; and exclusionary segregation via deed restrictions. But the New Metropolis is also the product of economic restructuring. Industry, office employment, and retailing had already been significantly decentralized by the 1980s as bigger and faster trucks, larger factories and warehouses, flexible specialization, and just-in-time delivery systems called for a new economic geography that could be accommodated only by larger and less expensive parcels of suburban land. Although heavy industry was usually tied to older outlying industrial sites with access to rail sidings, most other industry had become entirely footloose, able to set up wherever local municipalities could be persuaded to zone land and offer incentives. Corridors of land along major highways quickly became a common setting for footloose industries; industrial parks in campus-like settings became another; while the areas around the junction of major highways and around airports became the preferred location for the burgeoning bioscience and information-based industries, the biggest growth sectors of the modern economy. Meanwhile, along with decentralized population and industry came retailing, service, and office functions: strip malls, big-box discount stores, integrated shopping malls, specialized malls, power centers, fast-food franchises, hotel chains, family restaurants, corporate headquarters complexes, and office parks. By 2000, roughly three out of five jobs in American metropolitan areas were located in, for want of a better name, suburbs.

Globalization and the Quartering of Urban Space

Intensifying social polarization is another dimension of contemporary urbanization that is associated with globalization. The increased mobility of capital, together with the increasing span of control that is possible through advanced information and communications technologies, has resulted in an increasing concentration of wealth and an increasing gap between the fast world and the slow world (Chapter 2). In most of the world's metropolitan regions there has been an increase in the relative numbers of those who are rich and those who are poor, along with an increase in the financial distance between them. There has also been a greater differentiation among the intermediate groups, so that there is often a four- or five-part division of socioeconomic groups rather than a simple division into two. The United Nations Center for Human Settlements (UNCHS) has identified this social polarization as an indirect but crucial determinant of contemporary patterns of segregation of people and land uses around the world. The overall effect of globalization, according to the UNCHS *Global Report on Human Settlements 2001*,[8] is the "quartering" of cities into spatially partitioned, compartmentalized residential enclaves.

Although varying from city to city because of differences in national political and economic structures and in cities' roles in the international economy, their historical development, and their demographic composition, there are some basic features that characterize this quartering of urban space. Geographer Ronald van Kempen and planner Peter Marcuse describe six **socio-spatial formations**—residential patterns that, they suggest, add up to a new spatial order that characterizes globalizing cities:[9]

- protected enclaves of the rich—"citadels" or exclusionary enclaves that generally consist of expensive apartments in favorable locations;
- gentrified areas occupied by young professionals and managers, typically located in the inner districts of central cities;
- middle-class suburbs and medium- and high-rise apartment and condominium complexes;
- working-class neighborhoods, often of rented tenements, that are differentiated according to income, occupation, and ethnicity, sometimes with hardened spatial boundaries between them;
- ethnic enclaves—a specific form of tenement area where ethnic segregation is particularly pronounced;
- excluded ghettos—the slums inhabited by the very poor, the excluded, the never employed and permanently unemployed, and the homeless.

These socio-spatial formations represent another aspect of splintering urbanism. The premium spaces within the quartered city are plugged in to the networked infrastructures of information and communications technologies that underpin the fast world. They are partitioned off from the slow world and its spaces of perceived danger, difference, and poverty through processes of segregation and congregation, assisted by the practices of designers and developers who deploy and market security as a key feature of the built environment. Urban designers have become adept at programming **defensible space** into the layout of new developments, while architects have adopted bunker- and fortress-style elements for their buildings.

Developers, meanwhile, have learned to include access control and surveillance technologies in the packaging of the built environment. Gated communities are the most striking example of the way in which the quartering of urban space is being splintered into hardened security zones. In the United States, many of the private master-planned communities are gated; in all, about 8 million people live within the walls of gated communities. Gated communities are by no means confined to North American cities, however; they have become symptomatic of social polarization in globalizing cities everywhere.

---

[8]United Nations Center for Human Settlements, *Cities in a Globalizing World: Global Report on Human Settlements 2001*. London: Earthscan, 2001, p. 33.

[9]P. Marcuse and R. van Kampen, "Introduction," in P. Marcuse and R. van Kampen (eds.), *Globalizing Cities: A New Spatial Order?* Oxford: Blackwell, 2000, pp. 3–5.

CONCLUSION

Patterns of land use and the functional organization of economic and social subareas in cities are partly a product of economic, political, and technological conditions at the time of the city's growth, partly a product of regional cultural values and partly a product of processes of globalization. Geographers can draw on several perspectives in looking at patterns of land use within cities, including an economic perspective that emphasizes competition for space and a sociocultural perspective that emphasizes ethnic congregation and segregation. Nevertheless, urban structure varies considerably because of the influence of history, culture, and the different roles that cities have played within the world-system.

The evolution of the unintended metropolis of the periphery has been very different from the evolution of metropolitan areas in the world's core regions. Similarly, the problems they have faced are very different. In the core regions, the consequences of an economic transformation to a postindustrial economy have dominated urban change. Traditional manufacturing and related activities have been moved out of central cities, leaving decaying neighborhoods and a residual population of elderly and marginalized people. New, postindustrial activities have begun to cluster in redeveloped CBDs and in edge cities around metropolitan fringes. In a few cases, metropolitan growth has become so complex and extensive that 100-mile cities have begun to emerge, with half a dozen or more major commercial and industrial centers forming the nuclei of a series of interdependent urban realms.

In other parts of the world, traditional patterns of land use and the functional organization of economic and social subareas have been quite different, reflecting different historical legacies and different environmental and cultural influences. A basic trend affecting the cities of the world's periphery is demographic—the phenomenal rates of natural increase and in-migration that have given rise to overurbanization. The example of Lagos provides some sobering insights into the human consequences of overurbanization. An ever-growing informal sector of the economy, in which people seek economic survival, is reflected in extensive areas of shanty housing. High rates of unemployment, underemployment, and poverty generate acute social problems, which are overwhelming for city governments that are understaffed and underfunded. If present trends continue, such problems are likely to characterize increasing numbers of the world's largest settlements. Meanwhile, globalization processes are recasting metropolitan structure and intensifying social and economic inequalities between the fast world and the slow world. In the next chapter, we consider this question as part of a broader discussion of future geographies.

MAIN POINTS REVISITED

- **The internal structure of cities is shaped heavily by competition for territory and location. In general, all categories of land users—commercial and industrial, as well as residential—compete for the most convenient and accessible locations within the city.**

  An important exception is that wealthier households tend to trade off the convenience of accessibility for the greater utility of being able to consume larger amounts of (relatively cheap) suburban space. Poorer households, unable to afford the recurrent costs of transportation, must trade off living space for accessibility to jobs, thus ending up in high-density areas, at expensive locations, near their low-wage jobs.

- **Social patterns in cities are heavily influenced by territoriality, which provides a means of establishing and preserving group membership and identity.**

  Territoriality involves processes of congregation and discrimination, often resulting in segregation, the spatial separation of specific subgroups within a wider population. Segregation varies a great deal in both intensity and form, depending on the relative degrees of congregation and discrimination.

- **The typical North American city is structured around a central business district (CBD); a transitional zone; suburbs; secondary business districts and commercial strips; industrial districts; and, in larger metropolitan areas, edge cities.**

  This internal organization of cities reflects the way that they function, both to bring certain people and activities together and to sort them out into neighborhoods and functional subareas.

- **Traditionally, North American cities have experienced high rates of in-migration, which has resulted in their becoming structured into a series of concentric zones of neighborhoods of different ethnicity, demographic composition, and social status through processes of invasion and succession.**

  Within each concentric zone a mosaic of distinctive neighborhoods tends to develop—ecological niches where particular mixes of people have come to dominate a particular territory or geographic setting.

- **Urban structure varies a good deal from one region of the world to another because of the influence of history, culture, and the different roles that cities have played within the world-system.**

  European cities have evolved under circumstances very different from those faced by American cities, and consequently European cities exhibit some distinctive characteristics in urban form. European urban ideals have influenced many colonial cities, while the new cities of the world's peripheral regions are distinctive because of their explosive growth.

- **The most acute problems of the postindustrial cities of the world's core regions are localized in the central city areas that have borne the brunt of restructuring from an industrial to a postindustrial economy.**

  In these areas there are several interrelated problems: fiscal problems, infrastructure problems, and localized spirals of neighborhood decay and cycles of poverty.

- **The problems of the cities of the periphery stem from the way in which their demographic growth has outstripped their economic growth.**

  The result is high rates of long-term unemployment and underemployment, low and unreliable wages of informal-sector jobs, chronic poverty, and slum housing. Their low rates of economic growth reflect their dependent position in the global economy.

- **Traditional patterns of land use and spatial organization in many parts of the world are being transformed by the local effects of splintering urbanism.**

Economic and cultural globalization, together with the uneven evolution of networked infrastructures of information and communications technologies, is forging new landscapes of innovation, economic development, and cultural transformation, while at the same time intensifying social and economic inequalities between the fast world and the slow world.

KEY TERMS

Beaux Arts (p. 434)
central business district (CBD) (p. 425)
central cities (p. 427)
congregation (p. 424)
cycle of poverty (p. 429)

defensible space (p. 451)
dualism (p. 438)
edge cities (p. 450)
fiscal squeeze (p. 428)
gentrification (p. 426)

invasion and succession (p. 426)
isotropic surface (p. 424)
minority groups (p. 424)
Modern movement (p. 434)
redlining (p. 430)

segregation (p. 425)
socio-spatial formation (p. 451)
underclass (p. 430)
underemployment (p. 438)
zone in transition (p. 425)

ADDITIONAL READING

Bell, D., and M. Jayne (eds.), *City of Quarters*. Burlington, VT: Ashgate, 2004.

Biswas, R. K. (ed.), *Metropolis Now!* Vienna: Springer-Verlag, 2000.

Brunn, S., and J. F. Williams (eds.), *Cities of the World*, 3rd ed. New York: Harper Collins, 2003.

Fyfe, N. R., and J. T. Kenny (eds.), *The Urban Geography Reader*. New York: Routledge, 2005.

Goodman, D., and C. Chant (eds.), *European Cities and Technology: Industrial to Post-Industrial City*. London: Routledge, 1999.

Graham, S., and S. Marvin, *Splintering Urbanism*. New York: Routledge, 2001.

Hall, P., *Cities in Civilization*. New York: Pantheon, 1998.

Hubbard, P., *City*. New York: Routledge, 2006.

Kazepov, Y. (ed.), *Cities of Europe: Changing Contexts, Local Arrangements, and the Challenge to Urban Cohesion*. Malden, MA: Blackwell, 2005.

Knox, P. L., *Metroburbia USA*. New Brunswick, NJ: Rutgers University Press, 2008.

Knox, P. L., and L. McCarthy, *Urbanization: An Introduction to Urban Geography*, 2nd ed. Englewood Cliffs, NJ: Prentice Hall, 2005.

Knox, P. L., and S. Pinch, *Urban Social Geography*, 5th ed. London: Longman Scientific, 2006.

Morris, A. E. J., *History of Urban Form Before the Industrial Revolutions*. London: Longman, 1994.

Pacione, M., *Urban Geography: A Global Perspective*. New York: Routledge, 2001.

Potter, R. B., and S. Lloyd-Evans, *The City in the Developing World*. Harlow, Essex: Addison Wesley Longman, 1998.

Robbins, E., and R. El-Khoury (eds.), *Shaping the City: Studies in History, Theory, and Urban Design*. New York: Spon Press, 2003.

Scott, A. J. (ed.), *Global City-Regions: Trends, Theory, Policy*. Oxford: Oxford University Press, 2001.

Simmonds, R., and G. Hack (eds.), *Global City Regions: Their Emerging Forms*. New York: Spon Press, 2000.

United Nations Human Settlements Programme, *The State of the World's Cities 2006/2007*. Sterling, VA: Earthscan, 2006.

U.S. Department of Housing and Urban Development, *The State of the Cities 1999*. Washington, DC: U.S. Department of Housing and Urban Development, 1999.

Wallach, B., *Understanding the Cultural Landscape*. New York: Guilford, 2005. See Chapters 19 ("American Cities") and 21 ("Cities Abroad").

EXERCISES

On the Internet

The Companion Website at www.mygeoscienceplace.com contains a number of tools designed to reinforce the geographic concepts presented in this chapter. Structured within a four-step learning path, these online resources include: chapter Key Concepts; a fully-searchable myeBook version of the chapter, which allows you to add notes and highlights; RSS news feeds from various media agencies; links to additional web resources; Human Geography Videos; a digital Glossary; Concept Review questions; Thinking Spatially mapping exercises; and Critical Thinking essay questions.

Unplugged

1. Collect a week's worth of local newspapers and review the coverage of urban problems. What kinds of problems are covered, and for what kinds of cities? Compile a list of such categories, and then carefully analyze the content of the week's coverage, calculating the number of column inches devoted to each category of problems.

2. On a tracing-paper overlay of a street map of your town or city, plot the distribution of houses and apartments for sale or rent in different price brackets. (You can obtain the information from the real estate pages of your city's local newspaper; in smaller cities you may have to gather data from several issues of the paper a week or more apart in order for a pattern to emerge—your local library will likely have back issues.) What can you say about the spatial distribution that is revealed?

3. Most cities consist of "ordinary" cityscapes that are strongly evocative because they are widely understood as being a particular kind of place. Write a brief essay (500 words, or two double-spaced, typed pages) describing an "ordinary" cityscape with which you are familiar. What are its principal features, and how might it be considered typical of a particular kind of place?

Future Geographies

12

Will the Internet bring about new patterns of human interaction? Will we be able to cope with the environmental stresses that increasing industrialization and rapid population growth will bring to many parts of the world? Will the United States retain its position as the world's most powerful and influential nation? Will more countries move up from peripheral status to join the semiperiphery and core of the future world-system? What kind of problems will the future bring for local, regional, and international development? What new technologies are likely to have the most impact in reshaping human geographies? Will globalization undermine regional cultures? Will technology and human determination be able to cope with the environmental stresses that industrialization in the periphery and rapid population increases will inevitably create? Will new regions emerge based on new types of connectivity such as trade, the Internet, or any number of political movements such as mobilizations against globalization or the human rights movement? These are just a few of the many questions that spring from the key themes in human geography that we have examined in Chapters 5 through 11. This chapter examines some scenarios for future geographies, drawing on the principles and concepts established in Chapters 1 through 4.

Main Points

- In some ways, the future is already here, embedded in the world's institutional structures and in the dynamics of its populations.

- New and emerging technologies that are likely to have the most impact in reshaping human geographies include advanced transportation technologies, biotechnology, materials technologies, and information technologies.

- The changes involved in shaping future geographies will inevitably bring with them some critical issues, conflicts, and threats, including important geographical issues that center on scale, boundaries, and territories; on cultural dissonance; and on sustainability.

Tokyo waterfront

MAPPING OUR FUTURES

The most effective way to approach the questions listed above is to try to get a sense of how different aspects of globalization are changing the world and how they might continue to do so. As we discussed in Chapter 2, the globalization of the capitalist world-system involves processes that have been occurring for at least 500 years. But since World War II, world integration and transformation have been remarkably accelerated and dramatic. Among the forces driving integration and transformation are strengthening of regional alliances such as the European Union and the Organization of Petroleum Exporting Countries, the increasing connectivity of the most remote regions of the world due to telecommunications and transportation linkages, the emergence of the new economy in the core countries, and the rise of global institutions like the World Trade Organization.

How will the forces of broadening global connectivity—and the popular reactions to them—change the fates and fortunes of world regions whose current coherence owes more to eighteenth- and nineteenth-century European colonialism than to forces of integration or disintegration in the twenty-first century? To answer this and related questions, we need to understand what the experts think about the processes behind globalization and how they might affect its future potential. But we also need to first understand the very risky issue of predicting the future, how predictions are made, and how useful predictive exercises can be.

There is no shortage of visionary scenarios. Broadly speaking, futurists' projections can be divided into two kinds: optimistic and pessimistic. Optimistic futurists stress the potential for technological innovations to discover and harness new resources, to provide faster and more effective means of transportation and communication, and to make possible new ways of living. This sort of futurism is often characterized by science-fiction cities of mile-high skyscrapers and spaceship-style living pods, by bioecological harmony, and by unprecedented social and cultural progress through the information highways of cyberspace. It projects a world that will be stabilized and homogenized by supranational or even "world" governments. The sort of geography implied by such scenarios is rarely spelled out. Space and place, we are led to believe, will be transcended by technological fixes.

To pessimistic futurists, however, this is just "globaloney." They stress the finite nature of Earth's resources, the fragility of its environment, and population growth rates that exceed the capacity of peripheral regions to sustain them. Such doomsday forecasting is characterized by scenarios that include irretrievable environmental degradation, increasing social and economic polarization, and the breakdown of law and order. The sort of geography associated with these scenarios is rarely explicit, but it usually involves the probability of a sharp polarization between the haves and have-nots at every geographical scale.

Fortunately, we don't have to choose between the two extreme scenarios of optimism and pessimism. Using what we have learned from the study of human geography, we can suggest a more grounded outline of future geographies. To do so, we must first glance back at the past. Then, looking at present trends and using what we know about processes of geographic change and principles of spatial organization, we can begin to map out the kinds of geographies that the future most probably holds.

Looking back at the way that the geography of the world-system has unfolded, we can see now that a fairly coherent period of economic and geopolitical development occurred between the outbreak of World War I (in 1914) and the collapse of the Soviet Union (in 1989). Some historians refer to this period as the "short twentieth century." It was a period when the modern world-system developed its triadic core of the United States, Western Europe, and Japan, when geopolitics was based on an East-West divide, and when geoeconomics was based on a North-South divide. This was a time when the geographies of specific places and regions within these larger frameworks were shaped by the needs and opportunities of technology systems that were based on the internal combustion engine, oil and plastics, electrical engineering, aerospace industries, and electronics. In this short century the modern world was established, along with its now familiar landscapes and spatial structures: from the industrial landscapes of the core to the unintended metropolises of the periphery; from the voting blocs of the West to the newly independent nation-states of the South.

Looking around now, much of the established familiarity of the modern world and its geographies seems to be disappearing. We have entered a period of transition, triggered by the end of the Cold War in 1989 and rendered more complex by the geopolitical and cultural repercussions of the terrorist attacks of September 11, 2001, and the global financial "meltdown" of 2008. The result is a series of unexpected developments and unsettling juxtapositions. The United States, for example, has given economic aid to Russia; Eastern European countries have joined NATO and the European Union; Germany has unified, but Czechoslovakia and Yugoslavia have disintegrated; South Africa has been transformed, through an unexpectedly peaceful revolution, to black majority rule. Meanwhile, Islamist terrorists shoot up tourist buses, bomb office buildings, and sabotage aircraft; former communist Russian ultranationalists have become comradely with Austrian and German neo-Nazis; Hindus, Sikhs, and Muslims are in open warfare in South Asia; the United States has invaded Iraq; and Sudanese military factions steal food from aid organizations in order to sell it to the refugees for whom it was originally intended. Within Western countries, the cultural shifts and neoliberal political climate that have emerged over the same transitional period have given cause for concern about the fundamental pillars of civil society (see Box 12.1, "Dark Age Ahead?")

These examples show that we cannot simply project our future geographies from the landscapes and spatial structures of the past. Rather, we must map them out from a combination of existing structures and budding trends. We have to anticipate, in other words, how the shreds of tradition and the strands of contemporary change will be rewoven into new landscapes and new spatial structures.

Although this is certainly a speculative and tricky undertaking, we can draw with a good deal of confidence on what we know about processes of geographic change and principles of spatial organization. The study of human geography has taught us to understand spatial change as a composite of local place-making processes (see Chapter 6) that are subject to certain principles of spatial organization and that operate within the dynamic framework of the world-system (Chapter 2). It has also taught us that many important dimensions exist to spatial organization and spatial change, from the demographic dimension (Chapter 3) through to the urban (Chapter 11).

As we look to the future, we can appreciate that some dimensions of human geography are more certain than others (**Table 12.1**). In some ways, the future is already here, embedded in the world's institutional structures and in the dynamics of its populations. We know, for example, a good deal about the demographic trends of the next quarter century, given present populations, birth and death rates, and so on. We also know a good deal about the distribution of environmental resources and constraints, about the characteristics of local and regional economies, and about the legal and political frameworks within which geographic change will probably take place.

On the other hand, we can only guess at some aspects of the future. Two of the most speculative realms are those of politics and technology. While we can foresee some of the possibilities (maybe a spread and intensification of ethno-nationalism; perhaps a new railway era based on high-speed trains), politics and technology are both likely to spring surprises at any time. The September 11, 2001, terror attack in the United States was a painful example of the sort of political surprises that can occur. In a matter of about 100 minutes, terrorists who crashed two commercial jetliners into the World Trade Center towers and one into the Pentagon were able to deliver a blow to the U.S. economy, draw the country into full military alert, and profoundly transform daily life in the United States. Other possible events—such as political instability in Saudi Arabia, war between North Korea

TABLE 12.1 The 2020 Global Landscape

Relative Certainties	Key Uncertainties
A global multipolar system is emerging with the rise of China, India, and others.	Whether an energy transition away from oil and gas—supported by improved energy storage, biofuels, and clean coal—is completed during the 2025 time frame.
The relative power of nonstate actors—businesses, tribes, religious organizations, and even criminal networks—also will increase.	How quickly climate change occurs and the locations where its impact is most pronounced.
The unprecedented shift in relative wealth and economic power roughly from West to East now under way will continue.	Whether mercantilism stages a comeback and global markets recede.
The United States will remain the single most powerful country but will be less dominant.	Whether advances toward democracy occur in China and Russia.
Continued economic growth—coupled with 1.2 billion more people by 2025—will put pressure on energy, food, and water resources.	Whether regional fears about a nuclear-armed Iran trigger an arms race and greater militarization.
The number of countries with youthful populations in the "arc of instability" will decrease, but the populations of several youth-bulge states are projected to remain on rapid growth trajectories.	Whether the greater Middle East becomes more stable, especially whether Iraq stabilizes, and whether the Arab-Israeli conflict is resolved peacefully.
The potential for conflict will increase owing to rapid changes in parts of the greater Middle East and the spread of lethal capabilities.	Whether Europe and Japan overcome economic and social challenges caused or compounded by demography.
Terrorism is unlikely to disappear by 2025, but its appeal could lessen if economic growth continues in the Middle East and youth unemployment is reduced. For those terrorists that are active the diffusion of technologies will put dangerous capabilities within their reach.	Whether global powers work with multilateral institutions to adapt their structure and performance to the transformed geopolitical landscape.

Source: National Intelligence Council, *Global Trends 2025: A Transformed World.* Washington, DC: U.S. Government Printing Office, 2008, pp. iv–v.

12.1 GEOGRAPHY MATTERS

Dark Age Ahead?

Jane Jacobs, recognized in 2004 by *Business Week* magazine as one of the most influential public intellectuals in North America in the last 75 years, argues that the United States is slipping toward the beginnings of a new "Dark Age" as a result of the deterioration of five pillars of modern society: community and family, higher education, the application of science and technology, the integrity of the professions, and the role of government in relation to society's needs and potential.

Jacobs has long been a passionate advocate of cities and urban life, and her views have been both influential and controversial. But Jacobs's fears of an incipient Dark Age go well beyond her past concerns for urban development, centering as they do on several important aspects of higher education,[1] the application of knowledge, the role of the professions, and the role of government. "A culture is unsalvageable," writes Jacobs, "if stabilizing forces themselves become ruined." The roots of her concerns are based on evidence of:

- corporate immorality in the marketplace instead of entrepreneurship bonded to social justice;

- universities that serve employers and act as credential factories, stripping the music, art, ethics, idealism, and notion of the public good out of education;

- scientific research increasingly and immorally being bought by corporations or suppressed and ignored by governments; and

- a neoliberal political economy that is intent on abandoning the stewardship of urban and regional development.

She does not pull her punches. Here is what she has to say about universities: They have become credential factories, stripping the art, poetry, ethics, idealism and notion of the public good out of education. They are complicit in allowing scientific research to be controlled by corporations and directed—and sometimes suppressed—by governments. Rather than serving as one of the cultural pillars of society, universities now serve employers and act as "colleges of heraldry," awarding graduates a "coat of arms" (i.e., university diploma) to distinguish them from those without marketable credentials.

She is equally direct about the professions. This time it is not so much planners and design professionals that are the focus of her concern, but accountants, bankers, lawyers and other financial professions whose ethics and practices have serially been called into question in relation to the vast corporate scandals of the past two decades. Nevertheless, the design and public policy professions are implicated in her critique of urban trends. Cities—vital economic engines and crucibles of cultural change and innovation—are being starved of the money they need by national, state, and local governments. Neoliberal policies are responsible for sagging public transit and public education systems, increasing pollution, increasing social polarization, the erosion of community, and the burgeoning sullenness of citizens. Families, she argues, are rigged to fail by public policies that, unintentionally, force both parents to work to meet the financial needs of themselves and their children. Communities are rigged to fail by public policies that foster sprawling, placeless, and automobile-dependent suburbs—a reprise of her earlier critique in *The Death and Life of Great American Cities.*

It behooves us to reflect critically on Jacobs's extended essay. She confides that "I have written this cautionary book in hopeful expectations that time remains for corrective actions." If we accept that corrective actions are in fact warranted, then a foundation of geographic knowledge will be fundamental to the incisive analysis that will be necessary in order to contribute effectively to countervailing forces.

[1]J. Jacobs, *Dark Age Ahead: Caution.* New York: Random House, 2004.

and its neighbors, or unanticipated breakthroughs in biotechnology—would cause geographies to be rewritten suddenly and dramatically.

RESOURCES, TECHNOLOGY, AND SPATIAL CHANGE

As we saw in Chapter 2, technological breakthroughs and the availability of resources have had a profound influence on past patterns of development, and the same factors will certainly be a strong influence on future geographies. The expansion of the world economy and the globalization of industry will undoubtedly boost the overall demand for raw materials of various kinds, and this will spur the development of some previously underexploited but resource-rich regions in Africa, Eurasia, and East Asia.

Raw materials, however, will be only a fraction of future resource needs. The main issue, by far, will be energy resources. World energy consumption has been increasing steadily over the recent past, and as the periphery is industrialized and its population increases, the global demand for energy will expand rapidly. Basic industrial development tends to be highly energy-intensive. The International

Energy Agency, assuming (fairly optimistically) that energy in peripheral countries will be generated in the future as efficiently as it is today in core countries, estimates that developing-country energy consumption will more than double by 2015, lifting total world energy demand by almost 50 percent. By 2020, peripheral and semiperipheral countries will account for more than half of world energy consumption. Much of this will be driven by industrialization geared to meet the growing worldwide market for consumer goods, such as private automobiles, air conditioners, refrigerators, televisions, and household appliances.

Without higher rates of investment in exploration and extraction than at present, production will be slow to meet the escalating demand. As we saw in Chapter 4, many experts believe that current levels of production in fact represent "peak oil" and that by 2020 global oil production may be only 90 percent of its current peak. The result might well be a significant increase in energy prices. This would have important geographical ramifications: Companies would be forced to seriously reconsider their operations and force core-region households into a reevaluation of their residential preferences and commuting behavior, while peripheral-region households would be forced further into poverty. If the oil-price crisis of 1973 is anything to go by (after crude oil prices had been quadrupled by the OPEC cartel), the outcome could be a major revision of patterns of industrial location and a substantial reorganization of metropolitan form. Significantly higher energy costs may change the optimal location for many manufacturers, leading to deindustrialization in some regions and to new spirals of cumulative causation in others. Higher fuel costs will encourage some people to live nearer to their place of work; others will be able to take advantage of telecommuting to reduce personal transportation costs. It is also relevant to note that almost all of the increase in oil production over the next 15 or 20 years is likely to come from outside the core economies. This means that the world economy will become increasingly dependent on OPEC governments, which control over 70 percent of all proven oil reserves, most of them in the Middle East.

In countries that can afford the costs of research and development, new materials will reduce the growth of demand both for energy and for traditional raw materials such as aluminum, copper, and tin. Japan, for instance, may be able to reduce motor-vehicle fuel consumption by 15 percent (and thereby reduce its total fuel oil consumption by 3 percent) by using ceramics for major parts of engines. It may also be possible to substitute ceramics for expensive rare metals in creating heat-resistant materials. Improved engineering and product design will also reduce the need for the input of some resources. American cars, for example, were 22 percent lighter in 2004 than they were in 1974. In addition, of course, the future may well bring technological breakthroughs that dramatically improve energy efficiency or make renewable energy sources (such as wind, tidal, and solar power) commercially more viable. As with earlier breakthroughs that produced steam energy, electricity,

gasoline engines, and nuclear power, such advances would provide the catalyst for a major reorganization of the world's economic geographies.

Just what new technologies are likely to have the most impact in reshaping human geographies? Given what we know about past processes of geographic change and principles of spatial organization, it is clear that changes in transportation technology are of fundamental importance. Consider, for example, the impact of oceangoing steamers and railroads on the changing geographies of the nineteenth century, and the impact of automobiles and trucks on the changing geographies of the twentieth century (Chapter 2). Among the most important of the next generation of transportation technologies that will influence future geographies are high-speed rail systems, smart roads, and smart cars.

In addition, there are several emerging industrial technologies whose economic impact is likely to be so great that they will influence patterns of international, regional, and local development. The United States National Intelligence Council has identified six potentially disruptive technologies that could emerge in the coming years. A **disruptive technology** is defined as a technology with the potential to cause a—even if temporary—degradation or enhancement of one of the key elements of U.S. national power (geopolitical, military, economic, or social cohesion). The six disruptive technologies most likely to enhance or degrade U.S. national power to 2025 are:[1]

- energy storage technologies
- biofuels and bio-based chemicals
- clean coal technologies
- "biogerontechnology"
- service robotics
- an Internet of Things

Energy storage technologies have the potential to disrupt the way energy is stored and distributed for use in transportation and portable devices. These technologies include battery materials, ultracapacitors, and hydrogen storage materials, particularly for fuel cells. The biggest level of disruption that could occur, both in economic terms and in terms of global socioeconomic structure, would be the potential for one of these technologies (or a combination) to lead to a paradigm shift away from fossil fuels.

Biofuels and bio-based chemicals production technologies have the only potential near-term capability to provide alternatives to conventional gasoline and diesel-fuel and petrochemical energy sources. Crop-based biofuels are already in wide use, work in today's vehicles, and require no major investments in infrastructure for their use. Biofuels also help to address global-warming concerns by reducing net greenhouse gas emissions from vehicles. A large-scale

[1]National Intelligence Council, *Disruptive Civil Technologies. Conference Report*, April 2008, http://www.dni.gov/nic/confreports_disruptive_tech.html, accessed August 4, 2008.

move to energy-efficient biofuels could increase U.S. energy security and ease international competition for world oil supplies and reserves. Conversely, as we have seen, the growth of production in crop-based biofuels has unpredictable effects on patterns of food crop production and can cause sharp increases in food prices.

Clean coal technologies and an array of related technologies offer the potential to improve electrical generation efficiency, lower emissions of harmful pollutants, and provide fuels and chemicals. The development of clean coal technologies is gaining momentum in coal-rich nations, but it is not certain to succeed. Failure to successfully develop clean coal technology may result in environmental damage with major adverse economic impacts; conversely, a successful and rapid deployment of clean coal technology could pose a major challenge to other (predominantly oil) energy markets; and the resulting geopolitical instability could also be a major challenge to U.S. interests.

"Biogerontechnology" offers the means to accomplish control over and improve the human condition and promises improvements in people's life span. The advancement of the science and technology underlying the biological aging process has the potential to not only extend the average natural life span but also to simultaneously postpone many if not all of the costly and disabling conditions that humans experience in later life. The disruptive potential comes in the form of new treatment modalities and shifts in the cost and resulting allocation and use of health care resources. Countries will also be challenged as a result of changing demographic structures, new activity patterns of aging yet healthy citizens, and the resulting requirement to formulate new national economic and social policies.

Robotics and enabling technologies have the potential to replace humans in a variety of applications, with far-reaching implications. For example, the use of unmanned systems for terrorist activities could happen because of the increasing availability of commercial civilian robot platforms. On the other hand, unmanned military systems with a much greater level of autonomy could enhance the performance of soldiers. The development and implementation of robots in elder-care applications mean that robots could be working alongside humans in looking after and rehabilitating people; but a change in domestic and social responsibilities and a change in domestic employment requirements could adversely affect lower-income, service-oriented workers.

By 2025, Internet nodes may reside in everyday things—food packages, furniture, paper documents, and so on. Today's developments point to future opportunities and risks that will arise when people can remotely control, locate, and monitor even the most mundane devices and articles. Popular demand combined with technology advances could drive widespread diffusion of an Internet of Things that could, like the present Internet, contribute invaluably to economic development and military capability. Streamlining—or revolutionizing—supply chains and logistics could slash costs, increase

efficiencies, and reduce dependence on human labor. The ability to fuse sensor data from many distributed objects could deter crime and terrorism. Ubiquitous positioning technology could locate missing and stolen goods. However, to the extent that everyday objects become information-security risks, an Internet of Things could distribute those risks far more widely than the Internet has to date.

REGIONAL PROSPECTS

For many years now, organizations such as the United Nations and the World Bank have prepared forecasts of the world economy. These forecasts are based on economic models that take data on macroeconomic variables (for example, trends in countries' gross domestic product, imports and exports, economic structure, investment and saving performance, and demographic dynamism) and use known relationships between and among these variables to predict future outcomes. The problem is that economic projection is an inexact science. Economic models are not able to take into account the changes brought about by major technological innovations, significant geopolitical shifts, governments' willingness and ability to develop strong economic policies, or the rather mysterious longer-term ups and downs that characterize the world economy. What is clear is that in *overall* terms the global economy is vastly richer, more productive, and more dynamic than it was just 15 or 20 years ago, and there is every possibility that, in the longer term, the world economy will continue to expand, in spite of the instability and recession triggered by the 2008 global financial crisis.

Uneven Development

Any expansion, of course, will be uneven. The world in 2020 will have changed radically as the world recovers from the current recession. Emerging powers—China, India, and perhaps others, such as Brazil and Indonesia—have the potential to render obsolete the old categories of East and West, North and South, aligned and nonaligned, developed and developing. Traditional geographic groupings will increasingly lose salience in international relations. A state-bound world and a world of megacities, linked by flows of telecommunications, trade, and finance, will coexist. Competition for allegiances will be more open, less fixed than in the past.

In parallel with the world-system categorization of core, semiperipheral, and peripheral regions, we can think of large sectors of the world's population as being in one of three categories: the elite, the embattled, and the marginalized. The elite are participants in—and beneficiaries of—the fast world of new transport and communications technologies, globalized production networks, and global consumer culture. The embattled are also participants, but in dependent roles, with fewer benefits and limited opportunities: assembly-line workers, for example, in offshore commodity chains. The marginalized have to survive within the slow world, largely disconnected from formal economies and the dynamics of globalization.

The most significant point to keep in mind about the stratification of the world economy through contemporary globalization is that the elite, the marginalized, and the embattled are likely to be less concentrated than they once were in particular regions. It is no longer accurate to see Nigeria, for instance, as a wholly embattled region or its population as exclusively embattled. Instead, Nigeria, like the United States and most other countries, contains a range of stratified regions and groups within its national boundaries such that elite, embattled, and marginalized social groups are all part of the larger whole. Moreover, the elite regions and social groups have more in common with elite regions and groups in other parts of the world than they do with the embattled and marginalized groups within their own national boundaries (see **Figure 12.1**).

Nevertheless, an even greater gap between the haves and have-nots of the world seems likely to structure future geographies. The gap between the world's core areas and

Figure 12.1 Global social hierarchy These photographs underscore the new social hierarchy that has emerged around globalization. Pictured here are (top) members of the marginalized stratum in Haiti; (middle) members of the middle stratum in Mexico; and (bottom) members of the elite stratum in the United Kingdom.

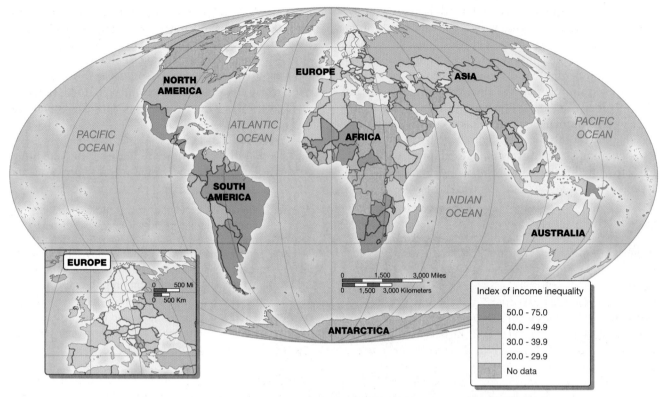

Figure 12.2   Index of income inequality This map shows the degree of inequality of income distributions within countries using the Gini coefficient, a commonly used indicator that compares actual income distributions to a hypothetical distribution of perfect equality. The higher the index score, the greater the degree of inequality within a country. (*Source:* United Nations Development Programme, *Cities in a Globalizing World.* Sterling, VA: Earthscan Publications, 2001, p. 18.)

the periphery has already begun to widen significantly. Even within the United States, the degree of income inequality ranks with that in China, Bolivia, Malaysia, Senegal, and Russia (**Figure 12.2**). There is also evidence of a growing disparity. U.S. Census Bureau data, for example, show that the gap between the rich and poor in the United States is now the widest it has been since World War II. This is partly a result of changing labor markets but also a result of government policy. As a result of the tax cuts enacted as part of the neoliberal policies of the Bush administration, for example, U.S. households in the bottom fifth of the income spectrum received tax cuts between 2001 and 2006 averaging $20, raising their after-tax incomes by an average of just 0.3 percent. Meanwhile, households in the middle fifth of the income spectrum received tax cuts averaging $740, raising their after-tax incomes an average of 2.5 percent. The top 1 percent of households received tax cuts averaging $44,200 that increased their after-tax income by an average of 5.4 percent. At the global level, the United Nations has calculated that the ratio of GDP per capita (measured at constant prices and exchange rates) between the developed and developing areas of the world increased from 10:1 in 1970 to 12:1 in 1985 and 13:1 in 2005. Little hope exists that any future boom in the overall world economy will reverse this trend. In spite of the globalization of the world-system (and in many ways *because*

of it), much of the world has been all but written off by the bankers and corporate executives of the core.

A New World Order?

The same factors that will consolidate the advantages of the core of the world-system as a whole—the end of the Cold War, the availability of advanced telecommunications, the transnational reorganization of industry and finance, the liberalization of trade, and the emergence of a global culture—will also open the way for a new geopolitical and geoeconomic order. This is likely to involve some new relationships between places, regions, and countries. As we have suggested, the old order of the "short" twentieth century (1914–1989), dominated both economically and politically by the United States, is rapidly disappearing. In our present transitional phase, the new world order is up for grabs; we are coming to the end of a geopolitical leadership cycle. This does not necessarily mean that the United States will be unable to renew or extend its position as the world's dominant power. As we saw in Chapter 2, Britain had two consecutive stints as the dominant world power—the hegemony that was able to impose its political view on the world and set the terms for a wide variety of economic and cultural practices.

The United States

The United States is the reigning hegemon and its economy is the largest in the world, with a broad resource base; a large, well-trained, and very sophisticated workforce; a domestic market that has greater purchasing power than any other single country; and a high level of technological sophistication. The United States also has the most powerful and technologically sophisticated military apparatus; and it has the dominant voice and last word in international economic and political affairs. It is at least as well placed as its rivals to exploit the new technologies and new industries of a globalizing economy. It also has a distinctive message as a global leader: free markets, personal liberty, private property, electoral democracy, and mass consumption.

For the moment, however, the United States is a declining hegemon, at least in relative terms. Its economic dominance is no longer unquestioned in the way it was in the 1950s, 1960s, and 1970s. A relatively sluggish rate of growth has brought the prospect of the U.S. economy being caught and overtaken by China. On some measures of economic development, the European Union has already overtaken the United States. More important, the globalization of the economy has severely constrained the ability of the United States to translate its economic might into the firm control of international financial markets that it used to enjoy.

The end of the Cold War, while a victory for the United States, robbed it of its image as Defender of the Free World and weakened the legitimacy of its role as global policeman. For a decade or so, the absence of a Cold War enemy and the globalization of economic affairs made it much more difficult for the United States to identify and define the national interest. The hesitancy in U.S. policies toward Kosovo, East Timor, Somalia, Sudan, Bosnia, and Haiti in the 1990s was symptomatic of such problems. The terrorist al-Qaeda attacks of 2001 gave a new focus to U.S. geopolitical strategy. But the invasion of Iraq in the absence of evidence of weapons of mass destruction, together with U.S. refusal to participate in high-profile international agreements such as the Kyoto Protocol and its lack of cooperation with the International Criminal Court and the United Nations Organization, has resulted in a significant weakening of the political, cultural, and moral leadership of the United States in global affairs, even from the perspective of elite groups (see Box 12.2, "How the World Sees the United States").

Growing numbers of ordinary people around the world, especially in the Middle East and the broader Muslim world, believe the United States is bent on regional domination—or direct political and economic domination of other states and their resources. In the future, growing distrust could prompt governments to take a more hostile approach to the United States, including resistance to supporting U.S. interests in multinational forums and the development of military capabilities as a hedge against the United States. Meanwhile, America's self-perception as a global leader and champion of benign universal values is contradicted by a deep well of national-greatness conservatism that is overwhelmingly self-referential. As geographer Neil Smith has argued, this nationalism has undercut the ability of the United States to impose its vision of globalization in the past and may well do so again in the future. The effects of the shock of the 2008 global credit crunch that originated in the United States (see Box 1.2) are difficult to predict in any detail, but in general it is likely to have undermined the country's ability to sustain its economic dominance and leadership in the world economy.

In summary, while the United States must be considered the strongest contender, it is by no means a foregone conclusion that it will, in fact, be so in the next cycle of economic and political leadership. The United States has choices not dissimilar to those faced by the British at the end of the nineteenth century. It can either oppose its rivals or accommodate them. It can oppose by pressing for a seamless global system that remains under its own hegemony. Or it can try to accommodate by coaxing the others into a global sharing of power, with some mix of regional spheres of interest and collective world responsibilities.

The European Union

It is not too much of a stretch to imagine a successful European Union as the main contender for world leadership. In geopolitical terms, the collapse of the Soviet Union advanced the prospects of the European Union significantly. The European Union, which began in 1952 as a trading bloc (the European Economic Community, or EEC), is now an economic union (with integrated economic policies among member states) and has moved a long way toward its goal of becoming a supranational political union (with a single set of institutions and policies). With the addition of Cyprus, the Czech Republic, Estonia, Hungary, Latvia, Lithuania, Malta, Slovakia, and Slovenia in 2004, and Bulgaria and Romania in 2007, the European Union now has a total population of 491 million (the U.S. population is 304 million) and an overall economy in 2007 of $16.8 trillion (the GDP of the United States in 2007 was $13.8 trillion). If and when it achieves full political union, its size will enable it to outvote the United States in the International Monetary Fund and the World Bank.

The EU has evolved dramatically since its start with six member nations in 1952. Now there is a common EU passport and a single currency, the euro, is used by many of its members. The EU regulates trade as well as coordinating energy, communications, and transportation. It has a president, a parliament, foreign policy powers, and a court whose decisions are binding on member countries and individuals. The EU is the world's largest internal market and largest exporting power. Of the world's 20 largest commercial banks, 10 are European (the rest: 6 American, 3 Chinese, and 1 Japanese). The EU also has one of the world's most vibrant and sophisticated industrial core regions that stretches from Southeast England to Northern Italy; its economic policies and

Figure 12.3 European growth axes Most of Europe's major cities and advanced manufacturing regions lie along a crescent-shaped axis that runs from Southeast England through Southwest Germany to Northern Italy and along emerging axes that extend to Southern France, Spain, eastern Austria, and Southern Italy.

infrastructure investments have fostered emerging axes of economic growth that stretch east-west and north-south (**Figure 12.3**); and European industries are world leaders in chemicals, insurance, engineering, construction, and aerospace industries. The EU also offers a distinctive message. In contrast to the American Dream,

> The European Dream emphasizes community relationships over individual autonomy, cultural diversity over assimilation, quality of life over the accumulation of wealth, sustainable development over unlimited material growth, deep play over unrelenting toil, universal human rights and the rights of nature over property rights, and global cooperation over the unilateral exercise of power.[2]

Successful enlargement, combined with already-successful monetary union and economic integration, would see Europe poised either to challenge U.S. hegemony or to become a senior partner among superpowers. EU citizens, however, have expressed reservations about establishing a constitution to underpin a supranational political union. There are also concerns that the extension of the EU to 30 countries would risk destroying its own internal balance and cohesion. Europe's future international role also depends greatly on whether it undertakes major structural economic and social reforms to deal with the problem of its aging workforce. This will demand more immigration and better integration of workers, most of whom are likely to be coming from North Africa and the Middle East. Even if more migrant workers are not allowed in, Western Europe

will have to integrate a growing Muslim population. Barring increased legal entry may only lead to more illegal migrants who will be harder to integrate, posing a long-term problem. It is possible to imagine European nations successfully adapting their workforces and social welfare systems to these new realities, but it is harder to see some countries—Germany, for example—successfully assimilating millions of new immigrant workers in a short period of time.

China and India According to the U.S. National Intelligence Council (NIC), the greatest benefits of globalization will accrue to countries and groups that can access and adopt new technologies. The council's report on *Mapping the Global Future* concludes that China and India are well positioned to become technology leaders and that the expected next revolution in high technology—involving the convergence of nano-, bio-, information, and materials technology—could further bolster China and India's prospects. Both countries are investing in basic research in these fields and are well placed to be leaders in a number of key fields. As the NIC report puts it:

> The likely emergence of China and India, as well as others, as new major global players—similar to the advent of a united Germany in the 19th century and a powerful United States in the early 20th century—will transform the geopolitical landscape, with impacts potentially as dramatic as those in the previous two centuries. In the same way that commentators refer to the 1900s as the "American Century," the

[2]J. Rifkin, *The European Dream*. Los Angeles: J. P. Tarcher, 2004, p. 3.

21st century may be seen as the time when Asia, led by China and India, comes into its own. A combination of sustained high economic growth, expanding military capabilities, and large populations will be at the root of the expected rapid rise in economic and political power for both countries . . . . Barring an abrupt reversal of the process of globalization or any major upheavals in these countries, the rise of these new powers is a virtual certainty.[3]

Most forecasts indicate that by 2020 China's GNI will exceed that of individual Western economic powers except the United States. India's GNI will have overtaken or be on the threshold of overtaking the larger European economies. Because of the sheer size of China's and India's populations—projected by the U.S. Census Bureau to be 1.5 billion and almost 1.4 billion, respectively, by 2025—their standard of living need not approach Western levels for these countries to become important economic powers. How China and India exercise their growing power and whether they relate cooperatively or competitively to other powers in the international system, however, are key uncertainties.

So, although China is not currently part of the core of the world-system, many observers predict a "Pacific Destiny" for the twenty-first century. In this scenario, China will be the hub of a world economy whose center of gravity is around the rim of the Pacific rather than the North Atlantic. China certainly has the potential to be a contender. It has a vast territory with a comprehensive resource base and a long history of political, cultural, and economic integration. It has the largest population of any country in the world (1.6 billion in 2008), and an econ-

[3]National Intelligence Council, *Mapping the Global Future: Report of the National Intelligence Council's 2020 Project.* Washington, DC: U.S. Government Printing Office, 2004, p. 11.

omy that has been growing very rapidly. Beginning in 1978, under the leadership of Deng Xiaoping, China completely reorganized and revitalized its economy. Agriculture was decollectivized, with communist collective farms modified to allow a degree of private profit taking. State-owned industries were closed or privatized, and centralized state planning was dismantled in order to foster private entrepreneurship. Since 1992, China has extended its "open-door" policy (that is, allowing trade with the rest of the world) beyond its special economic zones and allowed foreign investment aimed at Chinese domestic markets. In the 1980s and early 1990s, when the world economy was sluggish, China's manufacturing sector grew by almost 15 percent each year, and has continued into the 2000s with double-digit growth. China's increased participation in the world economy has created an entirely new situation, causing a deflationary trend in world prices for manufactures (**Figure 12.4**). Not only does the Chinese economy's size make it a major producer but its huge labor force ensures that its wage levels will not approach Western levels for a long time, thus guaranteeing a key competitive advantage. Overall, China's economy is already the fourth largest in the world after the European Union, the United States, and Japan.

Nevertheless, China's economy will remain largely agrarian for some time yet. The task of feeding, clothing, and housing its enormous population will also constrain its ability to modernize its economy to the point where it can dominate the world economy. Meanwhile, a great deal of social and political reform remains to be accomplished before free enterprise can really flourish. Finally, before China can emerge as a hegemonic power it must resolve a major feature of its contemporary human geography: the dramatically uneven development that has been a consequence of its economic reforms. The positive spiral of cumulative causation has affected only the larger cities and coastal regions, while

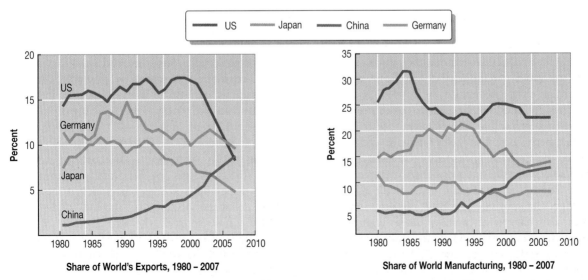

Figure 12.4 China's changing role in the world economy China's emergence as a significant player in the world economy is reflected in trends of manufacturing output and exports.

How the World Sees the United States

In 2004, the U.S. National Intelligence Council consulted with experts in six regional conferences, asking participants about their views of the role of the United States as a driver in shaping developments in their regions and globally.

It is impossible to predict how the emerging policies of the Obama administration and the recent economic crisis will affect these different regional and global views. For instance, China's economic growth rate is expected to continue to decline from a high point of 12.2 percent in the first half of 2008. This slowed growth is already leading to widespread disinvestment in housing, jobs, and infrastructure. How these changed conditions will be affected by the U.S. economic recovery efforts will not be known for a while. The Middle East provides another possible alternative to the one described below as Arab countries' populations were mostly jubilant over the election of Barack Obama and thus may be willing to extend more support with respect to U.S. efforts toward resolving regional conflicts, particularly in Iraq and Israel, than they might have otherwise. The important point to remember is that this report largely reflects the world that the new U.S. president has inherited and must seek to understand and change.

Asia

Participants felt that U.S. preoccupation with the war on terrorism is largely irrelevant to the security concerns of most Asians. The key question that the United States needs to ask itself is whether it can offer Asian states an appealing vision of regional security and order that will rival and perhaps exceed that offered by China. U.S. disengagement from what matters to its Asian allies would increase the likelihood that they would climb on Beijing's bandwagon and allow China to create its own regional security order that excludes the United States. Participants felt that the rise of China need not be incompatible with a U.S.-led international order. The critical question is whether or not the order is flexible enough to adjust to a changing distribution of power on a global level. An inflexible order would increase the likelihood of political conflict between emerging powers and the United States. If the order is flexible, it may be possible to forge an accommodation with rising powers and strengthen the order in the process.

Sub-Saharan Africa

Sub-Saharan African leaders worry that the United States and other advantaged countries will "pull up the drawbridge" and abandon the region. Participants opined that the United States and other Western countries may not continue to accept Africa's most successful "export," its people. The new African diaspora is composed overwhelmingly of economic migrants rather than political migrants as in previous eras. Some participants felt that Africans worry that Western countries will see some African countries as "hopeless" over the next 15 years because of prevailing economic conditions, ecological problems, and political circumstances. Participants feared that the United States will focus only on those African countries that are successful.

Latin America

Conference participants acknowledged that the United States is the key economic, political, and military player in the hemisphere. At the same time, Washington was viewed as traditionally not paying sustained

the vast interior regions of the country have become increasingly impoverished. This, ironically, is the very geographical pattern of spatial polarization that motivated Chinese communists to revolt in the 1940s. Today, it is a source of potential unrest and of political instability among the central government, the provincial governments of the interior regions, and the provincial governments of the coastal regions.

In the short term, China mostly needs to make room inside itself: to develop its internal market and moderate its growth. In some respects, China has been following such a strategy with a big program of infrastructure investment. Recently, it has also been trying to boost domestic demand by easing credit. But it remains to be seen whether the Chinese government can continue to control the situation. Managing things is not easy when capitalist energy is turning life upside down, creating vast new wealth amid a poor society based on different principles.

Alternatively, we may not see the same kind of hegemonic power in the new world order of the twenty-first century—there may not be a new hegemony at all.

Instead, the globalization of economics and culture may result in a polycentric network of nations, regions, and world cities bound together by flows of goods and capital. Order may come not from military strength rooted in national economic muscle, but from a mutual dependence on *trans*national production and marketing, with stability and regulation provided by powerful international institutions (such as the World Bank, the IMF, the World Trade Organization, the European Union, NATO, and the United Nations).

At the Margins Whatever the outcome of any new world order, the prospects for peripheral countries and marginalized populations are very bad indeed. Notwithstanding the debt-relief packages of the IMF and proclamations of the G8 group of the world's most affluent countries, peripheral countries and marginalized populations face formidable obstacles to significant progress. The worst-off regions face unprecedented levels of demographic, environmental, economic, and societal stress, so that the next 15–20 years are going to be played out from attention to the region and as reacting primarily to crises instead of responding to systemic problems. Participants saw a fundamentalist trend in Washington that would lead to isolation and unilateralism and undercut cooperation. Most shared the view that the U.S. "war on terrorism" had little to do with Latin America's security concerns. Latin American migrants are a stabilizing force in relations with the United States. An important part of the U.S. labor pool, migrants also remit home needed dollars along with new views on democratic governance and individual initiative that will have a positive impact on the region.

U.S. policies also can have a positive impact. Some participants said the region would benefit from U.S. application of regional mechanisms to resolve problems rather than punitive measures against regimes not to its liking, such as that of Fidel Castro.

Middle East

Participants felt that the role of U.S. foreign policy in the region will continue to be crucial. The perceived propping up of corrupt regimes by the United States in exchange for secure oil sources has helped to promote continued stagnation. Disengagement is highly unlikely but would in itself have an incalculable effect. Regarding the prospects for democracy in the region, participants felt that the West placed too much emphasis on the holding of elections, which, while important, is only one element of the democratization process. There was general agreement that if the United States and Europe can engage with and encourage reformers rather than confront and hector, genuine democracy would be achieved sooner.

Some Middle East experts argued that Washington has reinforced zero-sum politics in the region by focusing on top Arab rulers and not cultivating ties with emerging leaders in and outside the government. Although the Middle East has a lot to gain economically from globalization, it was agreed that Arabs/Muslims are nervous that certain aspects of globalization, especially the pervasive influence of Western, particularly American, values and morality are a threat to traditional cultural and religious values.

Europe and Eurasia

Participants engaged in a lively debate over whether a rift between the United States and Europe is likely to occur over the next 15 years, with some contending that a collapse of the U.S.-EU partnership would occur as part of the collapse of the international system. Several participants contended that if the United States shifts its focus to Asia, the U.S.-EU relationship could be strained to breaking point. They were divided over whether China's rise would draw the United States and Europe closer or not. They also differed over the importance of common economic, environmental, and energy problems to the alliance. Participants agreed that the United States has only limited influence on the domestic policies of the Central Asian states, although U.S. success or failure in Iraq would have spillover effects in Central Asia. Countries in western Eurasia, they believed, will continue to seek a balance between Russia and the West. In their view, Ukraine almost certainly will continue to seek admission to NATO and the European Union, while Moldova probably will maintain its orientation toward Russia.

Source: National Intelligence Council, *Mapping the Global Future.* Washington, DC: U.S. Government Printing Office, 2004, pp. 114–15.

---

a starting point of scarce basic resources, serious environmental degradation, overpopulation, disease, unprovoked crime, refugee migrations, and criminal anarchy.

While some African countries are now beginning to benefit from foreign investment (particularly from China, but also from Europe and the United States) and are developing an educated and entrepreneurial middle class based on growing numbers of service-sector jobs, continuing dependence on commodity exports means that many peripheral countries, especially in Africa, have and will probably continue to have reduced purchasing power in the global marketplace because of the relative decline in the value of their exports compared to imported goods from core countries. Continuing to disable the periphery's full participation in the global economy are the combined effects of external debt crisis, dwindling amounts of foreign aid, insufficient resources to purchase technology or develop indigenous technological innovations, and the high costs of marketing and transporting commodities.

Some governments, including those of Liberia, Sudan (**Figure 12.5**), and Sierra Leone, have lost control of large

Figure 12.5 Darfur In the Sudan, the brutal oppression, ethnic cleansing, and genocide sponsored by the central government allowed Arab rebels known as the Janjaweed to slaughter thousands of people in the Darfur region, causing the mass displacement of an estimated 1 million refugees. The people of the Darfur region resorted to popular political and military resistance for purposes of survival, forming a political movement known as the Sudan Liberation Movement. Five members of its military wing, the Sudan Liberation Army (SLA), stand in front of their transport in 2004.

parts of their territories: groups of unemployed youths plunder travelers; tribal groups war with one another; refugees trudge from war zones to camps and back again; environmental degradation proceeds unchecked (**Figure 12.6**). Meanwhile, health remains a serious issue. Malaria and tuberculosis are endemic to much of sub-Saharan Africa, and AIDS is truly epidemic. In Uganda, where annual spending on health is less than $5 per person (compared with international debt repayments of almost $20 per person), one in five children die before their fifth birthday, and HIV/AIDS is epidemic among young adults. More than 22 million people in sub-Saharan Africa in 2007 were infected with HIV, or 1 in every 12 people. Altogether, Africa, with just 10 percent of the world's total population, is home to more than two-thirds of the world's 33 million HIV-infected people. Given all this, it is not surprising that much of Africa is unattractive to the globalizing world economy, and it should not be surprising if it remains so.

CRITICAL ISSUES AND THREATS

We must add to our review of future prospects for spatial change the interdependent and less predictable dynamics of politics, culture, society, and environment. From what we have discussed already, it is clear that the immediate future will be characterized by a phase of geopolitical and geoeconomic transition, by the continued overall expansion of the world economy, and by the continued globalization of industry, finance, and culture. The processes of change involved in shaping this kind of future will inevitably bring forth critical issues, changes, conflicts, and threats. We can identify what several of these might involve: fault lines of cultural dissonance, complex security issues, and sustainability.

Globalizing Culture and Cultural Dissonance

At one level, globalization has brought a homogenization of culture through the language of consumer goods. This is the material culture of the fast world, enmeshed by Airbus jets, CNN, music video channels, cell phones, and the Internet and swamped by Coca-Cola, Budweiser, McDonald's, GAP clothing, Nikes, iPods, PlayStations, Toyotas, Disney franchising, and formula-driven Hollywood movies. Furthermore, sociologists have recognized that a distinctive culture of "global metropolitanism" is emerging among the transnational elite. This is simply homogenized culture at a higher plane of consumption (French wines instead of Budweiser, molecular gastronomy and international cuisine instead of McDonald's, Hugo

Figure 12.6 Wild zones Parts of Africa, the Middle East, and South Asia have become "wild zones," places where national governments have lost control over economic development, ethnic conflict, and environmental degradation. In 2007, according to estimates supplied by the United Nations High Commissioner for Refugees, there were more than 25 million refugees, asylum seekers, and displaced persons in the world. The map shows the areas of concern listed by the UNHCR in its Report for 2007. (*Source:* United Nations High Commissioner for Refugees, *2007 Global Trends: Refugees, Asylum-seekers, Returnees, Internally Displaced and Stateless Persons.* UNHCR, 2008. See http://www.unhcr. org/statistics/2007Global-Trends.zip, accessed August 15, 2008.)

Boss clothes instead of Levis, BMWs instead of Toyotas, and so on). The members of this new culture are people who make international conference calls, who make decisions and transact investments that are transnational in scope, who edit the news, design and market international products, and travel the world for business and pleasure.

These trends are transcending some of the traditional cultural differences around the world. We can, perhaps, more easily identify with people who use the same products, listen to the same music, and appreciate the same sports stars that we do. At the same time, however, sociocultural cleavages are opening up between the elites and the marginalized. Focusing people's attention on material consumption also obscures the emergence of new fault lines—between previously compatible cultural groups and between ideologically divergent civilizations.

Several reasons account for the appearance of these new fault lines. One is the release of pressure brought about by the end of the Cold War. The evaporation of external threats has allowed people to focus on other perceived threats and intrusions. Another is the globalization of culture itself. As people's lives are homogenized through their jobs and their material culture, many of them want to revive subjectivity, reconstruct we/us feelings, and reestablish a distinctive cultural identity. For the marginalized, a different set of processes is at work, however. The juxtaposition of poverty, environmental stress, and crowded living conditions alongside the materialism of the fast world creates a fertile climate for gangsterism. The same juxtaposition also provides the ideal circumstance for the spread and intensification of religious fundamentalism and for fundamentalist-inspired terrorism. This, perhaps more than anything else, represents a source of serious potential cultural dissonance.

The overall result is that cultural fault lines are opening up at every geographical scale. This poses some very

problematic dimensions for future regional geographies, from the scale of the city to the scale of the globe. The prospect at the metropolitan scale is one of fragmented and polarized communities, with outright cultural conflict suppressed only through electronic surveillance (**Figure 12.7**) and the "militarization" of urban space via security posts and "hardened" urban design using fences and gated streets. This, of course, presupposes a certain level of affluence in order to meet the costs of keeping the peace across economic and cultural fault lines. In the unintended metropolises of the periphery, where unprecedented numbers of migrants and refugees will be thrown together, the genuine possibility for anarchy and intercommunal violence exists—unless intergroup differences can be submerged in a common cause, such as religious fundamentalism.

At the subnational scale the prospect is one of increasing ethnic/racial rivalry, parochialism, and insularity. Examples of these phenomena can be found throughout the world. In North America, we see increasing ethnic rivalry represented by the secessionism of the French-speaking Quebeçois and the insistence of some Hispanic groups in the United States on the installation of Spanish as an alternative official language. In Europe, examples are the secessionism (from Spain) of the Basques, the separatist movement of the Catalans (also in Spain), the regional elitism of Northern Italy, and, recently, outright war between Serbs and Croats. In South Asia, examples are provided by the recurring hostility between Hindus and Muslims throughout the Indian subcontinent and between the Hindu Tamils and Buddhist Sinhalese of Sri Lanka. In Africa, ethnic rivalry, parochialism, and insularity are reflected in the continuing conflict between the Muslim majority in northern Sudan and the Christian minority in the south and widespread unrest in northeast Democratic Republic of Congo. Where the future brings prosperity, tensions and hostilities such as these will

Figure 12.7 Surveillance societies Social, economic, and ethnic polarization in the cities of the world's core countries has led to an increase in the electronic surveillance of both public and private spaces. Another significant change has been the increased presence of private security personnel in upscale settings: Private security officers now outnumber police officers in the United States. This photo shows the interior of Chicago's Office of Emergency Management, the center for Chicago's criminal database, "Citizen Law Enforcement and Analysis" (CLEAR).

probably be muted; where it brings economic hardship or decline, they will undoubtedly intensify.

The prospect at the global scale is of a rising consciousness of people's identities in terms of their broader historical, geographical, and racial "civilizations": Western, Latin American, Confucian, Japanese, Islamic, Hindu, and Slavic-Orthodox. According to some observers, deepening cleavages of this sort could replace the ideological differences of the Cold War era as the major source of tension and potential conflict in the world. And as many globalization experts have observed, they may also point the way toward new foundations for cooperation and unity as people see themselves less as citizens of a particular nation or state and more as world citizens sharing a common global cause (see Chapter 9).

Security

In addition to cultural fault lines, globalization will inevitably generate enormous economic, and, consequently, political upheavals that will translate into a complex set of security issues. With the gradual integration of China, India, and other emerging countries into the global economy, hundreds of millions of working-age adults will become available for employment in a more integrated world labor market. This enormous workforce—a growing portion of which will be well educated—will be an attractive, competitive source of low-cost labor at the same time that technological innovation is expanding the range of globally mobile occupations. The transition will hit the middle classes of the core regions in particular, bringing more rapid job turnover and requiring professional retooling, while outsourcing on an increasing scale will intensify the anti-globalization movement. Where these pressures lead will depend on how political leaders respond, how flexible labor markets become, and whether overall economic growth is sufficiently robust to absorb a growing number of displaced workers. In some regions, it is quite possible that weak governments, lagging economies, religious extremism, and youth bulges will align to create a perfect storm for internal conflict, with far-reaching repercussions for security elsewhere.

Such internal conflicts, particularly those that involve ethnic groups straddling national boundaries, risk escalating into regional conflicts and the failure of states, with expanses of territory and populations devoid of effective governmental control. These territories can become sanctuaries for transnational terrorists (such as al-Qaeda in Afghanistan) or for criminals and drug cartels (such as in Colombia). According to the U.S. National Intelligence Council, al-Qaeda itself will likely be superseded by similarly inspired Islamic extremist groups within the next 15 years, and there is a substantial risk that broad Islamic movements akin to al-Qaeda will merge with local separatist movements.[4] Information technology, allowing for instant connectivity, communication, and learning,

will enable the terrorist threat to become increasingly decentralized, evolving into an eclectic array of groups, cells, and individuals that do not need a stationary headquarters to plan and carry out operations. Training materials, targeting guidance, weapons know-how, and fund-raising will become virtual (i.e., online). As biotechnology information becomes more widely available, the number of people who can potentially misuse such information and wreak widespread loss of life will increase. Moreover, as biotechnology advances become more ubiquitous, stopping the progress of offensive biological warfare programs will become increasingly difficult. Over the next 10 to 20 years there is a risk that advances in biotechnology will allow the creation of advanced biological agents designed to target specific systems—human, animal, or crops.

Economic and political turbulence and instability will also be conducive to transnational crime. Transnational crime syndicates pose a considerable threat to global security. They distribute harmful materials, weapons, and drugs, exploit local communities, disrupt fragile ecosystems, and control significant economic resources. In 2007, transnational crime syndicates may have grossed up to $2 trillion—more than all national economies except those of the United States, Japan, and Germany. More than $520 billion of this comes from counterfeiting and piracy. The drug trade is the second-biggest earner, with an estimated $320 billion in takings. Trafficking in humans—for labor, sex work, and even for the removal of kidneys or other organs for transplant purposes—is another aspect of transnational crime. It is a small industry by comparison with drugs, worth under $44 billion but arguably the most pernicious aspect of organized crime. The U.S. State Department estimates that at least 600,000 to 800,000 people are sold internationally each year. According to the UN, up to 27 million people are now held in slavery, far more than at the peak of the African slave trade. The majority of the victims this time are Asian women. Other significant sources of crime include environmental products—everything from protected plants and animals to hazardous waste and banned chemicals. Weapons trafficking earns the syndicates comparatively little—estimated at less than $1 billion annually—but it contributes significantly to the complex security issues facing most countries. There are few international laws on arms trafficking, and UN arms embargoes often go unenforced.

Another dimension of the security issues associated with globalization concerns the environmental and human health issues that have emerged as a result of increases in global commerce and tourism. As people and goods move around the world, so do other species, pathogens, and zoonotic diseases. New plant and animal species can invade ecosystems, choke pastures, disrupt water systems, drive other species to extinction, and result in expensive and unanticipated consequences. In the United States, the European zebra mussel has invaded freshwater systems across almost half of the country, displacing other marine species and costing an estimated

[4]*Ibid.*, p. 9.

12.3 GEOGRAPHY MATTERS

The Asian Brown Cloud

At the heart of the World Summit on Sustainable Development in Johannesburg in September 2002 was the question: How can developing nations grow economically without overburdening Earth's environment and creating an uninhabitable planet for future generations? Emblematic of this issue is the Asian Brown Cloud, a blanket of air pollution 3 kilometers (nearly 2 miles) thick that hovers over most of the tropical Indian Ocean and South, Southeast, and East Asia, stretching from the Arabian Peninsula across India, Southeast Asia, and China almost to Korea. The brown haze was first identified by U.S. Air Force pilots but is now clearly visible in satellite photographs and from the Himalayas (**Figure 12.A**).

The Asian Brown Cloud consists of sulfates, nitrates, organic substances, black carbon, and fly ash, along with several other pollutants. It is an accumulated cocktail of contamination resulting from a dramatic increase in the burning of fossil fuels in vehicles, industries, and power stations in Asia's megacities, from forest fires used to clear land, and from the emissions from millions of inefficient cookers burning wood or cow dung. A study of the Asian Brown Cloud sponsored by the UN Environment Program and involving more than 200 scientists suggests that the Asian Brown Cloud not only influences local weather but also may have worldwide consequences.

The smog of the Asian Brown Cloud reduces the amount of solar radiation reaching the Earth's surface by 10 to 15 percent, with a consequent decline in the productivity of crops. But it can also trap heat, leading to warming of the lower atmosphere. It suppresses rainfall in some areas and increases it in others, while damaging forests and crops because of acid rain. The haze is also believed to be responsible for hundreds of thousands of premature deaths from respiratory diseases.

Figure 12.A   Asian Brown Cloud This view shows a dense haze over eastern China, looking eastward across the Yellow Sea toward Korea.

$1 billion in eradication costs during the 1990s. The total costs of losses from all kinds of invasive species in the United States now exceeds $120 billion annually. The diffusion of contagious diseases such as influenza, and of zoonotic diseases such as anthrax, avian flu, Ebola, and the West Nile virus, also poses a much greater security risk as a result of the speed and intensity of global flows—another dimension of the "risk society" (see Chapter 1) associated with globalization.

Sustainability

As we saw in Chapter 4, the world currently faces a daunting list of environmental threats: the destruction of tropical rain forests and the consequent loss of biodiversity; widespread, health-threatening pollution; the degradation of soil, water, and marine resources essential to food production; stratospheric ozone depletion; acid rain; and so on. Most of these threats are greatest in the world's periphery, where daily environmental pollution and degradation amounts to a catastrophe that will continue to unfold, slow-motion, in the coming years (see Box 12.3, "The Asian Brown Cloud").

Future trends will only intensify the contrasts between rich and poor regions. We know that environmental problems will be inseparable from processes of demographic change, economic development, and human welfare. In addition, it is becoming clear that environmental problems are going to be increasingly enmeshed in matters of national security and regional conflict. The spatial *interdependence* of economic, environmental, and social problems means that some parts of the world are ecological time bombs. The prospect of civil unrest and mass migrations resulting from the pressures of rapidly growing populations, deforestation, soil erosion, water depletion, air pollution, disease epidemics, and intractable poverty is real. These specters are alarming not only for the peoples of the affected regions but also for the peoples of rich and faraway countries, whose continued prosperity will depend on processes of globalization that are not disrupted by large-scale environmental disasters, unmanageable mass migrations, or the breakdown of stability in the world-system as a whole.

We cannot simply wait to see what the future will hold. If we are to have a better future (and if we are to *deserve* a better future), we must use our understanding of the world—and of geographical patterns and processes—to work toward more desirable outcomes. One cause for optimism here is the forthcoming United Nations Climate Change Conference, to be held in Denmark in 2009, which has the attention and support of most countries, including the United States. "Green" initiatives and the goals of sustainable development—equity, efficiency, and environmental quality—are increasingly widespread. No discipline is more relevant to the ideal of sustainable development than geography. Where else, as British geographer W. M. Adams has observed, can the science of the environment (physical geography) be married with an understanding of economic, technological, social, political, and cultural change (human geography)? What other discipline offers insights into environmental change, and who but geographers can cope with the diversity of environments and the sheer range of scales at which it is necessary to manage global change?

Those of us in the richer countries of the world have a special responsibility for leadership in sustainable development, because our present affluence is based on a cumulative past (and present) exploitation of the world's resources that is disproportionate to our numbers. We also happen to have the financial, technical, and human resources to enable us to take the lead in developing cleaner, less resource-intensive technologies, in transforming our economies to protect and work with natural systems, in providing more equitable access to economic opportunities and social services, and in supporting the technological and political frameworks necessary for sustainable development in poor countries. We cannot do it all at once, but we will certainly deserve the scorn and resentment of future generations if we do not try.

CONCLUSION

The beginning of the twenty-first century is going to be a period of fluid and transitional relationships among places, regions, and nations. Nevertheless, we know enough about contemporary patterns and trends, as well as geographic processes, to map out some plausible scenarios for the future. Some aspects of the future, though, we can only guess at. Two of the most speculative realms are those of politics and technology.

The future of the worst-off peripheral regions could be very bad indeed. They face unprecedented levels of demographic, environmental, and societal stress, with the events of the next 50 years being played out from a starting point of scarce basic resources, serious environmental degradation, overpopulation, disease, unprovoked crime, refugee migrations, and criminal anarchy.

For the world-system core, however, the long-term question is one of relative power and dominance. The same factors that consolidate the advantages of the core as a whole—the end of the Cold War, the availability of advanced telecommunications, the transnational reorganization of industry and finance, the liberalization of trade, and the emergence of a global culture—will also open the possibility for a new geopolitical and geoeconomic order, within which the economic and political relationships among core countries might change substantially.

Many aspects of future geographies will depend on trends in the demand for resources and on the exploitation of new technologies. The expansion of the world economy and the globalization of industry will undoubtedly boost the overall demand for raw materials of various kinds, and this will spur the development of previously underexploited but resource-rich regions in Africa, Eurasia, and East Asia. Raw materials will be only a fraction of future resource needs, however; the main issue, by far, will be energy resources.

It also appears that the present phase of globalization has the potential to create such disparities between the haves and the have-nots (as well as between the core and the periphery) that social unrest will ensue. The evidence of increasing dissatisfaction with the contemporary distribution of wealth both within and between the core and the periphery is widespread. In Russia, where public demonstrations against the government had been outlawed for decades, people are taking to the streets to protest government policies that favor transnational economic development at the expense of workers' minimum wages. In India, farmers damaged a Kentucky Fried Chicken restaurant to protest the company's role in dislocating the domestic poultry industry. In Mexico, industrial workers in transnational plants are challenging the absence of health and safety regulations that leave some exposed to harmful chemicals.

At the same time that protests against globalization and new geographies are being waged, the products of a global economy and culture are being widely embraced. The market for blue contact lenses is growing in such unlikely places as Bangkok and Nairobi, and plastic surgery to make eyes more Western looking is increasing in many Asian countries. CNN as well as American TV soap operas are eagerly viewed in even the most remote corners of the globe. Highway systems, airports, and container facilities are springing up throughout the periphery.

In short, future geographies are being negotiated at this very moment—from the board rooms of transnational corporations to the huts of remote villagers. The outcomes of these negotiations are still in the making as we all, in our daily lives, make seemingly insignificant decisions about what to wear, what to eat, where to work, how to travel, and how to entertain ourselves. These decisions help to either support or undermine the larger forces at work in the global economy, such as where to build factories, what products to make, or how to package and deliver them to the consumer. In essence, future geographies can be very much shaped by us through our understanding of the relatedness of people, places, and regions in a globalized economy.

MAIN POINTS REVISITED

- **In some ways the future is already here, embedded in the world's institutional structures and in the dynamics of its populations.**

  We know, for example, a good deal about the demographic trends of the next quarter-century, given present populations and birth and death rates. We also know a good deal about the distribution of environmental resources and constraints, about the characteristics of local and regional economies, and the legal and political frameworks within which geographic change will probably take place.

- **New and emerging technologies that are likely to have the most impact in reshaping human geographies include advanced transportation technologies, biotechnology, materials technologies, and information technologies.**

  The evolution of the world's geographies has always been shaped by the opportunities and constraints presented to different places and regions by successive technology systems.

Many aspects of future geographies will depend on trends in the demand for particular resources and on the exploitation of these new technologies.

- **The changes involved in shaping future geographies will inevitably bring some critical issues, conflicts, and threats, including important geographical issues that center on scale, boundaries, and territories; on cultural dissonance; and on the sustainability of development.**

  Many of these issues stem from the globalization of the economy, which is undermining the status of the territorial nation-state as the chief regulating mechanism of both global and local dimensions of the world-system. The implications for peripheral places and regions are dismal: No matter how strong governments may be in their apparatus of domestic power, they will be next to helpless in the face of acute environmental stress, increased cultural friction, escalating poverty and disease, and growing migrations of refugees.

KEY TERM

disruptive technology (p. 459)

ADDITIONAL READING

Castells, M., *The Information Age: Economy, Society and Culture*. Vol. 3: *End of Millenium*. Oxford: Blackwell, 1998.

Coates, J. F., J. B. Mahaffie, and A. Hines, *2025: Scenarios of U.S. and Global Society Reshaped by Science and Technology*. Greensboro, NC: Oakhill Press, 1997.

Glenn, J. C., T. J. Gordon, and E. Florescu, *2008 State of the Future*. New York: World Federation of UN Associations, 2008.

Huntington, S., "The Clash of Civilizations?" *Foreign Affairs* 72, 1993, pp. 22–49.

Jacobs, J., *Dark Age Ahead: Caution*. New York: Random House, 2004.

Johnston, R. J., P. J. Taylor, and M. Watts (eds.), *Geographies of Global Change: Remapping the World*. Cambridge: Blackwell, 2002.

Kaplan, R. D., "The Coming Anarchy," *Atlantic Monthly*, February 1994, pp. 44–76.

National Intelligence Council, *Mapping the Global Future*. Washington, DC: U.S. Government Printing Office, 2004.

O'Meara, P., H. D. Mehlinger, and M. Krain (eds.), *Globalization and the Challenges of a New Century*. Bloomington, IN: Indiana University Press, 2000.

Pearson, I. (ed.), *The Macmillan Atlas of the Future*. New York: Macmillan, 1998.

Smith, N., *The Endgame of Globalization*. New York: Routledge, 2005.

Worldwatch Institute, *State of the World 2008: Innovations for a Sustainable Economy*. New York: W.W. Norton, 2008.

EXERCISES

On the Internet

The Companion Website at www.mygeoscienceplace.com contains a number of tools designed to reinforce the geographic concepts presented in this chapter. Structured within a four-step learning path, these online resources include: chapter Key Concepts; a fully-searchable myeBook version of the chapter, which allows you to add notes and highlights; RSS news feeds from various media agencies; links to additional web resources; Human Geography Videos; a digital Glossary; Concept Review questions; Thinking Spatially mapping exercises; and Critical Thinking essay questions.

Unplugged

1. Using census data, construct a population pyramid (see Chapter 3) for any county or city with which you are familiar. What does this tell you about the future population of the locality?

2. Drawing on what you know about the geography of this locality and its regional, national, and global contexts, construct two scenarios, about 200 words each, for the future, each based on different assumptions about resources and technology.

3. Write a short essay (500 words, or two double-spaced, typed pages) in which you outline the possible effects of economic globalization on a particular place or region with which you are familiar.

Appendix A
Maps and Geographic Information Systems

Maps are representations of the world. They are usually two-dimensional graphic representations that use lines and symbols to convey information or ideas about spatial relationships. Maps express particular interpretations of the world, and they affect how we understand the world and how we see ourselves in relation to others. As such, all maps are "social products." In general, maps reflect the power of the people who draw them up. Just including things on a map—literally "putting something on the map"—can be empowering. The design of maps—what they include, what they omit, and how their content is portrayed—inevitably reflects the experiences, priorities, interpretations, and intentions of their authors. The most widely understood and accepted maps—"normal" maps—reflect the view of the world that is dominant in universities and government agencies.

Maps that are designed to represent the *form* of Earth's surface and to show permanent (or at least long-standing) features such as buildings, highways, field boundaries, and political boundaries are called *topographic maps* (see, for example, **Figure A1.1**). The usual device for representing the form of Earth's surface is the *contour,* a line that connects points of equal vertical distance above or below a zero data point, usually sea level.

Maps that are designed to represent the spatial dimensions of particular conditions, processes, or events are called *thematic maps.* These can be based on any one of a number of devices that allow cartographers or map makers to portray spatial variations or spatial relationships. One of these is the *isoline,* a line (similar to a contour) that connects places of equal data value (for example, air pollution, as in **Figure A1.2**). Maps based on isolines are known as *isopleth maps.* Another common device used in thematic maps is the *proportional symbol.* Thus, for example, circles, squares, spheres, cubes, or some other shape can be drawn in proportion to the frequency of occurrence of some particular phenomenon or event at a given location. **Figure A1.3** shows an example using proportional circles. Symbols such as arrows or lines can also be drawn proportionally in order to portray flows of things between particular places. Simple distributions can be effectively portrayed through *dot maps,* in which a single dot or other symbol represents a specified number of occurrences of some particular phenomenon or event. Yet another device is the *choropleth map,* in which tonal shadings are graduated to reflect area variations in numbers, frequencies, or densities (see, for example, **Figures 7.1**, p. 253 and 10.1, p. 394). Finally, thematic maps can be based on *located charts,* in which graphs or charts are located by place or region. In this way, a tremendous amount of information can be conveyed in one single map (**Figure A1.4**).

MAP SCALES

A *map scale* is simply the ratio between linear distance on a map and linear distance on Earth's surface. It is usually expressed in terms of corresponding lengths, as in "one centimeter equals one kilometer," or as a *representative fraction* (in this case, 1/100,000) or ratio (1:100,000). *Small-scale* maps are maps based on small representative fractions (for example, 1/1,000,000 or 1/10,000,000). Small-scale maps cover a large part of the Earth's surface on the printed page. A map drawn on this page to the scale of 1:10,000,000 would cover about half of the United States; a map drawn to the scale of 1:16,000,000 would easily cover the whole of Europe. *Large-scale* maps are maps based on larger representative fractions (e.g., 1/25,000 or 1/10,000). A map drawn on this page to the scale 1:10,000 would cover a typical suburban subdivision; a map drawn to the scale of 1:1,000 would cover just a block or two of it.

MAP PROJECTIONS

A map projection is a systematic rendering on a flat surface of the geographic coordinates of the features found on Earth's surface. Because Earth's surface is curved and it is not a perfect sphere, it is impossible to represent on a flat plane, sheet of paper, or monitor screen without some distortion. Cartographers have devised a number of different techniques of projecting latitude and longitude (see Figure 1.13) onto a flat surface, and the resulting representations of Earth each have advantages and disadvantages. None of them can represent distance correctly in all directions, though many can represent compass bearings or area without distortion. The choice of map projection depends largely on the purpose of the map.

Projections that allow distance to be represented as accurately as possible are called *equidistant projections.* These projections can represent distance accurately in only one direction (usually north-south), although they usually provide accurate scale in the perpendicular direction (which in most cases is the equator). Equidistant projections are often more aesthetically pleasing for representing Earth as a whole, or large portions of it. An example is the polyconic projection (**Figure A1.5**).

Figure A1.1   Topographic maps Topographic maps represent the form of Earth's surface in both horizontal and vertical dimensions. This extract is from a Swiss map of Lugano at the scale of 1:25,000. The height of landforms is represented by contours (lines that connect points of equal vertical distance above sea level), which on this map are drawn every 20 meters. Features such as roads, power lines, built-up areas, and so on are shown by stylized symbols. Note how the closely spaced contours of the hill slopes are able to represent the shape and form of the land. (*Source:* Extract from Carta Nazionale Della Svizzera, Sheet 1353, 1: 25,000 series, Ufficio Federale di Topografia, 3084 Wabern, Switzerland. Edizione 1998.)

Projections on which compass directions are rendered accurately are known as *conformal projections*. On the Mercator projection (Figure A1.5) for example, a compass bearing between any two points is plotted as a straight line. As a result, the Mercator projection has been widely used in navigation for hundreds of years. The Mercator projection was also widely used as the standard classroom wall map of the world for many years, and its image of the world has entered deeply into general consciousness. Many Europeans and North Americans have an exaggerated sense of the size of the northern continents and are unaware of the true size of Africa. Some projections are designed such that compass directions are correct only from one central point. These are known as *azimuthal projections*. They can be equidistant, as in the Azimuthal Equidistant projection (Figure A1.5), which is sometimes used to show air-route distances from a specific location, or equal-area, as in the Lambert Azimuthal Equal-Area projection.

Projections that portray areas on Earth's surface in their true proportions are known as **equal-area** or *equivalent projections*. Such projections are used where the cartographer wishes to compare and contrast distributions on the Earth's surface: the relative area of different types of land use, for example. Examples of equal-area projections include the Eckert IV projection, Bartholomew's Nordic projection (used in Figure 2.8) and the Mollweide projection (Figure A1.5). Equal-area projections such as the Mollweide projection are especially useful for thematic maps showing economic, demographic, or cultural data (see, for example, Figure 12.3). Unfortunately, preserving accuracy in terms of area tends to result in world maps on which many locations appear squashed and have unsatisfactory outlines.

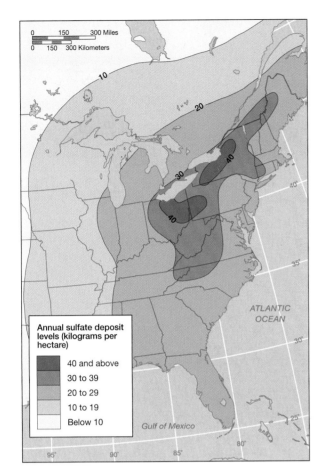

Figure A1.2   Isoline maps Isoline maps portray spatial information by connecting points of equal data value. Contours on topographic maps (see Figure A1.1) are isolines. This map shows air pollution in the eastern United States. (*Source:* Reprinted with permission of Prentice Hall, from J. M. Rubenstein, *The Cultural Landscape: An Introduction to Human Geography*, 1996, p. 584. Adapted from William K. Stevens, "Study of Acid Rain Uncovers Threat to Far Wider Area," *New York Times*, January 16, 1990, p. 21, map.)

Figure A1.3   An example of proportional symbols in thematic mapping Proportional circles showing the number of refugees originating in Southeast Asian countries in 1999, and the number of refugees granted asylum in Southeast Asian countries in 1999. (Adapted from S. Marston, P. Knox, and D. Liverman, *World Regions in Global Context.* Upper Saddle River, NJ: Prentice Hall, 2002, p. 497.)

For some applications, aesthetic appearance is more important than conformality, equivalence, or equidistance, so cartographers have devised a number of other projections. Examples include the Times projection, which is used in many world atlases, and the Robinson projection, which is used by the National Geographic Society in many of its publications. The Robinson projection (**Figure A1.6**) is a compromise projection that distorts both area and directional relationships but provides a general-purpose world map. There are also political considerations. Countries may appear larger and so more "important" on one projection rather than another. The Peters projection, for example (**Figure A1.7**), is a deliberate attempt to give prominence to the underdeveloped countries of the equatorial regions and the Southern Hemisphere. As such, it was officially adopted by the World Council of Churches and by numerous agencies of the United Nations and other international institutions. Its unusual shapes give it a shock value that gets people's attention. For some, however, its unusual shapes are ugly: It has been likened to laundry hung out to dry.

In this book we sometimes use another striking projection, the Dymaxion projection devised by Buckminster Fuller (**Figure A1.8**). Fuller was a prominent modernist architect and industrial designer who wanted to produce a map of the world with no significant distortion to any of the major landmasses. The Dymaxion projection does this, though it produces a world that at first may seem disorienting. This is not necessarily such a bad thing, for it can force us to take a fresh look at the world and at the relationships between places. Because Europe, North America, and Japan are all located toward the center of this map projection, it is particularly useful for illustrating two of the central themes of this book: the relationships among these prosperous regions, and the relationships between this prosperous core group and the less prosperous, peripheral countries of the world. Fuller's projection shows the economically peripheral countries of the world as being cartographically peripheral, too.

One particular kind of map projection that is sometimes used in small-scale thematic maps is the *cartogram*. In this kind of projection, space is transformed according to statistical factors, with the largest mapping units representing the greatest statistical values. **Figure A1.9a** shows a cartogram of the world in which countries are

Figure A1.4 Located charts By combining graphs, charts, or symbols with base maps, a great deal of information can be conveyed in a single figure. This example illustrates the profits generated through European plunder of global minerals, spices, and human beings over a 300-year period. (Adapted from S. Marston, P. Knox, and D. Liverman, *World Regions in Global Context*. Upper Saddle River, NJ: Prentice Hall, 2002, p. 84.)

represented as proportional to their population. This sort of projection is particularly effective in helping to visualize relative inequalities among the world's populations. **Figure A1.9b** shows a cartogram of the world in which the cost of telephone calls has been substituted for linear distance as the basis of the map. The deliberate distortion of the shapes of the continents in this sort of projection dramatically emphasizes spatial variations.

Finally, the advent of computer graphics has made it possible for cartographers to move beyond the use of maps as two-dimensional representations of Earth's surface. Computer software that renders three-dimensional statistical data on to the flat surface of a monitor screen or a piece of paper facilitates the *visualization* of many aspects of human geography in innovative and provocative ways (**Figure A1.10**).

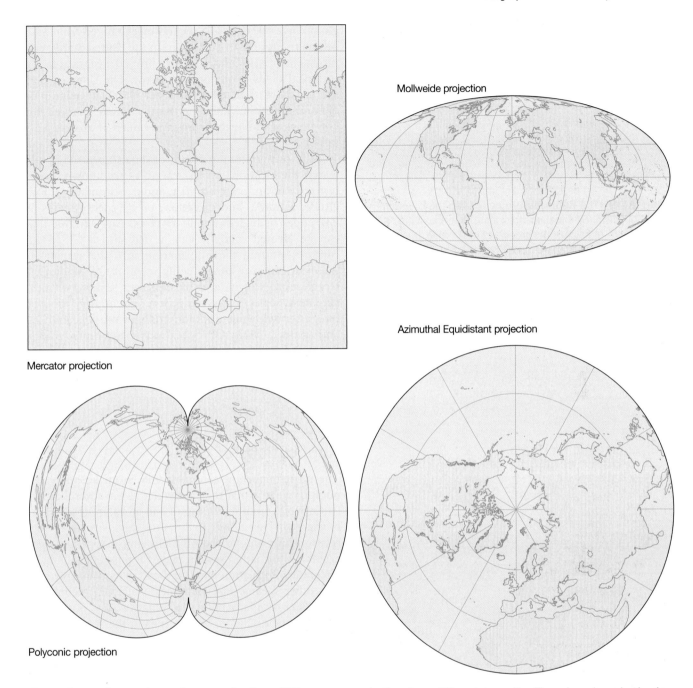

Mollweide projection

Mercator projection

Azimuthal Equidistant projection

Polyconic projection

Figure A1.5   Comparison of map projections Different map projections have different properties. The polyconic projection is true to scale along each east-west parallel and along the central north-south meridian. It is neither conformal nor equal-area, and it is free of distortion only along the central meridian. On the Mercator projection, compass directions between any two points are true, and the shapes of landmasses are true, but their relative size is distorted. On the azimuthal equidistant projection, distances measured from the center of the map are true, but direction, area, and shape are increasingly distorted with distance from the center point. On the Mollweide projection, relative sizes are true, but shapes are distorted.

GEOGRAPHIC INFORMATION SYSTEMS

Geographic information systems (GIS)—organized collections of computer hardware, software, and geographic data that are designed to capture, store, update, manipulate, and display geographically referenced information— have rapidly grown to become one of the most important methods of geographic analysis, particularly in the military and commercial worlds. The software in GIS incorporates programs to store and access spatial data, to manipulate those data, and to draw maps.

Between 2000 and 2005, GIS services grew at a rate of around 10 percent per year. In 2002, estimates of

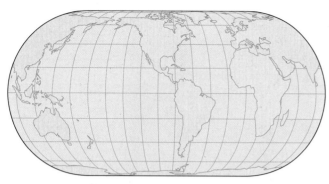

Figure A1.6 The Robinson projection On the Robinson projection, distance, direction, area, and shape are all distorted in an attempt to balance the properties of the map. It is designed purely for appearance and is best used for thematic and reference maps at the world scale. (*Source:* After E. F. Bergman, *Human Geography: Cultures, Connections, and Landscapes,* © 1995 by Prentice Hall, p. 12.)

global investment in GIS technologies ranged from $3.3 billion to more than $8 billion. In the United States, employment in GIS is now one of the 10 fastest-growing technical fields in the private sector. The United States spends more than $4 billion per year on geographic data acquisition, and the total annual market for GIS services in North America is valued at around $2.5 billion.

The primary requirement for data to be used in GIS is that the locations for the variables are known. Location may be annotated by x, y, and z coordinates of longitude, latitude, and elevation, or by such systems as ZIP codes or highway mile markers. Any variable that can be located spatially can be fed into a GIS. Data capture—putting the information into the system—is the most time-consuming component of GIS work. Different sources of data, using different systems of measurement, scales, and systems of representation, must be integrated with one another; changes must be tracked and updated. Many GIS operations in the United States, Europe, Japan, and Australia have begun to contract out such work to firms in countries where labor is cheaper. India has emerged as a major data-conversion center for GIS.

Applications of GIS

The most important aspect of GIS, from an analytical point of view, is that they allow data from several different sources, on different topics and at different scales, to be merged. This allows analysts to emphasize the spatial relationships among the objects being mapped. A geographic information system makes it possible to link, or integrate, information that is difficult to associate through any other means. For example, using GIS technology and water-company billing information, it is possible to simulate the discharge of materials in the septic systems in a neighborhood upstream from a wetland. The bills show how much water is used at each address. Because the amount of water a customer uses will roughly predict the amount of material that will be discharged into the septic systems, areas of heavy septic discharge can be located using a GIS.

GIS technology can render visible many aspects of geography that were previously unseen. GIS can, for example, produce incredibly detailed maps based on millions of pieces of information—maps that could never have been drawn by human hands. One example of such a map is the satellite image reconstruction of the vegetation cover of the United States shown in **Figure A1.11**. At the other extreme of spatial scale, GIS can put places under the microscope, creating detailed new insights using huge databases and effortlessly browsable media (**Figure A1.12**).

Many advances in GIS have come from military applications. GIS allows infantry commanders to calculate

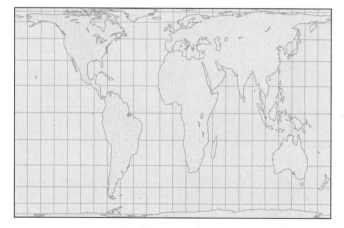

Figure A1.7 The Peters projection This equal-area projection was an attempt to offer an alternative to traditional projections which, Arno Peters argued, exaggerated the size and apparent importance of the higher latitudes—that is, the world's core regions—and so promoted the "Europeanization" of Earth. While it has been adopted by the World Council of Churches, the Lutheran Church of America, and various agencies of the United Nations and other international institutions, it has been criticized by cartographers in the United States on the grounds of aesthetics: One of the consequences of equal-area projections is that they distort the shape of landmasses. (*Source:* Adapted from E. F. Bergman, *Human Geography: Cultures, Connections, and Landscapes,* © 1995 by Prentice Hall, p. 13.)

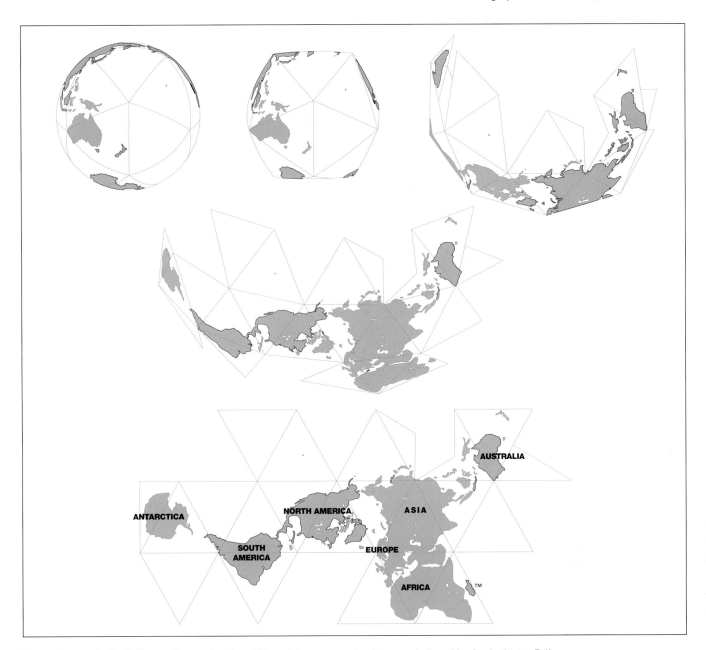

Figure A1.8   Fuller's Dymaxion projection This striking map projection was designed by Buckminster Fuller (1895–1983). As this figure shows, Fuller achieved his objective of creating a map with the minimum of distortion to the shape of the world's major landmasses by dividing the globe into triangular areas. Those areas not encompassing major landmasses were cut away, allowing the remainder of the globe to be "unfolded" into a flat projection. (*Source:* Buckminster Fuller Institute and Dymaxion Map Design, Santa Barbara, CA. The word *Dymaxion* and the Fuller Projection Dymaxion™ Map design are trademarks of the Buckminster Fuller Institute, Santa Barbara, CA, © 1938, 1967 & 1992. All rights reserved.)

line of sight from tanks and defensive emplacements, allows cruise missiles to fly below enemy radar, and provides a comprehensive basis for military intelligence. Beyond the military, GIS technology allows an enormous range of problems to be addressed. For instance, it can be used to decide how to manage farmland, to monitor the spread of infectious diseases, to monitor tree cover in metropolitan areas, to assess changes in ecosystems, to analyze the impact of proposed changes in the boundaries of legislative districts, to identify the location of potential business customers, to identify the location of potential criminals, and to provide a basis for urban and regional planning. Some of the most influential applications of GIS have resulted from geodemographic research. *Geodemographic research* uses census data and commercial data (such as sales data and property records) about the populations of small districts in creating profiles of those populations for market research. The digital media used by GIS make such applications very flexible. With GIS it is possible to zoom in and out,

(a)

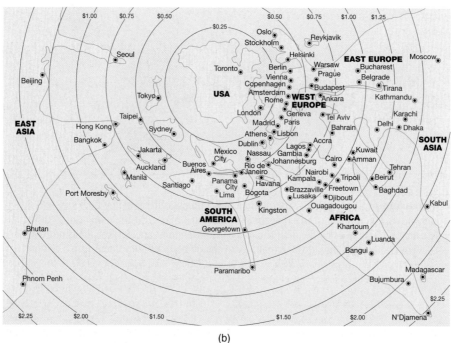

(b)

Figure A1.9   Example of a cartogram In a cartogram, space is distorted to emphasize a particular attribute of places or regions. (a) This example shows the relative size of countries based on their population rather than their area; the cartographers have maintained the shape of each country as closely as possible to make the map easier to read. As you can see, population-based cartograms are very effective in demonstrating spatial inequality. (b) In this example, the cost of telephone calls is substituted for linear distance as the basis of the map, thus deliberately distorting the shapes of the continents to dramatic effect. Countries are arranged around the United States according to the cost per minute of calls made from the United States in 1998. (Adapted from (a) M. Kidron and R. Segal (eds.), *The State of the World Atlas,* rev. 5th ed. London: Penguin Reference, 1995, pp. 28–29; (b) G. C. Staple (ed.), *TeleGeography 1999.* Washington, DC: TeleGeography, 1999, p. 82.)

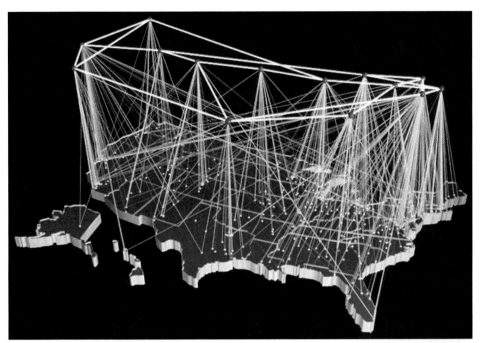

Figure A1.10   Visualization
This example shows the spatial structure of the Internet backbone and associated traffic flows within the United States. (After D. Cox and R. Patterson, http://www.ncsa. uiuc.edu/SCMS/DigLib/text/technology/ Visualization-Study-NSFNET-Cox.html)

evaluating spatial relationships at different spatial scales. Similarly, it is possible to vary the appearance and presentation of maps, using different colors and rendering techniques.

Critiques of GIS

Within the past five years applications of GIS have resulted in the creation of more maps than were created in all previous human history. One result is that as maps have become more commonplace, more people and more businesses have become more spatially aware. Nevertheless, some critics have argued that GIS represents no real advances in geographers' understanding of places and regions. The results of GIS, they argue, may be useful but are essentially mundane. This misses the point that however routine their subject may be, all maps constitute powerful and influential ways of representing the world. A more telling critique, perhaps, is that the real impact of GIS has been to increase the level of surveillance of the population by those who already possess power and control. The fear is that GIS may be helping to create a world in which people are not treated and judged by who they are and what they do but more by where they live. People's credit ratings, ability to buy insurance, and ability to secure a mortgage, for example, are all routinely judged in part by GIS-based analyses that take into account the attributes and characteristics of their neighbors.

Figure A1.11   Map of land cover This extract is from a map of land cover in the United States that was compiled from several data sets using GIS technology. These data sets included 1-kilometer resolution, Advanced Very High Resolution Radiometer (AVHRR) satellite imagery, and digital data sets on elevation, climate, water bodies, and political boundaries. Each of the 159 colors on the U.S. map represents a specific vegetation region. The purples and blues on this extract represent various subregions of western coniferous forests; the yellows are grasslands, and the reds are shrublands. The gray-brown region is the barren area of the Mojave Desert. (*Source:* United States Geological Survey, Map of Seasonal Land Cover Regions, 1993; see T. Loveland, J. W. Merchant, J. F. Brown, D. O. Ohlen, B. C. Reed, P. Olson, and J. Hutchinson, "Seasonal Land Cover Regions of the United States," *Annals, Association of American Geographers, 85* (1995), 339–355.)

Figure A1.12 GIS-Derived Planning Map This map of the Wachusett Reservoir Watershed in Massachusetts was prepared from a composite of 12 layers of GIS maps, each depicting a key aspect of land use and topography. The map is used to show in detail the parcels of land that the Commonwealth of Massachusetts needs to buy in order to keep development from damaging the watershed and the quality of the reservoir water. (*Source:* Adapted from R.W. Greene, *GIS in Public Policy*. Redlands, CA: ESRI Press, 2000, p. 67.)

ADDITIONAL READING

Dorling, D., and D. Fairbairn, *Mapping: Ways of Seeing the World.* London: Addison Wesley Longman, 1977.

Heywood, I, Cornelius, S and Carver, S. An Introduction to Geographical Information Systems, 3rd edition. Upper Saddle River, NJ: Prentice Hall, 2006.

Thrower, N. J. W., *Maps and Civilization: Cartography in Culture and Society,* 2nd ed. Chicago: University of Chicago Press, 1999.

Wilton, J. N., *The Mapmakers.* New York: Vintage Books, 2000.

Glossary

A

accessibility: the opportunity for contact or interaction from a given point or location, in relation to other locations.

acid rain: the wet deposition of acids upon Earth created by the natural cleansing properties of the atmosphere.

age-sex pyramid: a representation of the population based on its composition according to age and sex.

agglomeration diseconomies: the negative economic effects of urbanization and the local concentration of industry.

agrarian: referring to the culture of agricultural communities and the type of tenure system that determines access to land and the kind of cultivation practices employed there.

agribusiness: a set of economic and political relationships that organizes agro-food production from the development of seeds to the retailing and consumption of the agricultural product.

agricultural density: ratio between the number of agriculturists per unit of arable land and a specific area.

agricultural industrialization: process whereby the farm has moved from being the centerpiece of agricultural production to become one part of an integrated string of vertically organized industrial processes including production, storage, processing, distribution, marketing, and retailing.

agriculture: a science, art, and business directed at the cultivation of crops and the raising of livestock for sustenance and profit.

aquaculture: the cultivation of fish and shellfish under controlled conditions, usually in coastal lagoons.

autarky: insignificant contributions to the flows of imports and exports that constitute the geography of trade.

azimuthal projection: a map projection on which compass directions are correct only from one central point.

B

baby boom: population of individuals born between the years 1946 and 1964.

backwash effects: the negative impacts on a region (or regions) of the economic growth of some other region.

basic functions: economic activities that provide income from sales to customers beyond city limits.

Beaux Arts: a style of urban design that sought to combine the best elements of all of the classic architectural styles.

biofuels: renewable fuels derived from biological materials that can be regenerated.

biopharming: an application of biotechnology in which genes from other life forms (plant, animal, fungal, bacterial, or human) are inserted into a host plant.

Biorevolution: the genetic engineering of plants and animals with the potential to exceed the productivity of the Green Revolution.

biotechnology: technique that uses living organisms (or parts of organisms) to make or modify products, to improve plants and animals, or to develop microorganisms for specific uses.

bioterrorism: deliberate use of microorganisms or toxins from living organisms to induce death or disease.

Blue Revolution: the introduction of new production techniques, processing technology, infrastructure, and larger, motorized boats, as well as the application of transgenics into peripheral country fisheries.

boomburbs: rapidly growing suburban jurisdictions at the fringe of metropolitan areas.

Borlaug hypothesis: because global food demand is on the rise, restricting crop usage to traditional low-yield methods (such as organic farming) would require either the world population to decrease or the further conversion of forest land into cropland.

C

capitalism: a form of economic and social organization characterized by the profit motive and the control of the means of production, distribution, and the exchange of goods by private ownership.

carrying capacity: the maximum number of users that can be sustained, over the long term, by a given set of natural resources.

cartography: the body of practical and theoretical knowledge about making distinctive visual representations of Earth's surface in the form of maps.

census: count of the number of people in a country, region, or city.

central business district (CBD): the central nucleus of commercial land uses in a city.

central cities: the original, core jurisdictions of metropolitan areas.

central place: a settlement in which certain products and services are available to consumers.

central place theory: a theory that seeks to explain the relative size and spacing of towns and cities as a function of people's shopping behavior.

centrality: the functional dominance of cities within an urban system.

centrifugal forces: forces that divide or tend to pull the state apart.

centripetal forces: forces that strengthen and unify the state.

chemical farming: application of synthetic fertilizers to the soil—and herbicides, fungicides, and pesticides to crops—in order to enhance yields.

children's rights: the fundamental right of children to life, liberty, education, and health care codified by the United Nations Convention on the Rights of the Child in 1989.

citizenship: a category of belonging to a nation-state that includes civil, political, and social rights.

Clovis point: a flaked, bifaced projectile whose length is more than twice its width.

cognitive distance: the distance that people perceive to exist in a given situation.

cognitive images (mental maps): psychological representations of locations that are made up from people's individual ideas and impressions of these locations.

cognitive space: space defined and measured in terms of the nature and degree of people's values, feelings, beliefs, and perceptions about locations, districts, and regions.

cohort: a group of individuals who share a common temporal demographic experience.

colonial city: a city that was deliberately established or developed as an administrative or commercial center by colonial or imperial powers.

colonialism: the establishment and maintenance of political and legal domination by a state over a separate and alien society.

colonization: the physical settlement of a new territory of people from a colonizing state.

Columbian Exchange: interaction between the Old World—originating with the voyages of Columbus—and the New World.

commercial agriculture: farming primarily for sale, not direct consumption.

commodity chain: network of labor and production processes beginning with the extraction or production of raw materials and ending with the delivery of a finished commodity.

comparative advantage: principle whereby places and regions specialize in activities for which they have the greatest advantage in productivity relative to other regions—or for which they have the least disadvantage.

confederation: a group of states united for a common purpose.

conformal projection: a map projection on which compass bearings are rendered accurately.

conglomerate corporations: companies that have diversified into various economic activities, usually through a process of mergers and acquisitions.

congregation: the territorial and residential clustering of specific groups or subgroups of people.

conservation: the view that natural resources should be used wisely and that society's effects on the natural world should represent stewardship and not exploitation.

conventional farming: approach that uses chemicals in the form of plant protectants and fertilizers, or intensive, hormone-based practices in breeding and raising animals.

core regions: regions that dominate trade, control the most advanced technologies, and have high levels of productivity within diversified economies.

cosmopolitanism: an intellectual and aesthetic openness toward divergent experiences, images, and products from different cultures.

counterurbanization: the net loss of population from cities to smaller towns and rural areas.

creative destruction: the withdrawal of investments from activities (and regions) that yield low rates of profit in order to reinvest in new activities (and new places).

crop rotation: method of maintaining soil fertility in which the fields under cultivation remain the same but the crop being planted is changed.

crude birthrate (CBR): ratio of the number of live births in a single year for every thousand people in the population.

crude death rate (CDR): the number of deaths in a single year for every thousand people in the population.

crude density (arithmetic density): total number of people divided by the total land area.

cultural adaptation: the complex strategies human groups employ to live successfully as part of a natural system.

cultural complex: combination of traits characteristic of a particular group.

cultural ecology: study of the relationship between a cultural group and its natural environment.

cultural geography: how space, place, and landscape shape culture at the same time that culture shapes space, place, and landscape.

cultural hearths: the geographic origins or sources of innovations, ideas, or ideologies.

cultural landscape: a characteristic and tangible outcome of the complex interactions between a human group and a natural environment.

cultural nationalism: an effort to protect regional and national cultures from the homogenizing impacts of globalization, especially from the penetrating influence of U.S. culture.

cultural region: the areas within which a particular cultural system prevails.

cultural system: a collection of interacting elements that taken together shape a group's collective identity.

cultural trait: a single aspect of the complex of routine practices that constitute a particular cultural group.

culture: a shared set of meanings that are lived through the material and symbolic practices of everyday life.

cumulative causation: a spiral buildup of advantages that occurs in specific geographic settings as a result of the development of external economies, agglomeration effects, and localization economies.

cycle of poverty: the transmission of poverty and deprivation from one generation to another through a combination of domestic circumstances and local, neighborhood conditions.

D

debt trap: syndrome of always having to borrow in order to fund development.

decolonization: the acquisition, by colonized peoples, of control over their own territory.

deep ecology: approach to nature revolving around two key components: self-realization and biospherical egalitarianism.

defensible space: a physical setting that allows residents to identify with, survey, and exert a degree of social control over public space.

deforestation: the removal of trees from a forested area without adequate replanting.

deindustrialization: a relative decline in industrial employment in core regions.

democratic rule: a system in which public policies and officials are directly chosen by popular vote.

demographic collapse: phenomenon of near genocide of native populations.

demographic transition: replacement of high birth and death rates by low birth and death rates.

demography: the study of the characteristics of human populations.

dependency: high level of reliance by a country on foreign enterprises, investment, or technology.

dependency ratio: measure of the economic impact of the young and old on the more economically productive members of the population.

derelict landscapes: landscapes that have experienced abandonment, misuse, disinvestment, or vandalism.

desertification: the degradation of land cover and damage to the soil and water in grasslands and arid and semiarid lands.

dialects: regional variations in standard languages.

diaspora: spatial dispersion of a previously homogeneous group.

digital divide: inequality of access to telecommunications and information technology, particularly the Internet.

discourse: institutionalized ways of constituting knowledge.

disruptive technology: a technology with the potential to cause a significant change in key aspects of national military power, economic development, or social organization.

distance-decay function: the rate at which a particular activity or process diminishes with increasing distance.

division of labor: the specialization of different people, regions, or countries in particular kinds of economic activities.

domino theory: the theory that if one country in a region chooses or is forced to accept a communist political and economic system, then neighboring countries would be irresistibly susceptible to communism.

double cropping: practice used in the milder climates, where intensive subsistence fields are planted and harvested more than once a year.

doubling time: measure of how long it will take the population of an area to grow to twice its current size.

dualism: the juxtaposition in geographic space of the formal and informal sectors of the economy.

E

East/West divide: communist and noncommunist countries, respectively.

ecofeminism: view that patriarchal ideology is at the center of our present environmental malaise.

ecological footprint: measure of the human pressures on the natural environment from the consumption of renewable resources and the production of pollution indicating how much space a population needs compared to what is available.

ecological imperialism: introduction of exotic plants and animals into new ecosystems.

eco-migration: population movement caused by the degradation of land and essential natural resources.

economies of scale: cost advantages to manufacturers that accrue from high-volume production, since the average cost of production falls with increasing output.

ecosystem: community of different species interacting with each other and with the larger physical environment that surrounds it.

edge cities: nodal concentrations of shopping and office space situated on the outer fringes of metropolitan areas, typically near major highway intersections.

ejidos: communal lands given to groups of landless peasants who could farm collectively or as individuals but could not rent or sell the land.

elasticity of demand: degree to which levels of demand for a product or service change in response to changes in price.

emigration: move from a particular location.

environmental determinism: doctrine holding that human activities are controlled by the environment.

environmental ethics: philosophical perspective on nature that prescribes moral principles as guidance for our treatment of it.

environmental justice: movement reflecting a growing political consciousness, largely among the world's poor, that their immediate environs are far more toxic than those in wealthier neighborhoods.

equal-area (equivalent) projection: map projection that portrays areas on Earth's surface in their true proportions.

equidistant projection: map projection that allows distance to be represented as accurately as possible.

ethnicity: socially created system of rules about who belongs and who does not belong to a particular group based upon actual or perceived commonality.

ethnocentrism: attitude that one's own race and culture are superior to others'.

ethology: scientific study of the formation and evolution of human customs and beliefs.

existential imperative: tendency for people to define themselves in relation to their material world and their capacity to achieve a form of spiritual or psychic unity between themselves and their material worlds.

export-processing zones (EPZs): small areas within which especially favorable investment and trading conditions are created by governments in order to attract export-oriented industries.

external arena: regions of the world not yet absorbed into the modern world system.

external economies: cost savings that result from circumstances beyond a firm's own organization and methods of production.

F

famine: acute starvation associated with a sharp increase in mortality.

farm crisis: financial failure of and eventual foreclosure on thousands of family farms across the U.S. Midwest.

fast food: edibles that can be prepared and served very quickly, sold in a restaurant and served to customers in packaged form.

fast world: people, places, and regions directly involved, as producers and consumers, in transnational industry, modern telecommunications, materialistic consumption, and international news and entertainment.

federal state: form of government in which power is allocated to units of local government within the country.

fiscal squeeze: increasing limitations on city revenues, combined with increasing demands for expenditure.

flexible production systems: ability of manufacturers to shift quickly and efficiently from one level of output to another, or from one product configuration to another.

folk culture: traditional practices of small groups, especially rural people with a simple lifestyle who are seen to be homogeneous in their belief systems and practices.

food chain: five central and connected sectors (inputs, production, product processing, distribution, and consumption) with four contextual elements acting as external mediating forces (the state, international trade, the physical environment, and credit and finance).

food manufacturing: adding value to agricultural products through a range of treatments—such as processing, canning, refining, packing, and packaging—that occur off the farm and before products reach the market.

food regime: specific set of links that exists among food production and consumption and capital investment and accumulation opportunities.

food security: assured access by a person, household, or even a country to enough food at all times to ensure active and healthy lives.

food sovereignty: right of peoples, communities, and countries to define their own agricultural, labor, fishing, food, and land policies that are ecologically, socially, economically, and culturally appropriate to their unique circumstances.

forced migration: movement of an individual against his or her will.

Fordism: principles for mass production based on assembly-line techniques, scientific management, mass consumption based on higher wages, and sophisticated advertising techniques.

foreign direct investment: total of overseas business investments made by private companies.

formal region: groups of areal units that have a high degree of homogeneity in terms of particular distinguishing features.

friction of distance: deterrent or inhibiting effect of distance on human activity.

functional regions: regions with some variability in certain attributes but with an overall coherence to the structure and dynamics of economic, political, and social organization.

G

gateway city: serves as a link between one country or region and others because of its physical situation.

gender: social differences between men and women rather than the anatomical differences that are related to sex.

genetically modified organism (GMO): any organism that has had its DNA modified in a laboratory rather than through cross-pollination or other forms of evolution.

genre de vie: functionally organized way of life that is seen to be characteristic of a particular cultural group.

gentrification: invasion of older, centrally-located, working-class neighborhoods by higher-income households seeking the character and convenience of less expensive and well-located residences.

geodemographic analysis: practice of assessing the location and composition of particular populations.

geodemographic research: study of census data and commercial data (such as sales data and property records) about the populations of small districts to create profiles of those populations for market research.

geographic information system (GIS): organized collection of computer hardware, software, and geographic data that is designed to capture, store, update, manipulate, and display geographically referenced information.

geographical imagination: capacity to understand changing patterns, changing processes, and changing relationships among people, places, and regions.

geographical path dependence: historical relationship between the present activities associated with a place and the past experiences of that place.

geopolitics: state's power to control space or territory and shape the foreign policy of individual states and international political relations.

gerrymandering: practice of redistricting for partisan purposes.

global change: combination of political, economic, social, historical, and environmental problems at the world scale.

global civil society: set of institutions, organizations, and behaviors situated between the state, business world, and family including voluntary and non-profit organizations, philanthropic institutions, and social and political movements.

Global Positioning System (GPS): system of satellites that orbit Earth on precisely predictable paths, broadcasting highly accurate time and locational information.

globalization: increasing interconnectedness of different parts of the world through common processes of economic, environmental, political, and cultural change.

globalized agriculture: system of food production increasingly dependent upon an economy and set of regulatory practices that are global in scope and organization.

Green Revolution: export of a technological package of fertilizers and high-yielding seeds, from the core to the periphery, to increase global agricultural productivity.

gross domestic product (GDP): estimate of the total value of all materials, foodstuffs, goods, and services produced by a country in a particular year.

gross migration: total number of migrants moving into and out of a place, region, or country.

gross national income (GNI): similar to GDP, but also includes the value of income from abroad.

growth poles: economic activities that are deliberately organized around one or more high-growth industries.

guest workers: individuals who migrate temporarily to take up jobs in other countries.

H

hajj: religious pilgrimage.

hearth areas: geographic settings where new practices have developed and from which they have subsequently spread.

hegemony: domination over the world economy exercised by one national state in a particular historical epoch through a combination of economic, military, financial, and cultural means.

hinterland: sphere of economic influence of a town or city.

historical geography: geography of the past.

human geography: study of the spatial organization of human activity and of people's relationships with their environments.

human rights: people's individual rights to justice, freedom, and equality, considered by most societies to belong automatically to all people.

humanistic approach: point of view that places the individual—especially individual values, meaning systems, intentions, and conscious acts—at the center of analysis.

hunting and gathering: activities whereby people feed themselves through killing wild animals and fish and gathering fruits, roots, nuts, and other edible plants to sustain themselves.

I

identity: sense that people make of themselves through their subjective feelings based on their everyday experiences and wider social relations.

immigration: move to another location.

imperialism: extension of the power of a nation through direct or indirect control of the economic and political life of other territories.

import substitution: process by which domestic producers provide goods or services that formerly were bought from foreign producers.

infant mortality rate: annual number of deaths of infants under 1 year of age compared to the total number of live births for that same year.

inflation: increased supply of printed currency that leads to higher prices and international financial differentials.

informal sector: economic activities that take place beyond official record, not subject to formalized systems of regulation or remuneration.

infrastructure (or fixed social capital): underlying framework of services and amenities needed to facilitate productive activity.

initial advantage: critical importance of an early start in economic development; a special case of external economies.

intensive subsistence agriculture: practice that involves the effective and efficient use—usually through a considerable expenditure of human labor and application of fertilizer—of a small parcel of land in order to maximize crop yield.

internal migration: move within a particular country or region.

internally displaced persons: individuals who are uprooted within the boundaries of their own country because of conflict or human rights abuse.

international division of labor: specialization, by countries, in particular products for export.

international migration: move from one country to another.

international organization: group that includes two or more states seeking political and/or economic cooperation with each other.

international regime: orientation of contemporary politics around the international arena instead of the national.

intersubjectivity: shared meanings among people, derived from their lived experience of everyday practice.

intertillage: practice of mixing different seeds and seedlings in the same swidden.

intifada: uprising against Israel by the Palestinian people.

invasion and succession: process of neighborhood change whereby one social or ethnic group succeeds another.

irredentism: assertion by the government of a country that a minority living outside its formal borders belongs to it historically and culturally.

Islam: Arabic term that means submission to God's will.

Islamism: anticolonial, anti-imperial, and generally anticore political movement.

isotropic surface: hypothetical, uniform plain that is flat and has no variations in its physical attributes.

J

jihad: sacred struggle.

just-in-time production: manufacturing process in which daily or hourly delivery schedules of materials allow for minimal or zero inventories.

K

kinship: relationship based on blood, marriage, or adoption.

L

land reform: redistribution of land by the state with a goal of increasing productivity and reducing social unrest.

landscape as text: idea that landscapes can be read and written by groups and individuals.

language: communicating ideas or feelings by means of a conventionalized system of signs, gestures, marks, or articulate vocal sounds.

language branch: collection of languages that possess a definite common origin but have split into individual languages.

language family: collection of individual languages believed to be related in their prehistorical origin.

language group: collection of several individual languages that are part of a language branch, share a common origin, and have similar grammar and vocabulary.

latitude: angular distance of a point on Earth's surface, measured north or south from the equator, which is 0°.

law of diminishing returns: tendency for productivity to decline, after a certain point, with the continued application of capital and/or labor to a given resource base.

leadership cycles: periods of international power established by individual states through economic, political, and military competition.

life expectancy: average number of years a newborn infant can expect to live.

lifeworld: taken-for-granted pattern and context for everyday living through which people conduct their lives.

local food: food that is organically grown and produced within a fairly limited distance from where it is consumed.

localization economies: cost savings that accrue to particular industries as a result of clustering together at a specific location.

logistics: movement and storage of goods and the management of the entire supply chain, from purchase of raw materials through sale of the final product and back again, for the replenishment of goods as they are sold.

longitude: angular distance of a point on Earth's surface, measured east or west from the prime meridian (the line that passes through both poles and through Greenwich, England, and that has the value of 0°).

M

map projection: systematic rendering on a flat surface of the geographic coordinates of the features found on Earth's surface.

masculinism: assumption that the world is, and should be, shaped mainly by men, for men.

mechanization: replacement of human farm labor with machines.

medical geography: subarea of the discipline that specializes in understanding the spatial aspects of health and illness.

megacity: very large city characterized by both primacy and high centrality within its national economy.

middle cohort: members of the population 15 to 64 years of age who are considered economically active and productive.

migration: move beyond the same political jurisdiction, involving a change of residence—either as emigration or as immigration.

minisystem: society with a single cultural base and a reciprocal social economy.

minority groups: population subgroups that are seen—or that see themselves—as somehow different from the general population.

mobility: ability to move, either permanently or temporarily.

Modern movement: architectural movement based on the idea that buildings and cities should be designed and run like machines.

Modernity: forward-looking view of the world that emphasizes reason, scientific rationality, creativity, novelty, and progress.

Muslim: member of the Islamic community of believers whose duty is obedience and submission to the will of God.

N

nation: group of people often sharing common elements of culture, such as religion or language or a history or political identity.

nationalism: feeling of belonging to a nation as well as the belief that a nation has a natural right to determine its own affairs.

nation-state: ideal form consisting of a homogeneous group of people governed by their own state.

natural decrease: difference between CDR and CBR, which is the deficit of births relative to deaths.

natural increase: difference between the CBR and CDR, which is the surplus of births relative to deaths.

nature: social creation as well as the physical universe that includes human beings.

neocolonialism: economic and political strategies by which powerful states in core economies indirectly maintain or extend their influence over other areas or people.

neo-Fordism: economic principles in which the logic of mass production coupled with mass consumption is modified by the addition of more flexible production, distribution, and marketing systems.

neoliberal policies: economic policies that are predicated on a minimalist role for the state, assuming the desirability of free markets as the ideal condition not only for economic organization but also for political and social life.

neoliberalism: reduction in the role and budget of government, including reduced subsidies and the privatization of formerly publicly owned and operated concerns, such as utilities.

net migration: gain or loss in the total population of a particular area as a result of migration.

new world order: triumph of capitalism over communism, wherein the United States becomes the world's only superpower and therefore its policing force.

newly industrializing countries (NICs): countries formerly peripheral within the world system that have acquired a significant industrial sector, usually through foreign direct investment.

nontraditional agricultural exports (NTAEs): new export crops that contrast with traditional exports.

North/South divide: differentiation made between the colonizing states of the Northern Hemisphere and the formerly colonized states of the Southern Hemisphere.

nutritional density: ratio between the total population and the amount of land under cultivation in a given unit of area.

O

OECD: Organization for Economic Cooperation and Development.

offshore financial centers: islands or microstates that have become a specialized node in the geography of worldwide financial flows.

old-age cohort: members of the population 65 years of age and older who are considered beyond their economically active and productive years.

ordinary landscapes (vernacular landscapes): everyday landscapes that people create in the course of their lives.

organic farming: farming or animal husbandry done without commercial fertilizers, synthetic pesticides, or growth hormones.

Orientalism: discourse that positions the West as culturally superior to the East.

overurbanization: condition in which cities grow more rapidly than the jobs and housing they can sustain.

P

Paleolithic period: period when chipped-stone tools first began to be used.

pastoralism: subsistence activity that involves the breeding and herding of animals to satisfy the human needs of food, shelter, and clothing.

peripheral regions: regions with undeveloped or narrowly specialized economies with low levels of productivity.

physical geography: subarea of the discipline that studies Earth's natural processes and their outcomes.

place: specific geographic setting with distinctive physical, social, and cultural attributes.

plantation: large landholding that usually specializes in the production of one particular crop for market.

political ecology: approach to cultural geography that studies humans in their environment through the relationships of patterns of resource use to political and economic forces.

popular culture: practices and meaning systems produced by large groups of people whose norms and tastes are often heterogeneous and change frequently, often in response to commercial products.

population policy: official government policy designed to effect any or all of several objectives, including the size, composition, and distribution of population.

Postmodern urban design: style characterized by a diversity of architectural styles and elements, often combined in the same building or project.

Postmodernity: view of the world that emphasizes an openness to a range of perspectives in social inquiry, artistic expression, and political empowerment.

preservation: approach to nature advocating that certain habitats, species, and resources should remain off-limits to human use, regardless of whether the use maintains or depletes the resource in question.

primacy: condition in which the population of the largest city in an urban system is disproportionately large in relation to the second- and third-largest cities.

primary activities: economic activities that are concerned directly with natural resources of any kind.

producer services: services that enhance the productivity or efficiency of other firms' activities or that enable them to maintain specialized roles.

proxemics: study of the social and cultural meanings that people give to personal space.

pull factors: forces of attraction that influence migrants to move to a particular location.

push factors: events and conditions that impel an individual to move from a location.

Q

quaternary activities: economic activities that deal with the handling and processing of knowledge and information.

R

race: problematic classification of human beings based on skin color and other physical characteristics.

racialization: practice of categorizing people according to race, or of imposing a racial character or context.

rank-size rule: statistical regularity in size distributions of cities and regions.

reapportionment: process of allocating electoral seats to geographical areas.

redistricting: defining and redefining of territorial district boundaries.

redlining: practice whereby lending institutions delimit "bad-risk" neighborhoods on a city map and then use the map as the basis for determining loans.

refugee: individual who crosses national boundaries to seek safety and asylum.

region: larger-sized territory that encompasses many places, all or most of which share similar attributes in comparison with the attributes of places elsewhere.

regional geography: study of the ways unique combinations of environmental and human factors produce territories with distinctive landscapes and cultural attributes.

regionalism: feeling of collective identity based on a population's politico-territorial identification within a state or across state boundaries.

regionalization: classification of individual places or areal units.

religion: belief system and set of practices that recognize the existence of a power higher than humans.

remote sensing: collection of information about parts of Earth's surface by means of aerial photography or satellite imagery designed to record data on visible, infrared, and microwave sensor systems.

reurbanization: growth of population in metropolitan central cores, following a period of absolute or relative decline in population.

risk society: contemporary societies in which politics is increasingly about avoiding hazards.

rites of passage: ceremonial acts, customs, practices, or procedures that recognize key transitions in human life, such as birth, menstruation, and other markers of adulthood such as marriage.

romanticism: philosophy that emphasizes interdependence and relatedness between humans and nature.

S

sacred space: area recognized by individuals or groups as worthy of special attention as a site of special religious experiences or events.

secondary activities: economic activities that process, transform, fabricate, or assemble the raw materials derived from primary activities, or that reassemble, refinish, or package manufactured goods.

sectionalism: extreme devotion to local interests and customs.

segregation: spatial separation of specific population subgroups within a wider population.

self-determination: right of a group with a distinctive politico-territorial identity to determine its own destiny, at least in part, through the control of its own territory.

semiotics: practice of writing and reading signs.

semiperipheral regions: regions that are able to exploit peripheral regions but are themselves exploited and dominated by core regions.

sense of place: feelings evoked among people as a result of the experiences and memories that they associate with a place and the symbolism that they attach to it.

sexuality: set of practices and identities that a given culture considers related to each other and to those things it considers sexual acts and desires.

shifting cultivation: system in which farmers aim to maintain soil fertility by rotating the fields within which cultivation occurs.

shock city: city that is seen as the embodiment of surprising and disturbing changes in economic, social, and cultural life.

siltation: buildup of sand and clay in a natural or artificial waterway.

site: physical attributes of a location—its terrain, its soil, vegetation, and water sources, for example.

situation: location of a place relative to other places and human activities.

slash-and-burn: system of cultivation in which plants are cropped close to the ground, left to dry for a period, and then ignited.

slow food: attempt to resist fast food by preserving the cultural cuisine and the associated food and farming of an ecoregion.

slow world: people, places, and regions whose participation in transnational industry, modern telecommunications, materialistic consumption, and international news and entertainment is limited.

society: sum of the inventions, institutions, and relationships created and reproduced by human beings across particular places and times.

socio-spatial formation: specific combination of demographic groups, social classes, cultural values, and local institutions at a particular time and place.

sovereignty: exercise of state power over people and territory, recognized by other states and codified by international law.

spatial analysis: study of geographic phenomena in terms of their arrangement as points, lines, areas, or surfaces on a map.

spatial diffusion: way that things spread through space and over time.

spatial interaction: movement and flows involving human activity.

spatial justice: fairness of the distribution of society's burdens and benefits, taking into account spatial variations in people's needs and in their contribution to the production of wealth and social well-being.

splintering urbanism: fragmentation of the economic, social, and material fabric of cities as a result of the selective impact of new technologies and networked information and communications infrastructures.

spread effects: positive impacts on a region (or regions) of the economic growth of some other region.

squatter settlements: residential developments that occur on land that is neither owned nor rented by its occupants.

states: independent political units with territorial boundaries that are internationally recognized by other states.

strategic alliances: commercial agreements between transnational corporations, usually involving shared technologies, marketing networks, market research, or product development.

subsistence agriculture: farming for direct consumption by the producers; not for sale.

suburbanization: growth of population along the fringes of large metropolitan areas.

supranational organizations: collections of individual states with a common goal that may be economic and/or political in nature.

sustainable development: vision of development that seeks a balance among economic growth, environmental impacts, and social equity.

swidden: land that is cleared using the slash-and-burn process and is ready for cultivation.

symbolic landscapes: representations of particular values or aspirations that the builders and financiers of those landscapes want to impart to a larger public.

T

technology: physical objects or artifacts, activities or processes, and knowledge or know-how.

technology systems: clusters of interrelated energy, transportation, and production technologies that dominate economic activity for several decades at a time.

terms of trade: ratio of prices at which exports and imports are exchanged.

territorial organization: system of government formally structured by area, not by social groups.

territoriality: specific attachment of individuals or peoples to a specific location or territory.

territory: delimited area over which a state exercises control and which is recognized by other states.

terrorism: threat or use of force to bring about political change.

tertiary activities: economic activities involving the sale and exchange of goods and services.

threshold: minimum market size required to make the sale of a particular product or service profitable.

time-space convergence: rate at which places move closer together in travel or communication time or costs.

topological space: connections between, or connectivity of, particular points in space.

topophilia: emotions and meanings associated with particular places that have become significant to individuals.

total fertility rate (TFR): average number of children a woman will have throughout the years that demographers have identified as her childbearing years, approximately ages 15 through 49.

trading blocs: groups of countries with formalized systems of trading agreements.

transcendentalism: philosophy in which a person attempts to rise above nature and the limitations of the body to the point where the spirit dominates the flesh.

transhumance: movement of herds according to seasonal rhythms: warmer, lowland areas in the winter; cooler, highland areas in the summer.

transnational corporations: companies with investments and activities that span international boundaries and with subsidiary companies, factories, offices, or facilities in several countries.

transnational migrant: migrants who set up homes and/or work in more than one nation-state.

tribe: form of social identity created by groups who share a set of ideas about collective loyalty and political action.

U

underclass: subset of the poor, isolated from mainstream values and the formal labor market.

underemployment: situation in which people work less than full time even though they would prefer to work more hours.

undernutrition: inadequate intake of one or more nutrients and/or calories.

unitary state: form of government in which power is concentrated in the central government.

urban agriculture: establishment or performance of agricultural practices in or near an urban or citylike setting.

urban ecology: social and demographic composition of city districts and neighborhoods.

urban form: physical structure and organization of cities.

urban system: interdependent set of urban settlements within a specified region.

urbanism: way of life, attitudes, values, and patterns of behavior fostered by urban settings.

urbanization economies: external economies that accrue to producers because of the package of infrastructure, ancillary activities, labor, and markets typically associated with urban settings.

utility: usefulness of a specific place or location to a particular person or group.

V

vertical disintegration: evolution from large, functionally integrated firms within a given industry toward networks of specialized firms, subcontractors, and suppliers.

virgin soil epidemics: conditions in which the population at risk has no natural immunity or previous exposure to the disease within the lifetime of the oldest member of the group.

visualization: computer-assisted representation of spatial data, often involving three-dimensional images and innovative perspectives, in order to reveal spatial patterns and relationships more effectively.

vital records: information about births, deaths, marriages, divorces, and the incidence of certain infectious diseases.

voluntary migration: movement by an individual based on choice.

W

world city: city in which a disproportionate part of the world's most important business is conducted.

world empire: minisystems that have been absorbed into a common political system while retaining their fundamental cultural differences.

world music: musical genre defined largely in response to the sudden increase of non-English-language recordings released in the United Kingdom and the United States in the 1980s.

world region: large-scale geographic divisions based on continental and physiographic settings that contain major groupings of peoples with broadly similar cultural attributes.

world system: interdependent system of countries linked by economic and political competition.

Y

youth cohort: members of the population who are less than 15 years of age and generally considered to be too young to be fully active in the labor force.

Z

Zionism: movement for the establishment of a legally recognized home in Palestine for the Jewish people.

zone in transition: area of mixed commercial and residential land uses surrounding the CBD.

Photo Credits

T = top; M = middle; B = bottom; L = left; R = right

Chapter 1

Page ii: Wilfried Krecichwost/Getty Images. Page 3: Paul Knox. Page 4: Frederic J. Brown/Getty Images. Page 4: (B) Courtesy of Heike Mayer. Page 9: (T) Corbis/Bettmann. Page 9 (B): China Photos/Getty Images. Page 11: UN: Scott Olson/Getty Images. Page 14: Photo by Sean Smith, "Guardian Weekly", August 2002, p. 11. Page 15: Photo by Sean Smith, "Guardian Weekly", August 2002, p. 10. Page 16: Yapanchintsev Yevgeny/Corbis. Page 17 (T): Paul Knox. Page 17 (B): Christopher J. Morris/CORBIS All Rights Reserved. Page AP Wide World Photos. Page 19: Daniel Berehulak/Getty Images. Page 22: (T/L) Courtesy of Spaceimaging.com; (T/R) Marge Beaver/ Photography Plus; (M) NASA goddard Space Flight Center. Image by Reto Stockli (land surface, shallow water, clouds). Enhancements by Robert Simmon (ocean color, compositing, 3D globes, animation). Page 23: (M) Paul L. Knox. Page 30: Andrew Lichtenstein/Getty Images. Page 33: Frank Krahmer/Corbis. Page 34 (T): The Irish Image Collection/Corbis. Page 34 (B): Peter Adams/Getty Images. Page 35 (T): Paul Knox. Page 35 (B): Gabriel Bous/Getty Images. Page 36: Paul Knox. Page 37: Paul Knox

Chapter 2

Page 40: Keren Su/Corbis. Page 43 (B): Funai-Frente de Protecao Etno-Ambiental Envira/Reuters. Page 45 (C): Robert Estall/Corbis. Page 45 (B): Image Source/Corbis. Page 46: Courtesy of the Library of Congress. Page 49: Corbis. Page 50(T): The Bridgeman Art Library/Getty Images. Page 51: The Granger Collection. Page 54: Corbis. Page 59(R/B): Suzanne Brookens/Corbis. Page 61(T/L): The Bridgeman Art Library/Getty Images. Page 61 (C): The Bridgeman Art Library/Getty Images. Page 61 (R/B): Corbis. Page 62: The Bridgeman Art Library/Getty Images. Page 63: (M) Library of Congress; (B) Library of Congress; Page 65: (T/R) (c) Swim Ink 2, LLC/ CORBIS All Rights Reserved. Page 74 (B/R): Peter Ginter/Getty Images. Page 76: (c) Jim Hollander/Reuters/CORBIS All Rights Reserved. Page 77 (B/L): Nicholas Ratzenboeck/Getty Images.

Chapter 3

Page 82: Getty Images. Page 89 (T): Shutterstock. Page 89 (B): Jamie Kripke/Corbis. Page 96 (T): Roy Botterell/Corbis. Page 98: Grant Heilman Photography, Inc. Page 100: Photo Researchers, Inc. Page 109 (B): (c) Beatrice Mategwa/Reuters/CORBIS All Rights Reserved. Page 110: AP Wide World Photos. Page 115: (B) Corbis/ Bettmann; Page 116: Shutterstock. Page 117 (B): Siegfried Modola/ Getty Images. Page 121: Richard Vogel/Associated Press. Page 123: Larry Downing/Corbis.

Chapter 4

Page 126: Mel Curtis/Getty Images. Page 128: AP Wide World Photos. Page 129: (B/L) AP Wide World Photos; (T/R) New York, 2002 Photo by Martin Rowe, Lantern Books. Page 131 (T/R): Peter Menzel/ Menzel Photo. Page 131 (T/L): Peter Menzel/Menzel Photo. Page 135: AP Wide World Photos. Page 136: (c) Diego Lezama Orezzoli/ CORBIS All Rights Reserved. Page 137: Justin Sullivan/Getty Images. Page 140: (B) (c) Jean Clottes and French Ministry of Culture and

Communication, Regional Direcfton for Cultural Affairs - Rhone-Alps - Regional Department of Archaeology. Page 141 (T): AP Wide World Photos. Page 141 (B): (c) Warren Morgan/CORBIS All Rights Reserved; Page 142 (T): Arthur Lidov/National Geographic Society Image Collection. Page 142 (B): AP Wide World Photos. Page 143: AP Wide World Photos. Page 147 Sean Sprague/Mexicolore. Page 150: (T) (c) China Photo/Reuters/CORBIS All Rights Reserved; (B) (c) Norbert Schaefer/CORBIS All Rights Reserved. Page 151: (T) (c) Najlah Feanny/CORBIS SABA All Rights Reserved. Page 157: Xinhua Press/ Corbis. Page 161 (T): AP Wide World Photos. Page 161 (B): James Warwick/Getty Images. Page 162: AP Wide World Photos.

Chapter 5

Page 172: China Photos/Getty Images. Page 175 (T) Dave Hiser/ Corbis. Page 175: (B) (c) Paul A. Souders/CORBIS All Rights Reserved. Page 177: Michelly Rall/Getty Images. Page 179: (c) Gavriel Jecan/CORBIS All Rights Reserved. Page 180: (B) (c) Amet Jean Pierre/CORBIS Sygma All rights reserved. Page 181 (T): National Archives of the UK, ref. E31/2/1 f.56. Page 181: (B) (c) Nathan Benn/ CORBIS All Rights Reserved. Page 182: Shin Young-kyun Yonhap/ Associated Press. Page 189 (T): Charles Haire/Alamy. Page 189 (B): Brian Sokol/Getty Images. Page 190: John Gress/Corbis. Page 194: (c) Daniel Laine/CORBIS All Rights Reserved. Page 203: (L) AP Wide World Photos; (R) AP Wide World Photos. Page 206: AP Wide World Photos. Page 207: DEA Picture Library/Getty Images. Page 208: (c) Bob Krist/CORBIS All Rights Reserved. Page 210: AP Wide World Photos. Page 211: Aimee Corrigan.

Chapter 6

Page 214: Clive Rose/Getty Images. Page 217: AP Wide World Photos. Page 218 (B): AP Wide World Photos. Page 219: AP Wide World Photos. Page 221: AP Wide World Photos. Page 226 (T/L): Travelpix/Getty Images. Page 226 (T/R): Istockphoto. Page 226 (B/L): Fotolia. Page 226(B/R): Shutterstock. Page 227: (T) (c) Joseph Sohm/ CORBIS All Rights Reserved; (M) AP Wide World Photos; (B) AP Wide World Photos. Page 229: Martin Thomas Photography/Alamy Page 231: (T/L) (c) James Davis/Eye Ubiquitous/CORBIS All Rights Reserved; (B/L) Cristiano Mascaro/Getty Images, Inc. Page 232 (T): Richard Bickel/Corbis. Page 232 (M): Shutterstock. Page 235: Hanan Isachar/Corbis. Page 238: Paul Knox. Page 239: AP Wide World Photos. Page 240: LOOK Die Bildagentur der Fotografen GmbH/ Alamy. Page 241: Richard T. Nowitz/Corbis. Page 242: (M/L) Pepys Library, Magdafene College, Cambridge; (M/R) AP Wide World Photos; (B/L/R) AP Wide World Photos. Page 243: AP Wide World Photos. Page 244: Scott E Barbour/Getty Images. Page 245: AP Wide World Photos.

Chapter 7

Page 248: (c) Claro Cortes IV/Reuters/CORBIS All Rights Reserved. Page 252: Shaul Schwarz/Getty Images. Page 255: (T) Gideon Mendel/Corbis. Page 255: (B) Tom Stoddart/Getty Images. Page 256: David McNew/Getty Images. Page 261: David McNew/Getty Images. Page 262 Gideon Mendel/Corbis. Page 263 (M): Liu Jin/Getty Images. Page 263(B): Feng Li/Getty Images. Page 267: Peter Macdiarmid/ Getty Images. Page 268: Andy Aitchison/Corbis. Page 269: AP Wide World Photos. Page 275: (c) Vince Streano/CORBIS All Rights

INDEX

Carbon Emissions 2003

Internet Users 2002

Total Population

Wealth Growth

These maps from *Worldmapper.org* are equal area cartograms, otherwise known as density-equalizing maps. The cartograms resize each territory according to the variable being mapped.